FIFTH EDITION

UNIVERSE

The Solar System

UNIVERSE

The Solar System

FIFTH EDITION

Roger A. Freedman
University of California, Santa Barbara

Robert M. Geller
University of California, Santa Barbara

William J. Kaufmann III
San Diego State University

W.H. Freeman and Company

A Macmillan Higher Education Company

Publisher:	*Jessica Fiorillo*
Acquisitions Editor:	*Alicia Brady*
Development Editor:	*Brittany Murphy*
Senior Media and Supplements Editor:	*Amy Thorne*
Assistant Editor:	*Courtney Lyons*
Associate Director of Marketing:	*Debbie Clare*
Marketing Assistant:	*Samantha Zimbler*
Project Editor:	*Kerry O'Shaughnessy*
Production Manager:	*Julia DeRosa*
Cover and Text Designer:	*Victoria Tomaselli*
Illustration Coordinator:	*Janice Donnola*
Illustrations:	*Dragonfly Media Group, George Kelvin*
Photo Editors:	*Robin Fadool, Bianca Moscatelli*
Photo Researcher:	*Deborah Anderson*
Composition:	*Sheridan Sellers*
Printing and Binding:	*King Printing Co., Inc.*

Library of Congress Control Number: 2013950518
ISBN-13: 978-1-319-04246-2
ISBN-10: 1-319-04246-5

Fifth printing

W. H. Freeman and Company, 41 Madison Avenue, New York, NY 10010
Houndmills, Basingstoke RG21 6XS, England
www.whfreeman.com

About the Authors

Roger A. Freedman is on the faculty of the Department of Physics at the University of California, Santa Barbara. He grew up in San Diego, California, and was an undergraduate at the University of California campuses in San Diego and Los Angeles. He did his doctoral research in nuclear theory and its astrophysical applications at Stanford University under the direction of Professor J. Dirk Walecka. Dr. Freedman joined the faculty at UCSB in 1981 after three years of teaching and doing research at the University of Washington. Dr. Freedman holds a commercial pilot's license, and when not teaching or writing he can frequently be found flying with his wife, Caroline. He has flown across the United States and Canada. (Photo courtesy of Caroline J. Robillard)

Robert M. Geller teaches and conducts research in astrophysics at the University of California, Santa Barbara, where he also obtained his Ph.D. His doctoral research was in observational cosmology under Professor Robert Antonucci. Using data from the Hubble Space Telescope, he is currently involved in a search for bursts of light that are predicted to occur when a supermassive black hole consumes a star. His other project, in biomedicine, explores the use of magnetotactic bacteria to enhance the effectiveness of radiation therapy in treating cancer. Dr. Geller also has a strong emphasis on education, and he received the Distinguished Teaching Award at UCSB in 2003. His hobbies include rock climbing, and he built an unusual telescope with lenses made of water. (Photo courtesy of Richard Rouse)

William J. Kaufmann III was the author of the first four editions of *Universe*. Dr. Kaufmann earned his bachelor's degree magna cum laude in physics from Adelphi University in 1963, a master's degree in physics from Rutgers in 1965, and a Ph.D. in astrophysics from Indiana University in 1968. At 27, he became the youngest director of any major planetarium in the United States when he took the helm of the Griffith Observatory in Los Angeles. During his career, he also held positions at San Diego State University, UCLA, Caltech, and the University of Illinois. A prolific author, his many books include *Black Holes and Warped Spacetime, Relativity and Cosmology, Planets and Moons, Stars and Nebulas,* and *Galaxies and Quasars.* Dr. Kaufmann died in 1994.

Contents Overview

Chapters 17–26 can be found in Universe, 10th edition, of which this book is an abbreviated version.

Contents

III Stars and Stellar Evolution

Preface

Astronomy is one of the most dynamic and exciting areas of modern science. Its recent discoveries often capture widespread interest, and its subjects—the planets, stars, and the universe—inspire awe and inquiry. From observations of previously unseen worlds in the outer depths of our solar system to new evidence of how the most massive stars can collapse into black holes, astronomy provides a wide array of opportunities to show students the continuing process of science.

With so many possible concepts to explore, students can easily get overwhelmed. *Universe* has been developed to help students focus on and learn core concepts. Through clear, highlighted explanations of key topics, and multiple interactions with material, the text fosters an appreciative comprehension of the subject.

COSMIC CONNECTIONS figures summarize key ideas visually

Many students learn more through visual presentations than from reading long passages of text. To help these students, large figures, called **Cosmic Connections**, appear in most chapters of *Universe*. Cosmic Connections give an overview or summary of a particular important topic in a chapter. Subjects range from the variety of modern telescopes to the formation of the solar system; from the scale of distances in the universe to the life cycles of stars; and from the influence of gravitational tidal forces to the evolution of the universe after the Big Bang. Cosmic Connections convey the sense of excitement and adventure that lead students to study astronomy in the first place.

Enhanced, animated versions of the Cosmic Connections are available within the textbook's associated multimedia resources.

NEW! CONCEPTCHECK AND CALCULATIONCHECK QUESTIONS allow students to go beyond passive reading

Included at the end of each section, thought-provoking **ConceptCheck** questions provide immediate assessment by going beyond reading comprehension, often asking students to draw conclusions informed by the text, calling for applied thinking and synthesis of concepts.

> **CONCEPT**CHECK 4-1
>
> If Mars is moving retrograde, will it rise above the eastern horizon or above the western horizon?

> **CONCEPT**CHECK 11-4
>
> Microwaves are emitted from Venus's hot surface. What do 1.35-cm microwaves tell us about the presence of water on the surface of Venus and in its thick atmosphere?
>
> *Answer appears at the end of the chapter.*

Similar to Concept-Check questions and focusing on mathematics, **CalculationCheck** questions give students the opportunity to test themselves by solving different mathematical problems associated

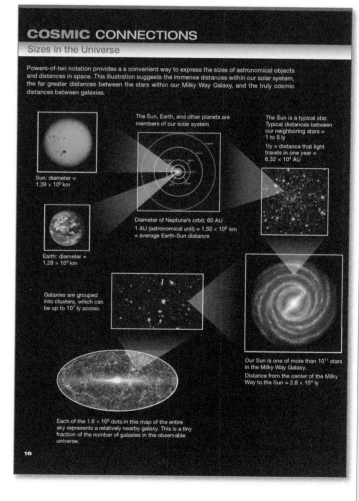

COSMIC CONNECTIONS

Sizes in the Universe

Powers-of-ten notation provides a a convenient way to express the sizes of astronomical objects and distances in space. This illustration suggests the immense distances within our solar system, the far greater distances between the stars within our Milky Way Galaxy, and the truly cosmic distances between galaxies.

Sun: diameter = 1.39×10^6 km

The Sun, Earth, and other planets are members of our solar system

The Sun is a typical star. Typical distances between our neighboring stars = 1 to 5 ly

1ly = distance that light travels in one year = 6.32×10^4 AU

Diameter of Neptune's orbit: 60 AU
1 AU (astronomical unit) = 1.50×10^8 km = average Earth-Sun distance

Earth: diameter = 1.28×10^4 km

Galaxies are grouped into clusters, which can be up to 10^7 ly across.

Our Sun is one of more than 10^{11} stars in the Milky Way Galaxy.
Distance from the center of the Milky Way to the Sun = 2.8×10^4 ly

Each of the 1.6×10^6 dots in this map of the entire sky represents a relatively nearby galaxy. This is a tiny fraction of the number of galaxies in the observable universe.

16

CALCULATIONCHECK 5-2

Which wavelength of light would our Sun emit most if its temperature were twice its current temperature of 5800 K?

Answers appear at the end of the chapter.

with the chapter conceps. They appear only within sections where relevant mathematical reasoning is present. Answers to all ConceptCheck and CalculationCheck questions are provided at the back of the book.

MOTIVATION STATEMENTS connect detailed discussion to the broader picture

4-5 Galileo's discoveries with a telescope strongly supported a heliocentric model

When Dutch opticians invented the telescope during the first decade of the seventeenth century, astronomy was changed forever. The scholar who used this new tool to amass convincing evidence that the planets orbit the Sun, not Earth, was the Italian mathematician and physical scientist Galileo Galilei (Figure 4-12).

Galileo used the cutting-edge technology of the 1600s to radically transform our picture of the universe

As students read a science text, they often wonder "What is the key idea? How does it relate to the discussion on previous pages? What motivated scientists to make these observations and deductions?" Free-standing **Motivation Statements** address these questions. These brief sentences show students how a section's material fits in with the larger picture of astronomy presented in the chapter. Some statements explore the scientific method. Others compare topics in one section with those in another section or chapter; still other statements give a quick summary of a section's conclusion.

CAUTIONS confront misconceptions

CAUTION! Note that the quantity c is the speed of light *in a vacuum*. Light travels more slowly through air, water, glass, or any other transparent substance than it does in a vacuum. In our study of astronomy, however, we will almost always consider light traveling through the vacuum (or near-vacuum) of space.

Many people think that Earth is closer to the Sun in summer than in winter and that the phases of the Moon are caused by Earth's shadow falling on the Moon. However, these "common sense" ideas are incorrect (as explained in Chapters 2 and 3). Throughout *Universe*, paragraphs marked by the **Caution** icon alert the reader to such common conceptual pitfalls.

ANALOGIES make astronomy relatable to students

When learning new astronomical ideas, it can be helpful to relate them to more familiar experiences on Earth. Throughout *Universe*, **Analogy** paragraphs make these connections. For example, the motions of the planets can be related to children on a merry-go-round, and the bending of light through a telescope lens is similar to the path of a car driving from firm ground onto sand.

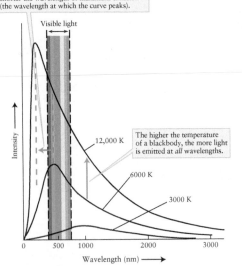

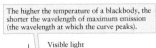

The higher the temperature of a blackbody, the shorter the wavelength of maximum emission (the wavelength at which the curve peaks).

Visible light

The higher the temperature of a blackbody, the more light is emitted at *all* wavelengths.

12,000 K

6000 K

3000 K

Intensity

0   500   1000   2000   3000

Wavelength (nm)

FIGURE 5-12
Blackbody Curves Each of these curves shows the intensity of light at every wavelength that is emitted by a blackbody (an idealized case of a dense object) at a particular temperature. The rainbow-colored band shows the range of visible wavelengths. The vertical scale has been compressed so that all three curves can be seen; the peak intensity for the 12,000-K curve is actually about 1000 times greater than the peak intensity for the 3000-K curve.

ANALOGY Figure skaters make use of the conservation of angular momentum. When a spinning skater pulls her arms and legs in close to her body, the rate at which she spins automatically increases (Figure 8-7). Even if you are not a figure skater, you can demonstrate this by sitting on a rotating office chair. Sit with your arms outstretched and hold a weight, like a brick or a full water bottle, in either hand. Now use your feet to start your body and the chair rotating, lift your feet off the ground, and then pull your arms inward. Your rotation will speed up quite noticeably.

EXPLANATORY ART helps students visualize concepts

Many **figures** in the text utilize balloon captions to help students interpret complex figures. Select **photos** also contain brief labels that point out the most important relevant features.

NEW! FOLD-OUT PHOTOS
highlight some of astronomy's most majestic sights

Throughout the text are special gatefold **photo spreads**, which take advantage of extra space to highlight awe-inspiring astronomy photography in full detail.

II Planets and Moons

The polar region in the north of Mars holds truly alien landscapes. Wind-blown dunes of dust form the mounds appearing somewhat pink in this color-enhanced image. In the Martian winter, the dunes are covered in carbon-dioxide snow. As summer approaches, the carbon dioxide evaporates, exposing the underlying dust. The black streaks occur on the steepest slopes where the exposed dust forms miniature avalanches. (NASA/JPL/University of Arizona) R I V U X G

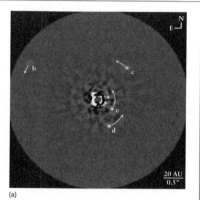

(a)

FIGURE 8-15   R I V U X G
Imaging Extrasolar Planets **(a)** Four planets can be seen orbiting the star HR 8799. The star is about 129 light-years from Earth and can be seen with the naked eye on a clear night. In capturing this image, special techniques make the star less visible to reduce its glare and reveal its planets. The white arrows indicate possible trajectories that each star might take over the next ten years. **(b)** About 170 light-years away, the star 2M1207 and

WAVELENGTH TABS astronomers observe the sky in many forms of light

Astronomers rely on telescopes that are sensitive to nonvisible forms of light. To help students appreciate these different kinds of observation, all of the photos in *Universe* appear with wavelength tabs. The highlighted letter on each tab indicates whether the image was made with **R**adio waves, **I**nfrared radiation, **V**isible light, **U**ltraviolet light, **X**-rays, or **G**amma rays.

TOOLS OF THE ASTRONOMER'S TRADE AND S.T.A.R. a problem-solving rubric

All the worked examples in *Universe* (found in boxes called **Tools of the Astronomer's Trade**) follow a logical and consistent sequence of steps called **S.T.A.R.**: assess the Situation, select the Tools, find the Answer, and Review the answer and explore its significance. Feedback from instructors indicates that when students follow these steps in their own work, they more rapidly develop important problem-solving skills.

BOX 4-4   TOOLS OF THE ASTRONOMER'S TRADE

Newton's Form of Kepler's Third Law

Kepler's original statement of his third law, $P^2 = a^3$, is valid only for objects that orbit the Sun. (Box 4-2 shows how to use this equation.) But Newton's form of Kepler's third law is much more general: It can be used in *any* situation where two objects of masses m_1 and m_2 orbit each other. For example, Newton's form is the equation to use for a moon orbiting a planet or a satellite orbiting Earth. This equation is

Newton's form of Kepler's third law:

$$P^2 = \left[\frac{4\pi^2}{G(m_1 + m_2)} \right] a^3$$

P = sidereal period of orbit, in seconds
a = semimajor axis of orbit, in meters
m_1 = mass of first object, in kilograms
m_2 = mass of second object, in kilograms
G = universal constant of gravitation = 6.67×10^{-11}

Notice that P, a, m_1, and m_2 *must* be expressed in these particular units. If you fail to use the correct units, your answer will be incorrect.

EXAMPLE: Io (pronounced "eye-oh") is one of the four large moons of Jupiter discovered by Galileo and shown in Figure 4-16. It orbits at a distance of 421,600 km from the center of Jupiter and has an orbital period of 1.77 days. Determine the combined mass of Jupiter and Io.

Situation: We are given Io's orbital period P and semimajor axis a (the distance from Io to the center of its circular orbit, which is at the center of Jupiter). Our goal is to find the sum of the masses of Jupiter (m_1) and Io (m_2).

Tools: Because this orbit is not around the Sun, we must use Newton's form of Kepler's third law to relate P and a. This relationship also involves m_1 and m_2, whose sum ($m_1 + m_2$) we are asked to find.

Answer: To solve for $m_1 + m_2$, we rewrite the equation in the form

$$m_1 + m_2 = \frac{4\pi^2 a^3}{GP^2}$$

To use this equation, we have to convert the distance a from kilometers to meters and convert the period P from days to seconds. There are 1000 meters in 1 kilometer and 86,400 seconds in 1 day, so

$$a = (421,600 \text{ km}) \times \frac{1000 \text{ m}}{1 \text{ km}} = 4.216 \times 10^8 \text{ m}$$

$$P = (1.77 \text{ days}) \times \frac{86,400 \text{ s}}{1 \text{ day}} = 1.529 \times 10^5 \text{ s}$$

We can now put these values and the value of G into the above equation:

$$m_1 + m_2 = \frac{4\pi^2 (4.216 \times 10^8)^3}{(6.67 \times 10^{-11})(1.529 \times 10^5)^2} = 1.90 \times 10^{27} \text{ kg}$$

Review: Io is very much smaller than Jupiter, so its mass is only a small fraction of the mass of Jupiter. Thus, $m_1 + m_2$ is very nearly the mass of Jupiter alone. We conclude that Jupiter has a mass of 1.90×10^{27} kg, or about 300 times the mass of Earth. This technique can be used to determine the mass of any object that has a second, much smaller object orbiting it. Astronomers use this technique to find the masses of stars, black holes, and entire galaxies of stars.

BOX 5-3 ASTRONOMY DOWN TO EARTH

Photons at the Supermarket

A beam of light can be regarded as a stream of tiny packets of energy called photons. The Planck relationships $E = hc/\lambda$ and $E = h\nu$ can be used to relate the energy E carried by a photon to its wavelength λ and frequency ν.

As an example, the laser bar-code scanners used at stores and supermarkets emit orange-red light of wavelength 633 nm. To calculate the energy of a single photon of this light, we must first express the wavelength in meters. A nanometer (nm) is equal to 10^{-9} m, so the wavelength is

$$\lambda = (633 \text{ nm}) \left(\frac{10^{-9} \text{ m}}{1 \text{ nm}} \right) = 633 \times 10^{-9} \text{ m} = 6.33 \times 10^{-7} \text{ m}$$

Then, using the Planck formula $E = hc/\lambda$, we find that the energy of a single photon is

$$E = \frac{hc}{\lambda} = \frac{(6.625 \times 10^{-34} \text{ J s})(3 \times 10^8 \text{ m/s})}{6.33 \times 10^{-7} \text{ m}} = 3.14 \times 10^{-19} \text{ J}$$

This amount of energy is very small. The laser in a typical bar-code scanner emits 10^{-3} joule of light energy per second, so the number of photons emitted per second is

$$\frac{10^{-3} \text{ joule per second}}{3.14 \times 10^{-19} \text{ joule per photon}} = 3.2 \times 10^{15} \text{ photons per second}$$

This number is so large that the laser beam seems like a continuous flow of energy rather than a stream of little energy packets.

ASTRONOMY DOWN TO EARTH reveals the universal applicability of the laws of nature

Astronomy Down to Earth boxes illustrate how the same principles astronomers use to explain celestial phenomena can also explain everyday behavior here on Earth, from the color of the sky to why diet soft drink cans float in water.

Up-to-date information shows the cutting edge of astronomy

Universe is up to date with the latest astronomical discoveries, ideas, and images. These include:

- New information on the formation of solar system (Chapter 8)
- *Kepler*'s discoveries of extrasolar planets (Chapter 8)
- The best evidence from *Curiosity* for the long-term flow of water on Mars (Chapter 11)
- Water geysers revealing a large subsurface ocean on Enceladus (Chapter 13)
- Cryovolcanism on Titan (Chapter 13)
- The history-making 2013 asteroid impact in Russia (Chapter 15)
- *Dawn*'s exploration of the asteroid Vesta (Chapter 15)
- New models of supernovae explosions (Chapter 20)
- Gamma-ray bubbles in the Milky Way (Chapter 22)
- Discovery of the Higgs particles (Chapter 26)
- Microtunnels found in a Martian meteorite consistent with microbial origins (Chapter 27)

FIGURE 8-20

Earth-Size Exoplanets Kepler-20 is a star with five known exoplanets. The exoplanets Kepler-20e and Kepler-20f are illustrated in comparison to Earth and Venus. Kepler-20e (at 0.05 AU) has a radius just 0.87 times smaller than Earth, while Kepler-20f (at 0.1 AU) is 1.03 times larger than Earth. However, these objects orbit extremely close to their star and are much too hot for liquid water on their surfaces. (Ames/JPL-Caltech/NASA)

Energy, Stellar Remnant Cores, and Active Galactic Nuclei

Energy is developed as an overarching concept in astronomy and is introduced in **Chapter 4, Gravitation and the Waltz of the Planets.** Our treatment begins with an understanding of kinetic and gravitational potential energy, aided by figures and ConceptCheck questions. Then, energy is considered with gravitational systems and provides an intuitive basis for a variety of orbital properties. In **Chapter 5, The Nature of Light,** we develop the connection between temperature and thermal energy, as well as the energy of light. With these powerful concepts, energy is considered throughout the text wherever it is the most natural description of phenomena.

4-7 Describing orbits with energy and gravity

TUTORIAL 4-3 Imagine a cannonball shot into the air; it arcs through the sky before it crashes back to the ground. If you want to shoot the cannonball into orbit around Earth (like a satellite), you might guess that the ball must be This guess is correct, and next orbits in terms of gravity and en

5-8 Spectral lines are produced when an electron jumps from one energy level to another within an atom

Niels Bohr began his study of the connection between atomic spectra and atomic structure by trying to understand the structure of hydrogen, the simplest and lightest of the elements. (As we dis-

Niels Bohr explained spectral lines with a radical new model of atom

Planetary Themes: Comparative Planetology and Systems Approach

The two chapters on comparative planetology—**Chapter 7, Our Solar System,** and **Chapter 8, The Origin of the Solar System**—have been revised and updated with the latest information. They contain explanations of the official distinctions between planets and dwarf planets, as well as discussions of new discoveries about objects in our own solar system. Mercury, Venus, and Mars are covered in a single chapter on the terrestrial planets, Chapter 11, to emphasize their similarities to Earth and to one another. Figures elucidating comparative planetology appear throughout the planetary coverage. In addition to its comparative planetology theme, *Universe* takes a unique *systems* approach to planetary surfaces and atmospheres: each planet's surface and atmosphere are discussed in tandem, where other texts address them separately. This underscores their interrelationships and builds a deeper understanding of the planets' overall characteristics.

End-of-chapter material provides students with opportunities to review and assess what they have learned and opportunities to extend their knowledge

Key Words

A list of **Key Words** appears at the end of each chapter, along with the number of the page where each term is introduced.

Key Ideas

Students can get the most benefit from these brief chapter summaries by using them in conjunction with the notes they take while reading.

Questions

Items from these sections are designed to be assigned as homework or used as jumping-off points for class discussion. Some questions ask students to analyze images in the text or evaluate how the mass media portray concepts in astronomy. Advanced questions are accompanied by a **Problem-Solving Tips and Tools** box to provide guidance. Web/e-Book questions challenge students to work with animations and interactive modules on the *Universe* Web site or e-Book (described below) or to research topics on the World Wide Web. Near the end of the book is a section of **Answers to Selected Questions.**

Scientific American Articles

Chosen by the textbook authors to illuminate core concepts in the text, these brief, relevant selections demonstrate the process of science and discovery. Revealing recent developments related to specific topics in the text, the articles also provide touchstones for classroom discussion.

Guest Essays

Several chapters in *Universe* end with essays written by scientists involved with some of the recent discoveries described in the text. These include essays by Scott Sheppard on Pluto and the Kuiper belt and Kevin Plaxco on astrobiology. For a full list of essays, see the Contents Overview on pages ix–x.

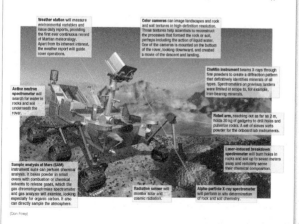

A revised selection of Observing Projects

The download code that comes free (bundled upon request) with this text grants 18 months of access to the *Starry Night*™ planetarium software from Simulation Curriculum Corp. Every chapter of *Universe* includes one or more **Observing Projects** that use this software package. Many projects are accompanied by an **Observing Tips and Tools** box to provide help and guidance.

> **60.** Use the *Starry Night*™ program to examine simulations of various features that appear on the surface of the Sun. Select **Favourites > Explorations > Sun** to show a simulated view of the visible surface of the Sun as it might appear from a spacecraft. **Stop** time flow and use the **Location Scroller** to examine this surface. **(a)** Which layer of the Sun's atmosphere is shown in this part of the simulation? **(b)** List the different features that are visible in this view of the Sun's surface.

Three versions of the text meet the needs of different instructors

In addition to the complete 27-chapter version of *Universe*, two shorter versions are also available. *Universe: The Solar System* includes Chapters 1–16 and 27; it omits the chapters on stars, stellar evolution, galaxies, and cosmology. *Universe: Stars and Galaxies* includes Chapters 1–8 and 16–27; it omits the detailed chapters on the solar system but includes the overview of the solar system in Chapters 7 and 8.

Multimedia

*U*niverse was developed to more deeply integrate learning resources, study aids, and multimedia tools into the use of the textbook. A variety of teaching and learning options provides an array of choices for students and instructors in their use of these materials, which were created based on input and contributions from a large number of faculty.

PREMIUM MULTIMEDIA RESOURCES

Electronic Versions

The *Universe* is offered in two electronic versions. One is an **Interactive e-Book**, available as part of LaunchPad, described below. The other is a PDF-based **e-Book from CourseSmart**, available through www.coursesmart.com. These options are provided to offer students and instructors flexibility in their use of course materials.

CourseSmart e-Book

Universe **CourseSmart e-Book** offers the complete text in an easy-to-use, flexible format. Students can choose to view the CourseSmart e-Book online or to download it to their computer or a mobile device, such as iPad, iPhone, or Android device. To help students study and mirror the experience of a printed textbook, CourseSmart e-Books feature notetaking, highlighting, and bookmark features.

ONLINE LEARNING OPTIONS

Universe supports instructors with a variety of online learning preferences. Its rich array of resources and platforms provides solutions according to each instructor's teaching method. Students can also access the resources through the **Companion Web Site.**

LaunchPad: Because Technology Should Never Get in the Way

At W. H. Freeman, we are committed to providing online instructional materials that meet the needs of instructors and students in powerful, yet simple ways—powerful enough to dramatically enhance teaching and learning, yet simple enough to use right away.

We have taken what we've learned from thousands of instructors and hundreds of thousands of students to create a new generation of technology: **LaunchPad**. LaunchPad offers our acclaimed content curated and organized for easy assignment in a breakthrough user interface in which power and simplicity go hand in hand.

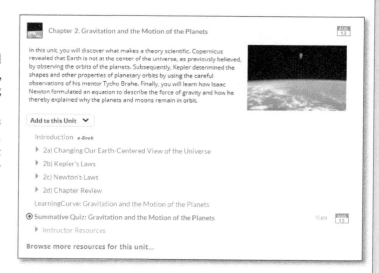

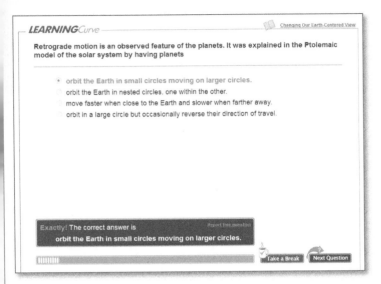

Curated LaunchPad Units make class prep a whole lot easier

Combining a curated collection of video, tutorials, animations, projects, multimedia activities and exercises, and e-Book content, LaunchPad's interactive units give instructors a building block to use as is, or as a starting point for their own learning units. An entire unit's worth of work can be assigned in seconds, drastically saving the amount of time it takes to have the course up and running.

LearningCurve

• Powerful adaptive quizzing, a gamelike format, direct links to the e-Book, instant feedback, and the promise of better grades make using **LearningCurve** a no-brainer.

• Customized quizzing tailored to the text adapts to students' responses and provides material at different difficulty levels and topics based on student performance. Students love the simple yet powerful system and instructors can access class reports to help refine lecture content.

Interactive e-Book

The **Interactive e-Book** is a complete online version of the textbook with easy access to rich multimedia resources that complete student understanding. All text, graphics, tables, boxes, and end-of-chapter resources are included in the e-Book, and the e-Book provides instructors and students with powerful functionality to tailor their course resources to fit their needs.

• Quick, intuitive navigation to any section or subsection

• Full-text search, including the Glossary and Index

• Sticky-note feature allows users to place notes anywhere on the screen, and choose the note color for easy categorization.

• "Top-note" feature allows users to place a prominent note at the top of the page to provide a more significant alert or reminder.

• Text highlighting, in a variety of colors

Astronomy Tutorials

• These self-guided, concept-driven experiential walkthroughs engage students in the process of scientific discovery as they make observations, draw conclusions, and apply their knowledge. Astronomy tutorials combine multimedia resources, activities, and quizzes.

Image Map Activities

• These activities use figures and photographs from the text to assess key ideas, helping students to develop their visual literacy. Students must click the appropriate section(s) of the image and answer corresponding questions.

Animations, Videos, Interactive Exercises, Flashcards

• Other LaunchPad resources highlight key concepts in introductory astronomy.

Assignments for Online Quizzing, Homework, and Self-Study

• Instructors can create and assign automatically graded homework and quizzes from the complete test bank, which is preloaded in LaunchPad. All quiz results feed directly into the instructors's gradebook.

Scientific American Newsfeed

• To demonstrate the continued process of science and the exciting new developments in astronomy, the *Scientific American* Newsfeed delivers regularly updated material from the well-known magazine. Articles, podcasts, news briefs, and videos on subjects related to astronomy are selected for inclusion by *Scientific American*'s editors. The newsfeed provides several updates per week, and instructors can archive or assign the content they find most valuable.

Gradebook

• The included gradebook quickly and easily allows instructors to look up performance metrics for a whole class, for individual students and for individual assignments. Having ready access to this information can help both in lecture prep and in making office hours more productive and efficient.

WebAssign Premium: Trusted Homework Management, Enhanced Learning Tools

www.webassign.net

For instructors interested in online homework management, **WebAssign Premium** features a time-tested, secure, online environment already used by millions of students worldwide. Featuring algorithmic problem-generation and supported by a wealth of astronomy-specific learning tools, WebAssign Premium for *Universe* presents instructors with a powerful assignment manager and study environment. WebAssign Premium provides the following resources:

• **Algorithmically generated problems:** Students receive homework problems containing unique values for computation, encouraging them to work out the problems on their own.

• Innovative **ImageActive** questions, which allow students to interact directly with figures to answer questions. These problems are most often critical thinking questions, which ask students to make connections across concepts and apply their knowledge.

• **Complete access to the interactive e-Book,** from a live table of contents as well as from relevant problem statements.

• Links to Tutorials, e-Book sections, animations, videos, and other interactive tools as hints and feedback to ensure a clearer understanding of the problems and the concepts they reinforce.

Sapling Learning

www.saplinglearning.com

Developed by educators with both online expertise and extensive classroom experience, Sapling Learning provides highly effective interactive homework and instruction that improve student learning outcomes for the problem-solving disciplines. Sapling Learning offers an enjoyable teaching and effective learning experience that is distinctive in three important ways:

• **Ease of use:** Sapling Learning's easy-to-use interface keeps students engaged in problem-solving, not struggling with the software.

• **Targeted instructional content:** Sapling Learning increases student engagement and comprehension by delivering immediate feedback and targeted instructional content.

• **Unsurpassed service and support:** Sapling Learning makes teaching more enjoyable by providing a dedicated Masters- and Ph.D.-level colleague to service instructors' unique needs throughout the course, including content customization.

We offer bundled packages with all versions of our texts that include Sapling Learning Online Homework.

Student Companion Web Site

The *Universe* **Book Companion Web site**, accessed at www.whfreeman.com/universe10e, provides a range of tools for student self-study and review. They include:

• **Online self-study quizzes** with instant feedback referring to specific sections in the text to help students study, review, and prepare for exams. Instructors can access results through an online database or they can have them e-mailed directly to their accounts.

• **Animations** of key figures

• **NASA videos** of important processes and phenomena

• **Vocabulary and concept-review flashcards**

• **Interactive exercises,** based on text illustrations, help students grasp the vocabulary in context.

Starry Night™ Planetarium Software

Starry Night™ is a brilliantly realistic planetarium software package. It is designed for easy use by anyone with an interest in the night sky. See the sky from anywhere on Earth or lift off and visit any solar system body or any location up to 20,000 light years away. View 2,500,000 stars along with more than 170 deep-space objects like galaxies, star clusters, and nebulae. You can travel 15,000 years in time, check out the view from the International Space Station, and see planets up close from any one of their moons. Included are stunning OpenGL graphics. You can also print handy star charts to explore the sky out of doors. This version of *Starry Night*™ also contains student exercises specific to the Freeman version of the software for use with *Universe*. *Starry Night*™ is available via online download by an access code packaged with the text at no extra charge upon instructor request.

Observing Projects Using *Starry Night*™

ISBN 1-4641-2502-3
by Marcel Bergman, T. Alan Clark, and William J. F. Wilson, University of Calgary
Available for packaging with the text and compatible with both PC and Mac, this book contains a variety of comprehensive lab activities for *Starry Night*™ planetarium software.

Test Bank CD-ROM

Windows and Mac versions on one disc, ISBN 1-4641-2496-5
by Thomas Krause, Towson University, T. Alan Clark, and William J. F. Wilson, University of Calgary
More than 3,500 multiple-choice questions are section-referenced. The easy-to-use CD-ROM provides test questions in a format that lets instructors add, edit, resequence, and print questions to suit their needs.

Online Course Materials

Blackboard, Moodle, Sakai, Canvas
As a service for adopters, we will provide content files in the appropriate online course

format, including the instructor and student resources for this text. The files can be used as is or can be customized to fit specific needs. Course outlines, prebuilt quizzes, links, activities, and a whole array of materials are included.

PowerPoint Lecture Presentations

A set of online lecture presentations created in PowerPoint allows instructors to tailor their lectures to suit their own needs using images and notes from the textbook. These presentations are available on the instructor portion of the companion Web site and within the LaunchPad.

Acknowledgments

We would like to thank our colleagues who have carefully scrutinized the manuscript of this edition. *Universe* is a stronger and better textbook because of their conscientious efforts.

Robert Antonucci, *University of California, Santa Barbara*
Nahum Arav, *Virginia Tech University*
Anca Constantin, *James Madison University*
John Cottle, *University of California, Santa Barbara*
Peter Detterline, *Kutztown University*
Tamara Davis, *University of Queensland*
Jacqueline Dunn, *Midwestern State University*
Martin Fisk, *Oregon State University*
Andy Howell, *University of California, Santa Barbara*
William Keel, *University of Alabama*
Charles Kerton, *Iowa State University*
Erik Kubik, *Providence College*
Peter Meinhold, *University of California, Santa Barbara*
Anatoly Miroshnichenko, *University of North Carolina, Greensboro*
Patrick Motl, *Indiana University, Kokomo*
Krishna Mukherjee, *Slippery Rock University*
Fritz Osell, *Northern Oklahoma College*
Stan Peale, *University of California, Santa Barbara*

Nicolas Pereyra, *University of Texas, Pan American*
Ylva Pihlstrom, *University of New Mexico*
Kevin Plaxco, *University of California, Santa Barbara*
Raghavan Rangarajan, *Physical Research Laboratory, Ahmedabad, India*
Sean Raymond, *Laboratoire d'Astrophysique de Bordeaux*
Ian Redmount, *Saint Louis University*
Adam Rengstorf, *Purdue University, Calumet*
Andrew Rivers, *Northwestern University*
Robert Rosner, *University of Chicago*
Alan Rubin, *University of California, Los Angeles*
Paul Schmidtke, *Arizona State University*
Frank Spera, *University of California, Santa Barbara*
Jack Sweeney, *Salem State University*
Nicole Vogt, *New Mexico State University*
Lauren White, *NASA/JPL*
Mark Whittle, *University of Virginia*
Michael E. Wysession, *Washington University, St. Louis*

We would also like to renew our thanks to those who have reviewed past editions of *Universe*:

Ann Bragg, *Marietta College*
Richard Bowman, *Bridgewater College*
Robert Braunstein, *Northern Virginia Community College, Loudoun*
Thomas Campbell, *Allen Community College*
Brian Carter, *Grossmont College*
Stephen Case, *Olivet Nazarene University*
Demian H. J. Cho, *Kenyon College*
Richard A. Christie, *Okanagan College*
Eric Collins, *California State University, Northridge*
George J. Corso, *DePaul University*
Jamie Day, *Transylvania University*
Steven Desch, *Arizona State University*
James Dickinson, *Clackamas Community College*
Jeanne Digel, *Canada College*
Edward Dingler, *Southwest Virginia Community College*
David Duluk, *Los Angeles City College/Glendale Community College*
Jean W. Dupon, *Menlo College*
Stephanie Fawcett, *McKendree University*
Kent Fisher, *Columbus State Community College*
Juhan Frank, *Louisiana State University*
Tony George, *Columbia Basin College*
Erika Gibb, *University of Missouri, St. Louis*
Daniel Greenberger, *City College of New York*
Wayne K. Guinn, *Lon Morris College*
Joshua Gundersen, *University of Miami*

Hunt Guitar, *New Mexico State University*
Kathleen A. Harper, *Denison University*
Lynn Higgs, *University of Utah*
Melinda Hutson, *Portland Community College*
Douglas R. Ingram, *Texas Christian University*
Adam G. Jensen, *University of Colorado, Boulder*
Darell Johnson, *Missouri Western State University*
Michael Joner, *Brigham Young University*
Lauren Jones, *Denison University*
Steve Kawaler, *Iowa State University*
Charles KeRton, *Iowa State University*
Kevin Kimberlin, *Bradley University*
H. S. Krawczynski, *Washington University, St. Louis*
Lauren Likkel, *University of Wisconsin, Eau Claire*
Dennis Machnik, *Plymouth State University*
Franck Marchis, *University of California, Berkeley*
Michele M. Montgomery, *University of Central Florida*
Edward M. Murphy, *University of Virginia*
Donna Naples, *University of Pittsburgh*
Gerald H. Newsom, *The Ohio State University*
Brian Oetiker, *Sam Houston State University*
Ronald P. Olowin, *Saint Mary's College*
Michael J. O'Shea, *Kansas State University*
J. Douglas Patterson, *Johnson County Community College*
Charles Peterson, *University of Missouri*

Richard Rand, *University of New Mexico*
Mike Reynolds, *Florida Community College*
Frederick Ringwald, *California State University, Fresno*
Todd Rimkus, *Marymount University*
Paul Robinson, *Westchester Community College*
Louis Rubbo, *Coastal Carolina University*
Kyla Scarborough, *Pittsburg State University*
Doug Showell, *University of Nebraska, Omaha*

Caroline Simpson, *Florida International University*
Glenn Spiczak, *University of Wisconsin, River Falls*
James D. Stickler, *Allegany College of Maryland*
Larry K. Smith, *Snow College*
Chris Taylor, *California State University, Sacramento*
Charles M. Telesco, *University of Florida*
Dale Trapp, *Concordia University, St. Paul*

Many others have participated in the preparation of this book and we thank them for their efforts. We are grateful for the guidance we have received from Alicia Brady, our acquisitions editor. We're also deeply grateful to our development editor on *Universe*, Brittany Murphy, and for the contributions of assistant editor, Courtney Lyons. Amy Thorne, the media and supplements editor, supervised the creation of the superb multimedia. Special thanks go to Kerry O'Shaughnessy, who skillfully kept all three versions of this book on track and on target. Vicki Tomaselli deserves credit for the design of the book. We also thank Janice Donnola and Dragonfly Media for coordinating and producing the excellent artwork, and extend our gratitude to photo editor Robin Fadool and photo researcher Deborah Anderson for finding just the right photographs. Louise Ketz deserves a medal for wading through the prose and copyediting it into proper English.

On a personal note, Roger Freedman would like to thank his father, Richard Freedman, for first cultivating his interest in space many years ago, and for his father's personal contributions to the exploration of the universe as an engineer for the Atlas and Centaur launch vehicle programs. Most of all, Roger thanks his charming wife, Caroline, for putting up with his long nights slaving over the computer!

Robert Geller would like to extend special thanks to Robert Antonucci (aka Ski). Ski has helped with numerous topics throughout this textbook, was Robert's Ph.D. thesis advisor, and—very conveniently—is his next door neighbor. While some discussions take the usual form of pointing to data plots on the computer, many discussions were literally over the fence. In addition to benefiting from Ski's seemingly unlimited expertise, he has made astronomy FUN from the first day Robert was a Teaching Assistant for his class.

Robert Geller would also like to thank his parents for nurturing a love of science. Although some home appliances that were taken apart in the name of science were never put back together, he always had their support. Most of all, Robert would like to thank his family for their support through years of long hours researching and writing for this textbook. His wife, Susanne, has even contributed scientific results to this book as part of the team that discovered the Higgs particle in 2012; if her next project is successful, she can add dark matter to her list. With the completion of this edition, Robert and his daughter, Zoe, will have some time to use their telescope, fish, and surf together.

Although we have made a concerted effort to make this edition error-free, some mistakes may have crept in unbidden. We would appreciate hearing from anyone who finds an error or wishes to comment on the text.

Roger A. Freedman
Department of Physics
University of California, Santa Barbara
Santa Barbara CA 93106
airboy@physics.ucsb.edu

Robert M. Geller
Department of Physics
University of California, Santa Barbara
Santa Barbara CA 93106
rhmg@physics.ucsb.edu

To the Student

HOW TO GET THE MOST FROM *UNIVERSE*

If you're like most students just opening this textbook, you're enrolled in one of the few science courses you'll take in college. As you study astronomy, you'll probably do relatively little reading compared to a literature or history course—at least in terms of the number of pages. But your readings will be packed with information, much of it new to you and (we hope) exciting. You can't read this textbook like a novel and expect to learn much from it. Don't worry, though. We wrote this book with you in mind. In this section, we'll suggest how Universe can help you succeed in your astronomy course, and take you on a guided tour of the book and media.

Apply these techniques to studying astronomy

• **Read before each lecture** You'll get the most out of your astronomy course if you read each chapter before hearing a lecture about its subject matter. That way, many of the topics will already be clear in your mind, and you'll understand the lecture better. You'll be able to spend more of your listening and note-taking time on the more challenging ideas presented in the lecture.

• **Take notes as you read and make use of office hours** Keep a notebook handy as you read, and write down the key points of each section so that you can review them later. If any parts of the section don't seem clear on first reading, make a note of them, too, including the page numbers. Once you've gone through the chapter, reread it with special emphasis on the ideas that gave you trouble the first time. If you're still unsure after the lecture, consult your instructor, either during office hours or after class. Bring your notes with you so your instructor can see which concepts are giving you trouble. Once your instructor has helped clarify things for you, revise your notes so you'll remember your newfound insights. You'll end up with a chapter summary in your own words. This will be a tremendous help when studying for exams!

• **Make use of your fellow students** Many students find it useful to form study groups for astronomy. You can hash out challenging topics with each other and have a good time while you're doing it. But make sure that you write up your homework by yourself, because the penalties for copying or plagiarizing other students' work can be severe in the extreme. Some students find individual assistance useful, too. If you think a tutor will be helpful, link up with one early. Getting a tutor late in the course, in the belief that you'll be able to catch up with what you missed earlier on, is almost always a lost cause.

• **Take advantage of the Web site and e-Book** Take some time to explore the Universe Web site (www.whfreeman.com/universe10e). There you'll find review materials, animations, videos, interactive exercises, flashcards, and many other features keyed to chapters in *Universe*. All of these features are designed to help you learn and enjoy astronomy, so make sure to take full advantage of them. You can also use the e-Book version of this textbook, which combines the complete content of the book and the Web site in a convenient online format.

• **Try astronomy for yourself with your star charts** At the back of this book you'll find a set of star charts for each month of the year in the northern hemisphere. (For a set of southern hemisphere star charts, see the *Universe* Web site.) Star charts can get you started with your own observations of the universe. Hold the chart overhead in the same orientation as the compass points, with southern horizon toward the south and western horizon toward the west. (To save strain on your arms, you may want to cut these pages out of the book.) Depending on the version of this textbook that your instructor requested, this book may also include access to the easy-to-use *Starry Night*™ planetarium program, which you can use to view the sky on any date and time as seen from anywhere on Earth.

Here's the most important advice of all

We haven't mentioned the most important thing you should do when studying astronomy: Have fun! Of all the different kinds of scientists, astronomers are among the most excited about what they do and what they study. Let some of that excitement about the universe rub off on you, and you'll have a great time with this course and with this textbook.

In preparing this edition of *Universe*, we've tried very hard to make it the kind of textbook that a student like you will find useful. We're very interested in your comments and opinions! Please feel free to e-mail or write to us and we will respond personally.

Best wishes for success in your studies!

Roger A. Freedman
Department of Physics
University of California, Santa Barbara
Santa Barbara CA 93106
airboy@physics.ucsb.edu

Robert M. Geller
Department of Physics
University of California, Santa Barbara
Santa Barbara CA 93106
rhmg@physics.ucsb.edu

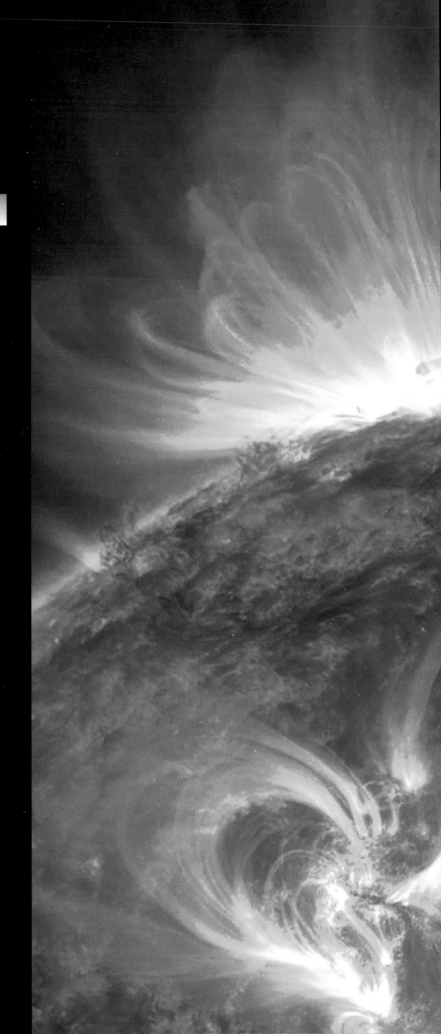

I Introducing Astronomy

Many details are revealed by looking at ultraviolet light emitted by the Sun. The bright arches result from magnetic fields poking out of the solar surface. Electrons accelerated by these magnetic fields then emit ultraviolet light. Part of these features can also be seen in visible light, where they are called solar flares. More than just emitting light, solar flares can hurl mountain-sized quantities of matter from the Sun. Sometimes the solar flares are directed at Earth, where they can damage communication satellites and pose risks to astronauts. (NASA)

R I V U X G

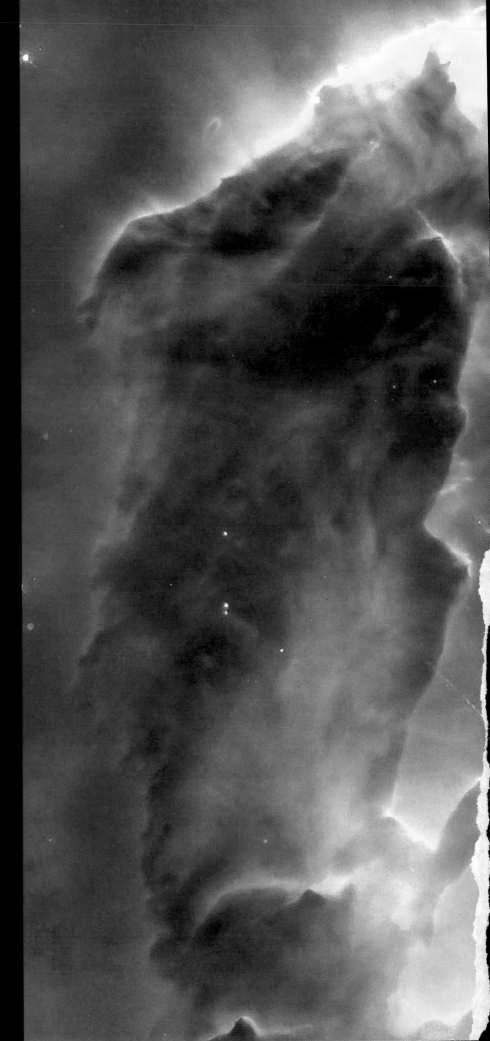

This tall pillar of hydrogen and dust extends a few light-years and lies within the Eagle Nebula about 7000 light-years away. There are several fingerlike structures near the top that contain newly forming solar systems within their tips. This dense column is slowly eroding as ultraviolet light from nearby stars breaks apart its gas and dust. There are several of these majestic star-forming structures next to each other, and they are often called the Pillars of Creation. (NASA)

R I V U X G

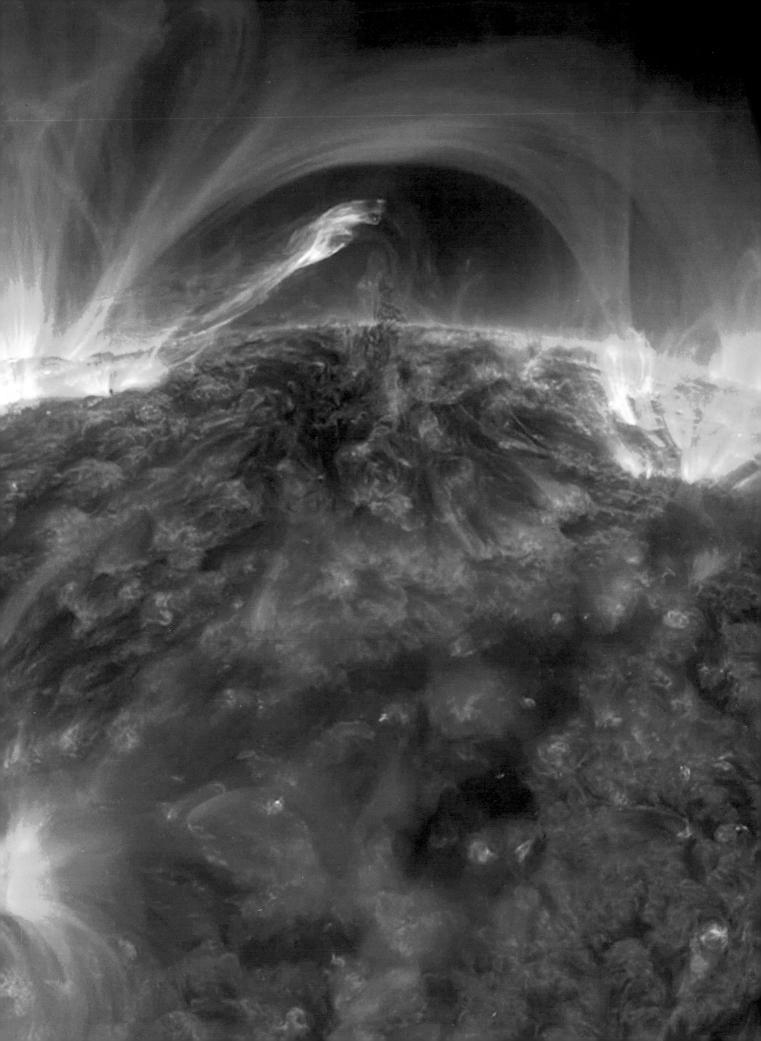

The night sky seen from Mauna Kea in Hawaii. The feature extending across the sky is the Milky Way, consisting of hundreds of billions of stars. The Milky Way is the galaxy in which we reside. (John Hook/Flickr/Getty Images)   R I V U X G

Astronomy and the Universe

LEARNING GOALS

By reading the sections of this chapter, you will learn

1-1 What distinguishes the methods of science from other human activities

1-2 How exploring other planets provides insight into the origins of the solar system and the nature of our Earth

1-3 Stars have a life cycle—they form, evolve over millions or billions of years, and die

1-4 Stars are grouped into galaxies, which are found throughout the universe

1-5 How astronomers measure position and size of a celestial object

1-6 How to express very large or very small numbers in convenient notation

1-7 Why astronomers use different units to measure distances in space

1-8 What astronomy can tell us about our place in the universe

Imagine yourself looking skyward on a clear, dark, moonless night, far from the glare of city lights. As you gaze upward, you see a panorama that no poet's words can truly describe and that no artist's brush could truly capture. Literally thousands of stars are scattered from horizon to horizon, many of them grouped into a luminous band called the Milky Way (which extends up and down across the middle of this photograph). As you watch, the entire spectacle swings slowly overhead from east to west as the night progresses.

For thousands of years people have looked up at the heavens and contemplated the universe. Like our ancestors, we find our thoughts turning to profound questions as we gaze at the stars. How was the universe created? Where did Earth, the Moon, and the Sun come from? What are the planets and stars made of? And how do we fit in? What is our place in the cosmic scope of space and time?

Wondering about the universe is a key part of what makes us human. Our curiosity, our desire to explore and discover, and, most important, our ability to reason about what we have discovered are qualities that distinguish us from other animals. The study of the stars transcends all boundaries of culture, geography, and politics. In a literal sense, astronomy is a universal subject—its subject is the entire universe.

1

1-1 To understand the universe, astronomers use the laws of physics to construct testable theories and models

Astronomy has a rich heritage that dates back to the myths and legends of antiquity. Centuries ago, the heavens were thought to be populated with demons and heroes, gods and goddesses. Astronomical phenomena were explained as the result of supernatural forces and divine intervention.

The course of civilization was greatly affected by a profound realization: *The universe is comprehensible.* For example, ancient Greek astronomers discovered that by observing the heavens and carefully reasoning about what they saw, they could learn something about how the universe operates. As we shall see in Chapter 3, ancient Greek astronomers measured the size of Earth and were able to understand and predict eclipses without appealing to supernatural forces. Modern science is a direct descendant of astronomy, which had many contributors from the Middle East, Africa, Asia, Central America, and, eventually, Greece.

The Scientific Method

Like art, music, or any other human creative activity, science makes use of intuition and experience. But the approach used by scientists to explore physical reality differs from other forms of intellectual endeavor in that it is based fundamentally on *observation, logic,* and *skepticism.* This approach, called the **scientific method,** requires that our ideas about the world around us be consistent with what we actually observe.

The scientific method goes something like this: A scientist trying to understand some observed phenomenon proposes a **hypothesis,** which is a collection of ideas that seems to explain what is observed. It is in developing hypotheses that scientists are at their most creative, imaginative, and intuitive. But their hypotheses must always agree with existing observations and experiments, because a discrepancy with what is observed implies that the hypothesis is wrong. (The exception is if the scientist thinks that the existing results are wrong and can give compelling evidence to show that they are wrong.) The scientist then uses logic to work out the implications of the hypothesis and to make predictions that can be tested. A hypothesis is on firm ground only after it has accurately forecast the results of new experiments or observations. (In practice, scientists typically go through these steps in a less linear fashion than we have described.)

Scientists describe reality in terms of **models,** which are hypotheses that have withstood observational or experimental tests. A model tells us about the properties and behavior of some object or phenomenon. A familiar example is a model of the atom, which scientists picture as electrons orbiting a central nucleus. Another example, which we will encounter in Chapter 18, is a model that tells us about physical conditions (for example, temperature, pressure, and density) in the interior of the Sun (Figure 1-1). A well-developed model uses mathematics—one of the most powerful tools for logical thinking—to make detailed predictions. For example, a successful model of the Sun's interior should describe what the values of temperature, pressure, and density are at each

> Hypotheses, models, theories, and laws are essential parts of the scientific way of knowing

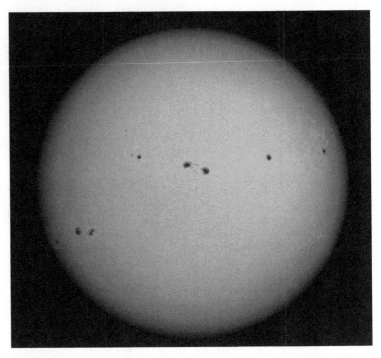

FIGURE 1-1   R I ☑ U X G

Our Star, the Sun The Sun is a typical star. Its diameter is about 1.39 million kilometers (roughly a million miles), and its surface temperature is about 5500°C (10,000°F). A detailed scientific model of the Sun tells us that it draws its energy from nuclear reactions occurring at its center, where the temperature is about 15 million degrees Celsius. (NSO/AURA/NSF)

depth within the Sun, as well as the relations between these quantities. For this reason, mathematics is one of the most important tools used by scientists.

A body of related hypotheses can be pieced together into a self-consistent description of nature called a **theory.** An example from Chapter 4 is the theory that the planets are held in their orbits around the Sun by the Sun's gravitational force (Figure 1-2). Without models and theories there is no understanding and no science, only collections of facts.

CAUTION! In everyday language the word "theory" is often used to mean an idea that looks good on paper, but has little to do with reality. In science, however, a good theory is one that explains reality very well and that can be applied to explain new observations. An excellent example is the theory of gravitation (Chapter 4), which was devised by the English scientist Isaac Newton in the late 1600s to explain the orbits of the six planets known at that time. When astronomers of later centuries discovered the planets Uranus and Neptune and the dwarf planet Pluto, they found that these planets also moved in accordance with Newton's theory. The same theory describes the motions of satellites around Earth as well as the orbits of planets around other stars.

An important part of a scientific theory is its ability to make predictions that can be tested by other scientists. If the predictions are verified by observation that lends support to the theory and

FIGURE 1-2

Planets Orbiting the Sun An example of a scientific theory is the idea that Earth and planets orbit the Sun due to the Sun's gravitational attraction. This theory is universally accepted because it makes predictions that have been tested and confirmed by observation. (The Sun and planets are actually much smaller than this illustration would suggest.) (Detlev Van Ravenswaay/Science Photo Library)

suggests that it might be correct. If the predictions are *not* verified, the theory needs to be modified or completely replaced. For example, an old theory held that the Sun and planets orbit around a stationary Earth. This theory led to certain predictions that could be checked by observation, as we will see in Chapter 4. In the early 1600s the Italian scientist Galileo Galilei used one of the first telescopes to show that these predictions were incorrect. As a result, the theory of a stationary Earth was rejected, eventually to be replaced by the modern model shown in Figure 1-2 in which Earth and other planets orbit the Sun.

An idea that *cannot* be tested by observation or experiment does not qualify as a scientific theory. An example is the idea that there is a little man living in your refrigerator who turns the inside light on or off when you open and close the door. The little man is invisible, weightless, and makes no sound, so you cannot detect his presence. While this is an amusing idea, it cannot be tested and so cannot be considered science.

Skepticism is an essential part of the scientific method. New hypotheses must be able to withstand the close scrutiny of other scientists. The more radical the hypothesis, the more skepticism and critical evaluation it will receive from the scientific community, because the general rule in science is that **extraordinary claims require extraordinary evidence.** That is why scientists as a rule do not accept claims that people have been abducted by aliens and taken aboard UFOs. The evidence presented for these claims is flimsy, secondhand, and unverifiable.

At the same time, scientists must be open-minded. They must be willing to discard long-held ideas if these ideas fail to agree with

new observations and experiments, provided the new data have survived critical review. (If an alien spacecraft really did land on Earth, scientists would be the first to accept that aliens existed—provided they could take a careful look at the spacecraft and its occupants.) That is why the validity of scientific knowledge can be temporary. With new evidence, some ideas only need modification, while occasionally, some ideas may need to be entirely replaced. As you go through this book, you will encounter many instances where new observations have transformed our understanding of Earth, the planets, the Sun and stars, and indeed the very structure of the universe.

Theories that accurately describe the workings of physical reality have a significant effect on civilization. For example, basing his conclusions in part on observations of how the planets orbit the Sun, Isaac Newton deduced a set of fundamental principles that describe how *all* objects move. These theoretical principles, which we will encounter in Chapter 4, work equally well on Earth as they do in the most distant corner of the universe. They represent our first complete, coherent description of how objects move in the physical universe. **Newtonian mechanics** had an immediate practical application in the construction of machines, buildings, and bridges. It is no coincidence that the Industrial Revolution followed hard on the heels of these theoretical and mathematical advances inspired by astronomy.

Newtonian mechanics and other physical theories have stood the test of time and been shown to have great and general validity. Proven theories of this kind are collectively referred to as the **laws of physics.** Thus, the most reliable theories with the broadest applicability can eventually be considered laws of physics. Astronomers use these laws to interpret and understand their observations of the universe. The laws governing light and its relationship to matter are of particular importance, because the only information we can gather about distant stars and galaxies is in the light that we receive from them. Using the physical laws that describe how objects absorb and emit light, astronomers have measured the temperature of the Sun and even learned what the Sun is made of. By analyzing starlight in the same way, they have discovered that our own Sun is a rather ordinary star and that the observable universe may contain 10 billion trillion stars just like the Sun.

Technology in Science

An important part of science is the development of new tools for research and new techniques of observation. As an example, until fairly recently everything we knew about the distant universe was based on visible light. Astronomers would peer through telescopes to observe and analyze visible starlight. By the end of the nineteenth century, however, scientists had begun to discover forms of light invisible to the human eye: X-rays, gamma rays, radio waves, microwaves, and ultraviolet and infrared radiation.

As we will see in Chapter 6, in recent years astronomers have constructed telescopes that can detect such nonvisible forms of light (Figure 1-3). These instruments give us views of the universe vastly different from anything our eyes can see. These new views have allowed us to see through the atmospheres of distant planets, to study the thin but incredibly violent gas that surrounds our Sun,

FIGURE 1-3   R I V U X G

A Telescope in Space Because it orbits outside Earth's atmosphere in the near-vacuum of space, the Hubble Space Telescope (HST) can detect not only visible light but also ultraviolet and near-infrared light coming from distant stars and galaxies. These forms of nonvisible light are absorbed by our atmosphere and hence are difficult or impossible to detect with a telescope on Earth's surface. This photo of HST was taken by the crew of the space shuttle *Columbia* after a servicing mission in 2002. (Courtesy of Scientific American/NASA/AAT)

and even to observe new solar systems being formed around distant stars. Aided by high-technology telescopes, today's astronomers carry on the program of careful observation and logical analysis begun thousands of years ago by their ancient Greek predecessors.

CONCEPTCHECK 1-1

Which is held in higher regard by professional astronomers: a hypothesis or a theory? Explain your answer.

Answer appears at the end of the chapter.

1-2 By exploring the planets, astronomers uncover clues about the formation of the solar system

The science of astronomy allows our intellects to voyage across the cosmos. We can think of three stages in this voyage: from Earth to other parts of the solar system, from the solar system to the stars, and from stars to galaxies and the grand scheme of the universe.

> Studying planetary science gives us a better perspective on our own unique Earth

The star we call the Sun and all the celestial bodies that orbit the Sun—including Earth, the other planets, all their various moons, and smaller bodies such as asteroids and comets—make up the **solar system.** Since the 1960s a series of unmanned spacecraft has been sent to explore each of the planets (Figure 1-4). Using the remote "eyes" of such spacecraft, we have flown over Mercury's cratered surface, peered beneath Venus's poisonous cloud cover, and discovered enormous canyons and extinct volcanoes on Mars. We have found active volcanoes on a moon of Jupiter, probed the atmosphere of Saturn's moon Titan, seen the rings of Uranus up close, and looked down on the active atmosphere of Neptune.

Along with rocks brought back by the *Apollo* astronauts from the Moon (the only world beyond Earth visited by humans), new information from spacecraft has revolutionized our understanding of the origin and evolution of the solar system. We have come to realize that many of the planets and their satellites were shaped by collisions with other objects. Craters on the Moon and on many other worlds are the relics of innumerable impacts by bits of interplanetary rock. The Moon may itself be the result of a catastrophic collision between Earth and a planet-sized object shortly after the solar system was formed. Such a collision could have torn sufficient material from the primordial Earth to create the Moon.

The oldest objects found on Earth are **meteorites,** chemically distinct bits of interplanetary debris that sometimes fall to our planet's surface. By using radioactive age-dating techniques, scientists have found that the oldest meteorites are 4.56 billion years old—older than any other rocks found on Earth or the Moon. The conclusion is that our entire solar system, including the Sun and planets, formed 4.56 billion years ago. The few thousand years of recorded human history is no more than the twinkling of an eye compared to the long history of our solar system.

The discoveries that we have made in our journeys across the solar system are directly relevant to the quality of human life on our own planet. Until recently, our understanding of geology, weather, and climate was based solely on data from Earth. Since

FIGURE 1-4

The Sun and Planets to Scale This montage of images from various spacecraft and ground-based telescopes shows the relative sizes of the planets and the Sun. The Sun is so large compared to the planets that only a portion of it fits into this illustration. The distances from the Sun to each planet are not shown to scale; the actual distance from the Sun to Earth, for instance, is 12,000 times greater than Earth's diameter. (Calvin J. Hamilton and NASA/JPL)

the advent of space exploration, however, we have been able to compare and contrast other worlds with our own. This new knowledge gives us valuable insight into our origins, the nature of our planetary home, and the limits of our natural resources.

1-3 By studying stars and nebulae, astronomers discover how stars are born, grow old, and die

The nearest of all stars to Earth is the Sun. Although humans have used the Sun's warmth since the dawn of our species, it was only in the 1920s and 1930s that physicists figured out how the Sun shines. At the center of the Sun, thermonuclear reactions—so

called because they require extremely high temperatures—convert hydrogen (the Sun's primary constituent) into helium. This violent process releases a vast amount of energy, which eventually makes its way to the Sun's surface and escapes as light (see Figure 1-1). Thus, through nuclear reactions, hydrogen acts as a "fuel" for stars. All the stars you can see in the nighttime sky also shine by nuclear reactions (Figure 1-5). By 1950 physicists could reproduce such nuclear reactions here on Earth in the form of a hydrogen bomb (Figure 1-6). In the future, converting hydrogen into helium might someday offer a cleaner method for the production of nuclear energy.

Because nuclear reactions consume the original material of which stars are made, the way an engine consumes its fuel, stars cannot last forever. Rather, they must form, evolve, and eventually die.

FIGURE 1-5 R I V U X G

Stars like Grains of Sand This Hubble Space Telescope image shows thousands of stars in the constellation Sagittarius. Each star shines because of thermonuclear reactions that release energy in its interior. Different colors indicate stars with different surface temperatures: stars with the hottest surfaces appear blue, while those with the coolest surfaces appear red. (The Hubble Heritage Team, AURA/STScI/NASA)

FIGURE 1-6 R I V U X G

A Thermonuclear Explosion A hydrogen bomb uses the same physical principle as the thermonuclear reactions at the Sun's center: the conversion of matter into energy by nuclear reactions. This thermonuclear detonation on October 31, 1952, had an energy output equivalent to 10.4 million tons of TNT (almost 1000 times greater than the nuclear bomb detonated over Hiroshima in World War II). This is a mere ten-billionth of the amount of energy released by the Sun in one second. (Defense Nuclear Agency)

CAUTION! Astronomers often use biological terms such as "birth" and "death" to describe stages in the evolution of inanimate objects like stars. Keep in mind that such terms are used only as *analogies,* which help us visualize these stages. They are not to be taken literally!

The Life Stories of Stars

The rate at which stars emit energy in the form of light tells us how rapidly they are consuming their nuclear "fuel" (hydrogen), and hence how long they can continue to shine before reaching the end of their life spans. More massive stars have more hydrogen, and thus, more nuclear "fuel," but consume it at such a prodigious rate that they live out their lives in just a few million years. Less massive stars have less material to consume, but their nuclear reactions proceed so slowly that their life spans are measured in billions of years. (Our own star, the Sun, is in early middle age: It is 4.56 billion years old, with a lifetime of 12.5 billion years.)

While no astronomer can watch a single star go through all of its life stages, we have been able to piece together the life stories of stars by observing many different stars at different points in their life cycles. Important pieces of the puzzle have been discovered by studying huge clouds of interstellar gas, called **nebulae** (singular **nebula**), which are found scattered across the sky. Within some nebulae, such as the Orion Nebula shown in Figure 1-7, stars are

FIGURE 1-8   R I **V** U X G

The Crab Nebula—Wreckage of an Exploded Star When a dying star exploded in a supernova, it left behind this elegant funeral shroud of glowing gases blasted violently into space. A thousand years after the explosion these gases are still moving outward at about 1800 kilometers per second (roughly 4 million miles per hour). The Crab Nebula is 6500 light-years from Earth and about 13 light-years across. (NASA, ESA, J. Hester and A. Loll/ Arizona State University)

born from the material of the nebula itself. Other nebulae reveal what happens when nuclear reactions stop and a star dies. Some stars that are far more massive than the Sun end their lives with a spectacular detonation called a **supernova** (plural **supernovae**) that blows the star apart. The Crab Nebula (Figure 1-8) is a striking example of a remnant left behind by a supernova.

> Studying the life cycles of stars is crucial for understanding our own origins

Dying stars can produce some of the strangest objects in the sky. Some dead stars become **pulsars,** which spin rapidly at rates of tens or hundreds of rotations per second. And some stars end their lives as almost inconceivably dense objects called **black holes,** whose gravity is so powerful that nothing—not even light—can escape. Even though a black hole itself emits essentially no radiation, a number of black holes have been discovered beyond our solar system by Earth-orbiting telescopes. This is done by detecting the X-rays emitted by gas falling toward a black hole.

During their death throes, stars return the gas of which they are made to interstellar space. (Figure 1-8 shows these expelled gases expanding away from the site of a supernova explosion.) This gas contains heavy elements—that is, elements heavier than hydrogen and helium—that were created during the star's lifetime by nuclear reactions in its interior. Interstellar space thus becomes enriched with newly manufactured atoms and molecules. The Sun and its planets were formed from interstellar material that was enriched in this way. This means that the atoms of iron and nickel that make up Earth, as well as the carbon in our bodies and the oxygen we

FIGURE 1-7   R I **V** U X G

The Orion Nebula—Birthplace of Stars This beautiful nebula is a stellar "nursery" where stars are formed out of the nebula's gas. Intense ultraviolet light from newborn stars excites the surrounding gas and causes it to glow. Many of the stars embedded in this nebula are less than a million years old, a brief interval in the lifetime of a typical star. The Orion Nebula is some 1500 light-years from Earth and is about 30 light-years across. (NASA, ESA, M. Robberto/STScI/ESA, and the Hubble Space Telescope Orion Treasury Project Team)

FIGURE 1-9   R I $\boxed{V}$ U X G

A Galaxy This spectacular galaxy, called M63, contains about a hundred billion stars. M63 has a diameter of about 60,000 light-years and is located about 35 million light-years from Earth. Along this galaxy's spiral arms you can see a number of glowing clumps. Like the Orion Nebula in our own Milky Way Galaxy (see Figure 1-7), these are sites of active star formation. (2004–2013 R. Jay GaBany, Cosmotography.com)

breathe, were created deep inside ancient stars. By studying stars and their evolution, we are really studying our own origins.

CONCEPTCHECK 1-2

If a star were twice as massive as our Sun, would it shine for a longer or shorter span of time? Why?

Answer appears at the end of the chapter.

1-4 By observing galaxies, astronomers learn about the origin and fate of the universe

Stars are not spread uniformly across the universe but are grouped together in huge assemblages called **galaxies.** Galaxies come in a wide range of shapes and sizes. Our Sun is just one star in a galaxy we call the Milky Way (see the figure that opens this chapter). A typical galaxy, like our Milky Way, contains several hundred billion stars. Some galaxies are much smaller, containing only a few million stars. Others are monstrosities that devour neighboring galaxies in a process called "galactic cannibalism."

Our Milky Way Galaxy has arching spiral arms like those of the galaxy shown in Figure 1-9. These arms are particularly active sites of star formation. In recent years, astronomers have discovered a mysterious object at the center of the Milky Way with a mass millions of times greater than that of our Sun. It now seems certain that this curious object is an enormous black hole.

Some of the most intriguing galaxies appear to be in the throes of violent convulsions and are rapidly expelling matter. The centers of these strange galaxies, which may harbor even more massive black holes, are often powerful sources of X-rays and radio waves.

Even more awesome sources of energy are found still deeper in space. Often located at distances so great that their light takes billions of years to reach Earth, we find the mysterious **quasars.** Although in some ways quasars look like nearby stars (Figure 1-10), they are among the most distant and most luminous objects in the sky. A typical quasar shines with the brilliance of a hundred galaxies. Detailed observations of quasars imply that they draw their energy from material falling into enormous black holes.

Galaxies and the Expanding Universe

The motions of distant galaxies reveal that they are moving away from us and from each other. In other words, the universe is *expanding.* Extrapolating into the past, we learn that the universe must have been born from an incredibly dense state some 13.7 billion years ago. A variety of evidence indicates that at that moment—the beginning of time—the universe began with a cosmic explosion, known as the **Big Bang,** which occurred throughout all space.

> The motions of distant galaxies motivate the ideas of the expanding universe and the Big Bang

Thanks to the combined efforts of astronomers and physicists, we are making steady advances in understanding the nature and history of the universe. This understanding may reveal the origin of some of the most basic properties of physical reality. Studying the most remote galaxies is also helping to answer questions about the

FIGURE 1-10   R I $\boxed{V}$ U X G

A Quasar The two bright starlike objects in this image look almost identical, but they are dramatically different. The object on the left is indeed a star that lies a few hundred light-years from Earth. But the "star" on the right is actually a quasar about 9 billion light-years away. To appear so bright even though they are so distant, quasars like this one must be some of the most luminous objects in the universe. The other objects in this image are galaxies like that in Figure 1-9. (Charles Steidel, California Institute of Technology; and NASA)

ultimate fate of the universe. Such studies suggest that the expansion of the universe will continue forever, and it is actually gaining speed.

The work of unraveling the deepest mysteries of the universe requires specialized tools, including telescopes, spacecraft, and computers. But for many purposes, the most useful device for studying the universe is the human brain itself. Our goal in this book is to help you use *your* brain to share in the excitement of scientific discovery.

In the remainder of this chapter we introduce some of the key concepts and mathematics that we will use in subsequent chapters. Study these carefully, for you will use them over and over again throughout your own study of astronomy.

1-5 Astronomers use angles to denote the positions and apparent sizes of objects in the sky

Whether they study planets, stars, galaxies, or the very origins of the universe, astronomers must know where to point their telescopes. For this reason, an important part of astronomy is keeping track of the positions of objects in the sky. A system for measuring angles is an essential part of this aspect of astronomy (Figure 1-11).

An **angle** measures the opening between two lines that meet at a point. A basic unit to express angles is the **degree**, designated by the symbol °. A full circle is divided into 360°, and a right angle measures 90° (Figure 1-11a). As Figure 1-11b shows, if you draw lines from your eye to each of the two "pointer stars" in the Big Dipper, the angle between these lines—that is,

> Angles are a tool that we will use throughout our study of astronomy

the **angular distance** between these two stars—is about 5°. (In Chapter 2 we will see that these two stars "point" to Polaris, the North Star.) The angular distance between the stars that make up the top and bottom of the Southern Cross, which is visible from south of the equator, is about 6° (Figure 1-11c).

Astronomers also use angles to describe the apparent size of a celestial object—that is, how wide the object appears in the sky. For example, the angle covered by the diameter of the full moon is about ½° (Figure 1-11a). We therefore say that the **angular diameter** (or **angular size**) of the Moon is ½°. Alternatively, astronomers say that the Moon **subtends**, or extends over, an angle of ½°. Ten full moons could fit side by side between the two pointer stars in the Big Dipper.

The average adult human hand held at arm's length provides a means of estimating angles, as **Figure 1-12** shows. For example, a fist covers an angle of about 10°, whereas a fingertip is about 1° wide. You can use various segments of your index finger extended to arm's length to estimate angles a few degrees across.

To talk about smaller angles, we subdivide the degree into 60 **arcminutes** (also called minutes of arc), which is commonly abbreviated as 60 arcmin or 60′. An arcminute is further subdivided into 60 **arcseconds** (or seconds of arc), usually written as 60 arcsec or 60″. Thus,

$$1° = 60 \text{ arcmin} = 60'$$
$$1' = 60 \text{ arcsec} = 60''$$

For example, on January 1, 2007, the planet Saturn had an angular diameter of 19.6 arcsec as viewed from Earth. That is a convenient, precise statement of how big the planet appeared in Earth's sky on that date. (Because this angular diameter is so small, to the naked eye Saturn appears simply as a point of light. To see any detail on Saturn, such as the planet's rings, requires a telescope.)

If we know the angular size of an object as well as the distance to that object, we can determine the actual linear size of the object

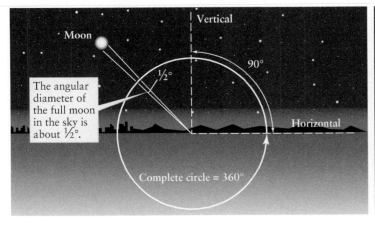

(a) Measuring angles in the sky

(b) Angular distances in the northern hemisphere

(c) Angular distances in the southern hemisphere

FIGURE 1-11

Measuring Angles **(a)** Angles are measured in degrees (°). There are 360° in a complete circle and 90° in a right angle. For example, the angle between the vertical direction (directly above you) and the horizontal direction (toward the horizon) is 90°. The angular diameter of the full moon in the sky is about ½°. **(b)** The seven bright stars that make up the Big Dipper can be seen from anywhere in the northern hemisphere. The angular distance between the two "pointer stars" at the front of the Big Dipper is about 5°. **(c)** The four bright stars that make up the Southern Cross can be seen from anywhere in the southern hemisphere. The angular distance between the stars at the top and bottom of the cross is about 6°.

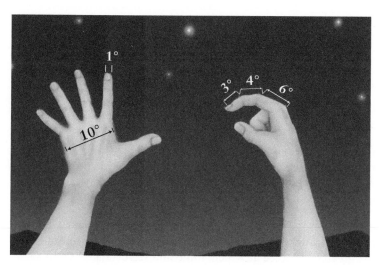

FIGURE 1-12

Estimating Angles with Your Hand The adult human hand extended to arm's length can be used to estimate angular distances and angular sizes in the sky.

(measured in kilometers or miles, for example). Box 1-1 describes how this is done.

CONCEPTCHECK **1-3**

Is it possible for a basketball to look bigger than the Moon? Smaller?

Answer appears at the end of the chapter.

1-6 Powers-of-ten notation is a useful shorthand system for writing numbers

Astronomy is a subject of extremes. Astronomers investigate the largest structures in the universe, including galaxies and clusters of galaxies. But they must also study atoms and atomic nuclei, among the smallest objects in the universe, in order to explain how and why stars shine. They also study conditions

> Learning powers-of-ten notation will help you deal with very large and very small numbers

BOX 1-1 TOOLS OF THE ASTRONOMER'S TRADE

The Small-Angle Formula

You can estimate the angular sizes of objects in the sky with your hand and fingers (see Figure 1-12). Using rather more sophisticated equipment, astronomers can measure angular sizes to a fraction of an arcsecond. Keep in mind, however, that *angular* size is not the same as *actual* size. As an example, if you extend your arm while looking at a full moon, you can completely cover the Moon with your thumb. That's because from your perspective, your thumb has a larger angular size (that is, it subtends a larger angle) than the Moon. But the actual size of your thumb (about 2 centimeters) is much less than the actual diameter of the Moon (more than 3000 kilometers).

The accompanying figure shows how the angular size of an object is related to its linear size. Part (**a**) of the figure shows that for a given angular size, the more distant the object, the larger its actual size. For example, your fingertip held at arm's length covers the full moon, but the Moon is much farther away and is far larger in linear size. Part (**b**) shows that for a given linear size, the angular size decreases the farther away the object. This is why a car looks smaller and smaller as it drives away from you.

We can put these relationships together into a single mathematical expression called the **small-angle formula**. Suppose that an object subtends an angle α (the Greek letter alpha) and is at a distance *d* from the observer, as in part (**c**) of the figure. If the angle α is small, as is almost always the case for objects in the sky, the width or linear size (*D*) of the object is given by the following expression:

The small-angle formula

$$D = \frac{\alpha d}{206{,}265}$$

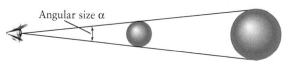

(a) For a given angular size α, the more distant the object, the greater its actual (linear) size

(b) For a given linear size, the more distant the object, the smaller its angular size

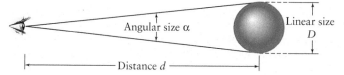

(a) Two objects that have the same angular size may have different linear sizes if they are at different distances from the observer. (b) For an object of a given linear size, the angular size is smaller the farther the object is from the observer. (c) The small-angle formula relates the linear size *D* of an object to its angular size α and its distance *d* from the observer.

D = width or linear size of an object

α = angular size of the object, in arcsec

d = distance to the object

The number 206,265 is required in the formula so that the units of α are in arcseconds. (206,265 is the number of

(continued on the next page)

BOX 1-1 *(continued)*

arcseconds in a complete 360° circle divided by the number 2π.) As long as the same units are used for D and d, any units for linear distance can be used (km, light-years, etc.).

The following examples show two different ways to use the small-angle formula. In both examples we follow a four-step process: Evaluate the *situation* given in the example, decide which *tools* are needed to solve the problem, use those tools to find the *answer* to the problem, and *review* the result to see what it tells you. Throughout this book, we'll use these same four steps in *all* examples that require the use of formulas. We encourage you to follow this four-step process when solving problems for homework or exams. You can remember these steps by their acronym: *S.T.A.R.*

EXAMPLE: On December 11, 2006, Jupiter was 944 million kilometers from Earth and had an angular diameter of 31.2 arcsec. From this information, calculate the actual diameter of Jupiter in kilometers.

Situation: The astronomical object in this example is Jupiter, and we are given its distance d and its angular size α (the same as angular diameter). Our goal is to find Jupiter's diameter D.

Tools: The equation to use is the small-angle formula, which relates the quantities d, α, and D. Note that when using this formula, the angular size α must be expressed in arcseconds.

Answer: The small-angle formula as given is an equation for D. Plugging in the given values $\alpha = 31.2$ arcsec and $d = 944$ million kilometers,

$$D = \frac{31.2 \times 944{,}000{,}000 \text{ km}}{206{,}265} = 143{,}000 \text{ km}$$

Because the distance d to Jupiter is given in kilometers, the diameter D is also in kilometers.

Review: Does our answer make sense? From Appendix 2 at the back of this book, the equatorial diameter of Jupiter measured by spacecraft flybys is 142,984 kilometers, so our calculated answer is very close.

EXAMPLE: Under excellent conditions, a telescope on Earth can see details with an angular size as small as 1 arcsec. What is the greatest distance at which you could see details as small as 1.7 meters (the height of a typical person) under these conditions?

Situation: Now the object in question is a person, whose linear size D we are given. Our goal is to find the distance d at which the person has an angular size α equal to 1 arcsec.

Tools: Again we use the small-angle formula to relate d, α, and D.

Answer: We first rewrite the formula to solve for the distance d, then plug in the given values $D = 1.7$ m and $\alpha = 1$ arcsec:

$$d = \frac{206{,}265D}{\alpha} = \frac{206{,}265 \times 1.7 \text{ m}}{1} = 350{,}000 \text{ m} = 350 \text{ km}$$

Review: This value is much less than the distance to the Moon, which is 384,000 kilometers. Thus, even the best telescope on Earth could not be used to see an astronaut walking on the surface of the Moon.

---

ranging from the incredibly hot and dense centers of stars to the frigid near-vacuum of interstellar space. To describe such a wide range of phenomena, we need an equally wide range of both large and small numbers.

Powers-of-Ten Notation: Large Numbers

Astronomers avoid such confusing terms as "a million billion billion" by using a standard shorthand system called **powers-of-ten notation**. All the cumbersome zeros that accompany a large number are consolidated into one term consisting of 10 followed by an **exponent**, which is written as a superscript. The exponent indicates how many zeros you would need to write out the long form of the number. Thus,

$$10^0 = 1 \text{ (one)}$$
$$10^1 = 10 \text{ (ten)}$$
$$10^2 = 100 \text{ (one hundred)}$$
$$10^3 = 1000 \text{ (one thousand)}$$
$$10^4 = 10{,}000 \text{ (ten thousand)}$$
$$10^6 = 1{,}000{,}000 \text{ (one million)}$$

$$10^9 = 1{,}000{,}000{,}000 \text{ (one billion)}$$
$$10^{12} = 1{,}000{,}000{,}000{,}000 \text{ (one trillion)}$$

and so forth. The exponent also tells you how many tens must be multiplied together to give the desired number, which is why the exponent is also called the **power of ten.** For example, ten thousand can be written as 10^4 ("ten to the fourth" or "ten to the fourth power") because $10^4 = 10 \times 10 \times 10 \times 10 = 10{,}000$.

In powers-of-ten notation, numbers are written as a figure between 1 and 10 multiplied by the appropriate power of ten. The approximate distance between Earth and the Sun, for example, can be written as 1.5×10^8 kilometers (or 1.5×10^8 km, for short). Once you get used to it, this is more convenient than writing "150,000,000 kilometers" or "one hundred and fifty million kilometers." (The same number could also be written as 15×10^7 or 0.15×10^9, but the preferred form is *always* to have the first figure be between 1 and 10.)

Calculators and Powers-of-Ten Notation

Most electronic calculators use a shorthand for powers-of-ten notation. To enter the number 1.5×10^8, you first enter 1.5, then press a key labeled "EXP" or "EE," then enter the exponent 8. (The EXP or EE key takes care of the "× 10" part of the expression.) The number will then appear on your calculator's display as "1.5 E 8," "1.5 8," or some variation of this; typically the "× 10" is not displayed as such. There are some variations from one kind of calculator to another, so you should spend a few minutes reading over your calculator's instruction manual to make sure you know the correct procedure for working with numbers in powers-of-ten notation. You will be using this notation continually in your study of astronomy, so this is time well spent.

CAUTION! Confusion can result from the way that calculators display powers-of-ten notation. Since 1.5×10^8 is displayed as "1.5 8" or "1.5 E 8," it is not uncommon to think that 1.5×10^8 is the same as 1.5^8. That is not correct, however; 1.5^8 is equal to 1.5 multiplied by itself 8 times, or 25.63, which is not even close to $150,000,000 = 1.5 \times 10^8$. Another, not uncommon, mistake is to write 1.5×10^8 as 15^8. If you are inclined to do this, perhaps you are thinking that you can multiply 1.5 by 10, then tack on the exponent later. Another process that does not work: 15^8 is equal to 15 multiplied by itself 8 times, or 2,562,890,625, which again is nowhere near 1.5×10^8. Reading over the manual for your calculator will help you to avoid these common errors.

Powers-of-Ten Notation: Small Numbers

You can use powers-of-ten notation for numbers that are less than one by using a minus sign in front of the exponent. A negative exponent tells you to *divide* by the appropriate number of tens. For example, 10^{-2} ("ten to the minus two") means to divide by 10 twice, so $10^{-2} = 1/10 \times 1/10 = 1/100 = 0.01$. This same idea tells us how to interpret other negative powers of ten:

$10^0 = 1$ (one)

$10^{-1} = 1/10 = 0.1$ (one tenth)

$10^{-2} = 1/10 \times 1/10 = 1/10^2 = 0.01$ (one hundredth)

$10^{-3} = 1/10 \times 1/10 \times 1/10 = 1/10^3 = 0.001$ (one thousandth)

$10^{-4} = 1/10 \times 1/10 \times 1/10 \times 1/10 = 1/10^4 = 0.0001$ (one ten-thousandth)

$10^{-6} = 1/10 \times 1/10 \times 1/10 \times 1/10 \times 1/10 \times 1/10 = 1/10^6 = 0.000001$ (one millionth)

$10^{-12} = 1/10 \times 1/10 \times 1/10 \times 1/10 \times 1/10 \times 1/10 \times 1/10 \times 1/10 \times 1/10 \times 1/10 \times 1/10 \times 1/10 = 1/10^{12} = 0.000000000001$ (one trillionth)

and so forth.

As these examples show, negative exponents tell you how many tenths must be multiplied together to give the desired number. For example, one ten-thousandth, or 0.0001, can be written as 10^{-4} ("ten to the minus four") because $10^{-4} = 1/10 \times 1/10 \times 1/10 \times 1/10 = 0.0001$.

A useful shortcut in converting a decimal to powers-of-ten notation is to notice where the decimal point is. For example, the decimal point in 0.0001 is four places to the left of the "1," so the exponent is –4, that is, $0.0001 = 10^{-4}$.

You can also use powers-of-ten notation to express a number like 0.00245, which is not a multiple of 1/10. For example, $0.00245 = 2.45 \times 0.001 = 2.45 \times 10^{-3}$. (Again, the standard for powers-of-ten notation is that the first figure is a number between 1 and 10.) This notation is particularly useful when dealing with very small numbers. A good example is the diameter of a hydrogen atom, which is much more convenient to state in powers-of-ten notation (1.1×10^{-10} meter, or 1.1×10^{-10} m) than as a decimal (0.00000000011 m) or a fraction (110 trillionths of a meter.)

Because it bypasses all the awkward zeros, powers-of-ten notation is ideal for describing the size of objects as small as atoms or as big as galaxies (Figure 1-13). Box 1-2 explains how powers-of-ten notation also makes it easy to multiply and divide numbers that are very large or very small.

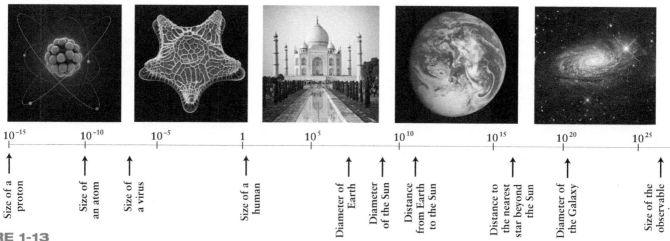

10^{-15} 10^{-10} 10^{-5} 1 10^5 10^{10} 10^{15} 10^{20} 10^{25}

Size of a proton | Size of an atom | Size of a virus | Size of a human | Diameter of Earth | Diameter of the Sun | Distance from Earth to the Sun | Distance to the nearest star beyond the Sun | Diameter of the Galaxy | Size of the observable

FIGURE 1-13

Examples of Powers-of-Ten Notation This scale gives the sizes of objects in meters, ranging from subatomic particles at the left to the entire observable universe at the right. The protons and neutrons in the illustration on the left form the nucleus of an atom; the electrons orbit around a hundred thousand times farther out. Second from left is the cell wall of a single-celled aquatic organism called a diatom, 10^{-4} meter (0.1 millimeter) in size. At the center is the Taj Mahal, about 60 meters tall and within reach of our unaided senses. On the right is the planet Earth, about 10^7 meters in diameter. At the far right is a galaxy, 10^{21} meters (100,000 light-years) in diameter. (Andrzej Wojcicki/Science Photo Library/Getty Images; Steve Gschmeissner/Science Photo Library/Getty; iStockphoto/Thinkstock; NASA/JPL; 2004–2013 R. Jay GaBany, Cosmography.com)

BOX 1-2 TOOLS OF THE ASTRONOMER'S TRADE

Arithmetic with Powers-of-Ten Notation

Using powers-of-ten notation makes it easy to multiply numbers. For example, suppose you want to multiply 100 by 1000. If you use ordinary notation, you have to write a lot of zeros:

$$100 \times 1000 = 100,000 \text{ (one hundred thousand)}$$

By converting these numbers to powers-of-ten notation, we can write this same multiplication more compactly as

$$10^2 \times 10^3 = 10^5$$

Because $2 + 3 = 5$, we are led to the following general rule for *multiplying* numbers expressed in terms of powers of ten: Simply *add* the exponents.

EXAMPLE: $10^4 \times 10^3 = 10^{4+3} = 10^7$.

To *divide* numbers expressed in terms of powers of ten, remember that $10^{-1} = 1/10$, $10^{-2} = 1/100$, and so on. The general rule for any exponent n is

$$10^{-n} = \frac{1}{10^n}$$

In other words, dividing by 10^n is the same as multiplying by 10^{-n}. To carry out a division, you first transform it into multiplication by changing the sign of the exponent, and then carry out the multiplication by adding the exponents.

EXAMPLE: $\dfrac{10^4}{10^6} = 10^4 \times 10^{-6} = 10^{4+(-6)} = 10^{4-6} = 10^{-2}$

Usually a computation involves numbers like 3.0×10^{10}, that is, an ordinary number multiplied by a factor of 10 with an exponent. In such cases, to perform multiplication or division, you can treat the numbers separately from the factors of 10^n.

EXAMPLE: We can redo the first numerical example from Box 1-1 in a straightforward manner by using exponents:

$$
\begin{aligned}
D &= \frac{31.2 \times 944,000,000 \text{ km}}{206,265} \\[2mm]
&= \frac{3.12 \times 10 \times 9.44 \times 10^8}{2.06265 \times 10^5} \text{ km} \\[2mm]
&= \frac{3.12 \times 9.44 \times 10^{1+8-5}}{2.06265} \text{ km} = 14.3 \times 10^4 \text{ km} \\[2mm]
&= 1.43 \times 10 \times 10^4 \text{ km} = 1.43 \times 10^5 \text{ km}
\end{aligned}
$$

1-7 Astronomical distances are often measured in astronomical units, light-years, or parsecs

Astronomers use many of the same units of measurement as do other scientists. They often measure lengths in meters (abbreviated m), masses in kilograms (kg), and time in seconds (s). (You can read more about these units of measurement, as well as techniques for converting between different sets of units, in Box 1-3.)

> Specialized units make it easier to comprehend immense cosmic distances

Like other scientists, astronomers often find it useful to combine these units with powers of ten and create new units using prefixes. As an example, the number 1000 ($= 10^3$) is represented by the prefix "kilo," and so a distance of 1000 meters is the same as 1 kilometer (1 km). Here are some of the most common prefixes, with examples of how they are used:

one-billionth meter	$= 10^{-9}$ m	$= 1$ nanometer
one-millionth second	$= 10^{-6}$ s	$= 1$ microsecond
one-thousandth arcsecond	$= 10^{-3}$ arcsec	$= 1$ milliarcsecond
one-hundredth meter	$= 10^{-2}$ m	$= 1$ centimeter
one thousand meters	$= 10^3$ m	$= 1$ kilometer
one million tons	$= 10^6$ tons	$= 1$ megaton

In principle, we could express all sizes and distances in astronomy using units based on the meter. Indeed, we will use kilometers to give the diameters of Earth and the Moon, as well as the Earth-Moon distance. But, while a kilometer (roughly equal to three-fifths of a mile) is an easy distance for humans to visualize, a megameter (10^6 m) is not. For this reason, astronomers have devised units of measurement that are more appropriate for the tremendous distances between the planets and the far greater distances between the stars.

When discussing distances across the solar system, astronomers use a unit of length called the **astronomical unit** (abbreviated **AU**). This is the average distance between Earth and the Sun:

$$1 \text{ AU} = 1.496 \times 10^8 \text{ km} = 92.96 \text{ million miles}$$

Thus, the average distance between the Sun and Jupiter can be conveniently stated as 5.2 AU.

To talk about distances to the stars, astronomers use two different units of length. The **light-year** (abbreviated **ly**) is the distance that light travels in one year. This is a useful concept because the speed of light in empty space always has the same value, 3.00×10^5 km/s (kilometers per second) or 1.86×10^5 mi/s (miles per second). In terms of kilometers or astronomical units, one light-year is given by

$$1 \text{ ly} = 9.46 \times 10^{12} \text{ km} = 63,240 \text{ AU}$$

This distance is roughly equal to 6 trillion miles.

FIGURE 1-14

A Parsec The parsec, a unit of length commonly used by astronomers, is equal to 3.26 light-years. The parsec is defined as the distance at which 1 AU perpendicular to the observer's line of sight subtends an angle of 1 arcsec.

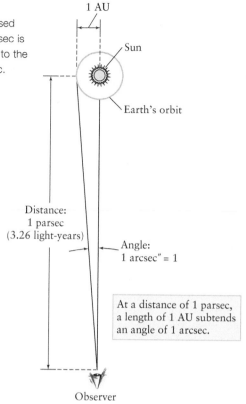

CAUTION! Keep in mind that despite its name, the light-year is a unit of distance and *not* a unit of time. As an example, Proxima Centauri, the nearest star other than the Sun, is a distance of 4.2 light-years from Earth. This means that light takes 4.2 years to travel to us from Proxima Centauri.

Physicists often measure interstellar distances in light-years because the speed of light is one of nature's most important numbers. But many astronomers prefer to use another unit of length, the *parsec*, because its definition is closely related to a method of measuring distances to the stars.

Imagine taking a journey far into space, beyond the orbits of the outer planets. As you look back toward the Sun, Earth's orbit subtends, or extends over, a smaller angle in the sky the farther you are from the Sun. As Figure 1-14 shows, the distance at which 1 AU subtends an angle of 1 arcsec is defined as 1 **parsec** (abbreviated **pc**):

$$1 \text{ pc} = 3.09 \times 10^{13} \text{ km} = 3.26 \text{ ly}$$

BOX 1-3 TOOLS OF THE ASTRONOMER'S TRADE

Units of Length, Time, and Mass

To understand and appreciate the universe, we need to describe phenomena not only on the large scales of galaxies but also on the submicroscopic scale of the atom. Astronomers generally use units that are best suited to the topic at hand. For example, interstellar distances are conveniently expressed in either light-years or parsecs, whereas the diameters of the planets are more comfortably presented in kilometers.

Most scientists prefer to use a version of the metric system called the International System of Units, abbreviated SI (after the French name Système International). In **SI units**, length is measured in meters (m), time is measured in seconds (s), and mass (a measure of the amount of material in an object) is measured in kilograms (kg). How are these basic units related to other measures?

When discussing objects on a human scale, sizes and distances can also be expressed in millimeters (mm), centimeters (cm), and kilometers (km). These units of length are related to the meter as follows:

$$1 \text{ millimeter} = 0.001 \text{ m} = 10^{-3} \text{ m}$$
$$1 \text{ centimeter} = 0.01 \text{ m} = 10^{-2} \text{ m}$$
$$1 \text{ kilometer} = 1000 \text{ m} = 10^{3} \text{ m}$$

A useful set of conversions for the English system is

$$1 \text{ in} = 2.54 \text{ cm}$$
$$1 \text{ ft} = 0.3048 \text{ m}$$
$$1 \text{ mi} = 1.609 \text{ km}$$

Each of these equalities can also be written as a fraction equal to 1. For example, you can write

$$\frac{0.3048 \text{ m}}{1 \text{ ft}} = 1 \quad \text{or} \quad \frac{1 \text{ ft}}{0.3048 \text{ m}} = 1$$

Fractions like this are useful for converting a quantity from one set of units to another. For example, the *Saturn V* rocket used to send astronauts to the Moon stands about 363 ft tall. How can we convert this height to meters? The trick is to remember that a quantity does not change if you multiply it by 1. Expressing the number 1 by the fraction (0.3048 m)/(1 ft), we can write the height of the rocket as

$$363 \text{ ft} \times 1 = 363 \text{ ft} \times \frac{0.3048 \text{ m}}{1 \text{ ft}} = 111 \frac{\text{ft} \times \text{m}}{\text{ft}} = 111 \text{ m}$$

(continued on the next page)

BOX 1-3 *(continued)*

EXAMPLE: The diameter of Mars is 6794 km. Let's try expressing this in miles.

CAUTION! You can get into trouble if you are careless in applying the trick of taking the number whose units are to be converted and multiplying it by 1. For example, if we multiply the diameter by 1 expressed as (1.609 km)/(1 mi), we get

$$6794 \text{ km} \times 1 = 6794 \text{ km} \times \frac{1.609 \text{ km}}{1 \text{ mi}} = 10{,}930 \frac{\text{km}^2}{\text{mi}}$$

The unwanted units of km did not cancel, so this answer cannot be right. Furthermore, a mile is larger than a kilometer, so the diameter expressed in miles should be a smaller number than when expressed in kilometers.

The correct approach is to write the number 1 so that the unwanted units *will* cancel. The number we are starting with is in kilometers, so we must write the number 1 with kilometers in the denominator ("downstairs" in the fraction). Thus, we express 1 as (1 mi)/(1.609 km):

$$6794 \text{ km} \times 1 = 6794 \text{ km} \times \frac{1 \text{ mi}}{1.609 \text{ km}}$$

$$= 4222 \text{ km} \times \frac{\text{mi}}{\text{km}} = 4222 \text{ mi}$$

Now the units of km cancel as they should, and the distance in miles is a smaller number than in kilometers (as it must be).

When discussing very small distances such as the size of an atom, astronomers often use the micrometer (μm) or the nanometer (nm). These are related to the meter as follows:

$$1 \text{ micrometer} = 1 \text{ μm} = 10^{-6} \text{ m}$$

$$1 \text{ nanometer} = 1 \text{ nm} = 10^{-9} \text{ m}$$

Thus, $1 \text{ μm} = 10^3 \text{ nm}$. (Note that the micrometer is often called the micron.)

The basic unit of time is the second (s). It is related to other units of time as follows:

$$1 \text{ minute (min)} = 60 \text{ s}$$

$$1 \text{ hour (h)} = 3600 \text{ s}$$

$$1 \text{ day (d)} = 86{,}400 \text{ s}$$

$$1 \text{ year (y)} = 3.156 \times 10^7 \text{ s}$$

In the SI system, speed is properly measured in meters per second (m/s). Quite commonly, however, speed is also expressed in km/s and mi/h:

$$1 \text{ km/s} = 10^3 \text{ m/s}$$

$$1 \text{ km/s} = 2237 \text{ mi/h}$$

$$1 \text{ mi/h} = 0.447 \text{ m/s}$$

$$1 \text{ mi/h} = 1.47 \text{ ft/s}$$

In addition to using kilograms, astronomers sometimes express mass in grams (g) and in solar masses ($M_\odot$), where the subscript $\odot$ is the symbol denoting the Sun. It is especially convenient to use solar masses when discussing the masses of stars and galaxies. These units are related to each other as follows:

$$1 \text{ kg} = 1000 \text{ g}$$

$$1 \text{ M}_\odot = 1.99 \times 10^{30} \text{ kg}$$

CAUTION! You may be wondering why we have not given a conversion between kilograms and pounds (lb). The reason is that these units do not refer to the same physical quantity! A kilogram is a unit of *mass,* which is a measure of the amount of material in an object. By contrast, a pound is a unit of *weight,* which tells you how strongly gravity pulls on that object's material. Consider a person who weighs 110 pounds on Earth, corresponding to a mass of 50 kg. Gravity is only about one-sixth as strong on the Moon as it is on Earth, so on the Moon this person would weigh only one-sixth of 110 pounds, or about 18 pounds. But that person's mass of 50 kg is the same on the Moon; wherever you go in the universe, you take all of your material along with you. We will explore the relationship between mass and weight in Chapter 4.

The distance from Earth to Proxima Centauri can be stated as 1.3 pc or as 4.2 ly. Whether you choose to use parsecs or light-years is a matter of personal taste.

For even greater distances, astronomers commonly use **kiloparsecs** and **megaparsecs** (abbreviated **kpc** and **Mpc**). As we saw before, these prefixes simply mean "thousand" and "million," respectively:

$$1 \text{ kiloparsec} = 1 \text{ kpc} = 1000 \text{ pc} = 10^3 \text{ pc}$$

$$1 \text{ megaparsec} = 1 \text{ Mpc} = 1{,}000{,}000 \text{ pc} = 10^6 \text{ pc}$$

For example, the distance from Earth to the center of our Milky Way Galaxy is about 8 kpc, and the galaxy shown in Figure 1-9 is about 11 Mpc away.

Some astronomers prefer to talk about thousands or millions of light-years rather than kiloparsecs and megaparsecs. Once again, the choice is a matter of personal taste. As a general rule, astronomers use whatever measuring sticks seem best suited for the issue at hand and do not restrict themselves to one system of measurement. For example, an astronomer might say that the supergiant star Antares has a diameter of 860 million kilometers and is located at a distance of 185 parsecs from Earth. The *Cosmic Connections* figure on the following page shows where these different systems are useful.

1-8 Astronomy is an adventure of the human mind

An underlying theme of this book is that the behavior of the universe is governed by underlying laws of nature. It is not a hodgepodge of unrelated things behaving in unpredictable ways. Rather, we find strong evidence that fundamental laws of physics govern the nature of the universe and the behavior of everything in it. These unifying concepts enable us to explore realms far removed from our earthly experience. Thus, a scientist can do experiments in a laboratory to determine the properties of light or the behavior of atoms and then use this knowledge to investigate the structure of the universe. Through careful testing, the laws of physics discovered on Earth have been found to apply throughout the universe at the most distant locations and in the distant past. Indeed, Albert Einstein expressed the view that **"the eternal mystery of the world is its comprehensibility."**

> Studying the universe benefits our lives on Earth

The discovery of fundamental laws of nature has had a profound influence on humanity. These laws have led to an immense number of practical applications that have fundamentally transformed commerce, medicine, entertainment, transportation, and other aspects of our lives. In particular, space technology has given us instant contact with any point on the globe through communication satellites, precise navigation to any point on Earth using signals from the satellites of the Global Positioning System (GPS), and accurate weather forecasts from meteorological satellites (Figure 1-15).

As important as the applications of science are, the pursuit of scientific knowledge for its own sake is no less important. We are fortunate to live in an age in which this pursuit is in full flower. Just as explorers such as Columbus and Magellan discovered the true size of our planet in the fifteenth and sixteenth centuries, astronomers of the twenty-first century are exploring the universe to an extent that is unparalleled in human history. Indeed, even the voyages into space imagined by such great science fiction writers as Jules Verne and H. G. Wells pale in comparison to today's reality. Over a few short decades, humans have walked on the Moon, sent robot spacecraft to dig into the Martian soil and explore the satellites of Saturn, and used the most powerful telescopes ever built to probe the limits of the observable universe. Never before has so much been revealed in so short a time.

As you proceed through this book, you will learn about the tools that scientists use to explore the natural world, as well as what they observe with these tools. But, most important, you will see how astronomers build from their observations an understanding of the universe in which we live. It is this search for understanding that makes science more than merely a collection of data and elevates it to one of the great adventures of the human mind.

FIGURE 1-15   R I V U X G

A Hurricane Seen from Space This image of Hurricane Frances was made on September 2, 2004, by GOES-12 (*Geostationary Operational Environmental Satellite 12*). A geostationary satellite like GOES-12 orbits around Earth's equator every 24 hours, the same length of time it takes the planet to make a complete rotation. Hence, this satellite remains over the same spot on Earth, from which it can monitor the weather continuously. By tracking hurricanes from orbit, GOES-12 makes it much easier for meteorologists to give early warning of these immense storms. The resulting savings in lives and property more than pay for the cost of the satellite. (NASA, NOAA)

It is an adventure that will continue as long as there are mysteries in the universe—an adventure we hope you will come to appreciate and share.

KEY WORDS

Terms preceded by an asterisk (*) are discussed in the Boxes.

angle, p. 8
angular diameter (angular size), p. 8
angular distance, p. 8
arcminute (', minute of arc), p. 8
arcsecond (", second of arc), p. 8
astronomical unit (AU), p. 12
Big Bang, p. 7
black hole, p. 6
degree (°), p. 8
exponent, p. 10
galaxy, p. 7
hypothesis, p. 2
kiloparsec (kpc), p. 14
laws of physics, p. 3
light-year (ly), p. 12

megaparsec (Mpc), p. 14
meteorite, p. 4
model, p. 2
nebula (*plural* nebulae), p. 6
Newtonian mechanics, p. 3
parsec (pc), p. 13
power of ten, p. 10
powers-of-ten notation, p. 10
pulsar, p. 6
quasar, p. 7
scientific method, p. 2
*SI units, p. 13
*small-angle formula, p. 9
solar system, p. 4
subtend (an angle), p. 8
supernova (*plural* supernovae), p. 6
theory, p. 2

Sizes in the Universe

Powers-of-ten notation provides a a convenient way to express the sizes of astronomical objects and distances in space. This illustration suggests the immense distances within our solar system, the far greater distances between the stars within our Milky Way Galaxy, and the truly cosmic distances between galaxies.

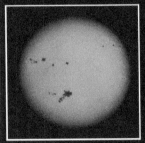

Sun: diameter = 1.39×10^6 km

The Sun, Earth, and other planets are members of our solar system

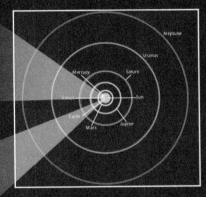

Diameter of Neptune's orbit: 60 AU

1 AU (astronomical unit) = 1.50×10^8 km = average Earth-Sun distance

The Sun is a typical star. Typical distances between our neighboring stars = 1 to 5 ly

1ly = distance that light travels in one year = 6.32×10^4 AU

Earth: diameter = 1.28×10^4 km

Galaxies are grouped into clusters, which can be up to 10^7 ly across.

Our Sun is one of more than 10^{11} stars in the Milky Way Galaxy.

Distance from the center of the Milky Way to the Sun = 2.8×10^4 ly

Each of the 1.6×10^6 dots in this map of the entire sky represents a relatively nearby galaxy. This is a tiny fraction of the number of galaxies in the observable

Astronomy, Science, and the Nature of the Universe: The universe is comprehensible. The scientific method is a procedure for formulating hypotheses about the universe. Hypotheses are tested by observation or experimentation in order to build consistent models or theories that accurately describe phenomena in nature.

Observations of the heavens have helped scientists discover some of the fundamental laws of physics. The laws of physics are in turn used by astronomers to interpret their observations.

The Solar System: Exploration of the planets provides information about the origin and evolution of the solar system, as well as about the history and resources of Earth.

Stars and Nebulae: Studying the stars and nebulae helps us learn about the origin and history of the Sun and the solar system.

Galaxies: Observations of galaxies tell us about the origin and history of the universe.

Angles: Astronomers use angles to denote the positions and sizes of objects in the sky. The size of an angle is measured in degrees, arcminutes, and arcseconds.

Powers-of-Ten Notation is a convenient shorthand system for writing numbers. It allows very large and very small numbers to be expressed in a compact form.

Units of Distance: Astronomers use a variety of distance units. These include the astronomical unit (the average distance from Earth to the Sun), the light-year (the distance that light travels in one year), and the parsec.

QUESTIONS

Review Questions

1. **TUTORIAL 1-1** What is the difference between a hypothesis and a theory?
2. What is the difference between a theory and a law of physics?
3. How are scientific theories tested?
4. Describe the role that skepticism plays in science.
5. Describe one reason why it is useful to have telescopes in space.
6. What caused the craters on the Moon?
7. What are meteorites? Why are they important for understanding the history of the solar system?
8. What makes the Sun and stars shine?
9. What role do nebulae like the Orion Nebula play in the life stories of stars?
10. What is the difference between a solar system and a galaxy?
11. What are degrees, arcminutes, and arcseconds used for? What are the relationships among these units of measure?
12. How many arcseconds equal 1°?
13. With the aid of a diagram, explain what it means to say that the Moon subtends an angle of ½°.

14. What is an exponent? How are exponents used in powers-of-ten notation?
15. What are the advantages of using powers-of-ten notation?
16. Write the following numbers using powers-of-ten notation: (a) ten million, (b) sixty thousand, (c) four one-thousandths, (d) thirty-eight billion, (e) your age in months.
17. How is an astronomical unit (AU) defined? Give an example of a situation in which this unit of measure would be convenient to use.
18. What is the advantage to the astronomer of using the light-year as a unit of distance?
19. What is a parsec? How is it related to a kiloparsec and to a megaparsec?
20. Give the word or phrase that corresponds to the following standard abbreviations: (a) km, (b) cm, (c) s, (d) km/s, (e) mi/h, (f) m, (g) m/s, (h) h, (i) y, (j) g, (k) kg. Which of these are units of speed? (*Hint:* You may have to refer to a dictionary. All of these abbreviations should be part of your working vocabulary.)
21. In the original (1977) *Star Wars* movie, Han Solo praises the speed of his spaceship by saying, "It's the ship that made the Kessel run in less than 12 parsecs!" Explain why this statement is obvious misinformation.
22. A reporter once described a light-year as "the time it takes light to reach us traveling at the speed of light." How would you correct this statement?

Advanced Questions

Questions preceded by an asterisk (*) are discussed in the Boxes.

> **Problem-solving tips and tools**
>
> The small-angle formula, given in Box 1-1, relates the size of an astronomical object to the angle it subtends. Box 1-3 illustrates how to convert from one unit of measure to another. An object traveling at speed v for a time t covers a distance d given by $d = vt$; for example, a car traveling at 90 km/h (v) for 3 h (t) covers a distance $d = (90 \text{ km/h})(3 \text{ h}) = 270 \text{ km}$. Similarly, the time t required to cover a given distance d at speed v is $t = d/v$; for example, if $d = 270$ km and $v = 90$ km/h, then $t = (270 \text{ km})/(90 \text{ km/h}) = 3 \text{ h}$.

23. What is the meaning of the letters **R I V U X G** that appear with some of the figures in this chapter? Why in each case is one of the letters highlighted? (*Hint:* See the Preface that precedes Chapter 1.)
24. The diameter of the Sun is 1.4×10^{11} cm, and the distance to the nearest star, Proxima Centauri, is 4.2 ly. Suppose you want to build an exact scale model of the Sun and Proxima Centauri, and you are using a ball 30 cm in diameter to represent the Sun. In your scale model, how far away would Proxima Centauri be from the Sun? Give your answer in kilometers, using powers-of-ten notation.

25. How many Suns would it take, laid side by side, to reach the nearest star? Use powers-of-ten notation. (*Hint:* See the preceding question.)

26. A hydrogen atom has a radius of about 5×10^{-9} cm. The radius of the observable universe is about 14 billion light-years. How many times larger than a hydrogen atom is the observable universe? Use powers-of-ten notation.

27. The Sun's mass is 1.99×10^{30} kg, three-quarters of which is hydrogen. The mass of a hydrogen atom is 1.67×10^{-27} kg. How many hydrogen atoms does the Sun contain? Use powers-of-ten notation.

28. The average distance from Earth to the Sun is 1.496×10^8 km. Express this distance (a) in light-years and (b) in parsecs. Use powers-of-ten notation. (c) Are light-years or parsecs useful units for describing distances of this size? Explain.

29. The speed of light is 3.00×10^8 m/s. How long does it take light to travel from the Sun to Earth? Give your answer in seconds, using powers-of-ten notation. (*Hint:* See the preceding question.)

30. When the *Voyager 2* spacecraft sent back pictures of Neptune during its flyby of that planet in 1989, the spacecraft's radio signals traveled for 4 hours at the speed of light to reach Earth. How far away was the spacecraft? Give your answer in kilometers, using powers-of-ten notation. (*Hint:* See the preceding question.)

31. The star Altair is 5.15 pc from Earth. (a) What is the distance to Altair in kilometers? Use powers-of-ten notation. (b) How long does it take for light emanating from Altair to reach Earth? Give your answer in years. (*Hint:* You do not need to know the value of the speed of light.)

32. The age of the universe is about 13.7 billion years. What is this age in seconds? Use powers-of-ten notation.

*33. Explain where the number 206,265 in the small-angle formula comes from.

*34. At what distance would a person have to hold a European 2-euro coin (which has a diameter of about 2.6 cm) in order for the coin to subtend an angle of (a) 1°? (b) 1 arcmin? (c) 1 arcsec? Give your answers in meters.

*35. A person with good vision can see details that subtend an angle of as small as 1 arcminute. If two dark lines on an eye chart are 2 millimeters apart, how far can such a person be from the chart and still be able to tell that there are two distinct lines? Give your answer in meters.

*36. The average distance to the Moon is 384,000 km, and the Moon subtends an angle of ½°. Use this information to calculate the diameter of the Moon in kilometers.

*37. Suppose your telescope can give you a clear view of objects and features that subtend angles of at least 2 arcsec. What is the diameter in kilometers of the smallest crater you can see on the Moon? (*Hint:* See the preceding question.)

*38. On April 18, 2006, the planet Venus was a distance of 0.869 AU from Earth. The diameter of Venus is 12,104 km. What was the angular size of Venus as seen from Earth on April 18, 2006? Give your answer in arcminutes.

*39. (a) Use the information given in the caption to Figure 1-7 to determine the angular size of the Orion Nebula. Give your answer in degrees. (b) How does the angular diameter of the Orion Nebula compare to the angular diameter of the Moon?

Discussion Questions

40. Scientists assume that "reality is rational." Discuss what this means and the thinking behind it.

41. All scientific knowledge is inherently provisional. Discuss whether this is a weakness or a strength of the scientific method.

42. How do astronomical observations differ from those of other sciences?

Web/eBook Questions

43. Use the links given in the *Universe* Web site or eBook, Chapter 1, to learn about the Orion Nebula (Figure 1-7). Can the nebula be seen with the naked eye? Does the nebula stand alone, or is it part of a larger cloud of interstellar material? What has been learned by examining the Orion Nebula with telescopes sensitive to infrared light?

44. Use the links given in the *Universe* Web site or eBook, Chapter 1, to learn more about the Crab Nebula (Figure 1-8). When did observers on Earth see the supernova that created this nebula? Does the nebula emit any radiation other than visible light? What kind of object is at the center of the nebula?

ACTIVITIES

Observing Projects

45. On a dark, clear, moonless night, can you see the Milky Way from where you live? If so, briefly describe its appearance. If not, what seems to be interfering with your ability to see the Milky Way?

46. Look up at the sky on a clear, cloud-free night. Is the Moon in the sky? If so, does it interfere with your ability to see the fainter stars? Why do you suppose astronomers prefer to schedule their observations on nights when the Moon is not in the sky?

47. Look up at the sky on a clear, cloud-free night and note the positions of a few prominent stars relative to such reference markers as rooftops, telephone poles, and treetops. Also note the location from where you make your observations. A few hours later, return to that location and again note the positions of the same bright stars that you observed earlier. How have their positions changed? From these changes, can you deduce the general direction in which the stars appear to be moving?

48. If you have access to the *Starry Night*™ planetarium software, install it on your computer. There are several guides to the use of this software. As an initial introduction, you can run through the step-by-step basics of the program by clicking the **Sky Guide** tab to the left of the main screen and then clicking the **Starry Night basics** hyperlink at the bottom of the **Sky Guide** pane. A more comprehensive guide is available by choosing the **Student Exercises** hyperlink and then the

Tutorial hyperlink. A User's Guide to this software is available under the **Help** menu. As a start, you can use this program to determine when the Moon is visible today from your location. If the viewing location in the *Starry Night*™ control panel is not set to your location, select **Set Home Location ...** in the **File** menu (on a Macintosh, this command is found under the **Starry Night** menu). Click the **List** tab in the **Home Location** dialog box; then select the name of your city or town and click the **Save As Home Location** button. Next, use the hand tool to explore the sky and search for the Moon by moving your viewpoint around the sky. (Click and drag the mouse to achieve this motion.) If the Moon is not easily seen in your sky at this time, click the **Find** tab at the top left of the main view. The **Find** pane that opens should contain a list of solar system objects. Ensure that there is no text in the edit box at the top of the **Find** pane. If the message "Search all Databases" is not displayed below this edit box, then click the magnifying glass icon in the edit box and select **Search All** from the drop-down menu that appears. Click the + symbol to the left of the listing for **Earth** to display **The Moon** and double-click on this entry in the list in order to center the view upon the Moon. (If a message is displayed indicating that "the Moon is not currently visible from your location," click on the **Best Time** button to advance to a more suitable time). You will see that the Moon can be seen in the daytime as well as at night. Note that the **Time Flow Rate** is set to **1x,** indicating that time is running forward at the normal rate. Note also the phase of the Moon. **(a)** Estimate how long it will take before the Moon reaches its full phase. Set the **Time Flow Rate** to **1 minutes. (b)** Find the time of moonset at your location. **(c)** Determine which, if any, of the following planets are visible tonight: Mercury, Venus, Mars, Jupiter, and Saturn. (*Hint:* Use the **Find** pane and click on each planet in turn to explore the positions of these objects.) Feel free to experiment with the many features of *Starry Night*™.

49. Use the *Starry Night*™ program to measure angular spacing between stars in the sky. Open **Favourites > Explorations > N Pole** to display the northern sky from Calgary, Canada, at a latitude of 51°. This view shows several asterisms, or groups of stars, outlined and labeled with their common names. The stars in the Big Dipper asterism outline the shape of a "dipper," used for scooping water from a barrel. The two stars in the Big Dipper on the opposite side of the scoop from the handle, Merak and Dubhe, can be seen to point to the brightest star in the Little Dipper, the Pole Star. The Pole Star is close to the North Celestial Pole, the point in the sky directly above the north pole of Earth. It is thus a handy aid to navigation for northern hemisphere observers because it indicates the approximate direction of true north. Measure the spacing between the two "pointer stars" in the Big Dipper and then the spacing between the Pole Star and the closest of the pointer stars, Dubhe. (*Hint:* These measurements are best made by activating the angular separation tool from the cursor selection control on the left side of the toolbar.) **(a)** What is the angular distance between the pointer stars Merak and Dubhe? **(b)** What is the angular spacing, or separation, between Dubhe (the pointer star at the end of the Big Dipper) and the Pole Star? **(c)** Approximately how many pointer-star spacings are there between Dubhe and the Pole Star?

Click the **Play** button in the toolbar. Notice that the Pole Star will appear to remain fixed in the sky as time progresses because it lies very close to the North Celestial Pole. Select **Edit > Undo Time Flow** or **File > Revert** from the menu to return to the initial view. Select **View > Celestial Guides > Celestial Poles** from the menu to indicate the position of the North Celestial Pole on the screen. Right-click on the Pole Star (Ctrl-click on a Macintosh) and select **Centre** from the drop down menu to center the view on the Pole Star. Zoom in and use the angular separation tool to measure the angular spacing between the Pole Star and the North Celestial Pole. **(d)** What is the angular separation between the Pole Star and the North Celestial Pole? **(e)** Select **File > Revert** from the menu and use the angular separation tool to measure the angle between the Pole Star and the horizon at Calgary. What is the relationship between this angle and the latitude of Calgary (51°)?

Collaborative Exercises

50. A scientific theory is fundamentally different than the everyday use of the word "theory." List and describe any three scientific theories of your choice and creatively imagine an additional three hypothetical theories that are not scientific. Briefly describe what is scientific and what is nonscientific about each of these theories.

51. Angles describe how far apart two objects appear to an observer. From where you are currently sitting, estimate the angular distance between the floor and the ceiling at the front of the room you are sitting in, the angular distance between the two people sitting closest to you, and the angular size of a clock or an exit sign on the wall. Draw sketches to illustrate each answer and describe how each of your answers would change if you were standing in the very center of the room.

52. Astronomers use powers of ten to describe the distances to objects. List an object or place that is located at very roughly each of the following distances from you: 10^{-2} m, 10^{0} m, 10^{1} m, 10^{3} m, 10^{7} m, 10^{10} m, and 10^{20} m.

ANSWERS

ConceptChecks

ConceptCheck 1-1: While a hypothesis is a testable idea that seems to explain an observation about nature, a scientific theory represents a set of well-tested and internally consistent hypotheses that are able to successfully and repeatedly predict the outcome of experiments and observations.

ConceptCheck 1-2: Shorter lifetime. The larger star has twice as much mass as the Sun. However, it consumes its nuclear fuel much more than twice as fast, so that its lifetime is shorter than the Sun.

ConceptCheck 1-3: Yes. The apparent size of an object is given by how wide it appears (in other words, the angle that it subtends). By moving a basketball near and far, it can appear very large or very small.

Why Astronomy? by Sandra M. Faber

As you study astronomy, you may ask, "Why am I studying this subject? What good is it for people in general and for me in particular?" Admittedly, astronomy does not offer the same practical benefits as other sciences, so how can it be important to your life?

On the most basic level, I think of astronomy as providing the ultimate background for human history. Recorded history goes back about 3000 years. For knowledge of the time before that, we consult archeologists and anthropologists about early human history and paleontologists, biologists, and geologists about the evolution of life and of our planet—altogether going back some five billion years. Astronomy tells us about the time before that, the ten billion years or so when the Sun, solar system, and Milky Way Galaxy formed, and even about the origin of the universe in the Big Bang. Knowledge of astronomy is part of a well-educated person's view of history.

Astronomy challenges our belief system and impels us to put our "philosophical house" in order. For example, the Bible says the world and everything in it were created in six days by the hand of God. However, according to the ancient Egyptians, Earth arose spontaneously from the infinite waters of the eternal universe, called Nun. Alaskan legends teach that the world was created by the conscious imaginings of a deity named Father Raven.

Modern astronomy, supported by physics and observations, differs from these stories of the creation of Earth. Astronomers believe that the Sun formed about five billion years ago by gravitational collapse from a dense cloud of interstellar gas and dust. At the same time, and over a period of several hundred thousand years, the planets condensed within the swirling solar nebula. Astronomers have actually seen young stars form in this way.

At issue here, really, is the question of how we are to gain information about the nature of the physical world—whether by revelation and intuition or by logic and observation. Where science stops and faith begins is a thorny issue for everyone, but particularly for astronomers—and for astronomy students.

Astronomy cultivates our notions about cosmic time and cosmic evolution. Given the short span of human life, it is all too easy to overlook that the universe is a dynamic place. This idea implies fragility—if something can change, it might even some day disappear. For instance, in another five billion years or so, the Sun will swell up and brighten to 1000 times its present luminosity, incinerating Earth in the process. This is far enough in the future that neither you nor I need to feel any personal responsibility for preparing to meet this challenge. However, other cosmic catastrophes will inevitably occur before then. Earth will be hit by a sizable piece of space debris—craters show that this happens every few million years or so. Enormous volcanic eruptions have occurred in the past and will certainly occur again. Another Ice Age is virtually certain to begin within the next 20,000 years, unless we first cook Earth ourselves by burning too much fossil fuel.

Such common notions as the inevitability of human progress, the desirability of endless economic growth, and Earth's ability to support its human population are all based on limited experience—they will probably not prove viable in the long run. Consequently, we must rethink who we are as a species and what is our proper activity on Earth. These long-term problems involve the whole human race and are vital to our survival and well-being. Astronomy is essential to developing a perspective on human existence and its relation to the cosmos.

Many astronomers believe that the ultimate, proper concept of "home" for the human race is our universe. It seems increasingly likely that a large number of other universes exist, with the vast majority incapable of harboring intelligent life as we know it. The parallel with Earth is striking. Among the solar system planets, only Earth can support human life. Among the great number of planets in our galaxy, only a small fraction may be such that we can call them home. The fraction of hospitable universes is likely to be smaller still. It seems, then, that our universe is the ultimate "home," a sanctuary in a vast sea of inhospitable universes.

I began this essay by talking about history and ended with issues that border on the ethical and religious. Astronomy is like that: It offers a modern-day version of Genesis—and of the Apocalypse, too. I hope that during this course you will be able to take time out to contemplate the broader implications of what you are studying. This is one of the rare opportunities in life to think about who you are and where you and the human race are going. Don't miss it.

Sandra M. Faber is professor of astronomy at the University of California, Santa Cruz, and astronomer at Lick Observatory. She chaired the now legendary group of astronomers called the Seven Samurai, who surveyed the nearest 400 elliptical galaxies and discovered a new mass concentration, called the Great Attractor. Dr. Faber received the Bok Prize from Harvard University in 1978, was elected to the National Academy of Sciences in 1985, and in 1986 won the coveted Dannie Heineman Prize from the American Astronomical Society, in recognition of her sustained body of especially influential astronomical research.

Earth's rotation makes stars appear to trace out circles in the sky. (Gemini Observatory) R I ☑ U X G

Knowing the Heavens

t is a clear night at the Gemini North Observatory atop Mauna Kea, a dormant volcano on the island of Hawaii. As you gaze toward the north, as in this time-exposure photograph, you find that the stars are not motionless. Rather, they move in counter-clockwise circles around a fixed point above the northern horizon. Stars close to this point never dip below the horizon, while stars farther from the fixed point rise in the east and set in the west. These motions fade from view when the Sun rises in the east and illuminates the sky. The Sun, too, arcs across the sky in the same manner as the stars. At day's end, when the Sun sets in the west, the panorama of stars is revealed for yet another night.

These observations are at the heart of *naked-eye astronomy*—the sort that requires no equipment but human vision. Naked-eye astronomy cannot tell us what the Sun is made of or how far away the stars are. For such purposes we need tools such as the Gemini Telescope, housed within the dome shown in the photograph. But by studying naked-eye astronomy, you will learn the answers to equally profound questions such as why there are seasons, why the night sky is different at different times of year, and why the night sky looks different in Australia than in North America. In discovering the answers to these questions, you will learn how Earth moves through space and will begin to understand our true place in the cosmos.

2-1 Naked-eye astronomy had an important place in civilizations of the past

Positional astronomy—the study of the positions of objects in the sky and how these positions change—has roots that extend far back in time. Four to five thousand years ago, the inhabitants of the British Isles erected stone structures, such as Stonehenge, that suggest a preoccupation with the motions of the sky. Alignments of these stones appear to show the changing locations where the Sun rose and set at key times during the year. Stoneworks of a different sort but with a similar astronomical purpose are found in the Americas, from the southwestern United States to the Andes of Peru. The ancestral Puebloans (also called Anasazi) of modern-day New Mexico, Arizona, Utah, and Colorado created stone carvings that were illuminated by the Sun on the first days of summer or winter (Figure 2-1). At the Incan city of Machu Picchu in Peru, a narrow window carved in a rock 2 meters thick looks out on the sunrise only on

> The astronomical knowledge of ancient peoples is the foundation of modern astronomy

December 21 each year. It is thought that the need to keep track of seasons for farming—to know when to sow seeds—provided at least one motivation for early astronomical alignments.

Other ancient peoples designed buildings with astronomical orientations. The great Egyptian pyramids, built around 3000 B.C.E., are oriented north-south and east-west with an accuracy much better than 1 degree. Similar alignments are found in the grand tomb of Shih Huang Ti (259 B.C.E.–210 B.C.E.), the first emperor of China.

Evidence of a highly sophisticated understanding of astronomy can be found in the written records of the Mayan civilization of Central America. Mayan astronomers deduced by observation that the apparent motions of the planet Venus follow a cycle that repeats every 584 days. They also developed a technique for calculating the position of Venus on different dates. The Maya believed that Venus was associated with war, so such calculations were important for choosing the most promising dates on which to attack an enemy.

These archaeological discoveries bear witness to an awareness of naked-eye astronomy by the peoples of many cultures. Many of the concepts of modern positional astronomy come to us from these ancients, including the idea of dividing the sky into constellations.

2-2 Eighty-eight constellations cover the entire sky

Looking at the sky on a clear, dark night, you might think that you can see millions of stars. Actually, the unaided human eye can detect only about 6000 stars. Because half of the sky

> The constellations provide a convenient framework for stating the position of an object in the heavens

is below the horizon at any one time, you can see at most about 3000 stars. When ancient peoples looked at these thousands of stars, they sometimes imagined that groupings of stars traced out pictures in the sky. Astronomers still refer to these groupings, called **constellations** (from the Latin for "group of stars").

You may already be familiar with some of these pictures or patterns in the sky, such as the Big Dipper, which is actually part of the large constellation Ursa Major (the Great Bear). Many constellations, such as Orion in Figure 2-2, have names derived from the myths and legends of antiquity. Although some star groupings vaguely resemble the figures they are supposed to represent (see Figure 2-2c), most do not.

The term "constellation" has a broader definition in present-day astronomy. On modern star charts, the entire sky is divided into 88 regions, each of which is called a constellation. For example, the constellation Orion is now defined to be an irregular patch of sky whose borders are shown in Figure 2-2b. When astronomers refer to the "Great Nebula" M42 in Orion, they mean that as seen from Earth this nebula appears to be within Orion's patch of sky. Some constellations cover large areas of the sky (Ursa Major being one of the biggest) and others very small areas (Crux, the Southern Cross, being the smallest). But because

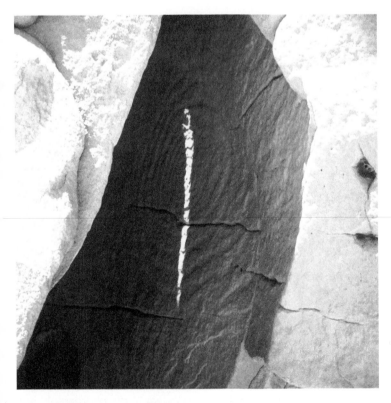

FIGURE 2-1   R I V U X G

The Sun Dagger at Chaco Canyon On the first day of winter, rays of sunlight passing between stone slabs bracket a spiral stone carving, or petroglyph, at Chaco Canyon in New Mexico. A single band of light strikes the center of the spiral on the first day of summer. This astronomically aligned petroglyph and others were carved by the ancestral Puebloan culture between 850 and 1250 C.E. (Courtesy Karl Kernberger)

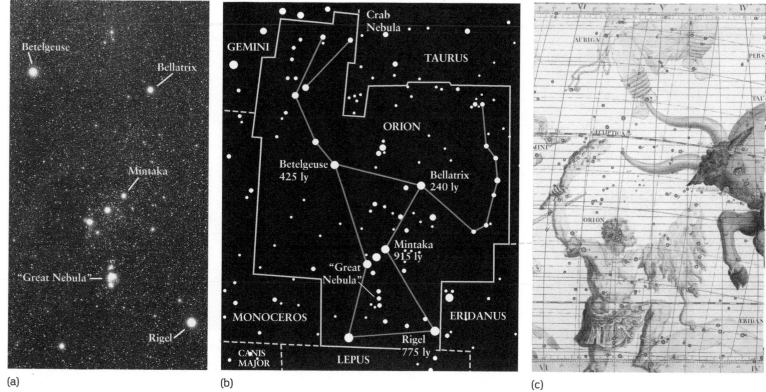

(a)　　　　　　　　　(b)　　　　　　　　　(c)

FIGURE 2-2　R I V U X G

Three Views of Orion The constellation Orion is easily seen on nights from December through March. **(a)** This photograph of Orion shows many more stars than can be seen with the naked eye. **(b)** A portion of a modern star atlas shows the distances in light-years (ly) to some of the stars in Orion. The yellow lines show the borders between Orion and its neighboring constellations (labeled in capital letters). **(c)** This fanciful drawing from a star atlas published in 1835 shows Orion the Hunter as well as other celestial creatures. (a: Eckhard Slawik/Science Source; c: Stapleton Collection/Corbis)

the modern constellations cover the entire sky, every star lies in one constellation or another.

CAUTION! When you look at a constellation's star pattern, it is tempting to conclude that you are seeing a group of stars that are all relatively close together. In fact, most of these stars are nowhere near one another. As an example, Figure 2-2b shows the distances in light-years from Earth to four stars in Orion. Although Bellatrix (Arabic for "the Amazon") and Mintaka ("the belt") appear to be close to each other, Mintaka is actually more than 600 light-years farther away from us. The two stars only *appear* to be close because they are in nearly the same direction as seen from Earth. The same illusion often appears when you see an airliner's lights at night. It is very difficult to tell how far away a single bright light is, which is why you can mistake an airliner a few kilometers away for a star trillions of times more distant.

The star names shown in Figure 2-2b are from the Arabic language. For example, Betelgeuse means "armpit," which makes sense when you look at the star atlas drawing in Figure 2-2c. Other types of names are also used for stars. For example, Betelgeuse is also known as α Orionis because it is the brightest star in Orion (α, or alpha, is the first letter in the Greek alphabet).

CAUTION! A number of unscrupulous commercial firms offer to name a star for you for a fee. The money that they charge you for this "service" is real, but the star names are not; none of these names is recognized by astronomers. If you want to use astronomy to commemorate your name or the name of a friend or relative, consider making a donation to your local planetarium or science museum. The money will be put to much better use!

CONCEPTCHECK **2-1**

If Jupiter is reported to be in the constellation of Taurus the Bull, does Jupiter need to be within the outline of the bull's body? Why or why not?

Answer appears at the end of the chapter.

2-3 The appearance of the sky changes during the course of the night and from one night to the next

Go outdoors soon after dark, find a spot away from bright lights, and note the patterns of stars in the sky. Do the same a few hours later. You will find that the entire pattern of stars (as well as the

Moon, if it is visible) has shifted its position. New constellations will have risen above the eastern horizon, and some will have disappeared below the western horizon. If you look again before dawn, you will see that the stars that were just rising in the east when

> By understanding the motions of Earth through space, we can understand why the Sun and stars appear to move in the sky

the night began are now low in the western sky. This daily motion, or **diurnal motion,** of the stars is apparent in time-exposure photographs (see the photograph that opens this chapter).

If you repeat your observations on the following night, you will find that the motions of the sky are almost but not quite the same. The same constellations rise in the east and set in the west, but a few minutes earlier than on the previous night. If you look again after a month, the constellations visible at a given time of night (say, midnight) will be noticeably different, and after six months you will see an almost totally different set of constellations. Only after a year has passed will the night sky have the same appearance as when you began.

Why does the sky go through diurnal motion? Why do the constellations slowly shift from one night to the next? As we will see, the answer to the first question is that Earth *rotates* once a day around an axis from the north pole to the south pole, while the answer to the second question is that Earth also *revolves* once a year around the Sun.

Diurnal Motion and Earth's Rotation

To understand diurnal motion, note that at any given moment it is daytime on the half of Earth illuminated by the Sun and nighttime on the other half (**Figure 2-3**). Earth rotates from west to east, making one complete rotation every 24 hours, which is why there is a daily cycle of day and night. Because of this rotation,

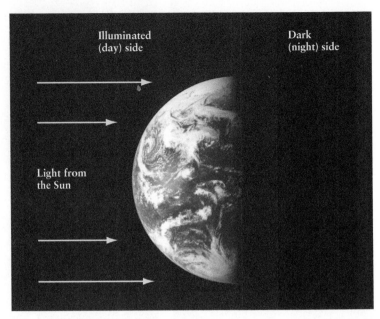

FIGURE 2-3 R I ☑ U X G

Day and Night on Earth At any moment, half of Earth is illuminated by the Sun. As Earth rotates from west to east, your location moves from the dark (night) hemisphere into the illuminated (day) hemisphere and back again. This image was recorded in 1992 by the *Galileo* spacecraft as it was en route to Jupiter. (JPL/NASA)

stars appear to us to rise in the east and set in the west, as do the Sun and Moon.

Figure 2-4 helps to further explain diurnal motion. It shows two views of Earth as seen from a point above the north pole.

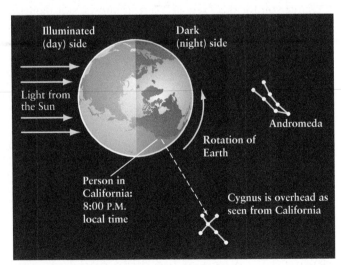

(a) Earth as seen from above the north pole

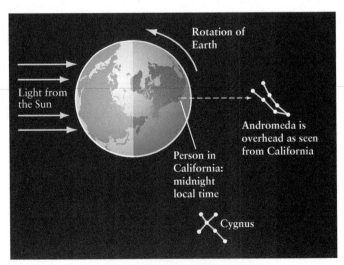

(b) 4 hours (one-sixth of a complete rotation) later

ANIMATION 2.1 **FIGURE 2-4**

Why Diurnal Motion Happens The diurnal (daily) motion of the stars, the Sun, and the Moon is a consequence of Earth's rotation. **(a)** This drawing shows Earth from a vantage point above the north pole. In this drawing, for a person in California the local time is 8:00 P.M. and the constellation Cygnus is directly overhead. **(b)** Four hours later, Earth has made one-sixth of a complete rotation to the east. As seen from Earth, the entire sky appears to have rotated to the west by one-sixth of a complete rotation. It is now midnight in California, and the constellation directly over California is Andromeda.

At the instant shown in Figure 2-4a, it is day in Asia but night in most of North America and Europe. Figure 2-4b shows Earth four hours later. Four hours is one-sixth of a complete 24-hour day, so Earth has made one-sixth of a rotation between Figures 2-4a and 2-4b. Europe is now in the illuminated half of Earth (the Sun has risen in Europe), while Alaska has moved from the illuminated to the dark half of Earth (the Sun has set in Alaska). For a person in California, in Figure 2-4a the time is 8:00 P.M. and the constellation Cygnus (the Swan) is directly overhead. Four hours later, the constellation over California is Andromeda (named for a mythological princess). Because Earth rotates from west to east, it appears to us on Earth that the entire sky rotates around us in the opposite direction, from east to west.

CONCEPTCHECK 2-2

People in which of the following cities in North America experience sunrise first: New York, San Francisco, Chicago, or Denver? Explain in terms of Earth's rotation.

CALCULATIONCHECK 2-1

If the constellation of Cygnus rises along the eastern horizon at sunset, at what time will it be highest above the southern horizon?

Answers appear at the end of the chapter.

Yearly Motion and Earth's Orbit

We described earlier that in addition to the diurnal motion of the sky, the constellations visible in the night sky also change slowly over the course of a year. This happens because Earth orbits, or revolves around, the Sun (Figure 2-5). Over the course of a year, Earth makes one complete orbit, and the darkened, nighttime side of Earth gradually turns toward different parts of the heavens. For example, as seen from the northern hemisphere, at midnight in late July the constellation Cygnus is close to overhead; at midnight in late September the constellation Andromeda is close to overhead; and at midnight in late November the constellation Perseus (commemorating a mythological hero) is close to overhead. If you follow a particular star on successive evenings, you will find that it rises approximately 4 minutes earlier each night, or 2 hours earlier each month.

CONCEPTCHECK 2-3

If Earth suddenly rotated on its axis 3 times faster than it does now, then how many times would the Sun rise and set each year?

Answer appears at the end of the chapter.

Constellations and the Night Sky

Constellations can help you find your way around the sky. For example, if you live in the northern hemisphere, you can use the Big Dipper in Ursa Major to find the north direction by drawing a straight line through the two stars at the front of the Big Dipper's bowl (Figure 2-6). The first moderately bright star you come to is Polaris, also called the North Star because it is located almost

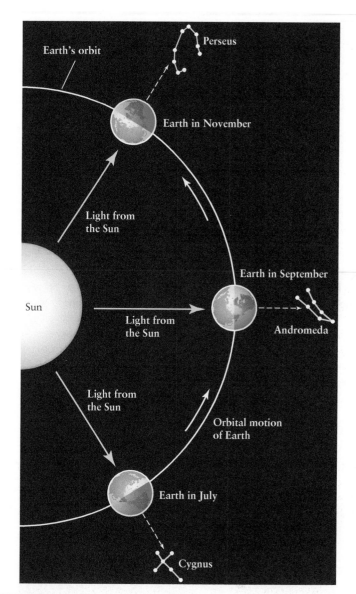

FIGURE 2-5

Why the Night Sky Changes During the Year As Earth orbits around the Sun, the nighttime side of Earth gradually turns toward different parts of the sky. Hence, the particular stars that you see in the night sky are different at different times of the year. This figure shows which constellation is overhead at midnight local time—when the Sun is on the opposite side of Earth from your location—during different months for observers at midnorthern latitudes (including the United States). If you want to view the constellation Andromeda, the best time of the year to do it is in late September, when Andromeda is nearly overhead at midnight.

directly over Earth's north pole. If you draw a line from Polaris straight down to the horizon, you will find the north direction.

As Figure 2-6 shows, by following the handle of the Big Dipper you can locate the bright reddish star Arcturus in Boötes (the Shepherd) and the prominent bluish star Spica in Virgo (the Virgin). The saying "Follow the arc to Arcturus and speed to Spica" may help you remember these stars, which are conspicuous in the evening sky during the spring and summer.

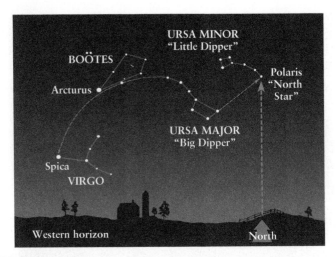

FIGURE 2-6

The Big Dipper as a Guide The North Star can be seen from anywhere in the northern hemisphere on any night of the year. This star chart shows how the Big Dipper can be used to point out the North Star as well as the brightest stars in two other constellations. The chart shows the sky at around 11 P.M. (daylight savings time) on August 1. Due to Earth's orbital motion around the Sun, you will see this same view at 1 A.M. on July 1 and at 9 P.M. on September 1. The angular distance from Polaris to Spica is 102°.

During winter in the northern hemisphere, you can see some of the brightest stars in the sky. Many of them are in the vicinity of the "winter triangle" (Figure 2-7), which connects bright stars in

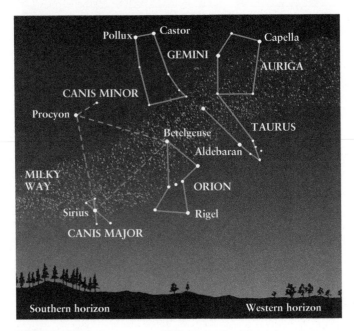

FIGURE 2-7

The "Winter Triangle" This star chart shows the view toward the southwest on a winter evening in the northern hemisphere (around midnight on January 1, 10 P.M. on February 1, or 8 P.M. on March 1). Three of the brightest stars in the sky make up the "winter triangle," which is about 26° on a side. In addition to the constellations involved in the triangle, the chart shows the prominent constellations Gemini (the Twins), Auriga (the Charioteer), and Taurus (the Bull).

FIGURE 2-8

The "Summer Triangle" This star chart shows the eastern sky as it appears in the evening during spring and summer in the northern hemisphere (around 1 A.M. daylight savings time on June 1, around 11 P.M. on July 1, and around 9 P.M. on August 1). The angular distance from Deneb to Altair is about 38°. The constellations Sagitta (the Arrow) and Delphinus (the Dolphin) are much fainter than the three constellations that make up the triangle.

the constellations of Orion (the Hunter), Canis Major (the Large Dog), and Canis Minor (the Small Dog).

A similar feature, the "summer triangle," graces the summer sky in the northern hemisphere. This triangle connects the brightest stars in Lyra (the Harp), Cygnus (the Swan), and Aquila (the Eagle) (Figure 2-8). A conspicuous portion of the Milky Way forms a beautiful background for these constellations, which are nearly overhead during the middle of summer at midnight.

A wonderful tool to help you find your way around the night sky is the planetarium program *Starry Night*™, which is on the CD-ROM that accompanies certain print copies of this book. (You can also obtain *Starry Night*™ separately.) In addition, at the end of this book you will find a set of selected star charts for the evening hours of all 12 months of the year. You may find stargazing an enjoyable experience, and *Starry Night*™ and the star charts will help you identify many well-known constellations.

Note that all the star charts in this section and at the end of this book are drawn for an observer in the northern hemisphere. If you live in the southern hemisphere, you can see constellations that are not visible from the northern hemisphere, and vice versa. In the next section we will see why this is so.

2-4 It is convenient to imagine that the stars are located on a celestial sphere

 Many ancient societies believed that all the stars are the same distance from Earth. They imagined the stars to be bits of fire imbedded in the inner surface of an immense hollow sphere, called the **celestial sphere,** with Earth at its center. In this picture of the universe, Earth was fixed and did not rotate. Instead, the entire celestial

> The concept of the imaginary celestial sphere helps us visualize the motions of stars in the sky

sphere rotated once a day around Earth from east to west, thereby causing the diurnal motion of the sky. The picture of a rotating celestial sphere fit well with naked-eye observations, and for its time was a useful model of how the universe works. (We discussed the role of models in science in Section 1-1).

Today's astronomers know that this simple model of the universe is not correct. Diurnal motion is due to the rotation of Earth, not the rest of the universe. Furthermore, as we learned when discussing the constellations in Section 2-2, the stars are not all at the same distance from Earth. Indeed, the stars that you can see with the naked eye range from 4.2 to more than 1000 light-years away, and telescopes allow us to see objects at distances of billions of light-years.

Thus, astronomers now recognize that the celestial sphere is an *imaginary* object that has no basis in physical reality. Nonetheless, the celestial sphere model remains a useful tool of positional astronomy. If we imagine, as did the ancients, that Earth is stationary and that the celestial sphere rotates around us, it is relatively easy to specify the directions to different objects in the sky and to visualize the motions of these objects. **Figure 2-9** depicts

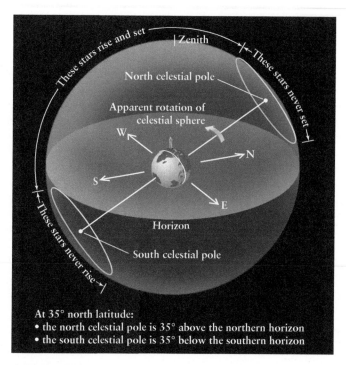

FIGURE 2-10

The View from 35° North Latitude To an observer at 35° north latitude (roughly the latitude of Los Angeles, Atlanta, Tel Aviv, and Tokyo), the north celestial pole is always 35° above the horizon. Stars within 35° of the north celestial pole are circumpolar; they trace out circles around the north celestial pole during the course of the night and are always above the horizon on any night of the year. Stars within 35° of the south celestial pole are always below the horizon and can never be seen from this latitude. Stars that lie between these two extremes rise in the east and set in the west.

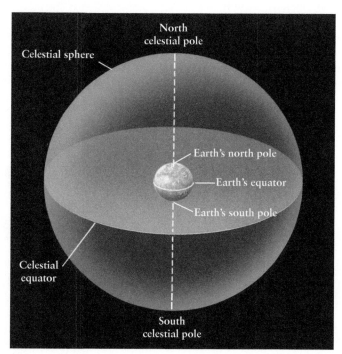

FIGURE 2-9

The Celestial Sphere The celestial sphere is the apparent sphere of the sky. The view in this figure is from the outside of this (wholly imaginary) sphere. Earth is at the center of the celestial sphere, so our view is always of the inside of the sphere. The celestial equator and poles are the projections of Earth's equator and axis of rotation out into space. The celestial poles are therefore located directly over Earth's poles.

the celestial sphere, with Earth at its center. (A truly proportional drawing would show the celestial sphere as being millions of times larger than Earth.) We picture the stars as points of light that are fixed on the inner surface of the celestial sphere.

Other features on the celestial sphere result from **projections.** The projection of a point on Earth is made by extending an imaginary line perpendicular to the surface of Earth until it intersects the celestial sphere. If we project Earth's north and south poles into space, we obtain the **north celestial pole** and the **south celestial pole.** Thus, the two celestial poles are where Earth's axis of rotation intersects the celestial sphere (see Figure 2-9). The star Polaris is less than 1° away from the north celestial pole, which is why it is called the North Star or the Pole Star. If we project Earth's equator out into space, we obtain the **celestial equator.** The celestial equator divides the sky into northern and southern hemispheres, just as Earth's equator divides Earth into two hemispheres.

The point in the sky directly overhead an observer anywhere on Earth is called that observer's **zenith.** The zenith and celestial sphere are shown in **Figure 2-10** for an observer located at 35° north latitude (that is, at a location on Earth's surface 35° north of the equator). The zenith is shown at the top of Figure 2-10, so Earth and the celestial sphere appear "tipped" compared to Figure 2-9. At any time, an observer can see only half of the celestial sphere; the other half is below the horizon, hidden by the body of

Earth. The hidden half of the celestial sphere is darkly shaded in Figure 2-10.

Motions of the Celestial Sphere

For an observer anywhere in the northern hemisphere, including the observer in Figure 2-10, the north celestial pole is always above the horizon. As Earth turns from west to east, it appears to the observer that the celestial sphere turns from east to west. Stars sufficiently near the north celestial pole revolve around the pole, never rising or setting. Such stars are called **circumpolar.** For example, as seen from North America or Europe, Polaris is a circumpolar star and can be seen at any time of night on any night of the year. The photograph that opens this chapter shows the circular trails of stars around the north celestial pole as seen from Hawaii (at 20° north latitude). Stars near the south celestial pole revolve around that pole but always remain below the horizon of an observer in the northern hemisphere. Hence, these stars can never be seen by the observer in Figure 2-10. Stars between those two limits rise in the east and set in the west.

CAUTION! Keep in mind that which stars are circumpolar, which stars never rise, and which stars rise and set depends on the latitude from which you view the heavens. As an example, for an observer at 35° south latitude (roughly the latitude of Sydney, Cape Town, and Buenos Aires), the roles of the north and south celestial poles are the opposite of those shown in Figure 2-10: Objects close to the *south* celestial pole are circumpolar, that is, they revolve around that pole and never rise or set. For an observer in the southern hemisphere, stars close to the *north* celestial pole are always below the horizon and can never be seen. Hence, astronomers in Australia, South Africa, and Argentina never see the North Star but are able to see other stars that are forever hidden from North American or European observers.

For observers at most locations on Earth, stars rise in the east and set in the west at an angle to the horizon. To see why this is so, notice that the rotation of the celestial sphere carries stars across the sky in paths that are parallel to the celestial equator (Figure 2-11).

CONCEPTCHECK 2-4

Where would you need to be standing on Earth for the celestial equator to pass through your zenith?

Answer appears at the end of the chapter.

How do we describe the location of a star? Using the celestial equator and poles, we can define a coordinate system based on angles to specify the position of any star on the celestial sphere. As Box 2-1 describes, the most commonly used coordinate system uses two angles, *right ascension* and *declination,* that are analogous to longitude and latitude on Earth. These coordinates tell us in what direction we should look to see the star. To locate the star's true position in three-dimensional space, we must also know the distance to the star.

2-5 The seasons are caused by the tilt of Earth's axis of rotation

TUTORIAL 2-2   In addition to rotating on its axis every 24 hours, Earth revolves around the Sun—that is, it *orbits* the Sun—in about 365¼ days (see Figure 2-5). As we travel with Earth around its orbit, we experience the annual cycle of seasons. But why *are* there seasons? Furthermore, the seasons are opposite in the northern and southern hemispheres. For

> Earth's tilt causes variations in the concentration of sunlight

(a) At middle northern latitudes

The stars move along paths parallel to the celestial equator…

To north celestial pole

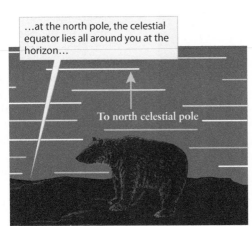

(b) At the north pole

…at the north pole, the celestial equator lies all around you at the horizon…

To north celestial pole

(c) At the equator

…and standing at the equator, the celestial equator is a semi-circle passing directly overhead.

To north celestial pole

ANIMATION 2-3   **FIGURE 2-11**   R I U X G
The Apparent Motion of Stars at Different Latitudes

As Earth rotates, stars appear to rotate around us along paths that are parallel to the celestial equator. **(a)** As shown in this long time exposure, at most locations on Earth the rising and setting motions are at an angle to the horizon that depends on the latitude. **(b)** At the north pole (latitude 90° north) the stars appear to move parallel to the horizon. **(c)** At the equator (latitude 0°) the stars rise and set along vertical paths. (a: David Miller/David Malin Images)

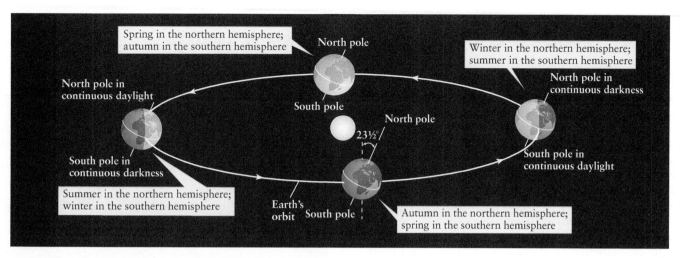

FIGURE 2-12

The Seasons Earth's axis of rotation is inclined 23½° away from the perpendicular to the plane of Earth's orbit. The north pole points in the same direction as it orbits the Sun, near the star Polaris. Consequently, the amount of solar illumination and the number of daylight hours at any location on Earth vary in a regular pattern throughout the year. This is the origin of the seasons.

example, February is midwinter in North America but midsummer in Australia. Why should this be?

The Origin of the Seasons

As we know from common experience, there are warmer temperatures in summer. This warmth is due to an increased amount of sunlight striking the ground beneath your feet and the surrounding areas. But what causes the concentration of sunlight to change through the seasons?

The reason why we have seasons, and why they are different in different hemispheres, is that Earth's axis of rotation is not perpendicular to the plane of Earth's orbit. Instead, as **Figure 2-12** shows, the axis is tilted about 23½° away from the perpendicular. Earth maintains this tilt as it orbits the Sun, with Earth's north pole always pointing in the same direction throughout the year. As we saw earlier, this direction just happens to have a star (named Polaris)

within 1°. (This stability is a hallmark of all rotating objects. A top will not fall over as long as it is spinning, and the rotating wheels of a motorcycle help to keep the rider upright.)

During part of the year, when Earth is in the part of its orbit shown on the left side of Figure 2-12, the northern hemisphere is tilted toward the Sun. As Earth spins on its axis, a point in the northern hemisphere spends more than 12 hours in the sunlight. Thus, the days there are long and the nights are short, and it is summer in the northern hemisphere. The summer is hot not only because of the extended daylight hours but also because the Sun is high in the northern hemisphere's sky. As a result, sunlight strikes the ground at a nearly perpendicular angle that heats the ground efficiently (**Figure 2-13a**). During this same time of year in the southern hemisphere, the days are short and the nights are long, because a point in this hemisphere spends fewer than 12 hours a day in the sunlight. The Sun is low in the sky, so sunlight strikes the

(a) The Sun in summer

(b) The Sun in winter

FIGURE 2-13

Solar Energy in Summer and Winter
At different times of the year, sunlight strikes the ground at different angles. **(a)** In summer, sunlight is concentrated and the days are also longer, which further increases the heating. **(b)** In winter, the sunlight is less concentrated, the days are short, and little heating of the ground takes place. This accounts for the low temperatures in winter.

BOX 2-1 TOOLS OF THE ASTRONOMER'S TRADE

Celestial Coordinates

Your *latitude* and *longitude* describe where on Earth's surface you are located. The latitude of your location denotes how far north or south of the equator you are, and the longitude of your location denotes how far west or east you are of an imaginary circle that runs from the north pole to the south pole through the Royal Observatory in Greenwich, England. In an analogous way, astronomers use coordinates called *declination* and *right ascension* to describe the position of a planet, star, or galaxy on the celestial sphere.

Declination is analogous to latitude. As the illustration shows, the **declination** of an object is its angular distance north or south of the celestial equator, measured along a circle passing through both celestial poles. Like latitude, it is measured in degrees, arcminutes, and arcseconds (see Section 1-5).

Right ascension is analogous to longitude. It is measured from a line that runs between the north and south celestial poles and passes through a point on the celestial equator called the *vernal equinox* (shown as a red dot in the illustration). This point is one of two locations where the Sun crosses the celestial equator during its apparent annual motion, as we discuss in Section 2-5. In Earth's northern hemisphere, spring officially begins when the Sun reaches the vernal equinox in late March. The **right ascension** of an object is the angular distance from the vernal equinox eastward along the celestial equator to the circle used in measuring its declination (see illustration). Astronomers measure right ascension in *time* units (hours, minutes, and seconds), corresponding to the time required for the celestial sphere to rotate through this angle. For example, suppose there is a star at your zenith right now

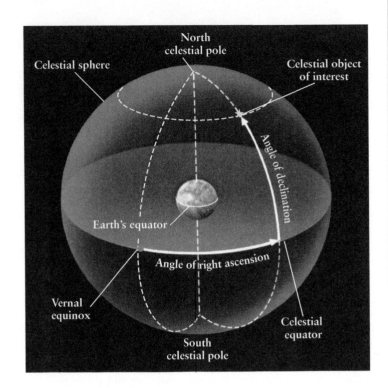

with right ascension $6^h 0^m 0^s$. Two hours and 30 minutes from now, there will be a different object at your zenith with right ascension $8^h 30^m 0^s$.

---

surface at a grazing angle that causes little heating (Figure 2-13b), and it is winter in the southern hemisphere.

Half a year later, Earth is in the part of its orbit shown on the right side of Figure 2-12. Now the situation is reversed, with winter in the northern hemisphere (which is now tilted away from the Sun) and summer in the southern hemisphere. During spring and autumn, the two hemispheres receive roughly equal amounts of illumination from the Sun, and daytime and nighttime are of roughly equal length everywhere on Earth.

CAUTION! A common misconception is that the seasons are caused by variations in the distance from Earth to the Sun. According to this idea, Earth is closer to the Sun in summer and farther away in winter. But in fact, Earth's orbit around the Sun is very nearly circular, and the Earth-Sun distance varies only about 3 percent over the course of a year. (Earth's orbit

only *looks* elongated in Figure 2-12 because this illustration shows the orbit from a side view; it would be circular if viewed overhead.) We are slightly closer to the Sun in January than in July, but this small variation has little influence on the cycle of the seasons. Also, if the seasons were really caused by variations in the Earth-Sun distance, the seasons would be the same in both hemispheres!

CONCEPTCHECK 2-5

If Earth's axis were not tilted, but rather was straight up and down compared to the plane created by the path of Earth's orbit, would observers near Earth's north pole still observe periods where the Sun never rises and the Sun never sets?

Answer appears at the end of the chapter.

The coordinates of the bright star Rigel for the year 2000 are R.A. = 5^h 14^m 32.2^s, Decl. = $-8°$ $12'$ $06''$. (R.A. and Decl. are abbreviations for right ascension and declination.) A minus sign on the declination indicates that the star is south of the celestial equator; a plus sign (or no sign at all) indicates that an object is north of the celestial equator. As we discuss in Box 2-2, right ascension helps determine the best time to observe a particular object.

It is important to state the year for which a star's right ascension and declination are valid. This is so because of precession, which we discuss in Section 2-6.

EXAMPLE: What are the coordinates of a star that lies exactly halfway between the vernal equinox and the south celestial pole?

Situation: Our goal is to find the right ascension and declination of the star in question.

Tools: We use the definitions depicted in the figure.

Answer: Since the circle used to measure this star's declination passes through the vernal equinox, this star's right ascension is R.A. = 0^h 0^m 0^s. The angle between the celestial equator and south celestial pole is $90°$ $0'$ $0''$, so the declination of this star is Decl. = $-45°$ $0'$ $0''$.

Review: The declination in this example is negative because the star is in the southern half of the celestial sphere.

EXAMPLE: At midnight local time you see a star with R.A. = 2^h 30^m 0^s at your zenith. When will you see a star at your zenith with R.A. = 21^h 0^m 0^s?

Situation: If you held your finger stationary over a globe of Earth, the longitude of the point directly under your finger would change as you rotated the globe. In the same way, the right ascension of the point directly over your head (the zenith) changes as the celestial sphere rotates. We use this concept to determine the time in question.

Tools: We use the idea that a change in right ascension of 24^h corresponds to an elapsed time of 24 hours and a complete rotation of the celestial sphere.

Answer: The time required for the sky to rotate through the angle between the stars is the difference in their right ascensions: 21^h 0^m 0^s − 2^h 30^m 0^s = 18^h 30^m 0^s. So the second star will be at your zenith 18½ hours after the first one, or at 6:30 P.M. the following evening.

Review: Our answer was based on the idea that the celestial sphere makes *exactly* one complete rotation in 24 hours. If this were so, from one night to the next each star would be in exactly the same position at a given time. But because of the way that we measure time, the celestial sphere makes slightly more than one complete rotation in 24 hours. (We explore the reasons for this in Box 2-2.) As a result, our answer is in error by about 3 minutes. For our purposes, this is a small enough error that we can ignore it.

How the Sun Moves on the Celestial Sphere

The plane of Earth's orbit around the Sun is called the **ecliptic plane** (Figure 2-14a). The ecliptic plane is easy to picture if you imagine hovering above the solar system, but is described quite differently when viewed from Earth. As a result of Earth's annual motion around the Sun, it appears to us (observing from Earth) that the Sun slowly changes its position on the celestial sphere over the course of a year. (Since we don't see the Sun and stars at the same time, this motion relative to the stars isn't readily familiar.) The circular path that the Sun appears to trace out against the background of stars is called the **ecliptic** (Figure 2-14b). The plane of this path is the same as the ecliptic plane. (The name *ecliptic* suggests that the path traced out by the Sun has something to do with eclipses. We will discuss the connection in Chapter 3.) Because there are 365¼ days in a year and 360° in a circle, the Sun appears to move along the ecliptic at a rate of about 1° per day. This motion is from west to east, that is, in the direction opposite to the apparent motion of the celestial sphere.

ANALOGY Note that at the same time that the Sun is making its yearlong trip around the ecliptic, the entire celestial sphere is rotating around us once per day. You can envision the celestial sphere as a merry-go-round rotating clockwise and the Sun as a restless child who is walking slowly around the merry-go-round's rim in the counterclockwise direction. During the time it takes the child to make a round trip, the merry-go-round rotates 365¼ times.

The ecliptic plane is *not* the same as the plane of Earth's equator, thanks to the 23½° tilt of Earth's rotation axis shown in Figure 2-12. As a result, the ecliptic and the celestial

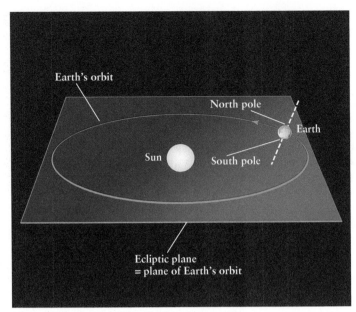

(a) In reality Earth orbits the Sun once a year

FIGURE 2-14

The Ecliptic Plane and the Ecliptic **(a)** The ecliptic plane is the plane in which Earth moves around the Sun. **(b)** As seen from Earth, the Sun appears to move around the celestial sphere along a circular path called the ecliptic.

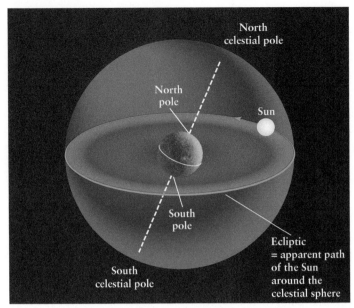

(b) It appears from Earth that the Sun travels around the celestial sphere once a year

Earth takes a year to complete one orbit around the Sun, so as seen by us the Sun takes a year to make a complete trip around the ecliptic.

equator are inclined to each other by that same 23½° angle (Figure 2-15).

CONCEPTCHECK 2-6

How long does the Sun take to move from being next to a bright star all the way around the celestial sphere and back to that same bright star?

Answer appears at the end of the chapter.

Equinoxes and Solstices

The ecliptic and the celestial equator intersect at only two points, which are exactly opposite each other on the celestial sphere. Each point is called an **equinox** (from the Latin for "equal night"), because when the Sun appears at either of these points, day and night are each about 12 hours long at all locations on Earth. The term "equinox" is also used to refer to the date on which the Sun passes through one of these special points on the ecliptic.

On about March 21 of each year, the Sun passes northward across the celestial equator at the **vernal equinox**. This marks the beginning of spring in the northern hemisphere ("vernal" is from the Latin for "spring"). On about September 22, the Sun moves southward across the celestial equator at the **autumnal equinox,** marking the moment when fall begins in the northern hemisphere (see Figure 2-15). Since the seasons are opposite in the northern and southern hemispheres, for Australians, South Africans, and South Americans the vernal equinox actually marks the beginning of autumn. The names of the equinoxes come from astronomers of the past who lived north of the equator.

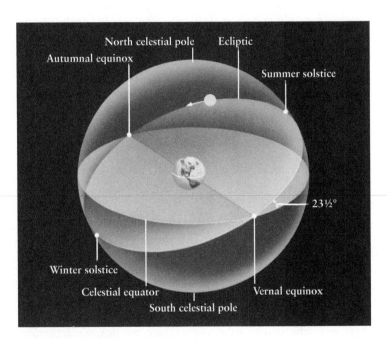

FIGURE 2-15

The Ecliptic, Equinoxes, and Solstices This illustration of the celestial sphere is similar to Figure 2-14b, but is drawn with the north celestial pole at the top and the celestial equator running through the middle. The ecliptic is inclined to the celestial equator by 23½° because of the tilt of Earth's axis of rotation. It intersects the celestial equator at two points, called equinoxes. The northernmost point on the ecliptic is the summer solstice, and the southernmost point is the winter solstice. The Sun is shown in its approximate position for August 1.

Between the vernal and autumnal equinoxes lie two other significant locations along the ecliptic. The point on the ecliptic farthest north of the celestial equator is called the **summer solstice.** "Solstice" is from the Latin for "solar standstill," and it is at the summer solstice that the Sun stops moving northward on the celestial sphere. At this point, the Sun is as far north of the celestial equator as it can get. It marks the location of the Sun at the moment summer begins in the northern hemisphere (about June 21). At the beginning of the northern hemisphere's winter (about December 21), the Sun is farthest south of the celestial equator at a point called the **winter solstice** (see Figure 2-15).

Because the Sun's position on the celestial sphere varies slowly over the course of a year, its daily path across the sky (due to Earth's rotation) also varies with the seasons (Figure 2-16). On the first day of spring or the first day of fall, when the Sun is at one of the equinoxes, the Sun rises directly in the east and sets directly in the west.

When the northern hemisphere is tilted away from the Sun and it is winter in the northern hemisphere, the Sun rises in the southeast. Daylight lasts for fewer than 12 hours as the Sun skims low over the southern horizon and sets in the southwest. Northern hemisphere nights are longest when the Sun is at the winter solstice.

The closer you get to the north pole, the shorter the winter days and the longer the winter nights. In fact, anywhere within 23½° of the north pole (that is, north of latitude 90° − 23½° = 66½° N) the Sun is below the horizon for 24 continuous hours at least one day of the year. The circle around Earth at 66½° north latitude is called the **Arctic Circle** (Figure 2-17). The corresponding region around the south pole is bounded by the **Antarctic Circle** at 66½° south latitude. At the time of the winter solstice,

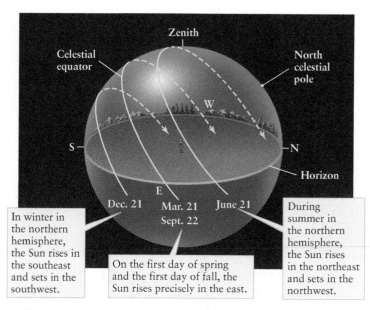

In winter in the northern hemisphere, the Sun rises in the southeast and sets in the southwest.

On the first day of spring and the first day of fall, the Sun rises precisely in the east.

During summer in the northern hemisphere, the Sun rises in the northeast and sets in the northwest.

ANIMATION 2-5 **FIGURE 2-16**

The Sun's Daily Path Across the Sky This drawing shows the apparent path of the Sun during the course of a day on four different dates. Like Figure 2-10, this drawing is for an observer at 35° north latitude.

explorers south of the Antarctic Circle enjoy "the midnight sun," or 24 hours of continuous daylight.

During summer in the northern hemisphere, when the northern hemisphere is tilted toward the Sun, the Sun rises in the northeast and sets in the northwest. The Sun is at its northernmost position at the summer solstice, giving the northern hemisphere

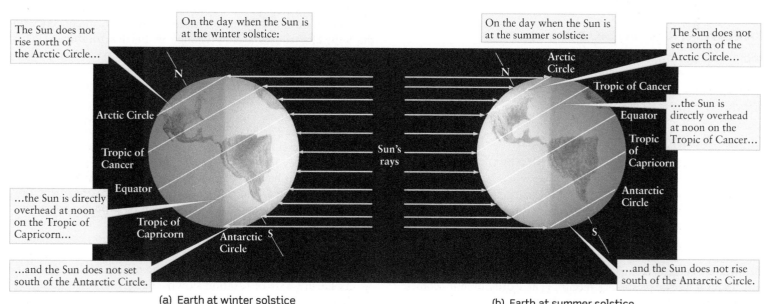

The Sun does not rise north of the Arctic Circle...

On the day when the Sun is at the winter solstice:

...the Sun is directly overhead at noon on the Tropic of Capricorn...

...and the Sun does not set south of the Antarctic Circle.

Sun's rays

On the day when the Sun is at the summer solstice:

The Sun does not set north of the Arctic Circle...

...the Sun is directly overhead at noon on the Tropic of Cancer...

...and the Sun does not rise south of the Antarctic Circle.

(a) Earth at winter solstice

(b) Earth at summer solstice

FIGURE 2-17

Tropics and Circles Four important latitudes on Earth are the Arctic Circle (66½° north latitude), Tropic of Cancer (23½° north latitude), Tropic of Capricorn (23½° south latitude), and Antarctic Circle (66½° south latitude).

These drawings show the significance of these latitudes when the Sun is (a) at the winter solstice and (b) at the summer solstice.

FIGURE 2-18 R I V U X G

The Midnight Sun This time-lapse photograph was taken on July 19, 1985, at 69° north latitude in northeast Alaska. At this latitude, the Sun is above the horizon continuously (that is, it is circumpolar) from mid-May to the end of July. (Doug Plummer/Science Photo Library)

the greatest number of daylight hours. At the summer solstice the Sun does not set at all north of the Arctic Circle (Figure 2-18) and does not rise at all south of the Antarctic Circle.

The variations of the seasons are much less pronounced close to the equator. Between the **Tropic of Capricorn** at 23½° south latitude and the **Tropic of Cancer** at 23½° north latitude, the Sun is directly overhead—that is, at the zenith—at high noon at least one day a year. Outside of the tropics, the Sun is never directly overhead, but is always either south of the zenith (as seen from locations north of the Tropic of Cancer) or north of the zenith (as seen from south of the Tropic of Capricorn).

CONCEPTCHECK 2-7

How often each year does an observer standing on Earth's equator experience no shadow during the noontime sun?

CALCULATIONCHECK 2-2

Approximately how many days are there between the summer solstice and the March equinox?

Answers appear at the end of the chapter.

2-6 The Moon helps to cause precession, a slow, conical motion of Earth's axis of rotation

The Moon is by far the brightest and most obvious naked-eye object in the nighttime sky. Like the Sun, the Moon slowly changes its position relative to the background stars; unlike the Sun, the Moon makes a complete trip around the celestial sphere in only about 4 weeks, or about a month. (The word "month" comes from the same Old English root as the word "moon.") Ancient astronomers realized that this motion occurs because the Moon orbits Earth in roughly 4 weeks. In 1 hour, the Moon moves on the celestial sphere by about ½°, or roughly its own angular size.

> Precession causes the apparent positions of the stars to slowly change over the centuries

The Moon's path on the celestial sphere is never far from the Sun's path (that is, the ecliptic). This is because the plane of the Moon's orbit around Earth is inclined only slightly from the plane of Earth's orbit around the Sun (the ecliptic plane shown in Figure 2-14a). The Moon's path varies somewhat from one month to the next, but always remains within a band called the **zodiac** that extends about 8° on either side of the ecliptic. Twelve famous constellations—Aries, Taurus, Gemini, Cancer, Leo, Virgo, Libra, Scorpius, Sagittarius, Capricornus, Aquarius, and Pisces—lie along the zodiac. The Moon is generally found in one of these 12 constellations. (Thanks to a redrawing of constellation boundaries in the mid-twentieth century, the zodiac actually passes through a thirteenth constellation—Ophiuchus, the Serpent Bearer—between Scorpius and Sagittarius.) As it moves along its orbit, the Moon appears north of the celestial equator for about two weeks and then south of the celestial equator for about the next two weeks. We will learn more about the Moon's motion, as well as why the Moon goes through phases, in Chapter 3.

The Moon not only moves around Earth but, in concert with the Sun, also causes a slow change in Earth's rotation. This is because both the Sun and the Moon exert a gravitational pull on Earth. We will learn much more about gravity in Chapter 4; for now, all we need is the idea that gravity is a universal attraction of matter for other matter.

The gravitational pull of the Sun and the Moon affects Earth's rotation because Earth is slightly fatter across the equator than it is from pole to pole: Its equatorial diameter is 43 kilometers (27 miles) larger than the diameter measured from pole to pole. Earth is therefore said to have an "equatorial bulge." Because of the gravitational pull of the Moon and the Sun on this bulge, the orientation of Earth's axis of rotation gradually changes, producing a motion called **precession.** Although the details are beyond the scope of this book, the combined actions of gravity and rotation cause Earth's axis to trace out a circle in the sky (Figure 2-19). As the axis precesses, it remains tilted about 23½° to the perpendicular.

As Earth's axis of rotation slowly changes its orientation, the north and south celestial poles—which are the projections of that axis onto the celestial sphere—change their positions relative to the stars. At present, the north celestial pole lies within 1° of the star Polaris, which is why Polaris is the North Star. But 5000 years ago, the north celestial pole was closest to the star Thuban in the constellation of Draco (the Dragon). Thus, that star and not

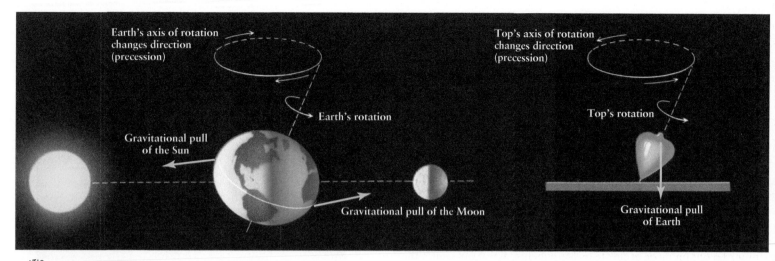

ANIMATION 2-6 **FIGURE 2-19**

Precession Because Earth's rotation axis is tilted, the gravitational pull of the Moon and the Sun on Earth's equatorial bulge together cause Earth to precess. As Earth precesses, its axis of rotation slowly traces out a circle in the sky, like the shifting axis of a spinning top.

Polaris was the North Star. And 12,000 years from now, the North Star will be the bright star Vega in Lyra (the Harp). It takes 26,000 years for the north celestial pole to complete one full precessional circle around the sky (Figure 2-20). The south celestial pole executes a similar circle in the southern sky.

Precession also causes Earth's equatorial plane to change its orientation. Because this plane defines the location of the celestial equator in the sky, the celestial equator precesses as well. The intersections of the celestial equator and the ecliptic define the equinoxes (see Figure 2-15), so these key locations in the sky also shift slowly from year to year. For this reason, the precession of Earth is also called the **precession of the equinoxes.** The first person to detect the precession of the equinoxes, in the second century B.C.E., was the Greek astronomer Hipparchus, who compared his own observations with those of Babylonian astronomers three centuries earlier. Today, the vernal equinox is located in the constellation Pisces (the Fishes). Two thousand years ago, it was in Aries (the Ram). Around the year 2600 C.E., the vernal equinox will move into Aquarius (the Water Bearer).

CAUTION! Astrological terms like the "Age of Aquarius" involve boundaries in the sky that are not recognized by astronomers and are generally not even related to the positions of the constellations. For example, most astrologers would call a person born on March 21, 1988, an "Aries" because the Sun was supposedly in the direction of that constellation on March 21. But due to precession, the Sun was actually in the constellation Pisces on that date! Indeed, astrology is not a science at all, but merely a collection of superstitions and hokum. Its practitioners use some of the terminology of astronomy but reject the logical thinking that is at the heart of science. James Randi has more to say about astrology and other pseudosciences in his essay "Why Astrology Is Not Science" at the end of this chapter.

The astronomer's system of locating heavenly bodies by their right ascension and declination, discussed in Box 2-1, is tied to the positions of the celestial equator and the vernal equinox.

Because of precession, these positions are changing, and thus the coordinates of stars in the sky are also constantly changing. These changes are very small and gradual, but they add up over the years. To cope with this difficulty, astronomers always make

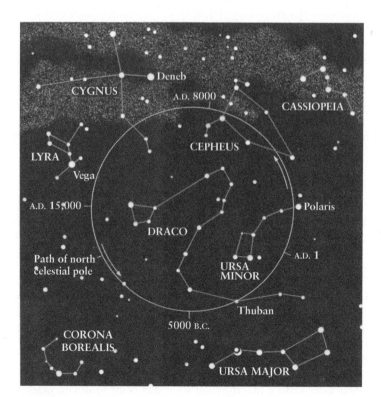

FIGURE 2-20

Precession and the Path of the North Celestial Pole As Earth precesses, the north celestial pole slowly traces out a circle among the northern constellations. At present, the north celestial pole is near the moderately bright star Polaris, which serves as the North Star. Twelve thousand years from now the bright star Vega will be the North Star.

note of the date (called the **epoch**) for which a particular set of coordinates is precisely correct. Consequently, star catalogs and star charts are periodically updated. Most current catalogs and star charts are prepared for the epoch 2000. The coordinates in these reference books, which are precise for January 1, 2000, will require very little correction over the next few decades.

2-7 Positional astronomy plays an important role in keeping track of time

Astronomers have traditionally been responsible for telling time. This is because we want the system of timekeeping used in everyday life to reflect the position of the Sun in the sky. Thousands of years ago, the sundial was invented to keep track of **apparent solar time**. To obtain more accurate measurements, astronomers use the **meridian**. As Figure 2-21 shows, this is a north-south circle on the celestial sphere that passes through the zenith (the point directly overhead) and both celestial poles. *Local noon* is defined as when the Sun crosses the **upper meridian**, which is the half of the meridian above the horizon. At *local midnight,* the Sun crosses the **lower meridian,** the half of the meridian below the horizon; this crossing cannot be observed directly.

> Ancient scholars developed a system of timekeeping based on the Sun

The crossing of the meridian by any object in the sky is called a **meridian transit** of that object. If the crossing occurs above the horizon, it is an *upper* meridian transit. When an object makes a meridian transit, it is not necessarily at the zenith directly overhead, but is at the *highest point above the horizon* that you will see it from your location.

An **apparent solar day** is the interval between two successive upper meridian transits of the Sun as observed from any fixed spot on Earth. Stated less formally, an apparent solar day is the time from one local noon to the next local noon, or from when the Sun is highest in the sky to when it is again highest in the sky.

The Sun as a Timekeeper

Unfortunately, the Sun is not a good timekeeper (Figure 2-22). The length of an apparent solar day (as measured by a device such as an hourglass) varies from one time of year to another. There are two main reasons why this is so, both having to do with the way in which Earth orbits the Sun.

The first reason is that Earth's orbit is not a perfect circle; rather, it is an ellipse, as Figure 2-22a shows in exaggerated form. As we will learn in Chapter 4, Earth moves more rapidly along its orbit when it is near the Sun than when it is farther away. Hence, the Sun appears to us to move more than 1° per day along the ecliptic in January, when Earth is nearest the Sun, and less than 1° per day in July, when Earth is farthest from the Sun. By itself, this effect would cause the apparent solar day to be longer in January than in July.

The second reason why the Sun is not a good timekeeper is the 23½° angle between the ecliptic and the celestial equator (see Figure 2-15). As Figure 2-22b shows, this causes a significant part of the Sun's apparent motion when near the equinoxes to be in

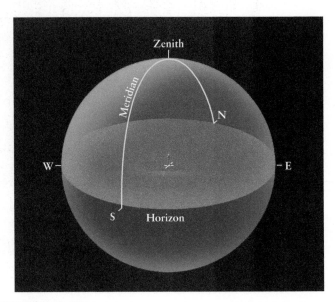

FIGURE 2-21

The Meridian The meridian is a circle on the celestial sphere that passes through the observer's zenith (the point directly overhead) and the north and south points on the observer's horizon. The passing of celestial objects across the meridian can be used to measure time. The upper meridian is the part above the horizon, and the lower meridian (not shown) is the part below the horizon.

a north-south direction. The net daily eastward progress in the sky is then somewhat foreshortened. At the summer and winter solstices, by contrast, the Sun's motion is parallel to the celestial equator. Thus, there is no comparable foreshortening around the beginning of summer or winter. This effect by itself would make the apparent solar day shorter in March and September than in June or December. Combining these effects with those due to Earth's noncircular orbit, we find that the length of the apparent solar day varies in a complicated fashion over the course of a year.

To avoid these difficulties, astronomers invented an imaginary object called the **mean sun** that moves along the celestial equator at a uniform rate. (In science and mathematics, "mean" is a synonym for "average.") The mean sun is sometimes slightly ahead of the real Sun in the sky, sometimes behind. As a result, mean solar time and apparent solar time can differ by as much as a quarter of an hour at certain times of the year.

Because the mean sun moves at a constant rate, it serves as a fine timekeeper. A **mean solar day** is the interval between successive upper meridian transits of the mean sun. It is exactly 24 hours long, the average length of an apparent solar day. One 24-hour day as measured by your alarm clock or wristwatch is a mean solar day.

Time zones were invented for convenience in commerce, transportation, and communication. In a time zone, all clocks and watches are set to the mean solar time for a meridian of longitude that runs approximately through the center of the zone. Time zones around the world are generally centered on meridians of longitude at 15° intervals. In most cases, going from one time zone to the next requires you to change the time on your wristwatch by exactly 1 hour. The time zones for most of North America are shown in Figure 2-23.

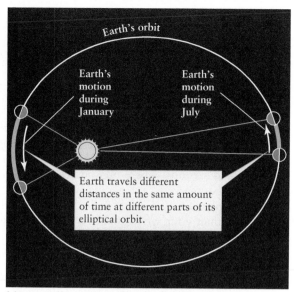

(a) A month's motion of Earth along its orbit

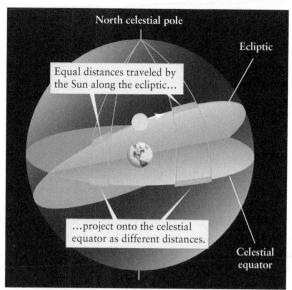

(b) A day's motion of the Sun along the ecliptic

FIGURE 2-22

Why the Sun Is a Poor Timekeeper There are two main reasons that the Sun is a poor timekeeper. **(a)** Earth's speed along its orbit varies during the year. It moves fastest when closest to the Sun in January and slowest when farthest from the Sun in July. Hence, the apparent speed of the Sun along the ecliptic is not constant. **(b)** Because of the tilt of Earth's rotation

axis, the ecliptic is inclined with respect to the celestial equator. Therefore, the projection of the Sun's daily progress (shown in orange) along the ecliptic onto the celestial equator (shown in blue) varies during the year. This causes further variations in the length of the apparent solar day.

In order to coordinate their observations with colleagues elsewhere around the globe, astronomers often keep track of time using Coordinated Universal Time, somewhat confusingly abbreviated UTC or UT. This is the time in a zone that includes Greenwich, England, a seaport just outside of London where the first internationally accepted time standard was kept. (UTC was formerly known as Greenwich Mean Time.) UTC is on a 24-hour system, with no A.M. or P.M. In North America, Eastern Standard Time (EST) is 5 hours different from UTC; 9:00 A.M. EST is 14:00 UTC. Coordinated Universal Time is also used by aviators and sailors, who regularly travel from one time zone to another.

Although it is natural to want our clocks and method of time-keeping to be related to the Sun, astronomers often use a system that is based on the apparent motion of the stars. This system, called **sidereal time,** is useful when aiming a telescope. Most observatories are therefore equipped with a clock that measures sidereal time, as discussed in Box 2-2.

2-8 Astronomical observations led to the development of the modern calendar

Just as the day is a natural unit of time based on Earth's rotation, the year is a natural unit of time based on Earth's revolution about the Sun. However, nature has not arranged things for our convenience. The year does not divide into exactly 365 whole days. Ancient astronomers realized that the length of a year is approximately 365¼ days, motivating the Roman emperor

> Our calendar is complex because a year does not contain a whole number of days

Julius Caesar to establish the system of "leap years" to account for this extra quarter of a day. By adding an extra day to the calendar every four years, he hoped to ensure that seasonal astronomical

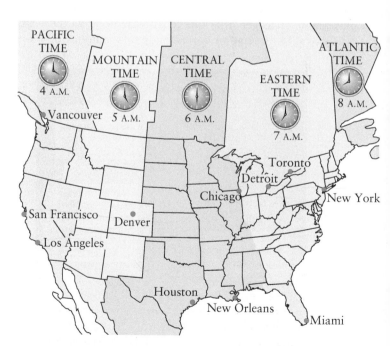

FIGURE 2-23

Time Zones in North America For convenience, Earth is divided into 24 time zones, generally centered on 15° intervals of longitude around the globe. There are four time zones across the continental United States, making for a 3-hour time difference between New York and California.

BOX 2-2 TOOLS OF THE ASTRONOMER'S TRADE

Sidereal Time

If you want to observe a particular object in the heavens, the ideal time to do so is when the object is high in the sky, on or close to the upper meridian.

Timing observations to occur when an object is high in the sky minimizes the distorting effects of Earth's atmosphere, which increase as you view objects closer to the horizon. For astronomers who study the Sun, this means making observations at local noon, which is not too different from noon as determined using mean solar time. For astronomers who observe planets, stars, or galaxies, however, the optimum time to observe depends on the particular object to be studied. Given the location of a given object on the celestial sphere, when will that object be on the upper meridian?

To answer this question, astronomers use *sidereal time* rather than solar time. It is different from the time on your wristwatch. In fact, a *sidereal clock* and an ordinary clock even tick at different rates, because they are based on different astronomical objects. Ordinary clocks are related to the position of the Sun, while sidereal clocks are based on the position of the vernal equinox, the location from which right ascension is measured. (See Box 2-1 for a discussion of right ascension.)

Regardless of where the Sun is, midnight sidereal time at your location is defined to be when the vernal equinox crosses your upper meridian. (Like solar time, sidereal time depends on where you are on Earth.) A **sidereal day** is the time between two successive upper meridian passages of the vernal equinox. By contrast, an apparent solar day is the time between two successive upper meridian crossings of the Sun. The illustration shows why these two kinds of day are not equal. Because Earth orbits the Sun, Earth must make one complete rotation plus about 1° to get from one local solar noon to the next. This extra 1° of rotation corresponds to 4 minutes of time,

which is the amount by which a solar day exceeds a sidereal day. To be precise:

$$1 \text{ sidereal day} = 23^h\ 56^m\ 4.091^s$$

where the hours, minutes, and seconds are in mean solar time.

One day according to your wristwatch is one mean solar day, which is exactly 24 hours of solar time long. A **sidereal clock** measures sidereal time in terms of sidereal hours, minutes, and seconds, where one sidereal day is divided into 24 sidereal hours.

This explains why a sidereal clock ticks at a slightly different rate than your wristwatch. As a result, at some times of the year a sidereal clock will show a very different time than an ordinary clock. (At local noon on March 21, when the Sun is at the vernal equinox, a sidereal clock will say that it is midnight. Do you see why?)

We can now answer the question in the opening paragraph. The vernal equinox, whose celestial coordinates are R.A. = $0^h\ 0^m\ 0^s$, Decl. = 0° 0′ 0″, crosses the upper meridian at midnight sidereal time (0:00). The autumnal equinox, which is on the opposite side of the celestial sphere at R.A. = $12^h\ 0^m\ 0^s$, Decl. = 0° 0′ 0″, crosses the upper meridian 12 sidereal hours later at noon sidereal time (12:00). As these examples illustrate, *any* object crosses the upper meridian when the sidereal time is equal to the object's right ascension. That is why astronomers measure right ascension in units of time rather than degrees, and why right ascension is always given in sidereal hours, minutes, and seconds.

EXAMPLE: Suppose you want to observe the bright star Spica (Figure 2-6), which has epoch 2000 coordinates R.A. = $13^h\ 25^m\ 11.6^s$, Decl. = −11° 9′ 41″. What is the best time to do this?

Situation: Our goal is to find the time when Spica passes through your upper meridian, where it can best be observed.

Tools: We use the idea that a celestial object is on your upper meridian when the sidereal time equals the object's right ascension.

Answer: Based on the given right ascension of Spica, it will be best placed for observation when the sidereal time at your location is about 13:25. (Note that sidereal time is measured using a 24-hour clock.)

Review: By itself, our answer doesn't tell you what time on your wristwatch (which measures mean solar time) is best for observing Spica. That's why most observatories are equipped with a sidereal clock.

While sidereal time is extremely useful in astronomy, mean solar time is still the best method of timekeeping for most earthbound purposes. All time measurements in this book are expressed in mean solar time unless otherwise stated.

To vernal equinox

Earth moves about 1° around its orbit in one day…

Sun

…so Earth must make a complete rotation plus 1° to bring this location to local solar noon on March 22.

Local solar noon on March 21 is at this location on Earth.

1°

1°

Earth on March 22

Earth on March 21

A month's motion of Earth along its orbit

events, such as the beginning of spring, would occur on the same date year after year.

Caesar's system would have been perfect if the year were exactly 365¼ days long and if there were no precession. Unfortunately, this is not the case. To be more accurate, astronomers now use several different types of years. For example, the **sidereal year** is defined to be the time required for the Sun to return to the same position with respect to the stars. It is equal to 365.2564 mean solar days, or 365^d 6^h 9^m 10^s.

The sidereal year is the orbital period of Earth around the Sun, but it is *not* the year on which we base our calendar. Like Caesar, most people want annual events—in particular, the first days of the seasons—to fall on the same date each year. For example, we want the first day of spring to occur on March 21. But spring begins when the Sun is at the vernal equinox, and the vernal equinox moves slowly against the background stars because of precession. Therefore, to set up a calendar we use the **tropical year**, which is equal to the time needed for the Sun to return to the vernal equinox. This period is equal to 365.2422 mean solar days, or 365^d 5^h 48^m 46^s. Because of precession, the tropical year is 20 minutes and 24 seconds shorter than the sidereal year.

Caesar's assumption that the tropical year equals 365¼ days was off by 11 minutes and 14 seconds. This tiny error adds up to about three days every four centuries. Although Caesar's astronomical advisers were aware of the discrepancy, they felt that it was too small to matter. However, by the sixteenth century the first day of spring was occurring on March 11.

The Roman Catholic Church became concerned because Easter kept shifting to progressively earlier dates. To straighten things out, Pope Gregory XIII instituted a calendar reform in 1582. He began by dropping 10 days (October 4, 1582, was followed by October 15, 1582), which brought the first day of spring back to March 21. Next, he modified Caesar's system of leap years.

Caesar had added February 29 to every calendar year that is evenly divisible by four. Thus, for example, 2008, 2012, and 2016 are all leap years with 366 days. But we have seen that this system produces an error of about three days every four centuries. To solve the problem, Pope Gregory decreed that only the century years evenly divisible by 400 should be leap years. For example, the years 1700, 1800, and 1900 (which would have been leap years according to Caesar) were not leap years in the improved Gregorian system, but the year 2000, which can be divided evenly by 400, *was* a leap year.

We use the Gregorian system today. It assumes that the year is 365.2425 mean solar days long, which is very close to the true length of the tropical year. In fact, the error is only one day in every 3300 years, which won't cause any problems for a long time.

KEY WORDS

Terms preceded by an asterisk (*) are discussed in the Boxes.

KEY IDEAS

Ideas preceded by an asterisk (*) are discussed in the Boxes.

Constellations and the Celestial Sphere: It is convenient to imagine the stars fixed to the celestial sphere with Earth at its center.

• The surface of the celestial sphere is divided into 88 regions called constellations.

Diurnal (Daily) Motion of the Celestial Sphere: The celestial sphere appears to rotate around Earth once in each 24-hour period. In fact, it is actually Earth that is rotating.

• The poles and equator of the celestial sphere are determined by extending the axis of rotation and the equatorial plane of Earth out to the celestial sphere.

• *The positions of objects on the celestial sphere are described by specifying their right ascension (in time units) and declination (in angular measure).

Seasons and the Tilt of Earth's Axis: Earth's axis of rotation is tilted at an angle of about 23½° from the perpendicular to the plane of Earth's orbit.

• The seasons are caused by the tilt of Earth's axis.

• Over the course of a year, the Sun appears to move around the celestial sphere along a path called the ecliptic. The ecliptic is inclined to the celestial equator by about 23½°.

• The ecliptic crosses the celestial equator at two points in the sky, the vernal and autumnal equinoxes. The northernmost point that the Sun reaches on the celestial sphere is the summer solstice, and the southernmost point is the winter solstice.

Precession: The orientation of Earth's axis of rotation changes slowly, a phenomenon called precession.

• Precession is caused by the gravitational pull of the Sun and Moon on Earth's equatorial bulge.

• Precession of Earth's axis causes the positions of the equinoxes and celestial poles to shift slowly.

• *Because the system of right ascension and declination is tied to the position of the vernal equinox, the date (or epoch) of observation must be specified when giving the position of an object in the sky.

Timekeeping: Astronomers use several different means of keeping time.

- Apparent solar time is based on the apparent motion of the Sun across the celestial sphere, which varies over the course of the year.

- Mean solar time is based on the motion of an imaginary mean sun along the celestial equator, which produces a uniform mean solar day of 24 hours. Ordinary watches and clocks measure mean solar time.

- *Sidereal time is based on the apparent motion of the celestial sphere.

The Calendar: The tropical year is the period between two passages of the Sun across the vernal equinox. Leap year corrections are needed because the tropical year is not exactly 365 days. The sidereal year is the actual orbital period of Earth.

QUESTIONS

Review Questions

1. Describe three structures or carvings made by past civilizations that show an understanding of astronomy.

2. How are constellations useful to astronomers? How many stars are not part of any constellation?

3. A fellow student tells you that only those stars in Figure 2-2b that are connected by blue lines are part of the constellation Orion. How would you respond?

4. Why is a particular star overhead at 10:00 P.M. on a given night rather than two hours later at midnight; how does this relate to Earth's orbit? Why are different stars overhead at midnight on June 1 rather than at midnight on December 1?

5. *TUTORIAL 2-1* What is the celestial sphere? Why is this ancient concept still useful today?

6. Imagine that someone suggests sending a spacecraft to land on the surface of the celestial sphere. How would you respond to such a suggestion?

7. What is the celestial equator? How is it related to Earth's equator? How are the north and south celestial poles related to Earth's axis of rotation?

8. Where would you have to look to see your zenith? Where on Earth would you have to be for the celestial equator to pass through your zenith? Where on Earth would you have to be for the south celestial pole to be at your zenith?

9. How many degrees is the angle from the horizon to the zenith? Does your answer depend on what point on the horizon you choose?

10. Why can't a person in Antarctica use the Big Dipper to find the north direction?

11. Is there any place on Earth where you could see the north celestial pole on the northern horizon? If so, where? Is there any place on Earth where you could see the north celestial pole on the western horizon? If so, where? Explain your answers.

12. How do the stars appear to move over the course of the night as seen from the north pole? As seen from the equator? Why are these two motions different?

13. *TUTORIAL 2-2* Using a diagram, explain why the tilt of Earth's axis relative to Earth's orbit causes the seasons as we orbit the Sun.

14. Give two reasons why it is warmer in summer than in winter.

15. What is the ecliptic plane? What is the ecliptic?

16. Why is the ecliptic tilted with respect to the celestial equator? Does the Sun appear to move along the ecliptic, the celestial equator, or neither? By about how many degrees does the Sun appear to move on the celestial sphere each day?

17. Where on Earth do you have to be in order to see the north celestial pole directly overhead? What is the maximum possible elevation of the Sun above the horizon at that location? On what date can this maximum elevation be observed?

18. What are the vernal and the autumnal equinoxes? What are the summer and winter solstices? How are these four points related to the ecliptic and the celestial equator?

19. At what point on the horizon does the vernal equinox rise? Where on the horizon does it set? (*Hint:* See Figure 2-16.)

20. How does the daily path of the Sun across the sky change with the seasons? Why does it change?

21. Where on Earth do you have to be in order to see the Sun at the zenith? As seen from such a location, will the Sun be at the zenith every day? Explain your reasoning.

22. What is precession of the equinoxes? What causes it? How long does it take for the vernal equinox to move 1° along the ecliptic?

23. What is the (fictitious) mean sun? What path does it follow on the celestial sphere? Why is it a better timekeeper than the actual Sun in the sky?

24. Why is it convenient to divide Earth into time zones?

25. Why is the time given by a sundial not necessarily the same as the time on your wristwatch?

26. What is the difference between the sidereal year and the tropical year? Why are these two kinds of year slightly different in length? Why are calendars based on the tropical year?

27. When is the next leap year? Was 2000 a leap year? Will 2100 be a leap year?

Advanced Questions

Questions preceded by an asterisk (*) are discussed in the Boxes.

Problem-solving tips and tools

To help you visualize the heavens, it is worth taking the time to become familiar with various types of star charts. These include the simple star charts at the end of this book, the monthly star charts published in such magazines as *Sky & Telescope* and *Astronomy*, and the more detailed maps of the heavens found in star atlases.

One of the best ways to understand the sky and its motions is to use the *Starry Night*™ computer program on the CD-ROM that accompanies certain printed copies of this book. This easy-to-use program allows you to view the sky on any date and at any time, as seen from any point on Earth, and to animate the sky to visualize its diurnal and annual motions.

You may also find it useful to examine a planisphere, a device consisting of two rotatable disks. The bottom disk shows all the stars in the sky (for a particular latitude), and the top one is opaque with a transparent oval window through which only some of the stars can be seen. By rotating the top disk, you can immediately see which constellations are above the horizon at any time of the year. A planisphere is a convenient tool to carry with you when you are out observing the night sky.

28. On November 1 at 8:30 P.M. you look toward the eastern horizon and see the bright star Bellatrix (shown in Figure 2-2b) rising. At approximately what time will Bellatrix rise one week later, on November 8?

29. Figure 2-4 shows the situation on September 21, when Cygnus is highest in the sky at 8:00 P.M. local time and Andromeda is highest in the sky at midnight. But as Figure 2-5 shows, on July 21 Cygnus is highest in the sky at midnight. On July 21, at approximately what local time is Andromeda highest in the sky? Explain your reasoning.

30. Figure 2-5 shows which constellations are high in the sky (for observers in the northern hemisphere) in the months of July, September, and November. From this figure, would you be able to see Perseus at midnight on May 15? Draw a picture to justify your answer.

31. Figure 2-6 shows the appearance of Polaris, the Little Dipper, and the Big Dipper at 11 P.M. (daylight saving time) on August 1. Sketch how these objects would appear on this same date at (**a**) 8 P.M. and (**b**) 2 A.M. Include the horizon in your sketches, and indicate the north direction.

32. Figure 2-6 shows the appearance of the sky near the North Star at 11 P.M. (daylight saving time) on August 1. Explain why the sky has this same appearance at 1 A.M. on July 1 and at 9 P.M. on September 1.

33. The time-exposure photograph that opens this chapter shows the trails made by individual stars as the celestial sphere appears to rotate around Earth. (**a**) For approximately what length of time was the camera shutter left open to take this photograph? (**b**) The stars in this photograph (taken in Hawaii, at roughly 20° north latitude) appear to rotate around one of the celestial poles. Which celestial pole is it? As seen from this location, do the stars move clockwise or counterclockwise around this celestial pole? (**c**) If you were at 20° south latitude, which celestial pole could you see? In which direction would you look to see it? As seen from this location, do the stars move clockwise or counterclockwise around this celestial pole?

34. (**a**) Redraw Figure 2-10 for an observer at the north pole. (*Hint:* The north celestial pole is directly above this observer.) (**b**) Redraw Figure 2-10 for an observer at the equator. (*Hint:*

The celestial equator passes through this observer's zenith.) (**c**) Using Figure 2-10 and your drawings from (a) and (b), justify the following rule, long used by navigators: The latitude of an observer in the northern hemisphere is equal to the angle in the sky between that observer's horizon and the north celestial pole. (**d**) State the rule that corresponds to (c) for an observer in the southern hemisphere.

35. The photograph that opens this chapter was taken next to the Gemini North Observatory atop Mauna Kea in Hawaii. The telescope is at longitude 155° 28' 09" west and latitude 19° 49' 26" north. (**a**) By making measurements on the photograph, find the approximate angular width and angular height of the photo. (**b**) How far (in degrees, arcminutes, and arcseconds) from the north celestial pole can a star be and still be circumpolar as seen from the Gemini North Observatory?

36. The Gemini North Observatory shown in the photograph that opens this chapter is located in Hawaii, roughly 20° north of the equator. Its near-twin, the Gemini South Observatory, is located roughly 30° south of the equator in Chile. Why is it useful to have telescopes in both the northern and southern hemispheres?

37. Is there any place on Earth where all the visible stars are circumpolar? If so, where? Is there any place on Earth where none of the visible stars is circumpolar? If so, where? Explain your answers.

38. This image of Earth was made by the *Galileo* spacecraft while en route to Jupiter. South America is at the center of the image and Antarctica is at the bottom of the image. (**a**) In which month of the year was this image made? Explain your reasoning. (**b**) When this image was made, was Earth relatively close to the Sun or relatively distant from the Sun? Explain your reasoning.

R I **V** U X G (NASA/JPL)

39. Figure 2-16 shows the daily path of the Sun across the sky on March 21, June 21, September 22, and December 21 for an observer at 35° north latitude. Sketch drawings of this kind

for (**a**) an observer at 35° south latitude; (**b**) an observer at the equator; and (**c**) an observer at the north pole.

40. Suppose that you live at a latitude of 40° N. What is the elevation (angle) of the Sun above the southern horizon at noon (**a**) at the time of the vernal equinox? (**b**) at the time of the winter solstice? Explain your reasoning. Include a drawing as part of your explanation.

41. In the northern hemisphere, houses are designed to have "southern exposure," that is, with the largest windows on the southern side of the house. But in the southern hemisphere houses are designed to have "northern exposure." Why are houses designed this way, and why is there a difference between the hemispheres?.

42. The city of Mumbai (formerly Bombay) in India is 19° north of the equator. On how many days of the year, if any, is the Sun at the zenith at midday as seen from Mumbai? Explain your answer.

43. Ancient records show that 2000 years ago, the stars of the constellation Crux (the Southern Cross) were visible in the southern sky from Greece. Today, however, these stars cannot be seen from Greece. What accounts for this change?

44. The Great Pyramid at Giza has a tunnel that points toward the north celestial pole. At the time the pyramid was built, around 2600 B.C.E., toward which star did it point? Toward which star does this same tunnel point today? (See Figure 2-20.)

45. The photo shows a statue of the Greek god Atlas. The globe that Atlas is holding represents the celestial sphere, with depictions of several important constellations and the celestial equator. Although the statue dates from around 150 C.E., it has been proposed that the arrangement of constellations depicts the sky as it was mapped in an early star atlas that dates from 129 B.C.E. Explain the reasoning that could lead to such a proposal.

R I ▐ U X G   (Scala/Art Resource, NY)

46. Unlike western Europe, Imperial Russia did not use the revised calendar instituted by Pope Gregory XIII. Explain why the Russian Revolution, which started on November 7, 1917, according to the modern calendar, is called the October Revolution in Russia. What was this date according to the Russian calendar at the time? Explain your answer.

*47. What is the right ascension of a star that is on the meridian at midnight at the time of the autumnal equinox? Explain your answer.

*48. The coordinates on the celestial sphere of the summer solstice are R.A. = 6^h 0^m 0^s, Decl. = $+23°$ $27'$. What are the right ascension and declination of the winter solstice? Explain your answer.

*49. Because 24 hours of right ascension takes you all the way around the celestial equator, $24^h = 360°$, what is the angle in the sky (measured in degrees) between a star with R.A. = 8^h 0^m 0^s, Decl. = $0°$ $0'$ $0''$ and a second star with R.A. = 11^h 20^m 0^s, Decl. = $0°$ $0'$ $0''$? Explain your answer.

*50. On a certain night, the first star in Advanced Question 49 passes through the zenith at 12:30 A.M. local time. At what time will the second star pass through the zenith? Explain your answer.

*51. At local noon on March 21, when the Sun is at the vernal equinox, a sidereal clock will say that it is midnight. Explain why.

*52. (**a**) What is the sidereal time when the vernal equinox rises? (**b**) On what date is the sidereal time nearly equal to the solar time? Explain your answer.

*53. How would the sidereal and solar days change (**a**) if Earth's rate of rotation increased, (**b**) if Earth's rate of rotation decreased, and (**c**) if Earth's rotation were retrograde (that is, if Earth rotated about its axis opposite to the direction in which it revolves about the Sun)?

Discussion Questions

54. Examine a list of the 88 constellations. Are there any constellations whose names obviously date from modern times? Where are these constellations located? Why do you suppose they do not have archaic names?

55. Describe how the seasons would be different if Earth's axis of rotation, rather than having its present 23½° tilt, were tilted (**a**) by 0° or (**b**) by 90°.

56. In William Shakespeare's *Julius Caesar* (act 3, scene 1), Caesar says:

> *But I am constant as the northern star,*
> *Of whose true-fix'd and resting quality*
> *There is no fellow in the firmament.*

Translate Caesar's statement about the "northern star" into modern astronomical language. Is the northern star truly "constant"? Was the northern star the same in Shakespeare's time (1564–1616) as it is today?

Web/eBook Questions

57. Search the World Wide Web for information about the national flags of Australia, New Zealand, and Brazil and the state flag of Alaska. Which stars are depicted on these flags? Explain any similarities or differences among these flags.

58. Some people say that on the date that the Sun is at the vernal equinox, and only on this date, you can stand a raw egg on end. Others say that there is nothing special about the vernal equinox, and that with patience you can stand a raw egg on end on any day of the year. Search the World Wide Web for information about this story and for hints about how to stand an egg on end. Use these hints to try the experiment yourself on a day when the Sun is *not* at the vernal equinox. What do you conclude about the connection between eggs and equinoxes?

59. Use the U.S. Naval Observatory Web site to find the times of sunset and sunrise on (**a**) your next birthday and (**b**) the date this assignment is due. (**c**) Are the times the same for the two dates? Explain why or why not.

ACTIVITIES

Observing Projects

> **Observing tips and tools**
>
> Moonlight is so bright that it interferes with seeing the stars. For the best view of the constellations, do your observing when the Moon is below the horizon. You can find the times of moonrise and moonset in your local newspaper or on the World Wide Web. Each monthly issue of the magazines *Sky & Telescope* and *Astronomy* includes much additional observing information.

60. On a clear, cloud-free night, use the star charts at the end of this book to see how many constellations of the zodiac you can identify. Which ones were easy to find? Which were difficult? Are the zodiacal constellations the most prominent ones in the sky?

61. Examine the star charts that are published monthly in such popular astronomy magazines as *Sky & Telescope* and *Astronomy*. How do they differ from the star charts at the end of this book? On a clear, cloud-free night, use one of these star charts to locate the celestial equator and the ecliptic. Note the inclination of the Milky Way to the ecliptic and celestial equator. The Milky Way traces out the plane of our galaxy. What do your observations tell you about the orientation of Earth and its orbit relative to the galaxy's plane?

62. Suppose you wake up before dawn and want to see which constellations are in the sky. Explain how the star charts at the end of this book can be quite useful, even though chart times are given only for the evening hours. Which chart most closely depicts the sky at 4:00 A.M. on the morning that this assignment is due? Set your alarm clock for 4:00 A.M. to see if you are correct.

63. Use *Starry Night*™ to observe the diurnal motion of the sky. First, set *Starry Night*™ to display the sky as seen from where you live, if you have not already done so. To do this, select **File > Set Home Location…** (**Starry Night > Set Home Location** on a Macintosh) and click on the **List** tab to find the name of your city or town. Highlight the name and note the latitude of your location as given in the list and click the **Save As Home Location** button. Select **Options > Other Options > Local Horizon…** from the menu. In the **Local Horizon Options** dialog box, click on the radio button labeled **Flat** in the **Horizon** style section and click **OK**. For viewers in the northern hemisphere, press the "N" key (or click the **N** button in the **Gaze** section of the toolbar) to set the gaze direction to the northern sky. If your location is in the southern hemisphere, press the "S" key (or click the **S** button in the Gaze section of the toolbar) to set the gaze direction to the south. Select **Hide Daylight** under the **View** menu to view the present sky without daylight. Select **View > Constellations > Astronomical** and **View > Constellations > Labels** to display the constellation patterns on the sky. In the toolbar, click on the **Time Flow Rate** control and set the time step to **1 minute**. Then click the **Play** button to run time forward. (The rapid motions of artificial Earth-orbiting satellites can prove irritating in this view. You can remove these satellites by clicking on **View > Solar System** and turning off **Satellites**). (**a**) Do the stars appear to rotate clockwise or counterclockwise? Explain this observation in terms of Earth's rotation. (**b**) Are any of the stars circumpolar, that is, do they stay above your horizon for the full 24 hours of a day? If some stars at your location are circumpolar, adjust time and locate a star that moves very close to the horizon during its diurnal motion. Click the **Stop** button and right-click (Ctrl-click on a Macintosh) on the star and then select **Show Info** from the contextual menu to open the **Info** pane. Expand the **Position in Sky** layer and note the star's declination (its N-S position on the sky with reference to a coordinate system whose zero value is the projection of Earth's equator). (**c**) How is this limiting declination, above which stars are circumpolar, related to your latitude, noted above?

    (i) The limiting declination is equal to the latitude of the observer's location.

    (ii) The limiting declination is equal to (90° − latitude).

    (iii) There is no relationship between this limiting declination and the observer's latitude.

(**d**) Now center your field of view on the southern horizon (if you live in the northern hemisphere) or the northern horizon (if you live in the southern hemisphere) and click **Play** to resume time flow. Describe what you see. Are any of these stars circumpolar?

64. Use the *Starry Night*™ program to observe the Sun's motion on the celestial sphere. Select **Favourites > Explorations > Sun** from the menu. The view shows the entire celestial sphere as if you were at the center of a transparent Earth

on January 1, 2010. The view is centered upon the Sun and shows the ecliptic, the celestial equator, and the boundary and name of the constellation in which the Sun is located. With the **Time Flow Rate** set to **8 hours**, click the **Play** button. Observe the Sun for a full year of simulated time. The motion of Earth in its orbit causes this apparent motion. (**a**) How does the Sun appear to move against the background stars? (**b**) What path does the Sun follow and does it ever change direction? (**c**) Through which constellations does the Sun appear to move over the course of a full year? In the toolbar, click the **Now** button to go to the current date and time. (**d**) In which constellation is the Sun located today? The Sun (and therefore this constellation) is high in the sky at midday. (**e**) Approximately how long do you think it will take for this constellation to be high in the sky at midnight?

65. Use *Starry Night*™ to demonstrate the reason for seasonal variations on Earth at mid-latitudes. A common misconception is that summertime is warmer because Earth is closer to the Sun in the summer. The real reason is that the tilt of the spin axis of Earth to its orbital plane places the Sun at a higher angle in the sky in the summer than in the winter. Thus, sunlight hits Earth's surface at a less oblique angle in summertime than in winter, thereby depositing greater heat. Open **Favourites > Explorations > Seasonal Variations** to view the southern sky in daylight from Calgary, Canada, at a latitude of 51°N, on December 21, 2013, at 12:38 P.M., local standard time, when the Sun is at its highest angle on that day. Form a table of values of Sun altitude and Sun–Earth distance as a function of date in the year. To find these values, move the cursor over the Sun to reveal the **Info** panel. (Ensure that the relevant information is displayed by opening **File > Preferences > Cursor Tracking (HUD)** and clicking on **Altitude** and **Distance from Observer**.) Note the date, Sun altitude, and distance in your table. Advance the **Date** to March 21, 2014, and then to June 21, 2014, noting the values of these parameters. (**a**) On which of the three dates, in winter, spring, and summer respectively, is the Sun at the highest altitude in the Calgary sky? (**b**) From the values in your table, what is the Sun altitude at midday in December in Calgary? (**c**) How does the Sun altitude at midday on March 21 relate to the latitude of Calgary?

    (i) The Sun's altitude is equal to the latitude of Calgary.

    (ii) The Sun's altitude is 90° minus the latitude of Calgary.

    (iii) The Sun's altitude is not related to the latitude of the location of the observer.

(**d**) On which of these three observing dates is Earth closest to the Sun? (**e**) On which of these three dates is Earth farthest from the Sun?

ANSWERS

ConceptChecks

ConceptCheck 2-1: No, Jupiter does not need to be one of the stars in the asterism that makes the outline of the bull's body. Jupiter would only need to be within the somewhat rectangular boundary of the constellation of Taurus.

ConceptCheck 2-2: Earth rotates from west to east. As locations on Earth rotate from the dark, nighttime side of the planet into the bright, daytime side of the planet, the easternmost cities experience sunrise first. New York is the farthest east of the cities listed, so the sun rises there first.

ConceptCheck 2-3: The Sun would still only rise and set once each day (since that defines what a day is), but each year would have 3 times as many days ($365 \times 3 = 1095$ days each year), and each day would be 3 times shorter.

ConceptCheck 2-4: The celestial equator is a projection, or an extension, of Earth's equator out into the sky. In order for this imaginary line to pass directly overhead, one would need to be standing somewhere on Earth's equator.

ConceptCheck 2-5: No, they would not; in this imaginary scenario, even near the north pole the Sun would rise and set every 12 hours all year long. Earth's tilted axis means that observers near Earth's north pole will experience six months when the Sun never rises (when it is tilted away from the Sun) and then six months when the Sun never sets (when it is tilted toward the Sun).

ConceptCheck 2-6: It takes one year. The Sun slowly moves through the sky a little each day, taking 365¼ days to return to the same place it was one year earlier.

ConceptCheck 2-7: To cast no shadow, the Sun must be directly overhead. As the Sun's position on the celestial sphere slowly moves back and forth between the northern and southern solstice points over the course of a year, the noontime Sun will be directly overhead (and will cast no shadow) for an observer at Earth's equator only twice each year, on the March and September equinoxes.

CalculationChecks

CalculationCheck 2-1: Earth rotates once in 24 hours, so when the stars of Cygnus rise in the east at sunset, they take about 12 hours to go from one side of the sky to the other. As a result, it takes about one half of that time, or 6 hours, for stars of Cygnus to move halfway across the sky. Thus, Cygnus will be highest in the sky around midnight.

CalculationCheck 2-2: The summer solstice occurs on about June 21 and the March equinox occurs about March 21, so there are about 9 months or 270 days between these two events.

Why Astrology Is Not Science by James Randi

'm involved in the strange business of telling folks what they should already know. I meet audiences who believe in all sorts of impossible things, often despite their education and intelligence. My job is to explain how science differs from the unproven, illogical assumptions of pseudoscience—and why it matters. Perhaps my best example is the difference between astronomy and astrology.

Both astrology and astronomy arose from the wonders of the night sky, from the stars to comets, planets, the Sun, and the Moon. Surely, humans have long reasoned, there must be some meaning in their motions. Surely the Moon's effect on tides hints at hidden "causes" for strange events. *Judiciary* (literally "judging") astrology therefore attempted to foretell the future—our earthly future. To serve it, *horary* (literally "hourly") astrology carefully tracked the heavens.

It is the latter that has become astronomy. Thanks to its process of careful measurement and testing, we now understand more about the true nature of the starry universe than astrologers could ever have imagined. With the birth of a new science, astronomers had a logical framework based on physical causes and systematic observations.

Astrology remains a popular delusion. Far too many believe today that patterns in the sky govern our lives. They accept the vague tendencies and portents of seers who cast horoscopes. They shouldn't. Just a glance at the tenets of astrology provides ample evidence of its absurdity.

An individual is said to be born under a sign. To the astrologer, the Sun was located "in" that sign at the moment of birth. (Stars are not seen in the daytime, but no matter—a calculation tells where the Sun is.) Each sign takes its name from a constellation, a totally imaginary figure invented for our convenience in referring to stars. Different cultures have different mythical figures up there, and so different schools of astrology assign different meanings to the signs they use.

In the spirit of equal-opportunity swindling, astrologers divide up the year fairly, ignoring variations in the size of constellations. Since Libra is tiny, while Virgo is huge, they chop some of the sky off Virgo and add it—along with bits of Scorpio—to bring Libra up to size. The Sun could well be declared "in" Libra when it is actually outside that constellation.

It gets worse. Science constantly challenges itself and changes. The rules of astrology could not, although they were made up thousands of years ago, and since then the "fixed" stars have moved. In particular, precession of the equinoxes has shifted objects in the sky relative to our calendar. The constellations have changed but astrology has not. If you were born August 7, you are said to be a Leo, but the Sun that day was really in the same part of the sky as the constellation Cancer.

With a theory like this to back it up, we should not be surprised at the bottom line: *A pseudoscience does not work.* Test after test has checked its predictions, and the result is always the same. One such investigator is Shawn Carlson of the University of California, San Diego. As he put it in *Nature* magazine, astrology is "a hopeless cause." Johannes Kepler, the pioneering astronomer, himself cast horoscopes, but they are little remembered today. Owen Gingerich, a historian of science at Harvard, puts it well: Kepler was the astrologer who destroyed astrology.

Astronomy works—it works very well indeed—which isn't easy. Because we humans tend to find what we want in any body of data, it takes science's careful process of observation, creative insight, and critical thinking to understand and predict changes in nature. As I write, a transit of Ganymede is due next Thursday at 21:47:20. At exactly that time, the satellite of Jupiter will cross in front of its planet as seen from Earth, and yet most of us will never know it. Still other moons of Jupiter may hold fresh clues to the formation of our entire solar system and the conditions for life elsewhere.

For most people, astronomy has too little fantasy or money in it, and they will never experience the beauty in its predictions. The dedicated labors of generations of scientists have enabled us to perform a genuine wonder.

James ("the Amazing") Randi works tirelessly to expose trickery so that others can relish the greater wonder of science. As a magician, he has had his own television show and an enormous public following. As a lecturer, he addresses teachers, students, and others worldwide. His newsletter and column for *The Skeptic* are key resources for educators. His many books include *Flim-Flam!, The Faith Healers,* and *The Mask of Nostradamus,* about a legendary con man with secrets of his own.

Mr. Randi is the founder of the James Randi Educational Foundation, and his one million dollar prize for "the performance of any paranormal event . . . under proper observing conditions" has gone unclaimed for more than 25 years. An amateur archeologist and astronomer as well, he lives in Florida with several untalented parrots and the occasional visiting magus.

The Sun in total eclipse, March 29, 2006.
(Stefan Seip)

R I V U X G

Eclipses and the Motion of the Moon

On March 29, 2006, a rare cosmic spectacle—a total solar eclipse—was visible along a narrow corridor that extended from the coast of Brazil through equatorial Africa and into central Asia. As shown in this digital composite taken in Turkey, the Moon slowly moved over the disk of the Sun. (Time flows from left to right in this image.) For a few brief minutes the Sun was totally covered, darkening the sky and revealing the Sun's thin outer atmosphere, or corona, which glows with an unearthly pearlescent light.

Such eclipses can be seen only on specific dates from special locations on Earth, so not everyone will ever see the Moon cover the Sun in this way. But anyone can find the Moon in the sky and observe how its appearance changes from night to night, from new moon to full moon and back again, and how the times when the Moon rises and sets differ noticeably from one night to the next.

In this chapter our subject is how the Moon moves as seen from Earth. We will explore why the Moon goes through a regular cycle of phases, and how the Moon's orbit around Earth leads to solar eclipses as well as lunar eclipses. We will also see how ancient astronomers used their observations of the Moon to determine the size and shape of Earth, as well as other features of the solar system. Thus, the Moon—which has always loomed large in the minds of poets, lovers, and dreamers—has also played a key role in the development of our modern picture of the universe.

3-1 The phases of the Moon are caused by its orbital motion

As seen from Earth, both the Sun and the Moon appear to move from west to east on the celestial sphere—that is, relative to the background of

> You can tell the Moon's position relative to Earth and the Sun by observing its phase

stars—but they move at very different rates. The Sun takes one year to make a complete trip around the imaginary celestial sphere along the path we call the *ecliptic* (Section 2-5). By comparison, the Moon takes only about four weeks. In the past, these similar motions led people to believe that both the Sun and the Moon orbit around Earth. We now know that only the Moon orbits Earth, while the Earth-Moon system as a whole (Figure 3-1) orbits the Sun. (In Chapter 4 we will learn how this was discovered.)

One key difference between the Sun and the Moon is the nature of the light that we receive from them. The Sun emits its own light. So do the stars, which are objects like the Sun but much farther away, and so does an ordinary lightbulb. By contrast, the light that we see from the Moon is reflected light. This is sunlight that has struck the Moon's surface, bounced off, and ended up in our eyes here on Earth.

CAUTION! You probably associate *reflection* with shiny objects like a mirror or the surface of a still lake. In science, however, the term refers to light bouncing off any object. You see most objects around you by reflected light. When you look at your hand, for example, you are seeing light from the Sun (or from a light fixture) that has been reflected from the skin of your hand and into your eyes. In the same way, moonlight is really sunlight that has been reflected by the Moon's surface.

Understanding the Moon's Phases

Figure 3-1 shows both the Moon and Earth as seen from a spacecraft. When this image was recorded, the Sun was far off to the right. Hence, only the right-hand hemispheres of both worlds were illuminated by the Sun; the left-hand hemispheres were in darkness and are not visible in the picture. In the same way, when we view the Moon from Earth, we see only the half of the Moon that faces the Sun and is illuminated. However, not all of the illuminated half of the Moon is necessarily facing us. As the Moon moves around Earth, from one night to the next, we see different amounts of the illuminated half of the Moon. These different appearances of the Moon are called **lunar phases.**

Figure 3-2 shows the relationship between the lunar phase visible from Earth and the position of the Moon in its orbit. For example, when the Moon is at position A, we see it in roughly the same direction in the sky as the Sun. Hence, the dark hemisphere of the Moon faces Earth. This phase, in which the Moon is barely visible for a day, is called **new moon.** Since a new moon can only be located near the Sun in the sky, it rises around sunrise, and sets around sunset.

As the Moon continues around its orbit from position A in Figure 3-2, more of its illuminated half becomes exposed to our view. The result, shown at position B, is a phase called **waxing crescent moon** ("waxing" is a synonym for "increasing"). About a week after new moon, the Moon is at position C; for a day we

FIGURE 3-1   R I V̄ U X G

Earth and the Moon This picture of Earth and the Moon was taken in 1992 by the *Galileo* spacecraft on its way toward Jupiter. The Sun, which provides the illumination for both Earth and the Moon, was far to the right and out of the camera's field of view when this photograph was taken. (NASA/JPL)

then see half of the Moon's illuminated hemisphere and half of the dark hemisphere. This phase is called **first quarter moon.**

As seen from Earth, a first quarter moon is one-quarter of the way around the celestial sphere from the Sun. It rises and sets about one-quarter of an Earth rotation, or 6 hours, after the Sun does: Moonrise occurs around noon, moonset occurs around midnight, and the moon is highest in the sky around 6 P.M.

CAUTION! Despite the name, a first quarter moon appears to be *half* illuminated, not one-quarter illuminated! The name means that this phase is one-quarter of the way through the complete cycle of lunar phases.

About four days later, the Moon reaches position D in Figure 3-2. Still more of the illuminated hemisphere can now be seen from Earth, giving us the phase called **waxing gibbous moon** ("gibbous" is another word for "swollen"). When you look at the Moon in this phase, as in the waxing crescent and first quarter phases, the illuminated part of the Moon is toward the west. Two weeks after new moon, when the Moon stands opposite the Sun in the sky (position E), we see the fully illuminated hemisphere. This phase is called **full moon.** Because a full moon is opposite the Sun on the celestial sphere, it rises at sunset and sets at sunrise.

Over the following two weeks, we see less and less of the Moon's illuminated hemisphere as it continues along its orbit, and the Moon is said to be *waning* ("decreasing"). While the Moon is waning, its illuminated side is toward the east. The phases are called **waning gibbous moon** (position F), **third quarter moon** (position G, also called *last quarter moon*), and **waning crescent moon** (position H). A third quarter moon appears one-quarter of the way around the celestial sphere from the Sun, but on the opposite side of the celestial sphere from a first quarter moon. Hence, a third quarter moon rises and sets about one-quarter

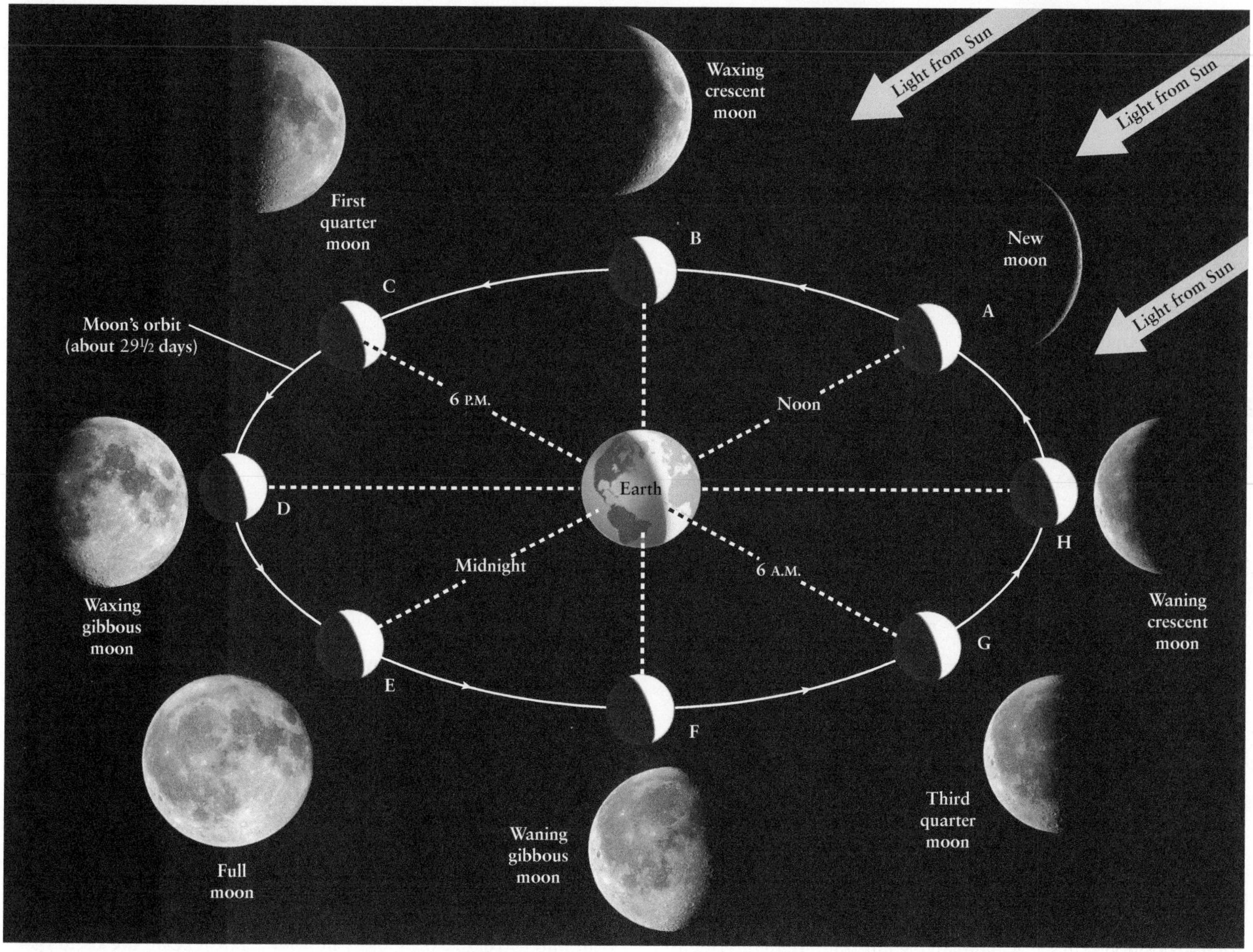

ANIMATION 3-1

FIGURE 3-2

Why the Moon Goes Through Phases This figure illustrates the Moon at eight positions on its orbit, along with photographs of what the Moon looks like at each position as seen from Earth. The changes in phase occur because light from the Sun illuminates one half of the Moon, and as the Moon orbits Earth we see varying amounts of the Moon's illuminated half. It takes about 29½ days for the Moon to go through a complete cycle of phases, and each of the phases shown is visible for one day during the complete cycle. The times given in the figure for each phase correspond to the time each phase is observed highest in the sky, halfway between its rise and set. (Photographs from Larry Landolfi/Science Source)

Earth rotation, or 6 hours, *before* the Sun: Moonrise is around midnight and moonset is around noon.

The Moon takes about four weeks to complete one orbit around Earth, so it likewise takes about four weeks for a complete cycle of phases from new moon to full moon and back to new moon. Since the Moon's position relative to the Sun on the celestial sphere is constantly changing, and since our system of timekeeping is based on the Sun (see Section 2-7), the times of moonrise and moonset are different on different nights. On average, the Moon rises and sets about an hour later each night.

Figure 3-2 also explains why the Moon is often visible in the daytime, as shown in Figure 3-3. From any location on Earth, about half of the Moon's orbit is visible at any time. For example, if it is midnight at your location, you are in the middle of the dark side of Earth that faces away from the Sun. At that time you can easily see the Moon if it is at position C, D, E, F, or G. If it is midday at your location, you are in the middle of Earth's illuminated side, and the Moon will be easily visible if it is at position A, B, C, G, or H. (The Moon is so bright that it can be seen even against the bright blue sky.) You can see that the Moon is prominent in the midnight sky for about half of its orbit, and prominent in the midday sky for the other half.

CAUTION! A very common misconception about lunar phases is that they are caused by the shadow of *Earth* falling on the Moon. As Figure 3-2 shows, this is not the case at all. Instead, phases are simply the result of our seeing the illuminated half of the Moon at different angles as the Moon moves around its orbit. To help you better visualize how this works, Box 3-1 describes how you can simulate the cycle shown in Figure 3-2 using ordinary objects on Earth. (As we will learn in Section 3-3, Earth's shadow does indeed fall on the Moon on rare occasions. When this happens, we see a lunar eclipse.)

FIGURE 3-3 R I V U X G

The Moon During the Day The Moon can be seen during the daytime as well as at night. The time of day or night when it is visible depends on its phase. (Karl Beath/Gallo Images)

BOX 3-1 ASTRONOMY DOWN TO EARTH

Phases and Shadows

Figure 3-2 shows how the relative positions of Earth, the Moon, and the Sun explain the phases of the Moon. You can visualize lunar phases more clearly by doing a simple experiment here on Earth. All you need are a small round object, such as an orange or a baseball, and a bright source of light, such as a street lamp or the Sun.

In this experiment, you play the role of an observer on Earth looking at the Moon, and the round object plays the role of the Moon. The light source plays the role of the Sun. Hold the object in your right hand with your right arm stretched straight out in front of you, with the object directly between you and the light source (position A in the accompanying illustration). In this orientation the illuminated half of the object faces away from you, like the Moon when it is in its new phase (position A in Figure 3-2).

Now, slowly turn your body to the left so that the object in your hand "orbits" around you (toward positions C, E, and G in the illustration). As you turn, more and more of the illuminated side of the "moon" in your hand becomes visible, and it goes through the same cycle of phases—waxing crescent, first quarter, and waxing gibbous—as does the real Moon. When you have rotated through half a turn so that the light source is directly behind you, you will be looking face on at the illuminated side of the object in your hand. This corresponds to a full moon (position E in Figure 3-2). Make sure your body does not cast a shadow on the "moon" in your hand—that would correspond to a lunar eclipse!

As you continue turning to the left, more of the unilluminated half of the object becomes visible as its phase moves through waning gibbous, third quarter, and waning crescent. When your body has rotated back to the same orientation that you were in originally, the unilluminated half of your handheld "moon" is again facing toward you, and its phase is again new. If you continue to rotate, the object in your hand repeats the cycle of "phases," just as the Moon does as it orbits around Earth.

The experiment works best when there is just one light source around. If there are several light sources, such as in a room with several lamps turned on, the different sources will create multiple shadows, and it will be difficult to see the phases of your handheld "moon." If you do the experiment outdoors using sunlight, you may find that it is best to perform it in the early morning or late afternoon, when shadows are most pronounced and the Sun's rays are nearly horizontal.

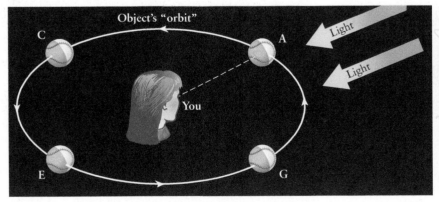

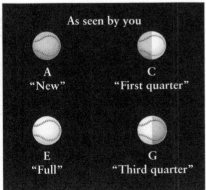

CONCEPTCHECK 3-1

If an observer on Earth sees just a tiny sliver of the crescent moon, how much of the Moon's total surface is being illuminated by the Sun?

CONCEPTCHECK 3-2

If the Moon appears in its waxing crescent phase, how will it appear in two weeks?

Answers appear at the end of the chapter.

3-2 The Moon always keeps the same face toward Earth

Although the phase of the Moon is constantly changing, one aspect of its appearance remains the same: It always keeps essentially the same hemisphere, or face, toward Earth. Thus, you will always see the same craters and mountains on the Moon, no matter when you look at it; the only difference will be the angle at which these surface features are illuminated by the Sun. (You can verify this by carefully examining the photographs of the Moon in Figure 3-2.)

> The Moon rotates in a special way: It spins exactly once per orbit

The Moon's Synchronous Rotation

Why is it that we only ever see one face of the Moon? You might think that it is because the Moon does not rotate (unlike Earth, which rotates around an axis that passes from its north pole to its south pole). To see that the Moon must rotate, consider Figure 3-4. This figure shows Earth and the orbiting Moon from a vantage point far above Earth's north pole. In this figure two craters on the lunar surface have been colored, one in red and one in blue. If the Moon did not rotate on its axis, as in Figure 3-4a, sometimes the red crater would be visible from Earth, while at other times the blue crater would be visible. Thus, we would see different parts of the lunar surface over time, which does not happen in reality.

In fact, the Moon always keeps the same face toward us because it *is* rotating, but in a very special way: It takes exactly as long for the Moon to rotate on its axis as it does to make one orbit around Earth. This situation is called **synchronous rotation.** As Figure 3-4b shows, this keeps the crater shown in red always facing Earth, so that we always see the same face of the Moon. In Chapter 4 we will learn why the Moon's rotation and orbital motion are in step with each other.

Is there a permanently "dark side of the Moon?" Not at all. To understand this, consider the red crater in Figure 3-4b. The red crater would spend two weeks (half of a lunar orbit) in darkness, and the next two weeks in sunlight. Thus, no part of the Moon

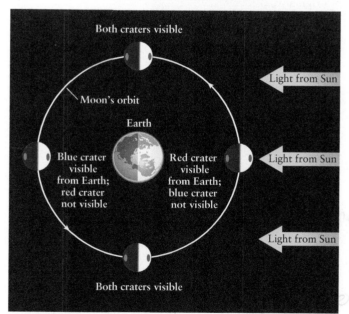

If the Moon did not rotate, we could see all sides of the Moon

(a) Wrong model

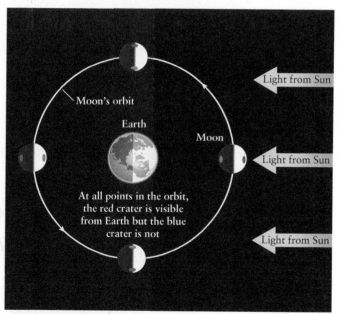

In fact, the Moon does rotate, and we see only one face of the Moon

(b) Correct picture

FIGURE 3-4

ANIMATION 3-2

The Moon's Rotation These diagrams show the Moon at four points in its orbit as viewed from high above Earth's north pole. **(a)** The wrong model: If the Moon did not rotate, then at various times the red crater would be visible from Earth while at other times the blue crater would be visible. Over a complete orbit, the entire surface of the Moon would be visible. **(b)** In reality, the Moon rotates on its north-south axis. Because the Moon makes one rotation in exactly the same time that it makes one orbit around Earth, we see only one face of the Moon.

is perpetually in darkness. The side of the Moon that constantly faces away from Earth is properly called the *far side*.

CONCEPTCHECK 3-3

Viewed from space, the far side of the Moon is fully dark at only one point during the lunar orbit. Can you identify this point in Figure 3-4b?

CONCEPTCHECK 3-4

If astronauts landed on the Moon near the center of the visible surface at full moon, how many Earth days would pass before the astronauts experienced darkness on the Moon?

Answers appear at the end of the chapter.

Sidereal and Synodic Months

It takes about four weeks for the Moon to complete one cycle of its phases as seen from Earth. This regular cycle of phases inspired our ancestors to invent the concept of a month. For historical reasons, the calendar we use today has months of differing lengths. Astronomers find it useful to define two other types of months, depending on whether the Moon's motion is measured relative to the stars or to the Sun. Neither corresponds exactly to the familiar months of the calendar.

The **sidereal month** is the time it takes the Moon to complete one full orbit of Earth, as measured *with respect to the stars*. The sidereal period is what you would observe while hovering in space (like the stars), watching the Moon orbit Earth. This true orbital period is equal to about 27.32 days. The **synodic month**, or *lunar month*, is the time it takes the Moon to complete one cycle of phases (that is, from new moon to new moon or from full moon to full moon) and thus is measured *with respect to the Sun* rather than the stars.

How long is a "day" on the Moon? On Earth, a 24-hour day measures the average time between successive sunrises (or sunsets). Therefore, a day on Earth is measured with respect to the Sun and is a synodic day. The lunar "day" is also the time from sunrise to sunrise as seen from the Moon's surface, and this is just what defines the Moon's synodic month; a lunar day is equal to a synodic month.

The synodic month is longer than the sidereal month because Earth is orbiting the Sun while the Moon goes through its phases. As **Figure 3-5** shows, the Moon must travel *more* than 360° along its orbit to complete a cycle of phases (for example, from one new moon to the next). Because of this extra distance, the synodic month is equal to about 29.53 days, about two days longer than the sidereal month.

Both the sidereal month and synodic month vary somewhat from one orbit to another, the latter by as much as half a day. The reason is that the Sun's gravity sometimes causes the

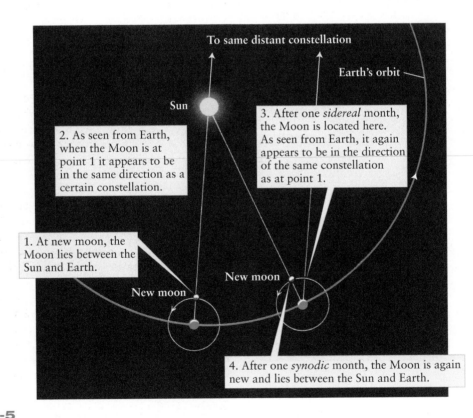

To same distant constellation

Earth's orbit

Sun

2. As seen from Earth, when the Moon is at point 1 it appears to be in the same direction as a certain constellation.

3. After one *sidereal* month, the Moon is located here. As seen from Earth, it again appears to be in the direction of the same constellation as at point 1.

1. At new moon, the Moon lies between the Sun and Earth.

New moon

New moon

4. After one *synodic* month, the Moon is again new and lies between the Sun and Earth.

ANIMATION 3-3   **FIGURE 3-5**

The Sidereal and Synodic Months The sidereal month is the time the Moon takes to complete one full revolution around Earth with respect to the background stars. However, because Earth is constantly moving along its orbit about the Sun, the Moon must travel through slightly more than 360° of its orbit to get from one new moon to the next. Thus, the synodic month—the time from one new moon to the next—is longer than the sidereal month.

Moon to speed up or slow down slightly in its orbit, depending on the relative positions of the Sun, Moon, and Earth. Furthermore, the Moon's orbit changes slightly from one month to the next.

CONCEPTCHECK 3-5

If Earth was orbiting the Sun much faster than it is now, would the length of time between full moons increase, decrease, or stay the same?

Answer appears at the end of the chapter.

3-3 Eclipses occur only when the Sun and Moon are both on the line of nodes

From time to time the Sun, Earth, and Moon all happen to lie along a straight line. When this occurs, the shadow of
Earth can fall on the Moon or the shadow of the Moon can fall on Earth. Such phenomena are called **eclipses.** They are perhaps the most dramatic astronomical events that can be seen with the naked eye.

> The tilt of the Moon's orbit makes lunar and solar eclipses rare events

A **lunar eclipse** occurs when the Moon passes through Earth's shadow. This occurs when the Sun, Earth, and Moon are in a straight line, with Earth between the Sun and Moon so that the Moon is at full phase (position E in Figure 3-2). At this point in the Moon's orbit, the face of the Moon seen from Earth would normally be fully illuminated by the Sun. Instead, it appears quite dim because Earth casts a shadow on the Moon.

A **solar eclipse** occurs when Earth passes through the Moon's shadow. As seen from Earth, the Moon moves in front of the Sun. Once again, this can happen only when the Sun, Moon, and Earth are in a straight line. However, for a solar eclipse to occur, the Moon must be between Earth and the Sun. Therefore, a solar eclipse can occur only at new moon (position A in Figure 3-2).

CAUTION! Both new moon and full moon occur at intervals of 29½ days. Hence, you might expect that there would be a solar eclipse every 29½ days, followed by a lunar eclipse about two weeks (half a lunar orbit) later. But in fact, there are only a few solar eclipses and lunar eclipses per year. Solar and lunar eclipses are so infrequent because the plane of the Moon's orbit and the plane of Earth's orbit are not exactly aligned, as Figure 3-6 shows. The angle between the plane of Earth's orbit and the plane of the Moon's orbit is about 5°. Because of this tilt, new moon and full moon usually occur when the Moon is either above or below the plane of Earth's orbit. When the Moon is not in the plane of Earth's orbit, the Sun, Moon, and Earth cannot align perfectly, and an eclipse cannot occur.

In order for the Sun, Earth, and Moon to be lined up for an eclipse, the Moon must lie in the same plane as Earth's orbit

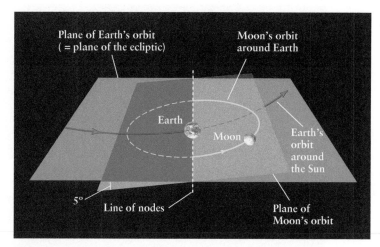

FIGURE 3-6

The Inclination of the Moon's Orbit This drawing shows the Moon's orbit around Earth (in yellow) and part of Earth's orbit around the Sun (in red). The plane of the Moon's orbit (shown in brown) is tilted by about 5° with respect to the plane of Earth's orbit, also called the plane of the ecliptic (shown in blue). These two planes intersect along a line called the line of nodes.

around the Sun. As we saw in Section 2-5, this plane is called the *ecliptic plane* because it is the same as the plane of the Sun's apparent path around the sky, or ecliptic (see Figure 2-14). Thus, when an eclipse occurs, the Moon appears from Earth to be on the ecliptic—this is how the ecliptic gets its name.

The planes of Earth's orbit and the Moon's orbit intersect along a line called the **line of nodes,** shown in Figure 3-6. The line of nodes passes through Earth and points in a particular direction in space. Eclipses can occur only if the line of nodes points toward the Sun—that is, if the Sun lies on or near the line of nodes—and if, at the same time, the Moon lies on or very near the line of nodes. Only then do the Sun, Earth, and Moon lie in a line straight enough for an eclipse to occur (Figure 3-7). Note that the alignment necessary for eclipses has nothing to do with either the solstices or the equinoxes.

Anyone who wants to predict eclipses must know the orientation of the line of nodes. But the line of nodes is gradually shifting because of the gravitational pull of the Sun on the Moon. As a result, the line of nodes rotates slowly westward. Astronomers calculate such details to fix the dates and times of upcoming eclipses.

There are at least two—but never more than five—solar eclipses each year. The last year in which five solar eclipses occurred was 1935. The fewest eclipses possible (two solar, zero lunar) happened in 1969. Lunar eclipses occur just about as frequently as solar eclipses, but the maximum possible number of eclipses (lunar and solar combined) in a single year is seven.

CONCEPTCHECK 3-6

Why don't lunar eclipses occur each time the Moon reaches full moon phase?

Answer appears at the end of the chapter.

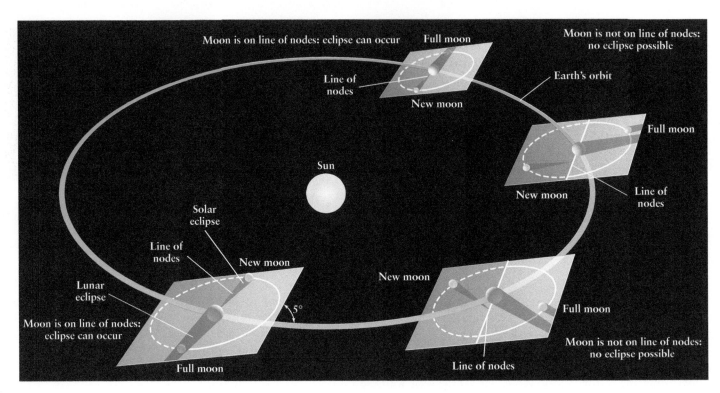

ANIMATION 3-2 **FIGURE 3-7**

Conditions for Eclipses Eclipses can take place only if the Sun and Moon are both very near to or on the line of nodes. Only then can the Sun, Earth, and Moon all lie along a straight line. A solar eclipse occurs only if the Moon is very near the line of nodes at new moon; a lunar eclipse occurs only if the Moon is very near the line of nodes at full moon. If the Sun and Moon are not near the line of nodes, the Moon's shadow cannot fall on Earth and Earth's shadow cannot fall on the Moon.

3-4 The character of a lunar eclipse depends on the alignment of the Sun, Earth, and Moon

The character of a lunar eclipse depends on exactly how the Moon travels through Earth's shadow. As Figure 3-8 shows, the shadow

> The Sun's tenuous outer atmosphere is revealed during a total solar eclipse

of Earth has two distinct parts. To picture how the Moon's surface would be illuminated, imagine an observer on the Moon as the Moon passes through the shadow. In the **umbra,** no portion of the Sun's surface can be seen from the Moon: This is the darkest part of the shadow. On the other hand, a portion of the Sun's surface is visible in the **penumbra,** which therefore is not quite as dark. Most people notice a lunar eclipse only if the Moon passes into Earth's

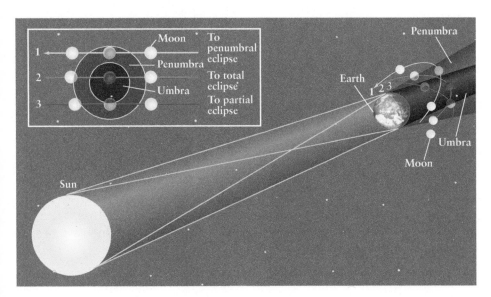

FIGURE 3-8

Three Types of Lunar Eclipse People on the nighttime side of Earth see a lunar eclipse when the Moon moves through Earth's shadow. In the umbra, the darkest part of the shadow, the Sun is completely covered by Earth. The penumbra is less dark because only part of the Sun is covered by Earth. The three paths show the motion of the Moon if the lunar eclipse is penumbral (Path 1, in yellow), total (Path 2, in red), or partial (Path 3, in blue). The inset shows these same paths, along with the umbra and penumbra, as viewed from Earth.

FIGURE 3-9   R I ☑ U X G

A Total Lunar Eclipse This sequence of nine photographs was taken over a 3-hour period during the lunar eclipse of January 20, 2000. The sequence, which runs from right to left, shows the Moon moving through Earth's umbra.

During the total phase of the eclipse (shown in the center), the Moon has a distinct reddish color. (Courtesy of Fred Espenak, www.mreclipse.com)

umbra. As this umbral phase of the eclipse begins, a bite appears to have been taken out of the Moon.

The inset in Figure 3-8 shows the different ways in which the Moon can pass into Earth's shadow. When the Moon passes through only Earth's penumbra (Path 1), we see a **penumbral eclipse.** During a penumbral eclipse, Earth blocks only part of the Sun's light and so none of the lunar surface is completely shaded. Because the Moon still looks full but only a little dimmer than usual, penumbral eclipses are easy to miss. If the Moon travels completely into the umbra (Path 2), **a total lunar eclipse** occurs. If only part of the Moon passes through the umbra (Path 3), we see a **partial lunar eclipse.**

If you were on the Moon during a total lunar eclipse, the Sun would be hidden behind Earth. But some sunlight would be visible through the thin ring of atmosphere around Earth, just as you can see sunlight through a person's hair if they stand with their head between your eyes and the Sun. As a result, a small amount of light reaches the Moon during a total lunar eclipse, and so the Moon does not completely disappear from the sky as

seen from Earth. Most of the sunlight that passes through Earth's atmosphere is red, and thus the eclipsed Moon glows faintly in reddish hues, as Figure 3-9 shows. (We'll see in Chapter 5 how our atmosphere causes this reddish color.)

Lunar eclipses occur at full moon, when the Moon is directly opposite the Sun in the sky. Hence, a lunar eclipse can be seen at any place on Earth where the Sun is below the horizon (that is, where it is nighttime). A lunar eclipse has the maximum possible duration if the Moon travels directly through the center of the umbra. The Moon's speed through Earth's shadow is roughly 1 kilometer per second (3600 kilometers per hour, or 2280 miles per hour), which means that **totality**—the period when the Moon is completely within Earth's umbra—can last for as long as 1 hour and 42 minutes.

On average, two or three lunar eclipses occur in a year. Table 3-1 lists all nine lunar eclipses from 2012 to 2015. Of all lunar eclipses, roughly one-third are total, one-third are partial, and one-third are penumbral.

TABLE 3-1 Lunar Eclipses, 2012–2015			
Date	Type	Where visible	Duration of totality (h = hours, m = minutes)
2012 June 4	Partial	Asia, Australia, Pacific, Americas	—
2012 Nov 28	Penumbral	Europe, eastern Africa, Asia, Australia, Pacific, North America	—
2013 Apr 25	Penumbral	Europe, Africa, Asia, Australia	—
2013 May 25	Penumbral	Americas, Africa	—
2013 Oct 18	Penumbral	Americas, Europe, Africa, Asia	—
2014 Apr 15	Total	Australia, Pacific, Americas	01h 18m
2014 Oct 08	Total	Asia, Australia, Pacific, Americas	59m
2015 Apr 04	Total	Asia, Australia, Pacific, Americas	5m
2015 Sep 28	Total	East Pacific, Americas, Europe, Africa, western Asia	1h 12m

Eclipse predictions by Fred Espenak, NASA/Goddard Space Flight Center.

Is it possible to have a total lunar eclipse without a penumbral eclipse?

Answer appears at the end of the chapter.

3-5 Solar eclipses also depend on the alignment of the Sun, Earth, and Moon

As seen from Earth, the angular diameter of the Moon is almost exactly the same as the angular diameter of the far larger but more distant Sun—about 0.5°. Thanks to this coincidence of nature, the Moon just "fits" over the Sun during a **total solar eclipse.**

Total Solar Eclipses

A total solar eclipse is a dramatic event. The sky begins to darken, the air temperature falls, and winds increase as the Moon gradually covers more and more of the Sun's disk. All nature responds: Birds go to roost, flowers close their petals, and crickets begin to chirp as if evening had arrived. As the last few rays of sunlight peek out from behind the edge of the Moon and the

> A lunar eclipse is most impressive when total, but can also be partial or penumbral

eclipse becomes total, the landscape around you is bathed in an eerie gray or, less frequently, in shimmering bands of light and dark. Finally, for a few minutes the Moon completely blocks out the dazzling solar disk (**Figure 3-10a**). The **solar corona**—the Sun's thin, hot outer atmosphere, which is normally too dim to be seen—blazes forth in the darkened daytime sky (Figure 3-10b). It is an awe-inspiring sight.

CAUTION! If you are fortunate enough to see a solar eclipse, keep in mind that the only time when it is safe to look at the Sun is during **totality,** when the solar disk is blocked by the Moon and only the solar corona is visible. Viewing this magnificent spectacle cannot harm you in any way. But you must *never* look directly at the Sun when even a portion of its intensely brilliant disk is exposed. *If you look directly at the Sun at any time without a special filter approved for solar viewing, you will suffer permanent eye damage or blindness.*

To see the remarkable spectacle of a total solar eclipse, you must be inside the darkest part of the Moon's shadow, also called the umbra, where the Moon completely blocks the Sun. Because the Sun and the Moon have nearly the same angular diameter as seen from Earth, only the tip of the Moon's umbra reaches Earth's surface (**Figure 3-11**). As Earth rotates, the tip of the umbra traces

(a)

(b)

FIGURE 3-10   R I **V** U X G

A Total Solar Eclipse **(a)** This photograph shows the total solar eclipse of August 11, 1999, as seen from Elâziğ, Turkey. The sky is so dark that the planet Venus can be seen to the left of the eclipsed Sun. **(b)** When the Moon completely covers the Sun's disk during a total eclipse, the faint solar corona is revealed. (a: Courtesy of Fred Espenak, www.mreclipse.com; b: © 2008 Miloslav Druckmüller, Martin Dietzel, Peter Aniol, Vojtech Rušin)

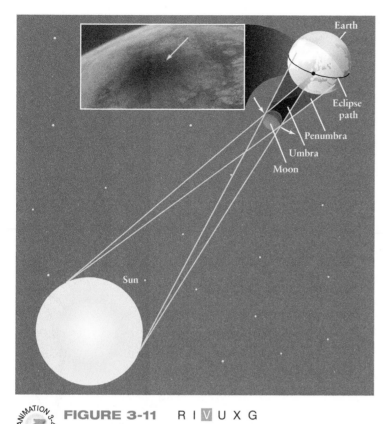

FIGURE 3-11   R I **V** U X G

The Geometry of a Total Solar Eclipse During a total solar eclipse, the tip of the Moon's umbra reaches Earth's surface. As Earth and the Moon move along their orbits, this tip traces an eclipse path across Earth's surface. People within the eclipse path see a total solar eclipse as the tip moves over them. Anyone within the penumbra sees only a partial eclipse. The inset photograph was taken from the *Mir* space station during the August 11, 1999, total solar eclipse (the same eclipse shown in Figure 3-10). The tip of the umbra appears as a black spot on Earth's surface. At the time the photograph was taken, this spot was 105 km (65 mi) wide and was crossing the cloud-covered English Channel at 3000 km/h (1900 mi/h). (Photograph by Jean-Pierre Haigneré, Centre National d'Etudes Spatiales, France/GSFS/NASA)

an **eclipse path** across Earth's surface. Only those locations within the eclipse path are treated to the spectacle of a total solar eclipse. The inset in Figure 3-11 shows the dark spot on Earth's surface produced by the Moon's umbra.

Partial Solar Eclipses

Immediately surrounding the Moon's umbra is the region of partial shadow called the penumbra. As seen from this area, the Sun's surface appears only partially covered by the Moon. During a solar eclipse, the Moon's penumbra covers a large portion of Earth's surface, and anyone standing inside the penumbra sees a **partial solar eclipse.** Such eclipses are much less interesting events than total solar eclipses, which is why astronomy enthusiasts strive to be inside the eclipse path. If you are within the eclipse path, you will see a partial eclipse before and after the brief period of totality (see the photograph that opens this chapter).

The width of the eclipse path depends primarily on the Earth-Moon distance during totality. The eclipse path is widest if the

Moon happens to be at **perigee,** the point in its orbit nearest Earth. In this case the width of the eclipse path can be as great as 270 kilometers (170 miles). In most eclipses, however, the path is much narrower.

CONCEPTCHECK 3-8

Why can a total lunar eclipse be seen by people all over the world whereas total solar eclipses can only be seen from a very limited geographic region?

Answer appears at the end of the chapter.

Annular Solar Eclipses

In some eclipses the Moon's umbra does not reach all the way to Earth's surface. This can happen if the Moon is at or near **apogee,** its farthest position from Earth. In this case, the Moon appears too small to cover the Sun completely. The result is a third type of solar eclipse, called an **annular eclipse.** During an annular eclipse, a thin ring of the Sun is seen around the edge of the Moon (Figure 3-12). The length of the Moon's umbra is

FIGURE 3-12   R I **V** U X G

An Annular Solar Eclipse This composite of six photographs taken at sunrise in Costa Rica shows the progress of an annular eclipse of the Sun on December 24, 1973. Note that at mideclipse the limb, or outer edge, of the Sun is visible around the Moon. (Photo Researchers/Getty Images)

nearly 5000 kilometers (3100 miles) less than the average distance between the Moon and Earth's surface. Thus, the Moon's shadow often fails to reach Earth even when the Sun, Moon, and Earth are properly aligned for an eclipse. Hence, annular eclipses are slightly more common—as well as far less dramatic—than total eclipses.

Even during a total eclipse, most people along the eclipse path observe totality for only a few moments. Earth's rotation, coupled with the orbital motion of the Moon, causes the umbra to race eastward along the eclipse path at speeds in excess of 1700 kilometers per hour (1060 miles per hour). Because of the umbra's high speed, totality never lasts for more than 7½ minutes. In a typical total solar eclipse, the Sun-Moon-Earth alignment and the Earth-Moon distance are such that totality lasts much less than this maximum.

The details of solar eclipses are calculated well in advance. They are published in such reference books as the *Astronomical Almanac* and are available on the World Wide Web. **Figure 3-13** shows the eclipse paths for all total solar eclipses from 1997 to 2020. **Table 3-2** lists all the total, annular, and partial eclipses from 2012 to 2015, including the maximum duration of totality for total eclipses.

Ancient astronomers achieved a limited ability to predict eclipses. In those times, religious and political leaders who were able to predict such awe-inspiring events as eclipses must have made a tremendous impression on their followers. One of three priceless manuscripts to survive the devastating Spanish Conquest shows that the Mayan astronomers of Mexico and Guatemala had

a fairly reliable method for predicting eclipses. The great Greek astronomer Thales of Miletus is said to have predicted the famous eclipse of 585 B.C.E, which occurred during the middle of a war. The sight was so unnerving that the soldiers put down their arms and declared peace.

In retrospect, it seems that what ancient astronomers actually produced were eclipse "warnings" of various degrees of reliability rather than true predictions. Working with historical records, these astronomers generally sought to discover cycles and regularities from which future eclipses could be anticipated. **Box 3-2** describes how you might produce eclipse warnings yourself.

CONCEPTCHECK 3-9

If you had a chance to observe a total solar eclipse and a total lunar eclipse, in general, how much longer would you expect one type to last than the other?

Answer appears at the end of the chapter.

3-6 Ancient astronomers measured the size of Earth and attempted to determine distances to the Sun and Moon

The prediction of eclipses was not the only problem attacked by ancient astronomers. More than 2000 years ago, centuries before

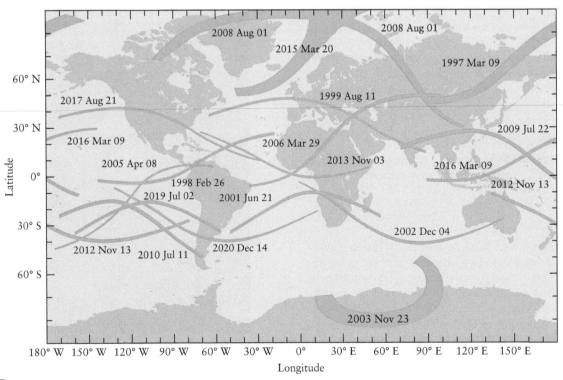

FIGURE 3-13

Eclipse Paths for Total Eclipses, 1997–2020 This map shows the eclipse paths for all 18 total solar eclipses occurring from 1997 through 2020. In each eclipse, the Moon's shadow travels along the eclipse path in a generally eastward direction across Earth's surface. (Courtesy of Fred Espenak, NASA/Goddard Space Flight Center)

TABLE 3-2 Solar Eclipses, 2012–2015

Date	Type	Where visible	Notes
2012 May 20	Annular	Asia, Pacific, North America	94% eclipsed
2012 Nov 13	Total	Australia, New Zealand, southern Pacific, southern South America	Maximum duration of totality 4m 02s
2013 May 10	Annular	Australia, New Zealand, central Pacific	Maximum duration of 06m 03s
2013 Nov 03	Total/Annular	Eastern Americas, southern Europe, Africa	Rare hybrid eclipse with maximum duration of 1m 40s
2014 Apr 29	Annular	Southern Indian Ocean, Australia, Antarctica	98% eclipsed
2015 Mar 20	Total	Iceland, Europe, northern Africa, northern Asia	Maximum duration of 02m 47s
2015 Sep 13	Partial	Southern Africa, southern Indian Ocean, Antarctica	79% eclipsed

Eclipse predictions by Fred Espenak, NASA/Goddard Space Flight Center.

BOX 3-2 TOOLS OF THE ASTRONOMER'S TRADE

Predicting Solar Eclipses

Suppose that you observe a solar eclipse in your hometown and want to figure out when you and your neighbors might see another eclipse. How would you begin?

First, remember that a solar eclipse can occur only if the line of nodes points toward the Sun at the same time that there is a new moon (see Figure 3-7). Second, you must know that it takes 29.53 days (one synodic month) to go from one new moon to the next. Because solar eclipses occur only during new moon, you must wait several whole lunar months for the proper alignment to occur again.

However, there is a complication: The line of nodes gradually shifts its position with respect to the background stars. It takes 346.6 days to move from one alignment of the line of nodes pointing toward the Sun to the next identical alignment. This period is called the **eclipse year.**

Therefore, to predict when you will see another solar eclipse, you need to know how many whole lunar months equal some whole number of eclipse years. This information will tell you how long you will have to wait for the next virtually identical alignment of the Sun, the Moon, and the line of nodes. By trial and error, you find that 223 lunar months is the same length of time as 19 eclipse years, because

$$223 \times 29.53 \text{ days} = 19 \times 346.6 \text{ days} = 6585 \text{ days}$$

This calculation is accurate to within a few hours. A more accurate calculation gives an interval, called the **saros,** that is about one-third of a day longer, or 6585.3 days (18 years, 11.3 days). Eclipses separated by the saros interval are said to form an *eclipse series.*

You might think that you and your neighbors would simply have to wait one full saros interval to go from one solar eclipse to the next. However, because of the extra one-third day, Earth will have rotated by an extra 120° (one-third of a complete rotation) when the next solar eclipse of a particular series occurs. The eclipse path will thus be one-third of the way around the world from you. Therefore, you must wait three full saros intervals (54 years, 34 days) before the eclipse path comes back around to your part of Earth. The illustration shows a series of six solar eclipse paths, each separated from the next by one saros interval.

There is evidence that ancient Babylonian astronomers knew about the saros interval. However, the discovery of the saros is more likely to have come from lunar eclipses than solar eclipses. If you are far from the eclipse path, there is a good chance that you could fail to notice a solar eclipse. Even if half the Sun is covered by the Moon, the remaining solar surface provides enough sunlight for the outdoor illumination not to be greatly diminished. By contrast, anyone on the nighttime side of Earth can see an eclipse of the Moon unless clouds block the view.

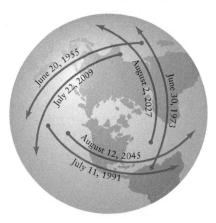

sailors of Columbus's era crossed the oceans, Greek astronomers were fully aware that Earth is not flat. They had come to this conclusion using a combination of observation and logical deduction, much like modern scientists. The Greeks noted that during lunar eclipses, when the Moon passes through Earth's shadow, the edge of the shadow is always circular. Because a sphere is the only shape that always casts a circular shadow from any angle, they concluded that Earth is spherical.

The ideas of geometry made it possible for Greek scholars to estimate cosmic distances

Eratosthenes and the Size of Earth

Around 200 B.C.E., the Greek astronomer Eratosthenes devised a way to measure the circumference of the spherical Earth. It was known that on the date of the summer solstice (the first day of summer; see Section 2-5) in the ancient town of Syene in Egypt, near present-day Aswan, the Sun shone directly down the vertical shafts of water wells. Hence, at local noon on that day, the Sun was at the zenith (see Section 2-4) as seen from Syene. Eratosthenes knew that the Sun never appeared at the zenith at his home in the Egyptian city of Alexandria, which is on the Mediterranean Sea almost due north of Syene. Rather, on the summer solstice in Alexandria, the position of the Sun at local noon was about 7° south of the zenith (Figure 3-14). This angle is about one-fiftieth of a complete circle, so he concluded that the distance from Alexandria to Syene must be about one-fiftieth of Earth's circumference.

In Eratosthenes's day, the distance from Alexandria to Syene was said to be 5000 stades. Therefore, Eratosthenes found Earth's circumference to be

$$50 \times 5000 \text{ stades} = 250,000 \text{ stades}$$

Unfortunately, no one today is sure of the exact length of the Greek unit called the stade. One guess is that the stade was about one-sixth of a kilometer, which would mean that Eratosthenes obtained a circumference for Earth of about 42,000 kilometers. This is remarkably close to the modern value of 40,000 kilometers.

Aristarchus and Distances in the Solar System

Eratosthenes was only one of several brilliant astronomers to emerge from the so-called Alexandrian school, which by his time had a distinguished tradition. One of the first Alexandrian astronomers, Aristarchus of Samos, had proposed a method of determining the relative distances to the Sun and Moon, perhaps as long ago as 280 B.C.E.

Aristarchus knew that the Sun, Moon, and Earth form a right triangle at the moment of first or third quarter moon, with the right angle at the location of the Moon (Figure 3-15). He estimated that, as seen from Earth, the angle between the Moon and the Sun at first and third quarters is 87°, or 3° less than a right angle. Using the rules of geometry, Aristarchus concluded that the Sun is about 20 times farther from us than is the Moon. We now know that Aristarchus erred in measuring angles and that the average distance to the Sun is about 390 times larger than the average distance to the Moon. It is nevertheless impressive that people were trying to measure distances across the solar system more than 2000 years ago.

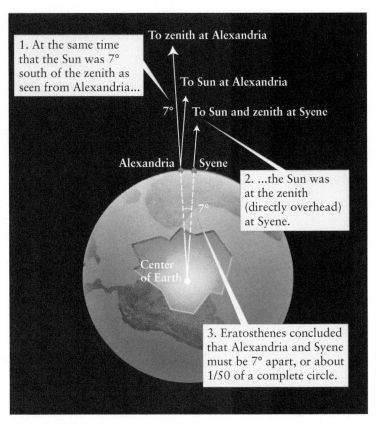

FIGURE 3-14

Eratosthenes's Method of Determining the Diameter of Earth Around 200 B.C.E., Eratosthenes used observations of the Sun's position at noon on the summer solstice to show that Alexandria and Syene were about 7° apart on the surface of Earth. This angle is about one-fiftieth of a circle, so the distance between Alexandria and Syene must be about one-fiftieth of Earth's circumference.

Aristarchus also made an equally bold attempt to determine the relative sizes of Earth, the Moon, and the Sun. From his observations of how long the Moon takes to move through Earth's shadow during a lunar eclipse, Aristarchus estimated the diameter of Earth to be about 3 times larger than the diameter of the Moon. To determine the diameter of the Sun, Aristarchus simply pointed out that the Sun and the Moon have the same angular size in the sky. Therefore, their diameters must be in the same proportion as their distances (see part a of the figure in Box 1-1). In other words, because Aristarchus thought the Sun to be 20 times farther from Earth than the Moon, he concluded that the Sun must be 20 times larger than the Moon. Once Eratosthenes had measured Earth's circumference, astronomers of the Alexandrian school could estimate the diameters of the Sun and Moon as well as their distances from Earth.

Table 3-3 summarizes some ancient and modern measurements of the sizes of Earth, the Moon, and the Sun and the distances between them. Some of these ancient measurements are far from the modern values. Yet the achievements of our ancestors still stand as impressive applications of observation and reasoning and important steps toward the development of the scientific method.

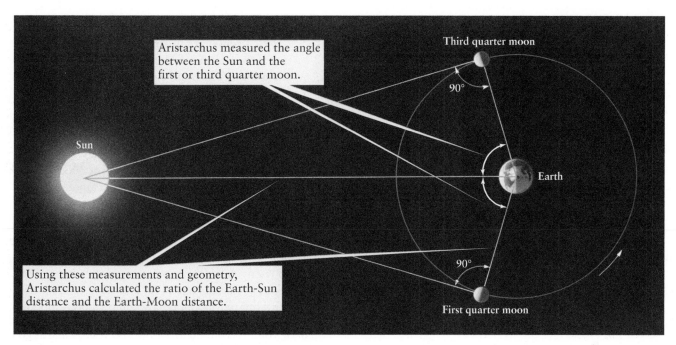

FIGURE 3-15

Aristarchus's Method of Determining Distances to the Sun and Moon Aristarchus knew that the Sun, Moon, and Earth form a right triangle at first and third quarter phases. Using geometrical arguments, he calculated the relative lengths of the sides of these triangles, thereby obtaining the distances to the Sun and Moon.

TABLE 3-3	Comparison of Ancient and Modern Astronomical Measurements	
	Ancient (km)	Modern (km)
Earth's diameter	13,000	12,756
Moon's diameter	4,300	3,476
Sun's diameter	9×10^4	1.39×10^6
Earth-Moon distance	4×10^5	3.84×10^5
Earth-Sun distance	10^7	1.50×10^8

CONCEPTCHECK 3-10

If Eratosthenes had found that the Sun's noontime summer solstice altitude at Alexandria was much closer to directly overhead, would he then assume that Earth was larger, smaller, or about the same size?

CALCULATIONCHECK 3-1

If Aristarchus had estimated the Sun to be 100 times farther from Earth than the Moon, how large would he have estimated the Sun to be?

Answers appear at the end of the chapter.

KEY IDEAS

Ideas preceded by an asterisk (*) are discussed in the Boxes.

Lunar Phases: The phases of the Moon occur because light from the Moon is actually reflected sunlight. As the relative positions of Earth, the Moon, and the Sun change, we see more or less of the illuminated half of the Moon.

Length of the Month: Two types of months are used in describing the motion of the Moon.

• With respect to the stars, the Moon completes one orbit around Earth in a sidereal month, averaging 27.32 days.

• The Moon completes one cycle of phases (one orbit around Earth with respect to the Sun) in a synodic month, averaging 29.53 days.

The Moon's Orbit: The plane of the Moon's orbit is tilted by about 5° from the plane of Earth's orbit, or ecliptic.

• The line of nodes is the line where the planes of the Moon's orbit and Earth's orbit intersect. The gravitational pull of the Sun gradually shifts the orientation of the line of nodes with respect to the stars.

Conditions for Eclipses: During a lunar eclipse, the Moon passes through Earth's shadow. During a solar eclipse, Earth passes through the Moon's shadow.

• Lunar eclipses occur at full moon, while solar eclipses occur at new moon.

• Either type of eclipse can occur only when the Sun and Moon are both on or very near the line of nodes. If this condition is not met, Earth's shadow cannot fall on the Moon and the Moon's shadow cannot fall on Earth.

Umbra and Penumbra: The shadow of an object has two parts: the umbra, within which the light source is completely blocked, and the penumbra, where the light source is only partially blocked.

Lunar Eclipses: Depending on the relative positions of the Sun, Moon, and Earth, lunar eclipses may be total (the Moon passes completely into Earth's umbra), partial (only part of the Moon passes into Earth's umbra), or penumbral (the Moon passes only into Earth's penumbra).

Solar Eclipses: Solar eclipses may be total, partial, or annular.

• During a total solar eclipse, the Moon's umbra traces out an eclipse path over Earth's surface as Earth rotates. Observers outside the eclipse path but within the penumbra see only a partial solar eclipse.

• During an annular eclipse, the umbra falls short of Earth, and the outer edge of the Sun's disk is visible around the Moon at mideclipse.

The Moon and Ancient Astronomers: Ancient astronomers such as Aristarchus and Eratosthenes made great progress in determining the sizes and relative distances of Earth, the Moon, and the Sun.

QUESTIONS

Review Questions

Questions preceded by an asterisk (*) are discussed in the Boxes.

1. Explain the difference between sunlight and moonlight.

2. *TUTORIAL 3.1* (a) Explain why the Moon exhibits phases. (b) A common misconception about the Moon's phases is that they are caused by Earth's shadow. Use Figure 3-2 to explain why this is not correct.

3. How would the sequence and timing of lunar phases be affected if the Moon moved around its orbit (a) in the same direction, but at twice the speed; (b) at the same speed, but in the opposite direction? Explain your answers.

4. At approximately what time does the Moon rise when it is (a) a new moon; (b) a first quarter moon; (c) a full moon; (d) a third quarter moon?

5. Astronomers sometimes refer to lunar phases in terms of the *age* of the Moon. This is the time that has elapsed since new moon phase. Thus, the age of a full moon is half of a 29½-day synodic period, or approximately 15 days. Find the approximate age of (a) a waxing crescent moon; (b) a third quarter moon; (c) a waning gibbous moon.

6. If you lived on the Moon, would you see Earth go through phases? If so, would the sequence of phases be the same as those of the Moon as seen from Earth, or would the sequence be reversed? Explain using Figure 3-2.

7. Is the far side of the Moon (the side that can never be seen from Earth) the same as the dark side of the Moon? Explain.

8. (a) If you lived on the Moon, would you see the Sun rise and set, or would it always be in the same place in the sky? Explain. (b) Would you see Earth rise and set, or would it always be in the same place in the sky? Explain using Figure 3-4.

9. What is the difference between a sidereal month and a synodic month? Which is longer? Why?

10. On a certain date the Moon is in the direction of the constellation Gemini as seen from Earth. When will the Moon next be in the direction of Gemini: one sidereal month later, or one synodic month later? Explain your answer.

11. *TUTORIAL 3.2* What is the difference between the umbra and the penumbra of a shadow?

12. Why doesn't a lunar eclipse occur at every full moon and a solar eclipse at every new moon?

13. What is the line of nodes? Why is it important to the subject of eclipses?

14. What is a penumbral eclipse of the Moon? Why do you suppose that it is easy to overlook such an eclipse?

15. Why is the duration of totality different for different total lunar eclipses, as shown in Table 3-1?

16. Can one ever observe an annular eclipse of the Moon? Why or why not?

17. 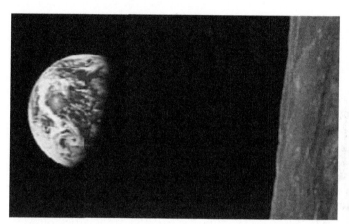 If you were looking at Earth from the side of the Moon that faces Earth, what would you see during (**a**) a total lunar eclipse; (**b**) a total solar eclipse? Explain your answers.

18. If there is a total eclipse of the Sun in April, can there be a lunar eclipse three months later in July? Why or why not?

19. Which type of eclipse—lunar or solar—do you think most people on Earth have seen? Why?

20. How is an annular eclipse of the Sun different from a total eclipse of the Sun? What causes this difference?

*21. What is the saros? How did ancient astronomers use it to predict eclipses?

22. How did Eratosthenes measure the size of Earth?

23. How did Aristarchus try to estimate the distance from Earth to the Sun and Moon?

24. How did Aristarchus try to estimate the diameters of the Sun and Moon?

Advanced Questions

Questions preceded by an asterisk (*) are discussed in the Boxes.

> **Problem-solving tips and tools**
>
> To estimate the average angular speed of the Moon along its orbit (that is, how many degrees around its orbit the Moon travels per day), divide 360° by the length of a sidereal month. It is helpful to know that the saros interval of 6585.3 days equals 18 years and 11⅓ days if the interval includes 4 leap years, but is 18 years and 10⅓ days if it includes 5 leap years.

25. The dividing line between the illuminated and unilluminated halves of the Moon is called the *terminator*. The terminator appears curved when there is a crescent or gibbous moon, but appears straight when there is a first quarter or third quarter moon (see Figure 3-2). Describe how you could use these facts to explain to a friend why lunar phases cannot be caused by Earth's shadow falling on the Moon.

26. What is the phase of the Moon if it rises at (**a**) midnight; (**b**) sunrise; (**c**) halfway between sunset and midnight; (**d**) halfway between noon and sunset? Explain your answers.

27. The Moon is highest in the sky when it crosses the meridian (see Figure 2-21), halfway between the time of moonrise and the time of moonset. At approximately what time does the Moon cross the meridian if it is (**a**) a new moon; (**b**) a first quarter moon; (**c**) a full moon; (**d**) a third quarter moon? Explain your answers.

28. The Moon is highest in the sky when it crosses the meridian (see Figure 2-21), halfway between the time of moonrise and the time of moonset. What is the phase of the Moon if it is highest in the sky at (**a**) midnight; (**b**) sunrise; (**c**) noon; (**d**) sunset? Explain your answers.

29. Suppose it is the first day of autumn in the northern hemisphere. What is the phase of the Moon if the Moon is located at (**a**) the vernal equinox; (**b**) the summer solstice; (**c**) the autumnal equinox; (**d**) the winter solstice? Explain your answers. (*Hint:* Make a drawing showing the relative positions of the Sun, Earth, and Moon. Compare with Figure 3-2.)

30. This photograph of Earth was taken by the crew of the *Apollo 8* spacecraft as they orbited the Moon. A portion of the lunar surface is visible at the right-hand side of the photo. In this photo, Earth is oriented with its north pole approximately at the top. When this photo was taken, was the Moon waxing or waning as seen from Earth? Explain your answer with a diagram.

R I **V** U X G   (NASA/JSC)

31. (**a**) The Moon moves noticeably on the celestial sphere over the space of a single night. To show this, calculate how long it takes the Moon to move through an angle equal to its own angular diameter (½°) against the background of stars. Give your answer in hours. (**b**) Through what angle (in degrees) does the Moon move during a 12-hour night? Can you notice an angle of this size? (*Hint:* See Figure 1-11.).

32. During an occultation, or "covering up," of Jupiter by the Moon, an astronomer notices that it takes the Moon's edge 90 seconds to cover Jupiter's disk completely. If the Moon's motion is assumed to be uniform and the occultation was "central" (that is, center over center), find the angular diameter of Jupiter. (*Hint:* Assume that Jupiter does not appear to move against the background of stars during this brief 90-second interval. You will need to convert the Moon's angular speed from degrees per day to arcseconds per second.)

33. The plane of the Moon's orbit is inclined at a 5° angle from the ecliptic, and the ecliptic is inclined at a 23½° angle from the celestial equator. Could the Moon ever appear at your zenith if you lived at (**a**) the equator; (**b**) the south pole? Explain your answers.

34. How many more sidereal months than synodic months are there in a year? Explain your answer.

35. Suppose Earth moved a little faster around the Sun so that it took a bit less than one year to make a complete orbit. If the speed of the Moon's orbit around Earth were unchanged, would the length of the sidereal month be the same, longer, or shorter than it is now? What about the synodic month? Explain your answers.

36. If the Moon revolved about Earth in the same orbit but in the opposite direction, would the synodic month be longer or shorter than the sidereal month? Explain your reasoning.

37. One definition of a "blue moon" is the second full moon within the same calendar month. There is usually only one full moon within a calendar month, so the phrase "once in a blue moon" means "hardly ever." Why are blue moons so rare? Are there any months of the year in which it would be impossible to have two full moons? Explain your answer.

38. You are watching a lunar eclipse from some place on Earth's night side. Will you see the Moon enter Earth's shadow from the east or from the west? Explain your reasoning.

39. The total lunar eclipse of October 28, 2004, was visible from South America. The duration of totality was 1 hour, 21 minutes. Was this total eclipse also visible from Australia, on the opposite side of Earth? Explain your reasoning.

40. During a total solar eclipse, the Moon's umbra moves in a generally eastward direction across Earth's surface. Use a drawing like Figure 3-11 to explain why the motion is eastward, not westward.

41. A total solar eclipse was visible from Africa on March 29, 2006 (see Figure 3-10b). Draw what the eclipse would have looked like as seen from France, to the north of the path of totality. Explain the reasoning behind your drawing.

42. Figures 3-11 and 3-13 show that the path of a total eclipse is quite narrow. Use this to explain why a glow is visible all around the horizon when you are viewing a solar eclipse during totality (see Figure 3-10a).

43. (a) Suppose the diameter of the Moon were doubled, but the orbit of the Moon remained the same. Would total solar eclipses be more common, less common, or just as common as they are now? Explain. (b) Suppose the diameter of the Moon were halved, but the orbit of the Moon remained the same. Explain why there would be *no* total solar eclipses.

44. Just as the distance from Earth to the Moon varies somewhat as the Moon orbits Earth, the distance from the Sun to Earth changes as Earth orbits the Sun. Earth is closest to the Sun at its *perihelion;* it is farthest from the Sun at its *aphelion.* In order for a total solar eclipse to have the maximum duration of totality, should Earth be at perihelion or aphelion? Assume that the Earth-Moon distance is the same in both situations. As part of your explanation, draw two pictures like Figure 3-11, one with Earth relatively close to the Sun and one with Earth relatively far from the Sun.

*45. On March 29, 2006, residents of northern Africa were treated to a total solar eclipse. (a) On what date and over what part of the world will the next total eclipse of that series occur? Explain. (b) On what date might you next expect a total eclipse of that series to be visible from northern Africa? Explain.

Discussion Questions

46. Describe the cycle of lunar phases that would be observed if the Moon moved around Earth in an orbit perpendicular to the plane of Earth's orbit. Would it be possible for both solar and lunar eclipses to occur under these circumstances? Explain your reasoning.

47. How would a lunar eclipse look if Earth had no atmosphere? Explain your reasoning.

48. In his 1885 novel *King Solomon's Mines,* H. Rider Haggard described a total solar eclipse that was seen in both South Africa and in the British Isles. Is such an eclipse possible? Why or why not?

49. Why do you suppose that total solar eclipse paths fall more frequently on oceans than on land? (You may find it useful to look at Figure 3-13.)

50. Examine Figure 3-13, which shows all of the total solar eclipses from 1997 to 2020. What are the chances that you might be able to travel to one of the eclipse paths? Do you think you might go through your entire life without ever seeing a total eclipse of the Sun?

Web/eBook Questions

51. Access the animation "The Moon's Phases" in Chapter 3 of the *Universe* Web site or eBook. This shows the Earth-Moon system as seen from a vantage point looking down onto the north pole. (a) Describe where you would be on the diagram if you are on the equator and the time is 6:00 P.M. (b) If it is 6:00 P.M. and you are standing on Earth's equator, would a third quarter moon be visible? Why or why not? If it would be visible, describe its appearance.

52. Search the World Wide Web for information about the next total lunar eclipse. Will the total phase of the eclipse be visible from your location? If not, will the penumbral phase be visible? Draw a picture showing the Sun, Earth, and Moon when the totality is at its maximum duration, and indicate your location on the drawing of Earth.

53. Search the World Wide Web for information about the next total solar eclipse. Through which major cities, if any, does the path of totality pass? What is the maximum duration of totality? At what location is this maximum duration observed? Will this eclipse be visible (even as a partial eclipse) from your location? Draw a picture showing the Sun, Earth, and Moon when the totality is at its maximum duration, and indicate your location on the drawing of Earth.

54. Access the animation "A Solar Eclipse Viewed from the Moon" in Chapter 3 of the *Universe* Web site or eBook. This shows the solar eclipse of August 11, 1999, as viewed from the Moon. Using a diagram, explain why the stars and the Moon's shadow move in the directions shown in this animation.

ACTIVITIES

Observing Projects

55. Observe the Moon on each clear night over the course of a month. On each night, note the Moon's location among the constellations and record that location on a star chart that also shows the ecliptic. After a few weeks, your observations will begin to trace the Moon's orbit. Identify the orientation of the line of nodes by marking the points where the Moon's orbit and the ecliptic intersect. On what dates is the Sun near the nodes marked on your star chart? Compare these dates with the dates of the next solar and lunar eclipses.

56. It is quite possible that a lunar eclipse will occur while you are taking this course. Look up the date of the next lunar eclipse in Table 3-1, on the World Wide Web, or in the current issue of a reference such as the *Astronomical Almanac* or *Astronomical Phenomena*. Then make arrangements to observe this lunar eclipse. You can observe the eclipse with the naked eye, but binoculars or a small telescope will enhance your viewing experience. If the eclipse is partial or total, note the times at which the Moon enters and exits Earth's umbra. If the eclipse is penumbral, can you see any changes in the Moon's brightness as the eclipse progresses?

57. Use *Starry Night™* to demonstrate the phases of the Moon. From the menu, select **Favourites > Explorations > Moon Phases.** In this view, you are looking down upon the southern hemisphere of Earth from a location 69,300 kilometers above Earth's surface with Earth centered in the view. The size of the Moon is greatly exaggerated in this view. The face of the Moon in this diagrammatic representation is configured to show its phase as seen from Earth. The inner green circle represents the Moon's orbit around Earth. The position of the Sun with respect to Earth is shown on the outer green circle, the ecliptic, near to the top of the view. In reality, the Sun would be in this direction from Earth but at a far larger distance. Click the **Play** button and observe the relative positions of the Sun, Earth, and Moon and the associated phases of the Moon. (**a**) Describe the geometry of the Sun, Moon, and Earth when the Moon is new. (**b**) Describe the geometry of the Sun, Moon, and Earth when the phase of the Moon is full. (**c**) When the phase of the Moon is at first or third quarter, what is the approximate angle at the Moon between the Moon-Sun and Moon-Earth lines? (**d**) What general term describes the phase of the Moon when the angle created by the Moon-Sun and Moon-Earth lines at the Moon is less than 90°? (**e**) What general term describes the phase of the Moon when the angle between the Moon-Sun and Moon-Earth lines at the Moon is greater than 90°?

58. Use *Starry Night™* to study the Moon's path in the sky and the phenomenon of eclipses, both lunar and solar. Select **Favourites > Explorations > Moon Motion** to display the sky as seen from the center of a transparent Earth on August 30, 2010, with daylight removed and the Moon centered in the view. As a guide to the positions and motions in the sky, the celestial equator and the associated equatorial grid are shown. The ecliptic, representing the plane of Earth's orbit extended onto the sky, is also shown. This line also represents the apparent path of the Sun across our sky in the course of a year. The **Time Flow Rate** is set to 1 hour but this can be adjusted if necessary to follow the Moon across the sky. (**a**) Is the Moon on the ecliptic at this time? (**b**) With the view locked onto the Moon, click the **Play** button. In which direction does the Moon move against the background stars? Keep in mind that the celestial equator runs in the east-west direction with east on the left. Note also that the ecliptic intersects this equator at an angle of 23½°. Does the Moon ever change its direction of motion relative to the stars? Why or why not? Describe its path relative to the ecliptic. (**c**) Determine how many days elapse between successive times that the Moon is on the ecliptic. (**d**) Set the **Time** and **Date** to 11 P.M. on July 21, 2009, and set the **Time Flow Rate** to 1 minute. Note that the Moon is almost on the ecliptic and that it is close to the position of the Sun in the sky. **Run time forward** at this 1-minute rate and stop time when the Moon is closest to the Sun. **Zoom** in to a field of view of about 3° and adjust the time in single steps to move the Moon to the position of maximum eclipse. What is the time of maximum eclipse? (**e**) Sketch the Sun and Moon on a piece of paper. Even though the Moon may not be directly over the Sun on the screen, an eclipse is occurring. What type of eclipse is it? Why is the Moon not directly over the Sun at this time? (**f**) Click on the **Events** tab, expand the **Event Filters** layer, and deselect all but the **Lunar and Solar Eclipse Events.** Expand the **Events Browser** layer, set the **Start** and **End** dates to cover the period from January 21, 2009, to January 21, 2010, and click **Find Events.** Right-click on each solar eclipse in turn, choose **View Event, Zoom** out to about 5°, and **Run time forward** and **backward** at a rate of 1 minute until the Moon is again closest to the Sun. Draw the Sun and Moon. Are they separated by the same amount as in part (d)? If not, why not? If there is a difference, what effect does it have on the eclipse?

59. Use *Starry Night™* to observe a lunar eclipse from a position in space near the Sun. From the main menu select **Favourites > Explorations > Lunar Eclipse.** This view from a position near to the Sun shows Earth and two tan-colored circles. The inner circle represents the umbra of Earth's shadow and the outer circle represents the boundary of the penumbra of Earth's shadow. (If these circles do not appear, click on **Options > Solar System > Planets-Moons…** and click on **Shadow Colour** to adjust the brightness of the umbral and penumbral outlines.) (**a**) What is the Moon's phase? Click the **Play** button and watch as time flows forward and the Moon passes through Earth's shadow. Select **File > Revert** from the menu, move the cursor over the Moon, right-click the mouse and click on **Centre** to center the view on the Moon. **Zoom** in to almost fill the field of view with the Moon and click the **Play** button again to watch the lunar eclipse in detail. Use the

Time controls to determine the duration of totality to the nearest minute. **(b)** What is the duration of totality for this eclipse? **(c)** Is it possible for a lunar eclipse to have a longer duration of totality than the one depicted in this simulation? Explain your answer.

Collaborative Exercise

60. Using a bright light source at the center of a darkened room, or a flashlight, use your fist held at arm's length to demonstrate the difference between a full moon and a lunar eclipse. (Use yourself or a classmate as Earth.) How must your fist "orbit" Earth so that lunar eclipses do not happen at every full moon? Create a simple sketch to illustrate your answers.

ANSWERS

ConceptChecks

ConceptCheck 3-1: Because the Sun is always shining on the Moon, the Moon is always half illuminated and half dark. Except in the rare events of eclipses, discussed later, the Moon's apparent changing phases are due to observers on Earth seeing differing amounts of the Moon's half illuminated surface.

ConceptCheck 3-2: The Moon goes through an entire cycle of phases in about four weeks, so if it is currently in the waxing gibbous phase, in one week it will be in the waxing gibbous phase, and after two weeks it will be in the waning gibbous phase.

ConceptCheck 3-3: The Moon is shown at four points of its orbit in Figure 3-4b. In the left position, the far side is fully dark.

ConceptCheck 3-4: If the Moon is full, then after one week, it would reach the third quarter phase, and the point that used to be in the center of the Moon's visible surface at full moon would now fall into darkness that would last for two weeks. This scenario can be seen in two figures: In Figure 3-4b, this scenario is shown by the red dot on the leftmost depiction of the Moon, and the Moon's position a week later at the bottom of the figure. In Figure 3-2, this scenario occurs between positions E and G.

ConceptCheck 3-5: Increase. If Earth was moving around the Sun faster than it is now, Earth would move farther around the Sun during the Moon's orbit and it would take longer for the Moon to reach the position where it was in line with the Sun and Earth, increasing its synodic period. This is illustrated (for the new moon) in Figure 3-5.

ConceptCheck 3-6: Lunar eclipses only occur when the Sun, Earth, and Moon all lie exactly on a line; this line is called the line of nodes. As shown in Figure 3-7, during a full moon the Sun, Earth, and Moon are not necessarily lined up because the Moon orbits in a different plane than the ecliptic. Usually, the full moon is above or below the ecliptic plane of Earth's orbit around the Sun and misses being covered by Earth's shadow.

ConceptCheck 3-7: No. Before the Moon passes into the full shadow (the umbra), where the Sun is completely blocked by Earth, the Moon must pass through the penumbra where only part of the Sun is blocked. This is illustrated by Path 2 in Figure 3-8.

ConceptCheck 3-8: Total solar eclipses are only observed in the small region where the Moon's tiny umbral shadow lands on Earth (see Figure 3-11), whereas when the Moon enters Earth's much larger umbral shadow, anyone on Earth who can see the Moon can observe it in Earth's shadow.

ConceptCheck 3-9: A total lunar eclipse can last for more than an hour whereas a total solar eclipse lasts only a few minutes.

ConceptCheck 3-10: If the noontime position of the Sun on the summer solstice was much closer to being directly overhead, it means that on the curved surface of Earth, the overhead directions at both Syene and Alexandria are pointing in similar directions. This scenario does not occur on Earth and would only do so if Earth were much larger. (Keep in mind that the distance between these two cities is assumed to be constant for this question, no matter what the size of Earth.)

CalculationCheck

CalculationCheck 3-1: Because their diameters must be in the same proportion as their distances, if Aristarchus assumed the Sun to be 100 times farther away when the Sun and Moon appeared to be the same diameter in the sky, he would have proposed that the Sun is 100 times wider than the Moon. Today we know that the Sun is about 400 times farther away from Earth and 400 times wider than the Moon.

Archaeoastronomy and Ethnoastronomy by Mark Hollabaugh

Many years ago I read astronomer John Eddy's *National Geographic* article about the Wyoming Medicine Wheel. A few years later I visited this archaeological site in the Bighorn Mountains and started on the major preoccupation of my career.

Archaeoastronomy combines astronomy and archaeology. You may be familiar with sites such as Stonehenge or the Mayan ruins of the Yucatán. You may not know that there are many archaeological sites in the United States that demonstrate the remarkable understanding a people, usually known as the Anasazi, had about celestial motions (see Figure 2-1). Chaco Canyon, Hovenweep, and Chimney Rock in the Four Corners area of the Southwest preserve ruins from this ancient Pueblo culture.

Moonrise at Chimney Rock in southern Colorado provides a good example of the Anasazi's knowledge of the lunar cycles. If you watched the rising of the Moon for many, many years, you would discover that the Moon's northernmost rising point undergoes an 18.6-year cycle. Dr. McKim Malville of the University of Colorado discovered that the Anasazi who lived there knew of the lunar standstill cycle and watched the northernmost rising of the Moon between the twin rock pillars of Chimney Rock.

My own specialty is the ethnoastronomy of the Lakota, or Teton Sioux, who flourished on the Great Plains of what are now Nebraska, the Dakotas, and Wyoming. Ethnoastronomy combines ethnography with astronomy. As an ethnoastronomer, I am less concerned with physical evidence in the form of ruins and more interested in myths, legends, religious belief, and current practices. In my quest to understand the astronomical thinking and customs of the nineteenth-century Lakota, I have traveled to museums, archives, and libraries in Nebraska, Colorado, Wyoming, South Dakota, and North Dakota. I frequently visit the Pine Ridge and Rosebud Reservations in South Dakota. The Lakota, and other Plains Indians, had a rich tradition of understanding celestial motions and developed an even richer explanation of why things appear the way they do in the sky.

My first professional contribution to ethnoastronomy was in 1996 at the Fifth Oxford International Conference on Astronomy in Culture held in Santa Fe, New Mexico. I had noticed that images of eclipses often appeared in Lakota winter counts—their method of making a historic record of events. The great Leonid meteor shower of 1833 appears in almost every Plains Indian winter count. As I looked at the hides or in ledger books recording these winter counts, I wondered why the Sun, the Moon, and the stars are so common among the Lakota of 150 years ago. My curiosity led me to look deeper: What did the Lakota think about eclipses? Why does their central ritual, the Sun Dance, focus on the Sun? Why do so many legends involve the stars?

The Lakota observed lunar and solar eclipses. Perhaps they felt they had the power to restore the eclipsed Sun or Moon. In August 1869, an Indian agency physician in South Dakota told the Lakota there would be an eclipse. When the Sun disappeared from view, the Lakota began firing their guns in the air. In their minds, they were more powerful than the white doctor because the result of their action was to restore the Sun.

The Lakota used a lunar calendar. Their names for the months came from the world around them. October was the moon of falling leaves. In some years, there actually are 13 new moons, and they often called this extra month the "Lost Moon." The lunar calendar often dictated the timing of their sacred rites. Although the Lakota were never dogmatic about it, they preferred to hold their most important ceremony, the Sun Dance, at the time of the full moon in June, which is when the summer solstice occurs.

Why did the Lakota pay attention to the night sky? Lakota elder Ringing Shield's statement about Polaris, recorded in the late nineteenth century, provides a clue: "One star never moves and it is *wakan*. Other stars move in a circle about it. They are dancing in the dance circle." For the Lakota, a driving force in their culture was a quest to understand the nature of the sacred, or *wakan*. Anything hard to understand or different from the ordinary was *wakan*.

Reaching for the stars, as far away as they are, was a means for the Lakota to bring the incomprehensible universe a bit closer to Earth. Their goal was the same as what Dr. Sandra M. Faber says in her essay "Why Astronomy?" at the end of Chapter 1, "a perspective on human existence and its relation to the cosmos."

Mark Hollabaugh taught physics and astronomy at Normandale Community College, St. Olaf College, Augsburg College, and the U.S. Air Force Academy. He majored in physics at St. Olaf College, earned an M.S. in astronomy at the University of Denver, and a Ph.D. in science education at the University of Minnesota. As a boy he watched the dance of the northern lights and the flash of meteors. Meeting *Apollo 13* commander Jim Lovell and going for a ride in Roger Freedman's airplane are as close as he came to his dream of being an astronaut.

An astronaut orbits Earth attached to the International Space Station's manipulator arm. (NASA) R I V U X G

Gravitation and the Waltz of the Planets

Sixty years ago the idea of humans orbiting Earth or sending spacecraft to other worlds was regarded as science fiction. Today, science fiction has become commonplace reality. Literally thousands of artificial satellites orbit our planet to track weather, relay signals for communications and entertainment, and collect scientific data about Earth and the universe. Humans live and work in Earth orbit (as in the accompanying photograph), have ventured as far as the Moon, and have sent dozens of robotic spacecraft to explore all the planets of the solar systemsphere, or corona, which glows with an unearthly pearlescent light.

While we think of spaceflight as an innovation of the twentieth century, we can trace its origins to a series of scientific revolutions that began in the 1500s. The first of these revolutions overthrew the ancient idea that Earth is an immovable object at the center of the universe, around which the Sun, the Moon, and the planets move. In this chapter we will learn how Nicolaus Copernicus, Tycho Brahe, Johannes Kepler, and Galileo Galilei helped us understand that Earth is itself one of several planets orbiting the Sun.

We will learn, too, about Isaac Newton's revolutionary discovery of why the planets move in the way that they do. This was just one aspect of Newton's immense body of work, which included formulating the fundamental laws of physics and developing a precise mathematical description of the force of gravitation—the force that holds the planets in their orbits.

Gravitation proves to be a truly universal force: It guides spacecraft as they journey across the solar system, and keeps satellites and astronauts in their orbits. In this chapter you will learn more about this important force of nature, which plays a central role in all parts of astronomy.

4-1 Ancient astronomers invented geocentric models to explain planetary motions

Since the dawn of civilization, scholars have attempted to explain the nature of the universe. The ancient Greeks were the first to use the principle that still guides scientists today: *The universe can be described and understood logically.* For example, more than 2500 years ago Pythagoras and his followers put forth the idea that nature can be described with mathematics. About 200 years later, Aristotle asserted that the universe is governed by physical laws. One of the most important tasks before the scholars of ancient Greece was to create a model (see Section 1-1) to explain the motions of objects in the heavens.

The Greek Geocentric Model

Most Greek scholars thought that the Sun, the Moon, the stars, and the planets revolve about a stationary Earth. A model of this kind, in which Earth is at the center of the universe, is called a **geocentric model.** Similar ideas were held by the scholars of ancient China.

> The scholars of ancient Greece imagined that the planets follow an ornate combination of circular paths

Today we recognize that the stars are not merely points of light on an immense celestial sphere. But in fact this is how the ancient Greeks regarded the stars in their geocentric model of the universe. To explain the diurnal motions of the stars, they assumed that the celestial sphere was *real,* and that it rotated around the stationary Earth once a day.

The Sun and Moon both participated in this daily rotation of the sky, which explained their rising and setting motions. To explain why the Sun and Moon both move slowly with respect to the stars, the ancient Greeks imagined that both of these objects orbit around Earth.

ANALOGY Imagine a merry-go-round that rotates clockwise as seen from above, as in **Figure 4-1a**. As it rotates, two children walk slowly counterclockwise at different speeds around the merry-go-round's platform. Thus, the children rotate along with the merry-go-round and also change their positions with respect to the merry-go-round's wooden horses. This scene is analogous to the way the ancient Greeks pictured the motions of the stars, the Sun, and the Moon. In their model, the celestial sphere rotated to the west around a stationary Earth (Figure 4-1b). The stars rotate along with the celestial sphere just as the wooden horses rotate along with the merry-go-round in Figure 4-1a. The Sun and Moon are analogous to the two children; they both turn westward with the celestial sphere, making one complete turn each day, and also move slowly eastward at different speeds with respect to the stars.

The geocentric model of the heavens also had to explain the motions of the planets. The ancient Greeks and other cultures of that time knew of five planets: Mercury, Venus, Mars, Jupiter, and Saturn, each of which is a bright object in the night sky. For example, when Venus is at its maximum brilliancy, it is 16 times brighter than the brightest star. (By contrast, Uranus and Neptune are quite dim and were not discovered until after the invention of the telescope.)

Like the Sun and Moon, all of the planets rise in the east and set in the west once a day. And like the Sun and Moon, from night to night the planets slowly move on the celestial sphere, that is, with respect to the background of stars. However, the character of this motion on the celestial sphere is quite different for the planets. Both the Sun and the Moon always move from west to east on the celestial sphere, that is, opposite the direction in which the celestial sphere appears to rotate. The Sun follows the path called the ecliptic

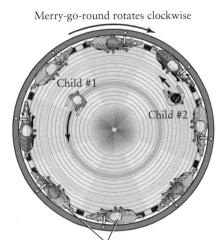

Merry-go-round rotates clockwise

Celestial sphere rotates to the west

Child #1

Child #2

Moon

Sun

Earth

Wooden horses fixed on merry-go-round

Stars fixed on celestial sphere

(a) A rotating merry-go-round

(b) The Greek geocentric model

ANIMATION 4-1 **FIGURE 4-1**
A Merry-Go-Round Analogy **(a)** Two children walk at different speeds around a rotating merry-go-round with its wooden horses. **(b)** In an analogous way, the ancient Greeks imagined that the Sun and Moon move around the rotating celestial sphere with its fixed stars. Thus, the Sun and Moon move from east to west across the sky every day and also move slowly eastward from one night to the next relative to the background of stars.

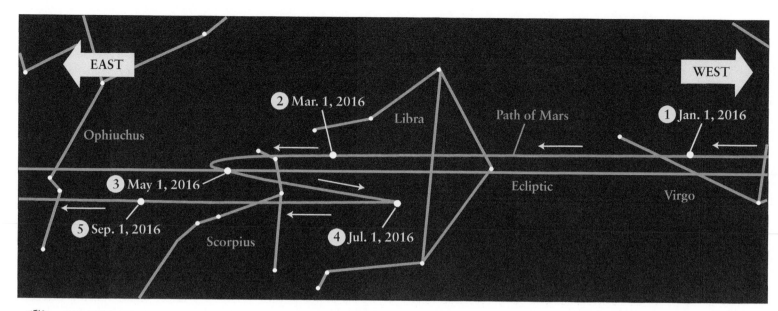

ANIMATION 4-2

FIGURE 4-2

The Path of Mars in 2016 From January 2016 through September 2016, Mars traverses zodiacal constellations from Virgo to Ophiuchus. Mars's motion is direct (from west to east, or from right to left in this figure) most of the time but is retrograde (from east to west, or from left to right in this figure) during May to July 2016. Notice that the speed of Mars relative to the stars is not constant: The planet travels farther across the sky from January 1 to March 1 than it does from March 1 to May 1.

(see Section 2-5), while the Moon follows a path that is slightly inclined to the ecliptic (see Section 3-3). Furthermore, the Sun and the Moon each move at relatively constant speeds around the celestial sphere. (The Moon's speed is faster than that of the Sun: It travels all the way around the celestial sphere in about a month while the Sun takes an entire year.) The planets, too, appear to move along paths that are close to the ecliptic. The difference is that each of the planets appears to wander back and forth on the celestial sphere with varying speed. As an example, Figure 4-2 shows the wandering motion of Mars with respect to the background of stars during 2016. (This figure shows that the name *planet* is well deserved; it comes from a Greek word meaning "wanderer.")

CAUTION! On a map of Earth with north at the top, west is to the left and east is to the right. Why, then, is *east* on the left and *west* on the right in Figure 4-2? The answer is that a map of Earth is a view looking downward at the ground from above, while a star map like Figure 4-2 is a view looking upward at the sky. You can see how east and west are flipped if you first draw a set of north-south and east-west axes on paper, and then hold the paper up against the sky.

Most of the time planets move slowly eastward relative to the stars, just as the Sun and Moon do. This eastward progress is called **direct motion.** For example, Figure 4-2 shows that Mars will be in direct motion from January through April 2016 and from July through September 2016. Occasionally, however, the planet seems to stop and then back up for several weeks or months. This occasional westward movement is called **retrograde motion.** Mars will undergo retrograde motion from May to July 2016 (see Figure 4-2), and will do so again about every 22½ months. All the other planets go through retrograde motion, but at different intervals. In the language of the merry-go-round analogy in Figure 4-1, the Greeks imagined the planets as children walking around the rotating merry-go-round but who keep changing their minds about which direction to walk!

CAUTION! Whether a planet is in direct or retrograde motion, over the course of a single night you will see it rise in the east and set in the west. That is because both direct and retrograde motions are much slower than the apparent daily rotation of the sky. Hence, they are best detected by mapping the position of a planet against the background stars from night to night over a long period. Figure 4-2 is a map of just this sort.

CONCEPTCHECK 4-1

If Mars is moving retrograde, will it rise above the eastern horizon or above the western horizon?

CONCEPTCHECK 4-2

How fast is Earth spinning on its axis in the Greek geocentric model?

Answers appear at the end of the chapter.

The Ptolemaic System

Explaining the nonuniform motions of the five planets was one of the main challenges facing the astronomers of antiquity. The Greeks developed many theories to account for retrograde motion and the loops that the planets trace out against the background stars. One of the most successful and enduring models was originated by Apollonius of Perga and by Hipparchus in the second

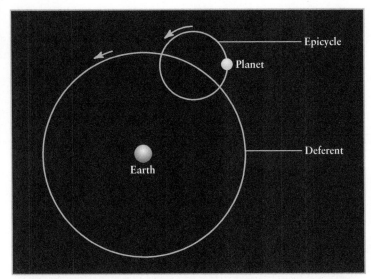

(a) Planetary motion modeled as a combination of circular motions

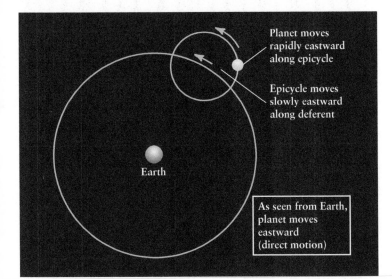

(b) Modeling direct motion

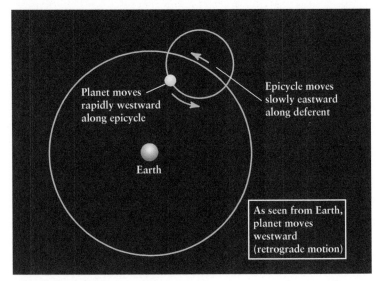

(c) Modeling retrograde motion

ANIMATION 4-3

FIGURE 4-3

A Geocentric Explanation of Retrograde Motion **(a)** The ancient Greeks imagined that each planet moves along an epicycle, which in turn moves along a deferent centered approximately on Earth. The planet moves along the epicycle more rapidly than the epicycle moves along the deferent. **(b)** At most times the eastward motion of the planet on the epicycle adds to the eastward motion of the epicycle on the deferent. Then the planet moves eastward in direct motion as seen from Earth. **(c)** When the planet is on the inside of the deferent, its motion along the epicycle is westward. Because this motion is faster than the eastward motion of the epicycle on the deferent, the planet appears from Earth to be moving westward in retrograde motion.

century B.C.E. and expanded upon by Ptolemy, the last of the great Greek astronomers, during the second century C.E. Figure 4-3a sketches the basic concept, usually called the **Ptolemaic system.** Each planet is assumed to move in a small circle called an **epicycle,** whose center in turn moves in a larger circle, called a **deferent,** which is centered approximately on Earth. Both the epicycle and deferent rotate in the same direction, shown as counterclockwise in Figure 4-3a.

As viewed from Earth, the epicycle moves eastward along the deferent. Most of the time the eastward motion of the planet on its epicycle adds to the eastward motion of the epicycle on the deferent (Figure 4-3b). Then the planet is seen to be in direct (eastward) motion against the background stars. However, when the planet is on the part of its epicycle nearest Earth, the motion of the planet along the epicycle is opposite to the motion of the epicycle along the deferent. The planet therefore appears to slow down and halt its usual eastward movement among the constellations, and actually goes backward in retrograde (westward) motion for a few weeks or months (Figure 4-3c). Thus, the concept of epicycles and deferents enabled Greek astronomers to explain the retrograde loops of the planets.

Using the wealth of astronomical data in the library at Alexandria, including records of planetary positions for hundreds of years, Ptolemy deduced the sizes and rotation rates of the epicycles and deferents needed to reproduce the recorded paths of the planets. After years of tedious work, Ptolemy assembled his calculations into 13 volumes, collectively called the *Almagest.* His work was used to predict the positions and paths of the Sun, Moon, and planets with unprecedented accuracy. In fact, the *Almagest* was so successful that it became the astronomer's bible, and for more than 1000 years, the Ptolemaic system endured as a useful description of the workings of the heavens.

A major problem posed by the Ptolemaic system was a philosophical one: It treated each planet independent of the others. There was no rule in the *Almagest* that related the size and rotation speed of one planet's epicycle and deferent to the corresponding sizes and speeds for other planets. This problem made the Ptolemaic system very unsatisfying to many Islamic and European astronomers of the Middle Ages. They thought that a correct model of the universe should be based on a simple set of underlying principles that applied to all of the planets.

With only limited observational data, the astronomers of the Middle Ages needed a way to judge if a scientific model might be correct. To help in this assessment, a guiding principle called

Occam's razor is often used to judge between competing scientific models. Named after the fourteenth-century English philosopher William of Occam, the principle states: *When two hypotheses can explain the available data, the hypothesis requiring the fewest new assumptions should be favored.* (The "razor" refers to shaving extraneous details or assumptions from an explanation.) In other words, the simplest explanation consistent with available data is most likely to be correct. The geocentric model required many parameters (or new assumptions) to characterize its epicycles and deferents, leaving open the possibility of a more simple theory. In the centuries that followed, Occam's razor would help motivate a simpler yet revolutionary view of the universe.

CONCEPTCHECK 4-3

Do the planets actually stop and change their direction of motion in the Ptolemaic model?

Answer appears at the end of the chapter.

4-2 Nicolaus Copernicus devised the first comprehensive heliocentric model

During the first half of the sixteenth century, a Polish lawyer, physician, canon of the church, and gifted mathematician named Nicolaus Copernicus (Figure 4-4) began to construct a new model of the universe. His model, which placed the Sun at the center, explained the motions of the planets in a more natural way than the Ptolemaic system. As we will see, it also helped lay the foundations of modern physical science. But Copernicus was not the first to conceive a Sun-centered model of planetary motion. In the third century B.C.E., the Greek astronomer Aristarchus suggested such a model as a way to explain retrograde motion.

A Heliocentric Model Explains Retrograde Motion

Imagine riding on a fast racehorse. As you pass a slowly walking pedestrian, he appears to move backward, even though he is traveling in the same direction as you and your horse. This sort of simple observation inspired Aristarchus to formulate a **heliocentric** (Sun-centered) **model** in which all the planets, including Earth, revolve about the Sun. Different planets take different lengths of time to complete an orbit, so from time to time one planet will overtake another, just as a fast-moving horse overtakes a person on foot. When Earth overtakes Mars, for example, Mars appears to move backward in retrograde motion, as Figure 4-5 shows. Thus, in the heliocentric picture, the occasional retrograde motion of a planet is merely the result of Earth's fast motion.

> The retrograde motion of the planets is a result of our viewing the universe from a moving Earth

As we saw in Section 3-6, Aristarchus demonstrated that the Sun is bigger than Earth (see Table 3-3). This made it sensible to imagine Earth orbiting the larger Sun. He also imagined that Earth rotated on its axis once a day, which explained the daily rising and setting of the Sun, Moon, and planets and the diurnal motions of the stars. To explain why the apparent motions of the planets never take them far from the ecliptic, Aristarchus proposed that

FIGURE 4-4
Nicolaus Copernicus (1473–1543) Copernicus was the first person to work out the details of a heliocentric system in which the planets, including Earth, orbit the Sun. (E. Lessing/Magnum)

the orbits of Earth and all the planets must lie in nearly the same plane. (Recall from Section 2-5 that the ecliptic is the projection onto the celestial sphere of the plane of Earth's orbit.)

This heliocentric model is conceptually much simpler than an Earth-centered system, such as that of Ptolemy, with all its "circles upon circles." In Aristarchus's day, however, the idea of an orbiting, rotating Earth seemed inconceivable, given Earth's apparent stillness and immobility. Nearly 2000 years would pass before a heliocentric model found broad acceptance.

CONCEPTCHECK 4-4

In the heliocentric model, could an imaginary observer on the Sun look out and see planets moving in retrograde motion?

Answer appears at the end of the chapter.

Copernicus and the Arrangement of the Planets

In the years after 1500, Copernicus came to realize that a heliocentric model has several advantages beyond providing a natural explanation of retrograde motion. In the Ptolemaic system, the arrangement of the planets—that is, which are close to Earth and which are far away—was chosen in large part by guesswork. By using a heliocentric model, Copernicus could determine the arrangement of the planets without ambiguity.

Copernicus realized that because Mercury and Venus are always observed fairly near the Sun in the sky, their orbits must be smaller than Earth's. Planets in such orbits are called **inferior planets**

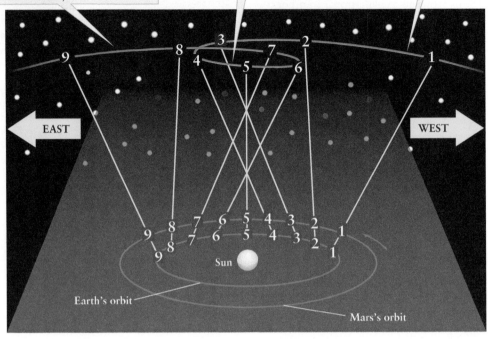

3. From point 6 to point 9, Mars again appears to move eastward against the background of stars as seen from Earth (direct motion).

2. As Earth passes Mars in its orbit from point 4 to point 6, Mars appears to move westward against the background of stars (retrograde motion).

1. From point 1 to point 4, Mars appears to move eastward against the background of stars as seen from Earth (direct motion).

EAST

WEST

Sun

Earth's orbit

Mars's orbit

ANIMATION 4-4

FIGURE 4-5

A Heliocentric Explanation of Retrograde Motion In the heliocentric model of Aristarchus, Earth and the other planets orbit the Sun. Earth travels around the Sun more rapidly than Mars. Consequently, as Earth overtakes and passes this slower-moving planet, Mars appears for a few months (from points 4 through 6) to fall behind and move backward with respect to the background of stars.

(Figure 4-6). The other visible planets—Mars, Jupiter, and Saturn— are sometimes seen on the side of the celestial sphere opposite the Sun, so these planets appear high above the horizon at midnight (when the Sun is far below the horizon). When this happens, Earth must lie between the Sun and these planets. Copernicus therefore concluded that the orbits of Mars, Jupiter, and Saturn must be larger than Earth's orbit. Hence, these planets are called **superior planets.**

Uranus and Neptune, as well as Pluto and a number of other small bodies called asteroids that also orbit the Sun, were discovered after the telescope was invented (and after the death of Copernicus). All of these can be seen at times in the midnight sky, so these also have orbits larger than that of Earth.

The heliocentric model also explains why planets appear in different parts of the sky on different dates. Both inferior planets (Mercury and Venus) go through cycles: The planet is seen in the west after sunset for several weeks or months, then for several weeks or months in the east before sunrise, and then in the west after sunset again.

Figure 4-6 shows the reason for this cycle. When Mercury or Venus is visible after sunset, it is near **greatest eastern elongation.** (The angle between the Sun and a planet as viewed from Earth is called the planet's **elongation.**) The planet's position in the sky is as far east of the Sun as possible, so it appears above the western horizon after sunset (that is, to the east of the Sun) and is often called an "evening star." At **greatest western elongation,** Mercury

or Venus is as far west of the Sun as it can possibly be. It then rises before the Sun, gracing the predawn sky as a "morning star" in the east. When Mercury or Venus is at **inferior conjunction,** it is between us and the Sun, and it moves from the evening sky into the morning sky over weeks or months. At **superior conjunction,** when the planet is on the opposite side of the Sun, it moves back into the evening sky.

A superior planet such as Mars, whose orbit is larger than Earth's, is best seen in the night sky when it is at **opposition.** At this point in its orbit the planet is in the part of the sky opposite the Sun and is highest in the sky at midnight. This is also when the planet appears brightest, because it is closest to us. But when a superior planet like Mars is located behind the Sun at **conjunction,** it is above the horizon during the daytime and thus is not well placed for nighttime viewing.

CONCEPTCHECK 4-5

How many times is Mars at inferior conjunction during one orbit around the Sun?

Answer appears at the end of the chapter.

Planetary Periods and Orbit Sizes

The Ptolemaic system has no simple rules relating the motion of one planet to another. But Copernicus showed that there *are* such

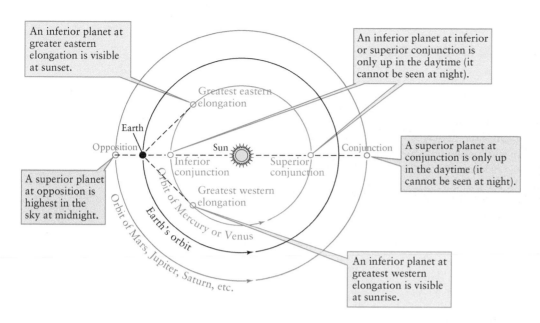

An inferior planet at greater eastern elongation is visible at sunset.

An inferior planet at inferior or superior conjunction is only up in the daytime (it cannot be seen at night).

A superior planet at conjunction is only up in the daytime (it cannot be seen at night).

A superior planet at opposition is highest in the sky at midnight.

An inferior planet at greatest western elongation is visible at sunrise.

FIGURE 4-6

Planetary Orbits and Configurations When and where in the sky a planet can be seen from Earth depends on the size of its orbit and its location on that orbit. Inferior planets have orbits smaller than Earth's, while superior planets have orbits larger than Earth's. (Note that in this figure you are looking down onto the solar system from a point far above Earth's northern hemisphere.)

rules in a heliocentric model. In particular, he found a correspondence between the time a planet takes to complete one orbit—that is, its **period**—and the size of the orbit.

Determining the period of a planet takes some care, because Earth, from which we must make the observations, is also moving. Realizing this, Copernicus was careful to distinguish between two different periods of each planet. The **synodic period** is the time that elapses between two successive identical configurations as seen from Earth—from one opposition to the next, for example, or from one conjunction to the next. The **sidereal period** is the true orbital period of a planet, the time it takes the planet to complete one full orbit of the Sun relative to the stars.

The synodic period of a planet can be determined by observing the sky from Earth, but the sidereal period has to be found by calculation. Copernicus figured out how to do this (Box 4-1). Table 4-1 shows the results for all of the planets.

To find a relationship between the sidereal period of a planet and the size of its orbit, Copernicus still had to determine the relative distances of the planets from the Sun. He devised a straightforward

TABLE 4-2	Average Distances of the Planets from the Sun	
Planet	Copernican value (AU*)	Modern value (AU)
Mercury	0.38	0.39
Venus	0.72	0.72
Earth	1.00	1.00
Mars	1.52	1.52
Jupiter	5.22	5.20
Saturn	9.07	9.55
Uranus	—	19.19
Neptune	—	30.07

*1 AU = 1 astronomical unit = average distance from Earth to the Sun.

geometric method of determining the relative distances of the planets from the Sun using trigonometry. His answers turned out to be remarkably close to the modern values, as shown in Table 4-2.

The distances in Table 4-2 are given in terms of the astronomical unit (AU), which is the average distance from Earth to the Sun (Section 1-7). The astronomical unit is 1 AU = 1.496×10^8 km (92.96 million miles). However, Copernicus did not know the precise value of this distance, so he could only determine the *relative* sizes of the orbits of the planets. (Using the astronomical unit, the average distance from the Sun to each of the planets in kilometers can be determined from Table 4-2. A table of these distances is given in Appendix 2.)

By comparing Tables 4-1 and 4-2, you can see the unifying relationship between planetary orbits in the Copernican model: The farther a planet is from the Sun, the longer it takes to travel around its orbit (that is, the longer its sidereal period). That is so for *two* reasons: (1) the larger the orbit, the farther a planet must

TABLE 4-1	Synodic and Sidereal Periods of the Planets	
Planet	Synodic period	Sidereal period
Mercury	116 days	88 days
Venus	584 days	225 days
Earth	—	1.0 year
Mars	780 days	1.9 years
Jupiter	399 days	11.9 years
Saturn	378 days	29.5 years
Uranus	370 days	84.1 years
Neptune	368 days	164.9 years

BOX 4-1 TOOLS OF THE ASTRONOMER'S TRADE

Relating Synodic and Sidereal Periods

We can derive a mathematical formula that relates a planet's sidereal period (the time required for the planet to complete one orbit) to its synodic period (the time between two successive identical configurations). To start with, let us consider an inferior planet (Mercury or Venus) orbiting the Sun as shown in the accompanying figure. Let P be the planet's sidereal period, S the planet's synodic period, and E Earth's sidereal period or sidereal year (see Section 2-8), which Copernicus knew to be nearly 365¼ days.

The rate at which Earth moves around its orbit is the number of degrees around the orbit divided by the time to complete the orbit, or 360°/E (equal to a little less than 1° per day). Similarly, the rate at which the inferior planet moves along its orbit is 360°/P.

During a given time interval, the angular distance that Earth moves around its orbit is its rate, (360°/E), multiplied by the length of the time interval. Thus, during a time S, or one synodic period of the inferior planet, Earth covers an angular distance of (360°/E)S around its orbit. In that same time, the inferior planet covers an angular distance of (360°/P)S. Note, however, that the inferior planet has gained one full lap on Earth, and hence has covered 360° more than Earth has (see the figure). Thus, (360°/P)S = (360°/E)S + 360°. Dividing each term of this equation by 360°S gives

For an inferior planet:

$$\frac{1}{P} = \frac{1}{E} + \frac{1}{S}$$

P = inferior planet's sidereal period
E = Earth's sidereal period = 1 year
S = inferior planet's synodic period

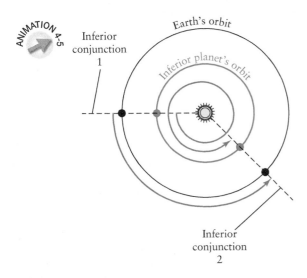

ANIMATION 4-5

Inferior conjunction 1

Earth's orbit

Inferior planet's orbit

Inferior conjunction 2

A similar analysis for a superior planet (for example, Mars, Jupiter, or Saturn) yields

For a superior planet:

$$\frac{1}{P} = \frac{1}{E} - \frac{1}{S}$$

P = superior planet's sidereal period
E = Earth's sidereal period = 1 year
S = superior planet's synodic period

Using these formulas, we can calculate a planet's sidereal period P from its synodic period S. Often astronomers express P, E, and S in terms of years, by which they mean Earth years of approximately 365.26 days.

EXAMPLE: Jupiter has an observed synodic period of 398.9 days, or 1.092 years. What is its sidereal period?

Situation: Our goal is to find the sidereal period of Jupiter, a superior planet.

Tools: Since Jupiter is a superior planet, we use the second of the two equations given above to determine the sidereal period P.

Answer: We are given Earth's sidereal period $E = 1$ year and Jupiter's synodic period $S = 1.092$ years. Using the equation $1/P = 1/E - 1/S$,

$$\frac{1}{P} = \frac{1}{1} - \frac{1}{1.092} = 0.08425,$$

so

$$P = \frac{1}{0.08425} = 11.87 \text{ years}$$

Review: Our answer means that it takes 11.87 years for Jupiter to complete one full orbit of the Sun. This sidereal period is greater than Earth's 1-year sidereal period because Jupiter's orbit is larger than Earth's orbit. Jupiter's synodic period of 1.092 years is the time from one opposition to the next, or the time that elapses from when Earth overtakes Jupiter to when it next overtakes Jupiter. This synodic period is so much shorter than the sidereal period because Jupiter moves quite slowly around its orbit. Earth overtakes it a little less often than once per Earth orbit, that is, at intervals of a little bit more than a year.

travel to complete an orbit; and (2) the larger the orbit, the slower a planet moves. For example, Mercury, with its small orbit, moves at an average speed of 47.9 km/s (107,000 mi/h). Saturn travels around its large orbit much more slowly, at an average speed of 9.64 km/s (21,600 mi/h). The older Ptolemaic model offers no such simple relations between the motions of different planets.

CONCEPTCHECK 4-6

What causes the planets to stop and change their direction of motion through the sky in the heliocentric model?

CONCEPTCHECK 4-7

Why is Jupiter's sidereal period longer than its synodic period? *Answers appear at the end of the chapter.*

The Shapes of Orbits in the Copernican Model

At first, Copernicus assumed that Earth travels around the Sun along a circular path. He found, however, that perfectly circular orbits could not accurately describe the paths of the other planets, so he had to add an epicycle to each planet. These epicycles were *not* added to explain retrograde motion, which Copernicus realized was because of the differences in orbital speeds of different planets, as shown in Figure 4-5. Rather, the small epicycles helped Copernicus account for slight variations in each planet's speed along its orbit.

Even though he clung to the old notion that orbits must be made up of circles, Copernicus had shown that a heliocentric model could explain the motions of the planets. He compiled his ideas and calculations into a book entitled *De revolutionibus orbium coelestium* (On the Revolutions of the Celestial Spheres), which was published in 1543, the year of his death.

For several decades after Copernicus, most astronomers saw little reason to change their allegiance from the older geocentric model of Ptolemy. The predictions that the Copernican model makes for the apparent positions of the planets are, on average, no better or worse than those of the Ptolemaic model. The test of Occam's razor does not really favor either model, because both use a combination of circles to describe each planet's motion.

More concrete evidence was needed to convince scholars to abandon the old, comfortable idea of a stationary Earth at the center of the universe. The story of how this evidence was accumulated begins nearly 30 years after the death of Copernicus, when a young Danish astronomer pondered the nature of a new star in the heavens.

CONCEPTCHECK 4-8

Why was Copernicus's model more accurate than Ptolemy's model?

Answer appears at the end of the chapter.

4-3 Tycho Brahe's astronomical observations disproved ancient ideas about the heavens

On November 11, 1572, a bright star suddenly appeared in the constellation Cassiopeia. At first, it was even brighter than Venus, but then it began to grow dim. After 18 months, it faded from view. Modern astronomers recognize this event as a supernova explosion, the violent death of a massive star.

> A supernova explosion and a comet revealed to Tycho that our universe is more dynamic than had been imagined

In the sixteenth century, however, the vast majority of scholars held with the ancient teachings of Aristotle and Plato, who had argued that the heavens are permanent and unalterable. Consequently, the "new star" of 1572 could not really be a star at all, because the heavens do not change; it must instead be some sort of bright object quite near Earth, perhaps not much farther away than the clouds overhead.

The 25-year-old Danish astronomer Tycho Brahe (1546–1601) realized that straightforward observations might reveal the distance to the new star. It is common experience that when you walk from one place to another, nearby objects appear to change position against the background of more distant objects. This phenomenon, whereby the apparent position of an object changes because of the motion of the observer, is called **parallax**. If the new star was nearby, then its position should shift against the background stars over the course of a single night because Earth's rotation changes our viewpoint. Figure 4-7 shows this predicted shift. (Actually, Tycho believed that the heavens rotate about Earth, as in the Ptolemaic model, but the net effect is the same.)

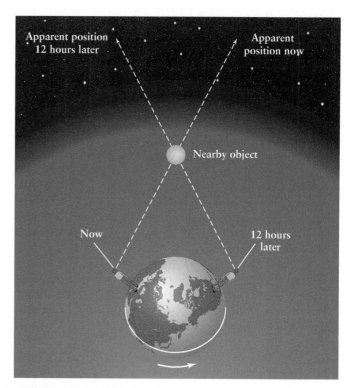

FIGURE 4-7

A Nearby Object Shows a Parallax Shift Tycho Brahe argued that if an object is near Earth, an observer would have to look in different directions to see that object over the course of a night and its position relative to the background stars would change. Tycho failed to measure such changes for a supernova in 1572 and a comet in 1577. He therefore concluded that these objects were far from Earth.

Tycho's careful observations failed to disclose any parallax. The farther away an object is, the less it appears to shift against the background as we change our viewpoint, and the smaller the parallax. Hence, the new star had to be quite far away, farther from Earth than anyone had imagined. This was the first evidence that the "unchanging" stars were in fact changeable.

Tycho also attempted to measure the parallax of a bright comet that appeared in 1577 and, again, found it too small to measure. Thus, the comet also had to be far beyond Earth. Furthermore, because the comet's position relative to the stars changed from night to night, its motion was more like a planet than a star. But most scholars of Tycho's time taught that the motions of the planets had existed in unchanging form since the beginning of the universe. If new objects such as comets could appear and disappear within the realm of the planets, the conventional notions of planetary motions needed to be revised.

Tycho's observations showed that the heavens are by no means pristine and unchanging. This discovery flew in the face of nearly 2000 years of astronomical thought. In support of these revolutionary observations, the king of Denmark financed the construction of two magnificent observatories for Tycho's use on the island of Hven, just off the Danish coast. The two observatories—Uraniborg ("heavenly castle") and Stjerneborg ("star castle")—allowed Tycho to design and have built a set of naked-eye astronomical instruments vastly superior in quality to any earlier instruments (Figure 4-8).

Tycho and the Positions of the Planets

With this state-of-the-art equipment, Tycho proceeded to measure the positions of stars and planets with unprecedented accuracy. In addition, he and his assistants were careful to make several observations of the same star or planet with different instruments, in order to identify any errors that might be caused by the instruments themselves. This painstaking approach to mapping the heavens revolutionized the practice of astronomy and is used by astronomers today.

A key goal of Tycho's observations during this period was to test the ideas Copernicus had proposed decades earlier about Earth going around the Sun. Tycho argued that if Earth was in motion, then nearby stars should appear to shift their positions with respect to background stars as we orbit the Sun. Tycho failed to detect any such parallax, and he concluded that Earth was at rest and the Copernican system was wrong.

On this point Tycho was in error, for nearby stars do in fact shift their positions as he had suggested. But even the nearest stars are so far away that the shifts in their positions are less than an arcsecond, too small to be seen with the naked eye. Tycho would have needed a telescope to detect the parallax that he was looking for, but the telescope was not invented until 1608, years after his death in 1601. Indeed, the first accurate determination of stellar parallax was not made until 1838.

Although he remained convinced that a stationary Earth was at the center of the universe, Tycho nonetheless made a tremendous contribution toward putting the heliocentric model on a solid foundation. From 1576 to 1597, he used his instruments to make comprehensive measurements of the positions of the planets with an accuracy of 1 arcminute. These measurements are as well as can be done with the naked eye and were far superior to any earlier measurements. Within the reams of data that Tycho

FIGURE 4-8

Tycho Brahe (1546–1601) Observing This contemporary illustration shows Tycho Brahe with some of the state-of-the-art measuring apparatus at Uraniborg, one of the two observatories that he built under the patronage of Frederik II of Denmark. (This magnificent observatory lacked a telescope, which had not yet been invented.) The data that Tycho collected were crucial to the development of astronomy in the years after his death. (Detlev van Ravenswaay/ Science Source)

compiled lay the truth about the motions of the planets. The person who would extract this truth was a German mathematician who became Tycho's assistant in 1600, a year before the great astronomer's death. His name was Johannes Kepler (Figure 4-9).

4-4 Johannes Kepler proposed elliptical paths for the planets about the Sun

TUTORIAL 4-1 The task that Johannes Kepler took on at the beginning of the seventeenth century was to find a model of planetary motion that agreed completely with Tycho's extensive and very accurate observations of planetary positions. To do this, Kepler found that he had to break with an ancient prejudice about planetary motions.

> Kepler's ideas apply not just to the planets, but to all orbiting celestial objects

Elliptical Orbits and Kepler's First Law

Astronomers had long assumed that heavenly objects move in circles, which were considered the most perfect and harmonious of all geometric shapes. They believed that if a perfect God resided in heaven along with the stars and planets, then the motions of these objects must be perfect too. Against this context, Kepler dared to try to explain planetary motions with noncircular curves. In

FIGURE 4-9

Johannes Kepler (1571–1630) By analyzing Tycho Brahe's detailed records of planetary positions, Kepler developed three general principles, called Kepler's laws, that describe how the planets move about the Sun. Kepler was the first to realize that the orbits of the planets are ellipses and not circles. (Erich Lessing/Art Resource)

particular, he found that he had the best success with a particular kind of curve called an **ellipse.**

You can draw an ellipse by using a loop of string, two thumbtacks, and a pencil, as shown in Figure 4-10a. Each thumbtack in the figure is at a **focus** (plural **foci**) of the ellipse; an ellipse has two foci. The longest diameter of an ellipse, called the **major axis,** passes through both foci. Half of that distance is called the **semimajor axis** and is usually designated by the letter a. A circle is a special case of an ellipse in which the two foci are at the same

point (this corresponds to using only a single thumbtack in Figure 4-10b). The semimajor axis of a circle is equal to its radius.

By assuming that planetary orbits were ellipses, Kepler found, to his delight, that he could make his theoretical calculations match precisely to Tycho's observations. This important discovery, first published in 1609, is now called **Kepler's first law:**

The orbit of a planet about the Sun is an ellipse with the Sun at one focus.

The semimajor axis a of a planet's orbit is the average distance between the planet and the Sun.

CAUTION! The Sun is at one focus of a planet's elliptical orbit, but there is *nothing* at the other focus. This "empty focus" has geometrical significance, because it helps to define the shape of the ellipse, but plays no other role.

Ellipses come in different shapes, depending on the elongation of the ellipse. The shape of an ellipse is described by its **eccentricity,** designated by the letter e. Figure 4-10b shows a few examples of ellipses with different eccentricities. The value of e can range from 0 (a circle) to just under 1 (nearly a straight line). The greater the eccentricity, the more elongated the ellipse. Because a circle is a special case of an ellipse, it is possible to have a perfectly circular orbit. But all of the objects that orbit the Sun have orbits that are at least slightly elliptical. The most circular of any planetary orbit is that of Venus, with an eccentricity of just 0.007; Mercury's orbit has an eccentricity of 0.206, and a number of small bodies called comets move in very elongated orbits with eccentricities of less than 1.

CONCEPTCHECK 4-9

Which orbit is closer to being circular—Venus's orbit, with $e = 0.007$, or Mars's orbit, with $e = 0.093$?

Answer appears at the end of the chapter.

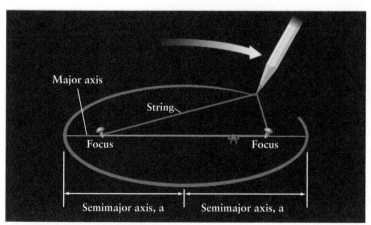

(a) The geometry of an ellipse

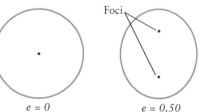

(b) Ellipses with different eccentricities

FIGURE 4-10

Ellipses (a) To draw an ellipse, use two thumbtacks to secure the ends of a piece of string, then use a pencil to pull the string taut. If you move the pencil while keeping the string taut, the pencil traces out an ellipse. The thumbtacks are located at the two foci of the ellipse. The major axis is the

greatest distance across the ellipse; the semimajor axis is half of this distance. (b) A series of ellipses with the same major axis but different eccentricities. An ellipse can have any eccentricity from $e = 0$ (a circle) to just under $e = 1$ (virtually a straight line).

Orbital Speeds and Kepler's Second Law

Once he knew the shape of a planet's orbit, Kepler was ready to describe exactly *how* it moves on that orbit. As a planet travels in an elliptical orbit, its distance from the Sun varies. Kepler realized that the speed of a planet also varies along its orbit. A planet moves most rapidly when it is nearest the Sun, at a point on its orbit called **perihelion**. Conversely, a planet moves most slowly when it is farthest from the Sun, at a point called **aphelion** (Figure 4-11). After much trial and error, Kepler found a way to describe just how a planet's speed varies as it moves along its orbit. Figure 4-11 illustrates this discovery, referred to as **Kepler's second law:**

A line joining a planet and the Sun sweeps out equal areas in equal intervals of time.

This relationship is also called the **law of equal areas**. In the idealized case of a circular orbit, a planet would have to move at a constant speed around the orbit in order to satisfy Kepler's second law. In the case of an elliptical orbit, a planet speeds up when it is closer to the Sun, and slows down when it is farther away.

CONCEPTCHECK 4-10

According to Kepler's second law, at what point in a communications satellite's elliptical orbit around Earth will it move the slowest?

Answer appears at the end of the chapter.

ANALOGY An analogy for Kepler's second law is a twirling ice skater holding weights in each hand. If the skater moves the weights closer to her body by pulling her arms straight in, her rate of spin increases and the weights move faster; if she extends her arms so the weights move away from her body, her rate of spin decreases and the weights slow down. Just like the weights, a planet in an elliptical orbit travels at a higher speed when it moves closer to the Sun (toward perihelion) and travels at a lower speed when it moves away from the Sun (toward aphelion).

Orbital Periods and Kepler's Third Law

Kepler's second law describes how the speed of a given planet changes as it orbits the Sun. Kepler also deduced from Tycho's data a relationship that can be used to compare the motions of *different* planets. Published in 1618 and now called **Kepler's third law**, it states a relationship between the size of a planet's orbit and the time the planet takes to go once around the Sun:

The square of the sidereal period of a planet is directly proportional to the cube of the semimajor axis of the orbit.

Kepler's third law says that the larger a planet's orbit—that is, the larger the semimajor axis, or average distance from the planet to the Sun—the longer the sidereal period, which is the time it takes the planet to complete an orbit. From Kepler's third law one can show that the larger the semimajor axis, the slower the average speed at which the planet moves around its orbit. (By contrast,

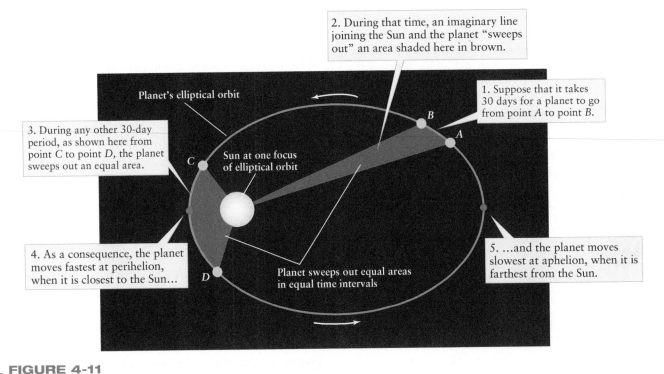

2. During that time, an imaginary line joining the Sun and the planet "sweeps out" an area shaded here in brown.

1. Suppose that it takes 30 days for a planet to go from point *A* to point *B*.

B

A

Planet's elliptical orbit

3. During any other 30-day period, as shown here from point *C* to point *D*, the planet sweeps out an equal area.

C

Sun at one focus of elliptical orbit

4. As a consequence, the planet moves fastest at perihelion, when it is closest to the Sun...

D

Planet sweeps out equal areas in equal time intervals

5. ...and the planet moves slowest at aphelion, when it is farthest from the Sun.

ANIMATION 4-6 **FIGURE 4-11**
Kepler's First and Second Laws According to Kepler's first law, a planet travels around the Sun along an elliptical orbit with the Sun at one focus. According to his second law, a planet moves fastest when closest to the Sun (at perihelion) and slowest when farthest from the Sun (at aphelion). As the planet moves, an imaginary line joining the planet and the Sun sweeps out equal areas in equal intervals of time (from A to B or from C to D). By using these laws in his calculations, Kepler found an excellent fit to the apparent motions of the planets.

Kepler's *second* law describes how the speed of a given planet is sometimes faster and sometimes slower than its average speed.) This qualitative relationship between orbital size and orbital speed is just what Aristarchus and Copernicus used to explain retrograde motion, as we saw in Section 4-2. Kepler's great contribution was to make this relationship a quantitative one.

It is useful to restate Kepler's third law as an equation. If a planet's sidereal period P is measured in years and the length of its semimajor axis a is measured in astronomical units (AU), where 1 AU is the average distance from Earth to the Sun (see Section 1-7), then Kepler's third law is

Kepler's third law

$$P^2 = a^3$$

P = planet's sidereal period, in years

a = planet's semimajor axis, in AU

If you know either the sidereal period of a planet or the semimajor axis of its orbit, you can find the other quantity using this equation. Box 4-2 gives some examples of how this is done.

We can verify Kepler's third law for all of the planets, including those that were discovered after Kepler's death, using data

BOX 4-2 TOOLS OF THE ASTRONOMER'S TRADE

Using Kepler's Third Law

Kepler's third law relates the sidereal period P of an object orbiting the Sun to the semimajor axis a of its orbit:

$$P^2 = a^3$$

You must keep two essential points in mind when working with this equation:

1. The period P *must* be measured in years, and the semimajor axis a *must* be measured in astronomical units (AU). Otherwise you will get nonsensical results.

2. This equation applies *only* to the special case of an object, like a planet, that orbits the Sun. If you want to analyze the orbit of the Moon around Earth, of a spacecraft around Mars, or of a planet around a distant star, you must use a different, generalized form of Kepler's third law. We discuss this alternative equation in Section 4-7 and Box 4-4.

EXAMPLE: The average distance from Venus to the Sun is 0.72 AU. Use this to determine the sidereal period of Venus.

Situation: The average distance from Venus to the Sun is the semimajor axis a of the planet's orbit. Our goal is to calculate the planet's sidereal period P.

Tools: To relate a and P we use Kepler's third law, $P^2 = a^3$.

Answer: We first cube the semimajor axis (multiply it by itself twice):

$$a^3 = (0.72)^3 = 0.72 \times 0.72 \times 0.72 = 0.373$$

According to Kepler's third law this is also equal to P^2, the square of the sidereal period. So, to find P, we have to "undo" the square, that is, take the square root. Using a calculator, we find

$$P = \sqrt{P^2} = \sqrt{0.373} = 0.61$$

Review: The sidereal period of Venus is 0.61 years, or a bit more than seven Earth months. This result makes sense: A planet with a smaller orbit than Earth's (an inferior planet) must have a shorter sidereal period than Earth.

EXAMPLE: A certain small asteroid (a rocky body a few tens of kilometers across) takes eight years to complete one orbit around the Sun. Find the semimajor axis of the asteroid's orbit.

Situation: We are given the sidereal period $P = 8$ years, and are to determine the semimajor axis a.

Tools: As in the preceding example, we relate a and P using Kepler's third law, $P^2 = a^3$.

Answer: We first square the period:

$$P^2 = 8^2 = 8 \times 8 = 64$$

From Kepler's third law, 64 is also equal to a^3. To determine a, we must take the *cube root* of a^3, that is, find the number whose cube is 64. If your calculator has a cube root function, denoted by the symbol $\sqrt[3]{\ }$, you can use it to find that the cube root of 64 is 4: $\sqrt[3]{64} = 4$. Otherwise, you can determine by trial and error that the cube of 4 is 64:

$$4^3 = 4 \times 4 \times 4 = 64$$

Because the cube of 4 is 64, it follows that the cube root of 64 is 4 (taking the cube root "undoes" the cube).

With either technique you find that the orbit of this asteroid has semimajor axis $a = 4$ AU.

Review: The period is greater than 1 year, so the semimajor axis is greater than 1 AU. Note that $a = 4$ AU is intermediate between the orbits of Mars and Jupiter (see Table 4-3). Many asteroids are known with semimajor axes in this range, forming a region in the solar system called the asteroid belt.

TABLE 4-3	A Demonstration of Kepler's Third Law ($P^2 = a^3$)			
Planet	Sidereal period P (years)	Semimajor axis a (AU)	P^2	a^3
Mercury	0.24	0.39	0.06	0.06
Venus	0.61	0.72	0.37	0.37
Earth	1.00	1.00	1.00	1.00
Mars	1.88	1.52	3.53	3.51
Jupiter	11.86	5.20	140.7	140.6
Saturn	29.46	9.55	867.9	871.0
Uranus	84.10	19.19	7,072	7,067
Neptune	164.86	30.07	27,180	27,190

Kepler's third law states that $P^2 = a^3$ for each of the planets. The last two columns of this table demonstrate that this relationship holds true to a very high level of accuracy.

from Tables 4-1 and 4-2. If Kepler's third law is correct, for each planet the numerical values of P^2 and a^3 should be equal. This result is indeed true to very high accuracy, as Table 4-3 shows.

CONCEPTCHECK 4-11

The space shuttle typically orbited Earth at an altitude of 300 km whereas the International Space Station orbits Earth at an altitude of 450 km. Although the space shuttle took less time to orbit Earth, which orbiter moved at a faster speed?

CALCULATIONCHECK 4-1

If Pluto's orbit has a semimajor axis of 39.5 AU, how long does it take Pluto to orbit the Sun once?

Answers appear at the end of the chapter.

The Significance of Kepler's Laws

Kepler's laws are a landmark in the history of astronomy. They made it possible to calculate the motions of the planets with better accuracy than any geocentric model ever had, and they helped to justify the idea of a heliocentric model. Kepler's laws also pass the test of Occam's razor, for they are simpler in every way than the schemes of Ptolemy or Copernicus, both of which used a complicated combination of circles.

But the significance of Kepler's laws goes beyond understanding planetary orbits. These same laws are also obeyed by spacecraft orbiting Earth, by two stars revolving about each other in a binary star system, and even by galaxies in their orbits about each other. Throughout this book, we shall use Kepler's laws in a wide range of situations.

As impressive as Kepler's accomplishments were, he did not prove that the planets orbit the Sun, nor was he able to explain *why* planets move in accordance with his three laws. These

advances were made by two other figures who loom large in the history of astronomy: Galileo Galilei and Isaac Newton.

CONCEPTCHECK 4-12

Do Kepler's laws of planetary motion apply only to the planets?

Answer appears at the end of the chapter.

4-5 Galileo's discoveries with a telescope strongly supported a heliocentric model

When Dutch opticians invented the telescope during the first decade of the seventeenth century, astronomy was changed forever. The scholar who used this new tool to amass convincing evidence that the planets orbit the Sun, not Earth, was the Italian mathematician and physical scientist Galileo Galilei (Figure 4-12).

> Galileo used the cutting-edge technology of the 1600s to radically transform our picture of the universe

While Galileo did not invent the telescope, he was the first to point one of these new devices toward the sky and to publish his observations. Beginning in 1610, he saw sights of which no one

FIGURE 4-12

Galileo Galilei (1564–1642) Galileo was one of the first people to use a telescope to observe the heavens. He discovered craters on the Moon, sunspots on the Sun, the phases of Venus, and four moons orbiting Jupiter. His observations strongly suggested that Earth orbits the Sun, not vice versa. (Erich Lessing/Art Resource)

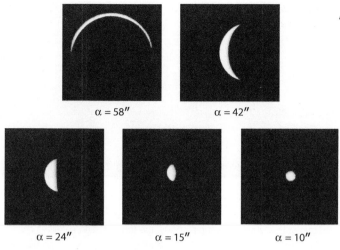

α = 58″   α = 42″

α = 24″   α = 15″   α = 10″

FIGURE 4-13   R I ⊻ U X G

The Phases of Venus This series of photographs shows how the appearance of Venus changes as it moves along its orbit. The number below each view is the angular diameter α of the planet in arcseconds. Venus has the largest angular diameter when it is a crescent, and the smallest angular diameter when it is gibbous (nearly full). (New Mexico State University Observatory)

had ever dreamed. He discovered mountains on the Moon, sunspots on the Sun, and the rings of Saturn, and he was the first to see that the Milky Way is not a featureless band of light but rather "a mass of innumerable stars."

The Phases of Venus

One of Galileo's most important discoveries with the telescope was that Venus exhibits phases like those of the Moon (Figure 4-13). Galileo also noticed that the apparent size of Venus as seen through his telescope was related to the planet's phase. Venus appears small at gibbous phase and largest at crescent phase. Galileo also noted a correlation between the phases of Venus and the planet's angular distance from the Sun.

Figure 4-14 shows that these relationships are entirely compatible with a heliocentric model in which Earth and Venus both go around the Sun. They are also completely *incompatible* with the Ptolemaic system, in which the Sun and Venus both orbit Earth. To explain why Venus is never seen very far from the Sun, the Ptolemaic model had to assume that the deferents of Venus and of the Sun move together in lockstep, with the epicycle of Venus centered on a straight line between Earth and the Sun (Figure 4-15). In this model, Venus was never on the opposite side of the Sun from Earth, and so it could never have shown the gibbous phases that Galileo observed.

CONCEPTCHECK 4-13

Which phase will Venus be in when it is at its maximum distance from Earth?

Answer appears at the end of the chapter.

The Moons of Jupiter

Galileo also found more unexpected evidence supporting the ideas of Copernicus. In 1610 Galileo discovered four moons, now

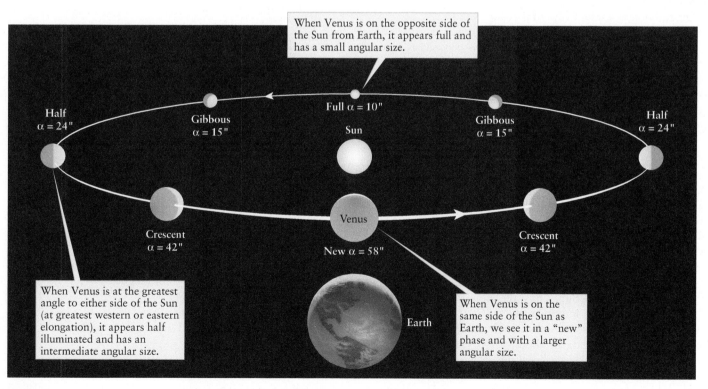

FIGURE 4-14

The Changing Appearance of Venus Explained in a Heliocentric Model A heliocentric model, in which Earth and Venus both orbit the Sun,

provides a natural explanation for the changing appearance of Venus shown in Figure 4-13.

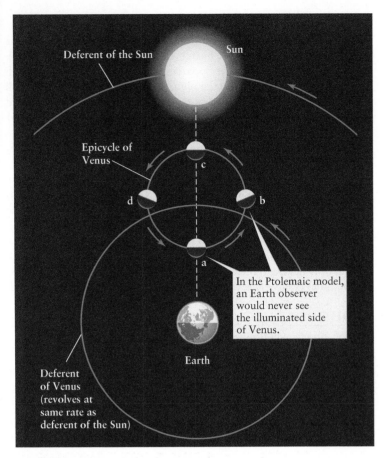

FIGURE 4-15

The Appearance of Venus in the Ptolemaic Model In the geocentric Ptolemaic model the deferents of Venus and the Sun rotate together, with the epicycle of Venus centered on a line (shown dashed) that connects the Sun and Earth. In this model an Earth observer would never see Venus as more than half illuminated. (At positions a and c, Venus appears in a "new" phase; at positions b and d, it appears as a crescent. Compare with Figure 3-2, which shows the phases of the Moon.) Because Galileo saw Venus in nearly fully illuminated phases, he concluded that the Ptolemaic model must be incorrect.

FIGURE 4-16   R I V U X G

Jupiter and Its Largest Moons This photograph, taken by an amateur astronomer with a small telescope, shows the four Galilean satellites alongside an overexposed image of Jupiter. Each satellite is bright enough to be seen with the unaided eye, were it not overwhelmed by the glare of Jupiter. (Rev. Ronald Royer/Science Source)

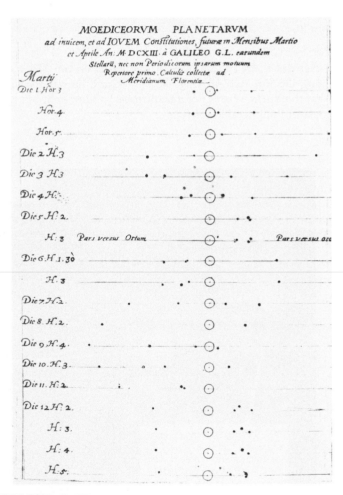

FIGURE 4-17

Early Observations of Jupiter's Moons In 1610 Galileo discovered four "stars" that move back and forth across Jupiter from one night to the next. He concluded that these are four moons that orbit Jupiter, much as our Moon orbits Earth. The circle represents Jupiter and the dots its moons. Compare the drawing numbered 7 with the photograph in Figure 4-16. (Royal Astronomical Society/Science Source)

called the Galilean satellites, orbiting Jupiter (Figure 4-16). He realized that they were orbiting Jupiter because they appeared to move back and forth from one side of the planet to the other. Figure 4-17 shows confirming observations made by Jesuit observers in 1620. Astronomers soon realized that the larger the orbit of one of the moons around Jupiter, the slower that moon moves and the longer it takes that moon to travel around its orbit. These are the same relationships that Copernicus deduced for the motions of the planets around the Sun. Thus, the moons of Jupiter behave like a Copernican system in miniature.

Galileo's telescopic observations constituted the first fundamentally new astronomical data in almost 2000 years. Contradicting prevailing opinion and religious belief, his discoveries strongly suggested a heliocentric structure of the universe. The Roman Catholic Church, which was a powerful political force in Italy and whose doctrine at the time placed Earth at the center of the universe, warned Galileo not to advocate a heliocentric model. He nonetheless persisted and was sentenced to spend the last years of his life under house arrest "for vehement suspicion of heresy." Nevertheless, there was no turning back. (The Roman Catholic Church lifted its ban against Galileo's heliocentric ideas in the 1700s.)

While Galileo's observations showed convincingly that the Ptolemaic model was entirely wrong and that a heliocentric model is the more nearly correct one, he was unable to provide a complete explanation of why Earth should orbit the Sun and not vice versa. The first person who was able to provide such an explanation was the Englishman Isaac Newton, born on Christmas Day of 1642, a dozen years after the death of Kepler and the same year that Galileo died. While Kepler and Galileo revolutionized our understanding of planetary motions, Newton's contribution was far greater: He deduced the basic laws that govern all motions on Earth as well as in the heavens.

CONCEPTCHECK 4-14

Why had Jupiter's moons not been observed prior to Galileo's time?

Answer appears at the end of the chapter.

4-6 Newton formulated laws of motion and gravity that describe fundamental properties of physical reality

TUTORIAL 4-2 Until the mid-seventeenth century, virtually all attempts to describe the motions of the heavens were *empirical,* or based directly on data and observations. From Ptolemy to Kepler, astronomers would adjust their ideas and calculations by trial and error until they ended up with answers that agreed with observation.

> The same laws of motion that hold sway on Earth apply throughout the universe

Isaac Newton (Figure 4-18) introduced a new approach. He began with three quite general statements, now called Newton's laws of motion. These laws, deduced from experimental observation, apply to all forces and all objects. Newton then showed that Kepler's three laws follow logically from these laws of motion and from a formula for the force of gravity that he derived from observation.

In other words, Kepler's laws are not just an empirical description of the motions of the planets, but a direct consequence of the fundamental laws of physical matter. Using this deeper insight into the nature of motions in the heavens, Newton and his successors were able to accurately describe not just the orbits of the planets but also the orbits of the Moon and comets.

Newton's First Law

Newton's laws of motion describe objects on Earth as well as in the heavens. Thus, we can understand each of these laws by considering the motions of objects around us. We begin with **Newton's first law of motion:**

An object remains at rest, or moves in a straight line at a constant speed, unless acted upon by a net outside force.

By **force** we mean any push or pull that acts on the object. An *outside* force is one that is exerted on the object by something other than the object itself. The net, or total, outside force is the

FIGURE 4-18
Isaac Newton (1642–1727) Using mathematical techniques that he devised, Isaac Newton formulated the law of universal gravitation and demonstrated that the planets orbit the Sun according to simple mechanical rules. (Corbis Images)

combined effect of all of the individual outside forces that act on the object.

Right now, you are demonstrating the first part of Newton's first law—remaining at rest. As you sit in your chair reading this passage, there are two outside forces acting on you: The force of gravity pulls you downward, and the chair pushes up on you. These two forces are of equal strength but of opposite direction, so their effects cancel one another—there is no *net* outside force. Hence, your body remains at rest as stated in Newton's first law. If you try to lift yourself out of your chair by grabbing your knees and pulling up, you will remain at rest because this force is not an outside force: It comes from your body.

CAUTION! It is easy to confuse the *net* (or total) outside force on an object (central to Newton's first law) with *individual* outside forces on an object. If you want to make a book move across the floor in a straight line at a *constant speed,* you must continually push on it. You might therefore think that your push is a net outside force. But another force also acts on the book— the force of friction as the book rubs across the floor. The force of your push and the force of friction combine to make the *net* outside force. If you push the book at a constant speed, then the force of your push exactly balances the force of friction, so there is no net outside force. If you stop pushing, there will be nothing to balance the effects of friction. Then the friction will be a net outside force and the book will slow to a stop.

Newton's first law tells us that if no net outside force acts on a moving object, it can only move in a straight line and at a constant speed. This means that a net outside force *must* be acting on the planets since they don't move in straight lines but instead move around elliptical paths. Another way to see that planetary orbits require a net outside force is that a planet would fly off into space at a constant speed along a straight line if there were no net outside force acting on it. Because planets don't fly off, Newton concluded that a force *must* act continuously on the planets to keep them in their elliptical orbits.

CONCEPTCHECK 4-15

Imagine a 10-kg rock speeding through empty space at 200 m/s, so far away from other objects that there is no gravitational force (or any other outside forces) exerted on the rock. Describe the rock's motion.

Answer appears at the end of the chapter.

Newton's Second Law

Newton's second law describes how the motion of an object *changes* if there is a net outside force acting on it. To appreciate Newton's second law, we must first understand three quantities that describe motion—speed, velocity, and acceleration.

Speed is a measure of how fast an object is moving. Speed and direction of motion together constitute an object's **velocity.** Compared with a car driving north at 100 km/h (62 mi/h), a car driving east at 100 km/h has the same speed but a different velocity. We can restate Newton's first law to say that an object has a constant velocity (its speed and direction of motion do not change) if no net outside force acts on the object.

Acceleration is the rate at which velocity changes. Because velocity involves both speed and direction, acceleration can result from changes in either. Contrary to popular use of the term, acceleration does not simply mean speeding up. A car is accelerating if it is speeding up, and it is also accelerating if it is slowing down or turning (that is, changing the direction in which it is moving).

You can verify these statements about acceleration if you think about the sensations of riding in a car. If the car is moving with a constant velocity (in a straight line at a constant speed), you feel the same, aside from vibrations, as if the car were not moving at all. But you can feel it when the car accelerates in any way: You feel thrown back in your seat if the car speeds up, thrown forward if the car slows down, and thrown sideways if the car changes direction in a tight turn. In Box 4-3 we discuss the reasons for these sensations, along with other applications of Newton's laws to everyday life.

An apple falling from a tree is a good example of acceleration that involves only an increase in speed. Initially, at the moment the stem breaks, the apple's speed is zero. After 1 second, its downward speed is 9.8 meters per second, or 9.8 m/s (32 feet per second, or 32 ft/s). After 2 seconds, the apple's speed is twice this, or 19.6 m/s. After 3 seconds, the speed is 29.4 m/s. Because the apple's speed increases by 9.8 m/s for each second of free fall, the rate of acceleration is 9.8 meters per second per second, or 9.8 m/s^2 (32 ft/s^2). Thus, Earth's gravity gives the apple a constant acceleration of 9.8 m/s^2 downward, toward the center of Earth.

A planet revolving about the Sun along a perfectly circular orbit is an example of acceleration that involves change of direction only. As the planet moves along its orbit, its speed remains constant. Nevertheless, the planet is continuously being accelerated because its direction of motion is continuously changing.

Newton's second law of motion says that in order to give an object an acceleration (that is, to change its velocity), a net outside force *must* act on the object. To be specific, this law says that the acceleration of an object is proportional to the net outside force acting on the object. That is, the harder you push on an object, the greater the resulting acceleration. This law can be succinctly stated as an equation. If a net outside force F acts on an object of mass m, the object will experience an acceleration a such that

Newton's second law

$$F = ma$$

F = net outside force on an object

m = mass of object

a = acceleration of object

The **mass** of an object is a measure of the total amount of material in the object. It is usually expressed in kilograms (kg) or grams (g). For example, the mass of the Sun is 2×10^{30} kg, the mass of a hydrogen atom is 1.7×10^{-27} kg, and the mass of an average adult is 75 kg. The Sun, a hydrogen atom, and a person have these masses regardless of where they happen to be in the universe.

CAUTION! It is important not to confuse the concepts of mass and weight. **Weight** is the force of gravity that acts on an object and, like any force, is usually expressed in pounds or newtons (1 newton = 0.225 pounds). For example, astronauts feel lighter on the Moon because they weigh less in the Moon's weaker gravity, but an astronaut's mass on the Moon is the same as her mass on Earth.

We can use Newton's second law to relate mass and weight. We have seen that the acceleration caused by Earth's gravity is 9.8 m/s^2. When a 50-kg swimmer falls from a diving board, the only outside force acting on her as she falls is her weight. Thus, from Newton's second law ($F = ma$), her weight is equal to her mass multiplied by the acceleration due to gravity:

$$50 \text{ kg} \times 9.8 \text{ m/s}^2 = 490 \text{ newtons} = 110 \text{ pounds}$$

Note that this answer is correct only when the swimmer is on Earth. She would weigh less on the Moon, where the pull of gravity is weaker, and more on Jupiter, where the gravitational pull is stronger. Floating deep in space, she would have no weight at all; she would be "weightless." Nevertheless, in all these circumstances, she would always have exactly the same mass, because mass is an inherent property of matter unaffected by details of the environment. Whenever we describe the properties of planets, stars, or galaxies, we speak of their masses, never of their weights.

We have seen that a planet is continually accelerating as it orbits the Sun. From Newton's second law, this means that there must be a net outside force that acts continually on each of the planets. As we will see in the next section, this force is the gravitational attraction of the Sun.

BOX 4-3 ASTRONOMY DOWN TO EARTH

Newton's Laws in Everyday Life

In our study of astronomy, we use Newton's three laws of motion to help us understand the motions of objects in the heavens. But you can see applications of Newton's laws every day in the world around you. By considering these everyday applications, we can gain insight into how Newton's laws apply to celestial events that are far removed from ordinary human experience.

Newton's *first* law, or principle of inertia, says that an object at rest naturally tends to remain at rest and that an object in motion naturally tends to remain in motion. This law explains the sensations that you feel when riding in an automobile. When you are waiting at a red light, your car and your body are both at rest. When the light turns green and you press on the gas pedal, the car accelerates forward but your body attempts to stay where it was. Hence, the seat of the accelerating car pushes forward into your body, and it feels as though you are being pushed back in your seat.

Once the car is up to cruising speed, your body wants to keep moving in a straight line at this cruising speed. If the car makes a sharp turn to the left, the right side of the car will move toward you. Thus, you will feel as though you are being thrown to the car's right side (the side on the outside of the turn). If you bring the car to a sudden stop by pressing on the brakes, your body will continue moving forward until the seat belt stops you. In this case, it feels as though you are being thrown toward the front of the car.

Newton's *second* law states that the net outside force on an object equals the product of the object's mass and its acceleration. You can accelerate a crumpled-up piece of paper to a pretty good speed by throwing it with a moderate force. But if you try to throw a heavy rock by using the same force, the acceleration will be much less because the rock has much more mass than the crumpled paper. Because of the smaller acceleration, the rock will leave your hand moving at only a slow speed.

Automobile airbags are based on the relationship between force and acceleration. It takes a large force to bring a fast-moving object suddenly to rest because this requires a large acceleration. In a collision, the driver of a car not equipped with airbags is jerked to a sudden stop and the large forces that act can cause major injuries. But if the car has airbags that deploy in an accident, the driver's body will slow down more gradually as it contacts the airbag, and the driver's acceleration will be less. (Remember that *acceleration* can refer to slowing down as well as to speeding up.) Hence, the force on the driver and the chance of injury will both be greatly reduced.

Newton's *third* law, the principle of action and reaction, explains how a car can accelerate at all. It is not correct to say that the engine pushes the car forward, because Newton's second law tells us that it takes a force acting from outside the car to make the car accelerate. Rather, the engine makes the wheels and tires turn, and the tires push backward on the ground. (You can see this backward force in action when a car drives through wet ground and sprays mud backward from the tires.) From Newton's third law, the ground must exert an equally large forward force on the car, and this is the force that pushes the car forward.

You use the same principles when you walk: You push backward on the ground with your foot, and the ground pushes forward on you. Icy pavement or a freshly waxed floor have greatly reduced friction. In these situations, your feet and the surface under you can exert only weak forces on each other, and it is much harder to walk.

Newton's Third Law

The last of Newton's general laws of motion is called **Newton's third law of motion:**

Whenever one object exerts a force on a second object, the second object exerts an oppositely directed force of equal strength on the first object.

Newton's third law is sometimes described in terms of actions and reactions: When two objects interact by exerting forces on each other, these action and reaction forces are equal in magnitude but opposite in direction.

For example, consider your weight; Earth pulls down on you with a gravitational force equal to your weight. But you also pull back up on Earth with a gravitational force of equal strength. In another example, consider a bat hitting a ball. Clearly, the bat exerts a large force on the ball when hitting a home run (this can be called an action force). However, the ball exerts a force of *equal strength* on the bat as well (this can be called a reaction force). It might seem totally counterintuitive that a small, passive ball could exert an equally large force on a bat that has someone's full swing behind it. However, imagine a spring between the ball and bat during the hit; a spring is a very good approximation for the deformed ball. Regardless of which object moves in order to compress the spring, the spring pushes outward with an equal force on both ends (outward on the ball and bat).

CONCEPTCHECK 4-16

Two sumo wrestlers push against each other during a match. One wrestler is much larger than the other. The larger wrestler's feet remain on the floor while the smaller wrestler's feet slip as he is accelerated in a push right out of the ring. Compare the force that the larger wrestler exerts on the smaller wrestler to the force the smaller wrestler exerts on the larger wrestler.

CONCEPTCHECK 4-17

In midair after stepping off a diving board, a diver is pulled down to the water by her weight. However, the diver pulls up on Earth with a force of equal strength, so why doesn't Earth move an equal amount as the diver?

Answers appears at the end of the chapter.

In these examples, you can think of each force in these pairs as a reaction to the other force, which is the origin of the phrase "action and reaction."

Newton realized that because the Sun is exerting a force on each planet to keep it in orbit, each planet must also be exerting an equal and opposite force on the Sun. However, the planets are much less massive than the Sun (for example, Earth has only 1/300,000 of the Sun's mass). Therefore, although the Sun's force on a planet is the same as the planet's force on the Sun, the planet's much smaller mass gives it a much larger acceleration, according to Newton's second law. This is why the planets circle the Sun instead of vice versa. Thus, Newton's laws reveal the reason for our heliocentric solar system.

The Law of Universal Gravitation

Tie a ball to one end of a piece of string, hold the other end of the string in your hand, and whirl the ball around in a circle. As the ball "orbits" your hand, it is continuously accelerating because its velocity is changing. (Even if its speed is constant, its direction of motion is changing.) In accordance with Newton's second law, this can happen only if the ball is continuously acted on by an outside force—the pull of the string. The pull is directed along the string toward your hand. In the same way, Newton saw, the force that keeps a planet in orbit around the Sun is a pull that always acts toward the Sun. That pull is **gravity,** or **gravitational force.**

> Newton's law of gravitation is truly universal: It applies to falling apples as well as to planets and galaxies

Newton's discovery about the forces that act on planets led him to suspect that the force of gravity pulling a falling apple straight down to the ground is fundamentally the same as the force on a planet that is always directed straight at the Sun. In other words, gravity is the force that shapes the orbits of the planets. What is more, he was able to determine how the force of gravity depends on the distance between the Sun and the planet. His result was a law of gravitation that could apply to the motion of distant planets as well as to the flight of a football on Earth. Using this law, Newton achieved the remarkable goal of deducing Kepler's laws from fundamental principles of nature.

To see how Newton reasoned, think again about a ball attached to a string. If you use a short string, so that the ball orbits in a small circle, and whirl the ball around your hand at a high speed, you will find that you have to pull fairly hard on the string (Figure 4-19a). But if you use a longer string, so that the ball moves in a larger orbit, and if you make the ball orbit your hand at a slow speed, you only have to exert a light tug on

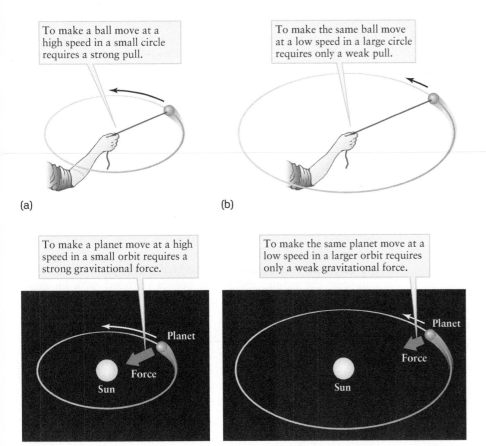

(a) (b)

(c) (d)

FIGURE 4-19

An Orbit Analogy **(a)** To make a ball on a string move at high speed around a small circle, you have to exert a substantial pull on the string. **(b)** If you lengthen the string and make the same ball move at low speed around a large circle, much less pull is required. **(c)** Similarly, a planet that orbits close to the Sun moves at high speed and requires a substantial gravitational force from the Sun, while **(d)** a planet in a large orbit moves at low speed and requires less gravitational force to stay in orbit.

the string (Figure 4-19b). The orbits of the planets behave in the same way: The larger the size of the orbit, the slower the planet's speed (Figure 4-19c, d). By analogy to the force of the string on the orbiting ball, Newton concluded that the force that attracts a planet toward the Sun must decrease with increasing distance between the Sun and the planet.

Using his own three laws and Kepler's three laws, Newton succeeded in formulating a general statement that describes the nature of the gravitational force. Newton's **law of universal gravitation** is as follows:

Two objects attract each other with a force that is directly proportional to the mass of each object and inversely proportional to the square of the distance between them.

This law states that *any* two objects exert gravitational pulls on each other. Normally, you notice only the gravitational force that Earth exerts on you, otherwise known as your weight. In fact, you are gravitationally attracted to *all* the objects around you. For example, a book exerts a gravitational force on you as you read it. But because the force exerted on you by this book is proportional to the book's mass, which is very small compared to Earth's mass, the force is too small to notice. (It can actually be measured with sensitive equipment.)

The farther apart two objects are, the weaker the gravitational force between them. Since the gravitational force weakens by the square of the distance between two objects, doubling their distance reduces their attraction by

$$\frac{1}{2^2} = \frac{1}{4}$$

and tripling their distance reduces their attraction by

$$\frac{1}{3^2} = \frac{1}{9}$$

Newton's law of universal gravitation can be stated as an equation. If two objects have masses m_1 and m_2 and are separated by a distance r, then the gravitational force F between these two objects is given by the following equation:

Newton's law of universal gravitation

$$F = G\left(\frac{m_1 m_2}{r^2}\right)$$

F = gravitational force between two objects
m_1 = mass of first object
m_2 = mass of second object
r = distance between objects
G = universal constant of gravitation

If the masses are measured in kilograms and the distance between them in meters, then the force is measured in newtons. In this formula, G is a number called the **universal constant of gravitation.** Laboratory experiments have yielded a value for G of

$$G = 6.67 \times 10^{-11} \text{ newton} \cdot \text{m}^2/\text{kg}^2$$

We can use Newton's law of universal gravitation to calculate the force with which any two objects attract each other. For example, to compute the gravitational force that the Sun exerts on Earth, we substitute values for Earth's mass ($m_1 = 5.98 \times 10^{24}$ kg), the Sun's mass ($m_2 = 1.99 \times 10^{30}$ kg), the distance between them ($r = 1$ AU $= 1.5 \times 10^{11}$ m), and the value of G into Newton's equation. We get

$$F_{\text{Sun-Earth}} = 6.67 \times 10^{-11}\left[\frac{(5.98 \times 10^{24}) \times (1.99 \times 10^{30})}{(1.50 \times 10^{11})^2}\right]$$

$$= 3.53 \times 10^{22} \text{ newtons}$$

If we calculate the force that Earth exerts on the Sun, we get exactly the same result. (Mathematically, we just let m_1 be the Sun's mass and m_2 be Earth's mass instead of the other way around. The product of the two numbers is the same, so the force is the same.) This is in accordance with Newton's third law: Any two objects exert *equal* gravitational forces on each other.

Your weight is just the gravitational force that Earth exerts on you, so we can calculate it using Newton's law of universal gravitation. Earth's mass is $m_1 = 5.98 \times 10^{24}$ kg, and the distance r to use is the distance between the *centers* of Earth and you. This distance is just the radius of Earth, which is $r = 6378$ km $= 6.378 \times 10^6$ m. If your mass is $m_2 = 50$ kg, your weight is

$$F_{\text{Earth-you}} = 6.67 \times 10^{-11}\left[\frac{(5.98 \times 10^{24}) \times (50)}{(6.378 \times 10^6)^2}\right]$$

$$= 490 \text{ newtons}$$

This value is the same as the weight of a 50-kg person that we calculated in Section 4-6. This example shows that your weight would have a different value on a planet with a different mass m_1 and a different radius r.

CONCEPTCHECK **4-18**

How much does the gravitational force of attraction change between two asteroids if the two asteroids drift 3 times closer together?

CALCULATIONCHECK **4-2**

How much would a 75-kg astronaut, weighing about 165 pounds on Earth, weigh in newtons and in pounds if he were standing on Mars, which has a mass of 6.4×10^{23} kg and a radius of 3.4×10^6 m?

Answers appear at the end of the chapter.

4-7 Describing orbits with energy and gravity

Imagine a cannonball shot into the air; it arcs through the sky before it crashes back to the ground. If you want to shoot the cannonball into orbit around Earth (like a satellite), you might guess that the ball must be shot faster to give it more energy. This guess is correct, and next we'll look at how to understand orbits in terms of gravity and energy.

Different Forms of Energy

Energy comes in a variety of familiar forms. For example, consider a car: The faster a car moves, the greater its energy. In general, the greater an object's speed, the more **kinetic energy** it has (Figure 4-20). The word *kinetic* refers to motion, so kinetic energy is the energy of motion. However, the mass of an object also contributes to its kinetic energy, and a speeding car therefore has more kinetic energy than a speeding bullet.

Now let's look at some other forms of energy. Both food and batteries store chemical energy, although they each contain very different chemicals. When energy is stored, we say it has the potential to be used, and this stored energy is often called *potential energy.* Consider a girl on the diving board of a swimming pool (Figure 4-21); gravity has the potential to pull her into the water (whenever she steps off the board). This is an example of stored gravitational energy, or **gravitational potential energy.** The higher the diving board, the more gravitational potential energy she has. After she steps off the board, she'll have both kinetic energy and gravitational potential energy. As she falls towards the water, her kinetic energy increases and her gravitational potential energy decreases.

While the concept of energy is easier to introduce with a diving board, the same principles apply to orbiting objects: Planets, moons, and satellites also have both kinetic and gravitational potential energy. For example, the faster a satellite orbits Earth, the greater its kinetic energy, and if it falls back to Earth, it loses gravitational potential energy.

To understand another aspect of energy, let's think more about the diver in Figure 4-21. Standing high atop the board, the diver has more gravitational potential energy than she would have while standing on the ground. In midair during the dive, some of that gravitational potential energy has transformed into kinetic energy. This illustrates a law called the **conservation of energy:** While energy can change from one form to another, energy cannot be created or destroyed.

There's another way to express the law of conservation of energy. First consider a collection of objects that are *isolated,* which means no energy can leave or enter this collection of objects. Second, find the total energy by taking into account all the forms of energy the collection of objects has. This total energy is conserved, meaning that the total energy stays constant over time, even as objects interact with each other and the various forms of energy can change.

CONCEPTCHECK 4-19

Consider a cannonball shot vertically, straight up into the air. At the ball's highest point, it momentarily comes to a complete stop before beginning to fall back down. At its highest point, what form of energy does the ball have, and where did the energy come from?

Answer appears at the end of the chapter.

As a feather falls gently to the ground it loses gravitational potential energy without gaining much kinetic energy. This is an example of **air drag**, where the kinetic energy of an object is transferred to the molecules in the surrounding air and the air gets hotter. Energy is still conserved—as it always is—and the nearby air molecules actually speed up. Since the kinetic energy of individual particles gives a gas its temperature and thermal energy, we can see that air drag converts kinetic energy into thermal energy (we will learn more about thermal energy in Section 5-3). Air drag can be good or bad, depending on the situation. Astronauts returning to Earth rely on this transfer of energy, as they use air drag during atmospheric reentry to reduce their speed for a safe landing. As we will see shortly, air drag is undesirable when it transfers energy away from an orbiting satellite, causing it to fall back to Earth.

Gravitational Force, Energy, and Orbits

Because there is a gravitational force between any two objects, Newton concluded that gravity is also the force that keeps the Moon in orbit around Earth. It is also the force that keeps artificial satellites in orbit. But if the force of gravity attracts two objects to each other, why don't satellites immediately fall to

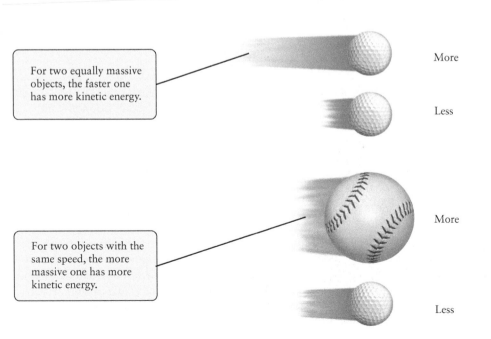

For two equally massive objects, the faster one has more kinetic energy.

More

Less

For two objects with the same speed, the more massive one has more kinetic energy.

More

Less

FIGURE 4-20
Kinetic Energy The kinetic energy of an object depends on both its speed and its mass.

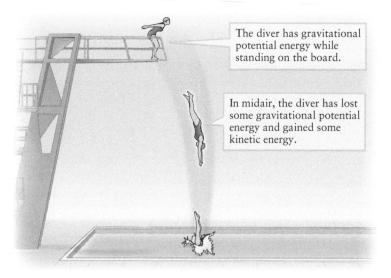

FIGURE 4-21

Gravitational Potential Energy The higher the diver, the more gravitational potential energy she has. As the diver descends, some of her gravitational potential energy has transformed into kinetic energy.

Earth? Why doesn't the Moon fall into Earth? And, for that matter, why don't the planets fall into the Sun?

To see the answer, imagine (as Newton did) dropping a ball from a great height above Earth's surface, as in **Figure 4-22**. After you drop the ball, it, of course, falls straight down (path A in Figure 4-22). But if you *throw* the ball horizontally, it travels some distance across Earth's surface before hitting the ground (path B). If you throw the ball harder, it travels a greater distance (path C). If you could throw or shoot it at just the right speed, the curvature of the ball's path will exactly match the curvature of Earth's surface (path E). Although Earth's gravity is making the ball fall, Earth's surface is falling away under the ball at the same rate.

Hence, the ball does not get any closer to the surface, and the ball is in circular orbit. So, the ball in path E is in fact falling, but it is falling *around* Earth rather than *toward* Earth.

As the hypothetical ball is launched faster and faster, you can also see that the paths (going from A to F) represent trajectories with more of both kinetic and gravitational potential energy; in other words, with increasing **orbital energy.** In general for an object orbiting the Sun or a planet:

The greater the orbital energy, the greater the average orbital distance (or the greater the semimajor axis, a).

A spacecraft is launched into orbit in just this way—by gaining enough speed and energy. Once the spacecraft is in orbit, no more rocket fuel is needed and, if not for air drag, the spacecraft could orbit indefinitely. However, air drag from the thin outer wisps of Earth's atmosphere slowly removes orbital energy from the closer satellites, slowing them down, and bringing them inward. Without occasional bursts of thrust, these satellites would fall back to Earth. The International Space Station is no exception, and air drag can decrease the station's altitude by a couple hundred feet each day!

CAUTION! An astronaut on board an orbiting spacecraft (like the one shown in the photograph that opens this chapter) feels "weightless." However, this is *not* because she is "beyond the pull of gravity." The astronaut is herself an independent satellite of Earth, and Earth's gravitational pull is what holds her in orbit. She feels "weightless" because she and her spacecraft are falling *together* around Earth, so there is nothing pushing her against any of the spacecraft walls. You feel the same "weightless" sensation whenever you are falling, such as when you jump off a diving board or ride the free-fall ride at an amusement park.

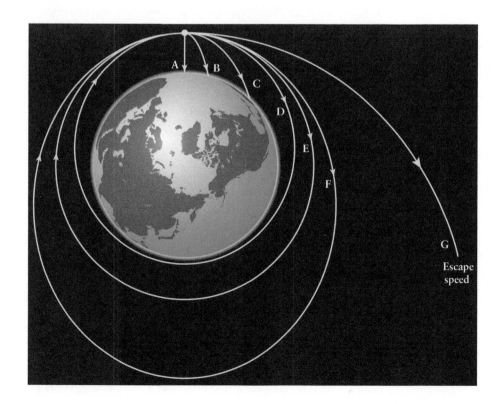

FIGURE 4-22

Orbits and the Escape Speed If a ball is dropped from a great height above Earth's surface, it falls straight down (A). If the ball is thrown with some horizontal speed, it follows a curved path before hitting the ground (B, C). If thrown with just the right speed (E), the ball goes into circular orbit; the ball's path curves but it never gets any closer to Earth's surface. If the ball is thrown with a speed that is slightly less (D) or slightly more (F) than the speed for a circular orbit, the ball's orbit is an ellipse. If thrown faster than the escape speed (G), the ball will leave Earth and never return.

CONCEPTCHECK 4-20

Suppose a spacecraft is initially shot into a circular orbit with path E in Figure 4-22. Which new path will the spacecraft likely take if air drag is present? Which new path will likely result if a rocket thruster is fired?

CONCEPTCHECK 4-21

As Earth orbits the Sun, does Earth's orbit decay due to air drag from its own atmosphere?

Answers appear at the end of the chapter.

The Escape Speed

If an object is hurled with enough speed, it can escape a planet altogether. For example, when a very large asteroid smashes into Earth, most of the debris ejected from the crater falls back to Earth's surface, but some rocks have enough speed to escape and roam the solar system. In Figure 4-22, the object on path G is shot with the **escape speed,** has more energy than any of the "bound" orbits, and never returns to Earth's orbit. The speed needed for escape depends on the size and mass of the planet, but *not* on the mass of the escaping object:

Escape Speed

$$v_{\text{escape}} = \sqrt{\frac{2GM}{R}}$$

V_{escape} = escape speed

M = mass of the planet

R = radius of the planet

G = universal constant of gravitation

For Earth, the escape speed is 11.2 km/s. That's about 25,000 mi/h, or 33 times the speed of sound. This speed is so fast that, except for the largest asteroid impacts, very little debris would have escaped Earth. However, the escape speed for Mars is only about 5 km/s, so compared to Earth, it is easier for impact debris from Mars to get ejected into the solar system. (In fact, about 100 rocks found on Earth were originally ejected from Mars by asteroid impacts.)

What about spacecraft; are they launched at the escape speed? The escape speed assumes a simple scenario where no air drag is present and where there is no rocket fuel to add energy during flight. However, even with fuel, spacecraft going to the Moon or beyond often leave Earth near the escape speed.

CONCEPTCHECK 4-22

Two objects, a large rock and a small rock, are shot with just enough speed to escape the Moon. Can both rocks be shot with equal speed? Can both rocks be shot with the same kinetic energy?

CONCEPTCHECK 4-23

New planets around other stars are constantly being discovered. Would it be possible to find a planet with the same mass as Earth but with a lower escape speed from its surface?

Answers appear at the end of the chapter.

FIGURE 4-23 R I V U X G

In Orbit Around the Moon This photograph taken from the spacecraft *Columbia* shows the lunar lander *Eagle* after returning from the first human landing on the Moon in July 1969. Newton's form of Kepler's third law describes the orbit of a spacecraft around the Moon, as well as the orbit of the Moon around Earth (visible in the distance). (Michael Collins, *Apollo 11,* NASA)

Gravitation and Kepler's Laws

Using his three laws of motion and his law of gravity, Newton found that he could derive Kepler's three laws mathematically. (Newton also invented calculus, which helped in this endeavor!) While Kepler's laws describe the planets in the solar system specifically, Newton showed that similar behavior results from any orbital system, such as a planet and its moons. Kepler's first law, concerning the elliptical shape of planetary orbits, proved to be a direct consequence of the $1/r^2$ factor in the law of universal gravitation. The law of equal areas, or Kepler's second law, turns out to be a consequence of the Sun's gravitational force on a planet being directed straight toward the Sun.

Newton also demonstrated that Kepler's third law follows logically from his law of gravity. Specifically, he proved that if two objects with masses m_1 and m_2 orbit each other, the period P of their orbit and the semimajor axis a of their orbit (that is, the average distance between the two objects) are related by an equation that we call **Newton's form of Kepler's third law:**

$$P^2 = \left[\frac{4\pi^2}{G(m_1 + m_2)} \right] a^3$$

Newton's form of Kepler's third law is valid whenever two objects orbit each other because of their mutual gravitational attraction (Figure 4-23). It is invaluable in the study of binary star systems, in which two stars orbit each other. If the orbital period P and semimajor axis a of the two stars in a binary system are known, astronomers can use this formula to calculate the sum $m_1 + m_2$ of the masses of the two stars. Within our own solar system, Newton's form of Kepler's third law makes it possible to learn about the masses of planets. By measuring the period and semimajor axis for a satellite, astronomers can determine the sum of the masses of the planet and the satellite. (The satellite can be a moon of the planet or a spacecraft that we place in orbit around the planet. Newton's laws apply in either case.) Box 4-4 gives an example of using Newton's form of Kepler's third law.

Newton also discovered new features of orbits around the Sun. For example, his equations soon led him to conclude that the orbit of an object around the Sun need not be an ellipse. It could be any one of a family of curves called conic sections.

BOX 4-4 TOOLS OF THE ASTRONOMER'S TRADE

Newton's Form of Kepler's Third Law

Kepler's original statement of his third law, $P^2 = a^3$, is valid only for objects that orbit the Sun. (Box 4-2 shows how to use this equation.) But Newton's form of Kepler's third law is much more general: It can be used in *any* situation where two objects of masses m_1 and m_2 orbit each other. For example, Newton's form is the equation to use for a moon orbiting a planet or a satellite orbiting Earth. This equation is

Newton's form of Kepler's third law:

$$P^2 = \left[\frac{4\pi^2}{G(m_1 + m_2)} \right] a^3$$

P = sidereal period of orbit, in seconds

a = semimajor axis of orbit, in meters

m_1 = mass of first object, in kilograms

m_2 = mass of second object, in kilograms

G = universal constant of gravitation = 6.67×10^{-11}

Notice that P, a, m_1, and m_2 *must* be expressed in these particular units. If you fail to use the correct units, your answer will be incorrect.

EXAMPLE: Io (pronounced "eye-oh") is one of the four large moons of Jupiter discovered by Galileo and shown in Figure 4-16. It orbits at a distance of 421,600 km from the center of Jupiter and has an orbital period of 1.77 days. Determine the combined mass of Jupiter and Io.

Situation: We are given Io's orbital period P and semimajor axis a (the distance from Io to the center of its circular orbit, which is at the center of Jupiter). Our goal is to find the sum of the masses of Jupiter (m_1) and Io (m_2).

Tools: Because this orbit is not around the Sun, we must use Newton's form of Kepler's third law to relate P and a. This relationship also involves m_1 and m_2, whose sum ($m_1 + m_2$) we are asked to find.

Answer: To solve for $m_1 + m_2$, we rewrite the equation in the form

$$m_1 + m_2 = \frac{4\pi^2 a^3}{GP^2}$$

To use this equation, we have to convert the distance a from kilometers to meters and convert the period P from days to seconds. There are 1000 meters in 1 kilometer and 86,400 seconds in 1 day, so

$$a = (421{,}600 \text{ km}) \times \frac{1000 \text{ m}}{1 \text{ km}} = 4.216 \times 10^8 \text{ m}$$

$$P = (1.77 \text{ days}) \times \frac{86{,}400 \text{ s}}{1 \text{ day}} = 1.529 \times 10^5 \text{ s}$$

We can now put these values and the value of G into the above equation:

$$m_1 + m_2 = \frac{4\pi^2 (4.216 \times 10^8)^3}{(6.67 \times 10^{-11})(1.529 \times 10^5)^2} = 1.90 \times 10^{27} \text{ kg}$$

Review: Io is very much smaller than Jupiter, so its mass is only a small fraction of the mass of Jupiter. Thus, $m_1 + m_2$ is very nearly the mass of Jupiter alone. We conclude that Jupiter has a mass of 1.90×10^{27} kg, or about 300 times the mass of Earth. This technique can be used to determine the mass of any object that has a second, much smaller object orbiting it. Astronomers use this technique to find the masses of stars, black holes, and entire galaxies of stars.

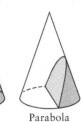
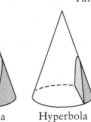
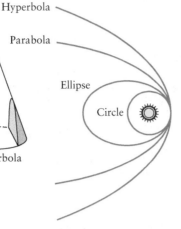

Hyperbola

Parabola

Ellipse

Circle

Circle   Ellipse   Parabola   Hyperbola

FIGURE 4-24

Conic Sections A conic section is any one of a family of curves obtained by slicing a cone with a plane. The orbit of one object about another can be any one of these curves: a circle, an ellipse, a parabola, or a hyperbola.

A **conic section** is any curve that you get by cutting a cone with a plane, as shown in Figure 4-24. You can get circles and ellipses by slicing all the way through the cone, and these represent the familiar circular and elliptical orbits. You can also get two types of open curves called **parabolas** and **hyperbolas**. An object traveling at the escape speed follows a parabolic curve, and for objects traveling even faster than the escape speed, the path is hyperbolic. Comets hurtling toward the Sun from the depths of space sometimes follow hyperbolic orbits and never return.

CONCEPTCHECK 4-24

A small celestial object on a hyperbolic orbit passes by a lonely moonless planet. Can the planet capture the object to acquire a moon?

Answer appears at the end of the chapter.

The Triumph of Newtonian Mechanics

Newton's ideas turned out to be applicable to an incredibly wide range of situations. Using his laws of motion, Newton himself proved that Earth's axis of rotation must precess because of the gravitational pull of the Moon and the Sun on Earth's equatorial bulge (see Figure 2-19). In fact, all the details of the orbits of the planets and their satellites could be explained mathematically with a body of knowledge built on Newton's work that is today called **Newtonian mechanics.**

Not only could Newtonian mechanics explain a variety of known phenomena in detail, but it could also predict new phenomena. For example, one of Newton's friends, Edmund Halley, was intrigued by three similar historical records of a comet that had been sighted at intervals of 76 years. Assuming these records to be accounts of the same comet, Halley used Newton's methods to work out the details of the comet's orbit and predicted its return in 1758. It was first sighted on Christmas night of 1757, a fitting memorial to Newton's birthday. To this day the comet bears Halley's name (Figure 4-25).

Another dramatic success of Newton's ideas was their role in the discovery of the eighth planet from the Sun. The seventh planet, Uranus, was discovered accidentally by William Herschel in 1781 during a telescopic survey of the sky. Fifty years later, however, it was clear that Uranus was not following its predicted orbit. John Couch Adams in England and Urbain Le Verrier in France independently calculated that the gravitational pull of a yet unknown, more distant planet could explain the deviations of Uranus from its orbit.

Le Verrier predicted that the planet would be found at a certain location in the constellation of Aquarius. A brief telescopic search on September 23, 1846, revealed the planet Neptune within 1° of the calculated position. Before it was sighted with a telescope, Neptune was actually predicted with pencil and paper.

Because it has been so successful in explaining and predicting many important phenomena, Newtonian mechanics has become the cornerstone of modern physical science. Even today, as we send astronauts into Earth orbit and spacecraft to the outer planets, Newton's equations are used to calculate the orbits and trajectories of these spacecraft. The *Cosmic Connections* figure on the next page summarizes some of the ways that gravity plays an important role on scales from apples to galaxies.

FIGURE 4-25   R I ⅤU X G

Comet Halley This most famous of all comets orbits the Sun with an average period of about 76 years. During the twentieth century, the comet passed near the Sun in 1910 and again in 1986 (when this photograph was taken). It will next be prominent in the sky in 2061. (Harvard College Observatory/Science Source)

COSMIC CONNECTIONS

Universal Gravitation

Gravity is one of the fundamental forces of nature. We can see its effects here on Earth as well as in the farthest regions of the observable universe.

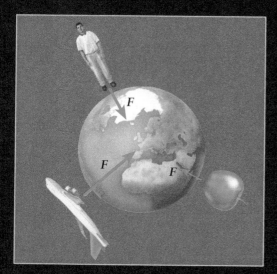

The weight of an ordinary object is just the gravitational force exerted on that object by Earth.

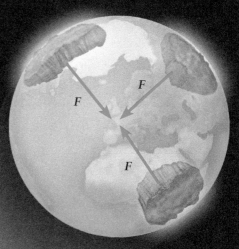

Earth is held together by the mutual gravitational attraction of its parts.

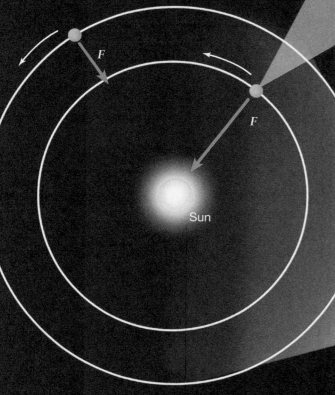

Gravitational forces exerted by the Sun keep the planets in their orbits. The farther a planet is from the Sun, the weaker the gravitational force that acts on the planet.

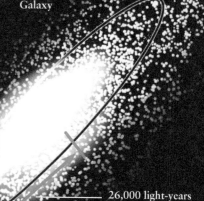

Milky Way Galaxy

Sun

26,000 light-years

The mutual gravitational attraction of all the matter in the Milky Way Galaxy holds it together. The gravitational force of the Galaxy on our Sun and solar system holds us in an immense orbit around the galactic center.

In the twentieth century, scientists found that Newton's laws do not apply in all situations. A new theory called *quantum mechanics* had to be developed to explain the behavior of matter on the very smallest of scales, such as within the atom and within the atomic nucleus. Albert Einstein developed the *theory of relativity* to explain what happens at very high speeds approaching the speed of light and in places where gravitational forces are very strong. For many purposes in astronomy, however, Newton's laws are as useful today as when Newton formulated them more than three centuries ago.

4-8 Gravitational forces between Earth and the Moon produce tides

We have seen how Newtonian mechanics explains why the Moon stays in orbit around Earth. It also explains why there are ocean tides, as well as why the Moon always keeps

> Tidal forces reveal how gravitation can pull objects apart rather than drawing them together

the same face toward Earth. Both of these are the result of *tidal forces*—a consequence of gravity that deforms planets and reshapes galaxies.

Tidal forces are differences in the gravitational pull at different points in an object. As an illustration, imagine that three billiard balls are lined up in space at some distance from a planet, as in **Figure 4-26a**. According to Newton's law of universal gravitation, the force of attraction between two objects is greater the closer the two objects are to each other. Thus, the planet exerts more force on the 3-ball (in red) than on the 2-ball (in blue) and exerts more force on the 2-ball than on the 1-ball (in yellow). Now, imagine that the three balls are released and allowed to fall toward the planet. Figure 4-26b shows the situation a short time later. Because of the differences in gravitational pull, a short time later the 3-ball will have moved farther than the 2-ball, which will in turn have moved farther than the 1-ball. But now imagine that same motion from the perspective of the 2-ball. From this perspective, it appears as though the 3-ball is pulled toward the planet while the 1-ball is pushed away (Figure 4-26c). These apparent pushes and pulls are called tidal forces.

Tidal Forces on Earth

The Moon has a similar effect on Earth as the planet in Figure 4-26 has on the three billiard balls. The arrows in **Figure 4-27a** indicate the strength and direction of the gravitational force of the Moon at several locations on Earth. The side of Earth closest to the Moon feels a greater gravitational pull than does Earth's center, and the side of Earth that faces away from the Moon feels less gravitational pull than does Earth's center. This means that just as for the billiard balls in Figure 4-26, there are tidal forces acting on Earth (Figure 4-27b). These tidal forces exerted by the Moon try to elongate Earth along a line connecting the centers of Earth and the Moon and try to squeeze Earth inward in the direction perpendicular to that line.

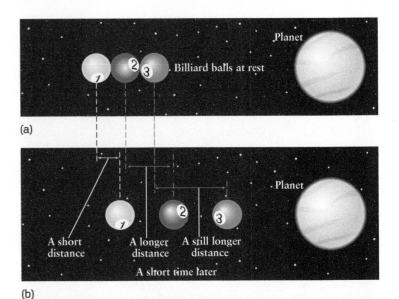

(a)

(b)

(c)

FIGURE 4-26

The Origin of Tidal Forces (a) Imagine three identical billiard balls placed some distance from a planet and released. (b) The closer a ball is to the planet, the more gravitational force the planet exerts on it. Thus, a short time after the balls are released, the blue 2-ball has moved farther toward the planet than the yellow 1-ball, and the red 3-ball has moved farther still. (c) From the perspective of the 2-ball in the center, it appears that forces have pushed the 1-ball away from the planet and pulled the 3-ball toward the planet. These forces are called tidal forces.

Because the body of Earth is largely rigid, it cannot deform very much in response to the tidal forces of the Moon. But the water in the oceans can and does deform into a football shape, as **Figure 4-28a** shows. As Earth rotates, a point on its surface goes from where the water is shallow to where the water is deep and back again. This is the origin of low and high ocean tides. (In this simplified description we have assumed that Earth is completely covered with water. The full story of the tides is much more complex, because the shapes of the continents and the effects of winds must also be taken into account.)

The Sun also exerts tidal forces on Earth's oceans. (The tidal effects of the Sun are about half as great as those of the Moon.) When the Sun, the Moon, and Earth are aligned, which happens at either new moon or full moon, the tidal effects of the Sun and Moon reinforce each other and the tidal distortion of the oceans is greatest. This produces large shifts in water level called **spring tides** (Figure 4-28b). At first quarter and last quarter, when the

Sun and Moon form a right angle with Earth, the tidal effects of the Sun and Moon partially cancel each other. Hence, the tidal distortion of the oceans is the least pronounced, producing smaller tidal shifts called **neap tides** (Figure 4-28c).

CAUTION! Note that spring tides have nothing to do with the season of the year called spring. Instead, the name refers to the way that the ocean level "springs up" to a greater than normal height. Spring tides occur whenever there is a new moon or full moon, no matter what the season of the year.

Tidal Forces on the Moon and Beyond

Just as the Moon exerts tidal forces on Earth, Earth exerts tidal forces on the Moon. Soon after the Moon formed some 4.56 billion years ago, it was molten throughout its volume. Earth's tidal forces deformed the molten Moon into a slightly elongated shape, with the long axis of the Moon pointed toward Earth. The Moon retained this shape and orientation when it cooled and solidified. To keep its long axis pointed toward Earth, the Moon spins once on its axis as it makes one orbit around Earth—that is, it is in synchronous rotation (Section 3-2). Hence, the same side of the Moon always faces Earth, and this is the side that we see. For

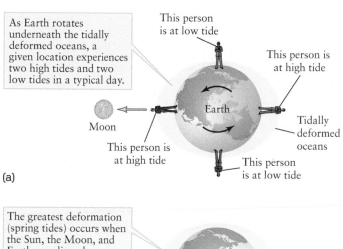

(a)

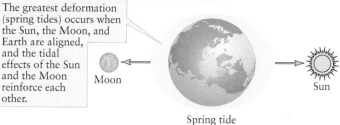

(b)

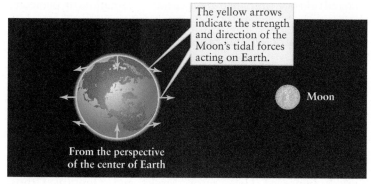

(c)

FIGURE 4-28

High and Low Tides (a) The gravitational forces of the Moon and the Sun deform Earth's oceans, giving rise to low and high tides. (b), (c) The strength of the tides depends on the relative positions of the Sun, the Moon, and Earth.

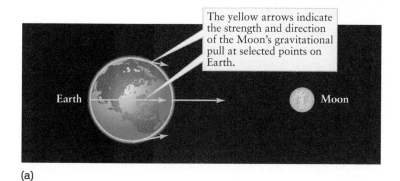

(a)

(b)

FIGURE 4-27

Tidal Forces on Earth (a) The Moon exerts different gravitational pulls at different locations on Earth. (b) At any location, the tidal force equals the Moon's gravitational pull at that location minus the gravitational pull of the Moon at the center of Earth. These tidal forces tend to deform Earth into a nonspherical shape.

the same reason, most of the satellites in the solar system are in synchronous rotation, and thus always keep the same side facing their planet.

Tidal forces are also important on scales much larger than the solar system. Figure 4-29 shows two spiral galaxies, like the one in Figure 1-9, undergoing a near-collision. During the millions of years that this close encounter has been taking place, the tidal forces of the larger galaxy have pulled an immense streamer of stars and interstellar gas out of the smaller galaxy.

Many galaxies, including our own Milky Way Galaxy, show signs of having been disturbed at some time by tidal interactions with other galaxies. By their effect on the interstellar gas from which stars are formed, tidal interactions can actually trigger the

FIGURE 4-29   R I V U X G

Tidal Forces on a Galaxy For millions of years the galaxies NGC 2207 and IC 2163 have been moving ponderously past each other. The larger galaxy's tremendous tidal forces have drawn a streamer of material a hundred thousand light-years long out of IC 2163. If you lived on a planet orbiting a star within this streamer, you would have a magnificent view of both galaxies. NGC 2207 and IC 2163 are, respectively, 143,000 light-years and 101,000 light-years in diameter. Both galaxies are 114 million light-years away in the constellation Canis Major. (NASA and the Hubble Heritage Team, AURA/STScI)

birth of new stars. Our own Sun and solar system may have been formed as a result of tidal interactions of this kind. Hence, we may owe our very existence to tidal forces. In this and many other ways, the laws of motion and of universal gravitation shape our universe and our destinies.

KEY WORDS

Terms preceded by an asterisk (*) are discussed in the Boxes.

KEY IDEAS

Apparent Motions of the Planets: Like the Sun and the Moon, the planets move on the celestial sphere with respect to the background of stars. Most of the time a planet moves eastward in direct motion, in the same direction as the Sun and the Moon, but from time to time it moves westward in retrograde motion.

The Ancient Geocentric Model: Ancient astronomers believed Earth to be at the center of the universe. They invented a complex system of epicycles and deferents to explain the direct and retrograde motions of the planets on the celestial sphere.

Copernicus's Heliocentric Model: Copernicus's heliocentric (Sun-centered) theory simplified the general explanation of planetary motions.

• In a heliocentric system, Earth is one of the planets orbiting the Sun.

• A planet undergoes retrograde motion as seen from Earth when Earth and the planet pass each other.

• The sidereal period of a planet, its true orbital period, is measured with respect to the stars. Its synodic period is measured with respect to Earth and the Sun (for example, from one opposition to the next).

Kepler's Improved Heliocentric Model and Elliptical Orbits: Copernicus thought that the orbits of the planets were combinations of circles. Using data collected by Tycho Brahe, Kepler deduced three laws of planetary motion: (1) the orbits are in fact ellipses; (2) a planet's speed varies as it moves around its elliptical orbit; and (3) the orbital period of a planet is related to the size of its orbit.

Evidence for the Heliocentric Model: The invention of the telescope led Galileo to new discoveries that supported a heliocentric model. These included his observations of the phases of Venus and of the motions of four moons around Jupiter.

Newton's Laws of Motion: Isaac Newton developed three principles, called the laws of motion, that apply to the motions of objects on Earth as well as in space. These are (1) the tendency of an object to maintain a constant velocity, (2) the relationship between the net outside force on an object and the object's acceleration, and (3) the principle of action and reaction. These laws and Newton's law of universal gravitation can be used to deduce Kepler's laws. They lead to extremely accurate descriptions of planetary motions.

• The mass of an object is a measure of the amount of matter in the object. Its weight is a measure of the force with which the gravity of some other object pulls on it.

• In general, the path of one object about another, such as that of a planet or comet about the Sun, is one of the curves called conic sections: circle, ellipse, parabola, or hyperbola.

Energy: There are different forms of energy an object can have. Kinetic and gravitational potential energy are the most important for orbiting objects.

• **Kinetic energy:** The greater an object's speed, the larger its kinetic energy. For two objects with equal speed, the more massive object has a larger kinetic energy.

• **Gravitational potential energy:** The farther an object is located from the surface of a planet, the greater its gravitational energy.

• **Conservation of energy:** Energy can change forms, and be transferred from one object to another, but energy cannot be created or destroyed.

Escape Speed: The escape speed of a planet is the ejection speed an object would need to have near the surface of the planet so that the object can "break free" from the planet and never return.

• The escape speed does not depend on the mass of the object, but it does depend on both the size and mass of the planet.

Tidal Forces: Tidal forces are caused by differences in the gravitational pull that one object exerts on different parts of a second object.

• The tidal forces of the Moon and the Sun produce tides in Earth's oceans.

• The tidal forces of Earth have locked the Moon into synchronous rotation.

QUESTIONS

Review Questions

1. How did the ancient Greeks explain why the Sun and the Moon slowly change their positions relative to the background stars?

2. In what direction does a planet move relative to the stars when it is in direct motion? When it is in retrograde motion? How do these compare with the direction in which we see the Sun move relative to the stars?

3. (**a**) In what direction does a planet move relative to the horizon over the course of one night? (**b**) The answer to (a) is the same whether the planet is in direct motion or retrograde motion. What does this tell you about the speed at which planets move on the celestial sphere?

4. What is an epicycle? How is it important in Ptolemy's explanation of the retrograde motions of the planets?

5. What is the significance of Occam's razor as a tool for analyzing theories?

6. How did the models of Aristarchus and Copernicus explain the retrograde motion of the planets?

7. How did Copernicus determine that the orbits of Mercury and Venus must be smaller than Earth's orbit? How did he determine that the orbits of Mars, Jupiter, and Saturn must be larger than Earth's orbit?

8. At what configuration (for example, superior conjunction, greatest eastern elongation, and so on) would it be best to observe Mercury or Venus with an Earth-based telescope? At what configuration would it be best to observe Mars, Jupiter, or Saturn? Explain your answers.

9. Is it ever possible to see Mercury at midnight? Explain your answer.

10. Which planets can never be seen at opposition? Which planets can never be seen at inferior conjunction? Explain your answers.

11. What is the difference between the synodic period and the sidereal period of a planet?

12. What is parallax? What did Tycho Brahe conclude from his attempt to measure the parallax of a supernova and a comet?

13. What observations did Tycho Brahe make in an attempt to test the heliocentric model? What were his results? Explain why modern astronomers get different results.

14. *TUTORIAL* What are the foci of an ellipse? If the Sun is at one focus of a planet's orbit, what is at the other focus?

15. What are Kepler's three laws? Why are they important?

16. At what point in a planet's elliptical orbit does it move fastest? At what point does it move slowest? At what point does it sweep out an area at the fastest rate?

17. A line joining the Sun and an asteroid is found to sweep out an area of 6.3 AU^2 during 2010. How much area is swept out during 2011? Over a period of five years?

18. The orbit of a spacecraft about the Sun has a perihelion distance of 0.1 AU and an aphelion distance of 0.4 AU. What is the semimajor axis of the spacecraft's orbit? What is its orbital period?

19. A comet with a period of 125 years moves in a highly elongated orbit about the Sun. At perihelion, the comet comes very close to the Sun's surface. What is the comet's average distance from the Sun? What is the farthest it can get from the Sun?

20. What observations did Galileo make that reinforced the heliocentric model? Why did these observations contradict the older model of Ptolemy? Why could these observations not have been made before Galileo's time?

21. Why does Venus have its largest angular diameter when it is new and its smallest angular diameter when it is full?

22. *TUTORIAL 4-2* What are Newton's three laws? Give an everyday example of each law.

23. How much force do you have to exert on a 3-kg brick to give it an acceleration of 2 m/s²? If you double this force, what is the brick's acceleration? Explain your answer.

24. What is the difference between weight and mass?

25. What is your weight in pounds and in newtons? What is your mass in kilograms?

26. *TUTORIAL 4-3* Suppose that Earth were moved to a distance of 3.0 AU from the Sun. How much stronger or weaker would the Sun's gravitational pull be on Earth? Explain your answer.

27. How far would you have to go from Earth to be completely beyond the pull of its gravity? Explain your answer.

28. A cannonball is shot horizontally from a barrel off a building. Name two forms of energy the cannonball has as it exits the barrel. How do these two forms of energy increase or decrease during the cannonball's flight? Before the cannonball was fired, where was the energy stored?

29. A satellite is in circular orbit. What two forms of energy are part of the satellite's orbital energy? Would its orbital energy need to increase or decrease in order to orbit at a larger distance from Earth?

30. If an object loses orbital energy through air drag, is energy still conserved? If so, where does the energy go?

31. Calculate the escape speed for Earth and show that it is 11.2 km/s. (Consult Appendix 2 for planetary data.)

32. Including the effects of Earth's atmosphere, would you expect the real escape speed for a hypothetical cannonball to be greater or less than 11.2 km/s? Why?

33. What are conic sections? In what way are they related to the orbits of planets in the solar system?

34. Why was the discovery of Neptune an important confirmation of Newton's law of universal gravitation?

35. What is a tidal force? How do tidal forces produce tides in Earth's oceans?

36. What is the difference between spring tides and neap tides?

Advanced Questions

Questions preceded by an asterisk (*) involve topics discussed in the Boxes.

> **Problem-solving tips and tools**
>
> Box 4-1 explains sidereal and synodic periods in detail. The semimajor axis of an ellipse is half the length of the long, or major, axis of the ellipse. For data about the planets and their satellites, see Appendices 1, 2, and 3 at the back of this book. If you want to calculate the gravitational force that you feel on the surface of a planet, the distance r to use is the planet's radius (the distance between you and the center of the planet). Boxes 4-2 and 4-4 show how to use Kepler's third law in its original form and in Newton's form.

37. Figure 4-2 shows the retrograde motion of Mars as seen from Earth. Sketch a similar figure that shows how Earth would appear to move against the background of stars during this same time period as seen by an observer on Mars.

*38. The synodic period of Mercury (an inferior planet) is 115.88 days. Calculate its sidereal period in days.

*39. Table 4-1 shows that the synodic period is *greater* than the sidereal period for Mercury, but the synodic period is *less* than the sidereal period for Jupiter. Draw diagrams like the one in Box 4-1 to explain why this is so.

*40. A general rule for superior planets is that the greater the average distance from the planet to the Sun, the more frequently that planet will be at opposition. Explain how this rule comes about.

41. In 2006, Mercury was at greatest western elongation on April 8, August 7, and November 25. It was at greatest eastern elongation on February 24, June 20, and October 17. Does Mercury take longer to go from eastern to western elongation, or vice versa? Explain why, using Figure 4-6.

42. Explain why the semimajor axis of a planet's orbit is equal to the average of the distance from the Sun to the planet at perihelion (the *perihelion distance*) and the distance from the Sun to the planet at aphelion (the *aphelion distance*).

43. A certain comet is 2 AU from the Sun at perihelion and 16 AU from the Sun at aphelion. (a) Find the semimajor axis of the comet's orbit. (b) Find the sidereal period of the orbit.

44. A comet orbits the Sun with a sidereal period of 64.0 years. (a) Find the semimajor axis of the orbit. (b) At aphelion, the comet is 31.5 AU from the Sun. How far is it from the Sun at perihelion?

45. One trajectory that can be used to send spacecraft from Earth to Mars is an elliptical orbit that has the Sun at one focus, its perihelion at Earth, and its aphelion at Mars. The spacecraft is launched from Earth and coasts along this ellipse until it reaches Mars, when a rocket is fired to either put the spacecraft into orbit around Mars or cause it to land on Mars. (a) Find the semimajor axis of the ellipse. (*Hint:* Draw a picture showing the Sun and the orbits of Earth, Mars, and the spacecraft. Treat the orbits of Earth and Mars as circles.) (b) Calculate how long (in days) such a one-way trip to Mars would take.

46. The mass of the Moon is 7.35×10^{22} kg, while that of Earth is 5.98×10^{24} kg. The average distance from the center of the Moon to the center of Earth is 384,400 km. What is the size of the gravitational force that Earth exerts on the Moon? What is the size of the gravitational force that the Moon exerts on Earth? How do your answers compare with the force between the Sun and Earth calculated in the text?

47. The mass of Saturn is approximately 100 times that of Earth, and the semimajor axis of Saturn's orbit is approximately 10 AU. To this approximation, how does the gravitational force that the Sun exerts on Saturn compare to the gravitational force that the Sun exerts on Earth? How do the accelerations of Saturn and Earth compare?

48. Suppose that you traveled to a planet with 4 times the mass and 4 times the diameter of Earth. Would you weigh more or less on that planet than on Earth? By what factor?

49. On Earth, a 50-kg astronaut weighs 490 newtons. What would she weigh if she landed on Jupiter's moon Callisto? What fraction is this of her weight on Earth? See Appendix 3 for relevant data about Callisto.

50. Except for some rare and exotic microbes, the Sun provides energy for life on Earth. Can you describe how energy in sunlight can end up in a kangaroo jumping through the air? In your description, can you name three forms of energy involved, other than sunlight?

51. You're talking with other students about how the energy of a satellite in circular orbit changes with altitude. One student says that lower orbits have more energy because Kepler's second law says that closer planets have higher speeds. Another student argues that orbits at larger distances must have more energy because rockets have to expend more energy to launch satellites farther into to space. Which student reaches the right conclusion?

52. Some argue that life might have started on either Mars or Earth and spread to the other planet by microbe-carrying debris ejected during very large asteroid impacts. Consult Appendix 2 for planetary data and calculate the escape speed for both Earth and Mars. Assuming all else being equal, for which planet is it easier for rocks to escape?

53. Imagine a planet like Earth orbiting a star with 4 times the mass of the Sun. If the semimajor axis of the planet's orbit (a) is 1 AU, what would be the planet's sidereal period? (*Hint:* Use Newton's form of Kepler's third law. Compared with the case of Earth orbiting the Sun, by what factor has the quantity $m_1 + m_2$ changed? Has a changed? By what factor must P^2 change?)

54. A satellite is said to be in a "geosynchronous" orbit if it appears always to remain over the exact same spot on rotating Earth. (**a**) What is the period of this orbit? (**b**) At what distance from the center of Earth must such a satellite be placed into orbit? (*Hint:* Use Newton's form of Kepler's third law.) (**c**) Explain why the orbit must be in the plane of Earth's equator.

55. Figure 4-23 shows the lunar module *Eagle* in orbit around the Moon after completing the first successful lunar landing in July 1969. (The photograph was taken from the command module *Columbia,* in which the astronauts returned to Earth.) The spacecraft orbited 111 km above the surface of the Moon. Calculate the period of the spacecraft's orbit. See Appendix 3 for relevant data about the Moon.

*56. In Box 4-4 we analyze the orbit of Jupiter's moon Io. Look up information about the orbits of Jupiter's three other large moons (Europa, Ganymede, and Callisto) in Appendix 3. Demonstrate that these data are in agreement with Newton's form of Kepler's third law.

*57. Suppose a newly discovered asteroid is in a circular orbit with synodic period 1.25 years. The asteroid lies between the orbits of Mars and Jupiter. (**a**) Find the sidereal period of the orbit. (**b**) Find the distance from the asteroid to the Sun.

58. The average distance from the Moon to the center of Earth is 384,400 km, and the diameter of Earth is 12,756 km. Calculate the gravitational force that the Moon exerts (**a**) on a 1-kg rock at the point on Earth's surface closest to the Moon, and (**b**) on a 1-kg rock at the point on Earth's surface farthest from the Moon. (**c**) Find the difference between the two forces you calculated in parts (**a**) and (**b**). This difference is the tidal force pulling these two rocks away from each other, like the 1-ball and 3-ball in Figure 4-26. Explain why tidal forces cause only a very small deformation of Earth.

Discussion Questions

59. Which planet would you expect to exhibit the greatest variation in apparent brightness as seen from Earth? Which planet would you expect to exhibit the greatest variation in angular diameter? Explain your answers.

60. Use two thumbtacks, a loop of string, and a pencil to draw several ellipses. Describe how the shape of an ellipse varies as the distance between the thumbtacks changes.

Web/eBook Questions

61. (**a**) Search the World Wide Web for information about Kepler. Before he realized that the planets move on elliptical paths, what other models of planetary motion did he consider? What was Kepler's idea of "the music of the spheres"? (**b**) Search the World Wide Web for information about Galileo. What were his contributions to physics? Which of Galileo's new ideas were later used by Newton to construct his laws of motion? (**c**) Search the World Wide Web for information about Newton. What were some of the contributions that he made to physics other than developing his laws of motion? What contributions did he make to mathematics?

62. **Monitoring the Retrograde Motion of Mars.** Watching Mars night after night reveals that it changes its position with respect to the background stars. To track its motion, access and view the animation "The Path of Mars in 2016" in Chapter 4 of the *Universe* Web site or eBook. (**a**) Through which constellations does Mars move? (**b**) On approximately what date does Mars stop its direct (west-to-east) motion and begin its retrograde motion? (*Hint:* Use the "Stop" function on your animation controls.) (**c**) Over how many days does Mars move retrograde?

ACTIVITIES

Observing Projects

63. It is quite probable that within a few weeks of your reading this chapter one of the planets will be near opposition or greatest eastern elongation, making it readily visible in the evening sky. Select a planet that is at or near such a configuration by searching the World Wide Web or by consulting a reference book, such as the current issue of the *Astronomical*

Almanac or the pamphlet entitled *Astronomical Phenomena* (both published by the U.S. government). At that configuration, would you expect the planet to be moving rapidly or slowly from night to night against the background stars? Verify your expectations by observing the planet once a week for a month, recording your observations on a star chart.

64. If Jupiter happens to be visible in the evening sky, observe the planet with a small telescope on five consecutive clear nights. Record the positions of the four Galilean satellites by making nightly drawings, just as the Jesuit priests did in 1620 (see Figure 4-17). From your drawings, can you tell which moon orbits closest to Jupiter and which orbits farthest? Was there a night when you could see only three of the moons? What do you suppose happened to the fourth moon on that night?

65. If Venus happens to be visible in the evening sky, observe the planet with a small telescope once a week for a month. On each night, make a drawing of the crescent that you see. From your drawings, can you determine if the planet is nearer or farther from Earth than the Sun is? Do your drawings show any changes in the shape of the crescent from one week to the next? If so, can you deduce if Venus is coming toward us or moving away from us?

66. Use the *Starry Night™* program to observe retrograde motion. Select **Favourites > Explorations > Retrograde** from the menu. The view from Earth is centered on Mars against the background of stars and the framework of star patterns within the constellations. The **Time Flow Rate** is set to 1 day. Click **Play** and observe Mars as it moves against the background constellations. An orange line traces Mars's path in the sky from night to night. Watch the motion of Mars for at least 2 years of simulated time. Since the view is centered on and tracks Mars in the view, the sky appears to move but the relative motion of Mars against this sky is obvious. (a) For most of the time, does Mars move generally to the left (eastward) or to the right (westward) on the celestial sphere? Select **File > Revert** from the menu to return to the original view. Use the time controls in the toolbar (**Play, Step time forward,** and **Step time backward**) along with the **Zoom** controls (+ and − buttons at the right of the toolbar or the mouse wheel) to determine when Mars's usual *direct* motion ends, when it appears that Mars comes to a momentary halt in the west-east direction, and *retrograde* motion begins. On what date does retrograde motion end and direct motion resume? (b) You have been observing the motion of Mars as seen from Earth. To observe the motion of Earth as seen from Mars, locate yourself on the north pole of Mars by selecting **Favourites > Explorations > Retrograde Earth** from the menu. The view is centered on and will track Earth as seen from the north pole of Mars, beginning on **June 23, 2010**. Click the **Play** button. As before, watch the motion for 2 years of simulated time. In which direction does Earth appear to move for most of the time? On what date does its motion change from direct to retrograde? On what date does its motion change from retrograde back

to direct? Are these roughly the same dates you found in part (a)? (c) To understand the motions of Mars as seen from Earth and vice versa, observe the motion of the planets from a point above the solar system. Select **Favourites > Explorations > Retrograde Overview** from the menu. This view, from a position 5 AU above the plane of the solar system, is centered on the Sun, and the orbits and positions of Mars and Earth on June 23, 2010, are shown. Click **Play** and watch the motions of the planets for 2 years of simulated time. Note that Earth catches up with and overtakes Mars as time proceeds. This relative motion of the two planets leads to our observation of retrograde motion. On what date during this 2-year period is Earth directly between Mars and the Sun? How does this date compare to the two dates you recorded in part (a) and the two dates you recorded in part (b)? Explain the significance of this.

67. Use *Starry Night™* to observe the phases of Venus and of Mars as seen from Earth. Select **Favourites > Explorations > Phases of Venus** and click the **Now** button in the toolbar to see an image of Venus if you were to observe it through a telescope from Earth right at this moment. (a) Draw the current shape (phase) of Venus. With the **Time Flow Rate** set to **30 days**, step time forward, drawing Venus to scale at each step. Make a total of 20 time steps and drawings. (b) From your drawings, determine when the planet is nearer or farther from Earth than is the Sun. (c) Deduce from your drawings when Venus is coming toward us or is moving away from us. (d) Explain why Venus goes through this particular cycle of phases. Select **Favourites > Explorations > Phases of Mars** and click the **Now** button in the toolbar. With the **Time Flow Rate** set to **30 days**, step time forward, and observe the changing phase of Mars as seen from Earth. (e) Compare this with the phases that you observed for Venus. Why are the cycles of phases as seen from Earth different for the two planets?

68. Use *Starry Night™* to observe the orbits of the planets of the inner solar system. Open **Favourites > Explorations > Kepler**. The view is centered on the Sun from a position in space 2.486 AU above the plane of the solar system and shows the Sun and the inner planets and their orbits, as well as many asteroids in the asteroid belt beyond the orbit of Mars. Click the **Play** button and observe the motions of the planets from this unique location. (a) Make a list of the planets visible in the view in the order of increasing distance from the Sun. (b) Make a list of the planets visible in the view in the order of increasing orbital period. (c) How do the lists compare? (d) What might you conclude from this observation? (e) Which of Kepler's laws accounts for this observation?

ANSWERS

ConceptChecks

ConceptCheck 4-1: When a planet is moving retrograde, the planet can be observed night after night to be slowly drifting from east to west compared to the very distant background stars (Figure 4-2). This is opposite to how a planet typically appears to move. However, regardless of this slow movement, all objects always appear to rise in the east and set in the west on a daily basis due to Earth's rotation.

ConceptCheck 4-2: The Greeks' ancient geocentric model used a nonspinning, stationary Earth where the stars, planets, and the Sun all moved around Earth.

ConceptCheck 4-3: In the Ptolemaic model, the planets are continuously orbiting in circles such that they appear to move backward for a brief time. However, the planets never actually stop and change their directions (Figure 4-3).

ConceptCheck 4-4: No. A planet only appears to move in retrograde motion if seen from another planet if the two planets move at different speeds and pass one another (Figure 4-5). An imaginary observer on the stationary Sun would only see planets moving in the same direction as they orbit the Sun.

ConceptCheck 4-5: Mars has an orbit around the Sun that is larger than Earth's orbit. As a result, Mars never moves to a position between Earth and the Sun, so Mars never is at inferior conjunction (Figure 4-6).

ConceptCheck 4-6: In Copernicus's heliocentric model, the more distant planets are moving slower than the planets closer to the Sun. As a faster-moving Earth moves past a slower-moving Mars, there is a brief time in which Mars appears to move backward through the sky. However, the planets never actually reverse their directions.

ConceptCheck 4-7: Slowly moving Jupiter does not move very far along its orbit in the length of time it takes for Earth to pass by Jupiter, move around the Sun, and pass by Jupiter again, giving Jupiter a synodic period similar to Earth's orbital period around the Sun. However, slow-moving Jupiter takes more than a decade to move around the Sun back to its original starting place as measured by the background stars, giving it a large sidereal period.

ConceptCheck 4-8: Initially, Copernicus's model was no more accurate at predicting the positions and motions of the planets than Ptolemy's model. However, Copernicus's model turned out to be more closely related to the actual motions of the planets around the Sun than was Ptolemy's model of planets orbiting Earth. With subsequent measurements, Ptolemy's model was proven wrong.

ConceptCheck 4-9: An ellipse with an eccentricity of zero is a perfect circle (see Figure 4-10b) and, compared to Mars's $e = 0.093$, the eccentricity of Venus' orbit is $e = 0.007$. Venus has a smaller eccentricity so its orbit is closer to a perfect circle in shape.

ConceptCheck 4-10: Kepler's second law says that objects are moving slowest when they are farthest from the object they are orbiting, so an Earth-orbiting satellite will move slowest when it is farthest from Earth. Just as the Sun is at one focus for a planet's elliptical orbit, Earth is at one focus for a satellite's elliptical orbit (Figure 4-11).

ConceptCheck 4-11: According to Kepler's third law, planets closer to the Sun move faster than planets farther from the Sun. For objects orbiting Earth, the object closer to Earth is also moving the fastest, which, in this case, is the space shuttle.

ConceptCheck 4-12: No. Kepler's laws of planetary motion apply to any objects in space that orbit around another object, including comets orbiting the Sun, man-made satellites and moons orbiting planets, and even stars orbiting other stars.

ConceptCheck 4-13: When Venus is on the opposite side of the Sun from Earth, it will be in a full or gibbous phase. The full phase can occasionally be observed because Earth and Venus do not orbit the Sun in the same exact plane.

ConceptCheck 4-14: Galileo was the first person to use and widely share what he learned from his telescope observations, and a telescope is necessary in order to observe Jupiter's tiny moons.

ConceptCheck 4-15: According to Newton's first law, the rock will continue to travel in the same direction, and with the same speed, as long as there is no net outside force. Note that this principle applies no matter how massive the rock is, or how fast it's traveling.

ConceptCheck 4-16: As described by Newton's third law, the forces the sumo wrestlers exert on each other are of equal strength, but in opposite directions. This is true even though they are different in size and even though one wrestler is pushed out of the ring. Frictional forces from the floor and their body masses influence which wrestler moves the most, but the forces they exert onto each other are of equal strength.

ConceptCheck 4-17: Expressing Newton's second law in terms of acceleration, the acceleration of an object is given by $a = F/m$. Since the gravitational forces on the diver and Earth are of equal strength, the smaller mass—the diver—will have the larger acceleration. Because Earth is so much more massive than the diver, Earth's acceleration in this case is nowhere near measurable.

ConceptCheck 4-18: According to Newton's universal law of gravitation, the gravitational attraction between two objects depends on the square of the distance between them. In this case, if the asteroids drift 3 times closer together, then the gravitational force of attraction between them increases 3^2 times, or, in other words, becomes 9 times greater.

ConceptCheck 4-19: At the cannonball's highest point, it is momentarily at rest and has gravitational potential energy, but no kinetic energy. Starting from the chemical energy in gunpowder, that chemical energy is transferred to the cannonball, giving it kinetic energy. As the ball rises, kinetic energy is transformed into gravitational potential energy until the ball reaches its peak.

ConceptCheck 4-20: Air drag *removes* orbital energy and could put the spacecraft on the elliptical path D. Firing the rocket thruster *adds* orbital energy and could put the spacecraft on the elliptical path F. A circular orbit requires special parameters, so it is not surprising that most moons and planets have at least slightly elliptical orbits.

ConceptCheck 4-21: In order for an object to feel air drag, it must pass through the air. Since Earth and our air orbit the Sun together, Earth feels no air drag and we can orbit the Sun indefinitely.

ConceptCheck 4-22: An object's escape speed from a planet does not depend on the object's mass, so both rocks can be shot at the same speed, which is the escape speed from the Moon. As illustrated in Figure 4-20, for two objects with the same speed, the more massive object has more kinetic energy, so the kinetic energy of the two rocks will not be equal, and more energy is required to shoot the larger rock off of the Moon.

ConceptCheck 4-23: Yes. The escape speed $v_{escape} = \sqrt{\frac{2GM}{R}}$ depends on both the mass and radius of the planet. If another Earth-mass planet had a larger radius (which is certainly possible), its escape speed would be less than that of Earth.

ConceptCheck 4-24: The object—on its high-energy hyperbolic orbit—would have to lose orbital energy in order to enter a lower-energy elliptical or circular orbit as a moon. Special circumstances are required to lose this much energy, which is why we do not think Earth simply captured our Moon as it passed by.

CalculationChecks

CalculationCheck 4-1: According to Kepler's third law, $P^2 = a^3$. So, if $P^2 = (39.5)^3$, then $P = 39.5^{3/2} = 248$ years.

CalculationCheck 4-2: Using Newton's universal law of gravitation, $F_{\text{Mars-astronaut}} = G(m_{\text{Mars}}) \times (m_{\text{astronaut}}) \div (\text{radius})^2 = 6.67 \times 10^{-11} \times 6.4 \times 10^{23} \times 75 \div (3.4 \times 10^6)^2 = 277$ newtons, which we can convert to pounds because 277 newtons $\times$ 0.255 lbs/N = 76 pounds.

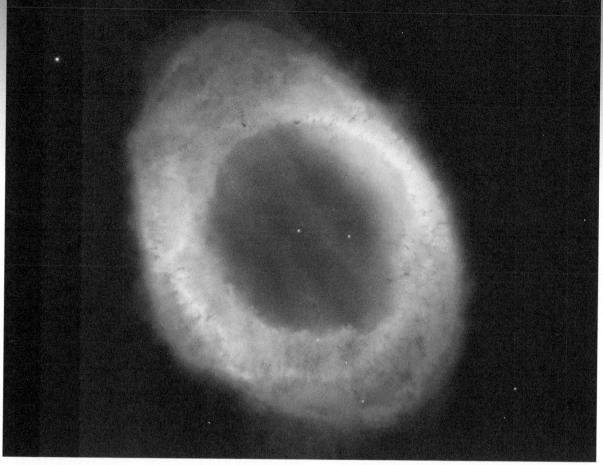

The Ring Nebula is a shell of glowing gases surrounding a dying star. The spectrum of the emitted light reveals which gases are present. (Hubble Heritage Team, AURA/STScI/NASA)) R I $\boxed{\text{V}}$ U X G

The Nature of Light

In the early 1800s, the French philosopher Auguste Comte argued that because the stars are so far away, humanity would never know their nature and composition. But the means to learn about the stars was already there for anyone to see—starlight. Just a few years after Comte's bold pronouncement, scientists began analyzing starlight to learn the very things that he had deemed unknowable.

We now know that atoms of each chemical element emit and absorb light at a unique set of wavelengths characteristic of that element alone. The red light in the accompanying image of a gas cloud in space is of a wavelength emitted by nitrogen and no other element; the particular green light in this image is unique to oxygen, and the particular blue light is unique to helium. The light from nearby planets, distant stars, and remote galaxies also has characteristic "fingerprints" that reveal the chemical composition of these celestial objects.

In this chapter we learn about the basic properties of light. Light has a dual nature: It has the properties of both waves and particles. The light emitted by an object depends upon the object's temperature; we can use this to determine the surface temperatures of stars. By studying the structure of atoms, we will learn why each element emits and absorbs light only at specific wavelengths and will see how astronomers determine what the atmospheres of planets and stars are made of. The motion of a light source also affects wavelengths, permitting us to deduce how fast stars and other objects are approaching or receding. These are but a few of the reasons why understanding light is a prerequisite to understanding the universe.

5-1 Light travels through empty space at a speed of 300,000 km/s

Galileo Galilei and Isaac Newton were among the first scientific thinkers to ask basic questions about light. Does light travel instantaneously from one

> The speed of light in a vacuum is a universal constant: It has the same value everywhere in the cosmos

place to another, or does it move with a measurable speed? Whatever the nature of light, it does seem to travel swiftly from a source to our eyes. We see a distant event before we hear the accompanying sound. (For example, we see a flash of lightning before we hear the thunderclap.)

In the early 1600s, Galileo tried to measure the speed of light. He and an assistant stood at night on two hilltops a known distance apart, each holding a shuttered lantern. First, Galileo opened the shutter of his lantern; as soon as his assistant saw the flash of light, he opened his own. Galileo used his pulse as a timer to try to measure the time between opening his lantern and seeing the light from his assistant's lantern. From the distance and time, he hoped to compute the speed at which the light had traveled to the distant hilltop and back.

Galileo found that the measured time failed to increase noticeably, no matter how far away the assistant was stationed. Galileo therefore concluded that the speed of light is too high to be measured by slow human reactions. Thus, he was unable to tell whether or not light travels instantaneously.

The Speed of Light: Astronomical Measurements

The first evidence that light does *not* travel instantaneously was presented in 1676 by Olaus Rømer, a Danish astronomer. Rømer had been studying the orbits of the moons of Jupiter by carefully timing the moments when they passed into or out of Jupiter's shadow. To Rømer's surprise, the timing of these eclipses of Jupiter's moons seemed to depend on the relative positions of Jupiter and Earth. When Earth was far from Jupiter (that is, near conjunction; see Figure 4-6), the eclipses occurred several minutes later than when Earth was close to Jupiter (near opposition).

Rømer realized that this puzzling effect could be explained if light needs time to travel from Jupiter to Earth. When Earth is closest to Jupiter, the image of one of Jupiter's moons disappearing into Jupiter's shadow arrives at our telescopes a little sooner than it does when Jupiter and Earth are farther apart (Figure 5-1). The range of variation in the times at which such eclipses are observed is about 16.6 minutes, which Rømer interpreted as the length of time required for light to travel across the diameter of Earth's orbit (a distance of 2 AU). The size of Earth's orbit was not accurately known in Rømer's day, and he never actually calculated the speed

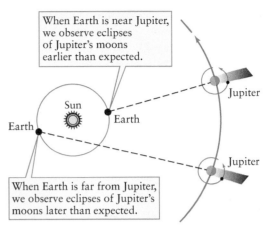

FIGURE 5-1
Rømer's Evidence That Light Does Not Travel Instantaneously
The timing of eclipses of Jupiter's moons as seen from Earth depends on the Earth-Jupiter distance. Rømer correctly attributed this effect to variations in the time required for light to travel from Jupiter to Earth.

of light. Today, using the modern value of 150 million kilometers for the astronomical unit, Rømer's method yields a value for the speed of light equal to roughly 300,000 km/s (186,000 mi/s).

The Speed of Light: A Universal Speed Limit

Almost two centuries after Rømer, the speed of light was measured very precisely using an ingenious experiment. In 1850, the French physicists Armand-Hippolyte Fizeau and Jean Foucault built the apparatus sketched in Figure 5-2. Light from a light source reflects from a rotating mirror toward a stationary mirror 20 meters away. The rotating mirror moves slightly while the light is making the round trip, so the returning light ray is deflected away from the source by a small angle. By measuring this angle and knowing the dimensions of their apparatus, Fizeau and

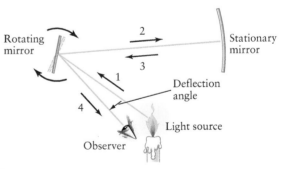

FIGURE 5-2
The Fizeau-Foucault Method of Measuring the Speed of Light
Light from a light source (1) is reflected off a rotating mirror to a stationary mirror (2) and from there back to the rotating mirror (3). The ray that reaches the observer (4) is deflected away from the path of the initial beam because the rotating mirror has moved slightly while the light was making the round trip. The speed of light is calculated from the deflection angle and the dimensions of the apparatus.

Foucault could deduce the speed of light. Once again, the answer was very nearly 300,000 km/s. Based on modern measurements, whether measuring the speed of light on Earth, in the solar system, or amongst galaxies, the speed of light appears to be the same throughout the known universe.

The speed of light in a vacuum (a space devoid of matter) is usually designated by the letter c (from the Latin *celeritas,* meaning "speed"). The modern value is $c = 299{,}792.458$ km/s $(186{,}282.397$ mi/s). In most calculations you can use

$$c = 3.00 \times 10^5 \text{ km/s} = 3.00 \times 10^8 \text{ m/s}$$

The most convenient set of units to use for c is different in different situations. The value in kilometers per second (km/s) is often most useful when comparing c to the speeds of objects in space, while the value in meters per second (m/s) is preferred when doing calculations involving the wave nature of light (which we will discuss in Section 5-2).

The speed of light is extremely fast. In one second, light travels a distance equal to about 7½ times around Earth's equator! Since there are more than 31 million seconds in a year, one light-year—the distance light travels in one year—is about 6 trillion miles.

CAUTION! Note that the quantity c is the speed of light *in a vacuum.* Light travels more slowly through air, water, glass, or any other transparent substance than it does in a vacuum. In our study of astronomy, however, we will almost always consider light traveling through the vacuum (or near-vacuum) of space.

The speed of light in empty space is one of the most important numbers in modern physical science. This value appears in many equations that describe atoms, gravity, electricity, and magnetism. According to Einstein's special theory of relativity, there is a universal speed limit: **Nothing can travel faster than the speed of light.**

CONCEPTCHECK 5-1

Why has the speed of light historically been so difficult to measure?

Answer appears at the end of the chapter.

5-2 Light is electromagnetic radiation and is characterized by its wavelength

TUTORIAL 5-1 Light is energy. This fact is apparent to anyone who has felt the warmth of the sunshine on a summer's day. But what exactly is light? How is it produced? What is it made of? How does it move through space? Scholars have struggled with these questions throughout history.

> Visible light, radio waves, and X-rays are all the same type of wave: They differ only in their wavelength

Newton and the Nature of Color

The first major breakthrough in understanding light came from a simple experiment performed by Isaac Newton around 1670. Newton was familiar with what he called the "celebrated Phenomenon of Colours," in which a beam of sunlight passing through a glass prism spreads out into the colors of the rainbow (**Figure 5-3**). This rainbow is called a **spectrum** (plural **spectra**).

Until Newton's time, it was thought that a prism somehow added colors to white light. To test this idea, Newton placed a second prism so that just one color of the spectrum passed through it: In **Figure 5-4**, only yellow light can pass through a slit to the second prism. According to the old theory, this should have caused a further change in the color of the light. But Newton found that each color of the spectrum was unchanged by the second prism; yellow remained yellow, blue remained blue, and so on. He concluded that a prism merely separates colors and does not add color. Hence, the spectrum produced by the first prism shows that *sunlight is a mixture of all the colors of the rainbow.* We often call sunlight "white light," but keep in mind that white light is just our brain's perception of sunlight's mixed colors.

Newton suggested that light is made of particles too small to detect individually. In 1678, however, the Dutch physicist and astronomer Christiaan Huygens proposed a rival explanation. He suggested that light travels in the form of waves rather than particles.

CONCEPTCHECK 5-2

The rear brake lights on a car emit white light, but are covered with plastic that only allows red light to pass through. What color of plastic would the red light have to pass through in order to emerge as green light?

Answer appears at the end of the chapter.

Young and the Wave Nature of Light

Around 1801, the English scientist Thomas Young carried out an experiment that convincingly demonstrated the wavelike aspect of light. He passed a beam of light through two thin, parallel slits in a plate, as shown in **Figure 5-5a**. On a white screen some distance beyond the slits, the light formed a pattern of alternating bright and dark bands. Young reasoned that if a beam of light was a stream of particles (as Newton had suggested), the two beams of light from the slits would simply form bright images of the slits on the white surface. The pattern of bright and dark bands he observed, however, is just what would be expected if light had wavelike properties. An analogy with water waves demonstrates why (Figure 5-5b).

Maxwell and Light as an Electromagnetic Wave

The discovery of the wave nature of light posed some obvious questions. What exactly is "waving" in light? That is, what is it about light that goes up and down like water waves on the ocean? Because we can see light from the Sun, planets, and stars, light waves must be able to travel across empty space. Hence, whatever is "waving" cannot be any material substance. What, then, is it?

The answer came from a seemingly unlikely source—a comprehensive theory that described electricity and magnetism. Numerous experiments during the first half of the nineteenth century demonstrated an intimate connection between electric and magnetic forces. A central idea to emerge from these experiments is the concept of a *field,* an immaterial yet measurable disturbance

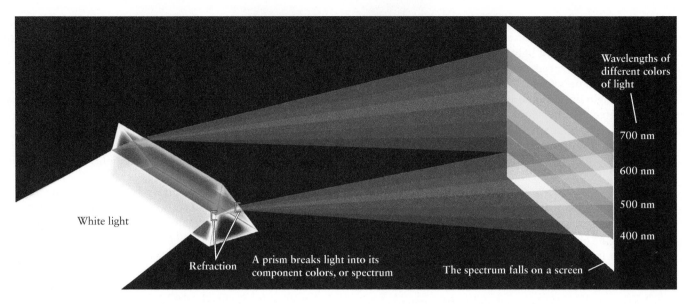

White light

Refraction

A prism breaks light into its
component colors, or spectrum

The spectrum falls on a screen

Wavelengths of
different colors
of light

700 nm

600 nm

500 nm

400 nm

FIGURE 5-3

A Prism and a Spectrum When a beam of sunlight passes through a
glass prism, the light is broken into a rainbow-colored band called a spectrum.

The wavelengths of different colors of light are shown on the right (1 nm =
1 nanometer = 10^{-9} m).

of any region of space in which electric or magnetic forces are
felt. Thus, an electric charge is surrounded by an electric field,
and a magnet is surrounded by a magnetic field. Experiments in
the early 1800s demonstrated that moving an electric charge produces a magnetic field; conversely, moving a magnet gives rise to
an electric field.

In the 1860s, the Scottish mathematician and physicist James
Clerk Maxwell succeeded in describing all the basic properties of
electricity and magnetism in four equations. This mathematical
achievement demonstrated that electric and magnetic forces are
really two aspects of the same phenomenon, which we now call
electromagnetism.

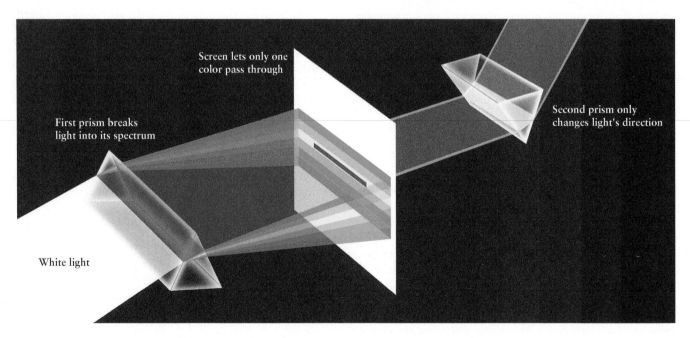

Screen lets only one
color pass through

First prism breaks
light into its spectrum

White light

Second prism only
changes light's direction

FIGURE 5-4

Newton's Experiment on the Nature of Light In a crucial experiment,
Newton took sunlight that had passed through a prism and sent it through
a second prism. Between the two prisms was a screen with a hole in it
that allowed only one color of the spectrum to pass through. This same

color emerged from the second prism. Newton's experiment proved that
prisms do not add color to light but merely bend different colors through
different angles. It also showed that white light, such as sunlight, is actually a
combination of all the colors that appear in its spectrum.

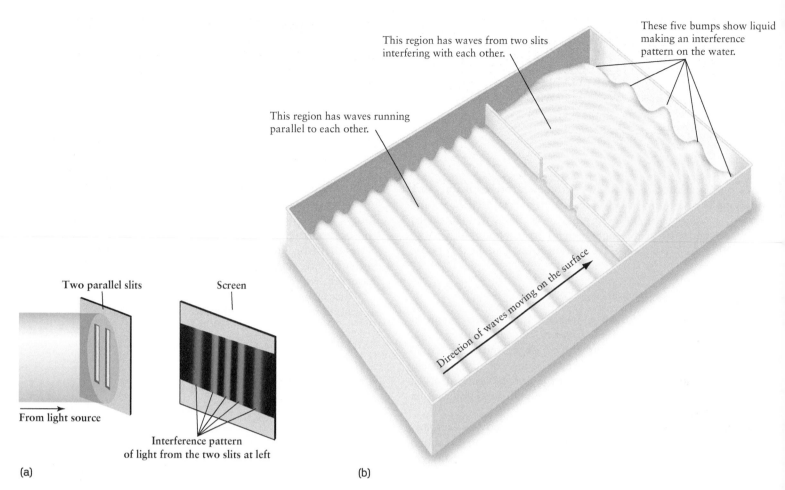

This region has waves running parallel to each other.

This region has waves from two slits interfering with each other.

These five bumps show liquid making an interference pattern on the water.

Two parallel slits

Screen

From light source

Direction of waves moving on the surface

Interference pattern of light from the two slits at left

(a)

(b)

FIGURE 5-5

Wave Interference (a) Electromagnetic radiation travels as waves. Thomas Young's interference experiment shows that light of a single color passing through a barrier with two slits behaves as waves that create alternating light and dark patterns on a screen. (b) Water waves passing through two slits creates two overlapping waves on the right side of a ripple tank; the result is an interference pattern. Constructive interference occurs where the crests (high points) from waves combine together. Destructive interference occurs where the crest from one wave combines with the trough (low point) of another, leaving the water's surface undisturbed. In the experiment with light, the bright bands result from constructive interference, and the dark bands, from destructive interference, which shows that light is made of waves. (© University of Colorado, Center for Intergrated Plasma Studies, Boulder, CO)

By combining his four equations, Maxwell showed that electric and magnetic fields should travel through space in the form of waves at a speed of 3.0×10^5 km/s—a value equal to the best available value for the speed of light. Maxwell's suggestion that these waves do exist and are observed as light was soon confirmed by experiments. Because of its electric and magnetic properties, light is also called **electromagnetic radiation.**

CAUTION! You may associate the term *radiation* with radioactive materials like uranium, but this term refers to anything that radiates, or spreads away, from its source. For example, scientists sometimes refer to sound waves as "acoustic radiation." Radiation does not have to be related to radioactivity!

Electromagnetic radiation consists of oscillating electric and magnetic fields, as shown in Figure 5-6. The distance between two successive wave crests is called the **wavelength** of the light, usually designated by the Greek letter λ (lambda). It is essential to keep in mind that no matter what the wavelength, all electromagnetic radiation travels at the same speed in a vacuum: $c = 3.0 \times 10^5$ km/s $= 3.0 \times 10^8$ m/s.

More than a century elapsed between Newton's experiments with a prism and the confirmation of the wave nature of light. One reason for this delay is that **visible light,** the light to which the human eye is sensitive, has extremely short wavelengths—less than a thousandth of a millimeter—that are not easily detectable. To express such tiny distances conveniently, scientists use a unit of length called the **nanometer** (abbreviated nm), where 1 nm $= 10^{-9}$ m. Experiments demonstrated that visible light has wavelengths covering the range from about 400 nm for violet light to about 700 nm for red light. Intermediate colors of the rainbow like yellow (550 nm) have intermediate wavelengths,

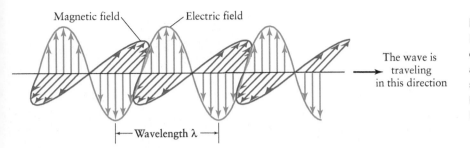

Magnetic field Electric field

The wave is traveling in this direction

|← Wavelength λ →|

FIGURE 5-6

Electromagnetic Radiation All forms of light consist of oscillating electric and magnetic fields that move through space at a speed of 3.00×10^5 km/s = 3.00×10^8 m/s. This figure shows a "snapshot" of these fields at one instant. The distance between two successive crests, called the wavelength of the light, is usually designated by the Greek letter λ (lambda).

as shown in Figure 5-7. (Some astronomers prefer to measure wavelengths in *angstroms*. One angstrom, abbreviated Å, is one-tenth of a nanometer: 1 Å = 0.1 nm = 10^{-10} m. In these units, the wavelengths of visible light extend from about 4000 Å to about 7000 Å. We will not use the angstrom unit in this book, however.)

Visible and Nonvisible Light

Maxwell's equations place no restrictions on the wavelength of electromagnetic radiation. Hence, electromagnetic waves could and should exist with wavelengths both longer and shorter than the 400- to 700-nm range of visible light. Consequently, researchers began to look for *invisible* forms of light. These are forms of electromagnetic radiation to which the cells of the human retina do not respond.

The first kind of invisible radiation to be discovered actually preceded Maxwell's work by more than a half century. Around 1800 the British astronomer William Herschel passed sunlight through a prism and held a thermometer just beyond the red end of the visible spectrum. The thermometer registered a temperature increase, indicating that it was being exposed to an invisible form of energy. This invisible energy, now called **infrared radiation,** was later realized to be electromagnetic radiation with wavelengths somewhat longer than those of visible light.

In experiments with electric sparks in 1888, the German physicist Heinrich Hertz succeeded in producing electromagnetic radiation with even longer wavelengths of a few centimeters or more. These are now known as **radio waves.** In 1895 another German physicist, Wilhelm Röntgen, invented a machine that produces electromagnetic radiation with wavelengths shorter than 10 nm, now known as **X-rays.** The X-ray machines in modern medical and dental offices are direct descendants of Röntgen's invention. Over the years radiation has been discovered with many other wavelengths.

Thus, visible light occupies only a tiny fraction of the full range of possible wavelengths, collectively called the **electromagnetic spectrum.** As Figure 5-7 shows, the electromagnetic spectrum stretches from the longest-wavelength radio waves to the shortest-wavelength gamma rays. Figure 5-8 shows some applications of nonvisible light in modern technology.

On the long-wavelength side of the visible spectrum, infrared radiation covers the range from about 700 nm to 1 mm. Astronomers interested in infrared radiation often express wavelength in *micrometers* or *microns,* abbreviated μm, where 1 μm = 10^{-3} mm = 10^{-6} m. **Microwaves** have wavelengths from roughly 1 mm to 10 cm, while radio waves have even longer wavelengths.

At wavelengths shorter than those of visible light, **ultraviolet radiation** extends from about 400 nm down to 10 nm. Next are X-rays, which have wavelengths between about 10 and 0.01 nm, and beyond them at even shorter wavelengths are **gamma rays.** Note that the rough boundaries between different types of radiation are simply arbitrary divisions in the electromagnetic spectrum.

Once again, in the vacuum of space, all light—from radio waves to gamma rays—travels at the speed of light, $c = 3 \times 10^8$ m/s.

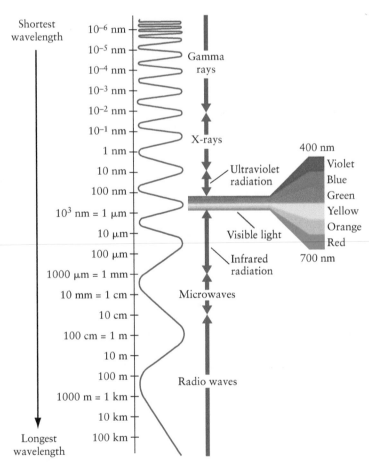

Shortest wavelength

10^{-6} nm
10^{-5} nm
10^{-4} nm — Gamma rays
10^{-3} nm
10^{-2} nm
10^{-1} nm
1 nm — X-rays
10 nm
100 nm — Ultraviolet radiation
10^3 nm = 1 μm
10 μm — Visible light
100 μm — Infrared radiation
1000 μm = 1 mm
10 mm = 1 cm — Microwaves
10 cm
100 cm = 1 m
10 m
100 m — Radio waves
1000 m = 1 km
10 km
100 km

Longest wavelength

400 nm
Violet
Blue
Green
Yellow
Orange
Red
700 nm

FIGURE 5-7

The Electromagnetic Spectrum The full array of all types of electromagnetic radiation is called the electromagnetic spectrum. It extends from the longest-wavelength radio waves to the shortest-wavelength gamma rays. Visible light occupies only a tiny portion of the full electromagnetic spectrum.

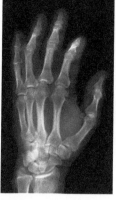

(a) Mobile phone: radio waves

(b) Microwave oven: microwaves

(c) TV remote: infrared light

(d) Tanning booth: ultraviolet light

(e) Medical imaging: X-rays

(f) Cancer radiotherapy: gamma rays

FIGURE 5-8

Uses of Nonvisible Electromagnetic Radiation **(a)** A mobile phone is actually a radio transmitter and receiver. The wavelengths used are in the range 16 to 36 cm. **(b)** A microwave oven produces radiation with a wavelength near 10 cm. The water in food absorbs this radiation, thus heating the food. **(c)** A remote control sends commands to a television using a beam of infrared light. **(d)** Ultraviolet radiation in moderation gives you a suntan, but in excess can cause sunburn or skin cancer. **(e)** X-rays can penetrate through soft tissue but not through bone, which makes them useful for medical imaging. **(f)** Gamma rays destroy cancer cells by breaking their DNA molecules, making them unable to multiply. (a: Maurizio Gambarini/dpa/Corbis; b: Michael Haegele/Corbis; c: Bill Lush/Taxi/Getty; d: Neil McAllister/Alamy; e: Ted Kinsman/Photo Researchers, Inc.; f: Will and Deni McIntyre/Science Source)

CONCEPTCHECK 5-3

Which form of electromagnetic radiation has a wavelength similar to the diameter of your finger?

Answer appears at the end of the chapter.

Frequency and Wavelength

Astronomers who work with radio telescopes often prefer to speak of *frequency* rather than wavelength. The **frequency** of a wave is the number of wave crests that pass a given point in one second. Equivalently, it is the number of complete *cycles* of the wave that pass per second (a complete cycle is from one crest to the next). Frequency is usually denoted by the Greek letter ν (nu). The unit of frequency is the cycle per second, also called the *hertz* (abbreviated Hz) in honor of Heinrich Hertz, the physicist who first produced radio waves. For example, if 500 crests of a wave pass you in one second, the frequency of the wave is 500 cycles per second or 500 Hz.

In working with frequencies, it is often convenient to use the prefix *mega-* (meaning "million," or 10^6, and abbreviated M) or *kilo-* (meaning "thousand," or 10^3, and abbreviated k). For example, AM radio stations broadcast at frequencies between 535 and 1605 kHz (kilohertz), while FM radio stations broadcast at frequencies in the range from 88 to 108 MHz (megahertz).

The relationship between the frequency and wavelength of an electromagnetic wave is a simple one. Because light moves at a constant speed $c = 3 \times 10^8$ m/s, if the wavelength (distance from one crest to the next) is made shorter, the frequency must increase (more of those closely spaced crests pass you each second). Thus, the shorter the wavelength of light, the higher the frequency, and likewise, the longer the wavelength of light, the lower the frequency. The frequency ν of light is related to its wavelength λ by the equation

Frequency and wavelength of an electromagnetic wave

$$\nu = \frac{c}{\lambda}$$

ν = frequency of an electromagnetic wave (in Hz)

c = speed of light = 3×10^8 m/s

λ = wavelength of the wave (in meters)

That is, the frequency of a wave equals the wave speed divided by the wavelength.

For example, hydrogen atoms in space emit radio waves with a wavelength of 21.12 cm. To calculate the frequency of this radiation, we must first express the wavelength in meters rather than centimeters: $\lambda = 0.2112$ m. Then we can use the above formula to find the frequency ν:

$$\nu = \frac{c}{\lambda} = \frac{3 \times 10^8 \text{ m/s}}{0.2112 \text{ m}} = 1.42 \times 10^9 \text{ Hz} = 1420 \text{ MHz}$$

Visible light has a much shorter wavelength and higher frequency than radio waves. You can use the above formula to show that for yellow-orange light of wavelength 600 nm, the frequency is 5×10^{14} Hz or 500 *million* megahertz!

While Young's experiment (Figure 5-5) showed convincingly that light has wavelike aspects, it was discovered in the early 1900s that light *also* has some of the characteristics of a stream of particles and waves. We will explore light's dual nature in Section 5-5.

CONCEPTCHECK 5-4

How do the frequencies of the longest wavelengths of light compare to the frequencies of the shortest wavelengths of light?

CALCULATIONCHECK 5-1

What is the wavelength of radio waves for the FM radio station 99.9 KTYD (which is 99.9 MHz)?

Answers appear at the end of the chapter.

5-3 An opaque object emits electromagnetic radiation according to its temperature

To learn about objects in the heavens, astronomers study the character of the electromagnetic radiation coming from those objects. Such studies can be very revealing because different kinds of electromagnetic radiation are typically produced in different ways. As an example, on Earth the most common way to generate radio waves is to make an electric

> As an object is heated, it glows more brightly and its peak color shifts to shorter wavelengths

current oscillate back and forth (as is done in the broadcast antenna of a radio station). By contrast, X-rays for medical and dental purposes are usually produced by bombarding atoms in a piece of metal with fast-moving particles extracted from other atoms. Our own Sun emits radio waves from near its glowing surface and X-rays from its corona (see the photo that opens Chapter 3). Hence, these observations indicate the presence of electric currents near the Sun's surface and of fast-moving particles in the Sun's outermost regions. (We will discuss the Sun at length in Chapter 18.)

Radiation from Heated Objects

TUTORIAL 5-2   The simplest and most common way to produce electromagnetic radiation, either on or off Earth, is to heat an object. The hot filament of wire inside an ordinary lightbulb emits white light, and a neon sign has a characteristic red glow because neon gas within the tube is heated by an electric current. In like fashion, almost all the visible light that we receive from space comes from hot objects like the Sun and the stars. The kind and amount of light emitted by a hot object tell us not only how hot it is but also about other properties of the object.

We can tell whether the hot object is made of relatively dense or relatively thin material. Consider the difference between a lightbulb

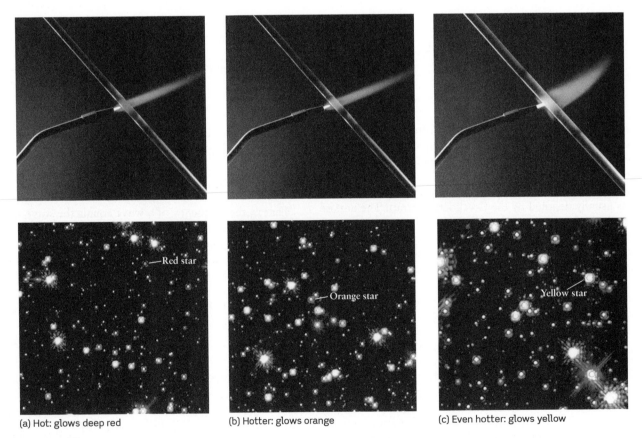

(a) Hot: glows deep red   (b) Hotter: glows orange   (c) Even hotter: glows yellow

FIGURE 5-9   R I V U X G

Objects at Different Temperatures Have Different Colors and Brightnesses This sequence of photographs shows the changing appearance of a piece of iron as it is heated. As the temperature increases, the amount of energy radiated by the bar increases, and so it appears brighter. The

apparent color of the bar also changes because, as the temperature increases, the dominant or peak wavelength of light emitted by the bar decreases. The stars shown have roughly the same temperatures as the bars above them.
(top row: © 1984 Richard Megna Fundamental Photographs; bottom row: NASA)

and a neon sign. The dense, solid filament of a lightbulb makes white light, which is a mixture of all different visible wavelengths, while the thin, transparent neon gas produces light of a rather definite red color and, hence, a rather definite wavelength. For now we will concentrate our attention on the light produced by dense, *opaque* objects. An **opaque** object does not allow light to travel through it without being scattered or absorbed—opaque is the opposite of transparent. Earth's atmosphere absorbs and scatters only some of the incoming sunlight, and we say that it is partially opaque, or partially transparent. Even though the Sun and stars are gaseous, not solid, it turns out that because they are so large, light from the inside scatters many times before emerging from the surface. Therefore, stars are opaque and emit light with many of the same properties as light emitted by a hot, glowing, solid object.

Imagine a welder or blacksmith heating a bar of iron. As the bar becomes hot, it begins to glow deep red, as shown in Figure 5-9a. (You can see this same glow from the coils of a toaster or from an electric range turned on "high.") While the bar emits a range of wavelengths, the dominant wavelength is red in color. As the temperature rises further, the bar begins to give off a brighter orange light (Figure 5-9b). At still higher temperatures, it shines with a brilliant yellow color (Figure 5-9c); the dominant wavelength is yellow light. If the bar could be prevented from melting and vaporizing, at extremely high temperatures it would emit a dazzling blue-white light. As shown in Figure 5-9, the temperature of a star determines its color, and the hottest stars are blue.

As this example shows, the amount of energy emitted by the hot, dense object and the dominant wavelength of the emitted radiation both depend on the temperature of the object. The hotter the object, the more energy it emits and the shorter the wavelength at which most of the energy is emitted. Colder objects emit relatively little energy, and this emission is primarily at long wavelengths.

These observations explain why you cannot see in the dark. The temperatures of people, animals, and furniture are much less than even that of the iron bar in Figure 5-9a. So, while these objects emit radiation even in a darkened room, most of this emission is at wavelengths greater than those of red light, in the infrared part of the spectrum (see Figure 5-7). Your eye is not sensitive to infrared, and you thus cannot see ordinary objects in a darkened room. But you can detect this radiation by using a camera that is sensitive to infrared light (Figure 5-10).

Temperature and Thermal Energy

To better understand the relationship between the temperature of a dense object and the radiation it emits, it is helpful to know just what "temperature" means. The temperature of a substance is directly related to the average speed of the tiny atoms or molecules that make up the substance. As shown in Figure 5-11, the hotter the substance, the faster its atoms or molecules are moving; the colder the substance, the slower the motion of its particles. The **thermal energy** of an object comes from the kinetic energy of its atoms and molecules. Therefore, the hotter a substance, the greater its thermal energy.

Scientists usually prefer to use the Kelvin temperature scale, on which temperature is measured in **kelvins** (K) upward from **absolute zero.** This is the coldest possible temperature, at which atoms move as slowly as possible (they can never quite stop completely). On the more familiar Celsius and Fahrenheit temperature scales, absolute

FIGURE 5-10   R I V U X G

An Infrared Portrait In this image made with a camera sensitive to infrared radiation, the different colors represent regions of different temperature. That this image is made from infrared radiation is indicated by the highlighted I in the wavelength tab. Red areas (like the man's face) are the warmest and emit the most infrared light, while blue-green areas (including the man's hands and hair) are at the lowest temperatures and emit the least radiation. (Dr. Arthur Tucker/Photo Researchers)

zero (0 K) is −273°C and −460°F. Ordinary room temperature is 293 K, 20°C, or 68°F. Box 5-1 discusses the relationships among the Kelvin, Celsius, and Fahrenheit temperature scales.

Figure 5-12 depicts quantitatively how the radiation from a dense object depends on its Kelvin temperature. Each curve in this figure shows the intensity of light emitted at each wavelength by a dense object at a given temperature: 3000 K (the temperature of a "cool" star, and at which molten gold boils), 6000 K (around the temperature of our Sun, and of an iron-welding arc), and 12,000 K (the temperature of a "hot" star, and also found in special industrial furnaces). In other words, the curves show the spectrum of light emitted by such an object. At any temperature, a hot, dense object emits at all wavelengths, so its spectrum is a smooth, continuous curve with no gaps in it.

As we will see shortly, this is an example of *blackbody radiation.* Figure 5-12 shows that *the higher the temperature, the greater the intensity of light at all wavelengths.* One way to understand this response to temperature is that the emitted light results from the motion of the material's atoms and molecules, which, as we saw, increases with temperature (see Figure 5-11).

The temperature not only indicates the total intensity of the light, but also the shape of its spectrum. The most important feature in the spectrum—the dominant wavelength—is called the **wavelength of maximum emission,** at which the curve has its peak

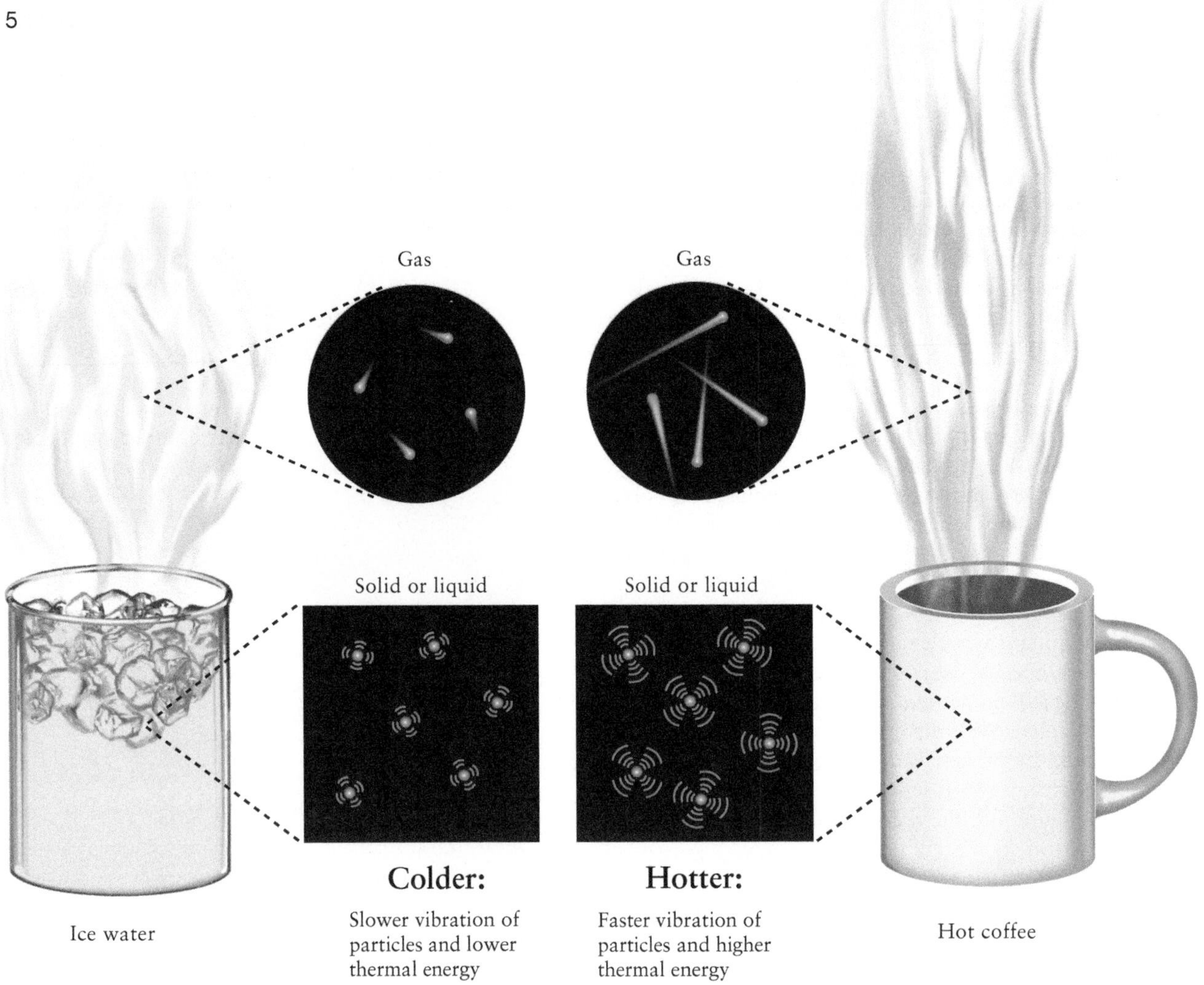

FIGURE 5-11

Thermal Energy The thermal energy of an object comes from the kinetic energy of its internal particles, and the hotter the material, the faster the particles move around. For hotter solids and liquids, particles vibrate faster and collide more intensely with their neighbors; for hotter gases the particles stream around at higher average speeds.

and the intensity of emitted energy is strongest (see Figure 5-12). The location of the peak also changes with temperature: *The higher the temperature, the shorter the wavelength of maximum emission.*

Figure 5-12 shows that for a dense object at a temperature of 3000 K, the wavelength of maximum emission is around 1000 nm (1 mm). Because this peak wavelength corresponds to the infrared range and is well outside the visible range, you might think that you cannot see the radiation from an object at this temperature. However, the glow from such an object *is* visible; the curve shows that this object emits light within the visible range, although about 90 percent of the emitted energy comes out at longer, invisible wavelengths (to the right of the visible spectrum). This invisible light is also emitted by lightbulbs. Incandescent lightbulbs create visible light by electrically heating a thin wire filament to about 3000 K, but because most of the energy is wasted at longer wavelengths, more efficient lights have become popular.

The 3000-K curve in Figure 5-12 is quite a bit higher at the red end of the visible spectrum than at the violet end, and a dense object at this temperature will appear yellowish in color. At even lower temperatures below 1000 K, the object would appear red (see Figure 5-9a). For higher temperatures, the 12,000-K curve has its wavelength of maximum emission in the ultraviolet part of the spectrum, at a wavelength shorter than visible light. But such a

hot, dense object also emits copious amounts of visible light (much more than at 6000 K or 3000 K, for which the curves are lower) and thus will have a very visible glow. The curve for this temperature is higher for blue light than for red light, and so the color of a dense object at 12,000 K is a brilliant blue or blue-white. The same principles apply to stars of different temperatures, which accounts for the different colors of stars in Figure 5-9. A star that looks blue has a high surface temperature, while an orange star has a relatively cool surface.

To summarize these observations:

The higher an object's temperature, the more intensely the object emits electromagnetic radiation and the shorter the wavelength at which it emits most strongly.

We will make frequent use of this general rule to analyze the temperatures of celestial objects such as planets and stars.

The curves in Figure 5-12 are drawn for an idealized type of dense object called a **blackbody**. A perfect blackbody does not reflect any light at all; instead, it absorbs all radiation falling on it. Because it reflects no electromagnetic radiation, the radiation that it does emit is entirely the result of its temperature. Ordinary objects, like tables, textbooks, and people, are not perfect blackbodies; they

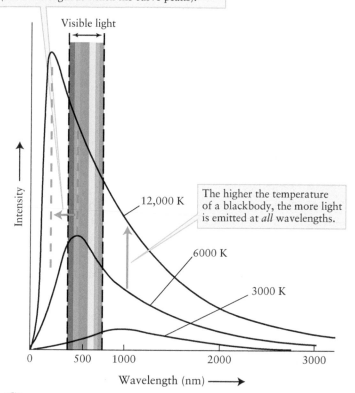

The higher the temperature of a blackbody, the shorter the wavelength of maximum emission (the wavelength at which the curve peaks).

Visible light

Intensity

12,000 K

The higher the temperature of a blackbody, the more light is emitted at *all* wavelengths.

6000 K

3000 K

0   500   1000   2000   3000

Wavelength (nm) ⟶

FIGURE 5-12

Blackbody Curves Each of these curves shows the intensity of light at every wavelength that is emitted by a blackbody (an idealized case of a dense object) at a particular temperature. The rainbow-colored band shows the range of visible wavelengths. The vertical scale has been compressed so that all three curves can be seen; the peak intensity for the 12,000-K curve is actually about 1000 times greater than the peak intensity for the 3000-K curve.

reflect light, which is why they are visible. A star such as the Sun, however, behaves very much like a perfect blackbody, because it absorbs almost completely any radiation falling on it from outside. The light emitted by a blackbody is called **blackbody radiation,** and the curves in Figure 5-12 are often called **blackbody curves.**

CAUTION! Despite its name, a blackbody does not necessarily *look* black. The Sun, for instance, does not look black because its temperature is high (around 5800 K), and so it glows brightly. But a room-temperature (around 300 K) blackbody would appear very black indeed. Even if a 300-K blackbody were as large as the Sun, it would emit only about 1/100,000 as much energy: Its blackbody curve is far too low to graph in Figure 5-12. Furthermore, most of the radiation from a room-temperature object is emitted at infrared wavelengths that are too long for our eyes to perceive.

Figure 5-13 shows the blackbody curve for a temperature of 5800 K. It also shows the intensity curve for light from the Sun, as measured from above Earth's atmosphere. (This is necessary because Earth's atmosphere absorbs certain wavelengths.) The peak of both curves is at a wavelength of about 500 nm, near the

middle of the visible spectrum. Note how closely the observed intensity curve for the Sun matches the blackbody curve. This is a strong indication that the temperature of the Sun's glowing surface is about 5800 K—a temperature that we can measure across a distance of 150 million kilometers! The close correlation between blackbody curves and the observed intensity curves for most stars is a key reason why astronomers are interested in the physics of blackbody radiation.

Blackbody radiation depends *only* on the temperature of the object emitting the radiation, not on the chemical composition of the object. The light emitted by molten gold at 2000 K is very nearly the same as that emitted by molten lead at 2000 K. Therefore, it might seem that analyzing the light from the Sun or from a star can tell astronomers the object's temperature but not what the star is made of. As Figure 5-13 shows, however, the intensity curve for the Sun (a typical star) is not precisely that of a blackbody. We will see later in this chapter that the *differences* between a star's spectrum and that of a blackbody allow us to determine the chemical composition of the star.

CONCEPTCHECK 5-5

An object is at 3000 K, emitting radiation that peaks at an infrared wavelength of around 1000 nm. Now the object is heated to 12,000 K and only emits a small fraction of its total energy at infrared wavelengths. At what temperature does the object emit more infrared radiation?

CONCEPTCHECK 5-6

Does a 0°C blackbody object give off any radiation?

Answers appear at the end of the chapter.

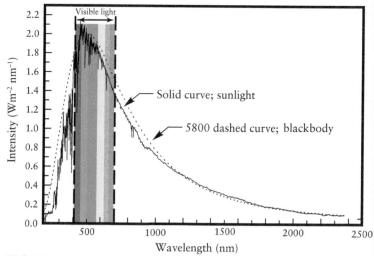

FIGURE 5-13

The Sun as a Blackbody This graph shows that the intensity of sunlight over a wide range of wavelengths (solid curve) is a remarkably close match to the intensity of radiation coming from a blackbody at a temperature of 5800 K (dashed curve). The measurements of the Sun's intensity were made above Earth's atmosphere (which absorbs and scatters certain wavelengths of sunlight). It is not surprising that the range of visible wavelengths includes the peak of the Sun's spectrum; the human eye evolved to take advantage of the most plentiful light available.

BOX 5-1 TOOLS OF THE ASTRONOMER'S TRADE

Temperatures and Temperature Scales

Three temperature scales are in common use. Throughout most of the world, temperatures are expressed in **degrees Celsius** (°C). The Celsius temperature scale is based on the behavior of water, which freezes at 0°C and boils at 100°C at sea level on Earth. This scale is named after the Swedish astronomer Anders Celsius, who proposed it in 1742.

Astronomers usually prefer the Kelvin temperature scale. This is named after the nineteenth-century British physicist Lord Kelvin, who made many important contributions to our understanding of heat and temperature. Absolute zero, the temperature at which atomic motion is at the absolute minimum, is –273°C in the Celsius scale but 0 K in the Kelvin scale. Atomic motion cannot be any less than the minimum, so nothing can be colder than 0 K; hence, there are no negative temperatures on the Kelvin scale. Note that we do *not* use degree (°) with the Kelvin temperature scale.

A temperature expressed in kelvins is always equal to the temperature in degrees Celsius plus 273. On the Kelvin scale, water freezes at 273 K and boils at 373 K. Water must be heated through a change of 100 K or 100°C to go from its freezing point to its boiling point. Thus, the "size" of a kelvin is the same as the "size" of a Celsius degree. When considering temperature changes, measurements in kelvins and Celsius degrees are the same. For extremely high temperatures the Kelvin and Celsius scales are essentially the same: For example, the Sun's core temperature is either 1.55×10^7 K or 1.55×10^7 °C.

The now-archaic Fahrenheit scale, which expresses temperature in **degrees Fahrenheit** (°F), is used only in the United States. When the German physicist Gabriel Fahrenheit introduced this scale in the early 1700s, he intended 100°F to represent the temperature of a healthy human body. On the Fahrenheit scale, water freezes at 32°F and boils at 212°F. There are 180 Fahrenheit degrees between the freezing and boiling points of water, so a degree Fahrenheit is only 100/180 = 5/9 as large as either a Celsius degree or a kelvin.

Two simple equations allow you to convert a temperature from the Celsius scale to the Fahrenheit scale and from Fahrenheit to Celsius:

$$T_F = \frac{9}{5} T_C + 32$$

$$T_C = \frac{5}{9} (T_F - 32)$$

T_F = temperature in degrees Fahrenheit

T_C = temperature in degrees Celsius

EXAMPLE: A typical room temperature is 68°F. We can convert this to the Celsius scale using the second equation:

$$T_C = \frac{5}{9} (68 - 32) = 20°C$$

To convert this to the Kelvin scale, we simply add 273 to the Celsius temperature. Thus,

$$68° = 20°C = 293 \text{ K}$$

The diagram displays the relationships among these three temperature scales.

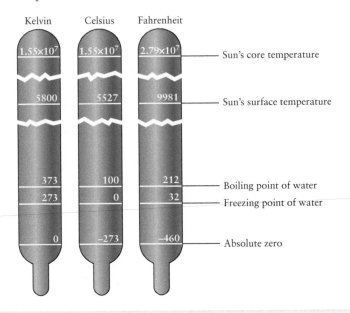

Kelvin	Celsius	Fahrenheit	
1.55×10^7	1.55×10^7	2.79×10^7	Sun's core temperature
5800	5527	9981	Sun's surface temperature
373	100	212	Boiling point of water
273	0	32	Freezing point of water
0	–273	–460	Absolute zero

5-4 Wien's law and the Stefan-Boltzmann law are useful tools for analyzing glowing objects like stars

The mathematical formula that describes the blackbody curves in Figure 5-12 is a rather complicated one. But there are two simpler formulas for blackbody radiation that prove to be very useful in many branches of astronomy. They are used by astronomers who investigate the stars as well as by those who study the planets (which are dense, relatively cool objects that emit infrared radiation). One of these formulas relates the temperature of a blackbody to its wavelength of maximum emission, and the other relates the temperature to the amount of energy that the blackbody emits. These formulas, which we will use throughout this book, restate in precise mathematical terms the qualitative relationships that we described in Section 5-3.

> Two simple mathematical formulas describing blackbodies are essential tools for studying the universe

Wien's Law

Figure 5-12 shows that the higher the temperature (T) of a blackbody, the shorter its wavelength of maximum emission (λ_{max}). In 1893 the German physicist Wilhelm Wien used ideas about both heat and electromagnetism to make this relationship quantitative. The formula that he derived, which today is called **Wien's law,** is

Wien's law for a blackbody

$$\lambda_{max} = \frac{0.0029 \text{ K m}}{T}$$

λ_{max} = wavelength of maximum emission of the object (in meters)

T = temperature of the object (in kelvins)

According to Wien's law, the wavelength of maximum emission of a blackbody is inversely proportional to its temperature in kelvins. In other words, if the temperature of the blackbody doubles, its wavelength of maximum emission is halved, and vice versa. For example, Figure 5-12 shows blackbody curves for temperatures of 3000 K, 6000 K, and 12,000 K. From Wien's law, a blackbody with a temperature of 6000 K has a wavelength of maximum emission $\lambda_{max} = (0.0029 \text{ K m})/(6000 \text{ K}) = 4.8 \times 10^{-7}$ m = 480 nm, in the visible part of the electromagnetic spectrum. At 12,000 K, or twice the temperature, the blackbody has a wavelength of maximum emission half as great, or $\lambda_{max} = 240$ nm, which is in the ultraviolet. At 3000 K, just half our original temperature, the value of λ_{max} is twice the original value—960 nm, which is an infrared wavelength. You can see that these wavelengths agree with the peaks of the curves in Figure 5-12.

CAUTION! Remember that Wien's law involves the wavelength of maximum emission in *meters*. If you want to convert the wavelength to nanometers, you must multiply the wavelength in meters by $(10^9 \text{ nm})/(1 \text{ m})$.

Wien's law is very useful for determining the surface temperatures of stars. It is not necessary to know how far away the star is, how large it is, or how much energy it radiates into space. All we need to know is the dominant wavelength of the star's electromagnetic radiation.

CONCEPTCHECK 5-7

What single piece of information do astronomers need in order to determine if a star is hotter than our Sun?

CALCULATIONCHECK 5-2

Which wavelength of light would our Sun emit most if its temperature were twice its current temperature of 5800 K?

Answers appear at the end of the chapter.

The Stefan-Boltzmann Law

The other useful formula for the radiation from a blackbody involves the total amount of energy the blackbody radiates at all wavelengths. (By contrast, the curves in Figure 5-12 show how much energy a blackbody radiates at each individual wavelength.)

Energy is usually measured in **joules** (J), named after the nineteenth-century English physicist James Joule. A joule is the amount of kinetic energy contained in the motion of a 2-kilogram mass moving at a speed of 1 meter per second. The joule is a convenient unit of energy because it is closely related to the familiar **watt** (W): 1 watt is 1 joule per second, or $1 \text{ W} = 1 \text{ J/s} = 1 \text{ J s}^{-1}$. (The superscript –1 means you are dividing by that quantity.) For example, a 100-watt lightbulb uses energy at a rate of 100 joules per second, or 100 J/s. The energy content of food is also often measured in joules; in most of the world, diet soft drinks are labeled as "low joule" rather than "low calorie."

The amount of energy emitted by a blackbody depends both on its temperature and on its surface area. These characteristics make sense: A large burning log radiates much more heat than a burning match, even though the temperatures are the same. To consider the effects of temperature alone, it is convenient to look at the amount of energy emitted from each square meter of an object's surface in a second. This quantity is called the **energy flux** (F). Flux means "rate of flow," and thus F is a measure of how rapidly energy is flowing out of the object. It is measured in joules per square meter per second, usually written as $J/m^2/s$ or $J \text{ m}^{-2} \text{ s}^{-1}$. Alternatively, because 1 watt equals 1 joule per second, we can express flux in watts per square meter (W/m^2, or $W \text{ m}^{-2}$).

The nineteenth-century Irish physicist David Tyndall performed the first careful measurements of the amount of radiation emitted by a blackbody. (He studied the light from a heated platinum wire, which behaves approximately like a blackbody.) By analyzing Tyndall's results, the Slovenian physicist Josef Stefan deduced in 1879 that the flux from a blackbody is proportional to the fourth power of the object's temperature (measured in kelvins). Five years after Stefan announced his law, Austrian physicist Ludwig Boltzmann showed how it could be derived mathematically from basic assumptions about atoms and molecules. For this reason, Stefan's law is commonly known as the **Stefan-Boltzmann law.** Written as an equation, the Stefan-Boltzmann law is

Stefan-Boltzmann law for a blackbody

$$F = \sigma T^4$$

F = energy flux, in joules per square meter of surface per second

σ = a constant = $5.67 \times 10^{-8} \text{ W m}^{-2} \text{ K}^{-4}$

T = object's temperature, in kelvins

The value of the Stefan-Boltzmann constant σ (the Greek letter sigma) is known from laboratory experiments.

The Stefan-Boltzmann law says that if you double the temperature of an object (for example, from 300 K to 600 K), then the energy emitted from the object's surface each second increases by a factor of $2^4 = 16$. If you increase the temperature by a factor of 10 (for example, from 300 K to 3000 K), the rate of energy emission increases by a factor of $10^4 = 10,000$. Thus, a chunk of iron at room temperature (around 300 K) emits very little

BOX 5-2 TOOLS OF THE ASTRONOMER'S TRADE

Using the Laws of Blackbody Radiation

The Sun and stars behave like nearly perfect blackbodies. Wien's law and the Stefan-Boltzmann law can therefore be used to relate the surface temperature of the Sun or a distant star to the energy flux and wavelength of maximum emission of its radiation. The following examples show how to do this.

EXAMPLE: The maximum intensity of sunlight is at a wavelength of roughly 500 nm $= 5.0 \times 10^{-7}$ m. Use this information to determine the surface temperature of the Sun

Situation: We are given the Sun's wavelength of maximum emission λ_{max}, and our goal is to find the Sun's surface temperature, denoted by $T_\odot$. (The symbol $\odot$ is the standard astronomical symbol for the Sun.)

Tools: We use Wien's law to relate the values of λ_{max} and $T_\odot$.

Answer: As written, Wien's law tells how to find λ_{max} if we know the surface temperature. To find the surface temperature from λ_{max}, we first rearrange the formula, then substitute the value of λ_{max}:

$$T_\odot = \frac{0.0029 \text{ K m}}{\lambda_{max}} = \frac{0.0029 \text{ K m}}{5.0 \times 10^{-7} \text{ m}} = 5800 \text{ K}$$

Review: This temperature is very high by Earth standards, about the same as an iron-welding arc.

EXAMPLE: Using detectors above Earth's atmosphere, astronomers have measured the average flux of solar energy arriving at Earth. This value, called the **solar constant,** is equal to 1370 W m^{-2}. Use this information to calculate the Sun's surface temperature. (This calculation provides a check on our result from the preceding example.)

Situation: The solar constant is the flux of sunlight as measured at Earth. We want to use the value of the solar constant to calculate $T_\odot$.

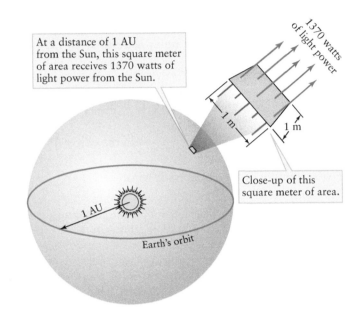

At a distance of 1 AU from the Sun, this square meter of area receives 1370 watts of light power from the Sun.

1370 watts of light power

1 m   1 m

Close-up of this square meter of area.

1 AU

Earth's orbit

Tools: It may seem that all we need is the Stefan-Boltzmann law, which relates flux to surface temperature. However, the quantity F in this law refers to the flux measured at the Sun's surface, *not* at Earth. Hence, we will first need to calculate F from the given information.

Answer: To determine the value of F, we first imagine a huge sphere of radius 1 AU with the Sun at its center, as shown in the figure. Each square meter of that sphere receives 1370 watts of power from the Sun, so the total energy radiated by the Sun per second is equal to the solar constant multiplied by the sphere's surface area. The result, called the **luminosity** of the Sun and denoted by the symbol $L_\odot$, is $L_\odot = 3.90 \times 10^{26}$ W. That is, in 1 second the Sun radiates 3.90×10^{26} joules of energy into space. Because we know the size of the Sun, we can compute the energy flux (energy emitted per square meter per second) at its surface. The radius of the Sun is

---

electromagnetic radiation (and essentially no visible light), but an iron bar heated to 3000 K glows quite intensely.

Box 5-2 gives several examples of applying Wien's law and the Stefan-Boltzmann law to typical astronomical problems.

CONCEPTCHECK 5-8

Is it ever possible for a cooler object to emit more energy than a warmer object?

Answer appears at the end of the chapter.

5-5 Light has properties of both waves and particles

At the end of the nineteenth century, physicists mounted a valiant effort to explain *how* a material can produce blackbody radiation. To this end they constructed theories based on Maxwell's description of light as electromagnetic waves. But while electromagnetic waves could explain many properties of light, it was not until the discovery of photons that all features of blackbody curves could be understood.

$R_\odot = 6.96 \times 10^8$ m, and the Sun's surface area is $4\pi R_\odot{}^2$. Therefore, its energy flux $F_\odot$ is the Sun's luminosity (total energy emitted by the Sun per second) divided by the Sun's surface area (the number of square meters of surface):

$$F_\odot = \frac{L_\odot}{4\pi R_\odot{}^2} = \frac{3.90 \times 10^{26} \text{ W}}{4\pi(6.96 \times 10^8 \text{ m})^2} = 6.41 \times 10^7 \text{ W m}^{-2}$$

Once we have the Sun's energy flux $F_\odot$ we can use the Stefan-Boltzmann law to find the Sun's surface temperature $T_\odot$:

$$T_\odot{}^4 = F_\odot/\sigma = 1.13 \times 10^{15} \text{ K}^4$$

Taking the fourth root (the square root of the square root) of this value, we find the surface temperature of the Sun to be $T_\odot = 5800$ K.

Review: Our result for $T_\odot$ agrees with the value we computed in the previous example using Wien's law. Notice that the solar constant of 1370 W m^{-2} is very much less than $F_\odot$ the flux at the Sun's surface. By the time the Sun's radiation reaches Earth, it is spread over a greatly increased area.

EXAMPLE: Sirius, the brightest star in the night sky, has a surface temperature of about 10,000 K. Find the wavelength at which Sirius emits most intensely.

Situation: Our goal is to calculate the wavelength of maximum emission of Sirius (λ_{max}) from its surface temperature T.

Tools: We use Wien's law to relate the values of λ_{max} and T.

Answer: Using Wien's law,

$$\lambda_{max} = \frac{0.0029 \text{ K m}}{T} = \frac{0.0029 \text{ K m}}{10,000 \text{ K}}$$

$$= 2.9 \times 10^{-7} \text{ m} = 290 \text{ nm}$$

Review: Our result shows that Sirius emits light most intensely in the ultraviolet. In the visible part of the spectrum, it emits more blue light than red light (like the curve for 12,000 K in Figure 5-11), so Sirius has a distinct blue color.

EXAMPLE: How does the energy flux from Sirius compare to the Sun's energy flux?

Situation: To compare the energy fluxes from the two stars, we want to find the *ratio* of the flux from Sirius to the flux from the Sun.

Tools: We use the Stefan-Boltzmann law to find the flux from Sirius and from the Sun, which from the preceding examples have surface temperatures 10,000 K and 5800 K, respectively.

Answer: For the Sun, the Stefan-Boltzmann law is $F_\odot = \sigma T_\odot{}^4$, and for Sirius we can likewise write $F_* = \sigma T_*{}^4$, where the subscripts $\odot$ and $*$ refer to the Sun and Sirius, respectively. If we divide one equation by the other to find the ratio of fluxes, the Stefan-Boltzmann constants cancel out and we get

$$\frac{F_*}{F_\odot} = \frac{T_*{}^4}{T_\odot{}^4} = \frac{(10,000 \text{ K})^4}{(5800 \text{ K})^4} = \left(\frac{10,000}{5800}\right)^4 = 8.8$$

Review: Because Sirius has such a high surface temperature, each square meter of its surface emits 8.8 times more energy per second than a square meter of the Sun's relatively cool surface. Sirius is actually a larger star than the Sun, so it has more square meters of surface area and, hence, its *total* energy output is *more* than 8.8 times that of the Sun.

The Discovery of Photons

In 1905, the great physicist Albert Einstein developed a radically new model for the nature of light. Central to this new picture is that light is made out of particles! Each particle of light is called a **photon**, which is a distinct packet of electromagnetic energy.

Photons have a dual nature in that they are both particlelike and wavelike. They are particlelike in the sense that they are the small packets of energy that make up light. As expected with particles, you can count the number of photons in an electromagnetic wave. But, photons are also wavelike because each one is itself a wave, where each photon has the same wavelength as the electromagnetic wave of which it is a small part. As expected with waves, photons exhibit interference patterns when passed through a double-slit experiment like the one in Figure 5-5.

The energy of each photon is related to the wavelength of light: the longer the wavelength, the lower the energy. Thus, a

> Light is made of smaller entities called photons, each one with a wavelength of its own

photon of red light (wavelength $\lambda = 700$ nm) has less energy than a photon of violet light ($\lambda = 400$ nm). Conversely, the shorter a photon's wavelength, the higher its energy.

It was quickly realized that photons explain more than just blackbody curves, and now the study of photons plays a key role in almost every branch of science. While scientific equipment is usually needed to detect individual photons, there is a somewhat familiar effect that photons can help to explain. When it comes to tanning, your skin must be exposed to ultraviolet light. For example, a low intensity exposure to ultraviolet light causes tanning, but even a high intensity exposure of visible light will not cause the skin to darken. (Recall that you do not get tanned with regular indoor lighting, which has negligible UV light). The reason is that tanning, at the molecular level, results from a chemical reaction in the skin that involves a single photon. A single high-energy, short-wavelength ultraviolet photon can trigger a tanning reaction. However, even many lower-energy, longer-wavelength photons of visible light cannot trigger the reaction. In this situation, we see that it is not the total energy of light that matters, but how the energy is bundled into packets—and that is a photon.

The Energy of a Photon

The relationship between the energy E of a single photon and the wavelength of the electromagnetic radiation can be expressed in a simple equation:

Energy of a photon (in terms of wavelength)

$$E = \frac{hc}{\lambda}$$

E = energy of a photon

h = Planck's constant

c = speed of light

λ = wavelength of light

The value of the constant h in this equation, now called *Planck's constant,* has been shown in laboratory experiments to be

$$h = 6.625 \times 10^{-34} \text{ J s}$$

The units of h are joules multiplied by seconds, called "joule-seconds" and abbreviated J s.

Because the value of h is so tiny, a single photon carries a very small amount of energy. For example, a photon of red light with wavelength 633 nm has an energy of only 3.14×10^{-19} J (Box 5-3), which is why we ordinarily do not notice that light comes in the form of photons. Even a dim light source emits so many photons per second that it seems to be radiating a continuous stream of energy.

The energies of photons are sometimes expressed in terms of a small unit of energy called the **electron volt** (eV). One electron volt is equal to 1.602×10^{-19} J, so a 633-nm photon has an energy of 1.96 eV. If energy is expressed in electron volts, Planck's constant is best expressed in electron volts multiplied by seconds, abbreviated eV s:

$$h = 4.135 \times 10^{-15} \text{ eV s}$$

Because the frequency v of light is related to the wavelength λ by $v = c/\lambda$, we can rewrite the equation for the energy of a photon as

Energy of a photon (in terms of frequency)

$$E = hv$$

E = energy of a photon

h = Planck's constant

v = frequency of light

The equations $E = hc/\lambda$ and $E = hv$ are together called **Planck's law.** Both equations express a relationship between a particlelike property of light (the energy E of a photon) and a wavelike property (the wavelength λ or frequency v).

BOX 5-3 ASTRONOMY DOWN TO EARTH

Photons at the Supermarket

A beam of light can be regarded as a stream of tiny packets of energy called photons. The Planck relationships $E = hc/\lambda$ and $E = hv$ can be used to relate the energy E carried by a photon to its wavelength λ and frequency v.

As an example, the laser bar-code scanners used at stores and supermarkets emit orange-red light of wavelength 633 nm. To calculate the energy of a single photon of this light, we must first express the wavelength in meters. A nanometer (nm) is equal to 10^{-9} m, so the wavelength is

$$\lambda = (633 \text{ nm}) \left(\frac{10^{-9} \text{ m}}{1 \text{ nm}} \right) = 633 \times 10^{-9} \text{ m} = 6.33 \times 10^{-7} \text{ m}$$

Then, using the Planck formula $E = hc/\lambda$, we find that the energy of a single photon is

$$E = \frac{hc}{\lambda} = \frac{(6.625 \times 10^{-34} \text{ J s})(3 \times 10^{8} \text{ m/s})}{6.33 \times 10^{-7} \text{ m}} = 3.14 \times 10^{-19} \text{ J}$$

This amount of energy is very small. The laser in a typical bar-code scanner emits 10^{-3} joule of light energy per second, so the number of photons emitted per second is

$$\frac{10^{-3} \text{ joule per second}}{3.14 \times 10^{-19} \text{ joule per photon}} = 3.2 \times 10^{15} \text{ photons per second}$$

This number is so large that the laser beam seems like a continuous flow of energy rather than a stream of little energy packets.

The photon picture of light is essential for understanding the detailed shapes of blackbody curves. As we will see, it also helps to explain how and why the spectra of the Sun and stars differ from those of perfect blackbodies.

CONCEPTCHECK 5-9

If a photon's wavelength is measured to be longer than the wavelength of a green photon, will it have a greater or lower energy than a green photon?

Answer appears at the end of the chapter.

5-6 Each chemical element produces its own unique set of spectral lines

In 1814 the German master optician Joseph von Fraunhofer repeated the classic experiment of shining a beam of sunlight through a prism (see Figure 5-3). But this time Fraunhofer subjected the resulting rainbow-colored spectrum to intense magnification. To his surprise, he discovered that the solar spectrum contains hundreds of fine, dark lines, now called **spectral lines** (Figure 5-14). A dark spectral line arises when light at a specific wavelength is at least partially *absorbed* so that the spectrum appears darker; it is also called an *absorption line*. By

> Spectroscopy is the key to determining the chemical composition of planets and stars

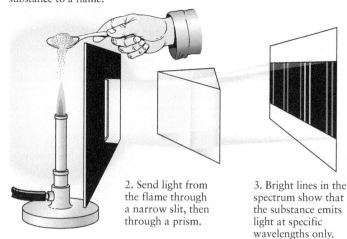

1. Add a chemical substance to a flame.

2. Send light from the flame through a narrow slit, then through a prism.

3. Bright lines in the spectrum show that the substance emits light at specific wavelengths only.

FIGURE 5-15

The Kirchhoff-Bunsen Experiment In the mid-1850s, Gustav Kirchhoff and Robert Bunsen discovered that when a chemical substance is heated and vaporized, the spectrum of the emitted light exhibits a series of bright spectral lines. They also found that each chemical element produces its own characteristic pattern of spectral lines. (In an actual laboratory experiment, lenses would be needed to focus the image of the slit onto the screen.)

FIGURE 5-14 R I **V** U X G

The Sun's Spectrum Numerous dark spectral lines are seen in this image of the Sun's spectrum. The spectrum is spread out so much that it had to be cut into segments to fit on this page. (N. A. Sharp, NOAO/NSO/Kitt Peak FTS/ AURA/NSF)

contrast, if the light from a perfect blackbody were sent through a prism, it would produce a smooth, continuous spectrum with no dark lines. Fraunhofer counted more than 600 dark lines in the Sun's spectrum; today we know of more than 30,000. The photograph of the Sun's spectrum in Figure 5-14 shows hundreds of these spectral lines.

Spectral Analysis

Half a century later, chemists discovered that they could produce spectral lines in the laboratory and use these spectral lines to analyze what kinds of atoms different substances are made of. Chemists had long known that many substances emit distinctive colors when sprinkled into a flame. To facilitate study of these colors, around 1857 the German chemist Robert Bunsen invented a gas burner (today called a Bunsen burner) that produces a clean flame with no color of its own. Bunsen's colleague, the Prussian-born physicist Gustav Kirchhoff, suggested that the colored light produced when substances were added to the flame might best be studied by passing the resulting light through a prism (Figure 5-15). The two scientists promptly discovered that the spectrum from the flame consists of a pattern of thin, bright spectral lines against a dark background. A bright spectral line arises because at least some additional light is being emitted at a specific wavelength; it is also called an *emission line*.

Kirchhoff and Bunsen then found that each chemical element produces its own unique pattern of spectral lines. Thus was born in 1859 the technique of **spectral analysis**: the identification of atoms and molecules by their unique patterns of spectral lines.

Atoms are the building blocks of substances on Earth and throughout the universe. Familiar examples are hydrogen, oxygen, carbon, iron, and gold. These atoms are also called chemical **elements** because a pure substance made from only one type of atom (such as pure gold) cannot be broken down into more basic chemicals. After Kirchhoff and Bunsen had recorded the prominent spectral lines of all the then-known elements, they soon began to discover other spectral lines in the spectra of vaporized mineral samples. In this way they discovered elements on Earth that had never been seen or inferred through any method.

Spectral analysis even allowed the discovery of new elements outside Earth. During the solar eclipse of 1868, astronomers found a new spectral line in light coming from the hot gases at the upper surface of the Sun while the main body of the Sun was hidden by the Moon. This line was attributed to a new element that was named helium (from the Greek *helios,* meaning "sun"). It took almost another 30 years to discover helium on Earth!

Atoms can also combine to form **molecules.** For example, two hydrogen atoms (symbol H) can combine with an oxygen atom (symbol O) to form a water molecule (symbol H_2O). Substances like water whose molecules include atoms of different elements are called **compounds.** Just as each type of atom has its own unique spectrum, so does each type of molecule. Figure 5-16 shows the spectra of several types of atoms and molecules. You can easily see in Figure 5-16 that each substance produces a unique pattern of spectral lines; each pattern can be thought of as a spectral "fingerprint" for identification. This is enormously important in

astronomy because it allows us to determine the detailed composition of distant planets and stars by analyzing the spectra of their light that reaches Earth.

Kirchhoff's Laws

The spectrum of the Sun, with its dark spectral lines superimposed on a bright background (see Figure 5-14), may seem to be unrelated to the spectra of bright lines against a dark background produced by substances in a flame (see Figure 5-15). But by the early 1860s, Kirchhoff's experiments had revealed a direct connection between these two types of spectra. His conclusions are summarized in three important statements about spectra that are today called **Kirchhoff's laws.** These laws, which are illustrated in Figure 5-17, are as follows:

Law 1 A hot opaque body, such as a perfect blackbody, or a hot, dense gas produces a **continuous spectrum**—a complete rainbow of colors without any spectral lines.

Law 2 A hot, transparent gas produces an **emission line spectrum**—a series of bright spectral lines against a dark background.

Law 3 A cool, transparent gas in front of a source of a continuous spectrum produces an **absorption line spectrum**—a series of dark spectral lines among the colors of the continuous spectrum. Furthermore, the dark lines in the absorption spectrum of a particular gas occur at exactly the *same* wavelengths as the bright lines in the emission spectrum of that same gas. As we will see in Section 5-8, absorption and emission lines correspond to distinct changes in an electron's orbit within an atom.

CAUTION! Figure 5-17 shows that light can either pass through a cloud of gas or be absorbed by the gas. But there is also a third possibility: The light can simply bounce off the atoms or molecules that make up the gas, a phenomenon called **light scattering.** In other words, photons passing through a gas cloud can miss the gas atoms altogether, be swallowed whole by the atoms (absorption), or bounce off the atoms like billiard balls colliding (scattering). Box 5-4 describes how light scattering explains the blue color of the sky and the red color of sunsets.

In general, a gas cloud can both emit and absorb light. Whether an emission line spectrum or an absorption line spectrum is observed from an intervening gas cloud depends on the relative temperatures of the gas cloud and its background. Absorption lines are seen if the background is hotter than the gas, and emission lines are seen if the background is cooler. This is why the gas cloud illustrating absorption (Law 3) in Figure 5-17 is cooler than the star.

Spectroscopy

Spectroscopy is the systematic study of spectra and spectral lines. As noted, spectral lines are tremendously important in astronomy, because they provide detailed evidence about the chemical composition of distant objects. As an example, the spectrum of the Sun shown in Figure 5-14 is an absorption line spectrum. The

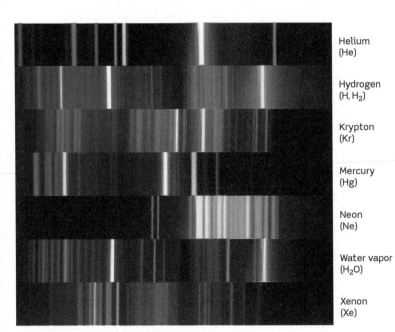

Helium (He)

Hydrogen (H, H_2)

Krypton (Kr)

Mercury (Hg)

Neon (Ne)

Water vapor (H_2O)

Xenon (Xe)

FIGURE 5-16 R I V U X G

Various Spectra These photographs show the spectra of different types of gases as measured in a laboratory on Earth. Each type of gas has a unique spectrum that is the same wherever in the universe the gas is found. Water vapor (H_2O) is a compound whose molecules are made up of hydrogen and oxygen atoms; the hydrogen molecule (H_2) is made up of two hydrogen atoms. For the hydrogen gas lamp, the bright lines are from single hydrogen atoms (H), and the fainter lines are emitted from hydrogen molecules (H_2). (Ted Kinsman/Science Photo Library)

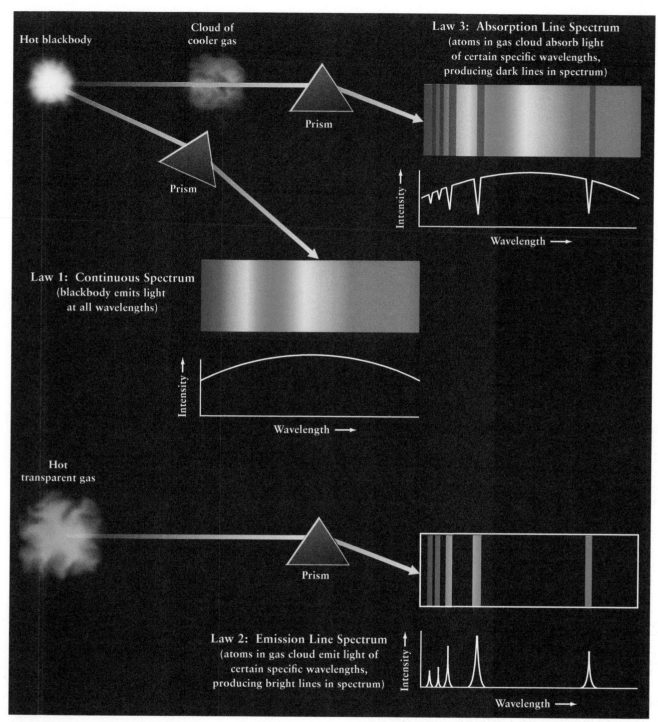

FIGURE 5-17

Kirchhoff's Laws for Continuous, Absorption Line, and Emission Line Spectra **Law 1:** A hot, opaque body emits a continuous spectrum of light. **Law 3:** If this light is passed through a cloud of a cooler gas, the cloud absorbs light of certain specific wavelengths, leaving the light that passes directly through the cloud with an absorption line spectrum. **Law 2:**

Atoms in a hot gas cloud collide energetically and emit light, creating an emission line spectrum that is unique to each type of atom. In each case the spectra is shown, as it would appear with a prism, and just below it, by a graph of intensity versus wavelength.

continuous spectrum comes from the hot surface of the Sun, which acts like a blackbody. The dark absorption lines are caused by this light passing through a cooler gas; this gas is the atmosphere that surrounds the Sun. Therefore, by identifying the spectral lines present in the solar spectrum, we can determine the chemical composition of the Sun's atmosphere.

BOX 5-4 ASTRONOMY DOWN TO EARTH

Why the Sky Is Blue

Light scattering is a process where photons bounce off particles, and change their direction of travel. The scattering particles can be atoms, molecules, or clumps of molecules. When you see regular objects like rocks or trees that do not produce their own intrinsic light, you see them because they scatter sunlight into your eyes.

An important fact about light scattering is that very small particles—ones that are smaller than a wavelength of visible light—are quite effective at scattering short-wavelength photons of blue light but less effective at scattering long-wavelength photons of red light. This fact explains a number of phenomena that you can see here on Earth.

The light that comes from the daytime sky is sunlight that has been scattered by the molecules that make up our atmosphere (see part a of the accompanying figure). Air molecules are less than 1 nm across, far smaller than the wavelength of visible light, so they scatter blue light more than red light—which is why the sky looks blue. In other words, the sky is blue because while incoming white sunlight contains all the colors, air molecules scatter blue light around more than red light, literally spreading blue light all throughout the atmosphere.

During the day, distant mountains often appear blue thanks to sunlight being scattered from the atmosphere between the mountains and your eyes. (The Blue Ridge Mountains, which extend from Pennsylvania to Georgia, and Australia's Blue Mountains derive their names from this effect.)

Light scattering also explains why sunsets are red. Again, sunlight contains all the colors, but as this light passes through our atmosphere the blue light is scattered away from the straight-line path from the Sun. On the other hand, red light undergoes relatively little scattering, so light coming directly from the Sun looks a bit redder once some blue has been removed. When you look toward the setting Sun (but, never look at the Sun!), the sunlight that reaches your eye has had to pass through a relatively thick layer of atmosphere (part b of the accompanying figure). Hence, a large fraction of the blue light from the Sun has been scattered, and the Sun appears quite red. If, at sunset, the sunlight shines onto clouds, the clouds also look red after much of the blue light has been scattered away.

The same effect also applies to sunrises, but sunrises seldom look as red as sunsets do. The reason is that dust is lifted into the atmosphere during the day by the wind (which, being driven by sunlight, typically blows stronger in the daytime), and dust particles in the atmosphere help to scatter even more blue light.

Finally, some scattered light appears white. If the small particles that scatter light are sufficiently concentrated, there will be almost as much scattering of red light as of blue light, and the scattered light will then appear white. This can occur when the scattering comes from dense ice crystals or water vapor and explains the white color of snow, clouds, and fog.

Light scattering has many applications to astronomy. For example, it explains why very distant stars in our Galaxy appear surprisingly red. The reason is that there are tiny dust particles throughout the space between the stars, and this dust scatters blue light. By studying how much scattering takes place, astronomers have learned about the tenuous material that fills interstellar space.

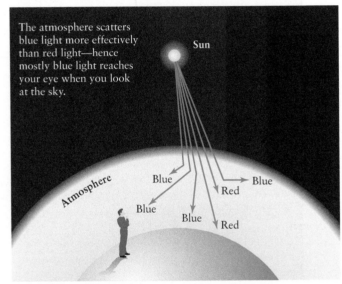

(a) Why the sky looks blue

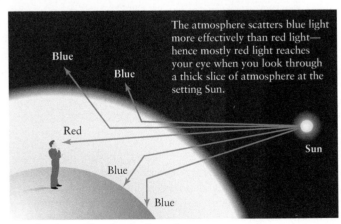

(b) Why the setting Sun looks red

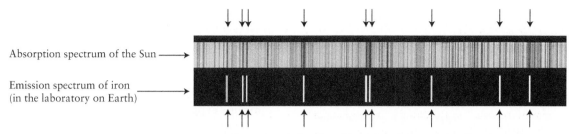

For each emission line of iron, there is a corresponding absorption
line in the solar spectrum; hence there must be iron in the Sun's atmosphere

FIGURE 5-18   R I V U X G

Iron in the Sun The upper part of this figure is a portion of the Sun's
spectrum at violet wavelengths, showing numerous dark absorption lines.
The lower part of the figure is a corresponding portion of the emission line
spectrum of vaporized iron. The iron lines coincide with some of the solar
lines, which proves that there is some iron (albeit a relatively small amount) in
the Sun's atmosphere. (Carnegie Observatories)

Even when the spectral lines from more than one type of atom
appear in the same spectrum, the different atoms can be identified.
Figure 5-18 shows both a portion of the Sun's absorption line
spectrum, which has numerous lines, and the emission line spectrum of iron vapor over the same wavelength range, measured on
Earth. This pattern of bright spectral lines in the lower spectrum
is iron's own distinctive "fingerprint," which no other substance
can imitate. Because some absorption lines in the Sun's spectrum
coincide with the iron lines, some vaporized iron must exist in the
Sun's atmosphere.

Spectroscopy can also help us analyze gas clouds in space,
such as the nebula surrounding the star cluster NGC 346 shown
in Figure 5-19. The particular shade of red that dominates the
color of this nebula is due to an emission line at a wavelength
of 656 nm. This is one of the characteristic wavelengths emitted
by hydrogen gas, so we can conclude that this nebula contains
hydrogen. More detailed analyses of this kind show that hydrogen
is the most common element in gaseous nebulae, and indeed in the
universe as a whole. The spectra of other nebulae, such as the Ring
Nebula shown in the image that opens this chapter, also reveal the
presence of nitrogen, oxygen, helium, and other gases.

What is truly remarkable about spectroscopy is that it can
determine chemical composition at any distance. The 656-nm red
light produced by a sample of heated hydrogen gas on Earth (the
bright red line in the hydrogen spectrum in Figure 5-16) is the
same as that observed coming from the nebula shown in Figure
5-19, located about 210,000 light-years away. Throughout this
book we will see many examples of how astronomers use spectra
to determine the nature of celestial objects.

Why does an atom absorb light of only particular wavelengths? And why does it then emit light of only these same wavelengths? Maxwell's theory of electromagnetism (see Section 5-2)
could not answer these questions. The answers did not come until
scientists began to discover the structure and properties of atoms.

FIGURE 5-19   R I V U X G

Analyzing the Composition of a Distant Nebula The glowing gas
cloud in this Hubble Space Telescope image lies 210,000 light-years away in
the constellation Tucana (the Toucan). The red light emitted by the nebula has
a wavelength of 656 nm, characteristic of an emission line from hydrogen gas.
Hot stars within the nebula emit high energy, ultra violet photons, which ionize
hydrogen (removing the electron). When the electron returns to the atom, it
jumps towards the center and gives off emission lines. (NASA, ESA, and A. Nota,
STScI/ESA)

CONCEPTCHECK **5-10**

Which one of Kirchhoff's laws describes the Sun's spectrum
in Figure 5-14? If helium gas is heated up enough to give the
spectrum at the top of Figure 5-16, which of Kirchhoff's laws
applies?

CONCEPTCHECK **5-11**

What type of spectra would result from a glowing field of hot,
dense lava as viewed by an orbiting satellite through Earth's
atmosphere?

Answers appear at the end of the chapter.

5-7 An atom consists of a small, dense nucleus surrounded by electrons

By the early twentieth century, it was known that **atoms** contain negatively charged particles called **electrons** and some material with positive charge, but the structure of atoms was still a mystery. The "plum pudding" model popular at that time proposed that the negative electrons were like plums uniformly mixed into a positively charged pudding. However, in 1910 this model was overturned by Ernest Rutherford, a gifted physicist from New Zealand.

> To decode the information in the light from immense objects like stars and galaxies, we must understand the structure of atoms

The Nucleus of an Atom

Rutherford discovered a more accurate model for the structure of an atom, shown in **Figure 5-20**. According to this model, a massive, positively charged **nucleus** at the center of the atom is orbited by tiny, negatively charged electrons. Rutherford concluded that at least 99.98% of the mass of an atom must be concentrated in its nucleus, whose diameter is only about 10^{-14} m. (The diameter of a typical atom is far larger, about 10^{-10} m.)

ANALOGY To appreciate just how tiny the nucleus is, imagine expanding an atom by a factor of 10^{12} to a diameter of 100 meters, about the length of a football field. On this scale, the nucleus would be just a centimeter across (no larger than your thumbnail) in the middle of the field, and the electrons would be orbiting at a distance near the goal posts.

We know today that the nucleus of an atom contains two types of particles, **protons** and **neutrons.** A proton has a positive electric charge, equal in magnitude to that of the negatively charged electron. As its name suggests, a neutron has no electric charge—it is electrically neutral. As an example, the helium atom has two protons and two neutrons. Protons and neutrons are held together in a nucleus by the so-called *strong nuclear force,* whose great strength

BOX 5-5 TOOLS OF THE ASTRONOMER'S TRADE

Atoms, the Periodic Table, and Isotopes

Each different chemical element is made of a specific type of atom. Each specific atom has a characteristic number of protons in its nucleus. For example, a hydrogen atom has 1 proton in its nucleus, an oxygen atom has 8 protons in its nucleus, and so on.

The number of protons in an atom's nucleus is the **atomic number** for that particular element. The chemical elements are most conveniently listed in the form of a **periodic table** (shown in the figure). Elements are arranged in the periodic table in order of increasing atomic number. With only a few exceptions, this sequence also corresponds to increasing average mass of the atoms of the elements. Thus, hydrogen (symbol H), with atomic number 1, is the lightest element. Iron (symbol Fe) has atomic number 26 and is a relatively heavy element.

All the elements listed in a single vertical column of the periodic table have similar chemical properties. For example, the elements in the far right column are all gases under the conditions of temperature and pressure found at Earth's surface, and they are all very reluctant to react chemically with other elements.

In addition to nearly 100 naturally occurring elements, the periodic table includes a number of artificially produced elements. Most of these elements are heavier than uranium (symbol U) and are highly radioactive, which means that they decay into lighter elements within a short time of being created in laboratory experiments. Scientists have succeeded in creating only a few atoms of elements 104 and above.

The number of protons in the nucleus of an atom determines which element that atom is. Nevertheless, the same element may have different numbers of neutrons in its nucleus. For example, oxygen (O) has atomic number 8, so every oxygen nucleus has exactly 8 protons. But oxygen nuclei can have 8, 9, or 10 neutrons. These three slightly different kinds of oxygen are called **isotopes.** The isotope with 8 neutrons is by far the most abundant variety. It is written as ^{16}O, or oxygen-16. The rarer isotopes with 9 and 10 neutrons are designated as ^{17}O and ^{18}O, respectively.

The superscript that precedes the chemical symbol for an element equals the total number of protons and neutrons in a nucleus of that particular isotope. For example, a nucleus of the most common isotope of iron, ^{56}Fe or iron-56, contains a total of 56 protons and neutrons. From the periodic table, the atomic number of iron is 26, so every iron atom has 26 protons in its nucleus. Therefore, the number of neutrons in an iron-56 nucleus is $56 - 26 = 30$. (Most nuclei have more neutrons than protons, especially in the case of the heaviest elements.)

It is extremely difficult to distinguish chemically between the various isotopes of a particular element. Ordinary chemical reactions involve only the electrons that orbit the atom, never the neutrons buried in its nucleus. But there are small differences in the wavelengths of the spectral lines for different isotopes of the same element. For example, the spectral line wavelengths of the hydrogen isotope ^{2}H are about 0.03% greater than the wavelengths for the most common hydrogen isotope, ^{1}H. Thus, different isotopes can be distinguished by careful spectroscopic analysis.

Isotopes are important in astronomy for a variety of reasons. By measuring the relative amounts of different isotopes of a given element in a Moon rock or meteorite, the age of that sample can be determined. The mixture of isotopes left behind when a star explodes into a supernova (see Section 1-3) tells astronomers about the processes that led to the explosion. And knowing the properties of different isotopes of hydrogen and helium is crucial to understanding the nuclear reactions that make the Sun shine. Look for these and other applications of the idea of isotopes in later chapters.

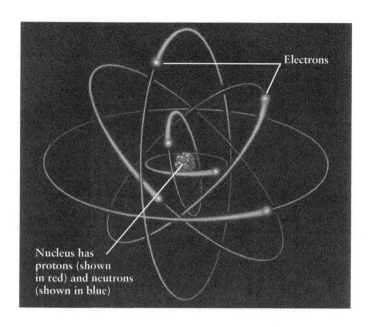

Electrons

Nucleus has protons (shown in red) and neutrons (shown in blue)

FIGURE 5-20
Rutherford's Model of the Atom In this model, electrons orbit the atom's nucleus, which contains most of the atom's mass. The nucleus contains two types of particles, protons and neutrons. The full behavior of electrons orbiting the nucleus is quite complicated and not represented in this simple illustration.

overcomes the electric repulsion between the positively charged protons. A proton and a neutron have almost the same mass, 1.7×10^{-27} kg, and each has about 2000 times as much mass as an electron (9.1×10^{-31} kg). In an ordinary atom there are as many positive protons as there are negative electrons, so the atom has no net electric charge. Because the mass of the electron is so small, the mass of an atom is not much greater than the mass of its nucleus.

While the solar system is held together by gravitational forces, electrons are held in atoms by electrical forces. Opposites attract: The negative charges on the orbiting electrons are attracted to the positive charges on the protons in the nucleus. Box 5-5 describes more about the connection between the structure of

Periodic Table of the Elements

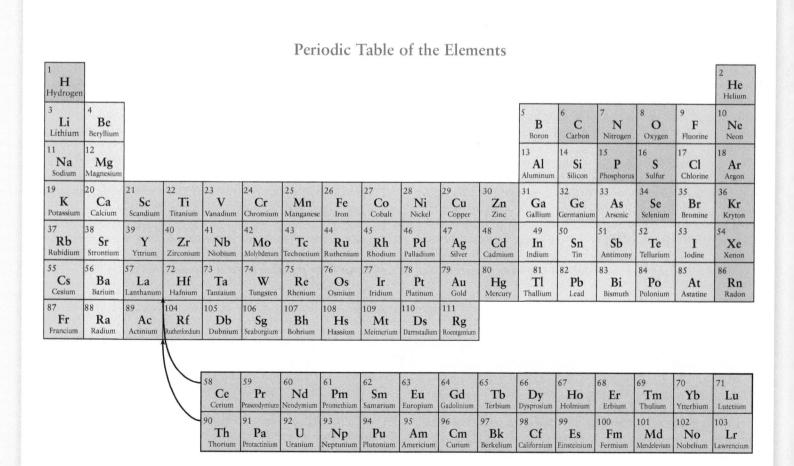

atoms and the chemical and physical properties of substances made of those atoms.

Rutherford's new model clarified the structure of the atom, but did not explain how these tiny particles within the atom give rise to spectral lines. The task of reconciling Rutherford's atomic model with Kirchhoff's laws of spectral analysis was undertaken by the young Danish physicist Niels Bohr, who joined Rutherford's laboratory in 1912.

5-8 Spectral lines are produced when an electron jumps from one energy level to another within an atom

Niels Bohr began his study of the connection between atomic spectra and atomic structure by trying to understand the structure of hydrogen, the simplest and lightest of the elements. (As we discussed in Section 5-6, hydrogen is also the most common element in the universe.) When Bohr was done, he had not only found a way to explain this atom's spectrum but had also found a justification for Kirchhoff's laws in terms of atomic physics.

> Niels Bohr explained spectral lines with a radical new model of the atom

Hydrogen and the Balmer Series

The most common type of hydrogen atom consists of a single electron and a single proton. Hydrogen atoms have a simple visible-light spectrum consisting of a pattern of lines that begins at a wavelength of 656.3 nm and ends at 364.6 nm. The first spectral line is called H_α (H-alpha), the second spectral line is called H_β (H-beta), the third is H_γ (H-gamma), and so forth (**Figure 5-21**). (These are the bright lines in the spectrum of hydrogen shown in Figure 5-16. The fainter lines between these appear when hydrogen atoms form into molecules.) The closer you get to the short-wavelength end of the spectrum at 364.6 nm, the more spectral lines you see.

The regularity in this spectral pattern was described mathematically in 1885 by Johann Jakob Balmer, a Swiss schoolteacher. In Figure 5-21, the spectral lines of hydrogen at visible wavelengths are today called **Balmer lines**, and the entire pattern from H_α onward is called the **Balmer series**. Stars in general, including the Sun, have Balmer absorption lines in their spectra, which shows they have atmospheres that contain hydrogen. Notice that the H_α line in Figure 5-21 is in the red portion of the visible spectrum,

and when seen as an emission line, it accounts for much of the red seen in beautiful nebulae (including Figure 5-19).

Using trial and error, Balmer discovered a formula from which the wavelengths (λ) of hydrogen's spectral lines can be calculated. Balmer's formula is usually written

$$\frac{1}{\lambda} = R\left(\frac{1}{4} - \frac{1}{n^2}\right)$$

In this formula R is the *Rydberg constant* ($R = 1.097 \times 10^7$ m^{-1}), named in honor of the Swedish spectroscopist Johannes Rydberg, and n can be any integer (whole number) greater than 2. The meaning of n will become clear shortly when we discuss electron orbits in the hydrogen atom. To get the wavelength λ_α of the spectral line H_α, you first put $n = 3$ into Balmer's formula:

$$\frac{1}{\lambda_\alpha} = (1.097 \times 10^7 \text{ m}^{-1})\left(\frac{1}{4} - \frac{1}{3^2}\right) = 1.524 \times 10^6 \text{ m}^{-1}$$

Then take the reciprocal:

$$\lambda_\alpha = \frac{1}{1.524 \times 10^6 \text{ m}^{-1}} = 6.563 \times 10^{-7} \text{ m} = 656.3 \text{ nm}$$

To get the wavelength of H_β, use $n = 4$, and to get the wavelength of H_γ use $n = 5$. If you use $n = \infty$ (the symbol ∞ stands for infinity), you get the short-wavelength end of the hydrogen spectrum at 364.6 nm. (Note that 1 divided by infinity equals zero.)

Bohr's Model of Hydrogen

Bohr realized that to fully understand the structure of the hydrogen atom, he had to be able to derive Balmer's formula using the laws of physics. He first made the rather wild assumption that the electron in a hydrogen atom can orbit the nucleus only in certain specific orbits. (This idea was a significant break with the ideas of Newton, in whose mechanics any orbit should be possible.) **Figure 5-22** shows the four smallest of these **Bohr orbits**, labeled by the numbers $n = 1$, $n = 2$, $n = 3$, and so on.

Although confined to one of these allowed orbits while circling the nucleus, an electron can jump from one Bohr orbit to another. For an electron to jump between orbits, the hydrogen atom must gain or lose a specific amount of energy.

An atom must absorb energy for the electron to go from an inner to an outer orbit; an atom must release energy for the electron to go from an outer to an inner orbit.

Shorter wavelength

FIGURE 5-21   R I V U X G

Balmer Lines in the Spectrum of a Star This portion of the spectrum of the star Vega in the constellation Lyra (the Harp) shows eight Balmer lines, from H_α at 656.3 nm through H_θ (H-theta) at 388.9 nm. The series converges at 364.6 nm, slightly to the left of H_θ. Parts of this image were made using ultraviolet radiation, which is indicated by the highlighted U in the wavelength tab. (NOAO)

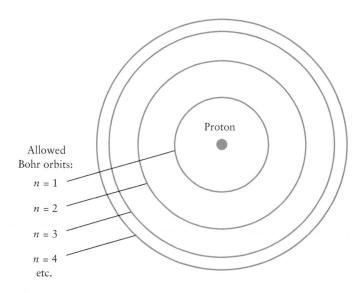

FIGURE 5-22

The Bohr Model of the Hydrogen Atom In this model, an electron circles the hydrogen nucleus (a proton) only in allowed orbits $n = 1, 2, 3$, and so forth. The first four Bohr orbits are shown here. This figure is not drawn to scale; in the Bohr model, the $n = 2, 3$, and 4 orbits are respectively 4, 9, and 16 times larger than the $n = 1$ orbit.

As an example, Figure 5-23 shows an electron jumping between the $n = 2$ and $n = 3$ orbits of a hydrogen atom as the atom absorbs or emits an H_{α} photon. Figure 5-23 shows absorption and emission of a single photon, but in a gas with many atoms, these same jumps produce absorption lines and emission lines.

Bohr's orbital model also explains Kirchhoff's observation (see Figure 5-17) that an atom can show the same spectral lines as either emission or absorption. Here's why: When the electron jumps from one orbit to another, the energy of the photon that is emitted or absorbed equals the difference in energy between these two orbits. This energy difference, and hence the photon energy, is the same whether the jump is from a low orbit to a high orbit (Figure 5-23a) or from the high orbit back to the low one (Figure 5-23b). If two photons have the same energy E, the relationship $E = hc/\lambda$ tells us that they must also have the same wavelength λ. Therefore, if an atom can emit photons of a given energy and wavelength, it can also absorb photons of precisely the same energy and wavelength.

The Bohr picture also helps us visualize how a hot, transparent gas produces an emission line spectrum (Kirchhoff's Law 2). When a gas is heated, its atoms move around rapidly and can collide forcefully with each other. These energetic collisions excite the atoms' electrons into various high orbits. The electrons then cascade back down to the innermost possible orbit, emitting photons whose energies are equal to the energy differences between different Bohr orbits. In this fashion, a hot gas produces an emission line spectrum with a variety of different wavelengths.

To produce an absorption line spectrum (Kirchhoff's Law 3), begin with a relatively cool gas, so that the electrons in most of the atoms are in inner, low-energy orbits. If a beam of light with a continuous spectrum is shone through the gas, most wavelengths will pass through undisturbed. *Only those photons will be absorbed whose energies are just right to excite an electron to an allowed outer orbit.* Hence, only certain wavelengths will be absorbed, and dark lines will appear in the spectrum at those wavelengths.

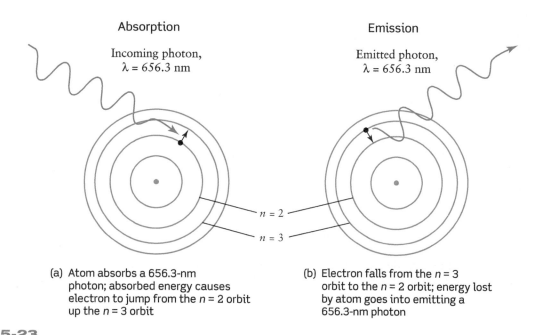

Absorption

Incoming photon,
$\lambda = 656.3$ nm

Emission

Emitted photon,
$\lambda = 656.3$ nm

$n = 2$

$n = 3$

(a) Atom absorbs a 656.3-nm photon; absorbed energy causes electron to jump from the $n = 2$ orbit up the $n = 3$ orbit

(b) Electron falls from the $n = 3$ orbit to the $n = 2$ orbit; energy lost by atom goes into emitting a 656.3-nm photon

ANIMATION 5-3 **FIGURE 5-23**

The Absorption and Emission of an H_{α} Photon This schematic diagram, drawn according to the Bohr model, shows what happens when a hydrogen atom (a) absorbs or (b) emits a photon whose wavelength is 656.3 nm.

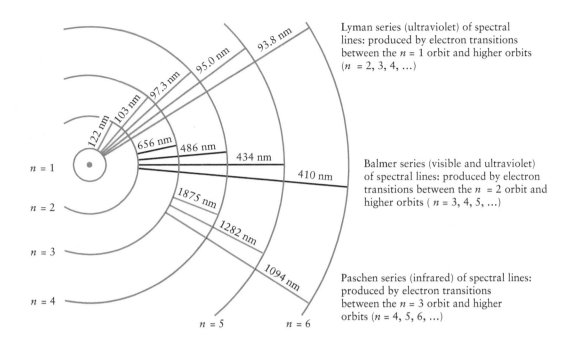

Lyman series (ultraviolet) of spectral lines: produced by electron transitions between the $n = 1$ orbit and higher orbits (n = 2, 3, 4, ...)

Balmer series (visible and ultraviolet) of spectral lines: produced by electron transitions between the $n = 2$ orbit and higher orbits (n = 3, 4, 5, ...)

Paschen series (infrared) of spectral lines: produced by electron transitions between the $n = 3$ orbit and higher orbits (n = 4, 5, 6, ...)

FIGURE 5-24

Electron Transitions in the Hydrogen Atom This diagram shows the photon wavelengths associated with different electron transitions in hydrogen. In each case, the same wavelength occurs whether a photon is emitted (when the electron drops from a high orbit to a low one) or absorbed (when the electron jumps from a low orbit to a high one). The orbits are not shown to scale.

Using his picture of allowed orbits and the formula $E = hc/\lambda$, Bohr was able to prove mathematically that the wavelength λ of the photon emitted or absorbed as an electron jumps between an inner orbit N and an outer orbit n is

Bohr formula for hydrogen wavelengths

$$\frac{1}{\lambda} = R\left(\frac{1}{N^2} - \frac{1}{n^2}\right)$$

N = number of inner orbit

n = number of outer orbit

R = Rydberg constant = 1.097×10^7 m^{-1}

λ = wavelength (in meters) of emitted or absorbed photon

If Bohr let $N = 2$ for the inner orbit in this formula, he got back the formula that Balmer discovered by trial and error. Hence, Bohr deduced the meaning of the Balmer series: All the Balmer lines are produced by electrons jumping between the second Bohr orbit ($N = 2$) and higher orbits ($n = 3, 4, 5$, and so on).

Bohr's formula also correctly predicts the wavelengths of other series of spectral lines that occur at nonvisible wavelengths (Figure 5-24). Using $N = 1$ for the inner orbit gives the **Lyman series,** which is entirely in the ultraviolet. All the spectral lines in this series involve electron transitions between the lowest Bohr orbit and all higher orbits ($n = 2, 3, 4$, and so on). This pattern of spectral lines begins with L$_\alpha$ (Lyman alpha) at 122 nm and converges on L$_\infty$ at 91.2 nm. Using $N = 3$ gives a series of infrared

wavelengths called the **Paschen series.** This series, which involves transitions between the third Bohr orbit and all higher orbits, begins with P$_\alpha$ (Paschen alpha) at 1875 nm and converges on P$_\infty$ at 822 nm. Additional series exist at still longer wavelengths.

Atomic Energy Levels

Today's view of the atom owes much to the Bohr model, but is different in certain ways. The modern picture is based on **quantum mechanics,** a branch of physics developed during the 1920s that deals with photons and subatomic particles. As a result of this work, physicists no longer picture electrons as moving in specific orbits about the nucleus. Instead, electrons are now known to have both particle and wave properties and are said to occupy only certain **energy levels** in the atom.

An extremely useful way of displaying the structure of an atom is with an **energy-level diagram.** Figure 5-25 shows such a diagram for hydrogen. The lowest energy level, called the **ground state,** corresponds to the $n = 1$ Bohr orbit. Higher energy levels, called **excited states,** correspond to successively larger Bohr orbits.

An electron can jump from the ground state ($n = 1$) up to the $n = 2$ level if the atom absorbs a Lyman-alpha photon with a wavelength of 122 nm. Such a photon has energy $E = hc/\lambda = 10.2$ eV (electron volts; see Section 5-5). That is why the energy level of $n = 2$ is shown in Figure 5-25 as having an energy 10.2 eV above that of the ground state (which is usually assigned a value of 0 eV). Similarly, the $n = 3$ level is 12.1 eV above the ground state, and so forth. Electrons can make transitions to higher energy levels by absorbing a photon or in a collision between atoms; they can make transitions to lower energy levels by emitting a photon.

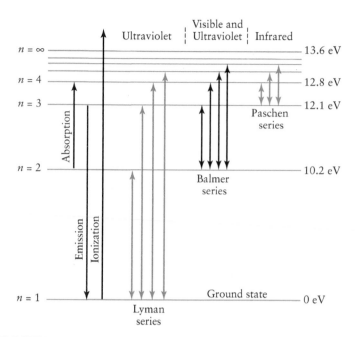

FIGURE 5-25

Energy-Level Diagram of Hydrogen A convenient way to display the structure of the hydrogen atom is in a diagram like this, which shows the allowed energy levels. The diagram shows a number of possible electron jumps, or transitions, between energy levels. An upward transition occurs when the atom absorbs a photon; a downward transition occurs when the atom emits a photon. (Compare with Figure 5-24.)

On the energy-level diagram for hydrogen, the $n = \infty$ level has an energy of 13.6 eV. (This corresponds to an infinitely large "orbit" in the Bohr model.) If the electron is initially in the ground state and the atom absorbs a photon of any energy greater than 13.6 eV, the electron will be removed completely from the atom. This process is called **ionization**. A 13.6-eV photon has a wavelength of 91.2 nm, equal to the shortest wavelength in the ultraviolet Lyman series (L_∞). So any photon with a wavelength of 91.2 nm or less can ionize hydrogen. (Recall that the Planck formula $E = hc/\lambda$ tells us that the shorter the wavelength, the higher the photon energy.)

As an example, the gaseous nebula shown in Figure 5-19 surrounds a cluster of hot stars that produce copious amounts of ultraviolet photons with wavelengths less than 91.2 nm. Hydrogen atoms in the nebula that absorb these photons become ionized and lose their electrons. When the electrons recombine with the nuclei, they cascade down the energy levels to the ground state and emit visible light in the process. This process is what makes the nebula glow. Note that this emission is due to the ionization of the atoms, not the temperature of the gas, and illustrates that Kirchhoff's Law 2 is not the only way to produce an emission line spectrum.

The Spectra of Other Elements

The same basic principles that explain the hydrogen spectrum also apply to the atoms of other elements. Electrons in each kind of atom can only be in certain energy levels, so only photons of certain wavelengths can be emitted or absorbed. Because each kind of atom has its own unique arrangement of electron energy levels, the pattern of spectral lines is likewise unique to that particular type of atom. These patterns are in general much more complicated than for the hydrogen atom.

The properties of energy levels explain the emission line spectra and absorption line spectra of gases. But what about the continuous spectra produced by dense objects like the glowing coils of a toaster or other blackbody object? These objects are made of atoms, so why don't they emit light with an emission line spectrum characteristic of the particular atoms of which they are made?

The reason is directly related to the difference between a gas on the one hand and a liquid or solid on the other. In a gas, atoms are widely separated and can emit photons without interference from other atoms. But in a liquid or a solid, atoms are so close that they almost touch, and thus these atoms interact strongly with each other. In addition, each of the many possible interactions in a liquid or solid have their own energy levels. As a result, the pattern of distinctive bright spectral lines that the atoms would emit in isolation becomes crowded with additional lines forming a continuous spectrum.

ANALOGY Think of atoms as being like tuning forks. If you strike a single tuning fork, it produces a sound wave with a single clear frequency and wavelength, just as an isolated atom emits light of definite wavelengths. But if you shake a box packed full of tuning forks, you will hear a clanging noise that is a mixture of sounds of all different frequencies and wavelengths. This is directly analogous to the continuous spectrum of light emitted by a dense object with closely packed atoms.

With the work of such people as Planck, Einstein, Rutherford, and Bohr, the interchange between astronomy and physics came full circle. Modern physics was born when Newton set out to understand the motions of the planets. Two and a half centuries later, physicists in their laboratories probed the properties of light and the structure of atoms. Their labors had immediate applications in astronomy. Armed with this new understanding of light and matter, astronomers were able to investigate in detail the chemical and physical properties of planets, stars, and galaxies.

CONCEPTCHECK 5-12

Two photons are emitted from a hydrogen atom. One comes from the electron jumping down from the $n = 3$ level to the $n = 1$ level. The other photon comes from an electron transition from $n = 2$ to $n = 1$. Which photon has the highest frequency?

CONCEPTCHECK 5-13

Hydrogen is, by far, the most abundant element in the universe. Are any hydrogen emission lines visible to humans? If so, what color is most common?

Answers appear at the end of the chapter.

5-9 The wavelength of a spectral line is affected by the relative motion between the source and the observer

In addition to telling us about temperature and chemical composition, the spectrum of a planet, star, or galaxy can also reveal something about that object's motion through space. This idea dates from 1842, when Christian Doppler, a professor of mathematics in Prague, pointed out that the observed wavelength of light must be affected by motion.

The Doppler Effect

In **Figure 5-26** a light source is moving from right to left; the circles represent the crests of waves emitted from the moving source at various positions. Each successive wave crest is emitted from a position slightly closer to the observer on the left, so she sees a shorter wavelength—the distance from one crest to the next—than she would if the source were stationary. What is the effect on a full spectrum of light? All the lines in the spectrum of an approaching source are shifted toward the short-wavelength (blue) end of the spectrum. This phenomenon is called a **blueshift**.

> The Doppler effect makes it possible to tell whether astronomical objects are moving toward us or away from us

The source is receding from the observer on the right in Figure 5-26. The wave crests that reach him are stretched apart, so that he sees a longer wavelength than he would if the source were stationary. All the lines in the spectrum of a receding source are shifted toward the longer-wavelength (red) end of the spectrum, producing a **redshift**. In general, the effect of relative motion on wavelength is called the **Doppler effect**. Police radar guns use the Doppler effect to check for cars exceeding the speed limit: The radar gun sends a radio wave toward the car, and measures the wavelength shift of the reflected wave. During reflection of the wave, the car acts like a moving source so the Doppler shift measures the speed of the car.

ANALOGY You have probably noticed a similar Doppler effect for sound waves. When a police car is approaching, the sound waves from its siren have a shorter wavelength and higher frequency than if the siren were at rest, and hence you hear a higher pitch. After the police car passes you and is moving away, you hear a lower pitch from the siren because the sound waves have a longer wavelength and a lower frequency.

Suppose that λ_0 is the wavelength of a particular spectral line from a light source that is not moving. It is the wavelength that you might look up in a textbook or determine in a laboratory experiment for this spectral line. If the source is moving, this particular spectral line is shifted to a different wavelength λ. The size of the wavelength shift is usually written as $\Delta\lambda$, where $\Delta\lambda = \lambda - \lambda_0$. Thus, $\Delta\lambda$ is the difference between the wavelength listed in textbooks and the wavelength that you actually observe in the spectrum of a moving star or galaxy.

Doppler proved that the wavelength shift ($\Delta\lambda$) is governed by the following simple equation:

Doppler shift equation

$$\frac{\Delta\lambda}{\lambda_0} = \frac{v}{c}$$

$\Delta\lambda$ = wavelength shift

λ_0 = wavelength if source is not moving

v = velocity of the source measured along the line of sight

c = speed of light = 3.0×10^5 km/s

CAUTION! The capital Greek letter Δ (delta) is commonly used as a symbol to denote change in the value of a quantity. Thus, $\Delta\lambda$ is the change in the wavelength λ due to the Doppler effect. It is *not* equal to a quantity Δ multiplied by a second quantity λ.

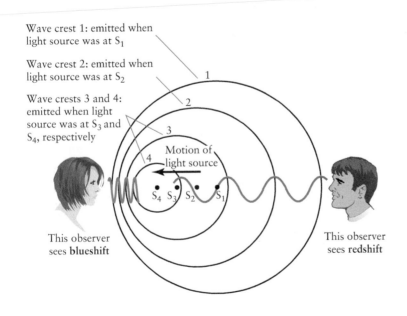

Wave crest 1: emitted when light source was at S_1

Wave crest 2: emitted when light source was at S_2

Wave crests 3 and 4: emitted when light source was at S_3 and S_4, respectively

Motion of light source

This observer sees **blueshift**

This observer sees **redshift**

FIGURE 5-26

The Doppler Effect The wavelength of light is affected by motion between the light source and an observer. The light source shown here is moving, so wave crests 1, 2, etc., emitted when the source was at points S_1, S_2, etc., are crowded together in front of the source but are spread out behind it. Consequently, wavelengths are shortened (blueshifted) if the source is moving toward the observer and lengthened (redshifted) if the source is moving away from the observer. Motion perpendicular to an observer's line of sight does not affect wavelength.

Interpreting the Doppler Effect

The velocity determined from the Doppler effect is called **radial velocity**, because v is the component of the star's motion parallel to our line of sight, or along a "radius" drawn from Earth to the star. Of course, a sizable fraction of a star's motion might be perpendicular to our line of sight, but this sideways or transverse movement across the sky does not cause Doppler shifts. Box 5-6 includes two examples of calculations with radial velocity using the Doppler formula.

CAUTION! The redshifts and blueshifts of stars visible to the naked eye, or even through a small telescope, are only a small fraction of a nanometer. These tiny wavelength changes are far too small to detect visually. (Astronomers were able to detect the tiny Doppler shifts of starlight only after they had developed highly sensitive equipment for measuring wavelengths. This was done around 1890, a half-century after Doppler's original proposal.) So, if you see a star with a red color, it means that the star really is red; it does *not* mean that it is moving rapidly away from us.

BOX 5-6   TOOLS OF THE ASTRONOMER'S TRADE

Applications of the Doppler Effect

Doppler's formula relates the radial velocity of an astronomical object to the wavelength shift of its spectral lines. Here are two examples that show how to use this remarkably powerful formula.

EXAMPLE: As measured in the laboratory, the prominent H_α spectral line of hydrogen has a wavelength $\lambda_0 = 656.285$ nm. But in the spectrum of the star Vega (Figure 5-21), this line has a wavelength $\lambda = 656.255$ nm. What can we conclude about the motion of Vega?

Situation: Our goal is to use the ideas of the Doppler effect to find the velocity of Vega toward or away from Earth.

Tools: We use the Doppler shift formula, $\Delta\lambda/\lambda_0 = v/c$, to determine Vega's velocity v.

Answer: The wavelength shift is

$$\Delta\lambda = \lambda - \lambda_0 = 656.255 \text{ nm} - 656.285 \text{ nm} = -0.030 \text{ nm}$$

The negative value means that we see the light from Vega shifted to shorter wavelengths—that is, there is a blueshift. (Note that the shift is very tiny and can be measured only using specialized equipment.) From the Doppler shift formula, the star's radial velocity is

$$v = c\frac{\Delta\lambda}{\lambda_0} = (3.00 \times 10^5 \text{ km/s}) \left(\frac{-0.030 \text{ nm}}{656.285 \text{ nm}}\right) = -14 \text{ km/s}$$

Review: The minus sign indicates that Vega is coming toward us at 14 km/s. *The star might also have some motion perpendicular to the line from Earth to Vega, but such motion produces no Doppler shift.*

By plotting the motions of stars such as Vega toward and away from us, astronomers have been able to learn how the Milky Way Galaxy (of which our Sun is a part) is rotating. From this knowledge, and aided by Newton's universal law of gravitation (see Section 4-6), they have made the surprising discovery that the Milky Way contains roughly 10 times more matter than had once been thought! The nature of this unseen *dark matter* is still one of the great unsolved mysteries in astronomy.

EXAMPLE: In the radio region of the electromagnetic spectrum, hydrogen atoms emit and absorb photons with a wavelength of 21.12 cm, giving rise to a spectral feature commonly called the *21-centimeter line*. The galaxy NGC 3840 in the constellation Leo (the Lion) is receding from us at a speed of 7370 km/s, or about 2.5% of the speed of light. At what wavelength do we expect to detect the 21-cm line from this galaxy?

Situation: Given the velocity of NGC 3840 away from us, our goal is to find the wavelength as measured on Earth of the 21-centimeter line from this galaxy.

Tools: We use the Doppler shift formula to calculate the wavelength shift $\Delta\lambda$, then use this to find the wavelength λ measured on Earth.

Answer: The wavelength shift is

$$\Delta\lambda = \lambda_0\left(\frac{v}{c}\right) = (21.12 \text{ cm})\left(\frac{7370 \text{ km/s}}{3.00 \times 10^5 \text{ km/s}}\right) = 0.52 \text{ cm}$$

Therefore, we will detect the 21-cm line of hydrogen from this galaxy at a wavelength of

$$\lambda = \lambda_0 + \Delta\lambda = 21.12 \text{ cm} + 0.52 \text{ cm} = 21.64 \text{ cm}$$

Review: The 21-cm line has been redshifted to a longer wavelength because the galaxy is receding from us. In fact, most galaxies are receding from us. This observation is one of the key pieces of evidence that the universe is expanding, and has been doing so since the Big Bang that took place almost 14 billion years ago.

The Doppler effect is an important tool in astronomy because it uncovers basic information about the motion of planets, stars, and galaxies. For example, the rotation of the planet Venus was deduced from the Doppler shift of radar waves reflected from its surface. Small Doppler shifts in the spectrum of sunlight have shown that the entire Sun is vibrating like an immense gong. The back-and-forth Doppler shifting of the spectral lines of certain stars reveals that these stars are being orbited by unseen companions; from this astronomers have discovered planets around other stars and massive objects that may be black holes. Astronomers also use the Doppler effect along with Kepler's third law to measure a galaxy's mass. These are but a few examples of how Doppler's discovery has empowered astronomers in their quest to understand the universe.

In this chapter we have glimpsed how much can be learned by analyzing light from the heavens. To analyze this light, however, it is first necessary to collect as much of it as possible, because most light sources in space are very dim. Collecting the faint light from distant objects is the key purpose of telescopes. In the next chapter we will describe both how telescopes work and how they are used.

CONCEPTCHECK 5-14

How is the spectrum changed when looking at the absorption spectrum from an approaching star as compared to how the star's spectrum would look if the star were stationary?

CALCULATIONCHECK 5-3

How fast and in what direction is a star moving if it has a line that shifts from 486.2 nm to 486.3 nm?

Answers appear at the end of the chapter.

KEY WORDS

Terms preceded by an asterisk (*) are discussed in the Boxes.

KEY IDEAS

The Nature of Light: Light is electromagnetic radiation. It has wavelike properties described by its wavelength λ and frequency ν, and travels through empty space at the constant speed $c = 3.0 \times 10^8$ m/s $= 3.0 \times 10^5$ km/s.

Thermal energy: The thermal energy of a material comes from the kinetic energy of its microscopic particles (atoms and molecules). The hotter a material, the faster its particles move, and the greater its thermal energy.

Blackbody Radiation: A blackbody is a hypothetical object that emits a continuous spectrum; the hotter the object, the greater the emission. Stars closely approximate the behavior of blackbodies, as do other hot, dense objects.

• The intensities of radiation emitted at various wavelengths by a blackbody at a given temperature are shown by a blackbody curve.

• Wien's law states that the dominant wavelength at which a blackbody emits electromagnetic radiation is inversely proportional to the Kelvin temperature of the object: λ_{max} (in meters) = (0.0029 K m)/T.

• The Stefan-Boltzmann law states that a blackbody radiates electromagnetic waves with a total energy flux F directly proportional to the fourth power of the Kelvin temperature T of the object: $F = \sigma T^4$.

Photons: Light is made of particles called photons. Each photon has a wavelength equal to the wavelength of the light that the photons make up.

• Planck's law relates the energy E of a photon to its frequency ν or wavelength λ: $E = h\nu = hc/\lambda$, where h is Planck's constant. The shorter the wavelength a photon has, the higher its frequency and the larger its energy.

Kirchhoff's Laws: Kirchhoff's three laws of spectral analysis describe conditions under which different kinds of spectra are produced.

• A hot, dense object such as a blackbody emits a continuous spectrum covering all wavelengths.

• A hot, transparent gas produces a spectrum that contains bright (emission) lines.

• A cool, transparent gas in front of a light source that itself has a continuous spectrum produces dark (absorption) lines in the continuous spectrum.

Atomic Structure: An atom has a small dense nucleus composed of protons and neutrons. The nucleus is surrounded by electrons that only occupy certain orbits or energy levels.

• When an electron jumps from one energy level to another, it emits or absorbs a photon of corresponding energy (and hence of a specific wavelength).

• The spectral lines of a particular atom correspond to various electron orbital transitions between energy levels in the atom. Each atom has a unique "spectral fingerprint."

The Doppler Shift: The Doppler shift enables us to determine the line-of-sight (radial) velocity of a light source from the displacement of its spectral lines.

• The spectral lines of an approaching light source are shifted toward short wavelengths (a blueshift); the spectral lines of a receding light source are shifted toward long wavelengths (a redshift).

• The size of a wavelength shift is proportional to the radial velocity of the light source relative to the observer.

QUESTIONS

Review Questions

1. When Jupiter is undergoing retrograde motion as seen from Earth, would you expect the eclipses of Jupiter's moons to occur several minutes early, several minutes late, or neither? Explain your answer.

2. Approximately how many times around Earth could a beam of light travel in one second?

3. How long does it take light to travel from the Sun to Earth, a distance of 1.50×10^8 km?

4. *TUTORIAL 5.1* How did Newton show that a prism breaks white light into its component colors but does not add any color to the light?

5. For each of the following wavelengths, state whether it is in the radio, microwave, infrared, visible, ultraviolet, X-ray, or gamma-ray portion of the electromagnetic spectrum. Explain your reasoning. (**a**) 2.6 μm, (**b**) 34 m, (**c**) 0.54 nm, (**d**) 0.0032 nm, (**e**) 0.620 μm, (**f**) 310 nm, (**g**) 0.012 m

6. What is meant by the frequency of light? How is frequency related to wavelength?

7. A cellular phone is actually a radio transmitter and receiver. You receive an incoming call in the form of a radio wave of frequency 880.65 MHz. What is the wavelength (in meters) of this wave?

8. A light source emits infrared radiation at a wavelength of 1150 nm. What is the frequency of this radiation?

9. (**a**) What is a blackbody? (**b**) In what way is a blackbody black? (**c**) If a blackbody is black, how can it emit light? (**d**) If you were to shine a flashlight beam on a perfect blackbody, what would happen to the light?

10. What does it mean for a material to be opaque? Can water molecules form an opaque object in some forms and a transparent substance in others? (*Hint:* Think of clouds.)

11. Describe how the thermal energy changes at a microscopic level as a blacksmith heats a piece of iron. If the iron is then removed from the furnace and glows bright red, what effect does this have on the thermal energy?

12. Why do astronomers find it convenient to use the Kelvin temperature scale in their work rather than the Celsius or Fahrenheit scales?

13. Explain why astronomers are interested in blackbody radiation.

14. In what way is the Sun's spectrum similar to a blackbody spectrum? In what way does it differ from a blackbody spectrum?

15. *TUTORIAL 5.2* Using Wien's law and the Stefan-Boltzmann law, explain the color and intensity changes that are observed as the temperature of a hot, glowing object increases.

16. If you double the Kelvin temperature of a hot piece of steel, how much more energy will it radiate per second?

17. The bright star Bellatrix in the constellation Orion has a surface temperature of 21,500 K. What is its wavelength of maximum emission in nanometers? What color is this star?

18. The bright star Antares in the constellation Scorpius (the Scorpion) emits the greatest intensity of radiation at a wavelength of 853 nm. What is the surface temperature of Antares? What color is this star?

19. (**a**) Describe an experiment in which light behaves like a wave. (**b**) Describe an experiment in which light behaves like a particle.

20. How is the energy of a photon related to its wavelength? What kind of photons carry the most energy? What kind of photons carry the least energy?

21. Which part of the electromagnetic spectrum contains light with a higher frequency: microwaves or radio waves?

22. To emit the same amount of light energy per second, which must emit more photons per second: a source of red light or a source of blue light? Explain your answer.

23. (**a**) Describe the spectrum of hydrogen at visible wavelengths. (**b**) Explain how Bohr's model of the atom accounts for the Balmer lines.

24. Why do different elements display different patterns of lines in their spectra?

25. What is the Doppler effect? Why is it important to astronomers?

26. If you see a blue star, what does its color tell you about how the star is moving through space? Explain your answer.

Advanced Questions

Questions preceded by an asterisk (*) involve topics discussed in the Boxes.

> ### Problem-solving tips and tools
>
> You can find formulas in Box 5-1 for converting between temperature scales. Box 5-2 discusses how a star's radius, luminosity, and surface temperature are related. Box 5-3 shows how to use Planck's law to calculate the energy of a photon. To learn how to do calculations using the Doppler effect, see Box 5-6.

27. Your normal body temperature is 98.6°F. What kind of radiation do you predominantly emit? At what wavelength (in nm) do you emit the most radiation?

28. What is the temperature of the Sun's surface in degrees Fahrenheit?

29. What wavelength of electromagnetic radiation is emitted with greatest intensity by this book? To what region of the electromagnetic spectrum does this wavelength correspond?

30. Can an object convert some of its orbital energy into thermal energy? If yes, describe a context in which this might occur.

31. Black holes are objects whose gravity is so strong that not even an object moving at the speed of light can escape from their surfaces. Hence, black holes do not themselves emit light. But it is possible to detect radiation from material falling *toward* a black hole. Calculations suggest that as this matter falls, it is compressed and heated to temperatures around 10^6 K. Calculate the wavelength of maximum emission for this temperature. In what part of the electromagnetic spectrum does this wavelength lie?

*32. Use the value of the solar constant given in Box 5-2 and the distance from Earth to the Sun to calculate the luminosity of the Sun.

*33. The star Alpha Lupi (the brightest in the constellation Lupus the Wolf) has a surface temperature of 21,600 K. How much more energy is emitted each second from each square meter of the surface of Alpha Lupi than from each square meter of the Sun's surface?

*34. Jupiter's moon Io has an active volcano named Pele whose temperature can be as high as 320°C. (a) What is the wavelength of maximum emission for the volcano at this temperature? In what part of the electromagnetic spectrum is this? (b) The average temperature of Io's surface is –150°C. Compared with a square meter of surface at this temperature, how much more energy is emitted per second from each square meter of Pele's surface?

*35. The bright star Sirius in the constellation of Canis Major (the Large Dog) has a radius of 1.67 $R_\odot$ and a luminosity of 25 $L_\odot$. (a) Use this information to calculate the energy flux at the surface of Sirius. (b) Use your answer in part (a) to calculate the surface temperature of Sirius. How does your answer compare to the value given in Box 5-2?

36. Instruments on board balloons and spacecraft detect 511-keV photons coming from the direction of the center of our Galaxy. (The prefix k means *kilo*, or thousand, so 1 keV = 10^3 eV.) What is the wavelength of these photons? To what part of the electromagnetic spectrum do these photons belong?

37. (a) Calculate the wavelength of P_Δ (P-delta), the fourth wavelength in the Paschen series. (b) Draw a schematic diagram of the hydrogen atom and indicate the electron transition that gives rise to this spectral line. (c) In what part of the electromagnetic spectrum does this wavelength lie?

38. (a) Calculate the wavelength of H_η (H-eta), the spectral line for an electron transition between the $n = 7$ and $n = 2$ orbits of hydrogen. (b) In what part of the electromagnetic spectrum does this wavelength lie? Use this to explain why Figure 5-21 is labeled R I V U X G.

39. Certain interstellar clouds contain a very cold, very thin gas of hydrogen atoms. Ultraviolet radiation with any wavelength shorter than 91.2 nm cannot pass through this gas; instead, it is absorbed. Explain why.

40. (a) Can a hydrogen atom in the ground state absorb an H-alpha (H_α) photon? Explain why or why not. (b) Can a hydrogen atom in the $n = 2$ state absorb a Lyman-alpha (L_α) photon? Explain why or why not.

41. An imaginary atom has just three energy levels: 0 eV, 1 eV, and 3 eV. Draw an energy-level diagram for this atom. Show all possible transitions between these energy levels. For each transition, determine the photon energy and the photon wavelength. Which transitions involve the emission or absorption of visible light?

42. The star cluster NGC 346 and nebula shown in Figure 5-19 are located within the Small Magellanic Cloud (SMC), a small galaxy that orbits our Milky Way Galaxy. The SMC and the stars and gas within it are moving away from us at 158 km/s. At what wavelength does the red H_α line of hydrogen (which causes the color of the nebula) appear in the nebula's spectrum?

43. The wavelength of H_β in the spectrum of the star Megrez in the Big Dipper (part of the constellation Ursa Major the Great Bear) is 486.112 nm. Laboratory measurements demonstrate that the normal wavelength of this spectral line is 486.133 nm. Is the star coming toward us or moving away from us? At what speed?

44. You are given a traffic ticket for going through a red light (wavelength 700 nm). You tell the police officer that because you were approaching the light, the Doppler effect caused a blueshift that made the light appear green (wavelength 500 nm). How fast would you have had to be going for this to be true? Would the speeding ticket be justified? Explain your answer.

Discussion Questions

45. The equation that relates the frequency, wavelength, and speed of a light wave, $\nu = c/\lambda$, can be rewritten as $c = \nu\lambda$. A friend who has studied mathematics but not much astronomy or physics might look at this equation and say: "This equation tells me that the higher the frequency ν, the greater the wave speed c. Since visible light has a higher frequency than radio waves, this means that visible light goes faster than radio waves." How would you respond to your friend?

46. (**a**) If you could see ultraviolet radiation, how might the night sky appear different? Would ordinary objects appear different in the daytime? (**b**) What differences might there be in the appearance of the night sky and in the appearance of ordinary objects in the daytime if you could see infrared radiation?

47. The accompanying visible-light image shows the star cluster NGC 3293 in the constellation Carina (the Ship's Keel). What can you say about the surface temperatures of most of the bright stars in this cluster? In what part of the electromagnetic spectrum do these stars emit most intensely? Are your eyes sensitive to this type of radiation? If not, how is it possible to see these stars at all? There is at least one bright star in this cluster with a distinctly different color from the others; what can you conclude about its surface temperature?

R I [V] U X G   (David Malin/Anglo-Australian Observatory)

48. The human eye is most sensitive over the same wavelength range at which the Sun emits the greatest intensity of radiation. Suppose creatures were to evolve on a planet orbiting a star somewhat hotter than the Sun. To what wavelengths would their vision most likely be sensitive?

49. Why do you suppose that ultraviolet light can cause skin cancer but ordinary visible light does not?

Web/eBook Question

50. Search the World Wide Web for information about rainbows. Why do rainbows form? Why do they appear as circular arcs? Why can you see different colors?

ACTIVITIES

Observing Projects

51. Turn on an electric stove or toaster oven and carefully observe the heating elements as they warm up. Relate your observations to Wien's law and the Stefan-Boltzmann law.

52. Use *Starry Night*™ to examine the celestial objects listed below. Select **Favourites > Explorations > Atlas** to show the whole sky as would be seen from the center of a transparent Earth. Ensure that deep space objects are displayed by selecting **View > Deep Space > Messier Objects** and **View > Deep Space > Bright NGC Objects** from the menu. Also, select **View > Deep Space > Hubble Images** and ensure that this option is turned **off**. To display each of the objects listed below, open the Find pane and then type the name of the selected object in the edit box and press the Enter (Return) key. The object will be centered in the view. Use the zoom controls to adjust your view until you can see the object in detail. For each object, decide whether you think it will have a continuous spectrum, an absorption line spectrum, or an emission line spectrum, and explain your reasoning. The objects to observe are (**a**) the Lagoon Nebula in Sagittarius (with a field of view of about $6° \times 4°$, you can compare and contrast the appearance of the Lagoon Nebula with the Trifid Nebula just to the north of it); (**b**) M31, the great galaxy in the constellation Andromeda (*Hint:* the light coming from this galaxy is the combined light of hundreds of billions of individual stars); (**c**) the Moon (*Hint:* moonlight is simply reflected sunlight).

53. Use the *Starry Night*™ program to compare the brightness of two similarly sized stars in the constellation Auriga. Select **Favourites > Explorations > Auriga**. The two stars Capella and Delta Aurigae are labeled in this view. Select **Preferences** from the **File** menu (Windows) or **Starry Night** menu (Mac) and set **Cursor Tracking (HUD)** options so that **Temperature** and **Radius** are shown in the HUD display. You will notice that these two stars have the same radius but differ in temperature. From these data, which of these stars is intrinsically brighter and by what proportion?

54. Use *Starry Night*™ to examine the temperatures of several relatively nearby stars. Select **Favourites > Explorations > Atlas**. Use the **File** menu (**Starry Night** menu on a Mac) and select **Preferences...** to open the **Preferences** dialog window. Click the box in the top left of this dialog window and choose **Cursor Tracking (HUD)**. Scroll through the **Show** list and click the checkbox next to **Temperature** to turn this option on. Then close the **Preferences** dialog window. Next, open the **Find** pane, click the magnifying glass icon in the edit box at the top of this pane, and select **Star** from the dropdown menu. To locate each of the following stars—Altair; Procyon; Epsilon Indi; Tau Ceti; Epsilon Eridani; Lalande 21185—type the name of the star in the edit box and then press the **Enter** (**Return**) key. Use the **HUD** to find and record the star's temperature. (**a**) Which of the stars have a longer wavelength of maximum emission λ_{max} than the Sun? (**b**) Which of the stars have a shorter λ_{max} than the Sun? (**c**) Which of the stars will have a reddish color?

55. You can use the *Starry Night*™ program to measure the speed of light by observing a particular event, in this case one of Jupiter's moons emerging from the planet's shadow, from two locations separated by a known distance. These two locations are at the north poles of Earth and the planet Mercury, respectively. Open **Favourites > Explorations > Io from Earth**. The view shows Jupiter as seen from the north pole of Earth at 9:12:00 P.M. standard time on September 25, 2010. You will see the label for Io to the left of the planet. Keep the field of view about 11 arcminutes wide. Click the **Play** button and observe Io suddenly brighten as it emerges from Jupiter's shadow. Depending upon your computer monitor, you may need to **Zoom out** slightly so that Io's transition from being invisible to visible occurs instantaneously (at high zoom levels, Io will brighten gradually). Next, use the **Time** controls to **Step time backward** and **forward** in 1-second intervals to determine the time to the nearest second at which Io brightens. Open the **Status** pane and expand the **Time** layer. Record the **Universal Time** for this event. Then open the **Info** pane and be sure that **Io Info** appears at the top of the **Info** pane. Expand the **Position in Space** layer and record the value given for **Distance from Observer**. Now select **Favourites > Explorations > Io from Mercury**. This view once again shows Io labeled to the left of Jupiter but in this instance you are viewing the scene from the north pole of the planet Mercury. With the field of view set to 11 arcminutes wide, click the **Play** button. **Stop** time flow as soon as you see Io suddenly brighten as it emerges from eclipse. Again, it may be necessary to adjust the zoom level to make this event appear instantaneous rather than gradual. Use the time controls to find the time to the nearest second at which Io brightens. Open the **Status** pane and record the **Universal Time** for this event as seen from Mercury. Then open the contextual menu for Io and select **Show Info**. Record the **Distance from Observer** of Io from the **Position in Space** layer. Use your observations to calculate the speed of light. First, calculate the difference in the time between the two observations in seconds. Then calculate the difference in the **Distance from Observer** in AU for each of the locations. Divide the difference in the distance by the difference in time to calculate the speed of light in AU per second. Finally, convert this value to kilometers per second by multiplying the result by 1.496×10^8 (the number of kilometers in 1 AU). How does your result compare to the accepted value of the speed of light

of 2.9979×10^5 kilometers per second? Explain the difference between your calculated value from these observations and the accepted value for the speed of light.

Collaborative Exercise

56. The Doppler effect describes how relative motion impacts wavelength. With a classmate, stand up and demonstrate each of the following: (**a**) a blueshifted source for a stationary observer; (**b**) a stationary source and an observer detecting a redshift; and (**c**) a source and an observer both moving in the same direction, but the observer is detecting a redshift. Create simple sketches to illustrate what you and your classmate did.

ANSWERS

ConceptChecks

ConceptCheck 5-1: The speed of light is incredibly fast, which made it very difficult to measure light's speed except over enormous distances.

ConceptCheck 5-2: As Newton found when passing sunlight through a series of prisms, when one color is isolated from white light, there are no longer any other colors present in the remaining light. As a result, you cannot turn pure red into any other color. White light contains all of the colors, but once any of those colors are absorbed, they cannot be recovered. The red plastic absorbs all the visible wavelengths other than red, so no green can be obtained.

ConceptCheck 5-3: The width of your finger is about 1 cm, which falls in the range of the wavelength of microwaves (1 mm–10 cm).

ConceptCheck 5-4: Because the relationship between wavelength and frequency is $c = \lambda \times f$, as one increases the other decreases (wavelength and frequency are inversely related). Thus, the longest wavelengths have the lowest frequencies and the shortest wavelengths have the highest frequencies.

ConceptCheck 5-5: As shown in Figure 5-12, at every wavelength, including infrared, the higher the temperature of a blackbody, the more energy it emits.

ConceptCheck 5-6: Yes. An object at 0°C still has a temperature of 237 K, and Kelvins are the temperature unit for working with blackbodies. At 273 K, the blackbody would peak in the infrared, although humans cannot see this light.

ConceptCheck 5-7: A star is hotter than our Sun if the star's peak wavelength of blackbody emission is a shorter wavelength than the peak from our Sun.

ConceptCheck 5-8: Yes. If the cooler object is much larger than the warmer object, the cooler object can emit a greater total energy. The energy flux F refers to the energy emitted per second for each square meter of surface area, so even for a cool object, the larger the object, the more energy it radiates. This is why some stars (called red giants) that are much larger than our Sun are actually brighter than our Sun, even though they are cooler.

ConceptCheck 5-9: Photons with longer wavelengths will have lower energy than those with shorter wavelengths because the greater the wavelength, the lower the energy of a photon associated with that wavelength.

ConceptCheck 5-10: While still hot, the Sun's outer atmosphere is cooler than deeper down where most of the Sun's continuous blackbody emission originates. Thus, Figure 5-14 is an absorption line spectrum described by Kirchhoff's third law. Kirchhoff's second law would describe the emission line spectrum from hot helium gas, as in Figure 5-16.

ConceptCheck 5-11: An absorption spectra results when the light from a hot, dense object passes through the cooler, transparent gas of our atmosphere. At visible wavelengths, the atmosphere does not produce a significant absorption spectrum, but at infrared wavelengths (which are also emitted by hot lava), the atmosphere has many absorption lines.

ConceptCheck 5-12: The greater a photon's energy, the higher its frequency. Since the change in energy for the $n = 3$ to $n = 1$ jump is the larger of the two transitions (see Figure 5-25), its emitted photon has the highest frequency. You can also compare the wavelengths of these two transitions in Figure 5-24, and the shorter the wavelength of light, the higher its frequency.

ConceptCheck 5-13: Yes. The most common visible-light hydrogen transition emits red photons of wavelength 656nm. It is an emission line (see Figure 5-24) with a transition from $n = 3$ to $n = 2$ and is referred to as H_α (H-alpha). This single transition accounts for most of the red seen in pretty nebulae such as Figure 5-19.

ConceptCheck 5-14: When the distance between an observer and a source is decreasing, the source's entire spectrum will be shifted toward shorter wavelengths; alternatively, when the distance between an object and a source is decreasing, the emissions lines will be shifted toward longer wavelengths. Thus, the star's absorption lines will be observed at shorter wavelengths.

CalculationCheck

CalculationCheck 5-1: This classic rock station in Santa Barbara, California, emits waves with a frequency of 99.9 MHz, but this can be rounded to 100 MHz. To calculate the wavelength of these radio waves, we rearrange the equation $c = \nu \div \lambda$ to get $\lambda = 3 \times 10^8$ m/s $\div 100 \times 10^6$ Hz = 3.00 m.

CalculationCheck 5-2: Wien's law can be rearranged to calculate the temperature of a star as $T = 0.0029$ K m $\div (5800$ K $\times 2) = 250$ nm, which is ultraviolet.

CalculationCheck 5-3: $\nu = c \times \Delta\lambda \div \lambda_0 = 3 \times 10^5$ km/s $\times (486.3$ nm $- 486.2$ nm$) \div 486.2$ nm = 61.7 km/s, and because it is moving toward longer wavelengths, the distance between the observer and the star must be increasing.

Visible-light image of galaxy M82

R I **V** U X G

Infrared image of galaxy M82

R **I** V U X G

Modern telescopes can view the universe in every range of electromagnetic radiation, although some must be placed above Earth's atmosphere. (NASA/JPL-Caltech/C. Engelbracht, University of Arizona)

Optics and Telescopes

There is literally more to the universe than meets the eye. As seen through a small telescope, the galaxy M82 (shown in the accompanying image) appears as a bright patch that glows with the light of its billions of stars. But when observed with a telescope sensitive to infrared light—with wavelengths longer than your eye can detect—M82 displays an immense halo that extends for tens of thousands of light-years.

The spectrum of this halo reveals it to be composed of tiny dust particles, which are ejected by newly formed stars. The halo's tremendous size shows that new stars are forming in M82 at a far greater rate than within our own Galaxy.

These observations are just one example of the tremendous importance of telescopes to astronomy. Whether a telescope detects visible or nonvisible light, its fundamental purpose is the same—to gather more light than the unaided human eye. Telescopes are used to gather the feeble light from distant objects to make bright, sharp images. Telescopes also produce finely detailed spectra of objects in space. These spectra reveal the chemical compositions of nearby planets as well as of distant galaxies and help astronomers understand the nature and evolution of astronomical objects of all kinds.

As the accompanying images of the galaxy M82 show, telescopes for nonvisible light reveal otherwise hidden aspects of the universe. In addition to infrared telescopes, radio telescopes have mapped out the structure of our Milky Way Galaxy; ultraviolet telescopes have revealed the workings of the Sun's outer atmosphere; and gamma-ray telescopes have detected the most powerful explosions in the universe. The telescope, in all its variations, is by far astronomers' most useful tool for collecting data about the universe.

6-1 A refracting telescope uses a lens to concentrate incoming light at a focus

The **optical telescope**—that is, a telescope designed for use with visible light—was invented in the Netherlands in the early seventeenth century. Soon after, Galileo used one of these new inventions for his groundbreaking astronomical observations (see Section 4-5). These first telescopes used carefully shaped pieces of glass, or **lenses,** to make distant objects appear larger and brighter. Telescopes of this same basic design are used today by many amateur astronomers. To understand telescopes of this kind, we need to understand how lenses work.

> Refracting telescopes gave humans the first close-up views of the Moon and planets

Refraction of Light

Lenses used in telescopes gather and focus light. Larger lenses can gather more light, which is why professional telescopes must be big. Focusing light requires that the light's path be bent. The ability of a lens to bend light is based on a universal fact: *Light travels at a slower speed in a dense substance.*

Thus, although the speed of light *in a vacuum* is 3.00×10^8 m/s, its speed in glass is less than 2×10^8 m/s. Just as a woman's walking pace slows suddenly when she walks from a boardwalk onto a sandy beach, so light slows abruptly as it enters a piece of glass. The same woman easily resumes her original pace when she steps back onto the boardwalk; in the same way, light resumes its original speed upon exiting the glass.

A material through which light travels is called a **medium** (*plural* **media**). As a beam of light passes from one transparent medium into another—say, from air into glass, or from glass back into air—the direction of the light can change. This path-bending phenomenon, called **refraction,** is caused by the change in the speed of light.

ANALOGY Imagine driving a car from a smooth pavement onto a sandy beach (**Figure 6-1a**). If the car approaches the beach head-on, it slows down when it enters the sand but keeps moving straight ahead. If the car approaches the beach at an angle, however, one of its front wheels will be slowed by the sand before the other is, and the car will veer from its original direction. In the same way, a beam of light changes direction when it enters a piece of glass at an angle (Figure 6-1b).

Figure 6-2a shows the refraction of a beam of light passing through a piece of flat glass. As the beam enters the upper surface of the glass, refraction takes place, and the beam is bent to a direction more nearly perpendicular to the surface of the glass. As the beam exits from the glass back into the surrounding air, a second refraction takes place, and the beam bends in the opposite sense.

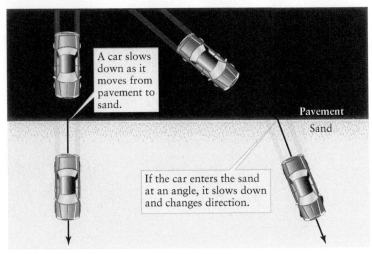

A car slows down as it moves from pavement to sand.

Pavement
Sand

If the car enters the sand at an angle, it slows down and changes direction.

(a) How cars behave

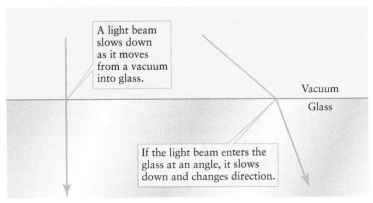

A light beam slows down as it moves from a vacuum into glass.

Vacuum
Glass

If the light beam enters the glass at an angle, it slows down and changes direction.

(b) How light beams behave

FIGURE 6-1

Refraction **(a)** When a car drives from smooth pavement into soft sand, it slows down. If it enters the sand at an angle, the front wheel on one side feels the drag of the sand before the other wheel, causing the car to veer to the side and change direction. **(b)** Similarly, light slows down when it passes from a vacuum into glass and changes direction if it enters the glass at an angle.

(The amount of bending depends on the speed of light in the glass, so different kinds of glass produce slightly different amounts of refraction.) Because the two surfaces of the glass are parallel, the beam emerges from the glass traveling in the same direction at which it entered.

Lenses and Refracting Telescopes

Something more useful happens if the glass is curved into a convex shape (one that is fatter in the middle than at the edges), like the lens in Figure 6-2b. When a beam of incoming light rays passes through the lens, refraction causes all the outgoing rays to converge at a point called the **focus.** If the light rays entering the lens are all parallel, the focus occurs at a special point called the **focal point.** The distance from the lens to the focal point is called the **focal length** of the lens.

The case of incoming parallel light rays, shown in Figure 6-2b, is not merely a theoretical ideal. The stars are so far away that light rays from them are essentially parallel, as Figure 6-3 shows. Consequently, a lens always focuses light from an astronomical object to the focal point. If the object has a very small angular

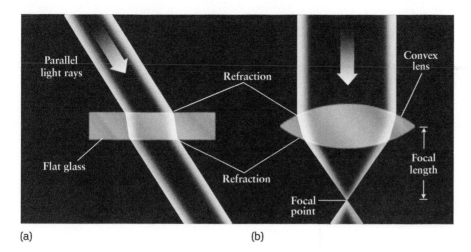

(a)   (b)

FIGURE 6-2

Refraction and Lenses **(a)** Refraction is the change in direction of a light ray when it passes into or out of a transparent medium such as glass. When light rays pass through a flat piece of glass, the two refractions bend the rays in opposite directions. There is no overall change in the direction in which the light travels. **(b)** If the glass is in the shape of a convex lens, parallel light rays converge to a focus at a special point called the focal point. The distance from the lens to the focal point is called the focal length of the lens.

size, like a distant star, all the light entering the lens from that object converges onto the focal point. The resulting image is just a single bright dot.

However, if the object is *extended*—that is, if it has a relatively large angular size, like the Moon or a nearby planet—then light coming from each point on the object is brought to a focus at its own individual point. The result is a clearly focused and extended image that lies in the **focal plane** of the lens (Figure 6-4), which is a plane that includes the focal point. You can use an ordinary magnifying glass in this way to make an image of the Sun on the ground.

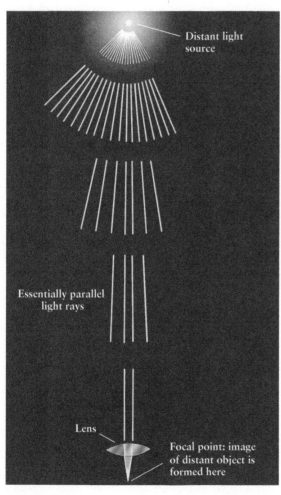

FIGURE 6-3

Light Rays from Distant Objects Are Parallel Light rays travel away in all directions from an ordinary light source. If a lens is located very far from the light source, only a few of the light rays will enter the lens, and these rays will be essentially parallel. This observation is why we drew parallel rays entering the lens in Figure 6-2b.

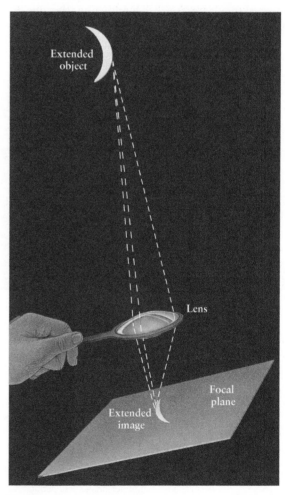

FIGURE 6-4

A Lens Creates an Extended Image of an Extended Object Light coming from each point on an extended object passes through a lens and produces an image of that point. All of these tiny images put together make an extended image of the entire object. The image of a very distant object is formed in a plane called the focal plane. The distance from the lens to the focal plane is called the focal length.

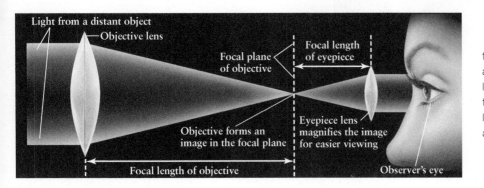

ANIMATION 6-1 **FIGURE 6-5**

A Refracting Telescope A refracting telescope consists of a large-diameter objective lens with a long focal length and a small eyepiece lens of short focal length. The eyepiece lens magnifies the image formed by the objective lens in its focal plane (shown as a dashed line). To take a photograph, the eyepiece is removed and an electronic detector is placed in the focal plane.

To use a lens to make a permanent picture of an astronomical object, you would place an electronic detector in the focal plane—the same type of detector used in digital cameras. In fact, an ordinary digital camera works in a very similar way for photographing objects here on Earth. However, many amateur astronomers want to view the image with their eye, not a camera, and so they add a second lens to magnify the image formed in the focal plane. Such an arrangement of two lenses is called a **refracting telescope,** or **refractor** (Figure 6-5). The large-diameter, long-focal-length lens at the front of the telescope, called the **objective lens,** forms the image; the smaller, shorter-focal-length lens at the rear of the telescope, called the **eyepiece lens,** magnifies the image for the observer.

Light-Gathering Power

In addition to the focal length, the other important dimension of the objective lens of a refractor is its diameter. Compared with a small-diameter lens, a large-diameter lens captures more light, produces

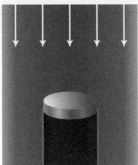

Small-diameter objective lens:
dimmer image, less detail

Large-diameter objective lens:
brighter image, more detail

FIGURE 6-6 R I V U X G

Light-Gathering Power These two photographs of the galaxy M31 in Andromeda were taken using the same exposure time and at the same magnification, but with two different telescopes with objective lenses of different diameters. The right-hand photograph is brighter and shows more detail because it was made using the large-diameter lens, which intercepts more starlight than a small-diameter lens. This same principle applies to telescopes that use curved mirrors rather than lenses to collect light (see Section 6-2). (Association of Universities for Research in Astronomy)

BOX 6-1 TOOLS OF THE ASTRONOMER'S TRADE

Magnification and Light-Gathering Power

The magnification of a telescope is equal to the focal length of the objective divided by the focal length of the eyepiece. Telescopic eyepieces are usually interchangeable, so the magnification of a telescope can be changed by using eyepieces of different focal lengths.

EXAMPLE: A small refracting telescope has an objective of focal length 120 cm. If the eyepiece has a focal length of 4.0 cm, what is the magnification of the telescope?

Situation: We are given the focal lengths of the telescope's objective and eyepiece lenses. Our goal is to calculate the magnification provided by this combination of lenses.

Tools: We use the relationship that the magnification equals the focal length of the objective (120 cm) divided by the focal length of the eyepiece (4.0 cm).

Answer: Using this relationship,

$$\text{Magnification} = \frac{120 \text{ cm}}{4.0 \text{ cm}} = 30 \text{ (usually written as 30×)}$$

Review: A magnification of 30× means that as viewed through this telescope, a large lunar crater that subtends an angle of 1 arcminute to the naked eye will appear to subtend an angle 30 times greater, or 30 arcminutes (one-half of a degree). This magnification makes the details of the crater much easier to see.

If a 2.0-cm-focal-length eyepiece is used instead, the magnification will be (120 cm)/(2.0 cm) = 60×. The shorter the focal length of the eyepiece, the greater the magnification.

The light-gathering power of a telescope depends on the diameter of the objective lens; it does not depend on the focal length. The light-gathering power is proportional to the square of the diameter. As an example, a fully dark adapted human eye has a pupil diameter of about 5 mm. By comparison, a small telescope whose objective lens is 5 cm in diameter has 10 times the diameter and $10^2 = 100$ times the light-gathering power

of the eye. (Recall that there are 10 mm in 1 cm.) Hence, this telescope allows you to see objects 100 times fainter than you can see without a telescope.

EXAMPLE: The same relationships apply to reflecting telescopes, discussed in Section 6-2. Each of the two Keck telescopes on Mauna Kea in Hawaii (discussed in Section 6-3; see Figure 6-16) uses a concave mirror 10 m in diameter to bring starlight to a focus. How many times greater is the light-gathering power of either Keck telescope compared to that of the human eye?

Situation: We are given the diameters of the pupil of the human eye (5 mm) and of the mirror of either Keck telescope (10 m). Our goal is to compare the light-gathering powers of these two optical instruments.

Tools: We use the relationship that light-gathering power is proportional to the square of the diameter of the area that gathers light. Hence, the *ratio* of the light-gathering powers is equal to the square of the ratio of the diameters.

Answer: We first calculate the ratio of the diameter of the Keck mirror to the diameter of the pupil. To determine this ratio, we must first express both diameters in the same units. Because there are 1000 mm in 1 meter, the diameter of the Keck mirror can be expressed as

$$10 \text{ m} \times \frac{1000 \text{ mm}}{1 \text{ m}} = 10,000 \text{ mm}$$

Thus, the light-gathering power of either of the Keck telescopes is greater than that of the human eye by a factor of

$$\frac{(10,000 \text{ mm})^2}{(5 \text{ mm})^2} = (2000)^2 = 4 \times 10^6 = 4,000,000$$

Review: Either Keck telescope can gather *4 million* times as much light as a human eye. When it comes to light-gathering power, the bigger the telescope, the better!

---

brighter images, and allows astronomers to detect fainter objects. (For the same reason, the iris of your eye opens when you go into a darkened room to allow you to see dimly lit objects.)

The **light-gathering power** of a telescope is directly proportional to the area of the objective lens, which in turn is proportional to the square of the lens diameter (Figure 6-6). Thus, if you double the diameter of the lens, the light-gathering power increases by a factor of $2^2 = 2 \times 2 = 4$. Box 6-1 describes how to compare the light-gathering power of different telescopes.

Because light-gathering power is so important for seeing faint objects, the lens diameter is almost always given when describing a telescope. For example, the Lick telescope on Mount Hamilton in California is a 90-cm refractor, which means that it is a refracting telescope whose objective lens is 90 cm in diameter. By comparison, Galileo's telescope of 1610 was a 3-cm refractor. The Lick telescope has an objective lens 30 times larger in diameter, and so has 30^2 or $30 \times 30 = 900$ times the light-gathering power of Galileo's instrument.

CONCEPTCHECK 6-1

If someone says they are using an 8-inch telescope, which dimension of the telescope—length or tube diameter—are they most likely referring to?

CONCEPTCHECK 6-2

If a thick lens is able to bend light more than a thin lens, which lens has a greater focal length?

Answers appear at the end of the chapter.

Magnification

In addition to their light-gathering power, telescopes are useful because they magnify distant objects. As an example, the angular diameter of the Moon as viewed with the naked eye is about 0.5°. But when Galileo viewed the Moon through his telescope, its apparent angular diameter was 10°, large enough so that he could identify craters and mountain ranges. The **magnification,** or **magnifying power,** of a telescope is the ratio of an object's angular diameter seen through the telescope to its naked-eye angular diameter. Thus, the magnification of Galileo's telescope was 10°/0.5° = 20 times, usually written as 20×.

The magnification of a refracting telescope depends on the focal lengths of both of its lenses:

$$\text{Magnification} = \frac{\text{focal length of objective lens}}{\text{focal length of eyepiece lens}}$$

This formula shows that using a long-focal-length objective lens with a short-focal-length eyepiece gives a large magnification. Box 6-1 illustrates how this formula is used.

CAUTION! Many people think that the primary purpose of a telescope is to magnify images. But in fact magnification is not the most important aspect of a telescope. The reason is that there is a limit to how sharp any astronomical image can be, due either to the blurring caused by Earth's atmosphere or to fundamental limitations imposed by the nature of light itself. (We will describe these effects in more detail in Section 6-3.) Magnifying a blurred image may make it look bigger but will not make it any clearer. Thus, beyond a certain point, there is nothing to be gained by further magnification. Astronomers place more priority on the light-gathering power of a telescope than in its magnification. Greater light-gathering power means brighter images, which makes it easier to see faint details.

Disadvantages of Refracting Telescopes

If you were to build a telescope like that in Figure 6-5 using only the instructions given so far, you would probably be disappointed with the results. The problem is that a lens bends different colors of light through different angles, just as a prism does (recall Figure 5-3). As a result, different colors do not focus at the same point, and stars viewed through a telescope that uses a simple lens are surrounded by fuzzy, rainbow-colored halos. **Figure 6-7a** shows this optical defect, called **chromatic aberration.**

One way to correct for chromatic aberration is to use an objective lens that is not just a single piece of glass. Different types of glass can be manufactured by adding small amounts of chemicals to the glass when it is molten. Because of the different properties of these added chemicals, the speed of light varies slightly from one kind of glass to another, and the refractive properties vary as well. If a thin lens is mounted just behind the main objective lens of a telescope, as shown in Figure 6-7b, and if the telescope

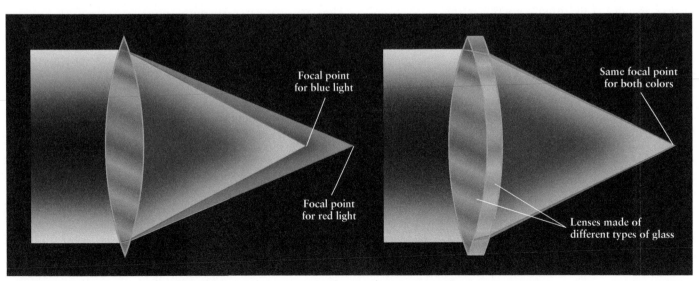

(a) The problem: chromatic aberration

(b) The solution: use two lenses

FIGURE 6-7

Chromatic Aberration **(a)** A single lens suffers from a defect called chromatic aberration, in which different colors of light are brought to a focus at different distances from the lens. While the aberration occurs for all colors, only red and blue are shown here. **(b)** This problem can be corrected by adding a second lens made from a different kind of glass.

Objective lens
(inside tube)

Light rays
enter here

Motor drive rotates telescope
to follow the motion of the
celestial sphere

Mount allows telescope
to be pointed to any
part of the sky

Floor can be raised or lowered to
keep the eyepiece within reach

(a)   Eyepiece   Astronomer

Objective lens made of two different
types of glass (see Figure 6-7b)

(b)

FIGURE 6-8   R I ☑ U X G

A Large Refracting Telescope **(a)** This giant refractor, built in 1897, is housed at Yerkes Observatory near Chicago. The telescope tube is 19.5 m (64 ft) long; it has to be this long because the focal length of the objective is just under 19.5 m (see Figure 6-5). As Earth rotates, the motor drive rotates the telescope in the opposite direction in order to keep the object being studied within the telescope's field of view. **(b)** This historical photograph shows the astronomer George van Biesbrock with the objective lens of the Yerkes refractor. This lens, the largest refracting lens still in use, is 102 cm (40 in.) in diameter. (a: Roger Ressmeyer/Corbis; b: Yerkes Observatory)

designer carefully chooses two different kinds of glass for these two lenses, different colors of light can be brought to a focus at the same point.

Chromatic aberration is only the most severe of a host of optical problems that must be solved in designing a high-quality refracting telescope. Master opticians of the nineteenth century devoted their careers to solving these problems, and several magnificent refractors were constructed in the late 1800s (Figure 6-8).

Unfortunately, there are several negative aspects of refractors that even the finest optician cannot overcome:

1. Because faint light must readily pass through the objective lens, the glass from which the lens is made must be totally free of defects, such as the bubbles that frequently form when molten glass is poured into a mold. Such defect-free glass is extremely expensive.

2. Glass is opaque to certain kinds of light. Ultraviolet light is absorbed almost completely, and even visible light is dimmed substantially as it passes through the thick slab of glass that makes up the objective lens.

3. It is impossible to produce a large lens that is entirely free of chromatic aberration.

4. Because a lens can be supported only around its edges, a large lens tends to sag and distort under its own weight as it tracks objects through the night. This distortion has negative effects on the image clarity.

For these reasons and more, few major refractors have been built since the beginning of the twentieth century. Instead, astronomers have avoided all of the limitations of refractors by building telescopes that use mirrors instead of lenses to form images.

CONCEPTCHECK **6-3**

How do the eyepieces with the largest focal length affect a telescope's overall magnification?

Answer appears at the end of the chapter.

6-2 A reflecting telescope uses a mirror to concentrate incoming light at a focus

Almost all modern telescopes form an image using the principle of **reflection.** To understand reflection, imagine drawing a dashed line perpendicular to the surface of a flat mirror at the point where a light ray strikes the mirror (Figure 6-9a). The angle *i* between the *incident* (arriving) light ray and the perpendicular is always equal to the angle *r* between the *reflected* ray and the perpendicular.

> All of the largest professional telescopes—and most amateur telescopes—are reflecting telescopes

In 1663, the Scottish mathematician James Gregory first proposed a telescope using reflection from a concave mirror—one that is fatter at the edges than at the middle. Such a mirror makes parallel light rays converge to a focus (Figure 6-9b). The distance between the reflecting surface and the focus is the focal length of the mirror. A telescope that uses a curved mirror to make an image of a distant object is called a **reflecting telescope,** or **reflector.** Using terminology similar to that used for refractors, the mirror that forms the image is called the **objective mirror** or **primary mirror.**

To make a reflector mirror, an optician grinds and polishes a large slab of glass into the appropriate concave shape. The glass is then coated with silver, aluminum, or a similar highly reflective

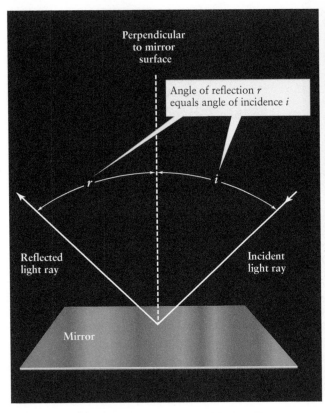

(a)

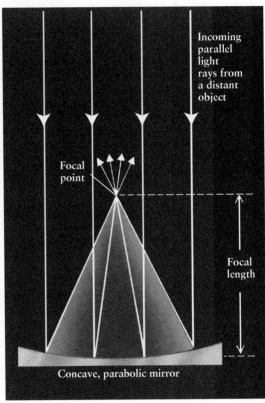

(b)

FIGURE 6-9

Reflection (a) The angle at which a beam of light approaches a mirror, called the angle of incidence (i), is always equal to the angle at which the beam is reflected from the mirror, called the angle of reflection (r).

(b) A concave mirror causes parallel light rays to converge to a focus at the focal point. The distance between the mirror and the focal point is the focal length of the mirror.

substance. Because light reflects off the surface of the coated glass rather than passing through it, defects within the glass—which would have very negative consequences for the objective lens of a refracting telescope—have no effect on the optical quality of a reflecting telescope.

Another advantage of reflector mirrors is that they do not suffer from the chromatic aberration that plagues refractors. This is because reflection is not affected by the wavelength of the incoming light (only the angle it hits the mirror), so all wavelengths are reflected to the same focus. (A small amount of chromatic aberration may arise if the image is viewed using an eyepiece lens.) Furthermore, the mirror can be fully supported by braces on its back, so that a large, heavy mirror can be mounted and well-supported without its shape warping much from gravity.

Designs for Reflecting Telescopes

Although a reflecting telescope has many advantages over a refractor, the arrangement shown in Figure 6-9a is not ideal. One problem is that the focal point is in front of the objective mirror. If you try to view the image formed at the focal point, your head will block part or all of the light from reaching the mirror.

To get around this problem, in 1668 Isaac Newton simply placed a small, flat mirror at a 45° angle in front of the focal point, as sketched in Figure 6-10a. This secondary mirror deflects the light

rays to one side, where Newton placed an eyepiece lens to magnify the image. A reflecting telescope with this optical design is appropriately called a **Newtonian reflector** (Figure 6-10b). The magnifying power of a Newtonian reflector is calculated in the same way as for a refractor: The focal length of the objective mirror is divided by the focal length of the eyepiece (see Box 6-1).

Later astronomers modified Newton's original design. The objective mirrors of some modern reflectors are so large that an astronomer could actually sit in an "observing cage" at the undeflected focal point directly in front of the objective mirror. (In practice, riding in this cage on a winter's night is a remarkably cold and uncomfortable experience.) This arrangement is called a **prime focus** (Figure 6-11a). It usually provides the highest-quality image, because there is no need for a secondary mirror (which might have imperfections).

Another popular optical design, called a **Cassegrain focus** after the French contemporary of Newton who first proposed it, also has a convenient, accessible focal point. A hole is drilled directly through the center of the primary mirror, and a convex secondary mirror placed in front of the original focal point reflects the light rays back through the hole (Figure 6-11b).

A fourth design is useful when there is optical equipment too heavy or bulky to mount directly on the telescope. Instead, a series of mirrors channels the light rays away from the telescope to a remote focal point where the equipment is located. This design is

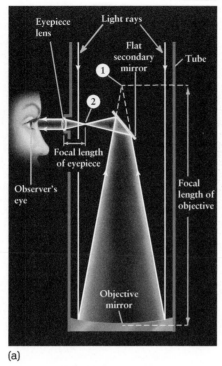

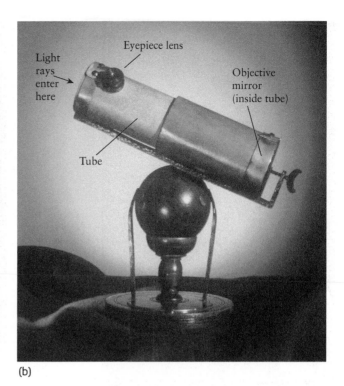

(a)   (b)

FIGURE 6-10   R I ☑ U X G

A Newtonian Telescope **(a)** In a Newtonian telescope, the image made by the objective is moved from point 1 to point 2 by means of a flat mirror called the secondary. An eyepiece magnifies this image, just as for a refracting

telescope (Figure 6-5). **(b)** This is a replica of a Newtonian telescope built by Isaac Newton in 1672. The objective mirror is 3 cm (1.3 inches) in diameter and the magnification is 40×. (Royal Greenwich Observatory/Science Photo Library)

Incoming parallel rays from a distant object

Secondary mirror

Secondary mirror

Focal point outside telescope tube

(a) Prime focus   **(b) Cassegrain focus**   **(c) Coudé focus**

called a **coudé focus,** from a French word meaning "bent like an elbow" (Figure 6-11c).

CAUTION! You might think that the secondary mirror in the Newtonian design and the Cassegrain and coudé designs shown in Figure 6-11 would cause a black spot or hole in the center of the telescope image. But this spot does not form. The reason is that light from every part of the object lands on every part of the primary, objective mirror. Hence, any portion of the mirror can itself produce an image of the distant object, as Figure 6-12 shows. The only effect of the secondary mirror is that it prevents part of the light from reaching the objective mirror, which reduces somewhat the light-gathering power of the telescope.

FIGURE 6-11

ANIMATION 6-2

Designs for Reflecting Telescopes Three common optical designs for reflecting telescopes are shown here. **(a)** Prime focus is used only on some large telescopes; an observer or instrument is placed directly at the focal point, within the barrel of the telescope. **(b)** The Cassegrain focus is used on reflecting telescopes of all sizes, from 90-mm (3.5-in.) reflectors used by amateur astronomers to giant research telescopes on Mauna Kea (see Figure 6-14). **(c)** The coudé focus is useful when large and heavy optical apparatus is to be used at the focal point. Light reflects off the objective mirror to a secondary mirror, then back down to a third, angled mirror. With this arrangement, the heavy apparatus at the focal point does not have to move when the telescope is repositioned.

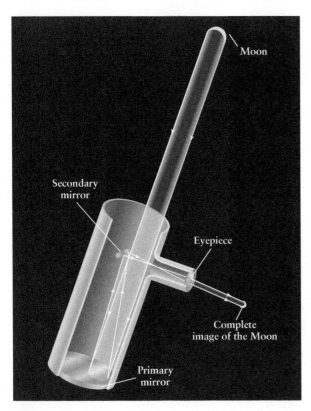

FIGURE 6-12

The Secondary Mirror Does Not Cause a Hole in the Image
This illustration shows how even a small portion of the primary (objective) mirror of a reflecting telescope can make a complete image of the Moon. Thus, the secondary mirror does not cause a black spot or hole in the image. (It does, however, make the image a bit dimmer by reducing the total amount of light that reaches the primary mirror.)

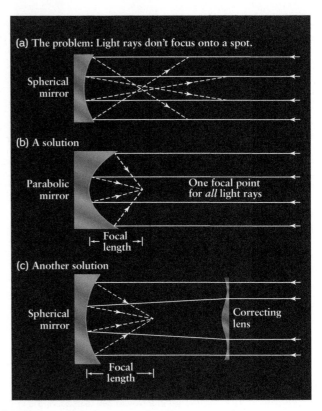

FIGURE 6-13

Spherical Aberration **(a)** Different parts of a spherically concave mirror reflect light to slightly different points. This effect, called spherical aberration, causes image blurring. This difficulty can be corrected by either **(b)** using a parabolic mirror or **(c)** using a correcting lens in front of the mirror.

Spherical Aberration

A reflecting telescope must be designed to minimize a defect called **spherical aberration** (Figure 6-13). At issue is the precise shape of a mirror's concave surface. A spherical surface is easy to grind and polish, but different parts of a spherical mirror focus light onto slightly different spots (Figure 6-13a). This results in a fuzzy image.

One common way to eliminate spherical aberration is to polish the mirror's surface to a parabolic shape, because a parabola reflects parallel light rays to a common focus (Figure 6-13b). Unfortunately, the astronomer then no longer has a wide-angle view. Furthermore, unlike spherical mirrors, parabolic mirrors suffer from a defect called **coma**, wherein star images far from the center of the field of view are elongated to look like tiny teardrops. A different approach is to use a spherical mirror, thus minimizing coma, and to place a thin correcting lens at the front of the telescope to eliminate spherical aberration (Figure 6-13c). This approach is only used on relatively small reflecting telescopes for amateur astronomers.

The Largest Reflectors

There are over a dozen optical reflectors in operation with primary mirrors between 8 meters (26.2 feet) and 11 meters (36.1 feet) in diameter. Figure 6-14a shows the objective mirror of one of the four Very Large Telescope (VLT) units in Chile, and Figure 6-14b shows the objective and secondary mirrors of the Gemini North telescope in Hawaii. A near-twin of Gemini North, called Gemini South, is in Cerro Pachón, Chile. These twins allow astronomers to observe both the northern and southern parts of the celestial sphere with essentially the same state-of-the-art instrument. Two other "twins" are the side-by-side 8.4-m objective mirrors of the Large Binocular Telescope in Arizona. Combining the light from these two mirrors gives double the light-gathering power, equivalent to a single 11.8-m mirror.

Several other reflectors around the world have objective mirrors between 3 and 6 meters in diameter, and dozens of smaller but still powerful telescopes have mirrors in the range of 1 to 3 meters. There are thousands of professional astronomers, each of whom has several ongoing research projects, and thus the demand for all of these telescopes is high. On any night of the year, nearly every research telescope in the world is being used to explore the universe.

CONCEPTCHECK 6-4

Does the angle of light reflected off a mirrored surface depend on wavelength? How does the answer to this question help or hurt a reflecting telescope?

Answer appears at the end of the chapter.

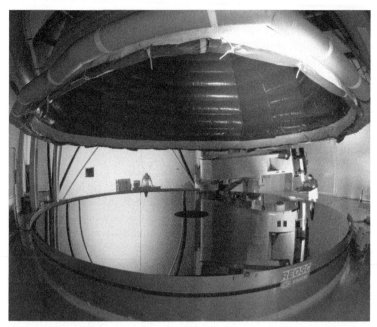

(a) A large objective mirror

FIGURE 6-14   R I V U X G

Reflecting Telescopes **(a)** This photograph shows technicians preparing an objective mirror 8.2 meters in diameter for the European Southern Observatory in Chile. The mirror was ground to a curved shape with a remarkable precision of 8.5 nanometers. **(b)** This view of the Gemini North telescope shows its 8.1-meter objective mirror (1). Light incident on this mirror is reflected toward the 1.0-meter secondary mirror (2), then through the hole in the objective mirror (3) to the Cassegrain focus (see Figure 6-11b). (a: SAGEM; b: NOAO/AURA/NSF)

(b) A large Cassegrain telescope

6-3 Telescope images are degraded by the blurring effects of the atmosphere and by light pollution

In addition to providing a brighter image, a large telescope also helps achieve a second major goal: It produces star images that are sharp and crisp. A quantity called **angular resolution** gauges how well fine details can be seen. Poor angular resolution causes star images to be fuzzy and blurred together.

To determine the angular resolution of a telescope, pick out two adjacent stars whose separate images are just barely discernible (Figure 6-15). The angle θ (the Greek letter theta) between these stars is the telescope's angular resolution; the *smaller* that angle, the finer the details that can be seen and the sharper the image.

When you are asked to read the letters on an eye chart, what's being measured is the angular resolution of your eye. If you have 20/20 vision, the angular resolution θ of your eye is about 1 arcminute, or 60 arcseconds. (You may want to review the definitions of angles and their units in Section 1-5.) Hence, with the naked eye it is impossible to distinguish two stars less than 1 arcminute apart or to see details on the Moon with an angular size smaller than this. All the planets have angular sizes (as seen from Earth) of

1 arcminute or less, which is why they appear as featureless points of light to the naked eye.

Limits to Angular Resolution

One factor limiting angular resolution is **diffraction,** which is the tendency of light waves to spread out when they are confined to a small area like the lens or mirror of a telescope. (A rough analogy is the way water exiting a garden hose sprays out in a wider angle when you cover part of the end of the hose with your thumb.) As a result of diffraction, a narrow beam of light tends to spread out within a telescope's optics, thus blurring the image. If diffraction were the only limit, the angular resolution of a telescope would be given by the formula

Diffraction-limited angular resolution

$$\theta = 2.5 \times 10^5 \frac{\lambda}{D}$$

θ = diffraction-limited angular resolution of a telescope, in arcseconds

λ = wavelength of light, in meters

D = diameter of telescope objective, in meters

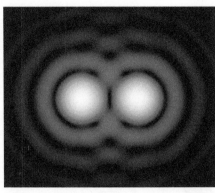

Two light sources with angular separation greater than angular resolution of telescope: two sources easily distinguished

(a)

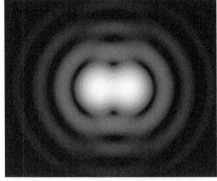

Light sources moved closer so that angular separation equals angular resolution of telescope: just barely possible to tell that there are two sources

(b)

FIGURE 6-15 R I U X G

Angular Resolution The angular resolution of a telescope indicates the sharpness of the telescope's images. **(a)** This telescope view shows two sources of light whose angular separation is greater than the angular resolution. **(b)** The light sources have been moved together so that their angular separation is equal to the angular resolution. If the sources were moved any closer together, the telescope image would show them as a single source. (Courtesy of John D. Monnier)

For a given wavelength of light, using a telescope with an objective of *larger* diameter D *reduces* the amount of diffraction and makes the angular resolution θ *smaller* (and hence better). For example, with red light with wavelength 640 nm, or 6.4×10^{-7} m, the diffraction-limited resolution of an 8-meter telescope (see Figure 6-15) would be

$$\theta = (2.5 \times 10^5)\frac{6.4 \times 10^{-7}\,\text{m}}{8\,\text{m}} = 0.02 \text{ arcsec}$$

In practice, however, ordinary optical telescopes cannot achieve such fine angular resolution. The problem is that turbulence in the air causes star images to jiggle around and twinkle. Even through the largest telescopes, a star still looks like a tiny blob rather than a pinpoint of light. A measure of the limit that atmospheric turbulence places on a telescope's resolution is called the **seeing disk.** This disk is the angular diameter of a star's image broadened by turbulence. The size of the seeing disk varies from one observatory site to another and from one night to another. At the observatories on Kitt Peak in Arizona and Cerro Tololo in Chile, the seeing disk is typically around 1 arcsec. Some of the very best conditions in the world can be found at the observatories atop Mauna Kea in Hawaii, where the seeing disk is often as small as 0.5 arcsec. These great conditions are one reason why so many telescopes have been built there (Figure 6-16).

CONCEPTCHECK 6-5

A small-diameter telescope collects less light than a larger-diameter telescope. If a telescope with a larger-diameter lens or mirror is used for observations of dim, barely illuminated features on Jupiter, does the larger diameter worsen or improve the image resolution compared to the smaller telescope?

Answer appears at the end of the chapter.

FIGURE 6-16 R I V U X G

The Telescopes of Mauna Kea The summit of Mauna Kea—an extinct Hawaiian volcano that reaches more than 4100 m (13,400 ft) above the waters of the Pacific— has nighttime skies that are unusually clear, still, and dark. To take advantage of these superb viewing conditions, Mauna Kea has become the home of many powerful telescopes. (AP Photo/University of Hawaii Institute of Astronomy, Richard Wainscoat)

Active Optics and Adaptive Optics

In many cases the angular resolution of a telescope is even worse than the limit imposed by the seeing disk. This occurs if the objective mirror deforms even slightly due to variations in air temperature or slight gravitational sagging of the telescope mount. To combat this, many large telescopes are equipped with an **active optics** system. Such a system slowly adjusts the mirror shape every few seconds to help keep the telescope in optimum focus and properly aimed at its target. These adjustments can be determined once during a testing phase for each orientation of the telescope and applied for all future observations. However, the adjustments discussed next are more challenging because they change rapidly throughout each observation.

Changing the mirror shape is also at the heart of a more refined technique called **adaptive optics**. The goal of this technique is to compensate for

> Adaptive optics produces sharper images by "undoing" atmospheric turbulence

atmospheric turbulence, so that the angular resolution can be smaller than the size of the seeing disk and can even approach the theoretical limit set by diffraction. Turbulence causes the image of a star to "dance" around erratically. While this turbulence produces twinkling stars seen with the naked eye, it produces blurry images in large telescopes. In an adaptive optics system, sensors monitor this dancing motion 10 to 100 times per second, and a powerful computer rapidly calculates the mirror shape needed to compensate. Fast-acting mechanical devices called *actuators* then deform the mirror accordingly, to produce a sharp, focused image. In some adaptive optics systems, the actuators deform a small secondary mirror rather than the large objective mirror.

One difficulty with adaptive optics is that a fairly bright star must be in or near the field of the telescope's view to serve as a "calibration target" for the sensors that track atmospheric turbulence. This is seldom the case, since the field of view of most telescopes is rather narrow, lowering the chances of viewing a sufficiently bright star. With no lack of creative problem-solving, astronomers get around this limitation by shining a laser beam toward a spot in the sky near the object to be observed (Figure 6-17). The laser beam causes atoms in Earth's upper atmosphere to glow, making an artificial "star." The light that comes down to Earth from this "star" travels through the same part of our atmosphere as the light from the object being observed, so its image in the telescope will "dance" around in the same erratic way as the image of a real star. By observing how the changing atmosphere distorts the artificial star, computers apply opposing adjustments to cancel the atmosphere's effects on the real star.

Figure 6-18 shows the dramatic improvement in angular resolution possible with adaptive optics. Images made with adaptive optics are nearly as sharp as if the telescope were in the vacuum of space, where there is no atmospheric distortion whatsoever and the only limit on angular resolution is diffraction. A number of large telescopes are now being used with adaptive optics systems.

CAUTION! The images in Figure 6-18 are **false color** images: They do not represent the true color of the stars shown. False color is often used when the image is made using wavelengths that the eye cannot detect, as with the infrared images in Figure 6-18. A different use of false color is to indicate the relative brightness of different parts of the image, as in the infrared image of a person in Figure 5-10. Throughout this book, we'll always point out when false color is used in an image.

FIGURE 6-17   R I V U X G

Creating an Artificial "Star" A laser beam shines upward from Yepun, an 8.2-meter telescope at the European Southern Observatory in the Atacama Desert of Chile. (Figure 6-14a shows the objective mirror for this telescope.) The beam strikes sodium atoms that lie about 90 km (56 miles) above Earth's surface, causing them to glow and make an artificial "star." Tracking the twinkling of this "star" makes it possible to undo the effects of atmospheric turbulence on telescope images. (European Southern Observatory)

Interferometry

Several large observatories are developing a technique called **interferometry** that promises to further improve the angular resolution of telescopes. The idea is to have two widely separated telescopes observe the same object simultaneously, then use fiber optic cables to "pipe" the light signals from each telescope to a central location where they "interfere" or blend together. This method makes the combined signal sharp and clear. The effective resolution of such

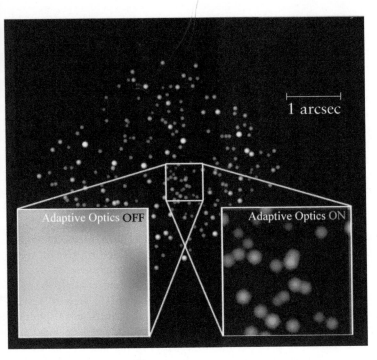

FIGURE 6-18 R ▮ V U X G

Using Adaptive Optics to "Unblur" Telescope Images The two false-color, inset images show the same 1-arcsecond-wide region of the sky at infrared wavelengths as observed with the 10.0-m Keck II telescope on Mauna Kea (see Figure 6-16). Without adaptive optics, it is impossible to distinguish individual stars in this region. With adaptive optics turned on, more than two dozen stars can be distinguished. (UCLA Galactic Center Group)

a combination of telescopes is equivalent to that of one giant telescope with a diameter equal to the **baseline,** or distance between the two telescopes. For example, the Keck I and Keck II telescopes atop Mauna Kea (Figure 6-16) are 85 meters apart, so when used as an interferometer the angular resolution is the same as a single 85-meter telescope.

Interferometry has been used for many years with radio telescopes (which we will discuss in Section 6-6), but is still under development with telescopes for visible light or infrared wavelengths. Astronomers are devoting a great deal of effort to this development because the potential rewards are great. For example, the Keck I and II telescopes used together should give an angular resolution as small as 0.005 arcsec, which corresponds to being able to read the bottom row on an eye chart 36 km (22 miles) away!

Light Pollution

Light from city street lamps and from buildings also degrades telescope images. This **light pollution** illuminates the sky, making it more difficult to see the stars. You can appreciate the problem if you have ever looked at the night sky from a major city. Only a few of the very brightest stars can be seen, as against the thousands that can be seen with the naked eye from in the desert or the mountains. To avoid light pollution, observatories are built in remote locations far from any city lights.

Unfortunately for astronomers, the expansion of cities has brought light pollution to observatories that in former times had none. As an example, the growth of Tucson, Arizona, has had deleterious effects on observations at the nearby Kitt Peak National Observatory. Efforts have been made to have cities adopt light fixtures that provide safe illumination for their citizens but produce little light pollution. These efforts have met with only mixed success.

One factor over which astronomers have absolutely no control is the weather. Optical telescopes cannot see through clouds, so it is important to build observatories where the weather is usually clear. One advantage of mountaintop observatories such as Mauna Kea is that most clouds form at altitudes below the observatory, giving astronomers a better chance of having clear skies.

In many ways the best location for a telescope is in orbit around Earth, where it is unaffected by weather, light pollution, or atmospheric turbulence. We will discuss orbiting telescopes in Section 6-7.

CONCEPTCHECK **6-6**

If astronomers are using an adaptive optics system on a night when the atmosphere is unusually turbulent, will the adaptive optics actuators deform the telescope's mirror more rapidly or less rapidly than on a typical night?

Answer appears at the end of the chapter.

6-4 A charge-coupled device is commonly used to record the image at a telescope's focus

Telescopes provide astronomers with detailed pictures of distant objects. The process of recording these pictures is called **imaging.**

Astronomical imaging really began in the nineteenth century with the invention of photography. It was soon realized that this new invention was a boon to astronomy. By mounting a camera at the focus of a telescope, long exposures can gather light from an object for an entire evening. Such long exposures can reveal details in galaxies, star clusters, and nebulae that would not be visible to an astronomer by simply looking through a telescope. Indeed, most large, modern telescopes do not have eyepieces at all.

Originally, cameras and telescopes used photographic film, but film has now been replaced by electronic detectors. Photographic film is not a very efficient light detector, with only 2% of the photons that strike photographic film actually triggering the chemical reaction needed to produce an image. And, of course, to analyze images on a computer we need them in some electronic form. For much greater efficiency, and to capture images digitally, we use charge-coupled devices (CCDs).

> The same technology that makes digital cameras possible has revolutionized astronomy

Charge-Coupled Devices

The most sensitive light detector currently available to astronomers is the **charge-coupled device (CCD).** At the heart of a CCD is a semiconductor wafer divided into an array of small,

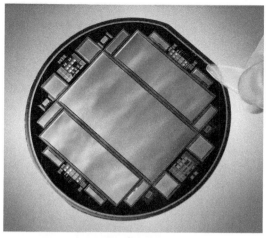

(a) Pan-STARRS detector CCDs

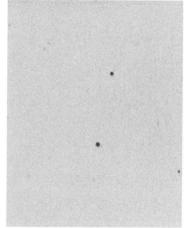

(b) An image made with photographic film

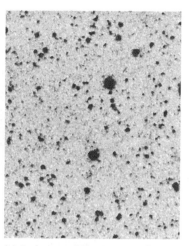

(c) An image of the same region of the sky made with a CCD

FIGURE 6-19 R I V U X G

Charge-Coupled Devices (CCDs) and Imaging **(a)** These CCDs are part of the Pan-STARRS detector developed at MIT's Lincoln Laboratory. The highlighted square contains 23 million pixels from 64 individual CCDs. The full detector is made from 60 of the 5 cm wide squares, has about 1.4 billion pixels, and is nearly 40 cm wide. **(b)** This negative print (black stars and white sky) shows a portion of the sky as imaged with a 4-meter telescope and photographic film. **(c)** This negative image of the same region of the sky was made with the same telescope, but with the photographic film replaced by a CCD. Many more stars and galaxies are visible. (a: LBNL/Science Source; b, c: Patrick Seitzer, National Optical Astronomy Observatories)

light-sensitive squares called picture elements or, more commonly, **pixels.** A personal digital camera uses a single CCD chip and might have around 10 million pixels. A large astronomical camera combines many CCDs to capture a wide field of view at high resolution (Figure 6-19a). For example, the square highlighted in Figure 6-19a contains 64 CCDs; the full detector combines 60 of these squares for about 1.4 billion pixels.

When an image from a telescope is focused on the CCD, an electric charge builds up in each pixel in proportion to the number of photons falling on that pixel. When the exposure is finished, the amount of charge on each pixel is read by a computer, where the resulting image can be analyzed and stored in digital form. Compared with photographic film, CCDs are some 35 times more sensitive to light, can record much finer details, and respond more uniformly to light of different colors. Figures 6-19b and 6-19c show the dramatic difference between photographic and CCD images. The great sensitivity of CCDs also makes them useful for **photometry,** which measures the brightness of a star or other astronomical object.

In the modern world of CCD astronomy, astronomers no longer need to spend the night in the unheated dome of a telescope. Instead, they operate the telescope electronically from a separate control room, where the electronic CCD images can be viewed on a computer monitor. By viewing the CCD's electronic image during the evening's observations, an astronomer gets immediate feedback and often makes changes to get the best results. The control room need not even be adjacent to the telescope. Although the Keck I and II telescopes (see Figure 6-16) are at an altitude of 4100 m (13,500 feet), astronomers can now make observations from a facility elsewhere on the island of Hawaii that is much closer to sea level. This saves the laborious drive to the summit of

Mauna Kea and eliminates the need for astronomers to acclimate to the high altitude.

Most of the images that you will see in this book were made with CCDs. Because of their extraordinary sensitivity and their ability to be used in conjunction with computers, CCDs have attained a role of central importance in astronomy.

CONCEPTCHECK 6-7

Why can CCDs more efficiently observe faint stars than photographic film or photographic plates?

Answer appears at the end of the chapter.

6-5 Spectrographs record the spectra of astronomical objects

We saw in Section 5-6 how the spectrum of an astronomical object provides a tremendous amount of information about that object, including its chemical composition and temperature. So, measuring spectra, or **spectroscopy,** is one of the most important uses of telescopes. Indeed, some telescopes are designed solely for measuring the spectra of distant, faint objects; they are never used for imaging.

> The spectrum of a planet, star, or galaxy can reveal more about its nature than an image

Spectrographs and Diffraction Gratings

An essential tool of spectroscopy is the **spectrograph,** a device that records spectra. This optical device is mounted at the focus

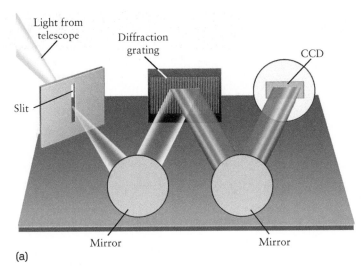

(a)

(b)

FIGURE 6-20 R I V U X G

A Grating Spectrograph **(a)** This optical device uses a diffraction grating to break up the light from a source into a spectrum. The spectrum is then recorded on a CCD. **(b)** A diffraction grating has a large number of parallel lines in its surface that reflect light of different colors in different direction. A compact disc, which stores information in a series of closely spaced pits, reflects light in a similar way. (Dale E. Boyer/Photo Researchers)

of a telescope. Figure 6-20a shows one design for a spectrograph, in which a **diffraction grating** is used to form the spectrum of a planet, star, or galaxy. A diffraction grating is a piece of glass on which thousands of very regularly spaced parallel lines have been cut. Some of the finest diffraction gratings have more than 10,000 lines per centimeter, which are usually cut by drawing a diamond back and forth across the glass. When light is shone on a diffraction grating, a spectrum is produced by the way in which light waves leaving different parts of the grating interfere with each other. (This same effect produces the rainbow of colors you see reflected from a compact disc or DVD, as shown in Figure 6-20b. Information is stored on the disc in a series of closely spaced pits, which acts as a diffraction grating.)

Older types of spectrographs used a prism rather than a diffraction grating to form a spectrum. Using a prism had several drawbacks. A prism does not disperse the colors of the rainbow evenly: Blue and violet portions of the spectrum are spread out more than the red portion. In addition, because the blue and violet wavelengths must pass through more of the prism's glass than do the red wavelengths (examine Figure 5-3), light is absorbed unevenly across the spectrum. Indeed, a glass prism is opaque to near-ultraviolet light. For these reasons, diffraction gratings are preferred in modern spectrographs.

In Figure 6-20a the spectrum of a planet, star, or galaxy formed by the diffraction grating is recorded on a CCD. Light from a hot gas (such as helium, neon, argon, iron, or a combination of these) is then focused on the spectrograph slit. The result is a *comparison spectrum* alongside the spectrum of the celestial object under study. The wavelengths of the bright spectral lines of the comparison spectrum are known from laboratory experiments and can therefore serve as reference markers. (See Figure 5-18, which shows the spectrum of the Sun and a comparison spectrum of iron.)

When the exposure is finished, electronic equipment measures the charge that has accumulated in each pixel. These data are used to graph light intensity versus wavelength. Dark absorption lines in the spectrum appear as depressions or valleys on the graph, while bright emission lines appear as peaks. Figure 6-21 compares two ways of exhibiting spectra with absorption lines and emission lines. Later in this book we shall see spectra presented in both these ways.

CONCEPTCHECK 6-8

A planet is viewed through a spectrograph like the one illustrated in Figure 6-20a. Does the CCD also record an image of the planet?

Answer appears at the end of the chapter.

6-6 A radio telescope uses a large concave dish to reflect radio waves to a focus

For thousands of years, all the information that astronomers gathered about the universe was based on ordinary visible light. In the twentieth century, however, astronomers first began to explore the nonvisible electromagnetic radiation coming from astronomical objects. In this way they have discovered aspects of the cosmos that are forever hidden to optical telescopes.

Astronomers have used ultraviolet light to map the outer regions of the Sun and the clouds of Venus, and used infrared radiation to see new stars and perhaps

> Observing at radio wavelengths reveals aspects of the universe hidden from ordinary telescopes

new planetary systems in the process of formation. By detecting radio waves from Jupiter and Saturn, they have mapped the intense magnetic fields that surround those giant planets; by detecting curious bursts of X-rays from space, they have learned about the utterly alien conditions in the vicinity of a black hole. It is no exaggeration to say that today's astronomers learn as much about the universe using telescopes for nonvisible wavelengths as they do using visible light.

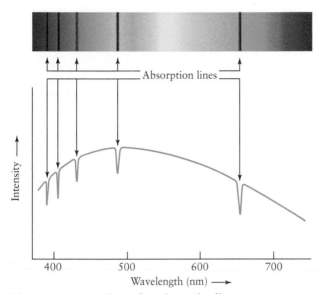

(a) Two representations of an absorption line spectrum

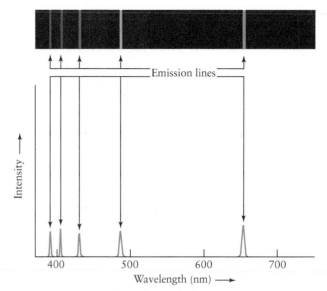

(b) Two representations of an emission line spectrum

FIGURE 6-21

Two Ways to Represent Spectra When a CCD is placed at the focus of a spectrograph, it records the rainbow-colored spectrum. A computer program can be used to convert the recorded data into a graph of intensity versus wavelength. **(a)** Absorption lines appear as dips on such a graph, while **(b)** emission lines appear as peaks. The dark absorption lines and bright emission lines in this example are the Balmer lines of hydrogen (see Section 5-8).

Radio Astronomy

Radio waves were the first part of the electromagnetic spectrum beyond the visible to be exploited for astronomy. The use of radio waves is a result of a research project seemingly unrelated to astronomy. In the early 1930s, Karl Jansky, a young electrical engineer at Bell Telephone Laboratories, was trying to locate the source of interference with the then-new transatlantic radio link. By 1932, he realized that one kind of radio noise is strongest when the constellation Sagittarius is high in the sky. The center of our Galaxy is located in the direction of Sagittarius, and Jansky concluded that he was detecting radio waves from an astronomical source.

At first, only Grote Reber, a radio engineer living in Illinois, took up Jansky's research. In 1936 Reber built in his backyard the first **radio telescope,** a radio-wave detector dedicated to astronomy. He modeled his design after an ordinary reflecting telescope, with a parabolic metal "dish" (reflecting antenna) measuring 10 m (31 ft) in diameter and a radio receiver at the focal point of the dish.

Reber spent the years from 1938 to 1944 mapping radio emissions from the sky at wavelengths of 1.9 m and 0.63 m. He found radio waves coming from the entire Milky Way, with the greatest emission from the center of the Galaxy. These results, together with the development of improved radio technology during World War II, encouraged the growth of radio astronomy and the construction of new radio telescopes around the world. Today, radio observatories are as common as major optical observatories.

Modern Radio Telescopes

Like Reber's prototype, a typical modern radio telescope has a large parabolic dish (Figure 6-22). An antenna tuned to the desired frequency is located at the focus (like the prime focus design for optical reflecting telescopes shown in Fig. 6-11a). The

incoming signal is relayed from the antenna to amplifiers and recording instruments, typically located in a room at the base of the telescope's pier.

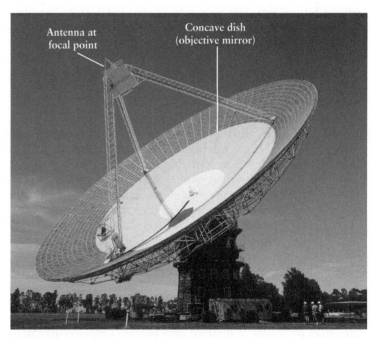

FIGURE 6-22   R I V U X G

A Radio Telescope The dish of the Parkes radio telescope in New South Wales, Australia, is 64 m (210 ft) in diameter. Radio waves reflected from the dish are brought to a focus and collected by an antenna at the focal point.

(David Nunuk/Photo Researchers)

CAUTION! The radio telescope in Figure 6-22 looks like a radar dish but is used in a different way. In radar, the dish is used to send out a narrow beam of radio waves. If this beam encounters an object like an airplane, some of the radio waves will be reflected back to the radar dish and detected by a receiver at the focus of the dish. Thus, a radar dish looks for radio waves *reflected* by distant objects. By contrast, a radio telescope is designed to receive radio waves *emitted* by objects in space.

Many radio telescope dishes, like the one in Figure 6-22, have visible gaps in them like a wire mesh. This does not affect their reflecting power because the holes are much smaller than the wavelengths of the radio waves; the radio waves reflect from the wire mesh as if from a solid surface. The same idea is used in the design of microwave ovens. The glass window in the oven door would allow the microwaves to leak out, so the window is covered by a metal screen with small holes. These holes are much smaller than the 12.2-cm (4.8-in.) wavelength of the microwaves, so the screen reflects the microwaves back into the oven.

Radio Telescopes: Limits to Angular Resolution

One great drawback of early radio telescopes was their very poor angular resolution. Recall from Section 6-3 that angular resolution is the smallest angular separation between two stars that can just barely be distinguished as separate objects. Unlike visible light, radio waves are only slightly affected by turbulence in the atmosphere, so the limitation on the angular resolution of a radio telescope is diffraction. The problem is that diffraction-limited angular resolution is directly proportional to the wavelength being observed: The longer the wavelength, the larger (and hence worse) the angular resolution and the fuzzier the image. (See the formula for angular resolution θ in Section 6-3.) As an example, a 1-m radio telescope detecting radio waves of 5-cm wavelength has an angular resolution 100,000 times poorer than a 1-m optical telescope. Because radio radiation has very long wavelengths, *small* radio telescopes can produce only blurry, indistinct images. VIDEO 6.7 A very large radio telescope can produce a somewhat sharper radio image, because as the diameter of the telescope increases, the angular resolution decreases. In other words, the bigger the dish, the better the resolution. For this reason, most modern radio telescopes have dishes more than 30 m (100 ft) in diameter. A large dish is also useful for increasing light-gathering power, because radio signals from astronomical objects are typically very weak in comparison with the intensity of visible light emitted by the same objects. But even the largest single radio dish in existence, the 305-m (1000-ft) Arecibo radio telescope in Puerto Rico, cannot come close to the resolution of the best optical instruments.

To improve angular resolution at radio wavelengths, astronomers combine observations from two or more widely separated radio telescopes using the interferometry technique that we described in Section 6-3. Interferometry using radio waves is much easier than for visible or infrared light because radio signals can be carried over electrical wires. Consequently, two radio telescopes observing the same astronomical object can be hooked together, even if they are separated by a large distance, or baseline, of many kilometers. The connected radio dishes can act as a single large-diameter dish with vastly improved angular resolution.

FIGURE 6-23 R I **V** U X G
The Very Large Array (VLA) The 27 radio telescopes of the VLA in central New Mexico are arranged along the arms of a Y. The north arm of the array is 19 km long; the southwest and southeast arms are each 21 km long. By spreading the telescopes out along the legs and combining the signals received, the VLA can give the same angular resolution as a single dish many kilometers in radius. (Courtesy of NRAO/AUI)

One of the largest arrangements of radio telescopes for interferometry is the Very Large Array (VLA), located in the desert near Socorro, New Mexico (Figure 6-23). The VLA consists of 27 parabolic dishes, each 25 m (82 ft) in diameter. These 27 telescopes are arranged along the arms of a gigantic Y that covers an area 27 km (17 mi) in diameter. By pointing all 27 telescopes at the same object and combining the 27 radio signals, this system can produce radio views of the sky with an angular resolution as small as 0.05 arcsec, comparable to that of the very best optical telescopes.

Dramatically better angular resolution can be obtained by combining the signals from radio telescopes at different observatories thousands of kilometers apart. This technique is called **very-long-baseline interferometry (VLBI)**. VLBI is used by a system called the Very Long Baseline Array (VLBA), which consists of ten 25-meter dishes at different locations between Hawaii and the Caribbean. Although the 10 dishes are not physically connected, they are all used to observe the same object at the same time. The data from each telescope are recorded electronically and processed later. By carefully synchronizing the 10 recorded signals, they can be combined just as if the telescopes had been linked together during the observation. With the VLBA, features smaller than 0.001 arcsec can be distinguished at radio wavelengths. This angular resolution is 100 times better than a large optical telescope with adaptive optics.

Even better angular resolution can be obtained by adding radio telescopes in space; the baseline is then the distance from the VLBA to the orbiting telescope. The first such space radio telescope, the Japanese HALCA spacecraft, operated from 1997 to 2003. Orbiting at a maximum distance of 21,400 km (13,300 mi) above Earth's surface, HALCA gave a baseline 3 times longer—and thus an angular resolution 3 times better—than can be obtained with Earthbound telescopes alone. A Russian-led mission launched in 2011, Spektr-R, continues the effort for space-based interferometry.

Figure 6-24 shows how optical and radio images of the same object can give different and complementary kinds of information. The visible-light image of Saturn (Figure 6-24a) shows clouds in

(a) R V U X G

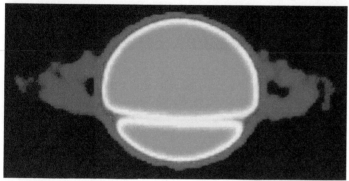

(b) R I V U X G

FIGURE 6-24

Optical and Radio Views of Saturn **(a)** This picture was taken by a spacecraft 18 million kilometers from Saturn. The view was produced by sunlight reflecting from the planet's cloudtops and rings. **(b)** The image is made by using radio waves, which is indicated by the highlighted R in the wavelength tab. This VLA image shows radio emission from Saturn at a wavelength of 2 cm. In this false-color image, the most intense radio emission is shown in red, the least intense in blue. Yellow and green represent intermediate levels of radio intensity; black indicates no detectable radio emission. Note the radio "shadow" caused by Saturn's rings where they lie in front of the planet. (NASA)

the planet's atmosphere and the structure of the rings. Like the visible light from the Moon, the light used to make this image is just reflected sunlight. By contrast, the false-color radio image of Saturn (Figure 6-24b) is a record of waves *emitted* by the planet and its rings. Analyzing such images provides information about the structure of Saturn's atmosphere and rings that could never be obtained from a visible-light image such as Figure 6-24a.

CONCEPTCHECK 6-9

What radio telescope would make an image with the better resolution: a single radio antenna dish with a diameter of 100 m, or two 25-m dishes connected together at a separation of 200 m?

Answer appears at the end of the chapter.

6-7 Telescopes in orbit around Earth detect radiation that does not penetrate the atmosphere

The many successes of radio astronomy show the value of observations at nonvisible wavelengths. But Earth's atmosphere is opaque to many

> Space telescopes make it possible to study the universe across the entire electromagnetic spectrum

wavelengths. Other than visible light and radio waves, very little radiation from space manages to penetrate the air we breathe. To overcome this, astronomers have placed a variety of telescopes in orbit around the planet.

Figure 6-25 shows the transparency of Earth's atmosphere to different wavelengths of electromagnetic radiation. The atmosphere is most transparent in two wavelength regions, the **optical window** (which includes the entire visible spectrum) and the **radio window** (which includes part, but not all, of the radio spectrum). There are also several relatively transparent regions at infrared wavelengths between 1 and 40 μm. Infrared radiation within these wavelength intervals can penetrate Earth's atmosphere somewhat and can be detected with ground-based telescopes. This wavelength range is

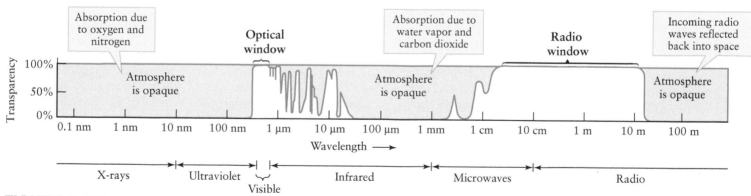

FIGURE 6-25

The Transparency of Earth's Atmosphere This graph shows the percentage of radiation that can penetrate Earth's atmosphere at different wavelengths. Regions in which the curve is high are called "windows," because the atmosphere is relatively transparent at those wavelengths.

There are also three wavelength ranges in which the atmosphere is opaque and the curve is near zero. At wavelengths where the atmosphere is opaque, ground-based telescopes cannot receive light from space because it is either absorbed or reflected by the atmosphere.

called the *near-infrared,* because it lies just beyond the red end of the visible spectrum.

CONCEPTCHECK 6-10

Look at Figure 6-25, which shows the transparency of Earth's atmosphere. Would astronomers most prefer to have a new ground-based telescope constructed that is most sensitive in the X-ray region, the ultraviolet wavelength region, or in the microwave region?

Answer appears at the end of the chapter.

Infrared Astronomy

Water vapor is the main absorber of infrared radiation from space, which is why infrared observatories are located at sites with exceptionally low humidity. The site must also be at high altitude to get above as much of the atmosphere's water vapor as possible. One site that meets both criteria is the summit of Mauna Kea in Hawaii, shown in Figure 6-16. (The complete lack of vegetation on the summit attests to its extreme dryness.) Some of the telescopes on Mauna Kea are designed exclusively for detecting infrared radiation. Others, such as the Keck I and Keck II telescopes, are used for both visible and near-infrared observations (see Figure 6-18).

Even at the elevation of Mauna Kea, water vapor in the atmosphere restricts the kinds of infrared observations that astronomers can make. This situation can be improved by carrying telescopes on board high-altitude balloons or aircraft. But the ultimate solution is to place a telescope in Earth's orbit and radio its data back to astronomers on the ground. The first such orbiting infrared observatory, the Infrared Astronomical Satellite (IRAS), was launched in 1983. During its nine-month mission, IRAS used its 57-cm (22-in.) telescope to map almost the entire sky at wavelengths from 12 to 100 μm.

The IRAS data revealed the presence of dust disks around nearby stars. Planets are thought to coalesce from disks of this kind, so this was the first observational indication that there might be planets orbiting other stars. The dust that IRAS detected is warm enough to emit infrared radiation but too cold to emit much visible light, so it remained undetected by ordinary optical telescopes. IRAS also discovered distant, ultraluminous galaxies that emit almost all their radiation at infrared wavelengths.

In 1995 the Infrared Space Observatory (ISO), a more advanced 60-cm reflector with better light detectors, was launched into orbit by the European Space Agency. During its two-and-one-half-year mission, ISO made a number of groundbreaking observations of very distant galaxies and of the thin, cold material between the stars of our own Galaxy. Like IRAS, ISO had to be cooled by liquid helium to temperatures just a few degrees above absolute zero. Had this not been done, the infrared blackbody radiation from the telescope itself would have outshone the infrared radiation from astronomical objects. The ISO mission came to an end when the last of the helium evaporated into space.

At the time of this writing the largest orbiting infrared observatory is the Herschel Space Observatory, a 3.5-m infrared telescope designed to survey the infrared sky with unprecedented resolution (Figure 6-26). Placed in orbit in 2009, the Herschel Space Observatory is being used to study galaxy formation in the early universe, star formation, and the chemical composition of atmospheres around planets, moons, and comets in our solar system.

FIGURE 6-26
Herschel Space Observatory This is an artist's concept of the Herschel Space Observatory. Herschel and the Planck observatory lifted off together on May 14, 2009 aboard an Ariane ECA rocket. (ESA, D. Ducros, 2009/NASA)

Ultraviolet Astronomy

Astronomers are also very interested in observing at ultraviolet wavelengths. These observations can reveal a great deal about hot stars, ionized clouds of gas between the stars, and the Sun's high-temperature corona (see Section 3-5), all of which emit copious amounts of ultraviolet light. The spectrum of ultraviolet sunlight reflected from a planet can also reveal the composition of the planet's atmosphere. However, Earth's atmosphere is opaque to ultraviolet light except for the narrow *near-ultraviolet* range, which extends from about 400 nm (the violet end of the visible spectrum) down to 300 nm.

To see shorter-wavelength *far-ultraviolet* light, astronomers must again make their observations from space. The first ultraviolet telescope was placed in orbit in 1962, and several others have since followed it into space. Small rockets have also lifted ultraviolet cameras briefly above Earth's atmosphere. Figure 6-27 shows an ultraviolet view of the constellation Orion, along with infrared and visible views.

The Far Ultraviolet Spectroscopic Explorer (FUSE), which went into orbit in 1999 and remained until 2007, specialized in measuring spectra at wavelengths from 90 to 120 nm. Highly ionized oxygen atoms, which can exist only in an extremely high-temperature gas, have a characteristic spectral line in this range. By looking for this spectral line in various parts of the sky, FUSE confirmed that our Milky Way Galaxy (Section 1-4) is surrounded by an immense "halo" of gas at temperatures in excess of 200,000 K. Only an ultraviolet telescope could have detected this "halo," which is thought to have been produced by exploding stars called supernovae (Section 1-3).

The Hubble Space Telescope

Infrared and ultraviolet satellites give excellent views of the heavens at selected wavelengths. But since the 1940s, astronomers had dreamed of having one large telescope that could be operated at

Immense clouds of dust are heated by hot, luminous, newly formed stars; the clouds glow at (a) ultraviolet and (b) infrared wavelengths

R I U̲ X G

(a) Ultraviolet Orion

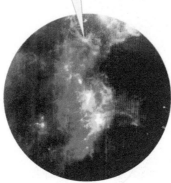

R I̲ V U X G

(b) Infrared Orion

At visible wavelengths the dust is opaque, obscuring the newly formed stars

R I V̲ U X G

(c) Visible Orion

Betelgeuse α   γ
   •*Bellatrix*

Mintaka •δ
Alnilam •ε
Alnitak •ζ

κ *Saiph*   β *Rigel*

(d) A star chart of Orion

FIGURE 6-27

Orion Seen at Ultraviolet, Infrared, and Visible Wavelengths

(a) An ultraviolet view of the constellation Orion was obtained during a brief rocket flight in 1975. This 100-s exposure covers the wavelength range 125–200 nm. **(b)** The false-color view from the Infrared Astronomical Satellite displays emission at different wavelengths in different colors: red for 100-μm radiation, green for 60-μm radiation, and blue for 12-μm radiation. Compare these images with **(c)** an ordinary visible-light photograph and **(d)** a star chart of Orion. (a: G. R. Carruthers, Naval Research Laboratory; b: NASA; c: mike black photography/Flickr/Getty Images)

any wavelength from the near-infrared through the visible range and out into the ultraviolet. This is the mission of the Hubble Space Telescope (HST), which was placed in a 600-km-high orbit by the space shuttle *Discovery* in 1990 (**Figure 6-28**). HST has a 2.4-m (7.9-ft) objective mirror and was designed to observe at wavelengths from 115 nm to 1 μm. Like most ground-based telescopes, HST uses a CCD to record images. (In fact, the development of HST helped drive advances in CCD technology.) The images are then radioed back to Earth in digital form.

The great promise of HST was that from its vantage point high above the atmosphere, its angular resolution would be limited only by diffraction. But soon after HST was placed in orbit, astronomers discovered that a manufacturing error had caused the telescope's objective mirror to suffer from spherical aberration. The mirror should have been able to concentrate 70% of a star's light into an image with an angular diameter of 0.1 arcsec. Instead, only 20% of the light was focused into this small area. The remainder was smeared out over an area about 1 arcsec wide, giving images little better than those achieved at major ground-based observatories.

On an interim basis, astronomers used only the 20% of incoming starlight that was properly focused and, with computer processing, discarded the remaining poorly focused 80%. This was practical only for brighter objects on which astronomers could afford to waste light. But many of the observing projects scheduled for HST involved extremely dim galaxies and nebulae.

These problems were resolved by a second space shuttle mission in 1993. Astronauts installed a set of small secondary mirrors whose curvature exactly compensated for the error in curvature of the primary mirror. Once these were in place, HST was able to make truly sharp images of extremely faint objects. Astronomers have used the repaired HST to make discoveries about the nature of planets, the evolution of stars, the inner workings of galaxies, and the expansion of the universe. You will see many HST images in later chapters.

The success of HST has inspired plans for its larger successor, the James Webb Space Telescope, or JWST (**Figure 6-29**). Planned for a 2018 launch, JWST will observe at visible and infrared wavelengths from 600 nm to 28 μm. With its 6.5-m objective mirror—2.5 times the diameter of the HST objective mirror, with 6 times the light-gathering power—JWST will study faint objects such as planetary systems forming around other stars and galaxies

FIGURE 6-28 R I V̲ U X G

The Hubble Space Telescope The largest telescope yet placed in orbit, HST is a joint project of NASA and the European Space Agency (ESA). HST has helped discover new moons of Pluto, probed the formation of stars, and found evidence that the universe is now expanding at a faster rate than several billion years ago. This photograph was taken from the space shuttle *Discovery* during a 1997 mission to service HST. (NASA)

near the limit of the observable universe. Unlike HST, which is in a relatively low-altitude orbit around Earth, JWST will orbit the Sun some 1.5 million km beyond Earth. In this orbit the telescope's view will not be blocked by Earth. Furthermore, by remaining far from the radiant heat of Earth it will be easier to keep JWST at the very cold temperatures required by its infrared detectors.

X-ray Astronomy

Space telescopes have also made it possible to explore objects whose temperatures reach the almost inconceivable values of 10^6 to 10^8 K. Atoms in such a high-temperature gas move so fast that when they collide, they emit X-ray photons of very high energy and very short wavelengths less than 10 nm. X-ray telescopes designed to detect these photons must be placed in orbit, because Earth's atmosphere is totally opaque at these wavelengths.

CAUTION! X-ray telescopes work on a very different principle from the X-ray devices used in medicine and dentistry. If you have your foot "X-rayed" to check for a broken bone, a piece of photographic film (or an electronic detector) sensitive to X-rays is placed under your foot and an X-ray beam is directed at your foot from above. The radiation penetrates through soft tissue but not as much through bone, so the bones cast an "X-ray shadow" on the film. A fracture will show as a break in the shadow. X-ray telescopes, by contrast, do *not* send beams of X-rays toward astronomical objects in an attempt to see inside them. Rather, these telescopes detect X-rays that the objects emit on their own.

Astronomers got their first quick look at the X-ray sky from brief rocket flights during the late 1940s. These observations confirmed that the Sun's corona (see Section 3-5) is a source of X-rays, and must therefore be at a temperature of millions of kelvins. In 1962 a rocket experiment revealed that objects beyond the solar system also emit X-rays.

Since 1970, a series of increasingly sensitive and sophisticated X-ray observatories have been placed in orbit, including NASA's Einstein Observatory, the European Space Agency's Exosat, and the German-British-American ROSAT. These telescopes have shown that other stars also have high-temperature coronae and have found hot, X-ray–emitting gas clouds so immense that hundreds of galaxies fit inside them. They also discovered unusual stars that emit X-rays in erratic bursts. These bursts are now thought to be coming from heated gas swirling around a small but massive object—possibly a black hole.

X-ray astronomy took a quantum leap forward in 1999 with the launch of NASA's Chandra X-ray Observatory and the European Space Agency's XMM-Newton. Named for the Indian-American Nobel laureate astrophysicist Subrahmanyan Chandrasekhar, Chandra can view the X-ray sky with an angular resolution of 0.5 arcsec (Figure 6-30a). This is comparable to the best ground-based optical telescopes and more than a thousand times better than the resolution of the first orbiting X-ray telescope. Chandra can also measure X-ray spectra 100 times more precisely than any previous spacecraft and can detect variations in X-ray emissions on time scales as short as 16 microseconds. This latter capability is essential for understanding how X-ray bursts are produced around black holes.

XMM-Newton (for *X-ray Multi-mirror Mission*) is actually three X-ray telescopes that all point in the same direction (Figure

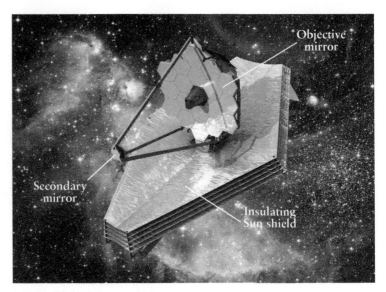

FIGURE 6-29
The James Webb Space Telescope (JWST) The successor to the Hubble Space Telescope, JWST will have an objective mirror 6.5 m (21 ft) in diameter. Like many Earthbound telescopes, JWST will be of Cassegrain design (see Figures 6-11b and 6-14b). Rather than using liquid helium to keep the telescope and instruments at the low temperatures needed to observe at infrared wavelengths, JWST will keep cool using a multilayer sunshield the size of two tennis courts. (ESA, C. Carreau)

6-30b). Their combined light-gathering power is 5 times greater than that of Chandra, which makes XMM-Newton able to observe fainter objects. (For reasons of economy, the mirrors were not ground as precisely as those on Chandra, so the angular resolution of XMM-Newton is only about 6 arcseconds.) It also carries a small but highly capable telescope for ultraviolet and visible observations. Hot X-ray sources are usually accompanied by cooler material that radiates at these longer wavelengths, so XMM-Newton can observe these hot and cool regions simultaneously.

The latest X-ray telescope in space is NuSTAR, with an incredible 10-m long mast (Figure 6-30c). NuSTAR will search for enormous black holes, study particles moving at almost the speed of light, and analyze how atoms are produced during supernova explosions.

Gamma-Ray Astronomy

Gamma rays, the shortest-wavelength photons of all, help us to understand phenomena even more energetic than those that produce X-rays. As an example, when a massive star explodes into a supernova, it produces radioactive atomic nuclei that are strewn across interstellar space. Observing the gamma rays emitted by these nuclei helps astronomers understand the nature of supernova explosions.

Like X-rays, gamma rays do not penetrate Earth's atmosphere, so space telescopes are required. One of the first gamma-ray telescopes placed in orbit was the Compton Gamma Ray Observatory (CGRO). One particularly important task for CGRO was the study of gamma-ray bursts, which are brief, unpredictable, and very intense flashes of gamma rays that are found in all parts of the sky. By analyzing data from CGRO and other orbiting observatories, astronomers have shown that the sources of these gamma-ray bursts are billions of light-years away. For these bursts to be visible across such great distances, their sources must be among the most energetic objects in the universe. By combining these gamma-ray

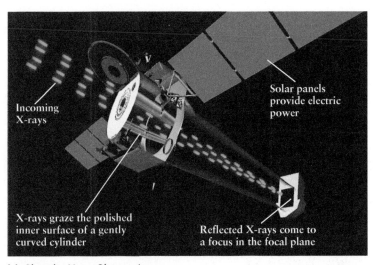

(a) Chandra X-ray Observatory

(b) XMM-Newton

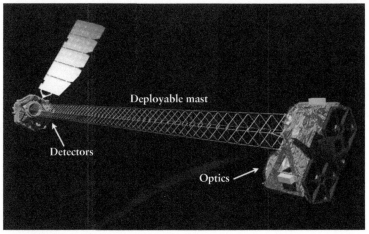

(c) NuSTAR

FIGURE 6-30

Three Orbiting X-Ray Observatories **(a)** X-rays are absorbed by
ordinary mirrors like those used in optical reflectors, but they can be reflected
if they graze the mirror surface at a very shallow angle. In the Chandra X-ray
Observatory, X-rays are focused in this way onto a focal plane 10 m (33 ft)
behind the mirror. **(b)** XMM-Newton is about the same size as Chandra, and
its three X-ray telescopes form images in the same way. **(c)** NuSTAR has
a 33-foot (10-meter) mast that deploys after launch to separate the optics
modules (right) from the detectors in the focal plane (left) (a: NASA/Chandra
X-ray Observatory Center/Smithsonian Astrophysical Observatory; b: D. Ducros/
European Space Agency; c: NASA)

observations with images made by optical telescopes, astronomers
have found that at least some of the gamma-ray bursts emanate
from stars that explode catastrophically. Launched in 2008, the
Fermi Gamma-ray Space Telescope (Figure 6-31) continues the
study of gamma-ray bursts and other highly energetic phenomena.

The *Cosmic Connections* figure shows the wavelengths at
which Earth-orbiting telescopes are particularly useful, as well as
summarizing the design of refracting and reflecting Earth-based
telescopes. The advantages and benefits of Earth-orbiting observatories cannot be overemphasized. We are no longer limited to the
narrow ranges of whatever wavelengths manage to leak through
our shimmering, hazy atmosphere (Figure 6-32). For the first time,
we are really *seeing* the universe.

CONCEPTCHECK **6-11**

What is the primary advantage of an orbiting space telescope,
compared to a ground-based telescope?

Answer appears at the end of the chapter.

FIGURE 6-31   R I V U X G

The Fermi Gamma-Ray Space Telescope This photograph shows
the telescope placed inside half of the launching rocket's nose cone. For
scale: Several engineers in white clean-room gowns are at the base of the
nose cone. (NASA)

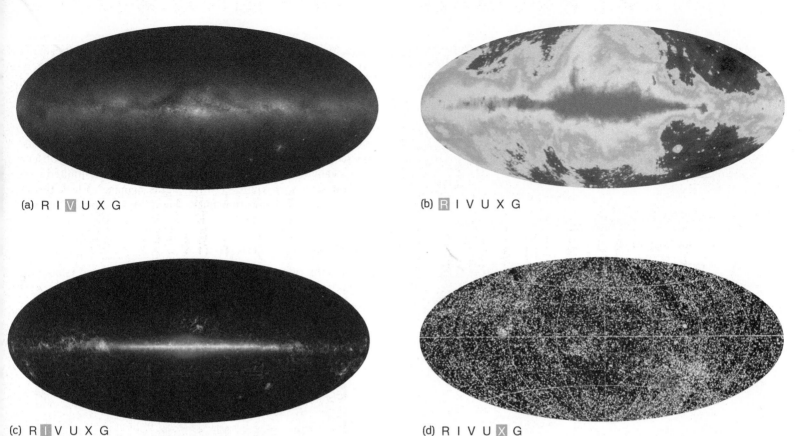

(a) R I [V] U X G

(b) [R] I V U X G

(c) R [I] V U X G

(d) R I V U [X] G

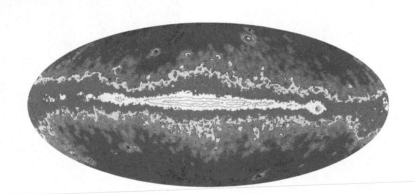

(e) R I V U X [G]

FIGURE 6-32

The Entire Sky at Five Wavelength Ranges These five views show the entire sky at visible, radio, infrared, X-ray, and gamma-ray wavelengths. The entire celestial sphere is mapped onto an oval, with the Milky Way stretching horizontally across the center. **(a)** In the visible view the constellation Orion is at the right, Sagittarius is in the middle, and Cygnus is toward the left. Many of the dark areas along the Milky Way are locations where interstellar dust is sufficiently thick to block visible light. **(b)** The radio view shows the sky at a wavelength of 21 cm. This wavelength is emitted by hydrogen atoms in interstellar space. The brightest regions (shown in red) are in the plane of the Milky Way, where the hydrogen is most concentrated. **(c)** This is a mosaic of the images covering the entire sky as observed by the Wide-field Infrared Survey Explorer (WISE). Most of the emission is from dust particles in the plane of the Milky Way that have been warmed by starlight. **(d)** The image is made by using X-rays, which is indicated by the highlighted X in the wavelength tab. The X-ray map from ROSAT shows about 50,000 X-ray sources whose intensity runs from red for the less intense sources through yellow and blue for the brightest ones. Extremely high temperature gas emits these X-rays. The white regions, which emit strongly at all X-ray wavelengths, are remnants of supernovae. **(e)** The image is made by using gamma rays, which is indicated by the highlighted G in the wavelength tab. The gamma-ray view from the Compton Gamma Ray Observatory includes all wavelengths less than about 1.2×10^{-5} nm (photon energies greater than 10^8 eV). The diffuse radiation from the Milky Way is emitted when fast-moving subatomic particles collide with the nuclei of atoms in interstellar gas clouds. The bright spots above and below the Milky Way are distant, extremely energetic galaxies. (a: Axel Mellinger; b: Max Planck Institute for Radio Astronomy/Science Photo Library/Science Source; c: NASA/JPL-Caltech/UCLA; d: Max Planck Institute for Extraterrestrial Physics/Science Photo Library/Science Source; e: NASA/Science Photo Library/Science Source)

COSMIC CONNECTIONS

Telescopes Across the EM Spectrum

Today, telescopes can view the universe in every range of electromagnetic radiation, although some must be above Earth's atmosphere to receive radiation without interference.

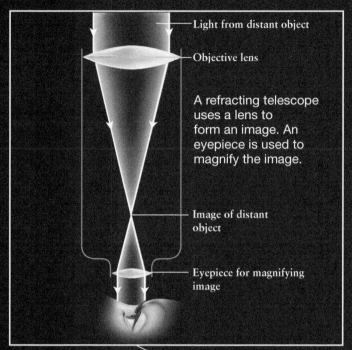

Light from distant object

Objective lens

A refracting telescope uses a lens to form an image. An eyepiece is used to magnify the image.

Image of distant object

Eyepiece for magnifying image

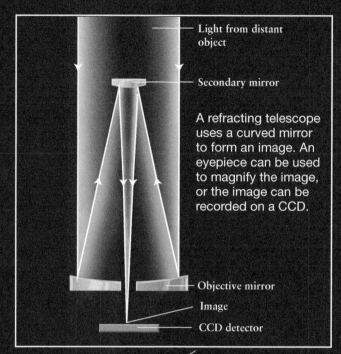

Light from distant object

Secondary mirror

A refracting telescope uses a curved mirror to form an image. An eyepiece can be used to magnify the image, or the image can be recorded on a CCD.

Objective mirror

Image

CCD detector

Earth's atmosphere is transparent to visible wavelengths, so visible-light telescopes (refracting or reflecting) can be used from Earth's surface.

Infrared telescopes are best placed in orbit since most infrared wavelengths do not penetrate the atmosphere.

Gamma-ray, X-ray, and ultraviolet telescopes must be placed in orbit because these wavelengths do not penetrate the atmosphere.

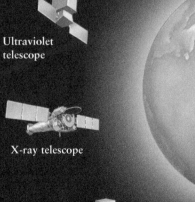

Ultraviolet telescope

Infrared telescope

X-ray telescope

Radio telescope

Gamma-ray telescope

The atmosphere is transparent to radio waves, so radio telescopes can be used from Earth's surface.

KEY IDEAS

Refracting Telescopes: Refracting telescopes, or refractors, produce images by bending light rays as they pass through glass lenses.

• Chromatic aberration is an optical defect whereby light of different wavelengths is bent in different amounts by a lens.

• Glass impurities, chromatic aberration, opacity to certain wavelengths, and structural difficulties make it inadvisable to build extremely large refractors.

Reflecting Telescopes: Reflecting telescopes, or reflectors, produce images by reflecting light rays to a focus point from curved mirrors.

• Reflectors are not subject to most of the problems that limit the useful size of refractors.

Angular Resolution: A telescope's angular resolution, which indicates ability to see fine details, is limited by two key factors.

• Diffraction is an intrinsic property of light waves. Its effects can be minimized by using a larger objective lens or mirror.

• The blurring effects of atmospheric turbulence can be minimized by placing the telescope atop a tall mountain with very smooth air. They can be dramatically reduced by the use of adaptive optics and can be eliminated entirely by placing the telescope in orbit.

Charge-Coupled Devices: Sensitive electronic light detectors called charge-coupled devices (CCDs) are often used at a telescope's focus to digitally record faint images.

Spectrographs: A spectrograph uses a diffraction grating to form the spectrum of an astronomical object.

Radio Telescopes: Radio telescopes use large reflecting dishes to focus radio waves onto a detector.

• Very large dishes provide reasonably sharp radio images. Higher resolution is achieved with interferometry techniques that link smaller dishes together.

Transparency of Earth's Atmosphere: Earth's atmosphere absorbs much of the radiation that arrives from space.

• The atmosphere is transparent chiefly in two wavelength ranges known as the optical window and the radio window. A few wavelengths in the near-infrared also reach the ground.

Telescopes in Space: For observations at wavelengths to which Earth's atmosphere is opaque, astronomers depend on telescopes carried above the atmosphere by rockets or spacecraft.

• Satellite-based observatories provide new information about the universe and permit coordinated observation of the sky at all wavelengths.

QUESTIONS

Review Questions

1. Describe refraction and reflection. Explain how these processes enable astronomers to build telescopes.

2. Explain why a flat piece of glass does not bring light to a focus while a curved piece of glass can.

3. Explain why the light rays that enter a telescope from an astronomical object are essentially parallel.

4. With the aid of a diagram, describe a refracting telescope. Which dimensions of the telescope determine its light-gathering power? Which dimensions determine the magnification?

5. What is the purpose of a telescope eyepiece? What aspect of the eyepiece determines the magnification of the image? In what circumstances would the eyepiece not be used?

6. Do most professional astronomers actually look through their telescopes? Why or why not?

7. Quite often advertisements appear for telescopes that extol their magnifying power. Is this a good criterion for evaluating telescopes? Explain your answer.

8. What is chromatic aberration? For what kinds of telescopes does it occur? How can it be corrected?

9. With the aid of a diagram, describe a reflecting telescope. Describe four different ways in which an astronomer can access the focal plane.

10. Explain some of the disadvantages of refracting telescopes compared to reflecting telescopes.

11. What kind of telescope would you use if you wanted to take a color photograph entirely free of chromatic aberration? Explain your answer.

12. Explain why a Cassegrain reflector can be substantially shorter than a refractor of the same focal length.

13. No major observatory has a Newtonian reflector as its primary instrument, whereas Newtonian reflectors are extremely popular among amateur astronomers. Explain why this is so.

14. What is spherical aberration? How can it be corrected?

15. *TUTORIAL 6-1* What is diffraction? Why does it limit the angular resolution of a telescope? What other physical phenomenon is often a more important restriction on angular resolution?

16. What is active optics? What is adaptive optics? Why are they useful? Would either of these be a good feature to include on a telescope to be placed in orbit?

17. Explain why combining the light from two or more optical telescopes can give dramatically improved angular resolution.

18. What is light pollution? What effects does it have on the operation of telescopes? What can be done to minimize these effects?

19. *TUTORIAL 6-2* What is a charge-coupled device (CCD)? Why have CCDs replaced photographic film for recording astronomical images?

20. What is a spectrograph? Why do many astronomers regard it as the most important device that can be attached to a telescope?

21. What are the advantages of using a diffraction grating rather than a prism in a spectrograph?

22. *TUTORIAL 6-3* Compare an optical reflecting telescope and a radio telescope. What do they have in common? How are they different?

23. Why can radio astronomers make observations at any time during the day, whereas optical astronomers are mostly limited to observing at night? (*Hint:* Does your radio work any better or worse in the daytime than at night?)

24. Why are radio telescopes so large? Why does a single radio telescope have poorer angular resolution than a large optical telescope? How can the resolution be improved by making simultaneous observations with several radio telescopes?

25. What are the optical window and the radio window? Why isn't there an X-ray window or an ultraviolet window?

26. Why is it necessary to keep an infrared telescope at a very low temperature?

27. How are the images made by an X-ray telescope different from those made by a medical X-ray machine?

28. Why must astronomers use satellites and Earth-orbiting observatories to study the heavens at X-ray and gamma-ray wavelengths?

Advanced Questions

> **Problem-solving tips and tools**
>
> You may find it useful to review the small-angle formula discussed in Box 1-1. The area of a circle is proportional to the square of its diameter. Data on the planets can be found in the appendices at the end of this book. Section 5-2 discusses the relationship between frequency and wavelength. Box 6-1 gives examples of how to calculate magnifying power and light-gathering power.

29. Show by means of a diagram why the image formed by a simple refracting telescope is upside down.

30. Ordinary photographs made with a telephoto lens make distant objects appear close. How does the focal length of a telephoto lens compare with that of a normal lens? Explain your reasoning.

31. The observing cage in which an astronomer can sit at the prime focus of the 5-m telescope on Palomar Mountain is about 1 m in diameter. Calculate what fraction of the incoming starlight is blocked by the cage.

32. (a) Compare the light-gathering power of the Keck I 10.0-m telescope with that of the Hubble Space Telescope (HST), which has a 2.4-m objective mirror. (b) What advantages does Keck I have over HST? What advantages does HST have over Keck I?

33. Suppose your Newtonian reflector has an objective mirror 20 cm (8 in.) in diameter with a focal length of 2 m. What magnification do you get with eyepieces whose focal lengths are (a) 9 mm, (b) 20 mm, and (c) 55 mm? (d) What is the telescope's diffraction-limited angular resolution when used with orange light of wavelength 600 nm? (e) Would it be possible to achieve this angular resolution if you took the telescope to the summit of Mauna Kea? Why or why not?

34. Several groups of astronomers are making plans for large ground-based telescopes. (a) What would be the diffraction-limited angular resolution of a telescope with a 40-meter objective mirror? Assume that yellow light with wavelength 550 nm is used. (b) Suppose this telescope is placed atop Mauna Kea. How will the actual angular resolution of the telescope compare to that of the 10-meter Keck I telescope? Assume that adaptive optics is not used.

35. The Hobby-Eberly Telescope (HET) at the McDonald Observatory in Texas has a spherical mirror, which is the least expensive shape to grind. Consequently, the telescope has spherical aberration. Explain why this does not affect the usefulness of HET for spectroscopy. (The telescope is not used for imaging.)

36. The four largest moons of Jupiter are roughly the same size as our Moon and are about 628 million (6.28×10^8) kilometers from Earth at opposition. What is the size in kilometers of the smallest surface features that the Hubble Space Telescope

(resolution of 0.1 arcsec) can detect? How does this compare with the smallest features that can be seen on the Moon with the unaided human eye (resolution of 1 arcmin)?

37. The Hubble Space Telescope (HST) has been used to observe the galaxy M100, some 70 million light-years from Earth. (a) If the angular resolution of the HST image is 0.1 arcsec, what is the diameter in light-years of the smallest detail that can be discerned in the image? (b) At what distance would a U.S. dime (diameter 1.8 cm) have an angular size of 0.1 arcsec? Give your answer in kilometers.

38. At its closest to Earth, Pluto is 28.6 AU from Earth. Can the Hubble Space Telescope distinguish any features on Pluto? Justify your answer using calculations.

39. The Institute of Space and Astronautical Science in Japan proposes to place a radio telescope into an even higher orbit than the HALCA telescope. Using this telescope in concert with ground-based radio-telescopes, baselines as long as 25,000 km may be obtainable. Astronomers want to use this combination to study radio emission at a frequency of 43 GHz from the molecule silicon monoxide, which is found in the interstellar clouds from which stars form. (1 GHz = 1 gigahertz = 10^9 Hz.) (a) What is the wavelength of this emission? (b) Taking the baseline to be the effective diameter of this radio-telescope array, what angular resolution can be achieved?

40. The mission of the Submillimeter Wave Astronomy Satellite (SWAS), launched in 1998, was to investigate interstellar clouds within which stars form. One of the frequencies at which it observed these clouds is 557 GHz (1 GHz = 1 gigahertz = 10^9 Hz), characteristic of the emission from interstellar water molecules. (a) What is the wavelength (in meters) of this emission? In what part of the electromagnetic spectrum is this? (b) Why was it necessary to use a satellite for these observations? (c) SWAS had an angular resolution of 4 arcminutes. What was the diameter of its primary mirror?

41. To search for ionized oxygen gas surrounding our Milky Way Galaxy, astronomers aimed the ultraviolet telescope of the FUSE spacecraft at a distant galaxy far beyond the Milky Way. They then looked for an ultraviolet spectral line of ionized oxygen in that galaxy's spectrum. Were they looking for an emission line or an absorption line? Explain.

42. A sufficiently thick interstellar cloud of cool gas can absorb low-energy X-rays but is transparent to high-energy X-rays and gamma rays. Explain why both Figure 6-32b and Figure 6-32d reveal the presence of cool gas in the Milky Way. Could you infer the presence of this gas from the visible-light image in Figure 6-32a? Explain.

Discussion Questions

43. If you were in charge of selecting a site for a new observatory, what factors would you consider important?

44. Discuss the advantages and disadvantages of using a small telescope in Earth's orbit versus a large telescope on a mountaintop.

Web/eBook Questions

45. Several telescope manufacturers build telescopes with a design called a Schmidt-Cassegrain. These use a correcting lens in an arrangement like that shown in Figure 6-13c. Consult advertisements on the World Wide Web to see the appearance of these telescopes and find out their cost. Why do you suppose they are very popular among amateur astronomers?

46. The Large Zenith Telescope (LZT) in British Columbia, Canada, uses a 6.0-m *liquid* mirror made of mercury. Use the World Wide Web to investigate this technology. How can a liquid metal be formed into the necessary shape for a telescope mirror? What are the advantages of a liquid mirror? What are the disadvantages?

47. Three of the telescopes shown in Figure 6-16—the James Clerk Maxwell Telescope (JCMT), the Caltech Submillimeter Observatory (CSO), and the Submillimeter Array (SMA)—are designed to detect radiation with wavelengths close to 1 mm. Search for current information about JCMT, CSO, and SMA on the World Wide Web. What kinds of celestial objects emit radiation at these wavelengths? What can astronomers see using JCMT, CSO, and SMA that cannot be observed at other wavelengths? Why is it important that they be at high altitude? How large are the primary mirrors used in JCMT, CSO, and SMA? What are the differences among the three telescopes? Which can be used in the daytime? What recent discoveries have been made using JCMT, CSO, or SMA?

48. In 2003 an ultraviolet telescope called GALEX (Galaxy Evolution Explorer) was placed into orbit. Use the World Wide Web to learn about GALEX and its mission. What aspects of galaxies was GALEX designed to investigate? Why is it important to make these observations using ultraviolet wavelengths?

49. At the time of this writing, NASA's plans for the end of the Hubble Space Telescope's mission were uncertain. Consult the Space Telescope Science Institute Web site to learn about plans for HST's final years of operation. Are future space shuttle missions planned to service HST? If so, what changes will be made to HST on such missions? What will become of HST at the end of its mission lifetime?

ACTIVITIES

Observing Projects

50. Obtain a telescope during the daytime along with several eyepieces of various focal lengths. If you can determine the telescope's focal length, calculate the magnifying powers of the eyepieces. Focus the telescope on some familiar object, such as a distant lamppost or tree. **DO NOT FOCUS ON THE SUN! Looking directly at the Sun can cause blindness.** Describe the image you see through the telescope. Is it upside down? How does the image move as you slowly and gently shift the telescope left and right or up and down? Examine the eyepieces, noting their focal lengths. By changing the eyepieces, examine the distant object under different

magnifications. How do the field of view and the quality of the image change as you go from low power to high power?

51. On a clear night, view the Moon, a planet, and a star through a telescope using eyepieces of various focal lengths and known magnifying powers. (To determine the locations in the sky of the Moon and planets, you may want to use the *Starry Night*™ program if you have access. You may also want to consult such magazines as *Sky & Telescope* and *Astronomy* or their Web sites.) In what way does the image seem to degrade as you view with increasingly higher magnification? Do you see any chromatic aberration? If so, with which object and which eyepiece is it most noticeable?

52. Many towns and cities have amateur astronomy clubs. If you are so inclined, attend a "star party" hosted by your local club. People who bring their telescopes to such gatherings are delighted to show you their instruments and take you on a telescopic tour of the heavens. Such an experience can lead to a very enjoyable, lifelong hobby.

53. Use the *Starry Night*™ program to compare the field of view, magnification, and quality of image provided by different optical instruments when observing various celestial objects. Open **Favourites > Explorations > Field of View** to examine several solar system objects. Click the **Find** tab on the left side of the **View** window. Remove any text from the edit box at the top of the **Find** pane, click on the magnifying glass icon at the left side of the edit box, and select the **Orbiting Objects** item from the dropdown menu that appears to bring up a list of solar system objects. Select each of the following objects in turn: the Moon, Jupiter, and Saturn. Double-click on the name of the selected object in the **Find** pane to center it in the view. Click the down arrow to the right of the **Zoom** panel in the toolbar and select first the **7 × 50 Binoculars** from the dropdown menu. Note the change in the quality of the view of the object under observation. Open the **Zoom** dropdown menu again and select the **25mm Plossl** eyepiece on a **Sample 4″ refractor**, followed by the **25mm Plossl** eyepiece on a **Sample 8″ Schmidt-Cassegrain** telescope, followed by the **10mm Plossl** eyepiece on an **8″ Schmidt-Cassegrain**. Before finding the next object, use the **Zoom** dropdown menu to return the field of view to **120°**. (a) Describe the change in the level of detail visible in the observed object as the field of view decreases. Open the **Find** pane again. Click on the magnifying glass icon on the left-hand side of the edit box at the top of the **Find** pane and select **Messier Objects** from the dropdown menu that appears. From the Messier Objects listed in the **Find** pane, double-click the entry for **Ring Nebula** to center this object in the view. If nothing appears in the view, open the **Options** side pane and click the checkbox to the left of **Messier Objects** under the **Deep Space** layer. Once more, use the **Zoom** dropdown menu to examine this object through the various instruments. (b) Assuming that the 8″ Schmidt-Cassegrain telescope has a focal length of 2000mm, what is the magnification produced by the 25mm and the 10mm eyepieces? (c) What is the ratio of the

magnification of the 10mm eyepiece to that of the 25mm eyepiece on the 8″ Schmidt-Cassegrain telescope? (d) What is the nominal field of view of the 10mm and 25mm eyepieces on the Schmidt-Cassegrain telescope (shown next to the eyepiece name in the **Zoom** dropdown menu)? (e) What is the ratio of the field of view of the 10mm eyepiece to the 25mm eyepiece? (Note that this is the inverse ratio to that which you calculated for the magnification yield of the two eyepieces.) (f) Describe the relationship between magnification and field of view for a particular telescope.

54. Use *Starry Night*™ to explore the difference between magnification and resolution. Select **Favourites > Explorations > Resolution**. The view is centered upon the full Moon as it might appear to the naked eye from Earth. Right-click the image of the Moon and select **Magnify** from the contextual menu to see the Moon as it might appear in good binoculars or a small telescope. (a) Describe the difference between the naked-eye image of the Moon and the magnified image. (b) What happens to the field of view when the image is magnified? (c) **Zoom** in to a field of view about **10′** wide (this is equivalent to increasing the magnification of the telescope). How does this further magnification affect the quality of the image, that is, the ability to distinguish close details on the lunar surface? (d) **Zoom** in to a field of view about **3′** wide. Does the clarity of the image improve, deteriorate, or remain unchanged with this further magnification? (e) Slowly **Zoom** in further to a field of view about **1′** wide. What happens to the quality of the image with this increase in magnification (zoom)? (e) Explain your observations based on the concept of resolution.

55. Use *Starry Night*™ to explore the effect of light pollution on the night sky. This exercise will also help you to determine the brightness of the faintest stars that are likely to be visible to the naked eye from your location under your present sky conditions and to estimate the fraction of possible stars that you can see with the unaided eye. This exercise is best done outdoors on a dark, clear night with a laptop computer and *Starry Night*™ set to function in night vision mode (select **Options > Night Vision**). Click the **Home** button to ensure that the view is correct for your present location and time. Open the **Options** side pane and expand the **Local View** layer. Turn the **Daylight** option off and then place the cursor over the words **Local Light Pollution** and click the **Local Light Pollution Options…** button that appears. This will open the **Local View Options** dialog window. Move this dialog window to the side of the view. In the **Local View Options** dialog window, click the checkbox to the left of the **Local Light Pollution** option to turn this feature on. You can now use the slide bar to adjust the **Local Light Pollution** level until the view matches your night sky. When you are satisfied that the view matches your sky, click on **OK** to dismiss the **Local View Options** dialog window. Be sure not to change the **Zoom** from the standard field of view, which is 100° wide. Move the cursor over some of the stars that appear on

your screen to display their properties in the **HUD** (Heads-up Display). (If necessary, open the **Preferences** dialog from the **File** menu in Windows or **Starry Night** menu on a Mac and add the **apparent magnitude** option to the **Cursor Tracking (HUD)** options.) (**a**) Make a note of the apparent magnitudes of some of the faintest stars in this view and compare these values to the faintest apparent magnitude of about +6 that can be seen under ideal conditions by the human eye. (*Reminder:* Magnitude values increase as stars become fainter.) (**b**) The second goal is to estimate what fraction of the visible stars you can see under these conditions, compared to the total number of stars you might see under ideal conditions. Open the **FOV** pane and click the **Add...** button in the **Other (This Chart)** layer. Select **Rectangular...** from the popup menu and use the **FOV Indicator** dialog window to create a rectangular indicator with a width and height of 10° and click **OK**. If necessary, expand the **Other (This Chart)** layer and click the checkbox for the 10° indicator you created. This will display the indicator in the center of the screen. Without changing the **Zoom** from the standard 100°–wide field of view, use the hand tool to drag the view so that a region of sky with a reasonable number of stars lies within the central 10° indicator. Count and record the number of stars in this square with your present setting. Now open the **Local View Options** dialog window again and set the **Local Light Pollution** to **less** (which essentially gives you ideal conditions) and repeat the count of visible stars. Divide the first number you count by the second. What fraction of the stars that would be visible under ideal conditions were you seeing? (**c**) To see how much light pollution occurs in large cities, adjust the slide bar for **Local Light Pollution** in the **Local View Options** dialog window to the far right for maximum light pollution and repeat the star-counting within the same limited sky region. Compared to viewing under ideal conditions, what fraction of the stars are observers in large urban centers seeing?

Collaborative Exercises

56. Stand up and have everyone in your group join hands, making as large a circle as possible. If a telescope mirror were built as big as your circle, what would be its diameter? What would be your telescope's diffraction-limited angular resolution for blue light? Would atmospheric turbulence have a noticeable effect on the angular resolution?

57. Are there enough students in your class to stand and join hands and make two large circles that recreate the sizes of the two Keck telescopes? Explain how you determined your answer.

ANSWERS

ConceptChecks

ConceptCheck 6-1: The diameter, because it is the width of the opening that determines how much light can be captured by the telescope.

ConceptCheck 6-2: The thin lens bends light less, so, according to Figure 6-5, it must focus light at a more distant location. Thus, the thinner lens has the greater focal length.

ConceptCheck 6-3: The larger the eyepiece focal length, the smaller the magnification.

ConceptCheck 6-4: Reflection off a mirror does not depend on wavelength. This is a key advantage for a reflecting telescope, because it prevents the chromatic aberration that blurs images in telescopes made from lenses.

ConceptCheck 6-5: The angular resolution is decreased for larger telescope diameters. This improves the sharpness of the image, by reducing the angle that light is spread out and blurred by the telescope. Therefore, a larger telescope is better in two ways: sharper images of even dimmer features.

ConceptCheck 6-6: Adaptive optics actuators slightly deform the telescope's mirror to match the apparent movement of a star due to distortions in Earth's atmosphere. The actuators must deform the mirror even more on nights when the atmosphere is fluctuating more rapidly.

ConceptCheck 6-7: Photographic film only captures 2% of the light it receives. For a faint star, there might be very little light captured by the film. CCDs, on the other hand, are 35 times more sensitive, capturing $35 \times 2\% = 70\%$ of the light they receive.

ConceptCheck 6-8: No. Instead of the CCD capturing a focused image of the planet, the spectrograph device spreads light from the planet out in wavelength to form a spectrum. A choice must be made to either look at an image using imaging equipment, or to look at a spectrum using a spectrograph.

ConceptCheck 6-9: If more than one radio dish is connected together, the distance between the two dishes—not the size of the dishes—determines the telescope's overall diameter when considering the angular resolution. Thus, two dishes at 200 m produce sharper images than a single 100-m dish.

ConceptCheck 6-10: Given these three choices, astronomers would much prefer to have a new telescope in the microwave region because X-rays and ultraviolet wavelengths rarely pass through Earth's atmosphere to the ground. While the atmosphere absorbs some portion of microwaves, not all microwaves are absorbed.

ConceptCheck 6-11: Orbiting space telescopes are placed far above most of Earth's atmosphere. Without Earth's atmosphere to absorb infrared, ultraviolet, X-ray, and gamma-ray light, astronomers can study the universe at these wavelengths.

March 10, 2006: The *Mars Reconnaissance Orbiter* spacecraft arrives at Mars (artist's impression). (JPL/NASA)

Comparative Planetology I: Our Solar System

LEARNING GOALS

By reading the sections of this chapter, you will learn

7-1 The important differences between the two broad categories of planets: terrestrial and Jovian

7-2 The similarities and differences among the large planetary satellites, including Earth's Moon

7-3 How the spectrum of sunlight reflected from a planet reveals the composition of its atmosphere and surface

7-4 Why some planets have atmospheres and others do not

7-5 The categories of the many small bodies that also orbit the Sun

7-6 How craters on a planet or satellite reveal the age of its surface and the nature of its interior

7-7 Why a planet's magnetic field indicates a fluid interior in motion

7-8 How the diversity of the solar system is a result of its origin and evolution

As recently as the mid-1900s, astronomers knew precious little about the other worlds that orbit the Sun. Even the best telescopes provided images of the planets that were frustratingly hazy and indistinct. Of asteroids, comets, and the satellites (or moons) of the planets, we knew even less.

Today, our knowledge of the solar system has grown exponentially, due almost entirely to robotic spacecraft. (The illustration shows an artist's impression of one such robotic explorer, the *Mars Reconnaissance Orbiter* spacecraft, as it approached Mars in 2006.) Spacecraft have been sent to fly past all the planets at close range, revealing details unimagined by astronomers of an earlier generation. We have landed spacecraft on the Moon, Venus, and Mars and dropped a probe into the immense atmosphere of Jupiter. This is truly the golden age of solar system exploration.

In this chapter we paint in broad outline our present understanding of the solar system. We will see that the planets come in a variety of sizes and chemical compositions. A rich variety also exists among the moons (or satellites) of the planets and among smaller bodies we call asteroids, comets, and trans-Neptunian objects. We will investigate the nature of craters on the Moon and other worlds of the solar system. And by exploring the magnetic fields of planets, we will be able to peer inside Earth and other worlds and learn about their interior compositions.

An important reason to study the solar system is to search for our own origins. In Chapter 8 we will see how astronomers have used evidence from the present-day solar system to understand how the Sun, Earth, and the other planets formed some four and a half billion years ago, and how they have evolved since then. But for now, we invite you to join us on a guided tour of the worlds that orbit our Sun.

CONCEPTCHECK 7-1

How many stars are in our solar system?

Answer appears at the end of the chapter.

7-1 The solar system has two broad categories of planets: Earthlike and Jupiterlike

Each of the planets that orbit the Sun is unique. Only Earth has liquid water and an atmosphere that humans can breathe; only Venus has a perpetual cloud layer made of sulfuric acid droplets; and only Jupiter has immense storm systems that persist for centuries. But there are also striking similarities among planets. Volcanoes are found not only on Earth but also on Venus and

Mars; rings encircle Jupiter, Saturn, Uranus, and Neptune; and impact craters dot the surfaces of Mercury, Venus, Earth, and Mars, showing that all of these planets have been bombarded by interplanetary debris.

How can we make sense of the many similarities and differences among the planets? An important first step is to organize our knowledge of the planets in a systematic way. We can organize

> We can understand the most important similarities and differences among the planets by comparing their orbits, masses, and diameters

this information in two ways. First, we can contrast the orbits of different planets around the Sun; and second, we can compare the planets' physical properties such as size, mass, average density, and chemical composition.

Comparing the Planets: Orbits

A planet falls naturally into one of two categories according to the size of its orbit. As **Figure 7-1** shows, the orbits of the four inner planets (Mercury, Venus, Earth, and Mars) are crowded in close to the Sun. In contrast, the orbits of the next four planets (Jupiter, Saturn, Uranus, and Neptune) are widely spaced at great distances from the Sun. **Table 7-1** lists the orbital characteristics of these eight planets.

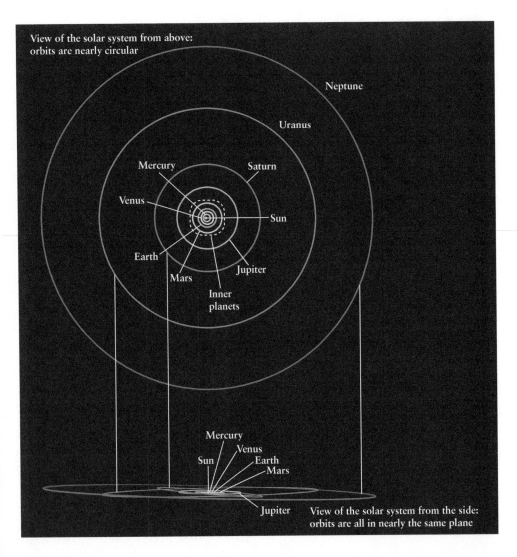

FIGURE 7-1

The Solar System to Scale This scale drawing shows the orbits of the planets around the Sun. The four inner planets are crowded in close to the Sun, while the four outer planets orbit the Sun at much greater distances. On the scale of this drawing, the planets themselves would be much smaller than the diameter of a human hair and too small to see.

TABLE 7-1   Characteristics of the Planets

	The Inner (Terrestrial) Planets			
	Mercury	**Venus**	**Earth**	**Mars**
Average distance from the Sun (10^6 km)	57.9	108.2	149.6	227.9
Average distance from the Sun (AU)	0.387	0.723	1.000	1.524
Orbital period (years)	0.241	0.615	1.000	1.88
Orbital eccentricity	0.206	0.007	0.017	0.093
Inclination of orbit to the ecliptic	7.00°	3.39°	0.00°	1.85°
Equatorial diameter (km)	4880	12,104	12,756	6794
Equatorial diameter (Earth = 1)	0.383	0.949	1.000	0.533
Mass (kg)	3.302×10^{23}	4.868×10^{24}	5.974×10^{24}	6.418×10^{23}
Mass (Earth = 1)	0.0553	0.8150	1.0000	0.1074
Average density (kg/m³)	5430	5243	5515	3934

R I V U X G
(NASA)

	The Outer (Jovian) Planets			
	Jupiter	**Saturn**	**Uranus**	**Neptune**
Average distance from the Sun (10^6 km)	778.3	1429	2871	4498
Average distance from the Sun (AU)	5.203	9.554	19.194	30.066
Orbital period (years)	11.86	29.46	84.10	164.86
Orbital eccentricity	0.048	0.053	0.043	0.010
Inclination of orbit to the ecliptic	1.30°	2.48°	0.77°	1.77°
Equatorial diameter (km)	142,984	120,536	51,118	49,528
Equatorial diameter (Earth = 1)	11.209	9.449	4.007	3.883
Mass (kg)	1.899×10^{27}	5.685×10^{26}	8.682×10^{25}	1.024×10^{26}
Mass (Earth = 1)	317.8	95.16	14.53	17.15
Average density (kg/m³)	1326	687	1318	1638

CAUTION! While Figure 7-1 shows the orbits of the planets, it does not show the planets themselves. The reason is simple: If Jupiter, the largest of the planets, were to be drawn to the same scale as the rest of this figure, it would be a dot just 0.0002 cm across—about $^{1}/_{300}$ of the width of a human hair and far too small to be seen without a microscope. The planets themselves are *very* small compared to the distances between them. Indeed, while an airplane traveling at 1000 km/h (620 mi/h) can fly around Earth in less than two days, at this speed it would take 17 *years* to fly from Earth to the Sun. The solar system is a very large and very empty place!

Most of the planets have orbits that are nearly circular. As we learned in Section 4-4, Kepler discovered in the seventeenth century that these orbits are actually ellipses. Astronomers denote the elongation of an ellipse by its *eccentricity* (see Figure 4-10b). The eccentricity of a circle is zero, and indeed most of the eight planets (with the notable exception of Mercury) have orbital eccentricities that are very close to zero.

If you could observe the solar system from a point several astronomical units (AU) above Earth's north pole, you would see that all the planets orbit the Sun in the same counterclockwise direction. Furthermore, the orbits of the eight planets all lie in nearly the same plane. In other words, these orbits are inclined at only slight angles to the plane of the ecliptic, which is the plane of Earth's orbit around the Sun (see Section 2-5). What's more, the plane of the Sun's equator is very closely aligned with the orbital planes of the planets. As we will see in Chapter 8, these near-alignments are not a coincidence. They provide important clues about the origin of the solar system.

Not included in Figure 7-1 or Table 7-1 is Pluto, which has an orbit that reaches beyond Neptune. Until the late 1990s, Pluto was generally regarded as the ninth planet. But in light of recent discoveries most astronomers now consider Pluto to be simply one member of a large collection of *trans-Neptunian objects* that orbit far from the Sun. Pluto is not even the largest of this new class of objects! Trans-Neptunian objects orbit the Sun in the same counterclockwise direction as the eight planets, though many of them have orbits that are steeply inclined to the plane of the ecliptic and have high eccentricities (that is, the orbits are quite elongated and noncircular). We will discuss trans-Neptunian objects, along with other small bodies that orbit the Sun, in Section 7-5.

CONCEPTCHECK 7-2

Is Mars classified as an inner planet or an outer planet? Is Mars a terrestrial planet?

CALCULATIONCHECK 7-1

Using Table 7-1, which of the planets has an orbital path that is most nearly a perfect circle in shape?

Answers appear at the end of the chapter.

Comparing the Planets: Physical Properties

When we compare the physical properties of the planets, we again find that they fall naturally into two classes— four small inner planets and four large outer ones. The four small inner planets are called **terrestrial planets** because they resemble Earth (in Latin, *terra*). They all have hard, rocky surfaces with mountains, craters, valleys, and volcanoes. You could stand on the surface of any one of them, although you would need a protective spacesuit on Mercury, Venus, or Mars. The four large outer planets are called **Jovian planets** because they resemble Jupiter. (Jove was another name for the Roman god Jupiter.) An attempt to land a spacecraft on the surface of any of the Jovian planets would be futile, because the materials of which these planets are made are mostly gaseous or liquid. The visible "surface" features of a Jovian planet are actually cloud formations in the planet's atmosphere. The photographs in Figure 7-2 show the distinctive appearances of the two classes of planets.

The most apparent difference between the terrestrial and Jovian planets is their *diameters*. Earth, with its diameter of about 12,756 km (7926 mi), is the largest of the four inner, terrestrial planets. In sharp contrast, the four outer, Jovian planets are much larger than the terrestrial planets. First place goes to Jupiter, whose

FIGURE 7-2   R I V U X G

The Planets to Scale This figure shows the planets from Mercury to Neptune to the same scale. The four terrestrial planets have orbits nearest Neptune the Sun, and the Jovian planets are the next four planets from the Sun. (Calvin J. Hamilton and NASA/JPL)

BOX 7-1 ASTRONOMY DOWN TO EARTH

Average Density

Average density—the mass of an object divided by that object's volume—is a useful quantity for describing the differences between planets in our solar system. This same quantity has many applications here on Earth.

A rock tossed into a lake sinks to the bottom, while an air bubble produced at the bottom of a lake (for example, by the air tanks of a scuba diver) rises to the top. These are examples of a general principle: An object sinks in a fluid if its average density is greater than that of the fluid, but rises if its average density is less than that of the fluid. The average density of water is 1000 kg/m^3, which is why a typical rock (with an average density of about 3000 kg/m^3) sinks, while an air bubble (average density of about 1.2 kg/m^3) rises.

At many summer barbecues, cans of soft drinks are kept cold by putting them in a container full of ice. When the ice melts, the cans of diet soda always rise to the top, while the cans of regular soda sink to the bottom. Why is this? The average density of a can of diet soda—which includes water, flavoring, artificial sweetener, and the trapped gas that makes the drink fizzy—is slightly less than the density of water, and so the can floats. A can of regular soda contains sugar instead of artificial sweetener, and the sugar is a bit heavier than the sweetener. The extra weight is just enough to make the average density of a can of regular soda slightly more than that of water, making the can sink. (You can test these statements for yourself by putting unopened cans of diet soda and regular soda in a sink or bathtub full of water.)

The concept of average density provides geologists with important clues about the early history of Earth. The average density of surface rocks on Earth, about 3000 kg/m^3, is less than Earth's average density of 5515 kg/m^3. The simplest explanation is that in the ancient past, Earth was completely molten throughout its volume, so that low-density materials rose to the surface and high-density materials sank deep into Earth's interior in a process called *chemical differentiation*. This series of events also suggests that Earth's core must be made of relatively dense materials, such as iron and nickel. A tremendous amount of other geological evidence has convinced scientists that this picture is correct.

---

equatorial diameter is more than 11 times that of Earth. On the other end of the scale, Mercury's diameter is less than two-fifths that of Earth. Figure 7-2 shows the Sun and the planets drawn to the same scale. The diameters of the planets are given in Table 7-1.

The *masses* of the terrestrial and Jovian planets are also dramatically different. If a planet has a moon, you can calculate the planet's mass from the moon's period and semimajor axis by using Newton's form of Kepler's third law (see Section 4-7 and Box 4-4). Astronomers have also measured the mass of each planet by sending a spacecraft to pass near the planet. The planet's gravitational pull (which is proportional to its mass) deflects the spacecraft's path, and the amount of deflection tells us the planet's mass. Using these techniques, astronomers have found that the four Jovian planets have masses that range from tens to hundreds of times greater than the mass of any of the terrestrial planets. Again, first place goes to Jupiter, whose mass is 318 times greater than Earth's.

Once we know the diameter and mass of a planet, we can learn something about what that planet is made of. The trick is to calculate the planet's **average density**, or mass divided by volume, measured in kilograms per cubic meter (kg/m^3). The average density of any substance depends in part on that substance's composition. For example, air near sea level on Earth has an average density of 1.2 kg/m^3, water's average density is 1000 kg/m^3, and a piece of concrete has an average density of 2000 kg/m^3. Box 7-1 describes some applications of the idea of average density to everyday phenomena on Earth.

The four inner, terrestrial planets have very high average densities (see Table 7-1); the average density of Earth, for example, is 5515 kg/m^3. By contrast, a typical rock found on Earth's surface has a lower average density, about 3000 kg/m^3. Thus, Earth must contain a large amount of material that is denser than rock. This information provides our first clue that terrestrial planets have dense iron cores.

In sharp contrast, the outer, Jovian planets have quite low densities. Saturn has an average density less than that of water. This information strongly suggests that the giant outer planets are composed primarily of light elements such as hydrogen and helium. All four Jovian planets probably have large cores of mixed rock and highly compressed water that are buried beneath low-density outer layers tens of thousands of kilometers thick.

We can conclude that the following general rule applies to the planets:

The terrestrial planets are made of rocky materials and have dense iron cores. These planets have solid surfaces and high average densities. The Jovian planets are composed primarily of light elements such as hydrogen and helium, resulting in low average densities. These planets have interiors made mostly of gas and liquid, and have no solid surface.

CONCEPTCHECK 7-3

A planet's average density can be estimated by measuring its size and how much the planet's gravity deflects a nearby spacecraft's path. If the density of rocks recovered from a planet's surface is lower than the planet's average density, what can one infer about the density of the planet's core?

CALCULATIONCHECK 7-2

If Earth's diameter is 12,756 km and Saturn's diameter is 120,536 km, how many Earths could fit across the diameter of Saturn?

Answers appear at the end of the chapter.

7-2 Seven large satellites are almost as big as the terrestrial planets

All the planets except Mercury and Venus have moons (also called satellites). At least 170 satellites are known: Earth has 1 (the Moon), Mars has 2, Jupiter has at least 66, Saturn at least 62, Uranus at least 27, and Neptune at least 13. Dozens of other small satellites probably remain to be discovered as our telescope technology continues to improve. Like the terrestrial planets, all of the satellites of the planets have solid surfaces.

You can see that there is a striking difference between the terrestrial planets, with few or no satellites, and the Jovian planets, each of which has so many moons that it resembles a miniature solar system. In Chapter 8 we will explore this evidence (as well as other evidence) that the Jovian planets formed in a manner similar to the solar system as a whole, but on a smaller scale.

> The various moons of the planets are not simply copies of Earth's Moon

Seven of the Jovian satellites are roughly as big as the planet Mercury! Table 7-2 lists these satellites and shows them to the same scale. Note that Earth's Moon and Jupiter's satellites Io and Europa have relatively high average densities, indicating that these moons are made primarily of rocky materials. By contrast, the average densities of Ganymede, Callisto, Titan, and Triton are all relatively low. Planetary scientists conclude that the interiors of these four moons also contain substantial amounts of water ice, which is less dense than rock. (In Section 7-4 we will learn about types of frozen "ice" made of substances other than water.)

CAUTION! Water ice may seem like a poor material for building a satellite, since the ice you find in your freezer can easily be cracked or crushed. But under high pressure, such as is found in the interior of a large satellite, water ice becomes as rigid as rock. (It also becomes denser than the ice found in ice cubes, although not as dense as rock.) Note that water ice is an important constituent only for satellites in the outer solar system, where the Sun is far away and temperatures are very low. For example, the surface temperature of Titan is a frigid 95 K (−178°C = −288°F). In Section 7-4 we will learn more about the importance of temperature in determining the composition of a planet or satellite.

The satellites listed in Table 7-2 are actually unusually large. Most of the known satellites have diameters less than 2000 km, and many are irregularly shaped and just a few kilometers across.

Interplanetary spacecraft have made many surprising and fascinating discoveries about the satellites of the solar system. We now know that Jupiter's satellite Io is the most geologically active world in the solar system, with numerous volcanoes that continually belch forth sulfur-rich compounds. The fractured surface of Europa, another of Jupiter's large satellites, suggests that a worldwide ocean of liquid water may lie beneath its icy surface. On Saturn's moon Titan, lakes of hydrocarbons, the primary source of energy on Earth, have been found near the poles—some the size of Lake Superior! In 2008, the space probe *Cassini* flew right through the plumes of geysers on Saturn's moon Enceladus as they ejected water-ice, dust, and gas. While Jupiter and Saturn lack a solid surface, their many moons contain a rich variety of features.

TABLE 7-2	The Seven Giant Satellites						
	Moon	**Io**	**Europa**	**Ganymede**	**Callisto**	**Titan**	**Triton**
Parent planet	Earth	Jupiter	Jupiter	Jupiter	Jupiter	Saturn	Neptune
Diameter (km)	3476	3642	3130	5268	4806	5150	2706
Mass (kg)	7.35×10^{22}	8.93×10^{22}	4.80×10^{22}	1.48×10^{23}	1.08×10^{23}	1.34×10^{23}	2.15×10^{22}
Average density (kg/m³)	3340	3530	2970	1940	1850	1880	2050
Substantial atmosphere?	No	No	No	No	No	Yes	No

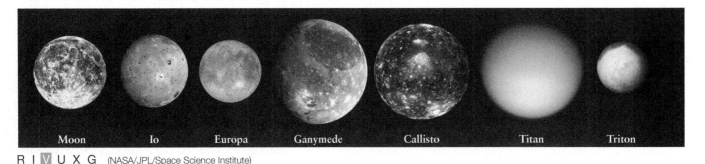

R I V U X G   (NASA/JPL/Space Science Institute)

How many moons in the solar system are larger than Earth's Moon (see Table 7-2)?

Answer appears at the end of the chapter.

7-3 Spectroscopy reveals the chemical composition of the planets

As we have seen, the average densities of the planets and satellites give us a crude measure of their **chemical compositions**—that is, what substances they are made of. For example, the low average density of the Moon (3340 kg/m^3) compared with Earth (5515 kg/m^3) tells us that the Moon contains relatively little iron or other dense metals. But to truly understand the nature of the planets and satellites, we need to know their chemical compositions in much greater detail than we can learn from average density alone.

> The light we receive from a planet or satellite is reflected sunlight—but with revealing differences in its spectrum

The most accurate way to determine chemical composition is by directly analyzing samples taken from a planet's atmosphere and soil. Unfortunately, of all the planets and satellites, we have such direct information only for Earth and the four worlds on which spacecraft have landed—Venus, the Moon, Mars, and Titan. In all other cases, astronomers must analyze sunlight reflected from the distant planets and their satellites. To do that, astronomers bring to bear one of their most powerful tools, **spectroscopy**, the systematic study of spectra and spectral lines. (We discussed spectroscopy in Sections 5-6 and 6-5.)

Determining Atmospheric Composition

Spectroscopy is a sensitive probe of the composition of a planet's *atmosphere*. If a planet has an atmosphere, then sunlight reflected from that planet must have passed through its atmosphere before coming back out. During this passage, some of the wavelengths of sunlight will have been absorbed. Hence, the spectrum of this reflected sunlight will have dark absorption lines. Astronomers look at the particular wavelengths absorbed and the amount of light absorbed at those wavelengths. Both of these depend on the kinds of chemicals present in the planet's atmosphere and the abundance of those chemicals.

For example, astronomers have used spectroscopy to analyze the atmosphere of Saturn's largest satellite, Titan (see **Figure 7-3a** and Table 7-2). The graph in Figure 7-3b shows the spectrum of visible sunlight reflected from Titan. (We first saw this method

(a) Saturn's satellite Titan

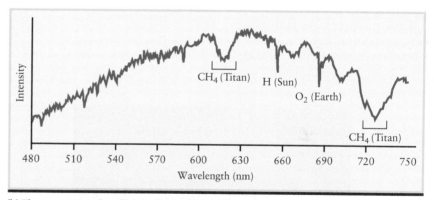

(b) The spectrum of sunlight reflected from Titan

FIGURE 7-3   R I **V** U X G

Analyzing a Satellite's Atmosphere through Its Spectrum

(a) Titan is the only satellite in the solar system with a substantial atmosphere. **(b)** The dips in the spectrum of sunlight reflected from Titan are due to absorption by hydrogen atoms (H), oxygen molecules (O$_2$), and methane molecules (CH$_4$). Of these, only methane is actually present in Titan's atmosphere. **(c)** This illustration shows the path of the light that reaches us from Titan. To interpret the spectrum of this light as shown in (b), astronomers must account for the absorption that takes place in the atmospheres of the Sun and Earth. (a: NASA/JPL/Space Science Institute)

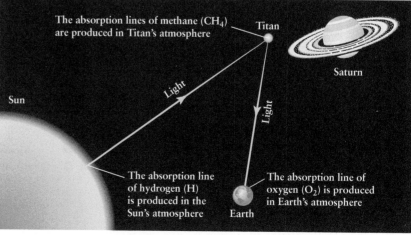

(c) Interpreting Titan's spectrum

of displaying spectra in Figure 6-21.) The dips in this curve of intensity versus wavelength represent absorption lines. However, not all of these absorption lines are produced in the atmosphere of Titan (Figure 7-3c). Before reaching Titan, light from the Sun's glowing surface must pass through the Sun's own hydrogen-rich atmosphere. This produces the hydrogen absorption line in Figure 7-3b at a wavelength of 656 nm. After being reflected from Titan, the light must pass through Earth's atmosphere before reaching the observing telescope; this is where the oxygen absorption line in Figure 7-3b is produced. Only the two dips near 620 nm and 730 nm are caused by gases in Titan's atmosphere.

These two absorption lines are caused not by individual atoms in the atmosphere of Titan but by atoms combined to form molecules. (We introduced the idea of molecules in Section 5-6.) Molecules, like atoms, also produce unique patterns of lines in the spectra of astronomical objects. The absorption lines in Figure 7-3b indicate the presence in Titan's atmosphere of methane molecules (CH_4, a molecule made of one carbon atom and four hydrogen atoms). Titan must be a curious place indeed, because on Earth, methane is a rather rare substance that is the primary ingredient in natural gas! When we examine other planets and satellites with atmospheres, we find that all of their spectra have absorption lines of molecules of various types.

In addition to visible-light measurements such as those shown in Figure 7-3b, it is very useful to study the *infrared* and *ultraviolet* spectra of planetary atmospheres. Many molecules have much stronger spectral lines in these nonvisible wavelength bands than in the visible. As an example, the ultraviolet spectrum of Titan shows that nitrogen molecules (N_2) are the dominant constituent of Titan's atmosphere. Furthermore, Titan's infrared spectrum includes spectral lines of a variety of molecules that contain carbon and hydrogen, indicating that Titan's atmosphere has a very complex chemistry. None of these molecules could have been detected by visible light alone. Because Earth's atmosphere is largely opaque to infrared and ultraviolet wavelengths (see Section 6-7), telescopes in space are important tools for these spectroscopic studies of the solar system.

It is a testament to the power of spectroscopy that when the robotic spacecraft *Huygens* landed on Titan in 2005, its onboard instruments confirmed the presence of methane and nitrogen in Titan's atmosphere—just as had been predicted years before by spectroscopic observations.

CONCEPTCHECK 7-5

If planets reflect some of the Sun's light rather than emitting visible light of their own, how can spectroscopy reveal information about a planet's atmosphere?

Answer appears at the end of the chapter.

Determining Surface Composition

Spectroscopy can also provide useful information about the *solid surfaces* of planets and satellites without atmospheres. When light shines on a solid surface, some wavelengths are absorbed while others are reflected. (For example, a plant leaf absorbs red and violet light but reflects green light—which is why leaves look green.) Unlike a gas, a solid illuminated by sunlight does not produce sharp, definite spectral lines. Instead, only broad absorption features appear in the spectrum. By comparing such a spectrum with the spectra of samples of different substances on Earth, astronomers can infer the chemical composition of the surface of a planet or satellite.

As an example, **Figure 7-4a** shows Jupiter's satellite Europa (see Table 7-2), and Figure 7-4b shows the infrared spectrum of

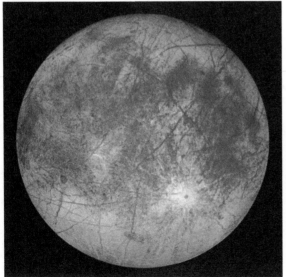

(a) Jupiter's moon Europa   R I **V** U X G

(b) The spectum of light reflected from Europa

FIGURE 7-4

Analyzing a Satellite's Surface from Its Spectrum **(a)** Unlike Titan (Figure 7-3a), Jupiter's satellite Europa has no atmosphere. **(b)** Infrared light from the Sun that is reflected from the surface of Europa has nearly the same spectrum as sunlight reflected from ordinary water ice.

light reflected from the surface of Europa. Because this spectrum is so close to that of water ice—that is, frozen water—astronomers conclude that water ice is the dominant constituent of Europa's surface. (We saw in Section 7-2 that water ice cannot be the dominant constituent of Europa's *interior,* because this satellite's density is too high.)

Unfortunately, spectroscopy tells us little about what the material is like just below the surface of a satellite or planet. For this purpose, there is simply no substitute for sending a spacecraft to a planet and examining its surface directly.

CONCEPTCHECK 7-6

How is the spectrum of reflected sunlight from a solid planetary surface different from the spectrum produced when sunlight passes through a planet's gaseous atmosphere?

Answer appears at the end of the chapter.

7-4 The Jovian planets are made of lighter elements than the terrestrial planets

Spectroscopic observations from Earth and spacecraft show that the outer layers of the Jovian planets are composed primarily of the lightest gases, hydrogen and helium (see Box 5-5). In contrast, chemical analysis of

> Differences in distance from the Sun, and hence in temperature, explain many distinctions between terrestrial and Jovian planets

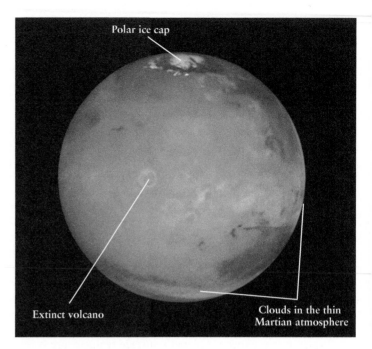

FIGURE 7-6   R I ☑ U X G

A Terrestrial Planet Mars is composed mostly of heavy elements such as iron, oxygen, silicon, magnesium, nickel, and sulfur. The planet's red surface can be seen clearly in this Hubble Space Telescope image because the Martian atmosphere is thin and nearly cloudless. Olympus Mons, the extinct volcano to the left of center, is nearly 3 times the height of Mount Everest. (Space Telescope Science Institute/JPL/NASA)

soil samples from Venus, Earth, and Mars demonstrate that the terrestrial planets are made mostly of heavier elements, such as iron, oxygen, silicon, magnesium, nickel, and sulfur. Spacecraft images such as Figure 7-5 and Figure 7-6 only hint at these striking differences in chemical composition, which are summarized in Table 7-3.

Temperature plays a major role in determining whether the materials of which planets are made exist as solids, liquids, or gases. Hydrogen (H_2) and helium (He) are gaseous except at

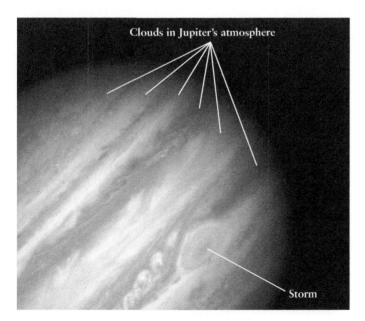

FIGURE 7-5   R I ☑ U X G

A Jovian Planet This Hubble Space Telescope image gives a detailed view of Jupiter's cloudtops. Jupiter is composed mostly of the lightest elements, hydrogen and helium, which are colorless; the colors in the atmosphere are caused by trace amounts of other substances. The giant storm at lower right, called the Great Red Spot, has been raging for more than 300 years. (Space Telescope Science Institute/JPL/NASA)

TABLE 7-3	Comparing Terrestrial and Jovian Planets	
	Terrestrial Planets	**Jovian Planets**
Distance from the Sun	Less than 2 AU	More than 5 AU
Size	Small	Large
Composition	Mostly rocky materials containing iron, oxygen, silicon, magnesium, nickel, and sulfur	Mostly light elements such as hydrogen and helium
Density	High	Low

extremely low temperatures and extraordinarily high pressures. By contrast, rock-forming substances such as iron and silicon are solids except at temperatures well above 1000 K. (You may want to review the discussion of temperature scales in Box 5-1.) Between these two extremes are substances such as water (H_2O), carbon dioxide (CO_2), methane (CH_4), and ammonia (NH_3), which solidify at low temperatures (from below 100 to 300 K) into solids called **ices.** (In astronomy, frozen water is just one kind of "ice.") At somewhat higher temperatures, they can exist as liquids or gases. For example, clouds of ammonia ice crystals are found in the cold upper atmosphere of Jupiter, but within Jupiter's warmer interior, ammonia exists primarily as a liquid.

CAUTION! The Jovian planets are sometimes called "gas giants." It is true that their primary constituents, including hydrogen, helium, ammonia, and methane, are gases under normal conditions on Earth. But in the interiors of these planets, pressures are so high that these substances are *liquids,* not gases. The Jovian planets might be better described as "liquid giants"!

As you might expect, a planet's surface temperature is related to its distance from the Sun. The four inner planets are quite warm. For example, midday temperatures on Mercury may climb to 700 K (= 427°C = 801°F), and during midsummer on Mars, it is sometimes as warm as 290 K (= 17°C = 63°F). The outer planets, which receive much less solar radiation, are cooler. Typical temperatures range from about 125 K (= −148°C = −234°F) in Jupiter's upper atmosphere to about 55 K (= −218°C = −360°F) at the tops of Neptune's clouds.

How Temperature Affects Atmospheres

The higher surface temperatures of the terrestrial planets help to explain the following observation: The atmospheres of the terrestrial planets contain virtually *no* hydrogen molecules or helium atoms. Instead, the atmospheres of Venus, Earth, and Mars are composed of heavier molecules such as nitrogen (N_2, 14 times more massive than a hydrogen molecule), oxygen (O_2, 16 times more massive), and carbon dioxide (22 times more massive). To understand the connection between surface temperature and the absence of hydrogen and helium, we need to know a few basic facts about gases.

The temperature of a gas is directly related to the speeds at which the atoms or molecules of the gas move: The higher the gas temperature, the greater the speed of its atoms or molecules. Furthermore, for a given temperature, lightweight atoms and molecules move more rapidly than heavy ones. On the four inner, terrestrial planets, where atmospheric temperatures are high, low-mass hydrogen molecules and helium atoms move so swiftly that they can escape from the relatively weak gravity of these planets. Hence, the atmospheres that surround the terrestrial planets are composed primarily of more massive, slower-moving molecules such as CO_2, N_2, O_2, and water vapor (H_2O). On the four Jovian planets, low temperatures and relatively strong gravity prevent even lightweight hydrogen and helium gases from escaping into space, and so their atmospheres are much more extensive. The combined mass of Jupiter's atmosphere, for example, is about a million (10^6) times greater than that of Earth's atmosphere. This

is comparable to the entire mass of Earth! Box 7-2 describes more about the ability of a planet's gravity to retain gases.

CONCEPTCHECK 7-7

If the average speed of hydrogen molecules in Earth's atmosphere is below Earth's escape speed, why do Earth's atmospheric hydrogen molecules slowly leak into space?

Answer appears at the end of the chapter.

7-5 Small chunks of rock and ice also orbit the Sun

In addition to the eight planets, many smaller objects orbit the Sun. These objects fall into three broad categories: asteroids, which are rocky objects found in the inner solar system; trans-Neptunian objects, which are found beyond Neptune in the outer solar system and contain both rock and ice; and comets, which are mixtures of rock and ice that originate in the outer solar system but can venture close to the Sun.

> The smaller bodies of the solar system contain important clues about its origin and evolution

Asteroids

Within the orbit of Jupiter are hundreds of thousands of rocky objects called **asteroids.** There is no sharp dividing line between planets and asteroids, which is why asteroids are also called **minor planets.** The largest asteroid, Ceres, has a diameter of about 900 km (around one-quarter our Moon's diameter). The next largest, Pallas and Vesta, are each about 500 km in diameter. Still smaller ones, like the asteroid shown in close-up in Figure 7-7, are more numerous. Hundreds of thousands of kilometer-sized

FIGURE 7-7 R I **V** U X G

An Asteroid The asteroid shown in this image, 433 Eros, is only 33 km (21 mi) long and 13 km (8 mi) wide—about the same size as the island of Manhattan. Because Eros is so small, its gravity is too weak to have pulled it into a spherical shape. This image was taken in 2000 by NEAR Shoemaker, the first spacecraft to orbit around and land on an asteroid. (NEAR Project, NLR, JHUAPL, Goddard SVS, NASA)

asteroids are known, and there are probably hundreds of thousands more that are boulder-sized or smaller. All of these objects orbit the Sun in the same direction as the planets.

Most (although not all) asteroids orbit the Sun at distances of 2 to 3.5 AU. This region of the solar system between the orbits of Mars and Jupiter is called the **asteroid belt.**

CAUTION! One common misconception about asteroids is that they are the remnants of an ancient planet that somehow broke apart or exploded, like the fictional planet Krypton in the comic book adventures of Superman. In fact, the combined mass of all the asteroids (including Ceres, Pallas, and Vesta) is much less than the mass of the Moon, and they were probably never part of any planet-sized body. The early solar system is thought to have been filled with asteroidlike objects, most of which were incorporated into the planets. The "leftover" objects that missed out on this process make up our present-day population of asteroids.

CONCEPTCHECK 7-8

Is the largest asteroid, Ceres, about the same size as a large city, a large U.S. state, a large country, or Earth itself?

Answer appears at the end of the chapter.

Trans-Neptunian Objects

While asteroids are the most important small bodies in the inner solar system, the outer solar system is the realm of the **trans-Neptunian objects.** As the name suggests, these are small bodies whose orbits lie beyond the orbit of Neptune. The first of these to be discovered (in 1930) was Pluto, with a diameter of only 2274 km. This is larger than any asteroid, but smaller than any planet or any of the moons listed in Table 7-2. The orbit of Pluto has a greater semimajor axis (39.54 AU), is more steeply inclined to the ecliptic (17.15°), and has a greater eccentricity (0.250) than that of any of the planets (**Figure 7-8**). In fact, Pluto's noncircular orbit sometimes takes it nearer the Sun than Neptune. (Happily, the orbits of Neptune and Pluto are such that they will never collide.) Pluto's density is only 2000 kg/m³, about the same as Neptune's moon Triton, shown in Table 7-2. Hence, its composition is thought to be a mixture of about 70% rock and 30% ice.

Since 1992 astronomers have discovered more than 900 other trans-Neptunian objects, and are discovering more each year. Like asteroids, the vast majority of trans-Neptunian objects orbit the Sun in the same direction as the planets. A handful of trans-Neptunian objects are comparable in size to Pluto; at least one, Eris, is even larger than Pluto, as well as being in an orbit that is much larger, more steeply inclined, and more eccentric (Figure 7-8).

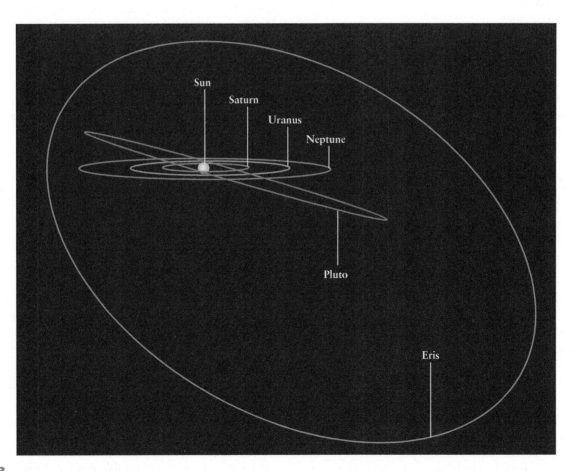

FIGURE 7-8

Trans-Neptunian Objects Pluto and Eris are the two largest trans-Neptunian objects, small worlds of rock and ice that orbit beyond Neptune. Unlike the orbits of the planets, the orbits of these two objects are steeply inclined to the ecliptic: Pluto's orbit is tilted by about 17°, and that of Eris is tilted by 44°.

BOX 7-2 TOOLS OF THE ASTRONOMER'S TRADE

Kinetic Energy, Temperature, and Whether Planets Have Atmospheres

A moving object possesses energy as a result of its motion. The faster it moves, the more energy it has (see Figure 4-20). Energy of this type is called **kinetic energy.** If an object of mass m is moving with a speed v its kinetic energy E_k is given by

Kinetic energy
$$E_k = \frac{1}{2} mv^2$$
of an object

E_k = kinetic energy of an object

m = mass of object

v = speed of object

This expression for kinetic energy is valid for all objects, both big and small, from atoms and molecules to planets and stars, as long as their speeds are slow in relation to the speed of light. If the mass is expressed in kilograms and the speed in meters per second, the kinetic energy is expressed in joules (J).

EXAMPLE: An automobile of mass 1000 kg driving at a typical freeway speed of 30 m/s (= 108 km/h = 67 mi/h) has a kinetic energy of

$$E_k = \frac{1}{2} (1000 \times 30^2) = 450{,}000 \text{ J} = 4.5 \times 10^5 \text{ J}$$

Consider a gas, such as the atmosphere of a star or planet. Some of the gas atoms or molecules will be moving slowly, with little kinetic energy, while others will be moving faster and have more kinetic energy. The temperature of the gas is a direct measure of the *average* amount of kinetic energy per atom or molecule. The hotter the gas, the faster atoms or molecules move, on average, and the greater the average kinetic energy of an atom or molecule (see Figure 5-11).

If the gas temperature is sufficiently high, typically several thousand kelvins, molecules move so fast that when they collide with one another, the energy of the collision can break the molecules apart into their constituent atoms. Thus, the Sun's atmosphere, where the temperature is 5800 K, consists primarily of individual hydrogen atoms rather than hydrogen molecules. By contrast, the hydrogen atoms in Earth's atmosphere (temperature 290 K) are combined with oxygen atoms into molecules of water vapor (H_2O).

The physics of gases tells us that in a gas of temperature T (in kelvins), the average kinetic energy of an atom or molecule is

Kinetic energy of a gas atom or molecule
$$E_k = \frac{3}{2} kT$$

E_k = average kinetic energy of a gas atom or molecule, in joules

$k = 1.38 \times 10^{-23}$ J/K

T = temperature of gas, in kelvins

The quantity k is called the Boltzmann constant. Note that the higher the gas temperature, the greater the average kinetic energy of an atom or molecule of the gas. This average kinetic energy becomes zero at absolute zero, or $T = 0$, the temperature at which molecular motion is at a minimum.

At a given temperature, all kinds of atoms and molecules will have the same average kinetic energy. But the average *speed* of a given kind of atom or molecule depends on the particle's mass. To see this dependence on mass, note that the average kinetic energy of a gas atom or molecule can be written in two equivalent ways:

$$E_k = \frac{1}{2} mv^2 = \frac{3}{2} kT$$

where v represents the average speed of an atom or molecule in a gas with temperature T. Rearranging this equation, we obtain

Average speed of a gas atom or molecule
$$v = \sqrt{\frac{3kT}{m}}$$

v = average speed of a gas atom or molecule, in m/s

$k = 1.38 \times 10^{-23}$ J/K

T = temperature of gas, in kelvins

m = mass of the atom or molecule, in kilograms

For a given gas temperature, the greater the mass of a given type of gas atom or molecule, the slower its average speed. (The value of v given by this equation is actually slightly higher than

Table 7-4 lists the seven largest trans-Neptunian objects known as of this writing. (Note that Charon is actually a satellite of Pluto.) Haumea and its two moons are shown. While Haumea's shape has not been observed directly, modeling of its fuzzy image suggests that it is unusually elongated, possibly from a collision.

Just as most asteroids lie in the asteroid belt, most trans-Neptunian objects orbit within a band called the **Kuiper belt** (pronounced "ki-per") that extends from 30 AU to 50 AU from the Sun and is centered on the plane of the ecliptic. Like asteroids, many more trans-Neptunian objects remain to be discovered:

the average speed of the atoms or molecules in the gas, but it is close enough for our purposes here. If you are studying physics, you may know that v is actually the root-mean-square speed.)

EXAMPLE: What is the average speed of the oxygen molecules that you breathe at a room temperature of 20°C (= 68°F)?

Situation: We are given the temperature of a gas and are asked to find the average speed of the gas molecules.

Tools: We use the relationship $v = \sqrt{3kT/m}$ where T is the gas temperature in kelvins and m is the mass of a single oxygen molecule in kilograms.

Answer: To use the equation to calculate the average speed v, we must express the temperature T in kelvins (K) rather than degrees Celsius (°C). As we learned in Box 5-1, we do this by adding 273 to the Celsius temperature, so 20°C becomes (20 + 273) = 293 K. The mass m of an oxygen molecule is not given, but you can easily find that the mass of an oxygen *atom* is 2.66×10^{-26} kg. The mass of an oxygen molecule (O_2) is twice the mass of an oxygen atom, or $2(2.66 \times 10^{-26}$ kg$)$ $= 5.32 \times 10^{-26}$ kg. Thus, the average speed of an oxygen molecule in 20°C air is

$$v = \sqrt{\frac{3(1.38 \times 10^{-23})(293)}{5.32 \times 10^{-26}}} = 478 \text{ m/s} = 0.478 \text{ km/s}$$

Review: This speed is about 1700 kilometers per hour, or about 1100 miles per hour. Hence, atoms and molecules move rapidly in even a moderate-temperature gas.

In some situations, atoms and molecules in a gas may be moving so fast that they can overcome the attractive force of a planet's gravity and escape into interplanetary space. The minimum speed that an object at a planet's surface must have in order to permanently leave the planet is called the planet's **escape speed** (see Figure 4-22). The escape speed for a planet of mass M and radius R is given by

$$v_{escape} = \sqrt{\frac{2GM}{R}}$$

where $G = 6.67 \times 10^{-11}$ N m²/kg² is the universal constant of gravitation.

The accompanying table gives the escape speed for the Sun, the planets, and the Moon. For example, to escape Earth, a cannon ball would have to be shot with an implausible speed greater than 11.2 km/s (25,100 mi/h).

A good rule of thumb is that a planet can retain a gas if the escape speed is at least 6 times greater than the average speed of the molecules in the gas. (Some molecules are moving slower than average, and others are moving faster, but very few are moving more than 6 times faster than average.) In such a case, very few molecules will be moving fast enough to escape from the planet's gravity.

EXAMPLE: Consider Earth's atmosphere. We saw that the average speed of oxygen molecules is 0.478 km/s at room temperature. The escape speed from Earth (11.2 km/s) is much more than 6 times the average speed of the oxygen molecules, so Earth has no trouble keeping oxygen in its atmosphere.

A similar calculation for hydrogen molecules (H_2) gives a different result, however. At 293 K, the average speed of a hydrogen molecule is 1.9 km/s. Six times this speed is 11.4 km/s, which is about the escape speed from Earth. Thus, Earth does not retain hydrogen in its atmosphere. Any hydrogen released into the air slowly leaks away into space. On Jupiter, by contrast, the escape speed is so high that even the lightest gases such as hydrogen are retained in its atmosphere. But on Mercury the escape speed is low and the temperature high (so that gas molecules move faster), and Mercury cannot retain any significant atmosphere at all.

Object	Escape speed (km/s)
Sun	618.0
Mercury	4.3
Venus	10.4
Earth	11.2
Moon	2.4
Mars	5.0
Jupiter	59.5
Saturn	35.5
Uranus	21.3
Neptune	23.5

Astronomers estimate that there are 35,000 or more such objects with diameters greater than 100 km. If so, the combined mass of all trans-Neptunian objects is comparable to the mass of Jupiter, and is several hundred times greater than the combined mass of all the asteroids found in the inner solar system.

Like asteroids, trans-Neptunian objects are thought to be debris left over from the formation of the solar system. In the inner regions of the solar system, rocky fragments have been able to endure continuous exposure to the Sun's heat, but any ice originally present would have evaporated. Far from the Sun, ice has survived for

TABLE 7-4 Seven Large Trans-Neptunian Objects

	Quaoar	Haumea	Sedna	Makemake	Pluto	Charon (satellite of Pluto)	Eris
Average distance from the Sun (AU)	43.54	43.34	489	45.71	39.54	39.54	67.67
Orbital period (years)	287	285	10,800	309	248.6	248.6	557
Orbital eccentricity	0.035	0.189	0.844	0.155	0.250	0.250	0.442
Inclination of orbit to the ecliptic	8.0°	28.2°	11.9°	29.0°	17.15°	17.15°	44.2°
Approximate diameter (km)	1250	1500	1600	1800	2274	1190	2900

R I **V** **U** X G

(Haumea: A. Field [STScI]/NASA; Charon: Lanthanum-138; all others: Alan Stern [Southwest Research Institute]/Marc Buie [Lowell Observatory]/NASA/ESA)

billions of years. Thus, *debris* in the solar system naturally divides into two families—rocky asteroids and partially icy trans-Neptunian objects—which can be arranged according to distance from the Sun, just like the two categories of planets (terrestrial and Jovian).

CONCEPTCHECK 7-9

Is Pluto an asteroid, planet, Kuiper belt object, or trans-Neptunian object?

CALCULATIONCHECK 7-3

Which has a larger average orbital radius, the asteroid belt or the Kuiper belt?

Answers appear at the end of the chapter.

Comets

Two objects in the Kuiper belt can collide if their orbits cross each other. When this happens, a fragment a few kilometers across can be knocked off one of the colliding objects and be diverted into an elongated orbit that brings it close to the Sun. Such small objects, each a combination of rock and ice, are called **comets**. When a comet comes close enough to the Sun, the Sun's radiation vaporizes some of the comet's ices, producing long flowing tails of gas and dust particles (**Figure 7-9**). Astronomers deduce the composition of comets by studying the spectra of these tails.

FIGURE 7-9 R I **V** U X G

A Comet This photograph shows Comet Hale-Bopp as it appeared in April 1997. The solid part of a comet like this is a chunk of dirty ice a few tens of kilometers in diameter. When a comet passes near the Sun, solar radiation vaporizes some of the icy material, forming a bluish tail of gas and a white tail of dust. Both tails can extend for tens of millions of kilometers. (Agencia el Universal/AP Images)

CAUTION! Science-fiction movies and television programs sometimes show comets tearing across the night sky like a rocket, which would be a pretty impressive sight. However, comets do not zoom across the sky. Like the planets, comets orbit the Sun. And like the planets, comets move hardly at all against the background of stars over the course of a single night (see Section 4-1). If you are lucky enough to see a bright comet, it will not zoom dramatically from horizon to horizon. Instead, it will seem to hang majestically among the stars, so you can admire it at your leisure.

Some comets appear to originate from locations far beyond the Kuiper belt. The source of these is thought to be a swarm of comets that forms a spherical "halo" around the solar system called the **Oort comet cloud** (also known as Oort cloud). This hypothesized "halo" extends to 50,000 AU from the Sun (about one-fifth of the way to the nearest other star). Because the Oort cloud is so distant, it has not yet been possible to detect objects in the Oort cloud directly.

CONCEPTCHECK 7-10

Consider a comet that has an orbital plane that is perpendicular to the plane of Earth's orbit around the Sun. Did this comet most likely originate from the Kuiper belt or the Oort cloud?

Answer appears at the end of the chapter.

7-6 Craters on planets and satellites are the result of impacts from interplanetary debris

One of the great challenges in studying planets and satellites is determining their internal structures. Are they solid or liquid inside? If there is liquid in the interiors, is the liquid calm or in agitated motion? Because planets are opaque, we cannot see directly into their interiors to answer these questions. But we can gather important clues about the interiors of terrestrial planets and satellites by studying the extent to which their surfaces are covered with craters (Figure 7-10). To see how information about the interiors is gathered, we first need to understand where craters come from.

> Scientists study craters on a planet or satellite to learn the age and geologic history of the surface

The Origin of Craters

The planets orbit the Sun in roughly circular orbits. But many asteroids and comets are in more elongated orbits. Such an elongated orbit can put these small objects on a collision course with a planet or satellite. If the object collides with a Jovian planet, it is swallowed up by the planet's thick atmosphere. (Astronomers actually saw an event of this kind in 1994, when a comet crashed into Jupiter.) But if the object collides with the solid surface of a terrestrial planet or a satellite, the result is an **impact crater** (see Figure 7-10). Such impact craters, found throughout the solar system, offer stark evidence of these violent collisions.

The easiest way to view impact craters is to examine the Moon through a telescope or binoculars. Some 30,000 lunar craters are visible, with diameters ranging from 1 km to several hundred km. Close-up photographs from lunar orbit have revealed millions of craters too small to be seen from Earth (Figure 7-10a). These smaller craters are thought to have been caused by impacts of relatively small objects called **meteoroids,** which range in size from a few hundred meters across to the size of a pebble or smaller. Meteoroids are the result of collisions between asteroids, whose

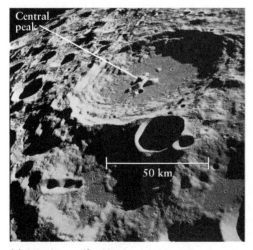

(a) A crater on the Moon

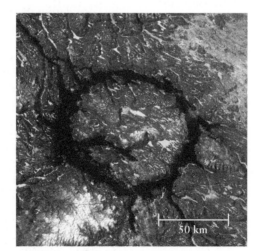

(b) A crater on Earth

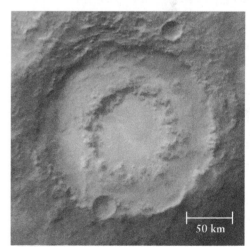

(c) A crater on Mars

FIGURE 7-10 R I V U X G

Impact Craters These images, all taken from spacecraft, show impact craters on three different worlds. **(a)** The Moon's surface has craters of all sizes. The large crater near the middle of this image is about 80 km (50 mi) in diameter, equal to the length of San Francisco Bay. **(b)** Manicouagan Reservoir in Quebec is the relic of a crater formed by an impact more than 200 million years ago. The crater was eroded over the ages by the advance and retreat of glaciers, leaving a ring lake 100 km (60 mi) across. **(c)** Lowell Crater in the southern highlands of Mars is 201 km (125 mi) across. Like the image of the Moon in part (a), there are craters on top of craters. Note the light-colored frost formed by condensation of carbon dioxide from the Martian atmosphere. (a: NASA; b: JSC/NASA; c: NASA/JPL/MSSS)

orbits sometimes cross. The chunks of rock that result from these collisions go into independent orbits around the Sun, which can lead them to collide with the Moon or another world.

When German astronomer Franz Gruithuisen proposed in 1824 that lunar craters were the result of impacts, a major sticking point was the observation that nearly all craters are circular. If craters were merely gouged out by high-speed rocks, a rock striking the Moon in any direction except straight downward would have created a noncircular crater. A century after Gruithuisen, it was realized that a meteoroid colliding with the Moon generates a shock wave in the lunar surface that spreads out from the point of impact. Such a shock wave produces a circular crater no matter what direction the meteoroid was moving. (In a similar way, the craters made by artillery shells are almost always circular.) Many of the larger lunar craters also have a central peak, which is characteristic of a high-speed impact (see Figure 7-10a). Craters made by other processes, such as volcanic action, would not have central peaks of this sort.

Comparing Cratering on Different Worlds

Not all planets and satellites show the same amount of cratering. The Moon is heavily cratered over its entire surface, with craters on top of craters, as shown in Figure 7-10a. On Earth, by contrast, craters are very rare. Geologists have identified fewer than 200 impact craters on our planet (Figure 7-10b). Our understanding is that both Earth and the Moon formed at nearly the same time and have been bombarded at comparable rates over their histories. Why, then, are craters so much rarer on Earth than on the Moon?

The answer is that Earth is a *geologically active* planet: Through **plate tectonics**—the motion of rocky plates over Earth's surface—the continents slowly change their positions over eons, new material flows onto the surface from the interior (as occurs in a volcanic eruption), and old surface material is pushed back into the interior (as occurs off the coast of Chile, where the ocean bottom is slowly being pushed beneath the South American continent). These processes, coupled with erosion from wind and water, cause craters on Earth to be erased over time. The few craters found on Earth today must be relatively recent, since there has not yet been time to erase them.

The Moon, by contrast, is geologically *inactive*. There are no volcanoes and no motion of continents (and, indeed, no continents). Furthermore, the Moon has neither oceans nor an atmosphere, so there is no erosion as we know it on Earth. With none of the processes that tend to erase craters on Earth, the Moon's surface remains pockmarked with the scars of billions of years of impacts.

In order for a planet to be geologically active, its interior must be at least partially molten. Even if the surface does not undergo plate tectonics—as in the case of Jupiter's moon Io—a partially molten interior is required to generate volcanoes with molten lava. Therefore, we would expect geologically inactive (and hence heavily cratered) worlds like the Moon to have less molten material in their interiors than does Earth. Investigation of these inactive worlds bears this out. But *why* is the Moon's interior less molten than Earth's?

ANALOGY To see one simple answer to this question, notice that a large turkey or roast taken from the oven will stay warm inside for hours, but a single meatball will cool off much more rapidly. The reason is that the meatball has more surface area relative to its volume, and so it can more easily lose heat to its surroundings. A planet or satellite also tends to cool down as it emits electromagnetic radiation into space (see Section 5-3); the smaller the planet or satellite, the greater its surface area relative to its volume, and the more readily it can radiate away heat (Figure 7-11). Both Earth and the Moon were probably completely molten when they first formed, but because the Moon (diameter 3476 km) is so much smaller than Earth (diameter 12,756 km), it has lost much of its internal heat and has a much more solid interior. There are two main sources for a planet's heat, which we will discuss in Chapter 8—gravity and radioactivity—and both of these sources result in more heat for larger planets.

Cratering Measures Geologic Activity

By considering these differences between Earth and the Moon, we have uncovered a general rule for worlds with solid surfaces:

The smaller the terrestrial world, the less internal heat it is likely to have retained, and, thus, the less geologic activity it will display on its surface. The less geologically active the world, the older and hence more heavily cratered its surface.

This rule means that we can use the amount of cratering visible on a planet or satellite to estimate the age of its surface and how geologically active it is. As an example, Mercury has a heavily cratered surface, which means that the surface is very old. This geologically inactive surface is in agreement with Mercury being the smallest of the terrestrial planets (see Table 7-1). Due to its small size, it has lost the internal heat required to sustain geologic activity. On Venus, by comparison, there are only about

Planet #1

Planet #2

Compared to planet #1, planet #2:
— has 1/2 the radius
— has 1/4 the surface area (so it can lose heat only 1/4 as fast)
— but has only 1/8 the volume (so it has only 1/8 as much heat to lose)

Hence compared to planet #1, planet #2:
— will cool off more rapidly
— will sustain less geologic activity
— will have more craters

FIGURE 7-11

Planet Size and Cratering Of these two hypothetical planets, the smaller one (#2) has less volume and less internal heat, as well as less surface area from which to radiate heat into space. But the ratio of surface area to volume is greater for the smaller planet. Hence, the smaller planet will lose heat faster, have a colder interior, and be less geologically active. It will also have a more heavily cratered surface, since it takes geologic activity to erase craters.

100 km

FIGURE 7-12   R I V U X G

A Martian Volcano Olympus Mons is the largest of the inactive volcanoes of Mars and the largest volcano in the solar system. The base of Olympus Mons measures 600 km (370 mi) in diameter, and the scarps (cliffs) that surround the base are 6 km (4 mi) high. The caldera, or volcanic crater, at the summit is approximately 70 km across, large enough to contain the state of Rhode Island. (Kees Veenenbos/Science Source)

a thousand craters larger than a few kilometers in diameter, many more than have been found on Earth but only a small fraction of the number on the Moon or Mercury. Venus is only slightly smaller than Earth, and it has enough internal heat to power the geologic activity required to erase most of its impact craters.

Mars is an unusual case, in that extensive cratering (Figure 7-10c) is found only in the higher terrain; the lowlands of Mars are remarkably smooth and free of craters. Thus, it follows that the Martian highlands are quite old, while the lowlands have a younger surface from which most craters have been erased. Considering the planet as a whole, the amount of cratering on Mars is intermediate between that on Mercury and Earth. This agrees with our general rule, because Mars is intermediate in size between Mercury and Earth. The interior of Mars was once hotter and more molten than it is now, so that geologic processes were able to erase some of the impact craters. A key piece of evidence that supports this picture is that Mars has a number of immense volcanoes (Figure 7-12). These volcanoes were active when Mars was young, but as this relatively small planet cooled down and its interior solidified, the supply of molten material to the volcanoes from the Martian interior was cut off. As a result, all of the volcanoes of Mars are now inactive.

As for all rules, there are limitations and exceptions to the rule relating a world's size to its geologic activity. One limitation is that the four terrestrial planets all have slightly different compositions, which affects the types and extent of geologic activity that can take place on their surfaces. The different compositions also complicate the relationship between the number of craters and the age of the surface. An important exception to our rule is Jupiter's satellite Io, which, despite its small size, is the most volcanic world in the solar system (see Section 7-2). Something must be supplying Io with energy to keep its interior hot; this energy comes from Jupiter, which exerts powerful tidal forces on Io as it moves in a relatively small orbit around its planet. These tidal forces cause Io to flex like a ball of clay being kneaded between your fingers, and this flexing heats up the satellite's interior. But despite these

limitations and exceptions, the relationships between a world's size, internal heat, geologic activity, and amount of cratering are powerful tools for understanding the terrestrial planets and satellites.

CONCEPTCHECK **7-11**

Io, which is a *moon,* is an exception to the rule that smaller worlds should have less geologic activity. Do you expect that some small *planets* might break this rule as well?

Answer appears at the end of the chapter.

7-7 A planet with a magnetic field indicates a fluid interior in motion

The amount of impact cratering on a terrestrial planet or satellite provides indirect evidence about whether the planet or satellite has a molten interior. But another, more direct tool for probing the interior of *any* planet or satellite is an ordinary compass, which senses the magnetic field outside the planet or satellite. Magnetic field measurements prove to be an extremely powerful way to investigate the internal structure of a world without having to actually dig into its interior. To illustrate how this works, consider the behavior of a compass on Earth.

> By studying the magnetic field of a planet or satellite, scientists can learn about that world's interior

The needle of a compass on Earth points north because it aligns with Earth's *magnetic field*. Such magnetic fields arise whenever electrically charged particles are in motion. For example, a loop of wire carrying an electric current generates a magnetic field in the space around it. The source of the magnetic field that surrounds an ordinary bar magnet (Figure 7-13a) is complicated, but part of it is created by the motion of negatively charged electrons within the iron atoms of which the magnet is made. Earth's magnetic field is similar to that of a bar magnet, as Figure 7-13b shows. The consensus among geologists is that this magnetic field is caused by the motion of the liquid portions of Earth's interior. Because this molten material (mostly iron) conducts electricity, these motions give rise to electric currents, which in turn produce Earth's magnetic field. Our planet's rotation helps to sustain these motions and hence the magnetic field. This process for producing a magnetic field is called a **dynamo**.

CAUTION! While Earth's magnetic field is similar to that of a giant bar magnet, you should not take this picture too literally. Earth is *not* simply a magnetized ball of iron. In an iron bar magnet, the electrons of different atoms orbit their nuclei in the same general direction, so that the magnetic fields generated by individual atoms add together to form a single, strong field. But at temperatures above 770°C (1418°F = 1043 K), the orientations of the electron orbits become randomized. The fields of individual atoms tend to cancel each other out, and the iron loses its magnetism. Geological evidence shows that almost all of Earth's interior is hotter than 770°C, so the iron there cannot be extensively magnetized. The correct picture is that Earth acts as a dynamo: The liquid iron carries electric currents, and these currents create Earth's magnetic field.

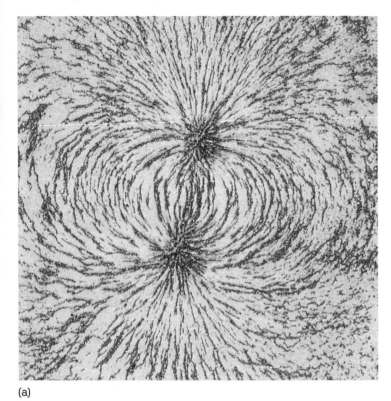

(a)

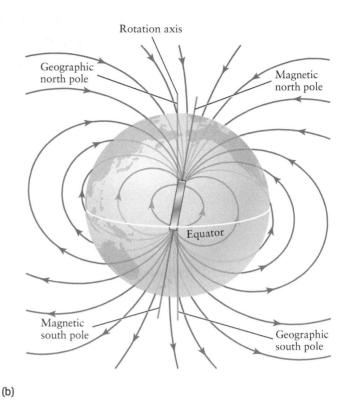

(b)

FIGURE 7-13

The Magnetic Fields of a Bar Magnet and of Earth **(a)** This picture was made by placing a piece of paper on top of a bar magnet, then spreading thin iron filings on the paper. Each elongated iron filing acts like a compass needle and aligns with the magnetic field at its location. The pattern of the filings show the magnetic field lines, which appear to stream from one of the magnet's poles to the other. **(b)** Earth's magnetic field lines have a similar pattern. Although Earth's field is produced in a different way—by electric currents in the liquid portion of our planet's interior—the field is much the same as if there were a giant bar magnet inside Earth. This "bar magnet" is not exactly aligned with Earth's rotation axis, which is why the magnetic north and south poles are not at the same locations as the true, or geographic, poles. A compass needle points toward the north magnetic pole, not the true north pole.

(a: Jules Bucher/Photo Researchers)

If a planet or satellite has a mostly solid interior, then the dynamo mechanism cannot work: Material in the interior cannot flow, there are no electric currents, and the planet or satellite does not generate a magnetic field. One example of this is the Moon. As we saw in Section 7-6, the extensive cratering of the lunar surface indicates that the Moon has no geologic activity and must therefore have a mostly solid interior. Measurements made during the *Apollo* missions, in which 12 humans visited the lunar surface between 1969 and 1972, showed that the present-day Moon indeed has no global magnetic field. However, careful magnetic measurements of lunar rocks returned by the *Apollo* astronauts indicate that the Moon *did* have a weak magnetic field when the rocks solidified. These rocks, like the rest of the lunar surface, are very old. Hence, in the distant past the Moon may have had a small amount of molten iron in its interior that acted as a dynamo. This material presumably solidified at least partially as the Moon cooled, so that the lunar magnetic field disappeared.

We have now identified another general rule about planets and satellites:

A planet or satellite with a global magnetic field has liquid material in its interior that conducts electricity and is in motion, generating the magnetic field.

Thus, by studying the magnetic field of a planet or satellite, we can learn about that world's interior. So, many spacecraft carry devices called **magnetometers** to measure magnetic fields. Magnetometers are often placed on a long boom extending outward from the body of the spacecraft (Figure 7-14). The boom isolates them from the magnetic fields produced by electric currents in the spacecraft's own circuitry.

Measurements made with magnetometers on spacecraft have led to a number of striking discoveries. For example, it has been found that Mercury has a planetwide magnetic field like Earth's, although it's only about 1% as strong as Earth's field. If Mercury's magnetic field was simply due to a magnetized crust, variations in the field due to crust-altering craters would have been detected, yet no such variations were observed. Instead, Mercury's smooth magnetic field—created beneath the surface—suggests that at least *some* of the planet's interior is in a liquid state to act as a magnetic dynamo. A liquid interior is quite surprising because Mercury's heavily cratered surface indicates a lack of geologic activity, which is usually associated with a solid interior. By contrast, Venus has no measurable planetwide magnetic field, even though the paucity of craters on its surface indicates the presence of geologic activity and hence a hot interior of the planet. One possible reason for the lack of a magnetic field on Venus is that the planet turns on its

FIGURE 7-14

Probing the Magnetic Field of Saturn This illustration depicts the *Cassini* spacecraft as it entered orbit around Saturn on July 1, 2004. In addition to telescopes for observing Saturn and its satellites, *Cassini* carries a magnetometer for exploring Saturn's magnetic field. The magnetometer is located on the long boom that extends down and to the right from the body of the spacecraft. (The glow at the end of the boom is the reflection of the Sun.) (JPL/NASA)

axis very slowly, taking 243 days for a complete rotation. Because of this slow rotation, the fluid material within the planet is hardly agitated at all, and so may not move in the fashion that generates a magnetic field.

Mars has a very interesting magnetic field that sheds light on its early history. The magnetometer aboard the *Mars Global Surveyor* spacecraft, which went into orbit around Mars in 1997, found that most of the planet's crust is magnetized (Figure 7-15): Mars has no planetwide magnetic field like Earth, but portions of its rocky surface are magnetized. This information actually tells us a lot about the Martian past. During the planet's first billion years or so, the interior was still hot and molten. Electric currents in the flowing molten material could then produce a planetwide magnetic field. As surface material cooled and solidified, the crust became magnetized by the planetwide field, and remained magnetized even after the internally generated planetwide field shut down.

The alternating stripes of magnetism on the Martian crust—seen most clearly in red and blue in the middle of Figure 7-15—tell us even more about the Martian past. Far from being "alien" features, these familiar stripes have been heavily studied in the places they occur on Earth, where they result from plate tectonics. Plate tectonics—the motion of rocky plates over a planet's surface—are driven by a planet's internal heat. Furthermore, on Earth plate tectonics stabilizes our atmosphere; Mars may have lost its atmosphere after its interior cooled too much to power further plate motion.

The most intense planetary magnetic field in the solar system is that of Jupiter: At the tops of Jupiter's clouds, the magnetic field is about 14 times stronger than the field at Earth's surface. It is thought that Jupiter's field, like Earth's, is produced by a dynamo acting deep within the planet's interior. Unlike Earth, however, Jupiter is composed primarily of hydrogen and helium, not substances like iron that conduct electricity. How, then, can Jupiter have a dynamo that generates such strong magnetic fields?

To answer this question, recall from Section 5-8 that a hydrogen atom consists of a single proton orbited by a single electron. Deep inside Jupiter, the pressure is so great and hydrogen atoms are squeezed so close together that electrons can hop from one atom to another. This hopping motion creates an electric current, just as the ordered movement of electrons in the copper wires of a flashlight constitutes an electric current. In other words, the highly compressed hydrogen deep inside Jupiter behaves like an electrically conducting metal; thus, it is called **liquid metallic hydrogen.**

Laboratory experiments show that hydrogen becomes a liquid metal when the pressure is more than about 1.4 million times ordinary atmospheric pressure on Earth. Recent calculations suggest that this transition occurs about 7000 km below Jupiter's cloudtops. Most of the planet's enormous bulk lies below this level, so there is a tremendous amount of liquid metallic hydrogen within Jupiter. Since Jupiter rotates rapidly—a "day" on Jupiter is just less than 10 hours long—this liquid metal moves rapidly, generating the planet's powerful magnetic field. Saturn also has

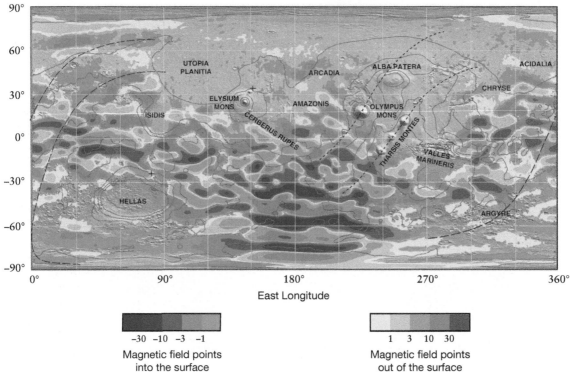

-30 -10 -3 -1
Magnetic field points
into the surface

1 3 10 30
Magnetic field points
out of the surface

FIGURE 7-15

Relic Magnetism on Mars For almost a billion years after Mars formed, molten material with electric currents generated a planetwide magnetic field. Most of the surface was magnetized during this time, which is the crustal field we measure today. Regions without magnetism probably experienced higher temperatures that can remove magnetism, or were formed by lava flows after the planetwide field shut down. Some magnetic stripes are visible in the lower middle of the image, where the field alternately points up and down. These stripes suggest that Mars went through a period with plate tectonics. (NASA)

a magnetic field produced by dynamo action in liquid metallic hydrogen. (The field is weaker than Jupiter's because Saturn is a smaller planet with less internal pressure, so there is less of the liquid metal available.)

Uranus and Neptune also have magnetic fields, but they cannot be produced in the same way: Because these planets are relatively small, the internal pressure is not great enough to turn liquid hydrogen into a metal. Instead, it is thought that both Uranus and Neptune have large amounts of liquid water in their interiors and that this water has molecules of ammonia and other substances dissolved in it. (The fluid used for washing windows has a similar chemical composition.) Under the pressures found in this interior water, the dissolved molecules lose one or more electrons and become electrically charged (that is, they become ionized; see Section 5-8). Water is a good conductor of electricity when it has such electrically charged molecules dissolved in it, and electric currents in this fluid would be the source of the magnetic fields of Uranus and Neptune. However, there is no direct evidence of currents in liquid water for these planets, and an alternative model proposes an electrically conductive mixture of hydrogen and rocky material.

CONCEPTCHECK 7-12

By comparing parts (a) and (b) of Figure 7-13, can you conclude that there is a large bar magnet inside Earth?

CONCEPTCHECK 7-13

Does every planet with a molten interior produce a magnetic field? Do any of the Jovian planets create their magnetic fields through electrically conductive molten iron?

Answers appear at the end of the chapter.

7-8 The diversity of the solar system is a result of its origin and evolution

Our brief tour of the solar system has revealed its almost dizzying variety. No two planets are alike, satellites come in all sizes, the extent of cratering varies from one terrestrial planet to another, and the magnetic fields of different planets vary dramatically in their strength and in how they are produced. (The *Cosmic Connections* that closes this chapter summarizes these properties of the planets.) All of this variety leads us to a simple yet profound question: *Why are the planets and satellites of the solar system so different from each other?*

> The similarities and differences among the planets can be logically explained by a model of the solar system's origin and evolution

Among humans, the differences from one individual to another result from heredity (the genetic traits passed on from an individual's parents) and environment (the circumstances under which the individual matures to an adult). As we will find in the following chapter, much the same is true for the worlds of the solar system.

In Chapter 8 we will see evidence that the entire solar system shares a common "heredity," in that the planets, satellites, comets, asteroids, and the Sun itself formed from the same cloud of interstellar gas and dust. The composition of this cloud was shaped by cosmic processes, including nuclear reactions that took place within stars that died long before our solar system was formed. We will see how different planets formed in different environments depending on their distance from the Sun and will discover how these environmental variations gave rise to the planets and satellites of our present-day solar system. And we will see how we can test these ideas of solar system origin and evolution by studying planetary systems orbiting other stars.

Our journey through the solar system is just beginning. In this chapter we have explored space to examine the variety of the present-day solar system; in Chapter 8 we will journey through time to see how our solar system came to be.

CALCULATIONCHECK 7-4

Using the *Cosmic Connections* figure, *Characteristics of the Planets,* how many Earth masses does it take to equal Jupiter's mass? How many Saturn masses does it take to equal Jupiter's mass?

Answer appears at the end of the chapter.

KEY WORDS

Terms preceded by an asterisk (*) are discussed in the Boxes.

asteroid, p. 180
asteroid belt, p. 181
average density, p. 175
chemical composition,
  p. 177
comet, p. 184
dynamo, p. 187
*escape speed, p. 183
ices, p. 180
impact crater, p. 185
Jovian planet, p. 174
*kinetic energy,
  p. 182
Kuiper belt, p. 182

liquid metallic hydrogen,
  p. 189
magnetometer, p. 188
meteoroid, p. 185
minor planet, p. 180
Oort comet cloud (Oort
  cloud), p. 185
plate tectonics, p. 186
spectroscopy, p. 177
terrestrial planet,
  p. 174
trans-Neptunian object,
  p. 181

KEY IDEAS

Properties of the Planets: All of the planets orbit the Sun in the same direction and in almost the same plane. Most of the planets have nearly circular orbits.

• The four inner planets are called terrestrial planets. They are relatively small (with diameters of 5000 to 13,000 km), have high average densities (4000 to 5500 kg/m^3), and are composed primarily of rocky materials.

• The four giant outer planets are called Jovian planets. They have large diameters (50,000 to 143,000 km) and low average densities (700 to 1700 kg/m^3) and are composed primarily of light elements such as hydrogen and helium.

Satellites and Small Bodies in the Solar System: Besides the planets, the solar system includes satellites of the planets, asteroids, comets, and trans-Neptunian objects.

• Seven large planetary satellites (one of which is the Moon) are comparable in size to the planet Mercury. The remaining satellites of the solar system are much smaller.

• Asteroids are small, rocky objects, while comets and trans-Neptunian objects are made of ice and rock. All are remnants left over from the formation of the planets.

• Most asteroids are found in the asteroid belt between the orbits of Mars and Jupiter, and most trans-Neptunian objects lie in the Kuiper belt outside the orbit of Neptune. Pluto is one of the largest members of the Kuiper belt.

Spectroscopy and the Composition of the Planets: Spectroscopy, the study of spectra, provides information about the chemical composition of objects in the solar system.

• The spectrum of a planet or satellite with an atmosphere reveals the atmosphere's composition. If there is no atmosphere, the spectrum indicates the composition of the surface.

• The substances that make up the planets can be classified as gases, ices, or rock, depending on the temperatures at which they solidify.

Impact Craters: When an asteroid, comet, or meteoroid collides with the surface of a terrestrial planet or satellite, the result is an impact crater.

• Geologic activity renews the surface and erases craters, so a terrestrial world with extensive cratering has an old surface and little or no geologic activity.

• Because geologic activity is powered by internal heat, and smaller worlds lose heat more rapidly, as a general rule smaller terrestrial worlds are more extensively cratered.

Magnetic Fields and Planetary Interiors: Planetary magnetic fields are produced by the motion of electrically conducting liquids inside the planet. This mechanism is called a dynamo. If a planet has no magnetic field, that is evidence that there is little such liquid material in the planet's interior or that the liquid is not in a state of motion.

• The magnetic fields of terrestrial planets are produced by metals such as iron in the liquid state. The stronger fields of the Jovian planets are generated by liquid metallic hydrogen or by water with ionized molecules dissolved in it.

QUESTIONS

Review Questions

1. Do all the planets orbit the Sun in the same direction? Are all of the orbits circular?

The Inner (Terrestrial) Planets	Close to the Sun – Small diameter, small mass – High density			
	Mercury	Venus	Earth	Mars
Average distance from the Sun (AU)	0.387	0.723	1.000	1.524
Equatorial diameter (Earth = 1)	0.383	0.949	1.000	0.533
Mass (Earth = 1)	0.0553	0.8150	1.0000	0.1074
Average density (kg/m^3)	5430	5243	5515	3934

Mercury

Venus

Earth

Mars

Atmosphere
None

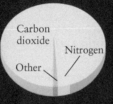

Atmosphere
Carbon dioxide

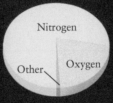

Atmosphere
Nitrogen, oxygen

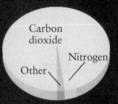

Atmosphere
Carbon dioxide

Magnetic field
Weak

Magnetic field
None

Magnetic field
Moderate, due to
liquid iron core

Magnetic field
Weak, due to
magnetized crust

Interior
Iron-nickel core,
rocky shell

Interior
Iron-nickel core,
rocky shell

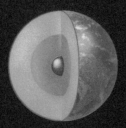

Interior
Iron-nickel core,
rocky shell

Interior
Iron-nickel core,
rocky shell

II Planets and Moons

The polar region in the north of Mars holds truly alien landscapes. Wind-blown dunes of dust form the mounds appearing somewhat pink in this color-enhanced image. In the Martian winter, the dunes are covered in carbon-dioxide snow. As summer approaches, the carbon dioxide evaporates, exposing the underlying dust. The black streaks occur on the steepest slopes where the exposed dust forms miniature avalanches. (NASA/JPL/University of Arizona)

R I **V** U X G

II Planets and Moons

The polar region in the north of Mars holds truly alien landscapes. Wind-blown dunes of dust form the mounds appearing somewhat pink in this color-enhanced image. In the Martian winter, the dunes are covered in carbon-dioxide snow. As summer approaches, the carbon dioxide evaporates, exposing the underlying dust. The black streaks occur on the steepest slopes where the exposed dust forms miniature avalanches. (NASA/JPL/University of Arizona)

R I **V** U X G

To attain this stunning view, the *Cassini* spacecraft was carefully lined up to watch Saturn eclipse the Sun. With the Sun directly behind Saturn, otherwise faint and diffuse ring particles appear bright in this color-enhanced image. The outermost ring contains Saturn's moon Enceladus, and the diffuse ring itself is produced by ejecta from that moon's ice-volcanoes. At about the "10 o'clock" position in the upper left, Earth can be seen just above the bright inner rings. (NASA/JPL/ Space Science Institute)

R I **V** U X G

The Outer (Jovian) Planets	Far from the Sun – Large diameter, large mass – Low density			
	Jupiter	Saturn	Uranus	Neptune
Average distance from the Sun (AU)	5.203	9.554	19.194	30.066
Equatorial diameter (Earth = 1)	11.209	9.449	4.007	3.883
Mass (Earth = 1)	317.8	95.16	14.53	17.15
Average density (kg/m³)	1326	687	1318	1638

Jupiter

Saturn

Uranus

Neptune

Atmosphere
Hydrogen, helium

Atmosphere
Hydrogen, helium

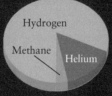

Atmosphere
Hydrogen, helium

Atmosphere
Hydrogen, helium

Magnetic field
Strong, due to
liquid metallic hydrogen

Magnetic field
Strong, due to
liquid metallic hydrogen

Magnetic field
Moderate, due to
dissolved ions

Magnetic field
Moderate, due to
dissolved ions

Interior
Rocky core, liquid
hydrogen and helium

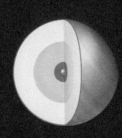

Interior
Rocky core, liquid
hydrogen and helium

Interior
Rocky core, liquid
water and ammonia

Interior
Rocky core, liquid
water and ammonia

2. What are the characteristics of a terrestrial planet?

3. What are the characteristics of a Jovian planet?

4. In what ways are the largest satellites similar to the terrestrial planets? In what ways are they different? Which satellites are largest?

5. *TUTORIAL 7.2* On March 16, 2007, Venus was 1.97×10^8 km from Earth and had an angular diameter of 12.7 arcsec. Using the small-angle formula from Box 1-1, calculate the diameter of Venus.

6. What is meant by the average density of a planet? What does the average density of a planet tell us?

7. *TUTORIAL 7.3* What are the differences in chemical composition between the terrestrial and Jovian planets?

8. The absorption lines in the spectrum of a planet or satellite do not necessarily indicate the composition of the planet or satellite's atmosphere. Why not?

9. Why are hydrogen and helium abundant in the atmospheres of the Jovian planets but present in only small amounts in Earth's atmosphere?

10. What is an asteroid? What is a trans-Neptunian object? In what ways are these minor members of the solar system like or unlike the planets?

11. What are the asteroid belt, the Kuiper belt, and the Oort cloud? Where are they located? How do the objects found in these three regions compare?

12. In what ways is Pluto similar to a terrestrial planet? In what ways is it different?

13. What is the connection between comets and the Kuiper belt? Between comets and the Oort cloud?

14. What is one piece of evidence that impact craters are actually caused by impacts?

15. What is the relationship between the extent to which a planet or satellite is cratered and the amount of geologic activity on that planet or satellite?

16. How do we know that the surface of Venus is older than Earth's surface but younger than the Moon's surface?

17. Why do smaller worlds retain less of their internal heat?

18. How does the size of a terrestrial planet influence the amount of cratering on the planet's surface?

19. How is the magnetic field of a planet different from that of a bar magnet? Why is a large planet more likely to have a magnetic field than a small planet?

20. Could you use a compass to find your way around Venus? Why or why not?

21. If Mars has no planetwide magnetic field, why does it have magnetized regions on its surface?

22. What is liquid metallic hydrogen? Why is it found only in the interiors of certain planets?

Advanced Questions

Questions preceded by an asterisk (*) involve topics discussed in the Boxes.

> **Problem-solving tips and tools**
>
> The volume of a sphere of radius r is $4\pi r^3/3$, and the surface area of a sphere of radius r is $4\pi r^2$. The surface area of a circle of radius r is πr^2. The average density of an object is its mass divided by its volume. To calculate escape speeds, you will need to review Box 7-2. Be sure to use the same system of units (meters, seconds, kilograms) in all your calculations involving escape speeds, orbital speeds, and masses. Appendix 6 gives conversion factors between different sets of units, and Box 5-1 has formulas relating various temperature scales.

23. Mars has two small satellites, Phobos and Deimos. Phobos circles Mars once every 0.31891 day at an average altitude of 5980 km above the planet's surface. The diameter of Mars is 6794 km. Using this information, calculate the mass and average density of Mars.

24. Figure 7-3 shows the spectrum of Saturn's largest satellite, Titan. Can you think of a way that astronomers can tell which absorption lines are due to Titan's atmosphere and which are due to the atmospheres of the Sun and Earth? Explain.

*25. (a) Find the mass of a hypothetical spherical asteroid 2 km in diameter and composed of rock with average density 2500 kg/m³. (b) Find the speed required to escape from the surface of this asteroid. (c) A typical jogging speed is 3 m/s. What would happen to an astronaut who decided to go for a jog on this asteroid?

*26. The hypothetical asteroid described in Question 25 strikes Earth with a speed of 25 km/s. (a) What is the kinetic energy of the asteroid at the moment of impact? (b) How does this energy compare with that released by a 20-kiloton nuclear weapon, like the device that destroyed Hiroshima, Japan, on August 6, 1945? (*Hint:* 1 kiloton of TNT releases 4.2×10^{12} joules of energy.)

*27. Suppose a spacecraft landed on Jupiter's moon Europa (see Table 7-2), which moves around Jupiter in an orbit of radius 670,900 km. After collecting samples from the satellite's surface, the spacecraft prepares to return to Earth. (a) Calculate the escape speed from Europa. (b) Calculate the escape speed from Jupiter at the distance of Europa's orbit. (c) In order to begin its homeward journey, the spacecraft must leave Europa with a speed greater than either your answer to (a) or your answer to (b). Explain why.

*28. A hydrogen atom has a mass of 1.673×10^{-27} kg, and the temperature of the Sun's surface is 5800 K. What is the average speed of hydrogen atoms at the Sun's surface?

*29. The Sun's mass is 1.989×10^{30} kg, and its radius is 6.96×10^8 m. (a) Calculate the escape speed from the Sun's surface (**b**) Using your answer to Question 28, explain why the Sun has lost very little hydrogen over its entire 4.56-billion-year history.

*30. Saturn's satellite Titan has an appreciable atmosphere, yet Jupiter's satellite Ganymede—which is about the same size and mass as Titan—has no atmosphere. Explain why there is a difference.

31. The distance from the asteroid 433 Eros (Figure 7-7) to the Sun varies between 1.13 and 1.78 AU. (**a**) Find the period of Eros's orbit. (**b**) Does Eros lie in the asteroid belt? How can you tell?

32. Imagine a trans-Neptunian object with roughly the same mass as Earth but located 50 AU from the Sun. (**a**) What do you think this object would be made of? Explain your reasoning. (**b**) On the basis of this speculation, assume a reasonable density for this object and calculate its diameter. How many times bigger or smaller than Earth would it be?

33. Consider a hypothetical trans-Neptunian object located 100 AU from the Sun. (**a**) What would be the orbital period (in years) of this object? (**b**) There are 360 degrees in a circle, and 60 arcminutes in a degree. How long would it take this object to move 1 arcminute across the sky? (**c**) Trans-Neptunian objects are discovered by looking for "stars" that move on the celestial sphere. Use your answer from part (b) to explain why these discoveries require patience. (**d**) Discovering trans-Neptunian objects also requires large telescopes equipped with sensitive detectors. Explain why.

34. The surfaces of Mercury, the Moon, and Mars are riddled with craters formed by the impact of space debris. Many of these craters are billions of years old. By contrast, there are only a few conspicuous craters on Earth's surface, and these are generally less than 500 million years old. What do you suppose explains the difference?

35. During the period of most intense bombardment by space debris, a new 1-km-radius crater formed somewhere on the Moon about once per century. During this same period, what was the probability that such a crater would be created within 1 km of a certain location on the Moon during a 100-year period? During a 10^6-year period? (*Hint:* If you drop a coin onto a checkerboard, the probability that the coin will land on any particular one of the board's 64 squares is $^1/_{64}$.)

36. When an impact crater is formed, material (called *ejecta*) is sprayed outward from the impact. (The accompanying photograph of the Moon shows light-colored ejecta extending outward from the crater Copernicus.) While ejecta are found surrounding the craters on Mercury, they do not extend as far from the craters as do ejecta on the Moon. Explain why, using the difference in surface gravity between the Moon (surface gravity = 0.17 that on Earth) and Mercury (surface gravity = 0.38 that on Earth).

Copernicus crater

Ejecta from Copernicus

100 km

R I V U X G   (NASA)

37. Mercury rotates once on its axis every 58.646 days, compared to 1 day for Earth. Use this information to argue why Mercury's magnetic field should be much smaller than Earth's.

38. As you can see in Figure 7-15, *Mars Global Surveyor* did not find a significant magnetic field in the northern region of Utopia Planitia, which is a large lava field. Based on this observation, would you expect the lava field to have formed before or after Mars ceased to have a molten core?

39. Liquid metallic hydrogen is the source of the magnetic fields of Jupiter and Saturn. Explain why liquid metallic hydrogen cannot be the source of Earth's magnetic field.

Discussion Questions

*40. There are no asteroids with an atmosphere. Discuss why not.

41. The *Galileo* spacecraft that orbited Jupiter from 1995 to 2003 discovered that Ganymede (Table 7-2) has a magnetic field twice as strong as that of Mercury. Does this discovery surprise you? Why or why not?

Web/eBook Question

42. Search the World Wide Web for information about impact craters on Earth. Where is the largest crater located? How old is it estimated to be? Which crater is closest to where you live?

43. **Determining Terrestrial Planet Orbital Periods.** Access the animation "Planetary Orbits" in Chapter 7 of the *Universe* Web site or eBook. Focus on the motions of the inner planets at the last half of the animation. Using the stop and start buttons, determine how many days it takes Mars, Venus, and Mercury to orbit the Sun once if Earth takes approximately 365 days.

ACTIVITIES

Observing Projects

44. Use a telescope or binoculars to observe craters on the Moon. Make a drawing of the Moon, indicating the smallest and largest craters that you can see. Can you estimate their sizes? For comparison, the Moon as a whole has a diameter of 3476 km. *Hint:* You can see craters most distinctly when the Moon is near first quarter or third quarter (see Figure 3-2). At these phases, the Sun casts long shadows across the portion of the Moon in the center of your field of view, making the variations in elevation between the rims and centers of craters easy to identify. You can determine the phase of the Moon by looking at a calendar or the weather page of the newspaper, by using the *Starry Night*™ program, or on the World Wide Web.

45. Use *Starry Night*™ to examine magnified images of the terrestrial major planets Mercury, Venus, Earth, and Mars, and the dwarf planet Ceres. Select each of these planets from **Favourites > Explorations** in turn. Use the location scroller cursor to rotate the image to see different views of the planet. (a) Describe each planet's appearance. From what you observe in each case, is there any way of knowing whether you are looking at a planet's surface or at complete cloud cover over the planet? (b) Which planet or planets have clouds? If a planet has clouds, open its contextual menu and choose **Surface Image/Model > Default** and use the location scroller to examine the planet's surface. (c) Which major planet shows the heaviest cratering? (d) Which of these terrestrial planets show evidence of liquid water? (e) What do you notice about Venus's rotation compared to the other planets?

46. Use *Starry Night*™ to examine the Jovian planets Jupiter, Saturn, Uranus, and Neptune. Select each of these planets from **Favourites > Explorations**. Use the location scroller cursor to examine each planet from different views. (a) Describe each planet's appearance. Which has the greatest color contrast in its cloud tops? (b) Which planet has the least color contrast in its cloud tops? (c) What can you say about the thickness of Saturn's rings compared to their diameter?

ANSWERS

ConceptChecks

ConceptCheck 7-1: Our solar system contains only one star, the Sun. The other stars are very far away.

ConceptCheck 7-2: Mars is classified as an inner planet, and all four inner planets are also terrestrial planets with hard surfaces (these are Mercury, Venus, Earth, and Mars). The inner planets orbit within 2 AU of the Sun, while the outer planets (Jupiter, Saturn, Uranus, and Neptune) all orbit farther out, more than 5 AU from the Sun.

ConceptCheck 7-3: If the rocks on the surface have a density lower than the planet's average density, then the planet's core has a density greater than the planet's average. This is not surprising, since gravity causes material of greater density to sink towards the center of a planet. (The concept of average density is discussed in Box 7-1.)

ConceptCheck 7-4: Table 7-2 lists four having diameters greater than our Moon's 3476 km: Io, Ganymede, Callisto, and Titan.

ConceptCheck 7-5: As sunlight passes through a planet's atmosphere (just before and after reflection off of the planet's surface), the atoms and molecules in the atmosphere absorb specific wavelengths of light unique to these atoms and molecules. By looking at this absorption spectrum in the reflected sunlight, astronomers can infer the composition of the atmosphere as illustrated in Figure 7-3.

ConceptCheck 7-6: The reflected spectrum from a solid surface shows broad absorption features, whereas the spectrum observed from light passing through a gaseous atmosphere shows sharper spectral lines.

ConceptCheck 7-7: Hydrogen molecules in a gaseous atmosphere have a range of speeds based on the temperature of the gas, with some moving about six times the average speed. These faster molecules exceed Earth's escape speed and leave Earth entirely as described in Box 7-2. Continuously, some of the remaining slower hydrogen molecules speed up through molecular collisions and steadily escape.

ConceptCheck 7-8: Ceres has a diameter of about 900 km. This is about the same size as a large U.S. state, such as the length of California.

ConceptCheck 7-9: Pluto orbits beyond Neptune. All such objects are called trans-Neptunian objects. One group of trans-Neptunian objects—including Pluto—orbits in the Kuiper belt, so Pluto is both a trans-Neptunian object and a Kuiper belt object. Pluto is neither a planet nor an asteroid.

ConceptCheck 7-10: Whereas the Kuiper belt lies in the same plane as Earth's orbit around the Sun, the Oort cloud is a spherical distribution of comets that completely surrounds the solar system. If a comet has an orbit that is considerably different from that of the flat plane of the solar system, it most likely came from the spherical Oort cloud that exists in all directions around our solar system.

ConceptCheck 7-11: No. Io's heat comes from tidal forces exerted by a very massive Jupiter. A planet, on the other hand, is not likely to experience strong interior-melting tidal forces from its smaller, orbiting moons.

ConceptCheck 7-12: No. Figure 7-13a shows the field created by a bar magnet. Figure 7-13b shows that these fields are similar in structure, but the source of the magnetic field is not a bar magnet. Instead, Earth's magnetic field is due to electric currents flowing in a molten iron interior.

ConceptCheck 7-13: The answer to both questions is no. Based on its similar size to Earth, Venus probably has a molten interior. However, it exhibits no magnetic field, probably due to its very slow rotation: Earth rotates about 243 times for each single rotation of Venus. As in the cases of Jupiter, Saturn, Uranus, and Neptune, some electrically conducting material other than iron is needed to explain the observed magnetic fields.

CalculationChecks

CalculationCheck 7-1: The shape of a planet's orbit is given by the value of its eccentricity. The closer this value is to zero, the closer the orbit's shape is to that of a perfect circle. According to the table, the orbit of Venus has the eccentricity closest to zero (0.007), making it the most circlelike of all planetary orbits.

CalculationCheck 7-2: If we divide Saturn's 120,536-km diameter by Earth's 12,756-km diameter—we find that 120,536 km ÷ 12,756 km = 9.449, so about 9½ Earth's would fit across Saturn's diameter.

CalculationCheck 7-3: The asteroid belt is located between Mars and Jupiter at about 3 AU from the Sun, whereas the much larger Kuiper belt is beyond the orbit of Neptune and is located between about 30 and 50 AU from the Sun.

CalculationCheck 7-4: The masses in this figure are all given as multiples of Earth's mass, so it would take just over 317 Earth masses to equal Jupiter's mass. The number of Saturn masses that would equal Jupiter's mass is 317.8/95.16 = 3.33.

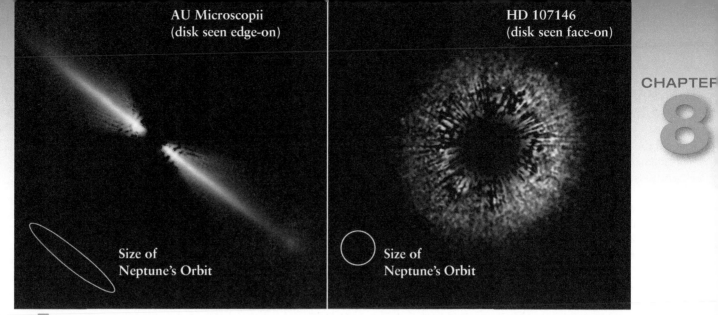

AU Microscopii
(disk seen edge-on)

HD 107146
(disk seen face-on)

Size of
Neptune's Orbit

Size of
Neptune's Orbit

R I V U X G

Planets are thought to form within the disks surrounding young stars such as these. Neptune's orbit, shown for scale, is about 60 AU across. (NASA, ESA, D. R. Ardila (JHU), D. A. Golimowski (JHU), J. E. Krist (STScI/JPL), M. Clampin (NASA/GSFC), J. P. Williams (UH/IfA), J. P. Blakeslee (JHU), H. C. Ford (JHU), G. F. Hartig (STScI), G. D. Illingworth (UCO-Lick) and the ACS Science Team)

Comparative Planetology II: The Origin of Our Solar System

LEARNING GOALS

By reading the sections of this chapter, you will learn

8-1 The key characteristics of the solar system that must be explained by any theory of its origins

8-2 How the abundances of chemical elements in the solar system and beyond explain the sizes of the planets

8-3 How we can determine the age of the solar system by measuring abundances of radioactive elements

8-4 Why scientists think the Sun and planets all formed from a cloud called the solar nebula

8-5 How the solar nebula model explains the formation of the terrestrial planets

8-6 How the Jovian planets formed and migrated in the early solar system

8-7 How astronomers search for planets around other stars

What did our solar system look like before the planets were fully formed? The answer may lie in these remarkable images from the Hubble Space Telescope. Each image shows an immense disk of gas and dust surrounding a young star. (In each image the light from the star itself was blocked out within the telescope to make the rather faint disk more visible.) Astronomers strongly suspect that in its infancy our own Sun was surrounded by a similar disk from which the planets of our solar system eventually coalesced.

In this chapter we will examine the evidence that led astronomers to this model of the origin of the planets. We will see how the abundances of different chemical elements in the solar system indicate that the Sun and planets formed from a thin cloud of interstellar matter some 4.54 billion years ago, an age determined by measuring the radioactivity of meteorites. We will learn how the nature of planetary orbits gives important clues to what happened as this cloud contracted and how evidence from meteorites reveals the chaotic conditions that existed within the cloud. And we will see how this cloud eventually evolved into the solar system that we see today.

In the past decade, astronomers have been able to test this picture of planetary formation by examining disks around young stars (like the ones in the accompanying images). Most remarkably, they have discovered planets in orbit around dozens of other stars. These recent observations provide valuable information about how our own system of planets came to be.

8-1 Any model of solar system origins must explain the present-day Sun and planets

How did the Sun and planets form? In other words, where did the solar system come from? This question has tantalized astronomers for centuries. Our goal in this chapter is to examine our current understanding of how the solar system came to be—that is, our current best *theory* of the origin of the solar system.

Recall from Section 1-1 that a theory is not merely a set of wild speculations, but a self-consistent collection of ideas that must pass the test of providing an accurate description of the real world. Since no humans were present to witness the formation of the planets, scientists must base their theories of solar system origins on their observations of the present-day solar system. (In an analogous way, paleontologists base their understanding of the lives of dinosaurs on the evidence provided by fossils that have survived to the present day.) In so doing, they are following the steps of the scientific method that we described in Section 1-1.

What key attributes of the solar system should guide us in building a theory of solar system origins? Among the many properties of the planets that we discussed in Chapter 7, three of the most important are listed in Table 8-1. Any theory that attempts to describe the origin of the solar system must be able to explain how these attributes came to be. We begin by considering what Property 1 tells us; we will return to Properties 2 and 3 and the orbits of the planets later in this chapter.

CONCEPTCHECK 8-1

Is Earth an exception to any of the three key properties of our solar system in Table 8-1?

Answer appears at the end of the chapter.

8-2 The cosmic abundances of the chemical elements are the result of how stars evolve

The small sizes of the terrestrial planets compared to the Jovian planets (Property 1 in Table 8-1) suggest that some chemical elements are quite common in our solar system, while others are quite rare. The tremendous masses of the Jovian planets—Jupiter alone has more mass than all of the other planets combined—means that the elements of which they are made, primarily hydrogen and helium, are very abundant. The Sun, too, is made almost entirely of hydrogen and helium. Its average density of 1410 kg/m^3 is in the same range as the densities of the Jovian planets (see Table 7-1), and its absorption spectrum (see Figure 5-14) shows the dominance of hydrogen and helium in the Sun's atmosphere. Hydrogen, the most abundant element, makes up nearly three-quarters of the combined mass of the Sun and planets. Helium is the second most abundant element. Together, hydrogen and helium account for about 98% of the mass of all the material in the solar system. All of the other chemical elements are relatively rare; combined, they make up the remaining 2% (Figure 8-1).

> The terrestrial planets are small because they are made of less abundant elements

The dominance of hydrogen and helium is not merely a characteristic of our local part of the universe. By analyzing the spectra of stars and galaxies, astronomers have found essentially the same pattern of chemical abundances out to the farthest distance attainable by the most powerful telescopes. Hence, the vast majority of the atoms in the universe are hydrogen and helium atoms. The elements that make up the bulk of Earth—mostly iron, oxygen, and silicon—are relatively rare in the universe as a whole, as are the elements of which living organisms are made—carbon, oxygen, nitrogen, and phosphorus, among others. (You may find it useful to review the periodic table of the elements, described in Box 5-5.)

The Origin of the Elements and Cosmic "Recycling"

There is a good reason for this overwhelming abundance of hydrogen and helium. A wealth of evidence has led astronomers to conclude that the universe began some 13.7 billion years ago with a violent event called the Big Bang (Chapter 26). Only the lightest elements—hydrogen and helium, as well as tiny amounts of lithium and perhaps beryllium—emerged from the enormously high temperatures following this cosmic event. All the heavier elements were later manufactured by stars, either by thermonuclear fusion reactions deep in their interiors or by the violent explosions that mark the end of massive stars. Were it not for these processes that take place only in stars, there would be no heavy elements in the universe, no planet like our Earth, and no humans to contemplate the nature of the cosmos.

TABLE 8-1	Three Key Properties of Our Solar System
Any theory of the origin of the solar system must be able to account for these properties of the planets.	
Property 1: Sizes and compositions of terrestrial planets versus Jovian planets	The terrestrial planets, which are composed primarily of rocky substances, are relatively small, while the Jovian planets, which are composed primarily of hydrogen and helium, are relatively large (see Sections 7-1 and 7-4).
Property 2: Directions and orientations of planetary orbits	All of the planets orbit the Sun in the same direction, and all of their orbits are in nearly the same plane (see Section 7-1).
Property 3: Sizes of terrestrial planet orbits versus Jovian planet orbits	The terrestrial planets orbit close to the Sun, while the Jovian planets orbit far from the Sun orbits versus Jovian planet orbits (see Section 7-1).

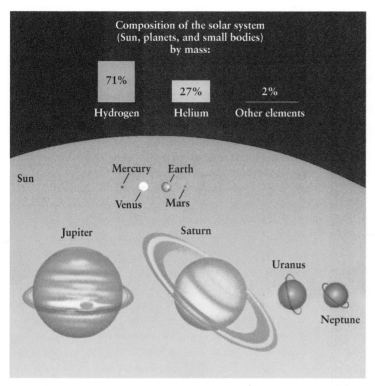

Composition of the solar system
(Sun, planets, and small bodies)
by mass:

71% Hydrogen   27% Helium   2% Other elements

Sun   Mercury   Earth
Venus   Mars

Jupiter   Saturn

Uranus

Neptune

FIGURE 8-1

Composition of the Solar System Hydrogen and helium make up almost all of the mass of our solar system. Other elements such as carbon, oxygen, nitrogen, iron, gold, and uranium constitute only 2% of the total mass.

Because our solar system contains heavy elements, it must be that at least some of its material was once inside other stars. But how did this material become available to help build our solar system? The answer is that near the ends of their lives, stars cast much of their matter back out into space. For most stars this process is a comparatively gentle one, in which a star's outer layers are gradually expelled. Figure 8-2 shows a star losing material in this fashion. This ejected material appears as the cloudy region, or **nebulosity** (from *nubes*, Latin for "cloud"), that surrounds the star and is illuminated by it. A few stars eject matter much more dramatically at the very end of their lives, in a spectacular detonation called a *supernova*, which blows the star apart (see Figure 1-8).

No matter how it escapes, the ejected material contains heavy elements dredged up from the star's interior, where they were formed. This material becomes part of the **interstellar medium,** a tenuous collection of gas and dust that pervades the spaces between the stars. As different stars die, they increasingly enrich the interstellar medium with heavy elements. Observations show that new stars form as condensations in the interstellar medium (Figure 8-3). Thus, these new stars have an adequate supply of heavy elements from which to develop a system of planets, satellites, comets, and asteroids. Our own solar system must have formed from enriched material in just this way. Thus, our solar system contains "recycled" material that was produced long ago inside a now-dead star. This "recycled" material includes all of the carbon in your body, all of the oxygen that you breathe, and all of the iron and silicon in the soil beneath your feet. Scientifically accurate, a rock song from 1970 proclaimed: "We are stardust."

The Abundances of the Elements

Stars create different heavy elements in different amounts. For example, oxygen (as well as carbon, silicon, and iron) is readily produced in the interiors of massive stars, whereas gold (as well as silver, platinum, and uranium) is created only under special circumstances. Consequently, gold is rare in our solar system and in the universe as a whole, while oxygen is relatively abundant (although still much less abundant than hydrogen or helium).

A convenient way to express the relative abundances of the various elements is to say how many atoms of a particular element are found for every trillion (10^{12}) hydrogen atoms. For example, for every 10^{12} hydrogen atoms in space, there are about 100 billion

Antares, an aging star, ejects gas and dust.

The dust ejected from Antares is visible because it reflects the star's light.

FIGURE 8-2 R I V U X G

A Mature Star Ejecting Gas and Dust The star Antares is shedding material from its outer layers, forming a thin cloud around the star. We can see the cloud because some of the ejected material has condensed into tiny grains of dust that reflect the star's light. (Dust particles in the air around you reflect light in the same way, which is why you can see them within a shaft of sunlight in a darkened room). Antares lies some 600 light-years from Earth in the constellation Scorpio. (David Malin/Anglo-Australian Observatory)

Gas and dust ejected from earlier generations of stars have coalesced to form new stars.

The dust reflects light emitted by the newly formed stars.

FIGURE 8-3   R I V U X G

New Stars Forming from Gas and Dust Unlike Figure 8-2, which depicts an old star that is ejecting material into space, this image shows young stars in the constellation Orion (the Hunter) that have only recently formed from a cloud of gas and dust. The bluish, wispy appearance of the cloud (called NGC 1973-1975-1977) is caused by starlight reflecting off interstellar dust grains within the cloud (see Box 5-4). The grains are made of heavy elements produced by earlier generations of stars. (David Malin/Anglo-Australian Observatory)

(10^{11}) helium atoms. From spectral analysis of stars and chemical analysis of Earth rocks, Moon rocks, and bits of interplanetary debris called meteorites, scientists have determined the relative abundances of the elements in our part of the Milky Way Galaxy today. Figure 8-4 shows the relative abundances of the 30 lightest elements, arranged in order of their **atomic numbers.** An element's atomic number is the number of protons in the nucleus of an atom of that element. It is also equal to the number of electrons orbiting

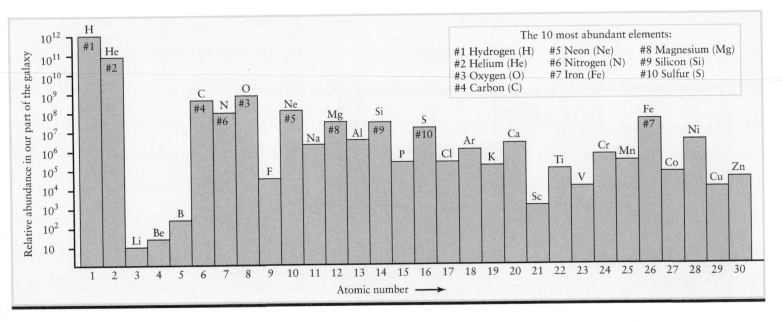

FIGURE 8-4

Abundances of the Lighter Elements This graph shows the abundances in our part of the Galaxy of the 30 lightest elements (listed in order of increasing atomic number) compared to a value of 10^{12} for hydrogen. The inset lists the 10 most abundant of these elements, which are also indicated in the graph. Notice that the vertical scale is not linear; each division on the scale corresponds to a tenfold increase in abundance. All elements heavier than zinc (Zn) have abundances of fewer than 1000 atoms per 10^{12} atoms of hydrogen.

the nucleus (see Box 5-5). In general, the greater the atomic number of an atom, the greater its mass.

CAUTION! Figure 8-4 shows that there is about 10 times more hydrogen than helium when comparing the *number of atoms*. Figure 8-1 shows that our solar system is made of 71% hydrogen versus 27% helium when comparing *mass*. The explanation of this seeming inconsistency lies in the difference between comparing the mass of atoms versus their numbers: Each helium atom has about 4 times the mass of a hydrogen atom, which makes helium's contribution to total mass larger than its contribution to the total number of atoms.

The box inset in Figure 8-4 lists the 10 most abundant elements. Note that even oxygen (chemical symbol O), the third most abundant element, is quite rare relative to hydrogen (H) and helium (He): There are only 8.5×10^8 oxygen atoms for each 10^{12} hydrogen atoms and each 10^{11} helium atoms. Expressed another way, for each oxygen atom in our region of the Milky Way Galaxy, there are about 1200 hydrogen atoms and 120 helium atoms.

In addition to the 10 most abundant elements listed in Figure 8-4, five elements are moderately abundant: sodium (Na), aluminum (Al), argon (Ar), calcium (Ca), and nickel (Ni). These elements have abundances in the range of 10^6 to 10^7 relative to the standard 10^{12} hydrogen atoms. Most of the other elements are much rarer. For example, for every 10^{12} hydrogen atoms in the solar system, there are only 6 atoms of gold.

The small cosmic abundances of elements other than hydrogen and helium help to explain why the terrestrial planets are so small (Property 1 in Table 8-1). Because the heavier elements required to make a terrestrial planet are rare, only relatively small planets can form out of them. By contrast, hydrogen and helium are so abundant that it was possible for these elements to form large Jovian planets.

CONCEPTCHECK 8-2

What is meant by the phrase "we are stardust"?

CONCEPTCHECK 8-3

From the abundances in Figure 8-4, do you expect that water (H_2O) might be a common substance in the galaxy?

Answers appear at the end of the chapter.

8-3 The abundances of radioactive elements reveal the solar system's age

The heavy elements can tell us even more about the solar system: They also help us determine its age. The particular heavy elements that provide us with this information are *radioactive*. Their atomic nuclei are unstable because they contain too many protons or too many neutrons. A radioactive nucleus therefore ejects particles until it becomes stable. In doing so, a nucleus may change from one element to another. Physicists refer to this transmutation as **radioactive decay.** For example, a radioactive form of the element rubidium (atomic

> Our solar system, which formed 9 billion years after the Big Bang, is a relative newcomer to the universe

number 37) decays into the element strontium (atomic number 38) when one of the neutrons in the rubidium nucleus decays into a proton and an electron (which is ejected from the nucleus).

Experiment shows that each type of radioactive nucleus decays at its own characteristic rate, which can be measured in the laboratory. Furthermore, the older a solid rock is, the less of its original radioactive nuclei remains. This behavior is the key to a technique called **radioactive dating**, which is used to determine how many years ago a rock cooled and solidified, or simply, to determine the "ages" of rocks. For example, if a rock contained a certain amount of radioactive rubidium when it first solidified, over time more and more of the atoms of rubidium within the rock will decay into strontium atoms. The ratio of the number of strontium atoms the rock contains to the number of rubidium atoms it contains then gives a measure of the age of the rock. Box 8-1 describes radioactive dating in more detail.

Dating the Solar System

Scientists have applied techniques of radioactive dating to rocks taken from all over Earth. The results show that most rocks are tens or hundreds of millions of years old, but that some rocks are as much as 4 billion (4×10^9) years old. These results confirm that geologic processes—for example, lava flows—have produced new surface material over Earth's history, as we concluded from the small number of impact craters found on Earth (see Section 7-6). They also show that Earth must be at least 4×10^9 years old.

Radioactive dating has also been applied to rock samples brought back from the Moon by the *Apollo* astronauts. The oldest *Apollo* specimen, collected from one of the most heavily cratered and hence most ancient regions of the Moon, is 4.5×10^9 years old. But the oldest rocks found anywhere in the solar system are **meteorites,** bits of interplanetary debris that survive passing through Earth's atmosphere and land on our planet's surface (Figure 8-5). Radioactive dating of meteorites reveals that they are all nearly

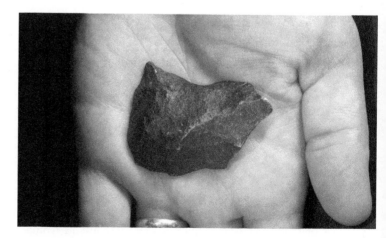

FIGURE 8-5   R I ▊ U X G

A Meteorite Although it resembles an ordinary Earth rock, this is actually a meteorite that fell from space. The proof of its extraterrestrial origin is the meteorite's composition and its surface. Searing heat melted the surface as the rock slammed into our atmosphere. Meteorites are the oldest objects in the solar system. (Ted Kinsman/Photo Researchers, Inc.)

BOX 8-1   TOOLS OF THE ASTRONOMER'S TRADE

Radioactive Dating

How old are the rocks found on Earth and other planets? Are rocks found at different locations the same age or different ages? How old are meteorites? Questions like these are important to scientists who wish to reconstruct the history of our solar system. The age of a rock is how long ago it solidified, but simply looking at a rock cannot tell us whether it was formed a million years, or a billion years ago. Fortunately, most rocks contain trace amounts of radioactive elements such as uranium. By measuring the relative abundances of various radioactive isotopes and their decay products within a rock, scientists can determine the rock's age.

As we saw in Box 5-5, every atom of a particular element has the same number of protons in its nucleus. However, different isotopes of the same element have different numbers of neutrons in their nuclei. For example, the common isotopes of uranium are ^{235}U and ^{238}U. Each isotope of uranium has 92 protons in its nucleus (correspondingly, uranium is element 92 in the periodic table; see Box 5-5). However, a ^{235}U nucleus contains 143 neutrons, whereas a ^{238}U nucleus has 146 neutrons.

A radioactive nucleus with too many protons or too many neutrons is unstable; to become stable, it *decays* by ejecting particles until it becomes stable. If the number of protons (the atomic number) changes in this process, the nucleus changes from one element to another.

Some radioactive isotopes decay rapidly, while others decay slowly. Physicists find it convenient to talk about the decay rate in terms of an isotope's **half-life.** The half-life of an isotope is the time interval in which one-half of the nuclei decay. For example, the half-life of ^{238}U is 4.5 billion (4.5×10^9) years. Uranium's half-life means that if you start out with 1 kg of ^{238}U, after 4.5 billion years, you will have only ½ kg of ^{238}U remaining; the other ½ kg will have turned into other elements. If you wait

another half-life, so that a total of 9.0 billion years has elapsed, only ¼ [0.25] kg of ^{238}U—one-half of one-half of the original amount—will remain. Several isotopes useful for determining the ages of rocks are listed in the accompanying table.

To see how geologists date rocks, consider the slow conversion of radioactive rubidium (^{87}Rb) into strontium (^{87}Sr). (The periodic table in Box 5-5 shows that the atomic numbers for these elements are 37 for rubidium and 38 for strontium, so in the decay a neutron is transformed into a proton. In this process an electron is ejected from the nucleus.) Over the years, the amount of ^{87}Rb in a rock decreases, while the amount of ^{87}Sr increases. Because the ^{87}Sr appears in the rock due to radioactive decay, this isotope is called *radiogenic*. Dating the rock is not simply a matter of measuring its ratio of rubidium to strontium, however, because the rock already had some strontium in it when it was formed. Geologists must therefore determine how much fresh strontium came from the decay of rubidium after the rock's formation.

To make this determination, geologists use as a reference another isotope of strontium whose concentration has remained constant. In this case, they use ^{86}Sr, which is stable and is not created by radioactive decay; it is said to be *nonradiogenic*. Dating a rock thus entails comparing the ratio of radiogenic and nonradiogenic strontium (^{87}Sr/^{86}Sr) in the rock to the ratio of radioactive rubidium to nonradiogenic strontium (^{87}Rb/^{86}Sr). Because the half-life for converting ^{87}Rb into ^{87}Sr is known, the rock's age can then be calculated from these ratios (see the table).

Radioactive isotopes decay with the same half-life no matter where in the universe they are found. Hence, scientists have used the same techniques to determine the ages of rocks from the Moon and of meteorites.

Original Radioactive Isotope	Final Stable Isotope	Half-Life (Years)	Range of Ages that Can Be Determined (Years)
Rubidium (^{87}Rb)	Strontium (^{87}Sr)	47.0 billion	10 million–4.54 billion
Uranium (^{238}U)	Lead (^{206}Pb)	4.5 billion	10 million–4.54 billion
Potassium (^{40}K)	Argon (^{40}Ar)	1.3 billion	50,000–4.54 billion
Carbon (^{14}C)	Nitrogen (^{14}N)	5730	100–70,000

the *same* age, about 4.54 billion years old. The absence of any younger or older meteorites indicates that these are all remnants of objects that formed around the same time when rocky material in the early solar system—which was initially hot—first cooled and solidified. We conclude that the age of the oldest meteorites, about 4.54×10^9 years, is the age of the solar system itself. Note that this almost inconceivably long span of time is only about one-third of the current age of the universe, 13.7×10^9 years.

Thus, by studying the abundances of radioactive elements, we are led to a remarkable insight: Some 4.54 billion years ago, a collection of hydrogen, helium, and a much smaller amount of heavy elements came together to form the Sun and all of the objects that orbit around it. All of those heavy elements, including the carbon atoms in your body and the oxygen atoms that you breathe, were created and cast off by stars that lived and died long before our solar system formed, during the first 9 billion years of the universe's existence. We are literally made of old star dust, and our solar system is relatively young.

CONCEPTCHECK 8-4

What is meant by the "age" of a rock? Is it the age of the rock's atoms?

Answer appears at the end of the chapter.

8-4 The Sun and planets formed from a solar nebula

We have seen how processes in the Big Bang and within ancient stars produced the raw ingredients of our solar system. But given these ingredients, how did they combine to make the Sun and planets? Astronomers have developed a variety of models for the origin of the solar system. The test of these models is whether they explain the properties of the present-day system of Sun and planets.

> Astronomers see young stars that may be forming planets today in the same way that our solar system did billions of years ago

The Failed Tidal Hypothesis

Any model of the origin of the solar system must explain why all the planets orbit the Sun in the same direction and in nearly the same plane (Property 2 in Table 8-1). One model that was devised explicitly to address this issue was the *tidal hypothesis,* proposed in the early 1900s. As we saw in Section 4-8, two nearby planets, stars, or galaxies exert tidal forces on each other that cause the objects to elongate. In the tidal hypothesis, another star happened to pass close by the Sun, and the star's tidal forces drew a long filament out of the Sun. The filament material would then go into orbit around the Sun, and all of it would naturally orbit in the same direction and in the same plane. From this filament the planets would condense. However, it was shown in the 1930s that the same tidal forces strong enough to pull a filament out of the Sun would also cause the filament to disperse before it could condense into planets. Hence, the tidal hypothesis cannot be correct.

The Successful Nebular Hypothesis

An entirely different model is now thought to describe the most likely series of events that led to our present solar system (**Figure 8-6**). The central idea of this model dates to the late 1700s, when the German philosopher Immanuel Kant and the French scientist Pierre-Simon de Laplace turned their attention to the manner in which the planets orbit the Sun. Both concluded that the arrangement of the orbits—all in the same direction and in nearly the same plane—could not be mere coincidence. To explain the orbits, Kant and Laplace independently proposed that our entire solar system, including the Sun as well as all of its planets and satellites, formed from a vast, rotating cloud of gas and dust called the **solar nebula** (Figure 8-6a). This model is called the **nebular hypothesis.**

The consensus among today's astronomers is that Kant and Laplace were exactly right. In the modern version of the nebular hypothesis, at the outset the solar nebula was similar in character

FIGURE 8-6

The Birth of the Solar System **(a)** A cloud of interstellar gas and dust begins to contract because of its own gravity. **(b)** As the cloud flattens and spins more rapidly around its rotation axis, a central condensation develops that evolves into a glowing protosun. The planets will form out of the surrounding disk of gas and dust.

to the nebulosity shown in Figure 8-3 and had a mass somewhat greater than that of our present-day Sun.

Each part of the nebula exerted a gravitational attraction on the other parts, and these mutual gravitational pulls tended to make the nebula contract. As it contracted, the greatest concentration of matter occurred at the center of the nebula, forming a relatively dense region called the **protosun.** As its name suggests, this part of the solar nebula eventually developed into the Sun. The planets formed from the much sparser material in the outer regions of the solar nebula. Indeed, the mass of all the planets together is only 0.1% of the Sun's mass.

Evolution of the Protosun

When you drop a ball, the gravitational attraction of Earth makes the ball fall faster and faster as it falls; in the same way, material falling inward toward the protosun would have gained speed as it approached the center of the solar nebula. As this fast-moving material ran into the protosun, the kinetic energy of the collision was converted into thermal energy, causing the temperature deep inside the solar nebula to climb. This process, in which the gravitational energy of a contracting gas cloud is converted into thermal energy, is called **Kelvin-Helmholtz contraction,** after the nineteenth-century physicists who first described it.

As the newly created protosun continued to contract and become denser, its temperature continued to climb as well. After about 10^5 (100,000) years, the protosun's surface temperature stabilized at about 6000 K, but the temperature in its interior kept increasing to ever higher values as the central regions of the protosun became denser and denser. Eventually, after perhaps 10^7 (10 million) years had passed since the solar nebula first began to contract, the gas at the center of the protosun reached a density of about 10^5 kg/m^3 (about 13 times denser than typical iron) and a temperature of a few million kelvins (that is, a few times 10^6 K). Under these extreme conditions, nuclear reactions that convert hydrogen into helium began in the protosun's interior. These nuclear reactions released energy that significantly increased the pressure in the protosun's core. When the pressure built up enough, it stopped further contraction of the protosun and a true star was born. In fact, the onset of nuclear reactions defines the end of a protostar and beginning of a star. Nuclear reactions continue to the present day in the interior of the Sun and are the source of all the energy that the Sun radiates into space.

The Protoplanetary Disk

If the solar nebula had not been rotating at all, everything would have fallen directly into the protosun, leaving nothing behind to form the planets. Instead, the solar nebula must have had an overall slight rotation, which caused its evolution to follow a different path. As the slowly rotating nebula collapsed inward, it would naturally have tended to rotate faster. This relationship between the size of an object and its rotation speed is an example of a general principle called the **conservation of angular momentum.**

ANALOGY Figure skaters make use of the conservation of angular momentum. When a spinning skater pulls her arms and legs in close to her body, the rate at which she spins automatically increases (Figure 8-7). Even if you are not a figure skater, you can demonstrate this by sitting on a rotating office chair. Sit with your arms outstretched and hold a weight, like a brick or a full water bottle, in either hand. Now use your feet to start your body and the chair rotating, lift your feet off the ground, and then pull your arms inward. Your rotation will speed up quite noticeably.

Astronomers see young stars that may be forming planets today in the same way that our solar system did billions of years ago. As the solar nebula began to rotate more rapidly, it also tended to flatten out (Figure 8-6b)—but why? From the perspective of a particle rotating along with the nebula, it felt as though there were a force pushing the particle away from the nebula's axis of rotation. (Likewise, passengers on a merry-go-round or spinning carnival ride seem to feel a force pushing them outward and away from the ride's axis of rotation.) This apparent force was directed opposite to the inward pull of gravity, and so it tended to slow the contraction of material toward the nebula's rotation axis. But there was no such effect opposing contraction in a direction parallel to the rotation axis. Some 10^5 (100,000) years after the solar nebula first began to contract, it had developed the structure shown in Figure 8-6b, with a rotating, flattened disk

(a) (b)

FIGURE 8-7 R I $\boxed{\text{V}}$ U X G

Conservation of Angular Momentum A figure skater who **(a)** spins slowly with her limbs extended will naturally speed up when **(b)** she pulls her limbs in. In the same way, the solar nebula spun more rapidly as its material contracted toward the center of the nebula. (AP Photo/Amy Sancetta)

surrounding what will become the protosun. This disk is called the **protoplanetary disk,** since planets formed from its material. This model explains why their orbits all lie in essentially the same plane and why they all orbit the Sun in the same direction.

There were no humans to observe these processes taking place during the formation of the solar system. But Earth astronomers have seen disks of material surrounding other stars that formed only recently. These, too, are called protoplanetary disks, because it is thought that planets can eventually form from these disks around other stars. Hence, these disks are planetary systems that are still "under construction." By studying these disks around other stars, astronomers are able to examine what our solar nebula may have been like some 4.5×10^9 years ago.

Figure 8-8 shows a number of protoplanetary disks in the Orion Nebula, a region of active star formation. A star is visible at the center of each disk, which reinforces the idea that our Sun began to shine before the planets were fully formed. (The images that open this chapter show even more detailed views of disks surrounding young stars.) A study of 110 young stars in the Orion Nebula detected protoplanetary disks around 56 of them, which suggests that systems of planets may form around a substantial fraction of stars. Later in this chapter we will see direct evidence for planets that have formed around stars other than the Sun.

CONCEPTCHECK 8-5

If the nebular hypothesis is correct, what must it explain about the planetary orbits?

CONCEPTCHECK 8-6

Was the solar nebula cold until nuclear reactions began powering the Sun?

Answers appear at the end of the chapter.

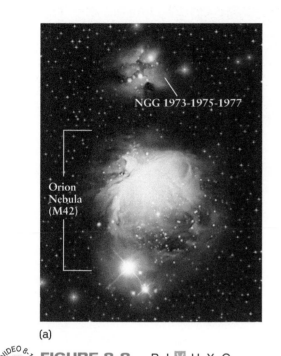

(a)

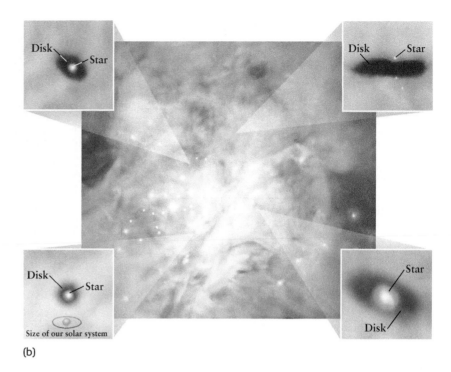

(b)

FIGURE 8-8   R I V U X G

Protoplanetary Disks **(a)** The Orion Nebula is a star-forming region located some 1500 light-years from Earth. It is the middle "star" in Orion's "sword" (see Figure 2-2a). The smaller, bluish nebula is the object shown in Figure 8-3. **(b)** This view of the center of the Orion Nebula is a mosaic of Hubble Space Telescope images. The four insets are false-color close-ups of four protoplanetary disks that lie within the nebula. A young, recently formed star is at the center of each disk. (The disk at upper right is seen nearly edge-on.) The inset at the lower left shows the size of our own solar system for comparison. (a: Anglo-Australian Observatory image by David Malin; b: C. R. O'Dell and S. K. Wong, Rice University; NASA)

8-5 The terrestrial planets formed by the accretion of planetesimals

We have seen how the solar nebula would have contracted to form a young Sun with a protoplanetary disk rotating around it. But how did the material in this disk form into planets? Why are the small terrestrial planets located in the inner solar system (Mercury, Venus, Earth, and Mars), while the giant Jovian planets are in the outer solar system (Jupiter, Saturn, Uranus, and Neptune)? In this section and the next we will see how the nebular hypothesis provides answers to these questions.

> Rocky planets formed in the inner solar nebula as a consequence of the high temperatures close to the protosun

The Condensation Temperature and the Snow Line

To understand how the planets, asteroids, and comets formed, we start by considering a cold and low-pressure solar nebula, before it was warmed by an emerging protosun. At the low pressures that prevailed, a substance does not form into a liquid state, but must exist as either a solid or a gas. An example of such a solid is in Figure 8-9, which shows a dust grain of the sort that would have been present throughout the solar nebula. Other substances

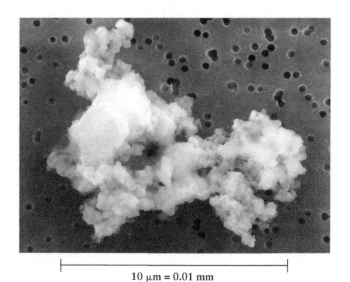

10 μm = 0.01 mm

FIGURE 8-9

A Grain of Cosmic Dust This highly magnified image shows a microscopic dust grain that came from interplanetary space. It entered Earth's upper atmosphere and was collected by a high-flying aircraft. Dust grains of this sort are abundant in star-forming regions like that shown in Figure 8-3. These tiny grains were also abundant in the solar nebula and served as the building blocks of the planets. (NASA)

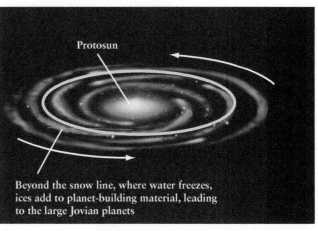

(a)

(b)

FIGURE 8-10

Temperature Distribution and Snow Line in the Solar Nebula **(a)** This graph shows how temperatures probably varied across the solar nebula as the planets were forming, and the present-day position of the planets (red arrows). Note the general decline in temperature with increasing distance from the center of the nebula. Between the present-day distances of Mars and Jupiter, the snow line marked where temperatures were low enough for water to condense and form ice; beyond about 16 AU, methane (CH_4) could also condense into ice. **(b)** Terrestrial planets formed inside the snow line, where the low abundance of solid dust grains kept these planets small. The Jovian planets formed beyond the snow line where solid ices of water, methane, and ammonia added their mass to build larger cores, and attract surrounding gas.

in the early solar nebula would have been in the form of small ice crystals (like snow), although at higher temperatures, these ices would evaporate to form a gas. Together, these solids—referred to as ices, dust grains, and ice-coated dust grains—were mixed with gaseous hydrogen and helium. But, then things began to heat up.

Due to Kelvin-Helmholtz contraction—the conversion of gravitational energy into thermal energy—a protosun began to form in the center of the nebula, and it was actually quite a bit more luminous than the present-day Sun. The temperature in the nebula varied significantly, rising above 2000 K closer to the hot protosun, and dropping below 50 K in the outermost regions (Figure 8-10a).

Our first goal in determining a model for the origin of the solar system is to understand why the inner four planets are small and rocky, whereas the outer Jovian planets are large and made of lighter elements (recall Properties 1 and 3 in Table 8-1). To answer this, we first need to understand the **condensation temperature,** which determines when a substance forms a solid or a gas.

If the temperature of a substance in the solar nebula is above its condensation temperature, the substance is a gas. On the other hand, if the temperature is below the condensation temperature, the substance solidifies (or condenses) into tiny specks of dust or icy frost. You can often see similar behavior on a cold morning. The morning air temperature can be above the condensation temperature of water, while the cold windows of parked cars may have temperatures below the condensation temperature. Thus, water molecules in the warmer air remain as a gas (water vapor) but form solid ice particles (frost) on the colder car windows.

With heat from the emerging protosun, the solar nebula was radically changed into two distinct regions. These regions—inner and outer—had very different properties and eventually formed very different planets:

- **Inner Region: rock and metal.** In the hot inner region of the nebula, only substances with high condensation temperatures could have remained solid. These rocky and metallic materials, in the form of solid dust grains (Figure 8-9), eventually formed much of the rock and metal of the terrestrial planets. Hydrogen compounds—such as water (H_2O), methane (CH_4), and ammonia (NH_3)—remained as gas; they could not condense into solids in the warm inner solar system. (Hydrogen and helium could not condense into solids anywhere in the solar system.) The terrestrial planets that formed in this inner region are Mercury, Venus, Earth, and Mars.

- **Outer Region: ices beyond the snow line.** In the area between the present-day orbits of Mars and Jupiter, the temperature dropped below around 170 K. In the low pressures of the solar nebula, this is the condensation temperature at which water vapor (H_2O) forms ice (see Figure 8-10a). One aspect of this **snow line**— the distance from the Sun at which water vapor solidifies into ice or frost—is that beyond this line, the solid ice particles can join with rock and metal grains resulting in more mass to build planets (Figure 8-10b). Farther out, at even cooler temperatures, methane (CH_4), and ammonia (NH_3) also form ices. (Recall from Section 7-4 that "ice" can refer to frozen carbon dioxide, methane, or ammonia as well as to frozen water.) To summarize: Beyond the snow line, rocky and metallic dust grains were coated with frost, which means that additional icy mass went into building even larger planets. The Jovian planets that formed in this outer region are Jupiter, Saturn, Uranus, and Neptune.

The abundances of material in the solar nebula provide the final insight about planet sizes. With a mixture similar to the Sun, the solar nebula was composed of about 98% hydrogen and helium, about 1.4% hydrogen compounds (H_2O, CH_4, and NH_3), and the remaining 0.6% of rocks and metals. Inside the snow line, the only solid material available to build planets consisted of rock and metal. Thus, with rock and metal having the lowest abundances, the inner terrestrial planets that formed from these dust grains are the smallest. On the other hand, beyond the snow line there was more solid mass available in the form of ice-coated dust grains to help build larger planets.

The phenomena described so far help us understand why smaller terrestrial planets formed close to the Sun, and larger Jovian planets formed farther out. However, there is more to this story, and some big mysteries remain. Next, we consider the process of building planets in more detail, beginning with the terrestrial planets.

Planetesimals, Protoplanets, and Terrestrial Planets

In the inner part of the solar nebula, the grains of rocks and metals would have collided and merged into small chunks. Initially, electric forces—that is, chemical bonds—held these chunks together, in the same way that chemical bonds hold an ordinary rock together. Over a few million years, these chunks coalesced into roughly a billion asteroidlike objects called **planetesimals,** with diameters of a kilometer or so. These larger planetesimals were massive enough to be held together by their own gravity, resulting from the mutual gravitational attraction of all the material within the planetesimal.

During the next stage, gravitational attraction between different planetesimals caused them to collide and accumulate into around a hundred still-larger objects called **protoplanets** (also called planetary embryos), each of which was roughly the size and mass of our Moon. This accumulation of material to form larger and larger objects is called **accretion.** For about a hundred million years, these Moon-sized protoplanets collided to form the inner planets. This final episode must have involved some truly spectacular, world-shattering collisions. In fact, detailed modeling suggests that one of these protoplanet collisions with the forming Earth probably created our Moon! Snapshots from a computer simulation of accreting planetesimals are shown in **Figure 8-11**.

In the inner solar nebula only materials with high condensation temperatures could form dust grains and hence protoplanets, so the result was a set of planets made predominantly of materials such as iron, silicon, magnesium, and nickel. In strong support for this model of solar system formation, these rocky and metallic materials match the composition of the present-day terrestrial planets. There is also supporting evidence from meteorites (dated from their radioactivity) that are thought to have come from smashed-up planetesimals during this early period. As expected for planetesimals from the inner solar system, these meteorites contain metallic grains mixed with rocky material (**Figure 8-12**).

At first the material that coalesced to form protoplanets in the inner solar nebula remained largely in solid form, despite the high temperatures close to the protosun. But as the protoplanets grew, they were heated by violent impacts as they collided with other planetesimals, as well as by the energy released from the decay of radioactive elements, and all this heat caused melting. Thus, the terrestrial planets began their existence as spheres of at least partially molten rocky materials. Material was free to move within these molten spheres, so the denser, iron-rich minerals sank

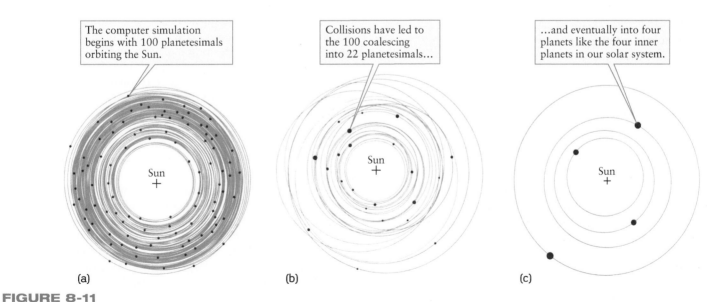

(a)   (b)   (c)

FIGURE 8-11

Accretion of the Inner Planets This computer simulation shows the formation of the inner planets over time. (Adapted from George W. Wetherill)

The computer simulation begins with 100 planetesimals orbiting the Sun.

Collisions have led to the 100 coalescing into 22 planetesimals...

...and eventually into four planets like the four inner planets in our solar system.

FIGURE 8-12   R I ☑ **U X G**

Primitive Meteorite This is a cross-section from a fragment of the Allende meteorite that landed in Chihuahua, Mexico, in 1969. A fireball was observed at 1:05 A.M. local time, and over the next 25 years about 3 tons of meteorite fragments were collected over an 8 km by 50 km area. Radioactive dating yields an age of 4.57 billion years, corresponding to the very early solar system. The meteorite contains rocks and metals that are consistent with the inner solar system and is thought to come from a smashed-up planetesimal. (ChinellatoPhoto/Shutterstock)

to the centers of the planets while the less dense silicon-rich rocky minerals floated to their surfaces. This process is called **chemical differentiation** (see Box 7-1). In this way the terrestrial planets developed their dense iron cores.

CONCEPTCHECK 8-7

How might the solar system be different if the location at which water could freeze in the solar nebula was much more distant from the Sun than it was in the solar nebula that formed our solar system?

CONCEPTCHECK 8-8

How does a rocky planet develop a core that has a higher density than rocks on its surface?

Answers appear at the end of the chapter.

8-6 Gases in the outer solar nebula formed the Jovian planets, and planetary migration reshaped the solar system

We have seen how the low abundance of solid material in the inner solar nebula led to the formation of the

> Low temperatures in the outer solar nebula made it possible for planets to grow to titanic size

small, rocky terrestrial planets. To explain the very different properties of the Jovian planets, we need to consider the conditions that prevailed in the relatively cool outer regions of the solar nebula.

The Core Accretion Model

The large Jovian planets *initially* formed through a similar process as the terrestrial planets—through the accretion of planetesimals. As discussed in Section 8-5, a key difference is that ices—in addition to rock and metal grains—were able to survive in the cooler outer regions of the solar nebula (see Figure 8-10). The elements of which ices are made are much more abundant than those that form rocky grains. Thus, more solid material would have been available to form planetesimals in the outer solar nebula than in the inner part. As a result, solid objects larger than any of the terrestrial planets could have formed in the outer solar nebula. Each such object could have become the core of a Jovian planet and served as a "seed" around which the rest of the planet eventually grew. For example, the mass of Jupiter's rock and metal core is estimated to equal about 10 Earth masses. However, while additional solid material beyond the snow line plays a role in forming the larger cores of Jovian planets, astronomers do not fully understand how their cores get as large as they do, and this is an active area of research.

For Jupiter, a large seed mass of rock, metal, and ice is only the beginning: Most of Jupiter's mass is hydrogen and helium. Recall that for a planet, retaining a gaseous atmosphere depends on both the planet's mass and on the gas temperature (see Box 7-2). Accordingly, due to the lower temperatures in the outer solar system, Jupiter's large seed mass could capture and retain hydrogen and helium gas. This picture—where a Jovian protoplanet core captures gas and grows by accretion—is called the **core accretion model.**

As Jupiter grew, its gravitational pull increased, allowing it to capture more gases and grow even larger until most of the available gas in its region had been captured. Because hydrogen and helium were so abundant (they are 98% of the solar nebula), Jupiter quickly grew to more than 300 Earth masses. You can see Jupiter forming before the other planets in **Figure 8-13**, which summarizes the formation of the solar system. Farther out in the solar nebula, Saturn would have gone through a similar process. About one-third the mass of Jupiter, Saturn's 95 Earth masses would also have taken longer to accumulate, forming a few million years after Jupiter. Uranus and Neptune formed well beyond the snow line, where temperatures were cold enough for additional ices of carbon dioxide, methane, and ammonia to form the bulk of these planets.

Like the protosun, each Jovian planet would have been surrounded by a disk—a solar nebula in miniature (see Figure 8-6). The large moons of the Jovian planets, including those shown in Table 7-2, are thought to have formed from ice particles and dust grains within these disks. Furthermore, since the moons formed from a rotating disk, these large satellites all orbit in the same direction. However, the Jovian planets also have smaller bodies—called irregular satellites—that orbit in the *opposite* direction, and these were probably captured after the planets formed.

Observations of protoplanetary disks around other protostars (such as those in Figure 8-8b) suggest that high-energy photons

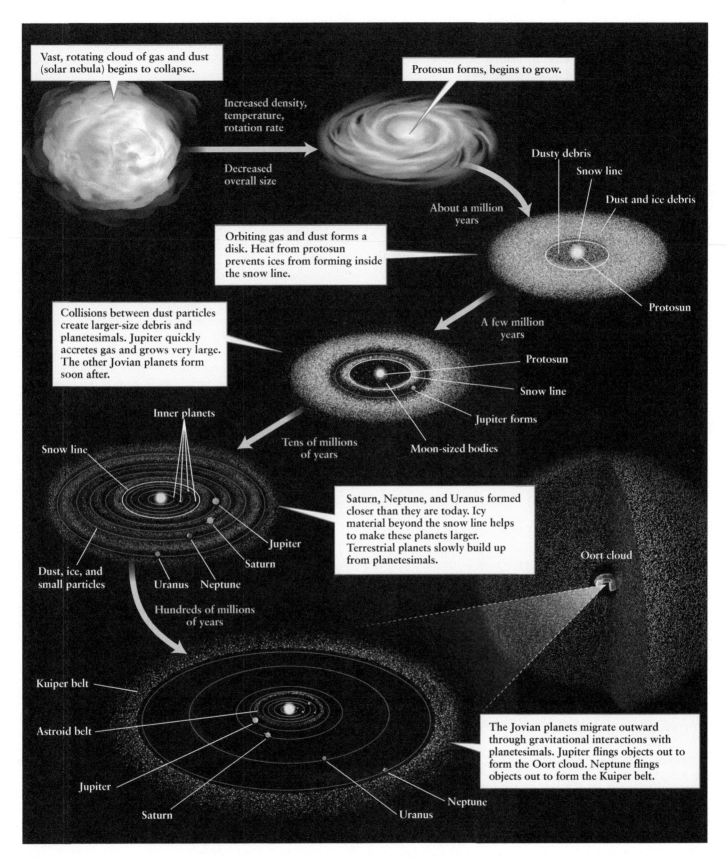

Vast, rotating cloud of gas and dust (solar nebula) begins to collapse.

Increased density, temperature, rotation rate

Decreased overall size

Protosun forms, begins to grow.

Dusty debris

Snow line

Dust and ice debris

About a million years

Orbiting gas and dust forms a disk. Heat from protosun prevents ices from forming inside the snow line.

Protosun

Collisions between dust particles create larger-size debris and planetesimals. Jupiter quickly accretes gas and grows very large. The other Jovian planets form soon after.

A few million years

Protosun

Snow line

Jupiter forms

Inner planets

Snow line

Moon-sized bodies

Tens of millions of years

Saturn, Neptune, and Uranus formed closer than they are today. Icy material beyond the snow line helps to make these planets larger. Terrestrial planets slowly build up from planetesimals.

Jupiter

Saturn

Dust, ice, and small particles

Uranus   Neptune

Oort cloud

Hundreds of millions of years

Kuiper belt

Astroid belt

Jupiter

Saturn

The Jovian planets migrate outward through gravitational interactions with planetesimals. Jupiter flings objects out to form the Oort cloud. Neptune flings objects out to form the Kuiper belt.

Uranus

Neptune

FIGURE 8-13

The Formation of the Solar System This sequence of drawings shows stages in the formation of the solar system.

would eventually disperse any remaining hydrogen and helium gas, and this gas ends up in the space between the stars. With hydrogen and helium gone from the disk, formation of the Jovian planets came to an end after a few million years. However, the accumulation of planetesimals that built the terrestrial planets did not stop when gas was expelled from the protoplanetary disk, and the terrestrial planets took much longer to form than the Jovian planets.

Early Migration of Jovian Planets Shapes the Inner Solar System

We now look at **migration,** which refers to changes in orbital distances of the planets. Migration of the planets results from interactions—gravitational tugs and pulls—between planets and different parts of the forming solar system. The details of planetary migration are far from certain, but the broad picture that follows is emerging from computer simulations.

Within the first few hundred thousand years, long before the terrestrial planets formed, Jupiter's interaction with the gaseous disk of hydrogen and helium led to an inward migration of Jupiter, followed by an outward migration. At its closest distance to the protosun, Jupiter migrated inward to about 1.5 AU, which is near the current orbit of Mars. With its large mass, Jupiter gravitationally deflected many of the planetesimals near the current Martian orbit. As a result, when Mars eventually formed in this region, it ended up with a low mass of about one-tenth of Earth's mass. In fact, it was the mysteriously low mass of Mars that prompted this analysis of Jupiter's inward migration. Even more, inward-then-outward migration of all the Jovian planets—called the **Grand Tack model**—also solves a long-standing mystery involving the asteroid belt.

The asteroid belt (Section 7-5) contains rocky objects typical of the inner solar system, but, surprisingly, also contains icy objects expected to have formed well beyond the asteroid belt. In the Grand Tack model, as both Jupiter and the other Jovian planets migrate outward, they deflect planetesimals inward to form the asteroid belt. Some of these planetesimals come from the inner solar system, but some also come from much farther out beyond the snow line, providing a very natural explanation for the icy objects in the asteroid belt.

This early migration, along with Jovian planet formation, was over within the solar system's first few million years or so. Another type of migration occurred over the next few hundred million years, and reshaped the outer solar system.

Late Migration of Jovian Planets Reshapes the Outer Solar System

Astronomers have long suspected that some planetary migration must have taken place in the outer solar system. To see why, consider Neptune. At Neptune's current large orbital distance, the timescale necessary to build a planet of Neptune's large size by core accretion is much longer than the time that the protoplanetary disk was around. However, planets can grow much faster closer to the protosun, where the protoplanetary disk is denser. Therefore, astronomers suspect that Neptune formed closer in, and then migrated outward.

The leading theory described here for late migration of the Jovian planets is based on the **Nice** (pronounced "niece") **model.**

This model was developed in Nice, France, around 2005 and results from detailed computer simulations. It is important to emphasize that the Nice model is a work in progress and has already evolved with new ideas and better simulations.

In the Nice model, the Jovian planets would have all originally formed within about 20 AU of the protosun, even though the outermost planet today, Neptune, is at 30 AU. Furthermore, the outermost planet during this early time could have been Uranus, at 20 AU. Gravitational interactions could have switched the orbital ordering of Neptune and Uranus; in the Nice model, both initial scenarios can lead to the arrangement of our present-day solar system.

There was also a disk of planetesimals beyond the outermost Jovian planet at 20 AU, and through gravitational encounters, these planetesimals, on average, were scattered inward. Furthermore, when a big planet knocks a small planetesimal inward, the planet itself is kicked slightly outward. Over a few hundred million years, as numerous planetesimals were knocked inward by Saturn, Neptune, and Uranus, these Jovian planets slowly migrated outward to their current locations.

Most of the inward-moving planetesimals even made it close to Jupiter. With its much greater mass, Jupiter did not migrate inward very much, but did gravitationally fling most of the planetesimals clear out of the solar system. However, a small fraction of these icy objects would not have made it all the way out of solar system and are thought to currently orbit at about 50,000 AU from the Sun—about 0.8 light-year away and one-fifth the distance to the nearest star. These planetesimals would be very loosely bound to our Sun and gravitational deflections from passing stars would spread many of their orbits into a spherical "halo." We call this hypothesized distribution of icy planetesimals the Oort cloud (Figure 8-14a), which formed beyond the objects of our next topic—the Kuiper belt.

Unstable Orbits, the Kuiper Belt, and the Late Heavy Bombardment

During the slow outward migration of the Jovian planets, their orbits remain mostly circular; in other words, they have very low eccentricity. The slowly expanding Jovian orbits are also quite stable and allow the migration to continue for about 600 million years or so. However, gravitational interactions with the planetesimals eventually destabilize planetary orbits, leading to elongated, or eccentric, orbits.

With elongated orbits, the planets gravitationally interact with each other more strongly. In many computer simulations, this nearly *doubles* the orbital distance of Neptune, sending Neptune out to its current distance of about 30 AU. As Neptune moves outward, its gravity flings nearby planetesimals to greater distances as well, creating an orbiting collection of icy objects called the Kuiper belt (Figure 8-14b). (The images that open this chapter show two young stars surrounded by dusty disks that resemble the Kuiper belt, one seen edge-on and the other seen face-on.) Taken together, the Oort cloud and Kuiper belt contain icy planetesimals that could not have formed at these great distances where matter was sparse, but instead were gravitationally deflected outward by Jupiter and then by Neptune.

The Nice model seems to solve another mystery of the solar system. Many astronomers think there was a cataclysmic period when the planets and moons of the inner solar system were subjected to a short but intense period of large impacts; this

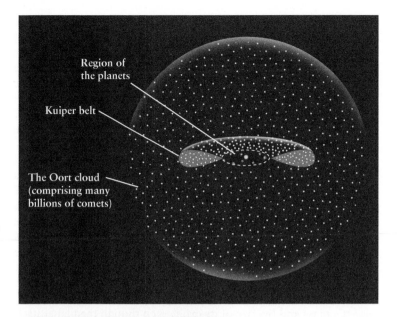

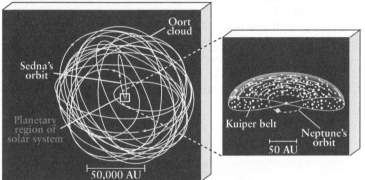

(a) Orbits of some Oort and Kuiper belt objects

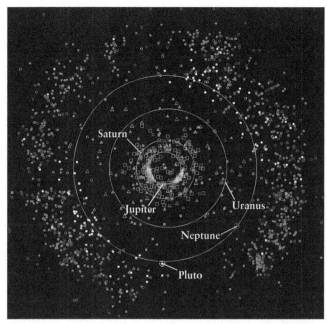

(b)

ANIMATION 8-4 **FIGURE 8-14**

The Kuiper Belt and Oort Cloud (a) The classical Kuiper belt of comets spreads from Neptune out to 50 AU from the Sun. Most of the estimated 200 million belt comets are believed to orbit in or near the plane of the ecliptic. The spherical Oort cloud extends from beyond the Kuiper belt. **(b)** Orbits of bodies in the Kuiper belt. (b: NASA and A. Field/Space Telescope Science Institute)

hypothesis is referred to as the **Late Heavy Bombardment.** The evidence comes from radioactive dating of craters on the Moon, but more lunar samples are needed to firmly establish a spike in the timing of impact events. This bombardment is considered "late" because the frequent planetesimal collisions of planet-building had long since passed.

So, what could cause a big jump in impacts? There are several ideas, but here is the most favored scenario. In the Nice model, the destabilized and elongated planetary orbits also led to deflections of planetesimals throughout the inner solar system. Very significantly, it takes about 600 million years before widespread deflection of the planetesimals, which is consistent with the surprisingly late timing of the Late Heavy Bombardment.

ANIMATION 8-3 To a *much* lesser extent, gravitational deflections continue today. Planetesimals from either the Kuiper belt or the Oort cloud are occasionally knocked into new orbits. If one of these icy objects is deflected on a path through the inner solar system, it begins to evaporate, producing a visible tail, and is seen as a comet (see Figure 7-9).

Prior to 1995 the only fully formed planetary system to which we could apply this model was our own. As we will see in the next section, astronomers can now further test this model on an ever-growing number of planets known to orbit other stars.

CONCEPTCHECK **8-9**

Why is it unlikely that Neptune formed by core accretion at its current location?

CONCEPTCHECK **8-10**

In the Nice model, how did Neptune get to its present location?

CONCEPTCHECK **8-11**

Does accretion refer to the accumulation of matter by gravitational attraction or the formation of chemical bonds? *Answers appear at the end of the chapter.*

8-7 A variety of observational techniques reveal planets around other stars

If planets formed around our Sun, have they formed around other stars? The answer is yes, and they are called **extrasolar planets,** or **exoplanets.** The discovery of exoplanets

> Many planets have been discovered around other stars, and some are Earth-size

brings up several questions that we will consider next. What are the properties of these exoplanets—are they similar to the planets of our solar system? What do they tell us about the formation of other solar systems? And perhaps the biggest question of all: Are there other Earthlike planets?

Since exoplanets were first detected in the 1990s, this field of astronomy has exploded with new discoveries. At the time of this writing, 834 exoplanets have been discovered with a wide range of properties, and some big surprises. (This number refers to confirmed exoplanets, and several thousand candidates are awaiting confirmation.) To appreciate how remarkable these discoveries are, we must look at the process that astronomers go through to search for exoplanets.

Searching for Exoplanets

It is very difficult to make direct observations of planets orbiting other stars. The problem is that planets are small and dim compared with stars; for example, the Sun is a billion times brighter than Jupiter at visible wavelengths. Exoplanets can be imaged directly, as in **Figure 8-15a**, but a star's glare makes this method the least productive method of detection. By the time the exoplanets in Figure 8-15b were imaged in 2010, hundreds had been detected through other means. One very powerful method of exoplanet detection is to search for stars that appear to "wobble" (**Figure 8-16**). If a star has a planet, it is not quite correct to say that the planet orbits the star. Rather, both the planet and the star move in elliptical orbits around a point called the **center of mass** (Figure 8-16a). Imagine the planet and the star as sitting at opposite ends of a very long seesaw; the center of mass is where you would have to place the support-point (or fulcrum) in order to balance the seesaw. Because of the star's much greater mass, the center of mass is much closer to the star than to the planet. Thus, while the planet may move in a very large orbit, the star will move in a much smaller orbit.

As an example, the Sun and Jupiter both orbit their common center of mass with an orbital period of 11.86 years. (Jupiter has more mass than the other seven planets put together, so it is a reasonable approximation to consider the Sun's wobble as being due to Jupiter alone.) The Sun's orbit around the common center of mass has a semimajor axis only slightly greater than the Sun's radius, so the Sun slowly wobbles around a point not far outside its surface. If astronomers elsewhere in the Galaxy could detect the Sun's wobbling motion, they could tell that there was a large planet (Jupiter) orbiting the Sun. They could even determine the planet's mass and the size of its orbit, even though the planet itself was unseen!

Detecting the wobble of other stars is not an easy task. One approach to the problem, called the **astrometric method,** involves making very precise measurements of a star's position in the sky relative to other stars. The goal is to find stars whose positions change in a cyclic way (Figure 8-16b). The measurements must be

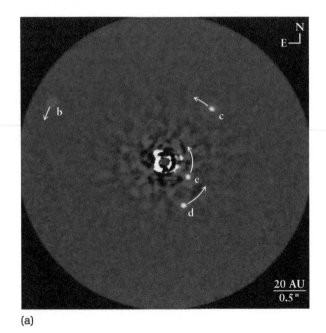

(a)

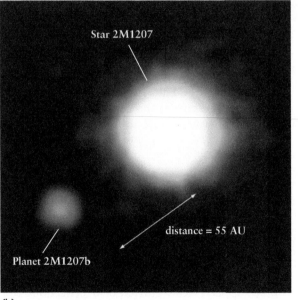

(b)

FIGURE 8-15   R ▮ V U X G

Imaging Extrasolar Planets (a) Four planets can be seen orbiting the star HR 8799. The star is about 129 light-years from Earth and can be seen with the naked eye on a clear night. In capturing this image, special techniques make the star less visible to reduce its glare and reveal its planets. The white arrows indicate possible trajectories that each star might take over the next ten years. **(b)** About 170 light-years away, the star 2M1207 and

a planet with about 1.5 times the diameter of Jupiter are captured in this infrared image. First observed in 2004, this extrasolar planet was the first to be visible in a telescopic image. At infrared wavelengths a star outshines a Jupiter-sized planet by only about 100 to 1, compared to 10^9 to 1 at visible wavelengths, making imaging possible in the infrared. (a: NASA/W. M. Keck Observatory; b: ESO/VLT/NACO)

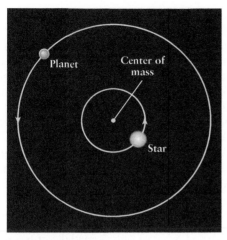

(a) A star and its planet

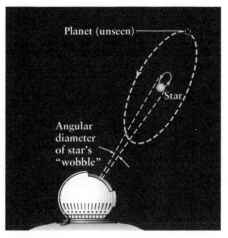

(b) The astrometric method

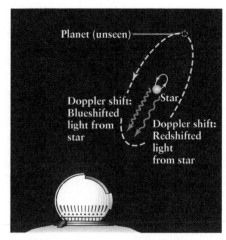

(c) The radial velocity method

FIGURE 8-16

Detecting a Planet by Measuring Its Parent Star's Motion
(a) A planet and its star both orbit around their common center of mass, always staying on *opposite sides* of this point. Even if the planet cannot be seen, its presence can be inferred if the star's motion can be detected.
(b) The astrometric method of detecting the unseen planet involves making direct measurements of the star's orbital motion. **(c)** In the radial velocity method, astronomers measure the Doppler shift of the star's spectrum as it moves alternately toward and away from Earth. Analyzing the Doppler shift of the star's light leads to the orbital period, orbital size (semimajor axis), and mass of the unseen planet. The radial velocity method works for circular and elliptical orbits and can even determine the orbital eccentricity.

made with very high accuracy (0.001 arcsec or better) and, ideally, over a long enough time to span an entire orbital period of the star's motion. The direct observation of a star's changing position has not yet played a significant role in detecting exoplanets, but as we will see, that will change in the near future.

A star's wobble can also be detected by the **radial velocity method** (Figure 8-16c). This method is based on the Doppler effect, which we described in Section 5-9. During repeated orbits, a wobbling star will alternately move toward and away from Earth. Only this motion along a radius or line between Earth and the star—in other words, the star's radial velocity—contributes to the Doppler shift. The star's movement causes the wavelengths of its entire spectrum to shift in a periodic fashion, and this can be measured in the star's dark absorption lines (see Figure 5-14). When the star is moving away from us, its spectrum will undergo a redshift to longer wavelengths. When the star is approaching, there will be a blueshift of the spectrum to shorter wavelengths (see Figure 5-26). These wavelength shifts are very small because the star's motion around its orbit is quite slow. As an example, the Sun moves around its small orbit at only 12.5 m/s (45 km/h, or 28 mi/h)—slow enough that someone could beat the Sun on a fast bike ride! If the Sun was moving directly toward an observer at this speed, the hydrogen absorption line at a wavelength of 656 nm in the Sun's spectrum would be shifted by only 2.6×10^{-5} nm, or about 1 part in 25 million. Detecting these tiny shifts requires extraordinarily careful measurements and painstaking data analysis.

In 1995, the radial velocity method was used to discover a planet orbiting the star 51 Pegasi, which is 47.9 light-years away. For the first time, solid evidence was found for a planet orbiting a star like our own Sun. Astronomers have since used the radial velocity method to discover about 500 exoplanets.

From radial velocity plots such as Figure 8-17, a planet's orbital distance and mass can be determined. To determine these quantities, the period of the radial velocity—this is measured

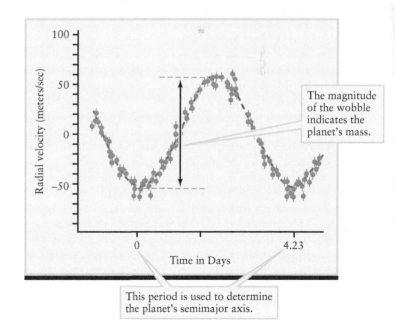

The magnitude of the wobble indicates the planet's mass.

This period is used to determine the planet's semimajor axis.

FIGURE 8-17

Discovering Planets from a Star's Wobble: 51 Pegasi As a star wobbles, it moves toward us and away from us, with a varying radial velocity that is measured through the Doppler shift. The shape of the curve indicates the planet's semimajor axis and also the planet's mass. However, the planetary mass is actually a minimum mass because an orbital plane tilted out of our line of sight would require an even larger mass to produce the observed Doppler shift.

from the star's Doppler shifting wobble—gives the period of the planet's orbit. Using Newton's form of Kepler's third law (see Box 4-4) and standard methods for estimating the mass of the star, we find the semimajor axis of the planet's orbit. Then, by knowing the planet's orbital distance and the gravitational force required to cause the star's observed radial velocity, we can determine the planet's mass.

Exoplanets: Surprising Masses and Orbits

What do we find in these other solar systems? The planet orbiting 51 Pegasi has a mass of at least 0.47 times that of Jupiter but orbits only 0.052 AU away from its star and with an orbital period of just 4.23 days. In our own solar system, this planet would orbit about *7 times closer to the Sun than Mercury!* In fact, as shown in **Figure 8-18**, most of the extrasolar planets discovered by the radial velocity method have large masses (comparable with Jupiter) and also orbit very close to their stars. However, this is an example of a **selection effect,** where an observational technique is more likely to detect some systems, and not others. Since a larger planet closer to its star creates a greater Doppler shift, the radial velocity method is more likely to detect this type of extrasolar system.

The difficulty of the radial velocity method to detect Earthlike planets in orbits similar to our own does not lessen the mystery of the so-called **hot Jupiters** that have been found. Orbiting so close to their stars, these planets would, in fact, be very hot; the temperatures at Mercury's distance can melt lead! Furthermore, according to the picture of our own solar system formation (Section 8-5), such Jovian planets would be expected to orbit relatively far from their stars, where temperatures were low enough to allow the buildup of a massive envelope of hydrogen and helium gas. But as Figure 8-18 shows, many extrasolar planets are in fact found orbiting very *close* to their stars.

Another surprising result is that many of the extrasolar planets found so far have noncircular orbits with very large eccentricities (see Figure 4-10b). As an example, the planet around the star 16 Cygni B has an orbital eccentricity of 0.67; its distance from the star varies between 0.55 AU and 2.79 AU. This eccentricity is quite unlike planetary orbits in our own solar system, where no planet has an orbital eccentricity greater than 0.2. Recall that solar system formation in the nebular hypothesis begins with matter mostly in circular motion. Therefore, to explain the highly eccentric exoplanet orbits requires some phenomena different than in our own solar system's past.

Do these observations mean that our picture of how planets form is incorrect? If Jupiterlike extrasolar planets such as that orbiting 51 Pegasi formed close to their stars, the mechanism of their formation might have been very different from that which operated in our solar system.

But another possibility is that extrasolar planets actually formed at large distances from their stars, just like our Jovian planets, and then migrated inward since their formation. Recall that in the Grand Tack model of our solar system formation (Section 8-6), Jupiter migrates inward, but to a much lesser extent than the extreme migration implied by these hot Jupiters. However, computer modeling indicates that Jupiter-size planets can migrate very close in to their stars due to interactions with a gaseous disk, if the migration occurs before the gas is dispersed.

Simulations even suggest that planets can migrate *all the way* inward and get swallowed by the central star.

One piece of evidence that young planets may spiral into their parent stars is the spectrum of the star HD 82943, which has at least two planets orbiting it. The spectrum shows that this star's atmosphere contains a rare form of lithium that is found in planets, but which is destroyed in stars by nuclear reactions within 30 million years. The presence of this exotic form of lithium, known as ^{6}Li, indicates that in the past HD 82943 was orbited by at least one other planet. It is thought that this planet was consumed by the star when it spiraled too close, leaving behind its ^{6}Li in the star's atmosphere.

Analyzing Extrasolar Planets with the Transit Method

The radial velocity method gives us only a partial picture of what extrasolar planets are like. It cannot give us precise values for a planet's mass, only a lower limit. (An exact determination of the mass would require knowing how the plane of the planet's orbit is inclined to our line of sight. Unfortunately, this angle is not known because the planet itself is unseen.) Furthermore, the radial velocity method by itself cannot tell us the diameter of a planet or what the planet is made of.

A technique called the **transit method** makes it possible to fill in these blanks about the properties of certain extrasolar planets. This method looks for the rare situation in which a planet comes between us and its parent star, in an event called a **transit** (**Figure 8-19**). As in a partial solar eclipse (Section 3-5), this causes a small but measurable dimming of the star's light. If a transit is seen, the orbit must be nearly edge-on to our line of sight. By knowing the orientation of the orbit due to the transit, and then using information obtained by additional radial velocity measurements of the star, we learn the true mass (not just a lower limit) of the orbiting planet. But learning the planet's mass is only the beginning.

There are three other benefits of the transit method. One, the amount by which the star is dimmed during the transit depends on how large the planet is, and so tells us the planet's diameter (Figure 8-19a). Two, during the transit the star's light passes through the planet's atmosphere, where certain wavelengths are absorbed by the atmospheric gases. This absorption affects the spectrum of starlight that we measure and therefore allows us to determine the composition of the planet's atmosphere (Figure 8-19b). Three, with an infrared telescope it is possible to detect a slight dimming when the planet goes *behind* the star (Figure 8-19c). That is because the planet emits infrared radiation due to its own temperature, and this radiation is blocked when the planet is behind the star. Measuring the amount of infrared dimming tells us the amount of radiation emitted by the planet, which in turn tells us the planet's surface temperature.

One example of what the transit method can tell us is the planet that orbits the star HD 209458 in Figure 8-19, which is 153 light-years from the Sun. The mass of this planet is 0.69 that of Jupiter, but its diameter is 1.32 times larger than that of Jupiter. This lower-mass, larger-volume planet has only one-quarter the density of Jupiter. It is also very hot: Because the planet orbits just 0.047 AU from the star, its surface temperature is a torrid 1130 K (860°C, or 1570°F), about twice the temperature needed to melt

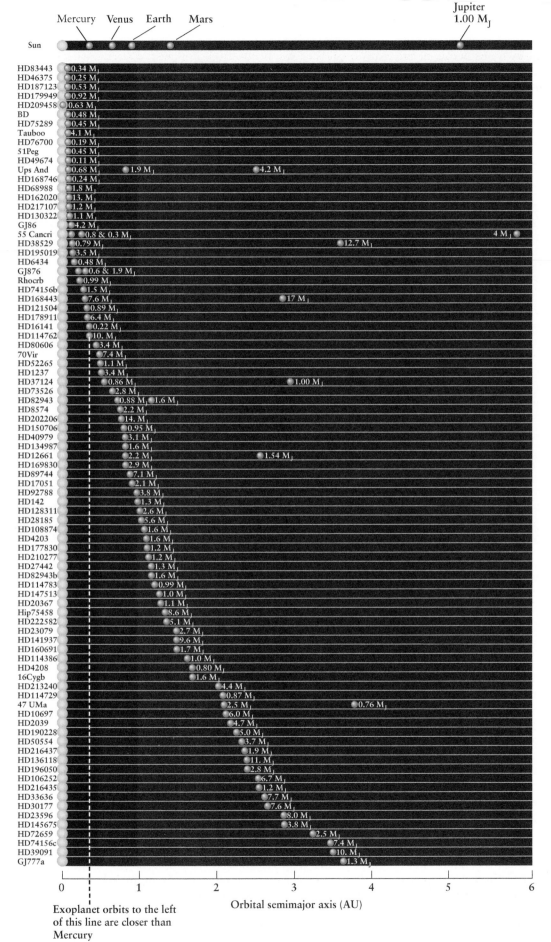

FIGURE 8-18

A Selection of Extrasolar
Planets This figure summarizes what we
know about planets orbiting a number of
other stars. The star name is given at the
left of each line. Each planet is shown at
its average distance from its star (equal to
the semimajor axis of its orbit), and some
stars are found with multiple planets. The
mass of each planet—actually a lower
limit—is given as a multiple of Jupiter's
mass (M_J), equal to 318 Earth masses.
Comparison with our own solar system (at
the top of the figure) shows how closely
many of these extrasolar planets orbit
their stars—many planets about the size
of Jupiter, orbit closer to their star than
Mercury does to our Sun! Due to higher
temperatures closer to the Sun, many of
these exoplanets are called "hot Jupiters."
However, the radial velocity method
detects large close-in planets most easily,
and if all planets could be detected, the
majority would be smaller and farther
away. (Adapted from the California and
Carnegie Planet Search)

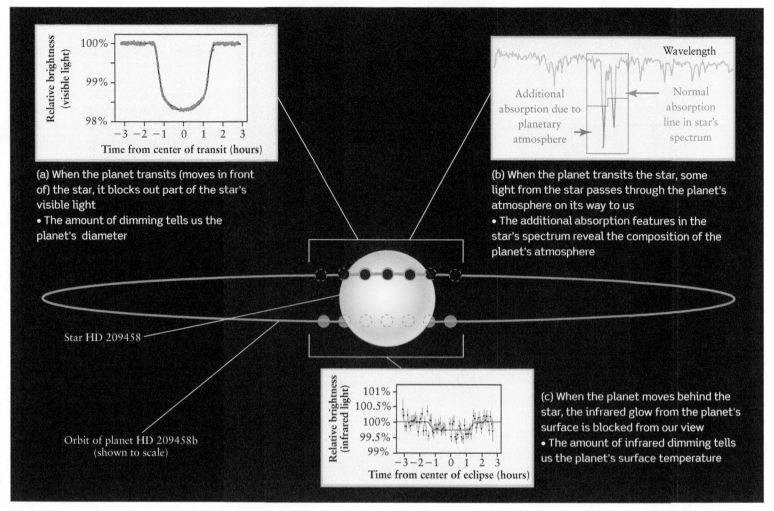

(a) When the planet transits (moves in front of) the star, it blocks out part of the star's visible light
• The amount of dimming tells us the planet's diameter

(b) When the planet transits the star, some light from the star passes through the planet's atmosphere on its way to us
• The additional absorption features in the star's spectrum reveal the composition of the planet's atmosphere

(c) When the planet moves behind the star, the infrared glow from the planet's surface is blocked from our view
• The amount of infrared dimming tells us the planet's surface temperature

FIGURE 8-19

A Transiting Extrasolar Planet If the orbit of an extrasolar planet is nearly edge-on to our line of sight, like the planet that orbits the star HD 209458, we can learn about the planet's (a) diameter, (b) atmospheric composition, and (c) surface temperature. (S. Seager and C. Reed, *Sky and* *Telescope*; H. Knutson, D. Charbonneau, R. W. Noyes (Harvard-Smithsonian CfA), T. M. Brown (HAO/NCAR), and R. L. Gilliland (STScI); A. Feild (STScI); NASA/JPL-Caltech/D. Charbonneau, Harvard-Smithsonian CfA)

lead! Observations of the spectrum during a transit of this hot Jupiter reveal the presence of hydrogen, carbon, oxygen, and sodium in a planetary atmosphere that is literally evaporating away in the intense starlight. Discoveries such as these give insight into the exotic circumstances that can be found in other planetary systems. Yet, we are also very interested in less exotic systems: the transit method can also discover terrestrial Earth-sized planets farther away from their star where temperatures might be just right for liquid water to form.

The Search for Earth-size Planets in the Habitable Zone

The ultimate goal of planetary searches is to find planets like Earth. An "Earth-size" planet is any rocky planet ranging in size from about half to double the size of Earth. We are eager to discover Earth-size planets with liquid water, which improves the chances that such a planet might harbor life. Why liquid water?

Even the most exotic forms of life on Earth require liquid water, and many biochemists argue that an alien life-form based on some alternative biochemistry might still require water.

A rocky planet in the so-called **habitable zone** of a solar system might have liquid water on its surface: In the habitable zone a planet is close enough to its star to melt water ice, yet not close enough to boil the water away. But the habitable zone is only a rough guide: Planetary greenhouse gases can warm up a cold planet or even boil off a planet's water (as in the case of Venus). Furthermore, as in the case of Jupiter's Europa—which is far beyond our habitable zone—tidal forces can warm up an icy moon to create liquid water.

Until recently, detecting Earth-size planets at Earth's orbital distance from a star—in other words, in the habitable zone—was impossible. To do this required a new telescope. Launched in 2009, the *Kepler* spacecraft was designed to detect Earth-size planets in habitable zones around Sunlike stars. *Kepler* uses the

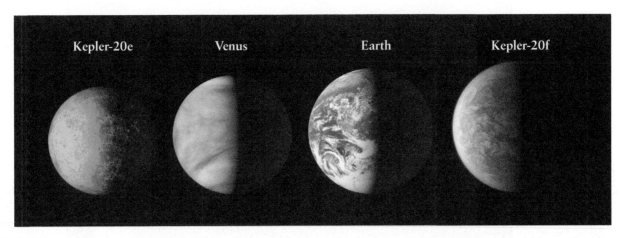

FIGURE 8-20

Earth-Size Exoplanets Kepler-20 is a star with five known exoplanets. The exoplanets Kepler-20e and Kepler-20f are illustrated in comparison to Earth and Venus. Kepler-20e (at 0.05 AU) has a radius just 0.87 times smaller than Earth, while Kepler-20f (at 0.1 AU) is 1.03 times larger than Earth. However, these objects orbit extremely close to their star and are much too hot for liquid water on their surfaces. (Ames/JPL-Caltech/NASA)

transit method by continuously monitoring the brightness of 145,000 stars in our Galaxy to search for a transit's dimming effect. As of this writing, *Kepler* has found 2321 candidates, and more than 90% of these will probably be confirmed as exoplanets. Of these, 207 candidates are roughly Earth-size (Figure 8-20) and 48 planets of various sizes were found in their stars' habitable zones. The *Kepler* team also estimates from this data set that about 5.4% of stars in our Galaxy host an Earth-size planet, and 17% of stars host multiple-planet systems.

Kepler is getting quite close to finding an Earth-size planet that is also in the habitable zone. One exoplanet only 2.4 times larger than Earth, Kepler-22b, is within the habitable zone, as illustrated in Figure 8-21. With more data on the way, habitable exoplanets might be just over the horizon.

Additional Searches for Extrasolar Planets

In 2004 astronomers began to use a property of space discovered by Albert Einstein as a tool for detecting extrasolar planets. As predicted by Einstein's general theory of relativity, a star's gravity can deflect the path of a light beam just as it deflects the path of a planet or spacecraft. If a star drifts through the line of sight between Earth and a more distant star, the closer star's gravity acts like a lens that focuses the more distant star's light (Figure 8-22). Such **microlensing** causes the distant star's image as seen in a telescope to become brighter. If the closer star has a planet, the planet's gravity will cause a secondary brightening whose magnitude depends on the planet's mass. Using this technique astronomers have detected a handful of extrasolar planets as far as 24,000 light-years away, and expect to find many more.

The most extensive search for extrasolar planets will be carried out by the European Space Agency's *Gaia* mission, scheduled for launch in 2013. *Gaia* will survey a billion (10^9) stars—1% of all the stars in the Milky Way Galaxy—and will be able to detect tiny stellar wobbles. To detect these wobbles, *Gaia* will measure minute changes in a star's position, as small as 2×10^{-6} arcsec, which is enough to see something move 1 cm on the Moon all the way from Earth! With such high-precision data, the astrometric method (Figure 8-16b) will finally come into its own. *Gaia* will also be able to follow up with radial velocity measurements on the

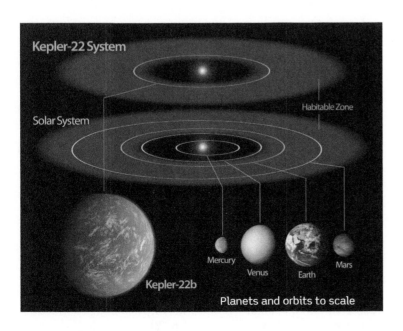

FIGURE 8-21

Exoplanets in the Habitable Zone The exoplanet Kepler-22b lies inside the habitable zone and is compared to our solar system in this illustration. This planet is 2.4 times the size of Earth, which raises hopes for a hard surface with at least some liquid water. Its mass is not yet known, but it could be around 10 times the mass of Earth. Each star determines the size of its surrounding habitable zone, and the two zones in this image are similar because the star, Kepler-22, is Sunlike. (NASA/Ames/JPL-Caltech)

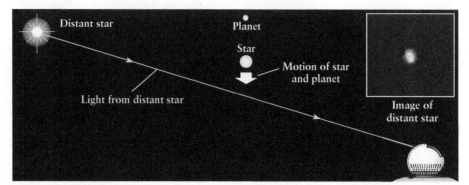

(a) No microlensing

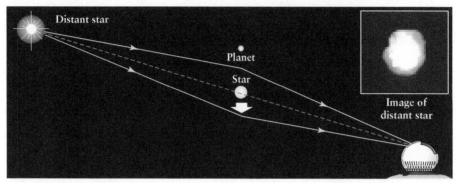

(b) Microlensing by star

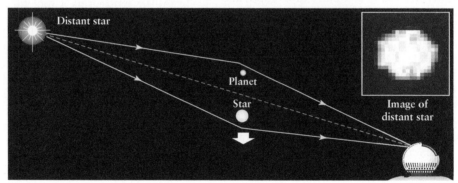

(c) Microlensing by star and planet

ANIMATION 8-9

FIGURE 8-22

Microlensing Reveals an Extrasolar Planet (a) A star with a planet drifts across the line of sight between a more distant star and a telescope on Earth. (b) The gravity of the closer star bends the light rays from the distant star, focusing the distant star's light and making it appear brighter. (c) The gravity of the planet causes a second increase in the distant star's brightness.

most promising candidates, as well as on transits. Astronomers estimate that *Gaia* will find about 10 exoplanets a day, leading to about 15,000 discoveries over the life of its mission. Over the next several years, the data from *Kepler* and *Gaia* will greatly advance our understanding of the formation and evolution of planetary systems, as well as find planets similar to Earth.

CONCEPTCHECK 8-12

Why will the radial velocity method fail to discover an extrasolar planet if the plane of the extrasolar planet's orbit is oriented perpendicular to the line between Earth and the wobbling star?

Answer appears at the end of the chapter.

KEY WORDS

Term preceded by an asterisk (*) is discussed in Box 8-1.

accretion, p. 209
astrometric method (for detecting extrasolar planets), p. 214
atomic number, p. 202
center of mass, p. 214
chemical differentiation, p. 210
condensation temperature, p. 208
conservation of angular momentum, p. 206

core accretion model, p. 210
exoplanet, p. 213
extrasolar planet, p. 213
Grand Tack model, p. 212
habitable zone, p. 218
*half-life, p. 204
hot Jupiter, p. 216
interstellar medium, p. 201
Kelvin-Helmholtz contraction, p. 206

Late Heavy Bombardment,
  p. 213
meteorite, p. 203
microlensing, p. 219
migration, p. 212
nebular hypothesis, p. 205
nebulosity, p. 201
Nice model, p. 212
planetesimal, p. 209
protoplanet, p. 209
protoplanetary disk, p. 206
protosun, p. 205

radial velocity method (for
  detecting extrasolar planets),
  p. 215
radioactive dating, p. 203
radioactive decay, p. 203
selection effect, p. 216
snow line, p. 208
solar nebula, p. 205
transit, p. 216
transit method (for detecting
  extrasolar planets), p. 216

KEY IDEAS

The Nebular Hypothesis: The most successful model of the origin of the solar system is called the nebular hypothesis. According to this hypothesis, the solar system formed from a cloud of interstellar material called the solar nebula. This occurred 4.54 billion years ago (as determined by radioactive dating).

The Solar Nebula and Its Evolution: The chemical composition of the solar nebula, by mass, was 98% hydrogen and helium (elements that formed shortly after the beginning of the universe) and 2% heavier elements (produced much later in the centers of stars and cast into space when the stars died). The heavier elements were in the form of ice and dust particles.

• The nebula flattened into a disk in which all the material orbited the center in the same direction, just as do the present-day planets.

Formation of the Planets and Sun: The terrestrial planets, the Jovian planets, and the Sun followed different pathways to formation.

• The four terrestrial planets formed through the accretion of rock and metal dust grains into planetesimals, then into larger protoplanets.

• Closer to the Sun—inside the snow line—water could not freeze into ice to help build larger planets. With rocks and metals making up only a small fraction of mass throughout the nebula, the terrestrial planets made from these materials in the inner solar system are small.

• Jovian planets form beyond the snow line, where ices of water and methane contribute to make larger cores as these planets form. These larger cores can then attract copious amounts of hydrogen and helium gas, becoming large Jovian planets.

• Gravitational interactions between the Jovian planets and a gaseous disk, and much later with a disk of planetesimals, leads to planetary migration. These migrations can explain the small size of Mars, the composition of the asteroid belt, the locations of the Jovian planets, aspects of the Kuiper belt and Oort cloud, and even account for a Late Heavy Bombardment.

• The Sun formed by gravitational contraction of the center of the nebula. After about 10^8 years, temperatures at the protosun's center became high enough to ignite nuclear reactions that convert hydrogen into helium, thus forming a true star.

Extrasolar Planets: Astronomers have discovered many planets orbiting other stars.

• Many of these planets are detected by the "wobble" of the stars around which they orbit. The radial velocity method detects this wobble through Doppler shifts. The astrometric method directly observes a star's change in position as it wobbles.

• Many extrasolar planets have been discovered by the transit method, and a lesser number by microlensing, and direct imaging.

• Most of the extrasolar planets discovered to date are quite massive and have orbits that are very different from planets in our solar system. This is expected because big planets close to stars are easier to detect. We are just beginning to detect Earth-size planets at distances that might allow liquid water on their surfaces.

QUESTIONS

Review Questions

1. Describe three properties of the solar system that are thought to be a result of how the solar system formed.

2. The graphite in your pencil is a form of carbon. Where were these carbon atoms formed?

3. What is the interstellar medium? How does it become enriched over time with heavy elements?

4. What is the evidence that other stars existed before our Sun was formed?

5. Why are terrestrial planets smaller than Jovian planets?

6. How do radioactive elements make it possible to determine the age of the solar system? What are the oldest objects that have been found in the solar system?

7. Consider a 4.54×10^9 year old meteorite. Are the atoms in that meteorite—which were created in stars—4.54×10^9 years old? Explain your reasoning.

8. What is the tidal hypothesis? What aspect of the solar system was it designed to explain? Why was this hypothesis rejected?

9. The half-life for uranium-238 in Box 8-1 is 4.5 billion years. There is plenty of uranium-238 in Earth, and much of Earth's heat comes from the radioactive decay of this isotope. Compared to today, about how much more uranium-238 was there when Earth formed? (*Hint:* See Box 8-1.)

10. What is the nebular hypothesis? Why is this hypothesis accepted?

11. What was the protosun? What caused it to shine? Into what did it evolve?

12. Why is it thought that a disk appeared in the solar nebula?

13. What are protoplanetary disks? What do they tell us about the plausibility of our model of the solar system's origin?

14. What is meant by a substance's condensation temperature? What role did condensation temperatures play in the formation of the planets?

15. At distances within the snow line, what is the state of water (solid, liquid, or gas)? How does this affect the formation of terrestrial planets?

16. Why are terrestrial planets smaller than Jovian planets?

17. What is a planetesimal? How did planetesimals give rise to the terrestrial planets?

18. (a) What is meant by accretion? (b) Why are the terrestrial planets denser at their centers than at their surfaces?

19. If hydrogen and helium account for 98% of the mass of all the atoms in the universe, why aren't Earth and the Moon composed primarily of these two gases?

20. Why did the terrestrial planets form close to the Sun while the Jovian planets formed far from the Sun?

21. How did the Jovian planets form?

22. Explain how our current understanding of the formation of the solar system can account for the following characteristics of the solar system: (a) All planetary orbits lie in nearly the same plane. (b) All planetary orbits are nearly circular. (c) The planets orbit the Sun in the same direction in which the Sun itself rotates.

23. In the Grand Tack model, why is Mars much smaller than Earth or Venus?

24. How does the Grand tack model help us understand the composition of the asteroid belt?

25. Why do we think that Neptune must have formed closer to the Sun, and later migrated outward to its present position?

26. In the Nice model of the Jovian planets, which planet migrates more, Jupiter or Neptune?

27. In the Nice model, how does the Oort cloud and the Kuiper belt form?

28. What is the Late Heavy Bombardment? How does the Nice model explain this event?

29. Explain why most of the satellites of Jupiter orbit that planet in the same direction that Jupiter rotates.

30. What is the radial velocity method used to detect planets orbiting other stars? Why is it difficult to use this method to detect planets like Earth?

31. What type of planets and orbits are easiest to detect in the radial velocity method? What are "hot Jupiters"? Explain your answer.

32. Summarize the differences between the planets of our solar system and those found orbiting other stars.

33. Is there evidence that planets have fallen into their parent stars? Explain your answer.

34. What does it mean for a planet to transit a star? What can we learn from such events?

35. What combination of methods are required to determine the average density of an exoplanet? Explain your reasoning. (*Hint:* The average density is the planet's mass divided by its volume.)

36. What is so special about the habitable zone? Is every planet in this zone like Earth?

37. What is microlensing? How does it enable astronomers to discover extrasolar planets?

38. A 1999 news story about the discovery of three planets orbiting the star Upsilon Andromedae ("Ups And" is near the top of stars listed in Figure 8-18) stated that "the newly discovered galaxy, with three large planets orbiting a star known as Upsilon Andromedae, is 44 light-years away from Earth." What is wrong with this statement?

Advanced Questions

Questions preceded by an asterisk (*) involve topics discussed in Box 4-4 or Box 8-1.

> **Problem-solving tips and tools**
>
> The volume of a disk of radius r and thickness t is $\pi r^2 t$. Box 1-1 explains the relationship between the angular size of an object and its actual size. An object moving at speed v for a time t travels a distance $d = vt$; Appendix 6 includes conversion factors between different units of length and time. To calculate the mass of 70 Virginis or the orbital period of the planet around 2M1207, review Box 4-4. Section 5-4 describes the properties of blackbody radiation.

39. Figure 8-4 shows that carbon, nitrogen, and oxygen are among the most abundant elements (after hydrogen and helium). In our solar system, the atoms of these elements are found primarily in the molecules CH_4 (methane), NH_3 (ammonia), and H_2O (water). Explain why you suppose this is.

40. (a) If Earth had retained hydrogen and helium in the same proportion to the heavier elements that exist elsewhere in the universe, what would its mass be? Give your answer as a multiple of Earth's actual mass. Explain your reasoning. (b) How does your answer to (a) compare with the mass of Jupiter, which is 318 Earth masses? (c) Based on your answer to (b), would you expect Jupiter's rocky core to be larger, smaller, or the same size as Earth? Explain your reasoning.

*41. If you start with 0.80 kg of radioactive potassium (^{40}K), how much will remain after 1.3 billion years? After 2.6 billion years? After 3.9 billion years? How long would you have to wait until there was *no* ^{40}K remaining?

*42. Three-quarters of the radioactive potassium (^{40}K) originally contained in a certain volcanic rock has decayed into argon (^{40}Ar). How long ago did this rock form?

43. Suppose you were to use the Hubble Space Telescope to monitor one of the protoplanetary disks shown in Figure 8-8b. Over the course of 10 years, would you expect to see planets forming within the disk? Why or why not?

44. The protoplanetary disk at the upper right of Figure 8-8b is seen edge-on. The diameter of the disk is about 700 AU. (a) Make measurements on this image to determine the thickness of the disk in AU. (b) Explain why the disk will continue to flatten as time goes by.

45. The accompanying infrared image shows IRAS 04302+2247, a young star that is still surrounded by a disk of gas and dust. The scale bar at the lower left of the image shows that at the distance of IRAS 04302+2247, an angular size of 2 arcseconds corresponds to a linear size of 280 AU. Use this information to find the distance to IRAS 04302+2247.

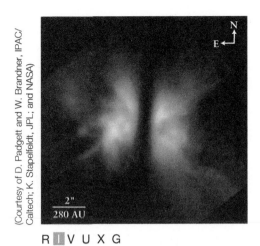

(Courtesy of D. Padgett and W. Brandner, IPAC/ Caltech; K. Stapelfeldt, JPL; and NASA)

2"
280 AU

R **I** V U X G

46. The image accompanying Question 45 shows a dark, opaque disk of material surrounding the young star IRAS 04302+2247. The disk is edge-on to our line of sight, so it appears as a dark band running vertically across this image. The material to the left and right of this band is still falling onto the disk. (a) Make measurements on this image to determine the diameter of the disk in AU. Use the scale bar at the lower left of this image. (b) If the thickness of the disk is 50 AU, find its volume in cubic meters. (c) The total mass of the disk is perhaps 2×10^{28} kg (0.01 of the mass of the Sun). How many atoms are in the disk? Assume that the disk is all hydrogen. A single hydrogen atom has a mass of 1.673×10^{-27} kg. (d) Find the number of atoms per cubic meter in the disk. Is the disk material thick or thin compared to the air that you breathe, which contains about 5.4×10^{25} atoms per cubic meter?

*47. The planet discovered orbiting the star 70 Virginis (70Vir is near the middle of the stars listed in Figure 8-18), 59 light-years from Earth, moves in an orbit with semimajor axis 0.48 AU and eccentricity 0.40. The period of the orbit is 116.7 days. Find the mass of 70 Virginis. Compare your answer with the mass of the Sun. (*Hint:* The planet has far less mass than the star.)

*48. Because of the presence of Jupiter, the Sun moves in a small orbit of radius 742,000 km with a period of 11.86 years. (a) Calculate the Sun's orbital speed in meters per second. (b) An astronomer on a hypothetical planet orbiting the star Vega, 25 light-years from the Sun, wants to use the astrometric method to search for planets orbiting the Sun. What would be the angular diameter of the Sun's orbit as seen by this alien astronomer? Would the Sun's motion be discernible if the alien astronomer could measure positions to an accuracy of 0.001 arcsec? (c) Repeat part (b), but now let the astronomer be located on a hypothetical planet in the Pleiades star cluster, 360 light-years from the Sun. Would the Sun's motion be discernible to this astronomer?

49. (a) Figure 8-19c shows how astronomers determine that the planet of HD 209458 has a surface temperature of 1130 K. Treating the planet as a blackbody, calculate the wavelength at which it emits most strongly. (b) The star HD 209458 itself has a surface temperature of 6030 K. Calculate its wavelength of maximum emission, assuming it to be a blackbody.

(c) If a high-resolution telescope were to be used in an attempt to record an image of the planet orbiting HD 209458, would it be better for the telescope to use visible or infrared light? Explain your reasoning.

*50. (a) The star 2M1207 shown in Figure 8-15b is 170 light-years from Earth. Find the angular distance between this star and its planet as seen from Earth. Express your answer in arcseconds. (b) The mass of 2M1207 is 0.025 that of the Sun; the mass of the planet is very much smaller. Calculate the orbital period of the planet, assuming that the distance between the star and planet shown in Figure 8-15b is the semimajor axis of the orbit. Is it possible that an astronomer could observe a complete orbit in one lifetime?

Discussion Questions

51. Propose an explanation why the Jovian planets are orbited by terrestrial-like satellites.

52. Suppose that a planetary system is now forming around some protostar in the sky. In what ways might this planetary system turn out to be similar to or different from our own solar system? Explain your reasoning.

53. Suppose astronomers discovered a planetary system in which the planets orbit a star along randomly inclined orbits. How might a theory for the formation of that planetary system differ from that for our own?

Web/eBook Question

54. Search the World Wide Web for information about recent observations of protoplanetary disks. What insights have astronomers gained from these observations? Is there any evidence that planets have formed within these disks?

55. In 2000, extrasolar planets with masses comparable to that of Saturn were first detected around the stars HD 16141 (also called 79 Ceti) and HD 46375. Search the World Wide Web for information about these "lightweight" planets. Do these planets move around their stars in the same kind of orbit as Saturn follows around the Sun? Why do you suppose this is? How does the discovery of these planets reinforce the model of planet formation described in this chapter?

56. In 2006 a planet called XO-1b was discovered using the transit method. Search the World Wide Web for information about this planet and how it was discovered. What unusual kind of telescope was used to make this discovery? Have other extrasolar planets been discovered using the same kind of telescope?

ACTIVITIES

Observing Projects

57. Use *Starry Night*™ to make observations of the solar system. Select **Favourites > Explorations > Solar System**. The view shows the names and orbits of the major planets of the solar system against the backdrop of the stars of the Milky Way Galaxy, from a location hovering 64 AU from the Sun. You may also see many smaller objects moving in the asteroid

belt between the orbits of Mars and Jupiter. (If not, select **View > Solar System** and click on the **Asteroids** box.) (**a**) Use the location scroller to look at the solar system from different angles, and observe the general distribution and motion of the major planets. Make a list of your observations. (**b**) How does the nebular hypothesis of solar system formation account for your observations?

58. Use *Starry Night*™ to examine stars that have planets. Select **Favourites > Explorations > Extrasolar Planets**. This is the view of nearby space looking back toward the location of Earth from a distance of about 90 light-years. Use the location scroller to look around. Stars marked with a light blue halo show evidence of having at least one planet. Right-click (Ctrl-click on a Mac) on a sample of these circled stars and select **Show Info** from the contextual menu to learn more about the star's properties. Under the **Other Data** layer, note the star's apparent magnitude, which is a measure of how bright each star appears as seen from Earth. Apparent magnitude uses an inverse scale: The greater the apparent magnitude, the dimmer the star. Most of the brighter stars you can see with the naked eye from Earth have apparent magnitudes between 0 and 1, while the dimmest star you can see from a dark location has apparent magnitude 6. Also note, under the **Other Data** layer info for your sample stars, the extrasolar mass (the mass of the extrasolar planet) and extrasolar semimajor axis (a measure of the planet's distance from its parent star). (**a**) Are most of the circled stars in your sample visible to the naked eye from Earth? List at least two stars that have extrasolar planets that are visible to the naked eye from Earth and include their apparent magnitudes. Expand your sample if necessary. (**b**) Are most of the planets in your sample larger than Jupiter? Offer a hypothesis to account for this. (**c**) You may have noted that a surprising number of extrasolar planets in your sample are both very massive and also orbit very close to their parent star. What hypothesis might account for this?

59. Use *Starry Night*™ to investigate stars that have planets orbiting them. Click the **Home** button in the toolbar. Open the **Options** pane and use the checkboxes in the **Local View** layer to turn off **Daylight** and the **Local Horizon**. Expand the **Stars** layer in the **Options** pane and then expand the **Stars** item and check the **Mark stars with extrasolar planets** option. Then use the **Find** pane to find and center each of the stars listed below. To do this, click the magnifying glass icon on the side of the edit box at the top of the **Find** pane and select **Star** from the dropdown menu; then type the name of the star in the edit box and press the **Enter** or **Return** key on the keyboard. Click on the **Info** tab for full information about the star. Expand the **Other Data** layer and note the luminosity of each of these four stars: 47 Ursae Majoris (3 known planets); 51 Pegasi (1 known planet); 70 Virginis (1 known planet); Rho Coronae Borealis (1 known planet). (**a**) Which stars are more luminous than the Sun? (**b**) Which are less luminous? (**c**) How do you think these differences

would have affected temperatures in the nebula in which each star's planets formed?

ANSWERS

ConceptChecks

ConceptCheck 8-1: No. Earth is not an exception to any of the three properties in Table 8-1, and in fact, all the planets are consistent with these properties.

ConceptCheck 8-2: The carbon atoms that make up much of our human bodies and the very oxygen atoms we breathe were made inside of stars. Stars have made most of the atoms other than hydrogen.

ConceptCheck 8-3: Yes. Hydrogen (H) is the first most abundant element, and oxygen (O) is the third most abundant element. Water (H_2O) is, in fact, quite common even though it is usually frozen.

ConceptCheck 8-4: No. The age of a rock refers to how much time has passed since the rock cooled and solidified. It is not the age of the rock's atoms, which were made in stars, and possibly different stars at different times.

ConceptCheck 8-5: To be correct, the nebular hypothesis must explain why all the planets orbit the Sun in the same direction and in nearly the same plane. This arrangement is unlikely to have arisen purely by chance.

ConceptCheck 8-6: No. Through Kelvin-Helmholtz contraction, gravitational energy is converted into heat—a lot of heat. The protosun, which formed before nuclear reactions began, had a surface temperature of 6000 K, slightly higher than our present-day Sun. Thus, the solar nebula was warmed by the protosun.

ConceptCheck 8-7: Frozen water helped to make Jupiter large enough to hold onto hydrogen and helium. The closest that a large, Jupiterlike planet could form would be much farther away in a much hotter solar nebula.

ConceptCheck 8-8: Called chemical differentiation, denser, iron-rich materials migrate to the center of a planet while the less dense silicon-rich minerals float to the outside surface before the initially molten planet solidifies in the early solar system.

ConceptCheck 8-9: At Neptune's present distance, the time to build a planet the size of Neptune is much longer than the time that the protoplanetary disk is around.

ConceptCheck 8-10: In the Nice model, once planetary orbits became unstable, Neptune was deflected to about twice its original orbital distance.

ConceptCheck 8-11: Accretion refers to the gravitational accumulation of matter. Chemical bonds might, or might not, be formed once matter has accumulated.

ConceptCheck 8-12: An extrasolar planet with an orbit that causes a star to wobble side to side will not exhibit any Doppler shifted spectra as seen from Earth because the star will not be moving alternately toward and away from Earth, and therefore the radial velocity method will fail to detect such an orbiting extrasolar planet.

The two hemispheres of Earth. (NASA Goddard Space Flight Center Image by Reto Stöckli)   R I V U X G

The Living Earth

When astronauts made the first journeys from Earth to the Moon between 1968 and 1972, they often reported that our planet is the most beautiful sight visible from space. Of all the worlds of the solar system, only Earth has the distinctive green color of vegetation, and only Earth has liquid water on its surface. Compared with the Moon, whose lifeless, dry surface has been ravaged by billions of years of impacts by interplanetary debris, Earth seems an inviting and tranquil place.

But the appearance of tranquility is deceiving. The seemingly solid surface of the planet is in a state of slow but constant motion, driven by the flow of hot, deformable rock within Earth's interior. Sunlight provides Earth's atmosphere with the energy that drives our planet's sometimes violent weather, including all the cloud patterns visible in the images shown here.

Unseen in these images are immense clouds of subatomic particles that wander around the outside of the planet in two giant belts, held in thrall by Earth's magnetic field. And equally unseen are trace gases in Earth's atmosphere, whose abundance may determine the future of our climate, and on which may depend the survival of entire species.

In this chapter our goal is to learn about Earth's components—its dynamic oceans and atmosphere, its ever-changing surface, and its hot, active interior—and how they interact to make up our planetary home. By learning about Earth, we also learn about some of the same processes shaping the planets and moons we will look at in Chapters 10 through 14.

9-1 Earth's atmosphere, oceans, and surface are extraordinarily active

The crew of an alien spacecraft exploring our solar system might overlook Earth altogether. Although it is the largest of the terrestrial planets, with a mass greater than that of Mercury, Venus, and Mars put together, Earth is far smaller than any of the giant Jovian planets (see Table 7-1 and Table 9-1). But a more careful inspection would reveal that Earth is unique among all the planets that orbit the Sun.

Dynamic Earth: Water, Air, and Land

Unlike the arid surfaces of Venus and Mars, Earth's surface is very wet. Indeed, nearly 71% of Earth's surface is covered with water (Figure 9-1). Water is also locked up into many of the minerals that

> To understand our dynamic planet, we must understand the energy sources that power its activity

make Earth's rocks, including those found in the driest deserts. Furthermore, Earth's liquid water is in constant motion. The oceans ebb and flow with the tide, streams and rivers flow down to the sea, and storms whip lake waters into a frenzy.

Earth's atmosphere is also in a state of perpetual activity. Winds blow at all altitudes, with speeds and directions that change from one hour to the next. The atmosphere is at its most dynamic when water evaporates to form clouds, then returns to the surface as rain or snow (Figure 9-2). While other worlds of the solar system have atmospheres, only Earth's contains the oxygen that animals (including humans) need to live.

The combined effects of water and wind cause erosion of mountains and beaches. But these are by no means the only forces reshaping the face of our planet. All across Earth we find evidence that the surface has been twisted, deformed, and folded (Figure 9-3). What is more, new material is continually being added to Earth's surface as lava pours forth from volcanoes and from immense cracks in the ocean floors. At the same time, other geologic activity slowly buries old surface material, or pulls

TABLE 9-1	EARTH DATA
Average distance from Sun:	1.000 AU $= 1.496 \times 10^8$ km
Maximum distance from Sun:	1.017 AU $= 1.521 \times 10^8$ km
Minimum distance from Sun:	0.983 AU $= 1.471 \times 10^8$ km
Eccentricity of orbit:	0.017
Average orbital speed:	29.79 km/s
Orbital period:	365.256 days
Rotation period:	23.9345 hours
Inclination of equator to orbit:	23.45°
Diameter (equatorial):	12,756 km
Mass:	5.974×10^{24} kg
Average density:	5515 kg/m³
Escape speed:	11.2 km/s
Albedo:	0.31
Surface temperature range:	Maximum: 60°C = 140°F = 333 K Mean: 14°C = 57°F = 287 K Minimum: −90°C = −130°F = 183 K
Atmospheric composition (by number of molecules):	78.08% nitrogen (N_2) 20.95% oxygen (O_2) 0.035% carbon dioxide (CO_2) about 1% water vapor

(NASA Goddard Space Flight Center Image by Reto Stöckli)

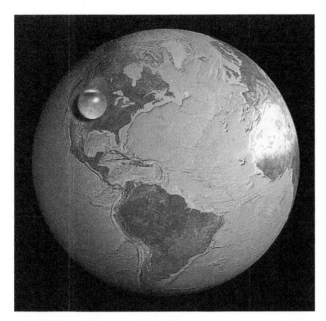

FIGURE 9-1   R I [V] U X G

Earth's Dynamic Oceans This image shows the relative size of Earth compared to a sphere containing all of Earth's water. The sphere is about 860 miles in diameter and includes fresh water, oceans, ice, and even water in the atmosphere. Nearly three-quarters of Earth's surface is covered with water, a substance that is essential to the existence of life. In contrast, there is no liquid water at all on Mercury, Venus, Mars, or the Moon. (Jack Cook, Woods Hole Oceanographic Institution, Howard Perlman, USGS)

FIGURE 9-2   R I [V] U X G

Earth's Dynamic Atmosphere This space shuttle image shows thunderstorm clouds over the African nation of Zaire. The tops of thunderstorm clouds frequently reach altitudes of 10,000 m (33,000 ft) or higher. At any given time, nearly 2000 thunderstorms are in progress over Earth's surface. (JSC/NASA)

material back into our planet's interior. These processes and others work together to renew and refresh our planet's exterior. Thus, although Earth is some 4.54 billion (4.54×10^9) years old, much of its surface is less than 200 million years old (Figure 9-4).

(We saw in Section 8-2 how scientists use radioactive dating to determine the ages of rocks and of Earth as a whole.)

Solar Energy and Convection

TUTORIAL 9-1 What powers all of this activity in Earth's oceans, atmosphere, and surface? There are three energy sources: radiation from the Sun, the tidal effects of the Moon and Sun, and Earth's own internal heat (Table 9-2).

The Sun is the principal source of energy for the atmosphere. Earth's surface is warmed by sunlight, which in turn warms the air next to the surface. Hot air is less dense than cool air and so it tends to rise (see the discussion of density in Box 7-1). As the air rises, it transfers heat to its surroundings. As a result, the rising air cools and becomes denser. It then sinks downward to be heated again, and the process starts over. This up-and-down motion is called **convection,** and the overall pattern of circulation is called a **convection current.** (You can see convection currents in action by

FIGURE 9-3   R I [V] U X G

Earth's Dynamic Surface These tan-colored ridges, or hogbacks, in Colorado's Rocky Mountains were once layers of sediment at the bottom of an ancient body of water. Forces within Earth folded this terrain and rotated the layers into a vertical orientation. The layers were revealed when wind and rain eroded away the surrounding material. (Mr. Klein/Shutterstock)

2. Over millions of years, layers of sediments built up over that rock. The most recent layer—the top—is about 250 million years old.

1. The rocks at the bottom of the Grand Canyon are 1.7-2.0 billion years old.

FIGURE 9-4   R I V U X G

Old and Young Rocks in the Grand Canyon The ages of rocks in Arizona's Grand Canyon demonstrate that geologic processes take place over very long time scales. (John Wang/PhotoDisc/Getty Images)

heating water on a stove, as Figure 9-5 shows.) Convection currents also play a key role in reshaping Earth's surface as warmer low-density material rises up from Earth's interior, cools and increases its density, and then sinks back down.

Solar energy also powers atmospheric activity by evaporating water from the surface. The energy in the water vapor is released later when it condenses to form water droplets, like those that make up clouds. In a typical thunderstorm (see Figure 9-2), the amount of energy released when this water condenses is as much as a city of 100,000 people uses in a month!

Solar energy also helps to power the oceans. Warm water from near the equator moves toward the poles, while cold polar water returns toward the equator. As we saw in Section 4-8, however, there is back-and-forth motion of the oceans due to the tidal forces of the Moon and Sun. Sometimes these two influences can reinforce each other, as when a storm (caused by solar energy) reaches a coastline at high tide (caused by tidal forces), producing waves strong enough to seriously erode beaches and sea cliffs.

TABLE 9-2   Earth's Energy Sources

Activity	Energy Sources
Motion of water in oceans, lakes, rivers	Solar energy, tidal forces
Motion of the atmosphere	Solar energy
Reshaping of surface	Earth's internal heat
Life	Solar energy (a few species that live on the ocean floor make use of Earth's internal heat)

Neither solar energy nor tidal forces can explain the reshaping of Earth's surface suggested by Figure 9-3. Rather, most geologic activity is powered by heat flowing from the interior of Earth itself. Most of the heat is caused by the nuclear decay of radioactive elements such as uranium and thorium deep inside Earth. There is even some heat still working its way out from the energetic collisions among planetesimals that created Earth (see Section 8-5).

The heat flow from Earth's interior to its surface is minuscule—just 1/6000 as great as the flow of energy we receive from the Sun—but it has a profound effect on the face of our planet. In Sections 9-2 and 9-3 we will explore the interior structure of Earth and discover that the heat-driven convection illustrated in Figure 9-5 also takes place in our Earth.

CONCEPTCHECK 9-1

Consider the convection illustrated in Figure 9-5. Is gravity necessary for convection to occur? Can convection occur if the temperature is the same above and below the fluid?

Answer appears at the end of the chapter.

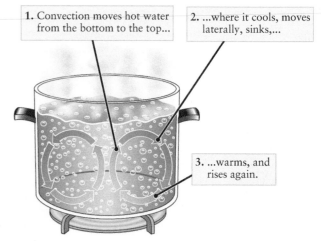

1. Convection moves hot water from the bottom to the top...

2. ...where it cools, moves laterally, sinks,...

3. ...warms, and rises again.

FIGURE 9-5

Convection in the Kitchen Heat supplied to the bottom of a pot warms the water there, making the water expand and lowering its average density. This low-density water rises and transfers heat to its cooler surroundings. The water that began at the bottom thus cools down, becomes denser, and sinks back to the bottom to repeat the process. (Adapted from F. Press, R. Siever, T. Grotzinger, and T. H. Jordan, *Understanding Earth,* 4th ed., W. H. Freeman, 2004)

Earth's Surface Temperature and the Greenhouse Effect

Since so little heat comes from inside our planet, the average surface temperature of Earth depends almost entirely on the amount of energy that reaches us from the Sun in the form of visible light and other wavelengths of electromagnetic radiation. (In an analogous way, whether you feel warm or cool outdoors on a summer day depends on whether you are in the sunlight, receiving lots of solar energy, or in the shade, receiving little of this energy.)

If Earth did nothing but *absorb* radiation from the Sun, it would get hotter and hotter until the surface temperature became high enough to melt rock. Happily, there are two reasons why this does not happen. One is that the clouds, snow, ice, and sand *reflect* about 31% of the incoming sunlight back into space. The fraction of incoming sunlight that a planet reflects is called its **albedo** (from the Latin for "whiteness"); thus, Earth's albedo is about 0.31. As a result, only 69% of the incoming solar energy is absorbed by Earth. A second reason is that Earth also *emits* radiation into space because of its temperature, in accordance with the laws that describe heated dense objects (see Section 5-3). Earth's average surface temperature is nearly constant, which means that on the whole it is neither gaining nor losing energy. Thus, the rate at which Earth emits energy into space must equal the rate at which it absorbs energy from the Sun.

To better understand this balance between absorbed and emitted radiation, remember that Wien's law tells us that the wavelength at which such an object emits most strongly (λ_{max}) is inversely proportional to its temperature (T) on the Kelvin scale (see Section 5-4 and Box 5-2). For example, the Sun's surface temperature is about 5800 K, and sunlight has its greatest intensity at a wavelength λ_{max} of 500 nm, which is green light in the middle of the visible spectrum. Earth's average surface temperature of 287 K is far lower than the Sun's, so Earth radiates most strongly at longer wavelengths in the infrared portion of the electromagnetic spectrum. The Stefan-Boltzmann law tells us that temperature also determines the *amount* of radiation that Earth emits: The higher the temperature, the more energy it radiates.

Given the amount of energy reaching us from the Sun each second as well as Earth's albedo, we can calculate the amount of solar energy that Earth should *absorb* each second. Since this amount must equal the amount of electromagnetic energy that Earth *emits* each second, which in turn depends on Earth's average surface temperature, we can calculate what Earth's average surface temperature should be. The result is a very chilly 254 K (−19°C = −2°F), so cold that oceans and lakes around the world should be frozen over. In fact, Earth's actual average surface temperature is 287 K (14°C = 57°F). What is wrong with our model? Why is Earth warmer than we would expect?

The explanation for this discrepancy is called the **greenhouse effect:** Our atmosphere prevents some of the radiation emitted by Earth's surface from escaping into space. Certain gases in our atmosphere called **greenhouse gases,** among them water vapor (H_2O) and carbon dioxide (CO_2), are transparent to visible light but not to infrared radiation. Consequently, visible sunlight has no trouble entering our atmosphere and warming the surface. But the infrared radiation coming from the heated surface is partially trapped by the atmosphere, thus raising the temperatures of both the atmosphere and the surface. As the surface and atmosphere

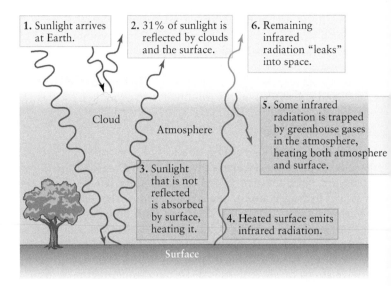

ANIMATION 9-1

FIGURE 9-6

The Greenhouse Effect Sunlight warms Earth's surface, which due to its temperature emits infrared radiation. Much of this radiation is absorbed by atmospheric water vapor and carbon dioxide, helping to raise the average temperature of the surface. Some infrared radiation does penetrate the atmosphere and leaks into space. In a state of equilibrium, the rate at which Earth loses energy to space in this way is equal to the rate at which it absorbs energy from the Sun.

become hotter, they both emit more infrared radiation, part of which is able to escape into space. The temperature levels off when the amount of infrared energy that escapes just balances the amount of solar energy reaching the surface (Figure 9-6). The result is that our planet's surface is some 33°C (59°F) warmer than it would be without the greenhouse effect, and water remains unfrozen over most of Earth.

The warming caused by the greenhouse effect gives our planet the moderate temperatures needed for the existence of life. For more than a century, however, our technological civilizations have been adding greenhouse gases to the atmosphere at an unprecedented rate. As we will see in Section 9-7, the likely consequences to our climate are extremely grave.

CONCEPTCHECK 9-2

What is the positive impact that the greenhouse effect has on our planet?

Answer appears at the end of the chapter.

9-2 Studies of earthquakes reveal Earth's layered interior structure

TUTORIAL 9-2 The kinds of rocks found on and near Earth's surface provide an important clue about our planet's interior. The densities of typical surface rocks, including the

> Like an exotic dessert, our planet's interior has layers of both solid and liquid material

rocky seafloor, are around 3000 kg/m³, but the average density of Earth as a whole (that is, its mass divided by its volume) is 5515 kg/m³. The interior of Earth must therefore be composed of substances much denser than those found near the surface. But what is this substance? Why is Earth's interior more dense than its crust? And is Earth's interior solid like rock or molten like lava?

An Iron-Rich Planet

Iron (chemical symbol Fe) is a good candidate for the substance that makes up most of Earth's interior for two reasons. First, iron atoms are quite massive (a typical iron atom has 56 times the mass of a hydrogen atom), and second, iron is relatively abundant. (Figure 8-4 shows that it is the seventh most abundant element in our part of the Milky Way Galaxy.) Other elements such as lead and uranium have more massive atoms, but these elements are quite rare. Hence, the solar nebula could not have had enough of these massive atoms to create Earth's dense interior. Furthermore, iron is common in meteoroids that strike Earth, which suggests that it was abundant in the planetesimals from which Earth formed.

Earth was almost certainly molten throughout its volume soon after its formation, about 4.54×10^9 years ago. Energy released by the violent impacts of numerous meteoroids and asteroids and by the decay of radioactive isotopes likely melted the solid material collected from the earlier planetesimals. Gravity caused abundant, dense iron to sink toward Earth's center, forcing less dense material to the surface. Figure 9-7a shows this process of *chemical differentiation*. (We discussed chemical differentiation in Section 8-5.) The result was a planet with the layered structure shown in Figure 9-7b—a central **core** composed of almost pure iron, surrounded by a **mantle** of dense, iron-rich minerals. The mantle, in turn, is surrounded by a thin **crust** of relatively light silicon-rich minerals. We live on the surface of this crust.

Seismic Waves as Earth Probes

How do we know that this layered structure is correct? The challenge in testing this model is that Earth's interior is as inaccessible as the most distant galaxies in space. The deepest wells go down only a few kilometers, barely penetrating the surface of our planet. Despite these difficulties, geologists have learned basic properties of Earth's interior by studying earthquakes and the seismic waves that they produce.

Over the centuries, forces build up in Earth's crust. Occasionally, these forces are relieved with a sudden motion called an **earthquake.** Most earthquakes occur within 20 km of Earth's surface. The point on Earth's surface directly over an earthquake's location is called the **epicenter.**

Earthquakes produce three different kinds of **seismic waves,** which travel around or through Earth in different ways and at different speeds. Geologists use sensitive instruments called **seismometer** to detect and record these vibratory motions. The first type of wave, which is analogous to ocean waves, causes the rolling motion that people feel around an epicenter. These waves are called **surface waves** because they travel only over Earth's surface. The two other kinds of waves, called **P waves** (for "primary") and **S waves** (for "secondary"), travel through the interior of Earth. P waves are called *longitudinal* waves because their oscillations are parallel to the direction of wave motion, like a spring that is alternately pushed and pulled. In contrast, S waves are called *transverse* waves because

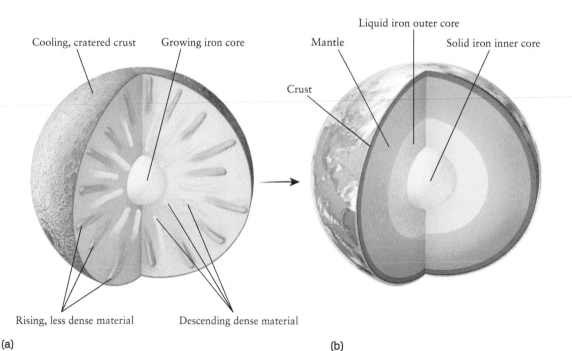

(a)

(b)

FIGURE 9-7

Chemical Differentiation and Earth's Internal Structure **(a)** The newly formed Earth was molten throughout its volume. Dense materials such as iron (shown in yellow) sank toward the center, while low-density materials (shown in orange) rose toward the surface. **(b)** The present-day Earth is no longer completely molten inside. A dense, solid iron core is surrounded by a less dense liquid core and an even less dense mantle. The crust, which includes the continents and ocean floors, is the least dense of all; it floats atop the mantle like the skin that forms on the surface of a cooling cup of cocoa.

their vibrations are perpendicular to the direction in which the waves move. S waves are analogous to waves produced by a person shaking a rope up and down (Figure 9-8).

What makes seismic waves useful for learning about Earth's interior is that they do not travel in straight lines. Instead, the paths that they follow through the body of Earth are bent because of the varying density and composition of Earth's interior. We saw in Section 6-1 that light waves behave in a very similar way. Just as light waves bend, or refract, when they pass from air into glass or vice versa (see Figure 6-2), seismic waves refract as they pass through different parts of Earth's interior. By studying how the paths of these waves bend, geologists can map out the general interior structure of Earth.

One key observation about seismic waves and how they bend has to do with the differences between S and P waves. When an earthquake occurs, seismometers relatively close to the epicenter record both S and P waves, but seismometers on the opposite side of Earth record only P waves. The absence of S waves was first explained in 1906 by British geologist Richard Dixon Oldham, who noted that transverse vibrations such as S waves cannot travel far through liquids. Oldham therefore concluded that our planet has a molten core. Furthermore, there is a region in which neither S waves nor P waves from a distant earthquake can be detected (Figure 9-9). This "shadow zone" results from the specific way in which P waves are deflected at the boundary between the solid mantle and the liquid iron outer core (the molten core). By measuring the size of the shadow zone, geologists have concluded that the radius of the molten core is about 3500 km (2200 mi), about 55% of our planet's overall radius but about double the radius of the Moon (1738 km = 1080 mi).

As the quality and sensitivity of seismometers improved, geologists discovered faint traces of P waves in an earthquake's shadow zone. In 1936, the Danish seismologist Inge Lehmann explained that some of the P waves passing through Earth are deflected into

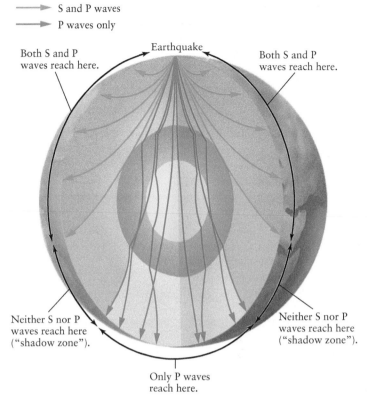

FIGURE 9-9

Earth's Internal Structure and the Paths of Seismic Waves
Seismic waves follow curved paths because of differences in the density and composition of the material in Earth's interior. The paths curve gradually where there are gradual changes in density and composition. Sharp bends occur only where there is an abrupt change from one kind of material to another, such as at the boundary between the outer core and the mantle. Only P waves can pass through Earth's liquid outer core.

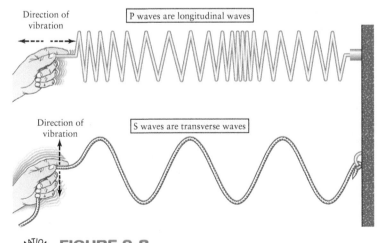

FIGURE 9-8

Seismic Waves Earthquakes produce two kinds of waves that travel through the body of our planet. One kind, called P waves, are longitudinal waves. They are analogous to those produced by pushing a spring in and out. The other kind, S waves, are transverse waves analogous to the waves produced by shaking a rope up and down.

the shadow zone by a small, solid **inner core** at the center of our planet. The radius of this solid iron inner core is about 1300 km (800 mi).

Earth's Major Layers

The seismic evidence reveals that our planet has a curious internal structure—a liquid **outer core** of iron sandwiched between a solid inner iron core and a solid mantle. Table 9-3 summarizes this structure. To understand this arrangement, we must look at how temperature and pressure inside Earth affect the melting point of rock.

Both temperature and pressure increase with increasing depth below Earth's surface. The temperature of Earth's interior rises steadily from about 14°C on the surface to nearly 5000°C at our planet's center (Figure 9-10).

Earth's outermost layer, the crust, is only about 5 to 35 km thick. It is composed of rocks for which the **melting point,** or temperature at which the rock changes from solid to liquid, is far higher than the temperatures actually found in the crust. Thus, the crust is solid.

TABLE 9-3	Earth's Internal Structure		
Region	**Depth Below Surface (km)**	**Distance From Center (km)**	**Average Density (kg/m³)**
Crust (solid)	0–5 (under oceans) 0–35 (under continents)	6343–6378	3500
Mantle (plastic, solid)	from bottom of crust to 2900	3500–6343	3500–5500
Outer core (liquid)	2900–5100	1300–3500	10,000–12,000
Inner core (solid)	5100–6400	0–1300	13,000

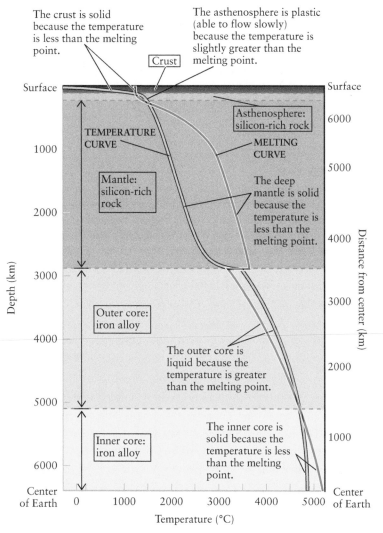

FIGURE 9-10

Temperature and Melting Point of Rock Inside Earth The temperature (yellow curve) rises steadily from Earth's surface to its center. The temperature at the center of Earth is about the same as the surface of the Sun! The melting point of Earth's material (red curve) is also shown on this graph. Where the temperature is below the melting point, as in the mantle and inner core, the material is solid; where the temperature is above the melting point, as in the outer core, the material is liquid. (Adapted from T. Grotzinger, T. H. Jordan, F. Press, and R. Siever, *Understanding Earth*, 5th ed., W. H. Freeman, 2007)

Earth's mantle, which extends to a depth of about 2900 km (1800 mi), is largely composed of substances rich in iron and magnesium. On Earth's surface, specimens of these substances have melting points slightly over 1000°C. However, the melting point of a substance depends on the pressure to which it is subjected—the higher the pressure, the higher the melting point. As Figure 9-10 shows, the actual temperatures throughout most of the mantle are less than the melting point of all the substances of which the mantle is made. Hence, the mantle is primarily solid. However, the upper levels of the mantle—called the **asthenosphere** (from the Greek *asthenia*, meaning "weakness")—are able to flow slowly and are therefore referred to as being **plastic.**

CAUTION! It may seem strange to think of a plastic material as one that is able to flow. Normally we think of plastic objects as being hard and solid, like the rigid plastic exterior of a computer or cell phone. But to make these objects, the plastic material is heated so that it can flow into a mold, then cooled so that it solidifies. Strictly speaking, the material only has a "plastic" property when it is able to flow. Maple syrup and tree sap are two other substances that flow at warm temperatures but become more solid when cooled. Furthermore, as with the asthenosphere, a material might seem solid to us if we could touch it, but actually flows when enormous forces act on it over millions of years.

At the boundary between the mantle and the outer core, there is an abrupt change in chemical composition, from silicon-rich materials to a mixture of iron (about 80%) with some nickel. Because this iron-nickel mixture has a lower melting point than the silicon-rich material above it, the melting curve in Figure 9-10 changes abruptly as it crosses from the mantle down into the outer core. As a result, the outer core is liquid.

At depths greater than about 5100 km, the pressure is more than 10^{11} newtons per square meter. This is about the same pressure that 3000 mid-size cars would make if stacked and balanced on a quarter! Because the melting point of the iron-nickel mixture under this pressure is higher than the actual temperature (see Figure 9-10), Earth's inner core is solid.

Until the 1980s, geologists knew little more about the inner core than that it is solid and dense. Since then, evidence from seismic waves has accumulated that suggests the inner core is a single **crystal** of iron—that is, iron atoms arranged in orderly rows, like carbon atoms are arranged within a diamond. If so, the inner core is the largest known crystalline object anywhere in the solar system. Furthermore, there are strong indications that the inner core is rotating at a slightly faster rate than the rest of Earth! These remarkable discoveries suggest that even more surprises may lurk deep within Earth's interior.

Heat naturally flows from where the temperature is high to where it is low. Figure 9-10 thus explains why heat flows from the center of Earth outward. We will see in the next section how this heat flow acts as the "engine" that powers our planet's geologic activity.

CONCEPTCHECK 9-3

Hypothetically, what would have changed about Earth's interior if S waves from earthquakes suddenly started being observed on the side of Earth opposite an earthquake?

Answer appears at the end of the chapter.

9-3 Plate movement produces earthquakes, mountain ranges, and volcanoes that shape Earth's surface

One of the most important geological discoveries of the twentieth century was the realization that Earth's crust is constantly changing. We have learned that the crust (which includes ocean seafloors) is divided into huge **plates** whose motions produce earthquakes, volcanoes, mountain ranges, and oceanic trenches. Furthermore, material in these plates is slowly but constantly replenished about every few hundred million years, so that most of Earth's surface is young compared to the planet's 4.54 billion year age. The motion of Earth's plates has come to be the central unifying theory of geology, much as the theory of evolution has become the centerpiece of modern biology. By understanding the effects of plate motion and what causes it, we also learn how to read the surface of other planets and moons to learn about their interiors.

> Evidence from the bottom of the ocean confirmed the theory of moving continents

Continental Drift

Anyone who carefully examines a map of Earth might come up with the idea of moving continents. South America, for example, would fit snugly against Africa were it not for the Atlantic Ocean. As Figure 9-11 shows, the fit between landmasses on either side of the Atlantic Ocean is quite remarkable. This observation inspired the German meteorologist Alfred Wegener to find evidence for "continental drift"—the idea that the continents on either side of the Atlantic Ocean have simply drifted apart. After much research, in 1915 Wegener published the theory that there had originally been a single gigantic supercontinent, which he called Pangaea (meaning "all lands"), that began to break up and drift apart roughly 200 million years ago. Other geologists refined this theory over time, arguing that Pangaea must have split up in stages to make the continents we see today (Figure 9-12). (Note: No humans were present to witness the break-up; the earliest humans did not appear on Earth until about 2 million years ago. At the

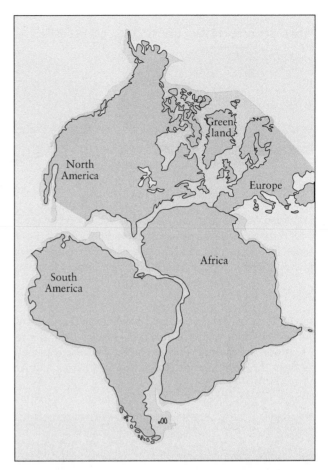

FIGURE 9-11
Fitting the Continents Together Africa, Europe, Greenland, North America, and South America fit together remarkably well. The fit is especially convincing if the edges of the continental shelves (shown in yellow) are used, rather than today's shorelines. This strongly suggests that these continents were in fact joined together at some point in the past. (Adapted from P. M. Hurley)

time that Pangaea broke up, *small* dinosaurs were the dominant land animals.)

Sadly, most geologists ridiculed Wegener's ideas and he died before they were accepted. Although it was generally accepted that the continents do "float" on the denser, somewhat plastic mantle beneath them, few geologists could accept the idea that entire continents could move around Earth at speeds as great as several centimeters per year. The "continental drifters" could not explain what forces could move the massive continents around.

Plate Tectonics

Beginning in the mid-1950s, however, geologists found evidence that material is being forced upward to the crust from deep within Earth. Completely submerged mountain ranges on the ocean floors were discovered, including the Mid-Atlantic Ridge, which is the longest mountain chain in the world and stretches all the way from Iceland to Antarctica (Figure 9-13). Through further analysis, it was found that rock from Earth's mantle is being melted and then forced upward along the Mid-Atlantic Ridge, which is in essence a long chain of underwater volcanoes that forms new seafloor.

(a) 237 million years ago: the supercontinent Pangaea

(b) 152 million years ago: the breakup of Pangaea

(c) The continents today

FIGURE 9-12

The Breakup of the Supercontinent Pangaea **(a)** The shapes of the continents led Alfred Wegener to conclude that more than 200 million (2×10^8) years ago, the continents were merged into a single supercontinent, which he called Pangaea. **(b)** Pangaea first split into two smaller land masses, Laurasia and Gondwana. **(c)** Over millions of years, the continents moved to their present-day locations. Among the evidence confirming this picture are nearly identical rock formations 200 million years in age that today are thousands of kilometers apart but would have been side by side on Pangaea. (Adapted from F. Press, R. Siever, T. Grotzinger, and T. H. Jordan, *Understanding Earth,* 4th ed., W. H. Freeman, 2004)

As newly risen material cools and hardens at the underwater ridge, this new seafloor material also *moves away* from the ridge, causing **seafloor spreading.** For example, because of seafloor spreading from the Mid-Atlantic Ridge, South America and Africa are moving apart at a speed of roughly 3 cm per year, with new seafloor forming in between. Working backward, these two continents would have been next to each other some 200 million years ago—just as Wegener suggested (see Figure 9-11).

In the early 1960s, geologists began to find additional evidence supporting the existence of large, moving plates. Thus was born the modern theory of crustal motion, which came to be known as **plate tectonics** (from the Greek *tekton,* meaning "builder").

Geologists today realize that earthquakes and volcanoes tend to occur at the boundaries of Earth's tectonic plates, where the plates are colliding, separating, or sliding against each other. The boundaries of the plates therefore stand out clearly when the epicenters of earthquakes are plotted on a map, as in Figure 9-14. The boundary in Figure 9-14 that surrounds the Pacific plate is called the Ring of Fire: More than 75% of the world's volcanoes are found here, along with 90% of the world's earthquakes.

Convection and Plate Motion

What makes the plates move? The answer is twofold. First, heat flows outward from Earth's hot interior to its cool crust, and second, this heat flows by convection. In Earth's convection, hotter interior temperatures cause rocky material to expand, *lowering* the material's density compared to its surroundings, which then causes it to rise. This material then cools off at or near the surface, *increasing* the material's density compared to its surroundings, which then causes it to sink. Figure 9-5 shows that convection can cause material to circulate in convection currents. While convection occurs at many layers within the mantle, it is the convection of Earth's outer layer that directly reshapes our planet's surface. This outer layer is called the **lithosphere** (from the Greek word *litho* for "rock"), which is physically divided into the plates described by plate tectonics. The lithosphere is a cooled rocky outer layer that contains the seafloor in the oceans, and the crust that forms continents.

Figure 9-15 shows the basic features of an oceanic seafloor plate. Hot, molten subsurface rock seeps upward along **oceanic rifts,** hardening to form the seafloor plates. (The Mid-Atlantic Ridge, shown in Figure 9-13, is an oceanic rift.) Where a seafloor

FIGURE 9-13

The Mid-Atlantic Ridge This artist's rendition shows the Mid-Atlantic Ridge, an immense 40,000 km–long mountain ridge that rises up from the floor of the North Atlantic Ocean. It is caused by lava seeping up from Earth's interior along a rift that extends from Iceland to Antarctica. (Courtesy of M. Tharp and B. C. Heezen)

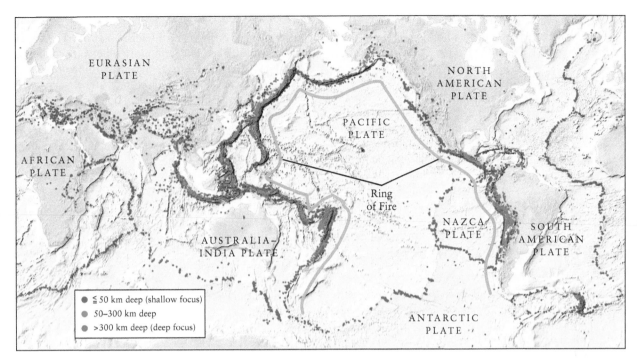

ANIMATION 9-3 **FIGURE 9-14**

Earth's Major Plates The boundaries of Earth's major plates are the scenes of violent seismic and geologic activity. Most earthquakes occur where plates separate, collide, or rub together. Plate boundaries are therefore easily identified simply by plotting earthquake epicenters (shown here as dots) on a map. The colors of the dots indicate the depths at which the earthquakes originate. Typical speeds of plates are between 2 cm and 10 cm per year, where some separate from each other and some move closer together. (Data from Harvard CMT catalog; plot by M. Boettcher and T. Jordan.)

plate meets a continental plate (on the right in Figure 9-15), the higher-density seafloor plate sinks back down into the mantle, creating a **subduction zone.** In most cases, the greatest force that moves seafloor plates comes from **slab pull** at the sinking edge of

the plate: Due to gravity, the cooler high-density seafloor slab falls into the warmer lower-density mantle.

Another force on the oceanic plate is **ridge push,** where molten rock emerging from the oceanic rift is pulled down the ridge's

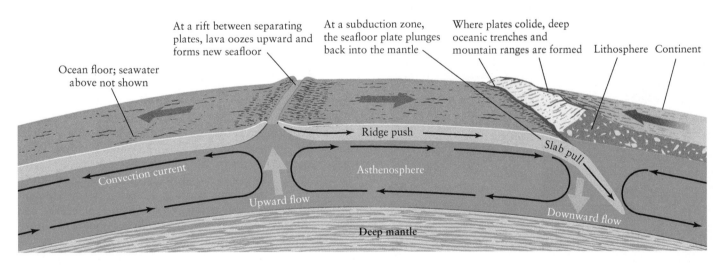

ANIMATION 9-4 **FIGURE 9-15**

The Mechanism of Plate Tectonics Convection currents transport rock to a rift where it oozes toward the surface as magma, hardens in the cold ocean, and spreads away from the rift to create new seafloor. Each side of the rift exhibits seafloor spreading and is a separate tectonic plate. When higher-density seafloor meets lower-density continental crust at a subduction zone, the seafloor plunges downward. Slab pull, due to gravity on this slab, is the dominant force moving the plate across Earth. Collision of plates at a subduction zone also leads to mountain-building, earthquakes, and volcanoes.

slopes by gravity, just as a ball rolls down a hill. Thus, gravity gives a push to the plate, contributing to its horizontal slide. With mountainous ridges typically a couple of miles high, this can be a large force.

CAUTION! The plates are not merely riding on top of deeper convection currents. Sometimes convection in lower layers drags the plates along, and other times the plates drag lower layers along, but both of these cases tend to create weaker forces than slab pull and ridge push. While convection currents also circulate deeper in the mantle, plate tectonics results from convection at Earth's surface. A further caution is that rocky material only becomes liquid (or magma) very near the surface at the rifts; most of the convection current is very slow plastic flow within mantle rock that is under great forces, and this mantle rock would feel very solid to the touch.

A rift between two plates is not always found at the ocean seafloor, although continental rifts occur much less often than oceanic rifts. As shown in Figure 9-16, the valley created by a continental rift can be filled with adjacent seawater. If the continental spreading persists it can slowly transition into a mid-ocean rift, as happened between the continental plates spreading from Pangaea. However, continental plates are thicker and tend to convect more

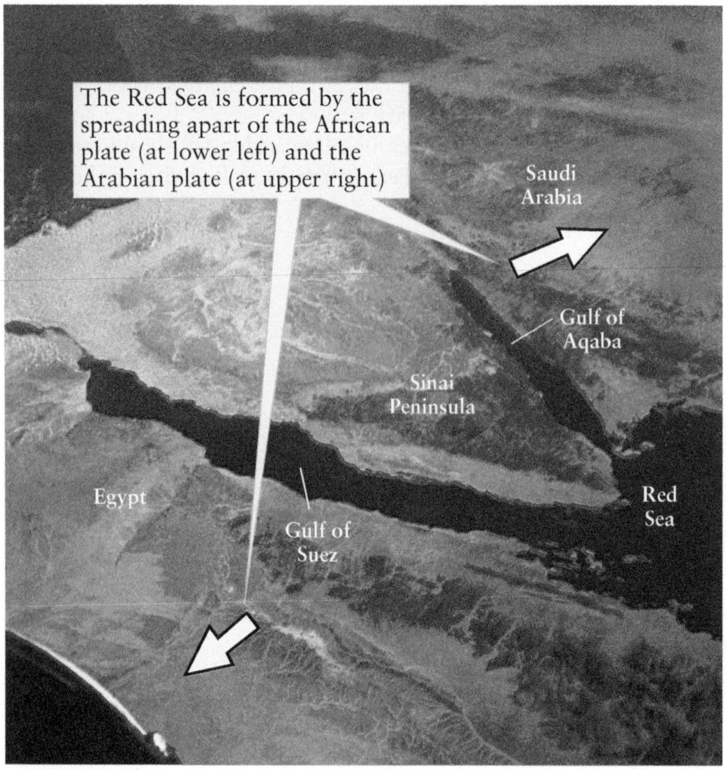

slowly than oceanic plates, and most continental plate motion comes from pushing by adjacent seafloor plates.

The boundaries between plates are the sites of some of the most impressive geological activity on our planet. Great mountain ranges, such as the Sierras and Cascades along the western coast of North America and the Andes along South America's west coast, are thrust up by ongoing collisions between continental plates and the plates of the ocean floor. The Himalayas, due to the collision of plates containing India and China, contain the highest mountains on Earth (Figure 9-17).

CONCEPTCHECK 9-4

What are two forces that move tectonic seafloor plates?

CALCULATIONCHECK 9-1

If seafloor spreading opens South America and Africa apart at a speed of 2 cm per year, how long would it have taken for the Atlantic to open to a current width of 600 million (6×10^8) cm?
Answers appear at the end of the chapter.

Plate Tectonics and the Varieties of Rocks

Rocks are solid material made from minerals. A **mineral** can be made out of a single type of atom, such as diamond made from carbon, but most minerals contain molecules made from several different types of atoms. For example, quartz—which can form sandstone and sand at the beach and desert—is made from molecules of silicon and oxygen (SiO_2). Plate tectonics play a major role in forming rocks and moving them around, as seen in the three major categories of rocks (Figure 9-18). **Igneous rocks** result when minerals cool from a molten state. (Molten rock is called **magma** when it is buried below the surface and **lava** when it flows out upon the surface, as in a volcanic eruption.) The ocean floor is made predominantly of a type of igneous rock called basalt (Figure 9-18a). Igneous rock is just what we expect for the spreading ocean seafloor, which is produced by material welling up from the mantle. Taken together, igneous rock in the seafloor and from volcanic flows on land comprise about 95% of Earth's crust.

Sedimentary rocks are produced by the accumulation of smaller particles. For example, winds and water can pile up layer upon layer of sand grains. Other minerals present amid the sand can gradually cement the grains together to produce sandstone (Figure 9-18b). In another process, the skeletal remains of tiny marine organisms can cover the ocean floor to form a sedimentary rock called limestone. The motion of tectonic plates can move such rocks to places far from where they were formed. This movement explains why the pyramids of Egypt are made of limestone that formed on the ocean floor. Plate tectonics can also upturn horizontally formed sedimentary rock, which explains the vertical layers of the hogback ridges in Colorado (see Figure 9-3).

Sometimes igneous or sedimentary rocks become buried far beneath the surface, where they are subjected to enormous pressure and high temperatures. These severe conditions can change the minerals and structure of the rocks, producing **metamorphic rock** (Figure 9-18c). The presence of metamorphic rocks at Earth's surface tells us that tectonic activity sometimes lifts up material from deep within the crust. This activity can happen when two plates collide, as shown in Figure 9-17.

The Red Sea is formed by the spreading apart of the African plate (at lower left) and the Arabian plate (at upper right)

Saudi Arabia

Gulf of Aqaba

Sinai Peninsula

Egypt

Gulf of Suez

Red Sea

FIGURE 9-16   R I V U X G
The Separation of Two Plates The plates that carry Africa and Arabia are moving apart, leaving a great rift that has been flooded to form the Red Sea. This view from orbit shows the northern Red Sea, which splits into the Gulf of Suez and the Gulf of Aqaba. (*Gemini 12*, NASA)

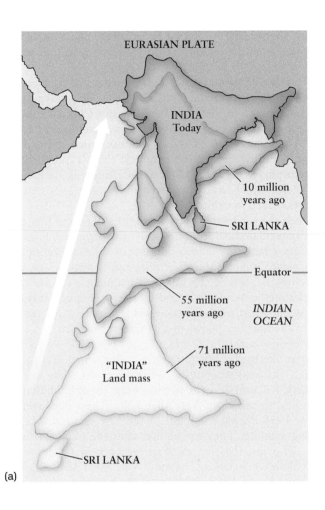

(a)

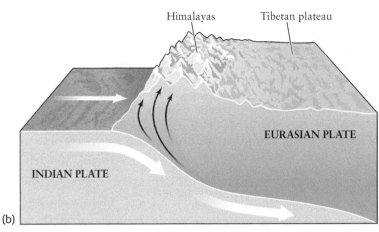

(b)

ANIMATION 9-8 **FIGURE 9-17**

The Collision of Two Plates Collisions between the plates that carry India and China have created Mount Everest and K2 in the Himalayas. **(a)** While plates typically move a few cm/yr, the Indian plate has sped along at up to 20 cm/yr. **(b)** Millions of years ago the ocean seafloor subducted neatly beneath the Eurasian plate, but now there is a buildup of crust on both plates that forms the Himalayas.

The Cycle of Supercontinents

Plate tectonic theory offers insight into geology on the largest of scales, that of an entire supercontinent. In recent years, geologists have uncovered evidence that points to a whole succession of supercontinents that once broke apart and then reassembled. Pangaea is only the most recent supercontinent in this cycle, which repeats about every 500 million years. As a result, intense episodes of mountain building have occurred at roughly 500-million-year intervals.

Apparently, a supercontinent sows the seeds of its own destruction because it blocks the flow of heat from Earth's interior. In one scenario, as soon as a supercontinent forms, temperatures beneath it rise, much as they do when a lid is placed on a heated cooking pot. As heat accumulates, the lithosphere domes upward and

(a) Igneous rock (basalt)   (b) Sedimentary rock (sandstone)   (c) Metamorphic rock (marble)

FIGURE 9-18   R I V U X G

Igneous, Sedimentary, and Metamorphic Rocks **(a)** Igneous rocks such as basalt are created when molten materials solidify. This example contains iron-rich minerals that give it its dark color. **(b)** Sedimentary rocks such as sandstone are typically formed when loose particles of soil or sand are fused into rock by the presence of other minerals, which act as a cement.

(c) Metamorphic rocks are produced when igneous or sedimentary rocks are subjected to high temperatures and pressures deep within Earth's crust. Marble is formed from sedimentary limestone; schist is formed from igneous rock. (a: Doug Martin/Science Source; b: Dorling Kindersley/Getty Images; c: Juniors/Superstock)

cracks. Molten rock from the overheated asthenosphere wells up to fill the resulting fractures, which continue to widen as pieces of the fragmenting supercontinent move apart.

The changes wrought by plate tectonics are very slow on the scale of a human lifetime, but have occurred on Earth's surface for more than 4 billion years. Ocean seafloor recycles back into the mantle every few hundred million years as continents are built up and broken apart. The lesson of plate tectonics is that the seemingly permanent face of Earth is in fact dynamic and ever-changing.

CONCEPTCHECK 9-5

What might cause a supercontinent to break apart?

CALCULATIONCHECK 9-2

If supercontinents are formed and break up in a 500-million-year cycle, about how many supercontinents could come and go over Earth's roughly 4.5-billion-year existence?

Answers appear at the end of the chapter.

9-4 Earth's magnetic field produces a magnetosphere and reverses direction

We saw in Section 7-7 that Earth's magnetic field provides evidence that our planet's interior is partially molten. While the magnetic field is generated deep inside Earth, it has effects on our surface, our atmosphere, and even in the solar system.

The Magnetosphere

Earth's magnetic field extends far above the atmosphere, where it interacts dramatically with charged particles from the Sun. This **solar wind** is a flow of mostly protons and electrons that streams constantly outward from the Sun's upper atmosphere. Near Earth, the particles in the solar wind move at speeds of roughly 450 km/s, or about a million miles per hour. Because this is considerably faster than sound waves can travel in the very thin gas between the

planets, the solar wind is said to be *supersonic*. (Because the gas between the planets is so thin, interplanetary sound waves carry too little energy to be heard by astronauts.)

If Earth had no magnetic field, we would be continually bombarded by the solar wind. But our planet does have a magnetic field, and the forces that this field can exert on charged particles are strong enough to deflect them away from us. The region of space around a planet in which the motion of charged particles is dominated by the planet's magnetic field is called the planet's **magnetosphere**. Figure 9-19 is a scale drawing of Earth's magnetosphere, which was discovered in the late 1950s by the first satellites placed in orbit.

When the supersonic particles in the solar wind first encounter Earth's magnetic field, they abruptly slow to subsonic speeds. The boundary where this sudden decrease in velocity occurs is called a **shock wave**. Still closer to Earth lies another boundary, called the **magnetopause**, where the outward magnetic pressure of Earth's field is exactly counterbalanced by the impinging pressure of the solar wind. Most of the particles of the solar wind are deflected around the magnetopause, just as water is deflected to either side of the bow of a ship.

Some charged particles of the solar wind manage to leak through the magnetopause. When they do, they are trapped by Earth's magnetic field in two huge, doughnut-shaped rings around Earth called the **Van Allen belts.** These belts were discovered in 1958 during the flight of the first successful U.S. Earth-orbiting satellite. They are named after the physicist James Van Allen, who insisted that the satellite carry a Geiger counter to detect charged particles. The inner Van Allen belt, which extends over altitudes of about 2000 to 5000 km, contains mostly protons. The outer Van Allen belt, about 6000 km thick, is centered at an altitude of about 16,000 km above Earth's surface and contains mostly electrons.

Aurorae

Sometimes the Van Allen belts become overloaded with particles. The particles then leak through the magnetic fields at their weakest points and cascade down into Earth's upper atmosphere, usually in

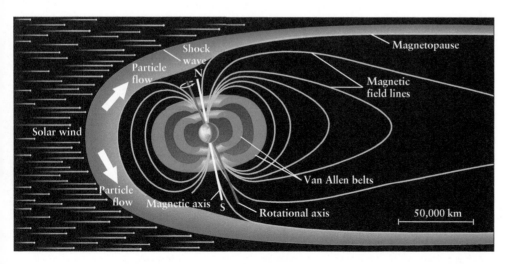

FIGURE 9-19

Earth's Magnetosphere Earth's magnetic field carves out a cavity in the solar wind, shown here in cross section. A shock wave marks the boundary where the supersonic solar wind is abruptly slowed to subsonic speeds. Most of the particles of the solar wind are deflected around Earth in a turbulent region (colored blue in this drawing). Earth's magnetic field also traps some charged particles in two huge, doughnut-shaped rings called the Van Allen belts (shown in red). This figure shows only a slice through the Van Allen belts.

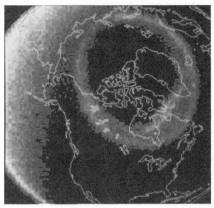  

(a) R I V U X G   (b) R I V U X G   (c) R I V U X G

VIDEO 9-1 **FIGURE 9-20**

The Aurora An increased flow of charged particles from the Sun can overload the Van Allen belts and cascade toward Earth, producing aurorae. **(a)** This false-color ultraviolet image from the *Dynamics Explorer 1* spacecraft shows a glowing oval of auroral emission about 4500 km in diameter centered on the north magnetic pole. **(b)** This photograph shows the aurora australis over Antarctica as seen from NASA's *IMAGE* satellite. **(c)** This view from Alaska shows aurorae at their typical altitudes of 100 to 400 km. The green color shows that the light is emitted by excited oxygen atoms in the upper atmosphere. (a: Courtesy of L. A. Frank and J. D. Craven, University of Iowa; b: NASA; c: J. Finch/Photo Researchers, Inc.)

a ring-shaped pattern (Figure 9-20a). As these high-speed charged particles collide with atoms in the upper atmosphere, they excite the atoms to high energy levels. The atoms then emit visible light as they drop down to their ground states, like the excited gas atoms in a neon light (see Section 5-8). The result is a beautiful, shimmering display called an **aurora** (*plural* **aurorae**). These are also called the **northern lights** (**aurora borealis**) or **southern lights** (**aurora australis**), depending on the hemisphere in which the phenomenon is observed. Figures 9-20b and 9-20c show the aurorae as seen from orbit and from Earth's surface. Aurorae have also been seen on Jupiter and Saturn, and detailed studies of aurorae on Earth have led to a clear interpretation of these distant light-rings on other worlds. For example, by spotting an aurora, we can quickly say that a planet probably has an electrically conductive fluid interior that is in motion and produces a global magnetic field!

Occasionally, a violent event on the Sun's surface sends a particularly intense burst of protons and electrons toward Earth. The resulting auroral display can be exceptionally bright and can often be seen over a wide range of latitudes. Such events also disturb radio transmissions and can damage communication satellites and transmission lines.

It is remarkable that Earth's magnetosphere, including its vast belts of charged particles, was entirely unknown until a few decades ago. Such discoveries remind us of how little we truly understand and how much remains to be learned even about our own planet.

Earth's Magnetic Field Reverses Direction

Earth's magnetic field is produced by a **dynamo.** While a full description is beyond the scope of this book, there are several key ingredients to a **magnetic dynamo.** A dynamo requires an electrically conductive rotating fluid. All electricity produces magnetic fields, and the rotation is necessary to keep the electricity flowing. Far from being solely of interest for Earth's geology, understanding the magnetic dynamo will help us understand the Sun, Mars, and moons around Jupiter.

In the dynamo, heat is required for convection in the liquid iron core. Recall (Section 9-1) that heat is released when water condenses from a vapor to a liquid, and this adds to a storm's energy. (Imagining the reverse is very easy: Heat is required to turn liquid water into steam.) Similarly, energy is released when material from our planet's liquid core cools and solidifies onto the solid core. This released energy powers convection of the liquid core, which combines with Earth's rotation to generate Earth's magnetic field. (While there is also fluid motion in the mantle, it does not produce an appreciable magnetic field: The silicon-rich rocks that comprise the mantle are poor conductors of electricity, and their motions are a million times slower than those in the outer core.)

> Like the motions of tectonic plates, Earth's magnetic field results from our planet's internal heat

If Earth was a permanent magnet, like the small magnets used to attach notes to refrigerators, it would be hard to imagine how our magnetic field could spontaneously reverse direction. But in a magnetic dynamo, computer simulations show that fields produced by electrical currents in Earth's liquid iron core would reverse direction from time to time (Figure 9-21). The flip of Earth's field is irregular, but occurs on average about every 300,000 years. The most recent reversal occurred 780,000 years ago—before that flip, a compass needle would have pointed south, not north! While Earth's present-day magnetic field varies, the evidence does not suggest we are heading toward another field flip anytime soon.

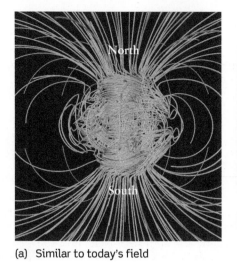

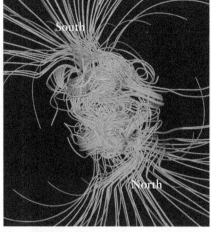

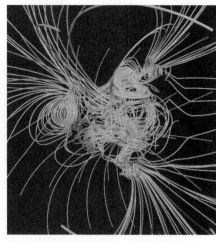

(a) Similar to today's field   (b) After a field reversal   (c) During a reversal

FIGURE 9-21

Simulations of Earth's Magnetic Dynamo Running for almost three months, a supercomputer was used to simulate the complex behavior of Earth's liquid iron core and the magnetic field it produces. During the simulation, the computed magnetic field spontaneously flipped, consistent with reversals by Earth's magnetic field. **(a)** Blue indicates magnetic field lines pointed toward Earth, as they do today at our north magnetic pole. **(b)** The field has reversed. **(c)** During field reversals, which can take a few thousand years to complete, the magnetic field structure can be very complicated. (a, b, and c: The computer model was developed and run by Gary A. Glatzmaier [University of California, Santa Cruz] and Paul H. Roberts [University of California, Los Angeles])

A historical record of Earth's magnetic field is found in the magnetized crust of the ocean seafloor. As illustrated in Figure 9-22, the seafloor indicates repeated reversals of Earth's magnetic field. As lava is cooled by seawater at a mid-ocean rift, the newly formed rocky crust is magnetized by Earth's magnetic field. The magma is too hot to hold a magnetic field before it solidifies, and upon cooling, the magnetic field it acquires is preserved in the oceanic crust. As seafloor spreading continues over time (Section 9-3), this magnetized rock is carried away, in a somewhat symmetric pattern on both sides of the rift. Now imagine what happens if Earth's magnetic field reverses direction. While the older magnetized rock is carried away, new lava rises and cools at the rift, getting magnetized by Earth's newly reversed field. The result is magnetic stripes, where the magnetic field alternately points mostly up and down (see the magnetic field penetrating Earth in Figure 7-13b).

Magnetic stripes—and all they imply about a molten interior and plate tectonics—are a compelling example of how phenomena discovered on Earth shed light on other worlds in the solar system. Consider the magnetic stripes on Mars shown in Figure 7-15. Generations of scientists have pieced together the story that these stripes tell on Earth, and with one orbiting magnetometer, a similar story might be revealed on Mars!

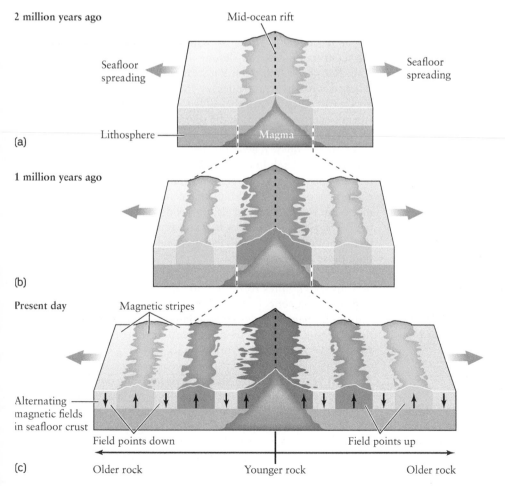

2 million years ago

Mid-ocean rift

Seafloor spreading

Seafloor spreading

Lithosphere — Magma

(a)

1 million years ago

(b)

Present day   Magnetic stripes

Alternating magnetic fields in seafloor crust

Field points down   Field points up

(c)   Older rock   Younger rock   Older rock

FIGURE 9-22

Magnetic Stripes on the Ocean Seafloor Magnetic measurements, illustrated here, clearly show that the seafloor preserves a record of past magnetic field reversals. New material is magnetized as it comes up at the mid-ocean rift and solidifies into the crust. Then, the newly magnetized crust is carried away by seafloor spreading as freshly solidified material repeats this process at the rift. Over millions of years, Earth's magnetic field flips several times, leaving stripes of alternating field directions in the seafloor. For simplicity, the magnetic field is drawn pointing up or down, but the actual direction depends on the latitude of the rift.

If Earth's iron-rich core solidified, would a magnetic compass still point north?

If the seafloor spreads away from a rift at 2 cm/yr, and Earth's magnetic field points in the same direction for 300,000 yr before flipping, how many km wide is the resulting magnetic stripe?

Answers appear at the end of the chapter.

9-5 Earth's atmosphere has changed substantially over our planet's history

Both plate tectonics and Earth's magnetic field paint a picture of a planet with a dynamic, evolving surface and interior. Our atmosphere also has changed and evolved substantially over Earth's 4.54-billion-year history. As we will see, this evolution explains why we have an atmosphere totally unlike that of any other world in the solar system.

> The composition of our atmosphere has been dramatically altered by the evolution of life

Earth's Early Atmosphere

When Earth first formed by accretion of planetesimals (see Section 8-5), gases were probably trapped within Earth's interior in the same proportions that they were present in the solar nebula. But since the early Earth was hot enough to be molten throughout its volume, most of these trapped gases were released. As we discussed in Section 7-4 and Box 7-2, Earth's gravity was too weak to prevent hydrogen and helium—the two most common kinds of atoms in the universe, but also the least massive—from leaking away into space. The atmosphere that remained still contained substantial amounts of hydrogen, but in the form of relatively massive molecules of water vapor (H_2O). In fact, water vapor was probably the dominant constituent of the early atmosphere, which is thought to have been about 100 times denser than our present-day atmosphere.

In addition to releasing water, intense volcanic activity of this early period would have released carbon dioxide (CO_2) and ammonia (NH_3). This process, in which volcanoes inject the atmosphere with gas, is called **outgassing.** As Earth cooled, much of the water condensed and fell as rain. But the rain lowered carbon dioxide levels as well.

Carbon dioxide dissolves in rainwater and the oceans, where it combines with other substances to form a class of minerals called *carbonates.* (Limestone and marble are examples of carbonate-bearing rock.) These form sediments on the ocean floor, which are eventually recycled into the crust by subduction. As an example, marble (Figure 9-18c) is a metamorphic rock formed deep within the crust from limestone, a carbonate-rich sedimentary rock.

At the same time that levels of water and carbon dioxide were decreasing, ultraviolet sunlight broke ammonia apart to create two gases in the atmosphere: hydrogen (H_2) and nitrogen (N_2). The hydrogen, being too light, escaped into space. Through the loss or relocation of water vapor, carbon dioxide, and hydrogen, the remaining nitrogen became the dominant component in the atmosphere. Today, nitrogen still forms the bulk of our atmosphere—about 78%—but it is the other main component that matters the most—oxygen.

Life's Impact on Earth's Atmosphere

The appearance of life on Earth set into motion a radical transformation of the atmosphere. Early single-cell organisms converted energy from sunlight into chemical energy using **photosynthesis,** a chemical process that consumes CO_2 and water and releases oxygen (O_2). Oxygen molecules are very reactive, so originally most of the O_2 produced by photosynthesis combined with other substances to form minerals called oxides. (Evidence for this can be found in rock formations of various ages. The oldest rocks have very low oxide content, while oxides are prevalent in rocks that formed after the appearance of organisms that used photosynthesis.) But as life proliferated, the amount of photosynthesis increased dramatically. Eventually, so much oxygen was being produced that it could not all be absorbed to form oxides, and O_2 began to accumulate in the atmosphere. Figure 9-23 shows how the amount of O_2 in the atmosphere has increased over the history of Earth.

About 2 billion (2×10^9) years ago, a new type of life evolved to take advantage of the newly abundant oxygen. These new organisms produced energy by consuming oxygen and releasing carbon dioxide—a process called **respiration** that is used by all modern animals, including humans. Such organisms thrived because photosynthetic plants continued to add even more oxygen

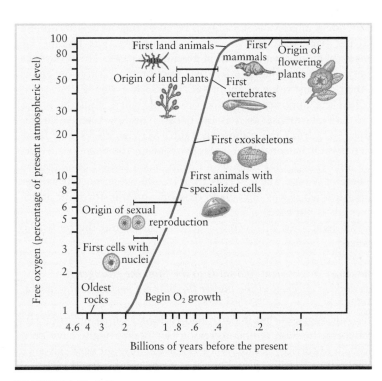

FIGURE 9-23

The Increase in Atmospheric Oxygen This graph shows how the amount of oxygen in the atmosphere (expressed as a percentage of its present-day value) has evolved with time. Note that the atmosphere contained essentially no oxygen until about 2 billion years ago. (Adapted from Preston Cloud, "The Biosphere," *Scientific American,* September 1983, p. 176)

TABLE 9-4	Chemical Compositions of Three Planetary Atmospheres		
	Venus	**Earth**	**Mars**
Nitrogen (N_2)	3.5%	78.08%	2.7%
Oxygen (O_2)	almost zero	20.95%	almost zero
Carbon dioxide (CO_2)	96.5%	0.035%	95.3%
Water vapor (H_2O)	0.003%	about 1%	0.03%
Other gases	almost zero	almost zero	2%

to the atmosphere. The abundance of atmospheric oxygen is due almost exclusively to the presence of life—a situation that has no parallel anywhere else in the solar system. For this reason, finding an oxygen atmosphere around a planet in another solar system would signal the possibility of life beyond Earth. And, as we saw in Section 8-7, determining the composition of atmospheres around extrasolar planets is now possible.

Comparing Atmospheres: Earth, Venus, and Mars

Table 9-4 shows the dramatic differences between Earth's atmosphere and those of Venus and Mars. (Mercury, the other terrestrial planet, is too small and has too little gravity to hold an appreciable atmosphere.) The greater intensity of sunlight on Venus caused higher temperatures, which boiled any liquid water and made it impossible for CO_2 to be taken out of the atmosphere and put back into rocks. Venus's atmosphere thus became far denser than our own and rich in greenhouse gases. The result was a very strong greenhouse effect that raised temperatures on Venus to their present value of about 460°C (733 K = 855°F).

Just the opposite happened on Mars, where sunlight is less than half as intense as it is on Earth. The lower temperatures drove CO_2 from the atmosphere into Martian rocks and froze groundwater to considerable depths beneath the planet's surface. The atmosphere that remains on Mars has a similar composition to that of Venus but is less than 1/10,000 as dense. On neither Venus nor Mars was life able to blossom and transform the atmosphere as it did here on Earth.

We have thus uncovered a general rule about the terrestrial planets:

The closer a terrestrial planet is to the Sun, the stronger the greenhouse effect, and the higher the planet's surface temperature.

This rule suggests that if Earth had formed a bit closer to or farther from the Sun, temperatures on our planet might have been too high or too low for life ever to evolve. Thus, our very existence is a result of Earth's special position in the solar system.

CONCEPTCHECK 9-7

Which, if any, of the two most abundant gases in Earth's present atmosphere came directly from volcanoes?

Answer appears at the end of the chapter.

9-6 Like Earth's interior, our atmosphere has a layered structure

While life has shaped our atmosphere's chemical composition, the structure of the atmosphere is controlled by the influence of sunlight. Scientists describe the structure of any atmosphere in terms of two properties: temperature and atmospheric pressure, which vary with altitude.

> Circulation in our atmosphere results from convection and Earth's rotation

Pressure, Temperature, and Convection in the Atmosphere

Atmospheric pressure at any height in the atmosphere is caused by the weight of all the air above that height. For a useful analogy, if 10 students form a vertical stack—one lying on top of another—it is the bottom student who feels all the weight from those above. It is similar for air, with the greatest air pressure at Earth's surface. The average atmospheric pressure at sea level is defined to be 1 **atmosphere** (1 atm), equal to 1.01×10^5 N/m^2 or 14.7 pounds per square inch. As you go higher in the atmosphere, there is less air above to weigh down on you. Hence, atmospheric pressure decreases smoothly with increasing altitude.

Unlike atmospheric pressure, temperature varies with altitude in a complex way. Figure 9-24 shows that temperature decreases with increasing altitude in some layers of the atmosphere, but in other layers, it actually *increases* with increasing altitude. These differences result from the individual ways in which each layer is heated. The lowest layer, called the **troposphere**, extends from the surface to an average altitude of 12 km (roughly 7.5 miles, or 39,000 ft). It is heated only indirectly by the Sun. Sunlight warms Earth's surface, which heats the lower part of the troposphere. By contrast, the upper part of the troposphere remains at cooler temperatures. This vertical temperature variation causes convection currents that move up and down through the

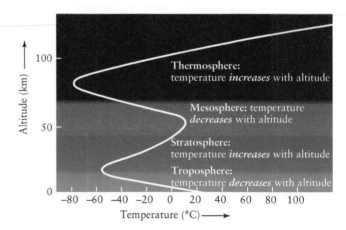

FIGURE 9-24

Temperature Profile of Earth's Atmosphere This graph shows how the temperature in Earth's atmosphere varies with altitude. In the troposphere and mesosphere, temperature decreases with increasing altitude; in the stratosphere and thermosphere, temperature actually increases with increasing altitude.

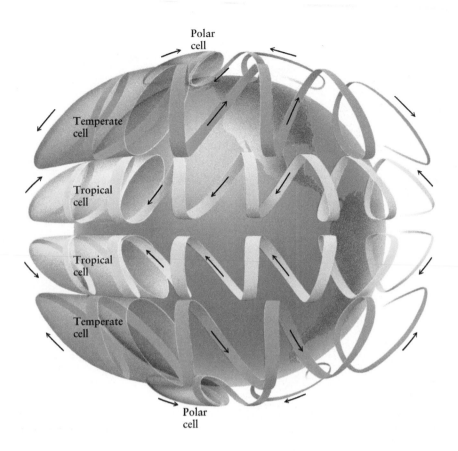

Polar
cell

Temperate
cell

Tropical
cell

Tropical
cell

Temperate
cell

Polar
cell

FIGURE 9-25

Circulation Patterns in Earth's Atmosphere The dominant circulation in our atmosphere consists of six convection cells, three in the northern hemisphere and three in the southern hemisphere. In the northern temperate region (including the continental United States), the prevailing winds at the surface are from the southwest toward the northeast. Farther south, within the northern tropical region (for example, Hawaii), the prevailing surface winds are from northeast to southwest.

troposphere (see Section 8-1). Much of Earth's weather is a consequence of this convection.

Convection on a grand scale is caused by the temperature difference between Earth's equator and its poles. If Earth did not rotate, heated air near the equator would rise upward and flow at high altitude toward the poles. There it would cool and sink to lower altitudes, at which it would flow back to the equator. However, Earth's rotation breaks up this simple convection pattern into a series of **convection cells.** In these cells, air flows east and west as well as vertically and in a north-south direction. The structure of these cells explains why the prevailing winds blow in different directions at different latitudes (**Figure 9-25**).

CONCEPTCHECK 9-8

Is the atmospheric layer called the troposphere heated from above or below?

Answer appears at the end of the chapter.

Upper Layers of the Atmosphere

Almost all the oxygen in the troposphere is in the form of O_2, a molecule made of two oxygen atoms. But in the **stratosphere,** which extends from about 12 to 50 km (about 7.5 to 31 mi) above the surface, an appreciable amount of oxygen is in the form of **ozone,** a molecule made of *three* oxygen atoms (O_3). Unlike O_2, ozone is very efficient at absorbing ultraviolet radiation from the Sun, which means that the stratosphere can directly absorb solar energy. The result is that the temperature actually increases as you move upward in the stratosphere. Convection requires that the

temperature must decrease, not increase, with increasing altitude, so there are essentially no convection currents in the stratosphere.

Above the stratosphere lies the **mesosphere.** Very little ozone is found there, so solar ultraviolet radiation is not absorbed within the mesosphere, and atmospheric temperature again declines with increasing altitude.

The temperature of the mesosphere reaches a minimum of about –75°C (= –103°F = 198 K) at an altitude of about 80 km (50 mi). This minimum marks the bottom of the atmosphere's thinnest and uppermost layer, the **thermosphere,** in which temperature once again rises with increasing altitude. This temperature increase is not due to the presence of ozone, because in this very low-density region oxygen and nitrogen are found as individual atoms rather than in molecules. Instead, the thermosphere is heated because these isolated atoms absorb very-short-wavelength solar ultraviolet radiation (which oxygen and nitrogen molecules cannot absorb).

CAUTION! At altitudes near 300 km the temperature of the thermosphere is about 1000°C (1800°F). This altitude is near the altitude at which the space shuttle and satellites orbit Earth. Nonetheless, a satellite in orbit does *not* risk being burned up as if in a hot oven. The reason is that the thermosphere is far less dense than the atmosphere at sea level. The high temperature simply means that an average atom in the thermosphere is moving very fast (see Box 7-2). But because the thermosphere is so thin (only about 10^{-11} as dense as the air at sea level), these fast-moving atoms are few and far between. Hence, the thermosphere contains very little thermal energy. When meteors burn up in the thermosphere it is because they move much faster than satellites and this generates heat regardless of the thermosphere's average temperature.

The *Cosmic Connections* figure compares the layered structure of our atmosphere with that of Earth's interior.

CONCEPTCHECK 9-9

What would happen to the temperature in the stratosphere if there were an absence of ozone?

Answer appears at the end of the chapter.

Comparing Earth's Atmosphere and Interior

Earth's atmosphere (which is a gas) and Earth's interior (which is partly solid, partly liquid) both decrease in density and pressure as you go farther away from Earth's center.

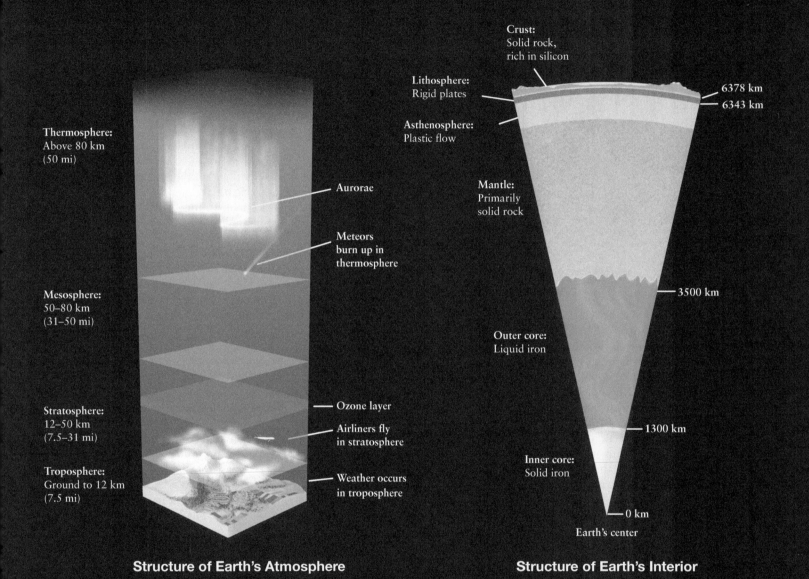

Thermosphere:
Above 80 km
(50 mi)

Aurorae

Meteors
burn up in
thermosphere

Mesosphere:
50–80 km
(31–50 mi)

Ozone layer

Stratosphere:
12–50 km
(7.5–31 mi)

Airliners fly
in stratosphere

Troposphere:
Ground to 12 km
(7.5 mi)

Weather occurs
in troposphere

Crust:
Solid rock,
rich in silicon

Lithosphere:
Rigid plates

6378 km

6343 km

Asthenosphere:
Plastic flow

Mantle:
Primarily
solid rock

3500 km

Outer core:
Liquid iron

1300 km

Inner core:
Solid iron

0 km

Earth's center

Structure of Earth's Atmosphere

Structure of Earth's Interior

9-7 A burgeoning human population is profoundly altering Earth's biosphere

One of the extraordinary characteristics of Earth is that it is covered with life, from the floors of the oceans to the tops of mountains and from frigid polar caps to blistering deserts.

> Global warming and its consequences pose a major challenge to our civilization

So far in this chapter we have hinted at how our planet and living organisms interact: The greenhouse effect has given Earth a suitable temperature for the evolution of life (Section 9-1), and over billions of years that evolution has transformed the chemical composition of the atmosphere (Section 9-5). Let's explore in greater depth how Earth and the organisms that live on it, especially humans, affect each other.

The Biosphere and Natural Climate Variation

All life on Earth subsists in a relatively thin layer called the **biosphere**, which includes the land, the oceans, the crust a kilometer beneath our feet, and the stratosphere 30 km overhead where microbes have been found. **Figure 9-26** is a portrait of Earth's biosphere based on NASA satellite data. The biosphere, which has taken billions of years to evolve to its present state, is a delicate, highly complex system in which plants and animals depend on each other for their mutual survival.

The state of the biosphere depends crucially on the temperatures of the oceans and atmosphere. Even small temperature changes can have dramatic consequences. An example that recurs every three to seven years is the El Niño phenomenon, in which temperatures at the surface of the equatorial Pacific Ocean rise by 2 to 3°C. Ordinarily, water from the cold depths of the ocean is able to well upward, bringing with it nutrients that are used by microscopic marine organisms called phytoplankton that live near the surface (see Figure 9-26). But during an El Niño, the warm surface water suppresses this upwelling, and the phytoplankton starve. This wreaks havoc on organisms such as mollusks that feed on phytoplankton, on the fish that feed on the mollusks, and on the birds and mammals that eat the fish. During the 1982–1983 El Niño, one-quarter of the adult sea lions off the Peruvian coast starved, along with all of their pups.

Many different factors can change the surface temperature of our planet. One is that the amount of energy radiated by the Sun can vary up or down by a few tenths of a percent. Reduced solar brightness may explain the period from 1450 to 1850, when European winters were substantially colder than they are today.

Even the gravitational influences of the Moon and the planets can affect the extent of ice that covers Earth. Thanks to these astronomical influences, the eccentricity of Earth's orbit varies with a period of 90,000 to 100,000 years, the tilt of its rotation axis varies between 22.1° and 24.5° with a 40,000-year period, and the orientation of its rotation axis changes due to precession (see Section 2-6) with a 26,000-year period. These effects can reinforce or cancel each other and lead to variations called Milankovitch cycles.

The Milankovitch cycles affect the prevalence of glaciers by altering the seasonal contrast between winter (when glaciers grow) and summer (when glaciers retreat). While incorrectly called ice ages, these times of periodically advancing ice sheets are actually

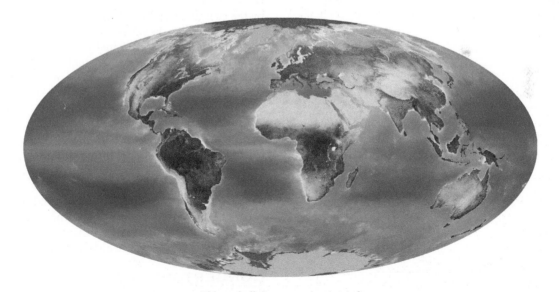

Chlorophyll Concentration (mg/m³)

| 0.01 | 0.1 | 1 | 10 | 50 |

Vegetation Index

| 0 | 0.2 | 0.4 | 0.6 | 0.8 |

FIGURE 9-26

Earth's Biosphere This image, based on data from the SeaWIFS spacecraft, shows the distribution of plant life over Earth's surface. Phytoplankton, which is a microorganism in fresh and ocean water, is a plant and contains the chlorophyll mapped in this image. Phytoplankton is more abundant where cold water brings nutrients to the surface. About equal parts of Earth's atmospheric oxygen come from land-based plants and phytoplankton. (NASA)

called *glacial periods* or *glaciations*. During the most recent glaciation, ice sheets covering what is now the state of Wisconsin were over 3 km thick, and their retreat from North America began only about 10,000 years ago.

As we saw in Sections 9-1 and 9-5, one of the most important factors affecting global temperatures is the abundance of greenhouse gases such as CO_2. Geologic processes can alter this abundance, either by removing CO_2 from the atmosphere (as happens when fresh rock is uplifted and exposed to water with CO_2 dissolved in it, which then forms carbonate minerals) or by supplying new CO_2. From time to time in our planet's history, natural events have caused dramatic increases in the amount of greenhouse gases in the atmosphere. One such event may have taken place 250 million years ago, when Siberia went through a period of intense volcanic activity. (The evidence for this is a layer of 250 million-year-old solidified lava in Siberia that covers an area about the size of Europe!) The tremendous amounts of CO_2 released by this activity could have elevated the global temperature by several degrees. Remarkably, the fossil record reveals that 95% of all species on Earth became extinct at this same time (this is called the Permian extinction).

CONCEPTCHECK 9-10

How can a tiny 3°C increase in ocean surface temperature disrupt the food chain?

Answer appears at the end of the chapter.

Human Effects on the Biosphere: Deforestation

Our species is having an increasing effect on the biosphere because our population is skyrocketing. **Figure 9-27** shows the sharp rise in the human population that began in the late 1700s with the

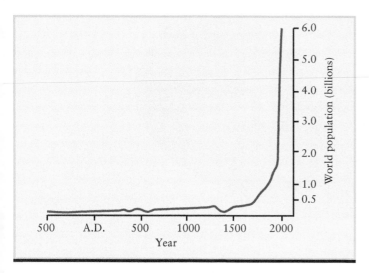

FIGURE 9-27

The Human Population Data and estimates from the U.S. Bureau of the Census, the Population Reference Bureau, and the United Nations Population Fund were combined to produce this graph showing the human population from 500 B.C.E. to 2000 C.E. The population began to rise in the eighteenth century and has been increasing at an astonishing rate since 1900.

FIGURE 9-28   R I 🗹 U X G

The Deforestation of Amazonia The Amazon, the world's largest rain forest, is being destroyed at a rate of 20,000 square kilometers per year in order to provide land for grazing and farming and as a source for lumber. About 80% of the logging is being carried out illegally. (Martin Wendler/Science Source)

Industrial Revolution and the spread of modern ideas about hygiene. This rise accelerated in the twentieth century thanks to medical and technological advances ranging from antibiotics to high-yield grains. In 1960 there were 3 billion people on Earth; in 1975, 4 billion; and in 1999, 6 billion. Projections by the United Nations Population Division show that there will be more than 8 billion people on Earth by the year 2030 and more than 9 billion by 2050.

Every human being has basic requirements: food, clothing, and housing. We all need fuel for cooking and heating. To meet these demands, we burn fossil fuel, cut down forests, and build sprawling cities. A striking example of this activity is occurring in the Amazon rain forest of Brazil. Tropical rain forests are vital to our planet's ecology because they absorb significant amounts of CO_2 and release O_2 through photosynthesis. Although rain forests occupy only 7% of the world's land areas, they are home to at least 50% of all plant and animal species on Earth. Nevertheless, to make way for farms and grazing land, people simply cut down the trees and set them on fire—a process called slash-and-burn. Such deforestation, along with extensive lumbering operations, is occurring in Malaysia, Indonesia, and Papua New Guinea. The rain forests that once thrived in Central America, India, and the western coast of Africa are almost gone (**Figure 9-28**).

CONCEPTCHECK 9-11

In what two ways does deforestation through slash-and-burn increase Earth's atmospheric carbon dioxide?

Answer appears at the end of the chapter.

Human Effects on the Biosphere: The Ozone Layer

Human activity is also having potentially disastrous effects on the upper atmosphere. Certain chemicals released into the air—in particular chlorofluorocarbons (CFCs), which have been used in

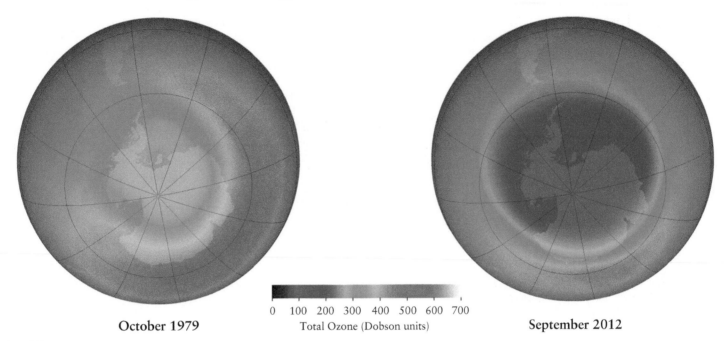

October 1979   Total Ozone (Dobson units)   September 2012

FIGURE 9-29

The Antarctic Ozone Hole These two false-color images show that there was a net decrease of 50% in stratospheric ozone over Antarctica between October 1979 and September 2012. The amount of ozone at midlatitudes, where most of the human population lives, decreased by 10 to 20% over the same period. (NASA)

refrigeration and electronics, and methyl bromide, which is used in fumigation—are destroying the ozone in the stratosphere. We saw in Section 9-6 that ozone molecules absorb ultraviolet radiation from the Sun. Without this high-altitude **ozone layer**, solar ultraviolet radiation would beat down on Earth's surface with greatly increased intensity. Such radiation breaks apart most of the delicate molecules that form living tissue. Hence, a complete loss of the ozone layer would lead to a catastrophic ecological disaster.

In 1985 scientists discovered an **ozone hole**, a region with an abnormally low concentration of ozone, over Antarctica. Since then this hole has generally expanded from one year to the next (Figure 9-29). Smaller but still serious effects have been observed in the stratosphere above other parts of Earth. As a result, there has been a worldwide increase in the number of deaths due to skin cancer caused by solar ultraviolet radiation. By international agreement, CFCs are being replaced by compounds that do not deplete stratospheric ozone, and sunlight naturally produces more ozone in the stratosphere. But even in the best of circumstances, the damage to the ozone layer is not expected to heal for many decades.

CONCEPTCHECK 9-12

If ozone is not actually leaking out of the ozone hole, what is moving through this hole?

Answer appears at the end of the chapter.

Human Effects on the Biosphere: Global Warming and Climate Change

The most troubling influence of human affairs on the biosphere is a consequence of burning fossil fuels (natural gas, petroleum, and coal) in automobiles, airplanes, and power plants as well as burning forests and brushland for agriculture and cooking (Figure 9-28). This burning releases carbon dioxide into Earth's atmosphere—and we are now adding CO_2 to the atmosphere faster than plants and geological processes can extract it. **Figure 9-30** shows

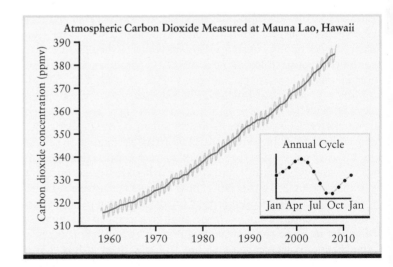

FIGURE 9-30

Atmospheric Carbon Dioxide Is Increasing This graph shows measurements of atmospheric carbon dioxide in parts per million (ppm). The sawtooth pattern results from plants absorbing more carbon dioxide during the spring and summer. The CO_2 concentration in the atmosphere has increased more than 20% since continuous observations started in 1958.

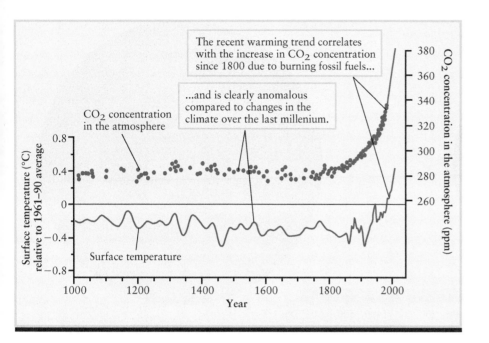

The recent warming trend correlates with the increase in CO_2 concentration since 1800 due to burning fossil fuels...

...and is clearly anomalous compared to changes in the climate over the last millenium.

CO_2 concentration in the atmosphere

Surface temperature

FIGURE 9-31

Atmospheric CO_2 and Changes in Global Temperature This figure shows how the carbon dioxide concentration in our atmosphere (upper curve) and Earth's average surface temperature (lower curve) have changed since 1000 C.E. The increase in CO_2 since 1800 due to burning fossil fuels has strengthened the greenhouse effect and caused a dramatic temperature increase. (Intergovernmental Panel on Climate Change and Hadley Centre for Climate Prediction and Research, UK Meteorological Office)

how the carbon dioxide content of the atmosphere has increased since 1958, when scientists began to measure this quantity on an ongoing basis.

To put the values shown in Figure 9-30 into perspective, we need to know the atmospheric CO_2 concentration in earlier eras. Scientists have learned this by analyzing air bubbles trapped at various depths in the ice that blankets the Antarctic and Greenland. Each winter a new ice layer is deposited, so the depth of a bubble indicates the year in which it was trapped. **Figure 9-31** uses data obtained in this way to show how the atmospheric CO_2 concentration has varied since 1000 C.E. While there has been some natural variation in the concentration, its value has skyrocketed since the beginning of the Industrial Revolution around 1800. Data from older, deeper bubbles of trapped air show that in the 650,000 years before the Industrial Revolution, the CO_2 concentration was never greater than 300 parts per million. The present-day CO_2 concentration is greater than this by 25% and has grown to its present elevated level in just over half a century. If there are no changes in our energy consumption habits, by 2050 the atmospheric CO_2 concentration will be greater than 600 parts per million.

Increasing atmospheric CO_2 is of concern because this strengthens the greenhouse effect and raises Earth's average surface temperature. Figure 9-31 also shows how this average temperature has varied since 1000 C.E. (Like the CO_2 data, the temperature data from past centuries comes from analyzing trapped air bubbles. Scientists measure the relative amounts of two different types of atmospheric oxygen called ^{16}O and ^{18}O, described in Box 5-5, in a bubble. The ratio of these is a sensitive measure of the temperature at that time the air was trapped inside the ice.) The recent dramatic increase in atmospheric CO_2 concentration has produced an equally dramatic increase in the average surface temperature (**Figure 9-32**). This temperature increase is called **global warming.** Other explanations for global warming have been proposed, such

as changes in the Sun's brightness, but only greenhouse gases produced by human activity can explain the steep temperature increase shown in Figure 9-31.

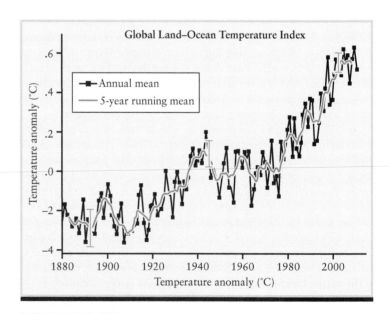

FIGURE 9-32

Global Warming Global temperatures have been rising. The annual temperature fluctuates quite a bit (the black line), and even the red plot of five-year averages shows periods where the temperature declines, but the overall trend of rising temperatures is firmly established. (The green error-bars show that uncertainties due to temperature variations are much smaller than the overall rise.) The temperature is the global mean over both land and ocean, and the temperature anomaly plotted on the vertical axis is just the temperature relative to the average that was measured during the period 1951 to 1980.

CONCEPTCHECK 9-13

Is the greenhouse effect the same as global warming?

CONCEPTCHECK 9-14

As shown in Figure 9-32, the average temperature from about 2002 to 2010 did not rise. Are these recent measurements consistent with a long-term rise in temperatures?

Answers appear at the end of the chapter.

Impact of Global Warming

The effects of global warming can be seen around the world. Glaciers worldwide are receding, the area of sea ice cover around the north pole has decreased by 20% since 1979, and a portion of the Antarctic ice shelf has broken off (Figure 9-33).

Unfortunately, global warming is predicted to intensify in the decades to come. The UN Intergovernmental Panel on Climate Change predicts that if nothing is done to decrease the rate at which we add greenhouse gases to our atmosphere, the average surface temperature will continue to rise by an additional 2 to 4°C during the twenty-first century. What is worse, the temperature increase is predicted to be greater at the poles than at the equator. The global pattern of atmospheric circulation (Figure 9-25) depends on the temperature difference between the warm equator and cold poles, so this entire pattern will be affected. The same is true for the circulation patterns in the oceans. As a result, temperatures will rise in some regions and decline in others, and the patterns of rainfall will be substantially altered. Agriculture depends on rainfall, so these changes in rainfall patterns can cause major disruptions in the world food supply. Studies suggest that the climate changes caused by a 3°C increase in the average surface temperature could cause a worldwide drop in wheat and corn crops of 20 to 400 million tons, putting hundreds of millions of people at risk of hunger.

The melting of polar ice due to global warming poses an additional risk to our civilization. When floating ice like that found near the north pole melts, the ocean level remains the same. (You can see this by examining a glass of water with an ice cube floating in it. The water level does not change when the ice melts.) But the ocean level rises when ice on land melts and runs off into the sea. The Greenland ice cap has been melting at an accelerating rate since 2000 and has the potential to raise sea levels by half a meter or more. Low-lying coastal communities such as Boston and New Orleans, as well as river cities such as London, will have a greater risk of catastrophic flooding. Some island nations of the Pacific will disappear completely beneath the waves.

While global warming is an unintended consequence of human activity, the solution to global warming will require concerted and thoughtful action. Global warming cannot be stopped completely: Even if we were to immediately halt all production of greenhouse gases, the average surface temperature would increase due to existing CO_2 still working its way through Earth's climate system. Instead, our goal is to minimize the effects of global warming by changing the ways in which we produce and use energy. Confronting global warming is perhaps the greatest challenge to face our civilization in the twenty-first century.

KEY WORDS

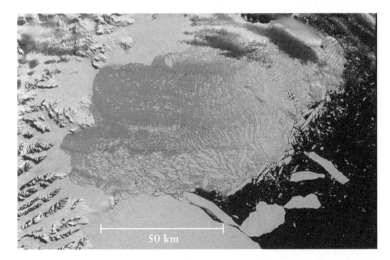

50 km

FIGURE 9-33 R I ☑ U X G

A Melting Antarctic Ice Shelf Global warming caused the Larsen B ice shelf to break up in early 2002. This ice shelf, which was about the size of Rhode Island, is thought to have been part of the Antarctic coast for the past 12,000 years. (NASA/GSFC/LaRC/JPL, MIST Team)

KEY IDEAS

Earth's Energy Sources: All activity in Earth's atmosphere, oceans, and surface is powered by three sources of energy.

• Solar energy is the energy source for the atmosphere. In the green-house effect, some of this energy is trapped by infrared-absorbing gases in the atmosphere, raising Earth's surface temperature.

• Tidal forces from the Moon and Sun help to power the motion of the oceans.

• The internal heat of Earth is the energy source for geologic activity.

Earth's Interior: Studies of seismic waves (vibrations produced by earthquakes) show that Earth has a small, solid inner core surrounded by a liquid outer core. The outer core is surrounded by the dense mantle, which in turn is surrounded by the thin low-density crust.

• Seismologists deduce Earth's interior structure by studying how longitudinal P waves and transverse S waves travel through Earth's interior.

• Earth's inner and outer cores are composed of almost pure iron with some nickel mixed in. The mantle is composed of iron-rich minerals.

• Both temperature and pressure steadily increase with depth inside Earth.

Plate Tectonics: Earth's crust and a small part of its upper mantle form a rigid layer called the lithosphere. The lithosphere is divided into huge plates that move about over the plastic layer called the asthenosphere in the upper mantle.

• Plate tectonics, or movement of the plates, is driven by convection within the asthenosphere. Molten material wells up at oceanic rifts, producing seafloor spreading, and is returned to the asthenosphere in subduction zones. As one end of a plate is subducted back into the asthenosphere, it helps to pull the rest of the plate along.

• Plate tectonics is responsible for most of the major features of Earth's surface, including mountain ranges, volcanoes, and the shapes of the continents and oceans.

• Plate tectonics is involved in the formation of the three major categories of rocks: igneous rocks (cooled from molten material), sedimentary rocks (formed by sediments in water), and metamorphic rocks (altered by extreme heat and pressure).

Earth's Magnetic Field and Magnetosphere: Electric currents in the liquid outer core generate a magnetic field. This magnetic field produces a magnetosphere that surrounds Earth.

• Most of the particles of the solar wind are deflected around Earth by the magnetosphere.

• Some charged particles from the solar wind are trapped in two huge, doughnut-shaped rings called the Van Allen belts. An excess of these particles can initiate an auroral display.

• Earth's magnetic field flips irregularly, but with an average of about 300,000 years between reversals. These field flips are preserved in magnetized rock on the ocean seafloor.

Earth's Atmosphere: Earth's atmosphere differs from those of the other terrestrial planets in its chemical composition, circulation pattern, and temperature profile.

• Earth's atmosphere evolved from being mostly water vapor to its present composition of 78% nitrogen and 21% oxygen.

• Earth's atmosphere is divided into layers called the troposphere, stratosphere, mesosphere, and thermosphere. Ozone molecules in the stratosphere absorb ultraviolet light.

• Because of Earth's rapid rotation, the circulation in its atmosphere is complex, with three circulation cells in each hemisphere.

The Biosphere: Human activity is changing Earth's biosphere, on which all living organisms depend.

• Industrial chemicals released into the atmosphere have damaged the ozone layer in the stratosphere.

• Deforestation and the burning of fossil fuels are increasing the greenhouse effect in our atmosphere and causing a global warming of the planet.

QUESTIONS

Review Questions

1. Describe the various ways in which Earth is unique among the planets of our solar system.

2. Describe the various ways in which Earth's surface is re-shaped over time.

3. Why are typical rocks found on Earth's surface much younger than Earth itself?

4. *TUTORIAL 9.1* What is convection? What causes convection in Earth's atmosphere?

5. Consider a variation of the pot of water in Figure 9-5. If the source of heat were placed at the top of the pot (instead of the bottom), would convection currents flow in the opposite direction? Explain your answer.

6. Describe how energy is transferred from Earth's surface to the atmosphere by both convection and radiation.

7. If heat flows to Earth's surface from both the Sun and Earth's interior, why do we say that the motions of the atmosphere are powered by the Sun?

8. How does solar energy help power the motion of water in Earth's oceans?

9. How does the greenhouse effect influence the temperature of the atmosphere? Which properties of greenhouse gases in the atmosphere cause this effect?

10. *TUTORIAL 9.2* How do we know that Earth was once entirely molten?

11. What are the different types of seismic waves? Why are seismic waves useful for probing Earth's interior structure?

12. Describe the interior structure of Earth.

13. What is a plastic material? Which parts of Earth's interior are described as being plastic?

14. The deepest wells and mines go down only a few kilometers. What then is the evidence that iron is abundant in Earth's core? That Earth's outer core is molten but the inner core is solid?

15. The inner core of Earth is at a higher temperature than the outer core. Why, then, is the inner core solid and the outer core molten instead of the other way around?

16. Describe the process of plate tectonics. Give specific examples of geographic features created by plate tectonics.

17. Explain how convection in Earth's interior drives the process of plate tectonics.

18. Why do you suppose that active volcanoes, such as Mount St. Helens in Washington State, are usually located in mountain ranges that border on subduction zones?

19. What is the difference between a rock and a mineral?

20. What are the differences among igneous, sedimentary, and metamorphic rocks? What do these rocks tell us about the sites at which they are found?

21. Why do geologists suspect that Pangaea was the most recent in a succession of supercontinents?

22. Why do geologists think that Earth's magnetic field is produced in the liquid outer core rather than in the mantle?

23. Describe Earth's magnetosphere. If Earth did not have a magnetic field, do you think aurorae would be more common or less common than they are today?

24. What are the Van Allen belts?

25. Explain how magnetic stripes are created around a rift on the ocean seafloor.

26. If the seafloor spreads from a rift at a speed of 2.5 cm/yr and a magnetic stripe is produced that is 20 km wide, for how many years is Earth's magnetic field in the same orientation without flipping?

27. Summarize the history of Earth's atmosphere. What role has biological activity played in this evolution?

28. Ozone and carbon dioxide each make up only a fraction of a percent of our atmosphere. Why, then, should we be concerned about small increases or decreases in the atmospheric abundance of these gases?

29. How are scientists able to measure what the atmospheric CO_2 concentration and average surface temperature were in the distant past?

30. Does global warming increase the surface temperature of all parts of Earth by equal amounts or by different amounts? What consequences does this have?

Advanced Questions

Problem-solving tips and tools

You can find most of the Earth data that you need in Tables 9-1 and 9-3. You will need to know that a sphere has surface area $4\pi r^2$ and volume $4\pi r^3/3$, where r is the sphere's radius. You may have to consult an atlas to examine the geography of the South Atlantic. Also, remember that the average density of an object is its mass divided by its volume. Sections 5-3 and 5-4 and Box 5-2 describe how to solve problems involving blackbody radiation, Section 2-6 discusses precession, and Section 4-4 describes the properties of elliptical orbits.

31. The total power in sunlight that reaches the top of our atmosphere is 1.75×10^{17} W. (a) How many watts of power are reflected back into space due to Earth's albedo? (b) If Earth had no atmosphere, all of the solar power that was not reflected would be absorbed by Earth's surface. In equilibrium, the heated surface would act as a blackbody that radiates as much power as it absorbs from the Sun. How much power would the entire Earth radiate? (c) How much power would one square meter of the surface radiate? (d) What would be the average temperature of the surface? Give your answer in both the Kelvin and Celsius scales. (e) Why is Earth's actual average temperature higher than the value you calculated in (d)?

32. On average, the temperature beneath Earth's crust increases at a rate of 20°C per kilometer. At what depth would water boil? (Assume the surface temperature is 20°C and ignore the effect of the pressure of overlying rock on the boiling point of water.)

33. What fractions of Earth's total volume are occupied by the core, the mantle, and the crust?

34. What fraction of the total mass of Earth lies within the inner core?

35. (a) Using the mass and diameter of Earth listed in Table 9-1, verify that the average density of Earth is 5500 kg/m³. (b) Assuming that the average density of material in Earth's mantle is about 3500 kg/m³, what must the average density of the core be? Is your answer consistent with the values given in Table 9-3?

36. Identical fossils of the reptile *Mesosaurus,* which lived 300 million years ago, are found in eastern South America and western Africa and nowhere else in the world. Explain how these fossils provide evidence for the theory of plate tectonics.

37. Measurements of the sea floor show that the Eurasian and North American plates have moved 60 km apart in the past 3.3 million years. How far apart (in millimeters) do they move in one year? (By comparison, your fingernails grow at a rate of about 50 mm/year.)

38. The oldest rocks found on the continents are about 4 billion years old. By contrast, the oldest rocks found on the ocean floor are only about 200 million years old. Explain why there is such a large difference in ages.

39. It was stated in Section 7-7 that iron loses its magnetism at temperatures above 770°C. Use this fact and Figure 9-10 to explain why Earth's magnetic field cannot be due to a magnetized solid core, but must instead be caused by the motion of electrically conducting material in the liquid core.

40. Most auroral displays (like those shown in Figure 9-20c) have a green color dominated by emission from oxygen atoms at a wavelength of 557.7 nm. (a) What minimum energy (in electron volts) must be imparted to an oxygen atom to make it emit this wavelength? (b) Why is your answer in (a) a *minimum* value?

41. Describe how the present-day atmosphere and surface temperature of Earth might be different (a) if carbon dioxide had never been released into the atmosphere; (b) if carbon

dioxide had been released, but life had never evolved on Earth.

42. The photograph below shows the soil at Bryce Canyon National Park, Utah. The color of the whitish layer is due to a lack of iron oxide. More recent soils typically contain iron oxide and have a darker color. Explain what this tells us about the history of Earth's atmosphere.

(Frank Vetere/Alamy)   R I V U X G

43. Earth's atmospheric pressure decreases by a factor of one-half for every 5.5-km increase in altitude above sea level. Construct a plot of pressure versus altitude, assuming the pressure at sea level is 1 atmosphere (1 atm). Discuss the characteristics of your graph. At what altitude is the atmospheric pressure equal to 0.001 atm?

44. Earth is at perihelion on January 3 and at aphelion on July 4. Because of precession, in 15,000 A.D. the amount of sunlight in summer will be more than at present in the northern hemisphere but less than at present in the southern hemisphere. Explain why.

45. Antarctica has an area of 13 million square kilometers and is covered by an ice cap that varies in thickness from 300 meters near the coast to 1800 meters in the interior. Estimate the volume of this ice cap. Assuming that water and ice have roughly the same density, estimate the amount by which the water level of the world's oceans would rise if Antarctica's ice were to melt completely (see Figure 9-33). What portions of Earth's surface would be inundated by such a deluge?

Discussion Questions

46. The human population on Earth is currently doubling about every 30 years. Describe the various pressures placed on Earth by uncontrolled human population growth. Can such growth continue indefinitely? If not, what natural and human controls might arise to curb this growth? It has been suggested that overpopulation problems could be solved by colonizing the Moon or Mars. Do you think this is a reasonable solution? Explain your answer.

47. In order to alleviate global warming, it will be necessary to dramatically reduce the amount of carbon dioxide that we release into the atmosphere by burning petroleum. What changes in technology and society do you think will be needed to bring this about?

Web/eBook Question

48. Search the World Wide Web for information about Rodinia, the supercontinent that predated Pangaea. When did this supercontinent form? When did it break apart? Do we have any evidence of supercontinents that predated Rodinia? Why or why not?

49. Because of plate tectonics, the arrangement of continents in the future will be different from today. Search the World Wide Web for information about "Pangaea Ultima," a supercontinent that may form in the distant future. When is it predicted to form? How will it compare to the supercontinent that existed 200 million years ago (see Figure 9-12a)?

50. Use the World Wide Web to investigate the current status of the Antarctic ozone hole. Is the situation getting better or worse? Is there a comparable hole over the north pole? Why do most scientists blame the chemicals called CFCs for the existence of the ozone hole?

51. Use the World Wide Web to research predictions of Earth's future surface temperature. What are some predictions for the surface temperature in 2050? In 2100? What effects may the increased temperature have on human health?

52. The Kyoto Protocol is an agreement made by a number of nations around the world to reduce their production of greenhouse gases. Use the World Wide Web to investigate the status of the Kyoto Protocol. How many nations have signed the protocol? What fraction of the world's greenhouse gas production comes from these nations? Have any developed nations failed to sign? How effective has the protocol been?

53. **Observing Mountain Range Formation.** Access the two animations "The Collision of Two Plates: South America" and "The Collision of Two Plates: The Himalayas" in Chapter 9 of the *Universe* Web site or eBook. (**a**) In which case are the mountains formed by tectonic uplift? In which case are the mountains formed by volcanoes from rising lava? (**b**) For each animation, describe which plate is moving in which direction.

ACTIVITIES

Observing Project

54. Use *Starry Night*™ to view Earth from space. Select **Favourites > Explorations > Earth** from the menu to place yourself about 12,238 km above Earth's surface. Use the location scroller to rotate Earth beneath you, allowing you to view different parts of its surface. Is there any evidence for the presence of an atmosphere on Earth? Explain.

55. Use *Starry Night*™ to view Earth from space. Open the view named **Favourites > Explorations > Earth's Surface**. The view shows the rotating Earth from a point in space about 12,000 km above the surface. The clouds and atmosphere have been removed from the image and the side of Earth facing away from the Sun is artificially brightened so that you

may use the location scroller and **Zoom** controls to inspect the entire surface of our home planet. (**a**) Can you see any evidence of plate tectonics on Earth? (**b**) Can you see any evidence of life or of man-made structures or objects?

ANSWERS

ConceptChecks

ConceptCheck 9-1: For convection to occur, gravity is necessary, as well as a cooler temperature on top of the fluid. Convection depends on lower-density material rising when surrounded by higher-density material, and this requires gravity. But, the density depends on temperature, and if the entire fluid were at the same temperature, there would be no changes in fluid density; thus, a cooler temperature on top is needed.

ConceptCheck 9-2: By trapping heat, the greenhouse effect keeps Earth's temperature warmer than it would otherwise be, allowing liquid water and hence life to exist on Earth.

ConceptCheck 9-3: Because longitudinal S waves are unable to travel through liquid, the sudden appearance of S waves would suggest that the planet's interior has solidified.

ConceptCheck 9-4: The largest force is slab pull, as the dense sinking portion of the plate pulls on the entire seafloor plate. Another force is due to the portion of the plate on the side of the underwater mountains forming the midocean ridge. The weight of this portion of the plate pulls it down the ridge and pushes the entire plate; this force is called ridge push.

ConceptCheck 9-5: In one scenario, a supercontinent traps heat beneath it, causing that region to expand and crack. This process allows fissures to be filled by rapidly upwelling material from underneath, breaking the continent apart.

ConceptCheck 9-6: No. Earth's magnetic field is predominantly created by a dynamo effect caused by the movement of iron-rich material in the liquid core.

ConceptCheck 9-7: Nitrogen and oxygen are the most abundant gases in our atmosphere, but neither came directly form volcanoes. Ammonia (NH_3) came directly from volcanoes, but after sunlight broke this gas apart, only nitrogen gas remained. Later, photosynthesizing microorganisms produced oxygen.

ConceptCheck 9-8: As sunlight warms Earth's surface, energy in the form of heat increases the temperature of the troposphere near Earth's surface such that the troposphere is heated from below.

ConceptCheck 9-9: Ozone in the stratosphere absorbs ultraviolet radiation from the Sun and helps warm the stratosphere. So, a deficit of ozone would make the stratosphere cooler in temperature.

ConceptCheck 9-10: A 3°C change in ocean surface temperature is sufficient to starve plankton, the base of the food chain, by preventing cooler, nutrient-rich water from moving to the surface.

ConceptCheck 9-11: Living plants remove CO_2 from the atmosphere through photosynthesis, and burning plants release the CO_2 they have captured back into the atmosphere.

ConceptCheck 9-12: Atmospheric ozone prevents much of the Sun's harmful ultraviolet light from entering Earth's atmosphere. The ozone hole is a region allowing more ultraviolet light to enter Earth's atmosphere.

ConceptCheck 9-13: No. Our planet's natural greenhouse effect has kept the planet warm for billions of years. Global warming refers to the sudden and significant rise in temperatures since the 1800s.

ConceptCheck 9-14: Yes. As can be seen in Figure 9-32, global temperatures do not rise every single year, or every decade, and global warming refers to the overall rising trend since the 1800s.

CalculationChecks

CalculationCheck 9-1: Using the relationship that distance equals rate times (6×10^8 cm) $\div$ 2 cm/year = 3×10^8 years or 300 million years.

CalculationCheck 9-2: (5×10^9 years) $\div$ (5×10^8 years/cycle) = 10 cycles.

CalculationCheck 9-3: The width of the stripe is (speed) $\times$ (time) = (2 cm/yr) $\times$ (300,000 yr) = 600,000 cm. 1 km = 10^5 cm, so the width is (6×10^5 cm) $\times$ (1 km/10^5 cm) = 6 km wide (or about 3.7 miles).

Walking on the lunar surface, April 1972. (John Young, *Apollo 16*, Science Source)   R I V̄ U X G

Our Barren Moon

Other than Earth itself, the most familiar world in the solar system is the Moon. It is so near to us that some of its surface features are visible to the naked eye. But the Moon is also a strange and alien place, with dramatic differences from our own Earth. It has no substantial atmosphere, no global magnetic field, and no liquid water of any kind. On the airless Moon, the sky is black even at midday. And unlike the geologically active Earth, the Moon has a surface that changes very little with time. When *Apollo 16* astronauts Charlie Duke (shown in the photograph) and John Young explored the region of the lunar highlands known as the Descartes Formation, the youngest rocks that they found were virtually the same age as the oldest rocks found on Earth.

Because the Moon has changed so little over the eons, it has preserved the early history of the inner solar system and the terrestrial planets. In this chapter we will see evidence that the Moon's formation and early evolution was violent and chaotic. The *Apollo* missions revealed that the Moon may have formed some 4.5 billion years ago as the result of a titanic collision between the young Earth and a rogue protoplanet the size of Mars. The young Moon was then pelted incessantly for hundreds of millions of years by large chunks of interplanetary debris, producing a cratered landscape. We will discover that the largest of these impacts formed immense circular basins that then flooded with molten lava. You can see the solidified lava as dark patches that cover much of the face of the Moon. In later chapters we will find other evidence that catastrophic collisions have played an important role in shaping the solar system.

10-1 The Moon's airless, dry surface is covered with plains and craters

Compared with the distances between planets, which are measured in tens or hundreds of millions of kilometers, the average distance from Earth to the Moon is just 384,400 km (238,900 mi). Despite being so close, the Moon has an angular diameter of just ½°, which tells us that it is a rather small world. The diameter of the Moon is 3476 km (2160 mi), just 27% of Earth's diameter and about the same as the distance from Las Vegas to Philadelphia or from London to Cairo (Figure 10-1). Table 10-1 lists some basic data about the Moon.

Motions of the Earth-Moon System

Although the Moon has a small mass (just 1.23% of Earth's mass), Newton's third law tells us that the Moon exerts just as much gravitational force on Earth as Earth does on the Moon (see Section 4-6). Hence, it is not strictly correct to say that the Moon orbits Earth; rather, they both orbit around a point between their centers called the **center of mass** of the Earth-Moon system. (Because Earth is so much more massive than the Moon, the center of mass is close to Earth's center: Indeed, it actually lies within Earth's interior.) The center of mass then follows an elliptical orbit around the Sun (Figure 10-2a). In a similar way, a spinning wrench sliding along a table actually rotates around its center of mass (Figure 10-2b).

The Moon's Rotation and Libration

Even without binoculars or a telescope, you can easily see that dark gray and light gray areas cover vast expanses of the Moon's rocky surface. If you observe the Moon over many nights, you will see the Moon go through its phases as the dividing line between the illuminated and darkened portion of the Moon (the **terminator**) shifts. But you will always see the same pattern of dark and light areas, because the Moon keeps the same side turned toward Earth; you never see the **far side** (or back side) of the Moon. We saw in Section 3-2 that this is the result of the Moon's **synchronous rotation**: It takes precisely the same amount of time for the Moon to rotate on its axis (one lunar "day") as it does to complete one orbit around Earth.

With patience, however, you can see slightly more than half the lunar surface, because the Moon appears to wobble slightly over the course of an orbit. It seems to rock back and forth around its north-south axis and to nod up and down in a north-south direction. This apparent periodic wobbling, called **libration**, permits us to view 59% of our satellite's surface.

We stress that the Moon only *appears* to be wobbling. In fact, its rotation axis keeps nearly the same orientation in space as the Moon goes around its orbit. The reason why the Moon seems to rock back and forth is that its orbit around Earth is slightly elliptical. As a result, the Moon's orbital motion does not keep pace with its rotation at all points around the orbit. Furthermore, because the Moon's rotation axis is not

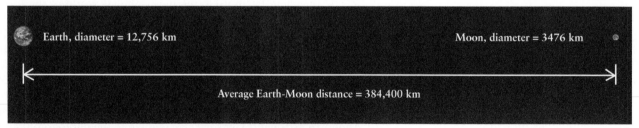

Earth, diameter = 12,756 km   Moon, diameter = 3476 km

Average Earth-Moon distance = 384,400 km

(a) The Earth-Moon system

Earth

Moon

(b) Earth and the Moon to scale, shown 10 times larger than in part (a)

FIGURE 10-1   R I V U X G

Comparing Earth and the Moon (a) The Moon is a bit more than one-quarter the diameter of Earth. The average Earth-Moon distance is about 30 times Earth's diameter. (b) Earth has blue oceans, an atmosphere streaked with white clouds, and continents continually being reshaped by plate tectonics. By contrast, the Moon has no oceans, no atmosphere, and no evidence of plate tectonics. (a: NASA; b: *Earth:* NASA Goddard Space Flight Center Image by Reto Stöckli; *Moon:* NASA/GSFC/Arizona State Univ./Lunar Reconnaissance Orbiter)

TABLE 10-1 Moon Data

Distance from Earth (center to center):	Average: 384,400 km = 238,900 mi
	Maximum (apogee): 405,500 km
	Minimum (perigee): 363,300 km
Eccentricity of orbit:	0.0549
Average orbital speed:	3680 km/h
Sidereal period (relative to fixed stars):	27.322 days
Synodic period (new moon to new moon):	29.531 days
Inclination of lunar equator to orbit:	6.68°
Inclination of orbit to ecliptic:	5.15°
Diameter (equatorial):	3476 km = 2160 mi =
	0.272 Earth diameter
Mass:	7.349×10^{22} kg =
	0.0123 Earth mass
Average density:	3344 kg/m³
Escape speed:	2.4 km/s
Surface gravity (Earth =1):	0.17
Albedo:	0.11
Average surface temperatures:	Day: 130°C = 266°F = 403 K
	Night: −180°C = −292°F = 93 K
Atmosphere:	Essentially none

R I V U X G
(Larry Landolfi, Science Source)

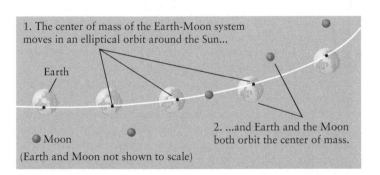

1. The center of mass of the Earth-Moon system moves in an elliptical orbit around the Sun...

Earth

Moon

2. ...and Earth and the Moon both orbit the center of mass.

(Earth and Moon not shown to scale)

(a)

FIGURE 10-2   R I V U X G

Motion of the Earth-Moon System (a) Earth and the Moon both orbit around their center of mass, which in turn follows an elliptical orbit around the Sun. (b) The illustration in part (a) is analogous to this time-lapse image of a spinning wrench sliding across a table. (b: Berenice Abbott/Photo Researchers)

1. The center of mass of this wrench (shown as a red cross) moves in a straight line...

2. ...and the wrench rotates around the center of mass.

(b)

exactly perpendicular to the plane of its orbit, it appears to us to nod up and down.

An Airless World

Clouds always cover some portion of Earth's surface. However, clouds are never seen on the Moon (see Figure 10-1), because the Moon is too small a world to have an atmosphere. Its surface gravity is low, only about one-sixth as great as that of Earth, and thus any gas molecules can easily escape into space (see Box 7-2). Because there is no atmosphere to scatter sunlight, the daytime sky on the Moon appears as black as the nighttime sky on Earth (see the photograph that opens this chapter).

The absence of an atmosphere means that the Moon can have no liquid water on its surface. Here on Earth, water molecules are kept in the liquid state by the pressure of the atmosphere pushing down on them. To see what happens to liquid water on the airless Moon, consider what happens when the pressure of the atmosphere is reduced. The air pressure in Denver, Colorado (elevation 1650 m, or 5400 ft), is only 83% as great as at sea level. Therefore, in Denver, less energy has to be added to the molecules in a pot of water to make them evaporate, and water boils at only 95°C (= 368 K = 203°F), as compared with 100°C (= 373 K = 212°F) at sea level. (This temperature dependence on pressure is why some foods require longer cooking times at high elevations, where the boiling water used for cooking is not as hot as at sea level.) On the Moon, the atmospheric pressure is essentially zero, and the boiling temperature is lower than the temperature of the lunar surface. Under these conditions, water can exist as a solid (ice) or a vapor but not as a liquid.

CAUTION! If astronauts were to venture out onto the lunar surface without wearing spacesuits, the lack of atmosphere would *not* cause their blood to boil. That's because our skin is strong enough to contain bodily fluids. In the same way, if we place an orange in a sealed container and remove all the air, the orange retains all of its fluid and does not explode. It even tastes the same afterward! The reasons why astronauts wore pressurized spacesuits on the Moon (see the photograph that opens this chapter) were to provide oxygen for breathing, to regulate body temperature, and to protect against ultraviolet radiation from the Sun. Since the Moon has no atmosphere, and thus no ozone layer like that of Earth to screen out ultraviolet wavelengths (Section 9-7), an unprotected astronaut would suffer a sunburn after just 10 seconds of exposure!

Lunar Craters

Since the Moon has no atmosphere to obscure its surface, you can get a clear view of several different types of lunar features with a small telescope or binoculars. These include *craters,* the dark-colored *maria,* and the light-colored *lunar highlands* or *terrae* (Figure 10-3). Taken together, these different features show the striking differences between the surfaces of Earth and the Moon.

With an Earth-based telescope, some 30,000 **craters** are visible, with diameters ranging from 1 km to several hundred km. Close-up photographs from lunar orbit have revealed millions of craters too small to be seen from Earth. Following a tradition established in the seventeenth century, the most prominent craters are named after famous philosophers, mathematicians, and

FIGURE 10-3 R I V U X G

The Near Side of the Moon The key features of the lunar surface can be seen with binoculars, a small telescope, or even the naked eye. This composite image was made with more than 1300 images from the *Lunar Reconnaissance Orbiter*. (NASA/GSFC/Arizona State Univ./Lunar Reconnaissance Orbiter)

scientists, such as Plato, Aristotle, Pythagoras, Copernicus, and Kepler. Virtually all lunar craters, both large and small, are the result of the Moon's having been bombarded by meteoroids and asteroids. (Meteoroids are just the smaller range of these space rocks.) For this reason, they are also called **impact craters** (see Section 7-6). High-speed impacts have been recreated in laboratories to study what happens for different impacting materials and different angles of impact. Based on experiments, two key pieces of evidence for the impact-origin of lunar craters are the central peak found in some craters (see Figure 7-10a), and that different angles of impact still produce circular craters.

An asteroid striking the lunar surface generates a shock wave over the surface outward from the point of impact. Such shock waves can spread over an area much larger than the size of the asteroid that generated the wave. Very large craters more than 100 km across, such as the crater Clavius shown in Figure 10-4, were probably created by the impact of a fast-moving piece of rock only a few kilometers in radius.

Maria: Dark and Relatively Young

The large dark areas on the lunar surface shown in Figure 10-3 are called **maria** (pronounced "MAR-ee-uh"). They form a pattern that looks crudely like a human face, sometimes called "the man in the Moon." The singular form of maria is **mare** (pronounced "MAR-ay"), which means "sea" in Latin. These terms date from the seventeenth century, when observers using early telescopes thought these dark features were large bodies of water. They gave the maria fanciful and romantic names such as Mare Tranquillitatis (Sea of Tranquility), Mare Nubium (Sea of Clouds), Mare Nectaris (Sea of

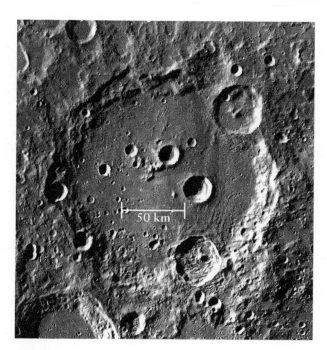

FIGURE 10-4   R I ◻ U X G

The Crater Clavius This crater is one of the largest on the Moon. Clavius has a diameter of 232 km (144 mi) and a depth of 4.9 km (16,000 ft), measured from the crater's floor to the top of the surrounding rim. Later impacts formed the smaller craters inside Clavius. (NASA)

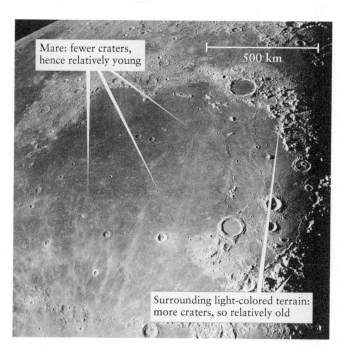

FIGURE 10-5   R I ◻ U X G

Mare Imbrium from Earth Mare Imbrium is the largest of 14 dark plains that dominate the Earth-facing side of the Moon. Its diameter of 1100 km (700 mi) is about the same as the distance from Chicago to Philadelphia or from London to Rome. The maria formed after the surrounding light-colored terrain, so they have not been exposed to meteoritic bombardment for as long and hence have fewer craters. (Carnegie Observatories)

Nectar), Mare Serenitatis (Sea of Serenity), and Mare Imbrium (Sea of Showers). While modern astronomers still use these names, they now understand that the maria are not bodies of water (which, as we have seen, could not exist on the airless Moon) but rather the remains of huge lava flows. The maria get their distinctive appearance from the dark color of the solidified lava.

On the whole, the maria have far fewer craters than the surrounding terrain. Craters are caused by meteoritic bombardment, so the maria have not been exposed to that bombardment for as long as the surrounding terrain. Hence, the maria must be relatively young, and the lava flows that formed them must have occurred at a later stage in the Moon's geologic history (Figure 10-5). The craters that are found in the maria were caused by the relatively few impacts that have occurred since the maria solidified (Figure 10-6).

Although they are much larger than craters, maria also tend to be circular in shape (see Figure 10-5). This suggests that, like craters, these depressions in the lunar surface were also created by large impacts. It is thought that very large asteroids some tens of kilometers across struck the Moon, forming basins. These depressions subsequently flooded with lava that flowed out from the Moon's interior through cracks in the lunar crust. The solidified lava formed the maria that we see today. However, the large basin-creating impacts probably did not directly cause the lava flows because the lava is typically dated to about 500 million years after the impacts. The *Cosmic Connections* figure depicts our understanding of how lunar craters and maria formed.

An impact large enough to create a mare basin must have thrust upward tremendous amounts of material around the impact site. This explains the mountain ranges that curve in circular arcs around mare basins. Analysis of lunar rocks as well as

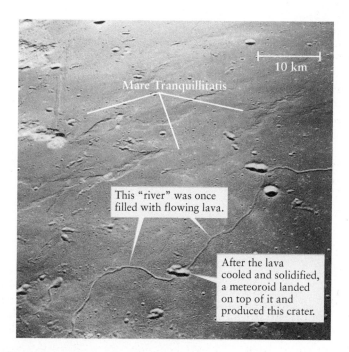

FIGURE 10-6   R I ◻ U X G

Details of Mare Tranquillitatis This color photograph from lunar orbit reveals numerous tiny craters in the surface of a typical mare. The curving "river," or rille, was carved out by flowing lava that later solidified; similar features are found in volcanic areas on Earth. The *Apollo 11* landing site is near the center of the top edge of this image. (*Apollo 10*, NASA)

COSMIC CONNECTIONS

The Formation of Craters and Maria on the Moon

The physical features of lunar craters and maria show that they were formed by asteroids impacting the Moon's surface at high speed. The crater labeled in the photograph, called Aristillus after a Greek astronomer who lived around 300 B.C.E., is 55 km (34 miles) across and 3600 m (12,000 ft) deep.

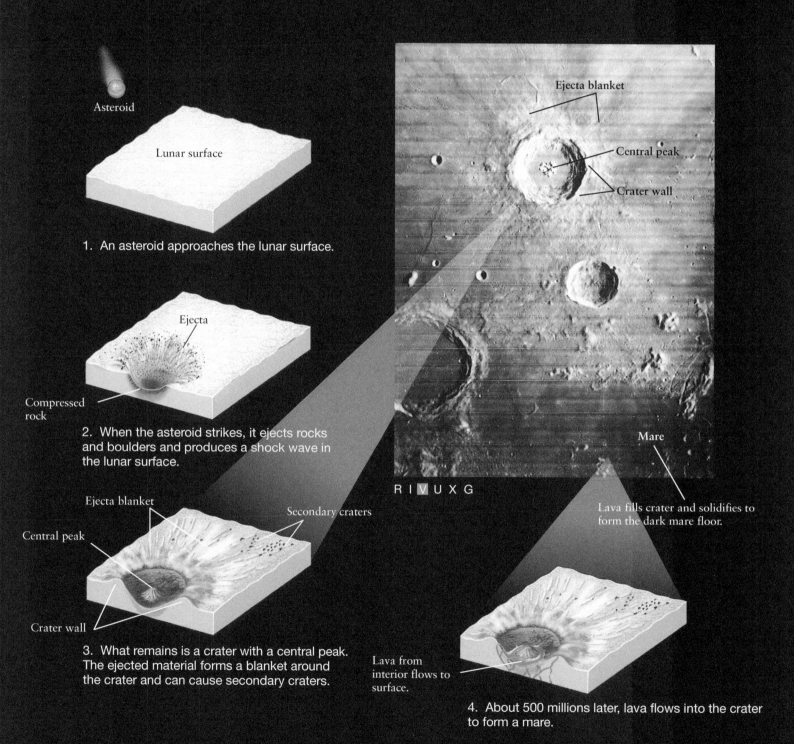

Asteroid

Lunar surface

1. An asteroid approaches the lunar surface.

Ejecta

Compressed rock

2. When the asteroid strikes, it ejects rocks and boulders and produces a shock wave in the lunar surface.

Ejecta blanket

Central peak

Secondary craters

Crater wall

3. What remains is a crater with a central peak. The ejected material forms a blanket around the crater and can cause secondary craters.

Ejecta blanket

Central peak

Crater wall

Mare

R I V U X G

Lava fills crater and solidifies to form the dark mare floor.

Lava from interior flows to surface.

4. About 500 millions later, lava flows into the crater to form a mare.

observations made from lunar orbit have validated this picture of the origin of lunar mountains.

Lunar Highlands: Light-Colored and Ancient

The light-colored terrain that surrounds the maria is called **terrae**, from the Latin for "lands." (Seventeenth-century astronomers thought that this terrain was the dry land that surrounded lunar oceans.) Detailed measurements by astronauts in lunar orbit demonstrated that the maria on the Moon's Earth-facing side are 2 to 5 km below the average lunar elevation, while the terrae extend to several kilometers above the average elevation. For this reason terrae are often called **lunar highlands** (see Figure 10-3). The flat, low-lying, dark gray maria cover only 15% of the lunar surface; the remaining 85% is light gray lunar highlands. Since the lunar highlands are more heavily cratered than the maria, we can conclude that the highlands are older.

Remarkably, when the Moon's far side was photographed for the first time by the Soviet *Luna 3* spacecraft in 1959, scientists were surprised to find almost no maria there. Lunar highlands predominate on the side of the Moon that faces away from Earth (Figure 10-7). Presumably, the crust is thicker on the far side, so even the most massive impacts there were unable to crack through the crust and let lava flood onto the surface. The Moon is not unique in having major portions with different crustal thicknesses: Earth's northern hemisphere has much more continental crust than its southern hemisphere, and the southern hemisphere of Mars has a thicker crust than the northern hemisphere.

Although the Moon's far side has fewer maria than the near side, the highlands on the far side are as heavily pockmarked by impacts as those on the near side. The most prominent feature on the far side is the South Pole–Aitken Basin, the largest known impact crater in the solar system. This basin is about 2500 km (1600 mi) across—about the same size as the continent of Europe. The Aitken Basin isn't as dark as the maria because very little lava has filled in the basin, which also leaves it with an extraordinary depth of 12 km.

A World Without Plate Tectonics

Something is entirely missing from our description of the Moon's surface—any mention of plate tectonics. Indeed, no evidence for plate tectonics can be found on the Moon. Recall from Section 9-3 that Earth's major mountain ranges were created by collisions between lithospheric plates. In contrast, it appears that the Moon is a *one-plate world*: Its entire lithosphere is a single, solid plate. There are no rifts where plates are moving apart, no mountain ranges of the sort formed by collisions between plates, and no subduction zones. Thus, while lunar mountain ranges bear the names of terrestrial ranges, such as the Alps, Carpathians, and Apennines, they are entirely different in character. They were formed by material ejected from impact sites, not by plate tectonics. Geologically, the Moon is a nearly dead world. In Section 7-6 we discussed the general

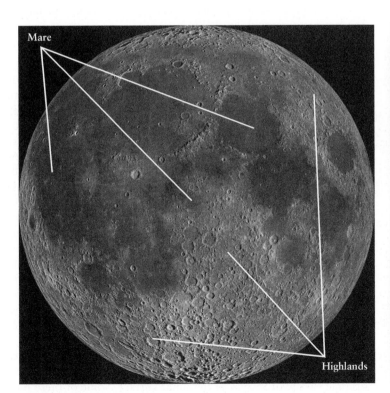

Near side: highlands and maria

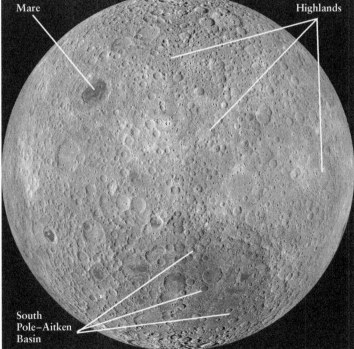

Far side: highlands but almost no maria

VIDEO 10-1 **FIGURE 10-7**   R I V U X G

The Near and Far Sides of the Moon The iron-rich darker maria are a younger surface made from ancient lava flows and are 2 to 5 km below the average surface elevation. The cratered highlands are an older surface several kilometers above the average elevation. The South Pole–Aitken Basin reaches depths of 12 km, but was never filled in by darker lava flows. These images were made by the *Lunar Reconnaissance Orbiter*. (a: NASA/GSFC/Arizona State Univ./Lunar Reconnaissance Orbiter; b: NASA)

rule that a world with abundant craters has an old, geologically inactive surface. The Moon is a prime example of this rule. Without an atmosphere to cause erosion and without plate tectonics, the surface of the Moon changes only very slowly. That is why the Moon's craters, many of which are now known to be billions of years old, remain visible today. On Earth, by contrast, the continents and seafloor are continually being reshaped by seafloor spreading, mountain building, and subduction. Thus, any craters that were formed by impacts on the ancient Earth's surface have long since been erased. Fewer than 200 impact craters are known to exist on Earth today; these were formed within the last few hundred million years and have not yet been erased.

> In contrast to Earth and its active geology, the Moon's surface is geologically dead

In 2010, evidence for some surprisingly recent geological activity on the Moon was found. The Moon has long cliffs called *scarps* that develop as the entire Moon cools and shrinks (Figure 10-8). A good analogy for the formation of these scarps is the wrinkling seen on a fruit's skin as the fruit dries out and contracts. It had been thought that the Moon cooled quickly after formation, and that all scarps were billions of years old. However, the craters disturbed by a scarp can indicate the scarp's age and suggest some of these geologic features might be as young as a few hundred million years old. The Moon has probably shrunk about 600 feet in diameter since it first formed, and this new evidence suggests that it might still be shrinking today.

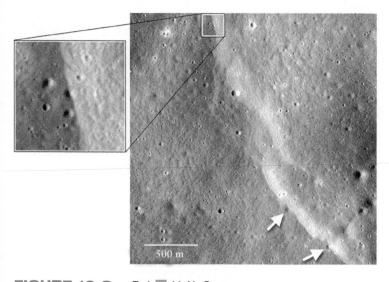

FIGURE 10-8   R I V U X G

Recent Geologic Activity of Scarps This image shows a long cliff called a scarp—formed as the Moon cools, shrinks, and wrinkles—that is cutting across several small craters (see the inset, and the two arrows). Until discovery in 2010, it was thought these scarps were billions of years old. But small craters allow for a unique way to estimate ages. Small craters (less than about 400 meters) typically do not last very long on the lunar surface because they are disturbed by ongoing micrometeorite impacts. Furthermore, when an image shows that a scarp cuts across a crater, it means the scarp is even younger than the crater. By analyzing scarps and their underlying craters, some scarps are thought to be less than a billion years old, and could be as young as a hundred million years old. (NASA)

CONCEPTCHECK 10-1

Why does darker rock appear at lower elevations?
Answer appears at the end of the chapter.

10-2 Human exploration of the Moon in the 1960s and 1970s has been continued by robotic spacecraft

For thousands of years, storytellers have invented fabulous tales of voyages to the Moon. The seventeenth-century French satirist Cyrano de Bergerac wrote of a spacecraft whose motive power was provided by spheres of morning dew. (After all, reasoned Cyrano, dew rises with the morning sun, so a spacecraft would rise along with it.) In his imaginative tale *Somnium*, Cyrano's contemporary Johannes Kepler (the same Kepler who deduced the laws of planetary motion) used the astronomical knowledge of his time to envision what it would be like to walk on the Moon's surface. In 1865 the French author Jules Verne published *From the Earth to the Moon*, a story of a spacecraft sent to the Moon from a launching site in Florida.

Almost exactly a hundred years later, Florida was indeed the starting point of the first voyages made by humans to other worlds. Between 1969 and 1972, 12 astronauts walked on the Moon, and fantasy became magnificent reality.

> The Moon is the only world other than Earth to be visited by humans

The First Spacecraft to the Moon

As in Verne's story, politics had much to do with the motivation for going to the Moon. But in fact, there were excellent scientific reasons to do so. Because Earth is a geologically active planet, typical terrestrial surface rocks are only a few hundred million years old, which is only a small fraction of Earth's age. Thus, all traces of Earth's origins have been erased. By contrast, rocks on the geologically inactive surface of the Moon have been essentially undisturbed for billions of years. Samples of lunar rocks have helped answer many questions about the Moon and have shed light on the origin of Earth and the birth of the entire solar system.

Lunar missions began in 1959, when the Soviet Union sent three small spacecraft—*Lunas 1, 2,* and *3*—toward the Moon. American attempts to reach the Moon began in the early 1960s with Project Ranger. Equipped with six television cameras, the *Ranger* spacecraft transmitted close-up views of the lunar surface taken in the few minutes before it crashed onto the Moon. These images showed far more detail than even the best Earth-based telescope.

In 1966 and 1967, five *Lunar Orbiter* spacecraft were placed in orbit around the Moon and made a high-resolution photographic survey of 99.5% of the lunar surface. Some of these photographs show features as small as 1 m across and were essential in helping NASA scientists choose suitable landing sites for the astronauts.

Even after the *Ranger* missions, it was unclear whether the lunar surface was a safe place to land. One leading theory was that billions of years of impacts by meteoroids and the solar wind had ground the Moon's surface to a fine powder. Perhaps a spacecraft attempting to land would simply sink into a quicksandlike sea of dust! Thus, before a manned landing on the Moon could be attempted, it was necessary to perform a trial landing of an unmanned spacecraft: The purpose of the *Surveyor* program was to land unmanned spacecraft on the Moon. Between June 1966 and January 1968, five *Surveyors* successfully touched down on the Moon, sending back pictures and data directly from the lunar surface. These missions demonstrated convincingly that the Moon's surface was as solid as that of Earth. Figure 10-9 shows an astronaut visiting *Surveyor 3,* two and a half years after it had landed on the Moon.

That's One Small Step for Man ...

VIDEO 10-2 The first of six manned lunar landings took place on July 20, 1969, when the *Apollo 11* lunar module *Eagle* set down in Mare Tranquillitatis (Sea of Tranquility) seen in Figure 10-6. Astronauts Neil Armstrong and Edwin "Buzz" Aldrin were the first humans to set foot on the Moon. *Apollo 12* also landed in a mare, Oceanus Procellarum (Ocean of Storms). The *Apollo 13* mission experienced a nearly catastrophic inflight explosion that made it impossible for it to land on the Moon. Fortunately, titanic efforts by the astronauts and ground personnel brought it safely home. The remaining four missions, *Apollo 14* through *Apollo 17,* were made in progressively more challenging terrain, chosen to permit the exploration of a wide variety of geologically interesting features. The first era of human exploration of the Moon culminated in December 1972, when *Apollo 17* landed in a rugged region of the lunar highlands. Figure 10-10 shows one of the *Apollo* bases.

FIGURE 10-10   R I V U X G

The *Apollo 15* Base This photograph shows astronaut James Irwin and the *Apollo 15* landing site at the foot of the Apennine Mountains near the eastern edge of Mare Imbrium. The hill in the background is about 5 km away and rises about 3 km high. At the right is a Lunar *Rover,* an electric-powered vehicle used by astronauts to explore a greater radius around the landing site. The last three *Apollo* missions carried Lunar *Rovers.* (Dave Scott, *Apollo 15,* NASA)

VIDEO 10-3 **FIGURE 10-9**   R I V U X G

Visiting an Unmanned Pioneer *Surveyor 3* landed on the Moon in April 1967. It was visited in November 1969 by *Apollo 12* astronaut Alan Bean, who took pieces of the spacecraft back to Earth for analysis. (The *Apollo 12* lunar module is visible in the distance.) The retrieved pieces showed evidence of damage by micrometeoroid impacts, demonstrating that small bits of space debris are still colliding with the Moon. (Pete Conrad, *Apollo 12,* NASA)

In a curious postscript to the *Apollo* missions, in the 1990s an "urban legend" arose that the Moon landings had never taken place and were merely an elaborate hoax by NASA. While amusing, this legend does not hold up under even the slightest scientific scrutiny. For example, believers in the "Moon hoax" cannot explain why the rocks brought back from the Moon are substantially different in their chemistry from Earth rocks, as we will see later in this chapter. Those arguing for a moon hoax simply say that the scientists, along with countless others involved in the lunar missions, are lying.

Between 1966 and 1976, a series of unmanned Soviet spacecraft also orbited the Moon and landed on its surface. The first landing, by *Luna 9,* came four months before the U.S. *Surveyor 1* arrived. The first lunar satellite, *Luna 10,* orbited the Moon four months before *Lunar Orbiter 1.* In the 1970s, robotic spacecraft named *Luna 17* and *Luna 21* landed vehicles that roamed the lunar surface, and *Lunas 16, 20,* and *24* brought samples of rock back to Earth. The *Luna* samples complement those returned by the *Apollo* astronauts because they were taken from different parts of the Moon's surface.

Recent Lunar Exploration and the Search for Water

While robots and probes can pursue scientific research in space more efficiently than humans, there might be other reasons for people to colonize space some day. Indeed, some countries have already outlined initial steps toward a lunar base in the coming decades. However, it is an expensive endeavor—it costs over $100,000 to transport the mass in a gallon of water to the Moon. Contemplation of a manned lunar base and the resources required to maintain it has added extra importance to the question, "Is there water on the moon?" To help answer this question, in 1994 a small NASA spacecraft named *Clementine* spent more than

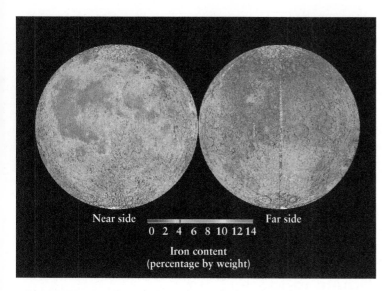

FIGURE 10-11

Iron on the Moon Images at different wavelengths from the *Clementine* spacecraft were used to make this map of the concentration of iron on the lunar surface. The areas of highest concentration (red) coincide closely with the maria on the near side (compare with Figure 10-7), confirming that the maria formed from iron-rich lavas. The lowest iron concentration (blue) is found in the lunar highlands. (No data were collected in the gray areas.) The large green region of intermediate iron concentration on the far side is the South Pole–Aitken Basin (see Figure 10-7). The impact that created this basin may have excavated all the way to the Moon's mantle, in which iron is more abundant than in the crust. (JSC/NASA)

two months observing the Moon from lunar orbit. It carried an array of very high-resolution cameras sensitive to ultraviolet and infrared light as well as to the visible spectrum. Different types of materials on the lunar surface reflect different wavelengths, so analysis of *Clementine* images made in different parts of the electromagnetic spectrum has revealed the composition of large areas of the Moon's surface. For example, the unique reflection signature of iron led to the iron-map in Figure 10-11.

One remarkable result of the *Clementine* mission was evidence for a patch of water-ice tens or hundreds of kilometers across at the Moon's southern pole, in locations where sunlight never reaches and where temperatures are low enough to prevent ice from evaporating. Using a remarkable technique, radio waves beamed from *Clementine* were reflected off these areas and were then detected by radio telescopes on Earth! The reflected radio waves showed the unique signature of reflection from water-ice. Then, data from the *Lunar Prospector* spacecraft, which went into orbit around the Moon in 1998, reinforced the *Clementine* findings and indicate that even more water-ice might lie in deep craters at the Moon's northern pole.

Encouragingly, the *Lunar Prospector* used a different measurement technique than *Clementine*, which strengthens the overall case for water. However, the *Lunar Prospector* observations show only that there is an excess of hydrogen atoms at both lunar poles. The assumption is that this hydrogen most likely exists within water molecules (H_2O) but the possibility exists that the hydrogen could be within other, more exotic minerals. Overall, this evidence for water on the moon is highly suggestive but not overwhelming.

At the end of the *Lunar Prospector* mission in July 1999, the spacecraft was intentionally crashed into a crater at the lunar south pole in the hope that a plume of water vapor might rise from the impact site. Although more than a dozen large Earth-based telescopes watched the Moon during the impact, no such plume or debris of any kind was observed. The advantage of seeing water kicked up in a plume is that you can detect the unique infrared spectrum of H_2O directly, not just hydrogen atoms, which would *confirm* the presence of water.

Designed specifically for impact and analysis, the *Lunar Crater Observation and Sensing Satellite* (LCROSS) tried again to detect water directly. The impact target was suspected to contain water in a permanently shadowed region of a crater only 100 meters from the lunar south pole, and should have been cold enough to retain water-ice for billions of years. In 2009, after ferrying the LCROSS probe from Earth, a spent fuel canister weighing more than 2 tons was used as a man-made meteoroid to crash into the suspected ice. Traveling at about 8 times the speed of sound, the impact created a large plume of debris (Figure 10-12). Then, a few minutes later the probe itself traveled through the plume, analyzing the debris, before smashing into the surface.

The results were spectacular and lunar water was finally confirmed! As shown in Figure 10-13, the characteristic absorption features due to H_2O were found in the plume's infrared spectrum. Ammonia and methane were also found, matching the abundances expected for comets, indicating that comets are the most likely source of the water.

About 155 kg (41 gallons) of water were ejected from LCROSS's 30-meter impact crater. However, the concentration of water in the crater is similar to that found in Earth's Sahara Desert. As a resource for supporting human activity on the Moon, the water discovered so far on the Moon is quite scarce. On the

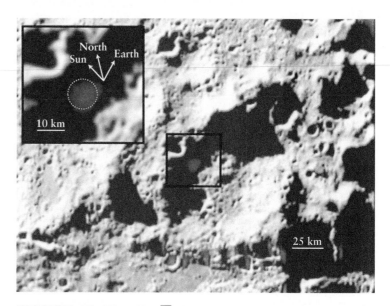

FIGURE 10-12   R I V U X G

Water in the Plume Created by LCROSS The permanent shadow in Cabeus crater is visible, in addition to the plume that rose 20 seconds after the impact of a spent fuel canister. About 5% of the ejected material is water. Other evidence suggests that the water may have originally come from comets. (NASA)

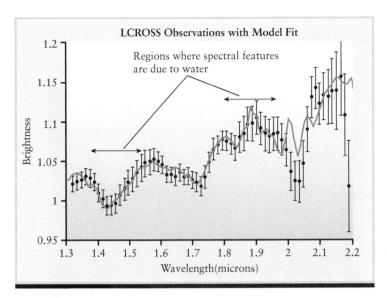

FIGURE 10-13

Evidence for Lunar Water The infrared spectrum of the plume shown in Figure 10-12 indicates both water vapor and water-ice. The collected data are shown in black with their vertical error bars. The red curve is a model based on the known spectrum of water vapor and ice. There is a very good fit between the model and data in the regions where the spectral features are due to water.

other hand, the ice has scientific value analogous to ice cores that have been obtained on Earth: By preserving material in ice for billions of years, craters may hold clues to the origins and evolution of our solar system.

CONCEPTCHECK 10-2

What provides better evidence of water on the Moon—a clear excess of hydrogen atoms or the spectrum shown in Figure 10-13?

Answer appears at the end of the chapter.

10-3 The Moon has no global magnetic field but has a small molten core

We saw in Section 7-6 and Section 7-7 that a small planet or satellite will have little remaining internal heat, and hence is likely to have a mostly solid interior. Because it lacks substantial amounts of moving molten material in its interior, a small world is unlikely to generate a global magnetic field. However, a small world can still have a small molten core, which is just what has been found for the Moon.

> Compared to earthquakes, moonquakes are few and feeble—but they reveal key aspects of the Moon's interior

No Global Magnetic Field

The Moon landings gave scientists an unprecedented opportunity to explore an alien world. The *Apollo* spacecraft were therefore packed with an assortment of apparatus that the astronauts deployed around the landing sites (see Figure 10-9). For example, all the missions carried magnetometers, which confirmed that the present-day Moon has no global magnetic field. However, careful magnetic measurements of lunar rocks returned by the *Apollo* astronauts indicated that the Moon did have a weak magnetic field when the rocks solidified billions of years ago. Hence, the Moon must originally have had a small, molten, iron-rich core. (We saw in Section 7-7 that Earth's magnetic field is generated by fluid motions in its interior.)

The Moon's Interior

The *Apollo* astronauts also set up four seismometers on the Moon's surface, which made it possible for scientists to investigate the Moon's interior using seismic waves just as is done for Earth (Section 9-2). It was found that the Moon exhibits far less seismic activity than Earth. Roughly 3000 **moonquakes** were detected per year, whereas a similar seismometer on Earth would record hundreds of thousands of earthquakes per year. Furthermore, typical moonquakes are very weak. A major terrestrial earthquake is in the range from 6 to 8 on the Richter scale, while a major moonquake measures 0.5 to 1.5 on that scale and would go unnoticed by a person standing near the epicenter.

Analysis of moonquakes reveals that the Moon has a thick mantle surrounding an iron-rich core (Figure 10-14). The Moon also has a crust with an average thickness of about 60 km on the Earth-facing side and up to 100 km on the far side. For comparison, the thickness of Earth's crust ranges from 5 km under the oceans to about 70 km under major mountain ranges on the

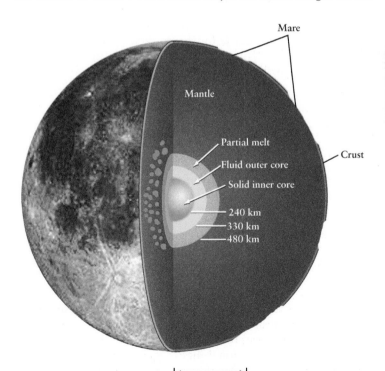

FIGURE 10-14

The Internal Structure of the Moon Like Earth, the Moon has a crust, a mantle, and a core. The red dots indicate deep earthquakes detected by seismometers left by the *Apollo* missions in the 1970s. Reanalysis of this data in 2011 uncovered a liquid outer core surrounded by a solid inner core. (NASA)

BOX 10-1 TOOLS OF THE ASTRONOMER'S TRADE

Calculating Tidal Forces

The gravitational force between any two objects decreases with increasing distance between the objects. This principle leads to a simple formula for estimating the tidal force that Earth exerts on parts of the Moon.

Consider two small objects, each of mass m, on opposite sides of the Moon. Because the two objects are at different distances from Earth's center, Earth exerts different forces F_{near} and F_{far} on them (see the accompanying illustration). The consequence of this difference is that the two objects tend to pull away from each other and away from the center of the Moon (see Section 4-8, especially Figure 4-26). The *tidal force* on these two objects, which is the force that tends to pull them apart, is the difference between the forces on the individual objects:

$$F_{tidal} = F_{near} - F_{far}$$

Both F_{near} and F_{far} are inversely proportional to the square of the distance from Earth (see Section 4-7). If this distance were doubled, both forces would decrease to $\frac{1}{2^2} = \frac{1}{4}$ of their original values. But the tidal force F_{tidal}, which is the *difference* between these two forces, decreases even more rapidly with distance: It is inversely proportional to the *cube* of the distance from Earth to the Moon. That is, doubling the Earth-Moon distance decreases the tidal force to $\frac{1}{2^3} = \frac{1}{8}$ of its original value. If we use the symbols M_{Earth} for Earth's mass (5.974×10^{24} kg), d for the Moon's diameter, r for the center-to-center distance from Earth to the Moon, and G for the universal constant of gravitation (6.67×10^{-11} newton m²/kg²), the tidal force is given approximately by the formula

$$F_{tidal} = \frac{2GM_{Earth}md}{r^3}$$

EXAMPLE: Calculate the tidal force for two 1-kg lunar rocks at the locations shown in the figure.

Situation: We are asked to calculate a tidal force. We know the masses of Earth (M_{Earth}) and of each of the two rocks

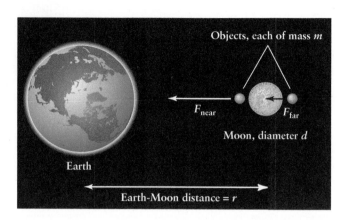

($m = 1$ kg). The Earth-Moon distance (r) and the diameter of the Moon (d, equal to the separation between the two rocks) are given in Table 10-1.

Tools: We use the equation given above for the tidal force F_{tidal}.

Answer: To use the given values in the equation, we must first convert the values for the average Earth-Moon distance r and the Moon's diameter d from kilometers to meters:

$$d = 3476 \text{ km} \times \frac{10^3 \text{ m}}{1 \text{ km}} = 3.476 \times 10^6 \text{ m}$$

$$r = 384{,}4000 \text{ km} \times \frac{10^3 \text{ m}}{1 \text{ km}} = 3.844 \times 10^8 \text{ m}$$

The tidal force is then

$$F_{tidal} = \frac{2(6.67 \times 10^{-11})(5.974 \times 10^{24})(1)(3.476 \times 10^6)}{(3.844 \times 10^8)^3}$$

$$= 4.88 \times 10^{-5} \text{ newton}$$

---

continents (see Table 9-3). This slightly lopsided distribution of mass between the near and far sides of the Moon helps to keep the Moon in synchronous rotation, so that we always see the same side of the Moon (see Section 4-8).

More than 30 years after the original *Apollo* data shed light on the Moon's internal structure, the seismic data were reanalyzed in 2011 and revealed a molten core. Like Earth, the Moon's iron-rich core has a solid inner core, surrounded by a liquid outer core. However, the liquid core is too small for the convection that is needed to generate a global magnetic field.

On the whole, seismic waves indicate that the Moon's interior is much more rigid and less fluid than the interior of Earth. Plate tectonics requires that there be fluid motion just below a planet's surface, so it is not surprising that there is no evidence for plate tectonics on the lunar surface.

The Origin of Moonquakes

Earthquakes tend to occur at the boundaries between tectonic plates, where plates move past each other, collide, or undergo

Review: To put our result into perspective, let us compare it to the *weight* of a 1-kg rock on the Moon. The weight of an object is equal to its mass multiplied by the acceleration due to gravity (see Section 4-6). This acceleration is equal to 9.8 m/s^2 on Earth, but is only 0.17 as great on the Moon. So, the weight of a 1-kg rock on the Moon is

$$0.17 \times 1 \text{ kg} \times 9.8 \text{ m/s}^2 = 1.7 \text{ newtons}$$

Thus, the tidal force on lunar rocks is *much* less than their weight, which is a good thing. If the tidal force (which tries to pull a rock away from the Moon's center) were greater than the weight (which pulls it toward the Moon's center), loose objects would levitate off the Moon's surface and fly into space!

The tidal force of Earth on the Moon deforms the solid body of the Moon, so that it acquires *tidal bulges* on its near and far sides. The mass of each bulge must be proportional to the tidal force that lifts lunar material away from the Moon's center. So, the above formula tells us that the mass of each bulge is inversely proportional to the cube of the Earth-Moon distance r:

$$m_{\text{bulge}} = \frac{A}{r^3}$$

(Determining the value of the constant A requires knowing how easily the Moon deforms, which is a complicated problem beyond our scope.) We can then use our formula for the tidal force F_{tidal} to tell us the *net* tidal force on the Moon. To make this calculation, we think of the objects of mass m on opposite sides of the Moon as being the tidal bulges. So, we substitute m_{bulge} for m in the tidal force formula

$$F_{\text{tidal-net}} = \frac{2GM_{\text{Earth}}m_{\text{bulge}}d}{r^3} = \frac{2GM_{\text{Earth}}}{r^3}\frac{A}{r^3}d$$

$$= \frac{2GM_{\text{Earth}}Ad}{r^6}$$

Thus, the net tidal force on the Moon is inversely proportional to the *sixth* power of the Earth-Moon distance. Even a small decrease in this distance can cause a substantial increase in the net tidal force.

EXAMPLE: Compare the net tidal force on the Moon at perigee (when it is closest to Earth) and at apogee (when it is farthest away).

Situation: We are asked to *compare* the tidal forces on the Moon at two different distances from Earth. To make this comparison, we will find the *ratio* of the tidal force at perigee to the tidal force at apogee.

Tools: We use the equation given above for the net tidal force on the Moon.

Answer: The Earth-Moon distances at perigee and apogee are given in Table 10-1. If we take the ratio of the tidal forces at these two distances, the constants (including the unknown quantity A) cancel out:

$$\frac{F_{\text{tidal-net-perigee}}}{F_{\text{tidal-net-apogee}}} = \frac{2GM_{\text{Earth}}Ad/r^6{}_{\text{perigee}}}{2GM_{\text{Earth}}Ad/r^6{}_{\text{apogee}}} = \left(\frac{r_{\text{apogee}}}{r_{\text{perigee}}}\right)^6$$

$$= \left(\frac{405{,}500 \text{ km}}{363{,}000 \text{ km}}\right)^6 = (1.116)^6 = 1.93$$

Review: Our result tells us that the net tidal force at perigee is 1.93 times as great (that is, almost double) as at apogee. Tidal forces on the Moon help create the stresses that generate moonquakes, so it is not surprising that these quakes are more frequent at perigee than at apogee.

---

subduction (see Figure 9-14). But there are no plate motions on the Moon. What, then, causes moonquakes? The answer turns out to be Earth's tidal forces. Just as the tidal forces of the Sun and Moon deform Earth's oceans and give rise to ocean tides (see Figure 4-26), Earth's tidal forces deform the solid body of the Moon. The amount of the deformation changes as the Moon moves around its elliptical orbit and the Earth-Moon distance varies. As a result, the body of the Moon flexes slightly, triggering moonquakes. The tidal stresses are greatest when the Moon is nearest Earth—that is, at perigee (see Section 3-5)—which is just

when the *Apollo* seismometers reported the most moonquakes. Box 10-1 has more information about how the tidal stresses on the Moon depend on its distance from Earth.

Without tectonics, and without the erosion caused by an atmosphere or oceans, the only changes occurring today on the lunar surface are those due to meteoroid impacts. These impacts were also monitored by the seismometers left on the Moon by the *Apollo* astronauts. These devices, which remained in operation for several years, were sensitive enough to detect a hit by a grapefruit-sized meteoroid anywhere on the lunar surface. The data show that every year the Moon is struck by 80 to 150 meteoroids having masses between 0.1 and 1000 kg (roughly ¼ lb to 1 ton).

CONCEPTCHECK 10-3

Does every world with a liquid iron-rich core contain a magnetic dynamo?

Answer appears at the end of the chapter.

10-4 Lunar rocks reveal a geologic history quite unlike that of Earth

One of the most important scientific goals of the *Apollo* missions was to learn as much as possible about the lunar surface. Although only one of the 12 astronauts who visited the Moon was a professional geologist, all of the astronauts received intensive training in field geology, and they were in constant communication with planetary scientists on Earth as they explored the lunar surface.

Weathering by Micrometeorites and the Solar Wind

The Moon has no atmosphere or oceans to cause "weathering" of the surface. But the *Apollo* astronauts found that the lunar surface has undergone a kind of erosion due to billions of years of relentless micrometeorite bombardment. When lunar astronauts retrieved equipment left on the moon two years earlier, micrometeorite impacts showed that this weathering is still taking place. Another significant form of weathering comes from bombardment by charged particles in the solar wind.

On longer timescales, craters fewer than 400 meters wide could be erased by weathering over a period of hundreds of millions of years to perhaps a billion years (this is used to date the scarps shown in Figure 10-8). This bombardment has pulverized the surface rocks into a layer of fine powder and rock fragments called the **regolith,** from the Greek for "blanket of stone" (Figure 10-15). This layer varies in depth from about 2 to 20 m. Just as porous foam material helps to absorb sound waves in a recording studio, the regolith absorbs most of the light waves, or sunlight, that falls on it. This absorption of light helps to explain why the Moon only reflects about 11% of incident sunlight (see albedo, Table 10-1), compared to a typical Earth desert that reflects about 25% of incident light.

> Although no wind or rain disturbs the lunar surface, it is "weathered" over the ages by the impact of tiny meteoroids

Research reported in 2012 suggests that the regolith might be dangerous for lunar explorers. As the *Apollo* astronauts found, even after wearing sealed space suits when venturing about, fine regolith dust eventually makes its way into living quarters. The dust is sharp like glass, and some of the particles are very small. Small particles are particularly dangerous for the lungs, where they can lodge deep into lung tissue and cause a range of serious health problems.

While the effects of weathering can be dangerous to the touch, the impacts that cause weathering might provide an intriguing visual display. Meteorites smaller than 1 mm in size are referred to as micrometeorites, but less frequent impacts by larger objects

FIGURE 10-15 R I V U X G

The Regolith Billions of years of bombardment by space debris have pulverized the uppermost layer of the Moon's surface into powdered rock. This layer, called the regolith, is utterly bone dry. It nonetheless sticks together like wet sand, as shown by the sharp outline of an astronaut's bootprint. (*Apollo 11*, NASA)

still occur. These larger impacts might solve a mystery that is more than 500 years old—a sparkling moon. Observers have long noted occasional flashes of light from the moon, although their origin and even their very existence, was in dispute. In 2012, researches found that when meteorites as small as 10 cm hit the lunar surface, enough heat is generated to produce light that is visible from Earth. Such events—called **transient lunar phenomena**—can occur hundreds of times a year, and are more frequent during meteor showers.

Moon Rocks

In addition to their observations of six different landing sites, the *Apollo* astronauts brought back 2415 individual samples of lunar material totaling 382 kg (842 lb). In addition, the unmanned Soviet spacecraft *Luna 16, 20,* and *24* returned 300 g (⅔ lb) from three other sites on the Moon. This collection of lunar material has provided important information about the early history of the Moon that could have been obtained in no other way.

All of the lunar samples are igneous rocks; there are no metamorphic or sedimentary rocks (see Section 9-3, especially Figure 9-18). The presence of only igneous rocks suggests that most or all of the Moon's surface was once molten. Indeed, Moon rocks are composed mostly of the same minerals that are found in volcanic rocks on Earth.

Astronauts who visited the maria discovered that these dark regions of the Moon are covered with basaltic rock quite similar to the dark-colored rocks formed by lava from volcanoes on Hawaii and Iceland. The rock of these low-lying lunar plains is called **mare basalt** (Figure 10-16).

In contrast to the dark maria, the lunar highlands are composed of a light-colored rock called **anorthosite** (Figure 10-17). On Earth,

FIGURE 10-16   R I ☑ U X G

Mare Basalt This 1.53-kg (3.38-lb) specimen of mare basalt was brought
back by the crew of *Apollo 15*. Gas must have dissolved under pressure in
the lava from which this rock solidified. When the lava reached the airless lunar
surface, the pressure on the lava dropped and the gas formed bubbles. Some
of the bubbles were frozen in place as the rock cooled, forming the holes on its
surface. (Figure 9-18a shows a sample of basalt from Earth.) (NASA)

FIGURE 10-17   R I ☑ U X G

Anorthosite The light-colored lunar terrae (highlands) are composed of this
ancient type of rock, which is thought to be the material of the original lunar
crust. Lunar anorthosites vary in color from dark gray to white; this sample from
the *Apollo 16* mission is a medium gray. It measures 18 × 16 × 7 cm. (NASA)

anorthositic rock is found only in such very old mountain ranges
as the Adirondacks in the eastern United States. Anorthosite is less
dense than basalt. As a result, during the period when the Moon's
surface was molten, the less dense anorthosite rose to the top, forming the majority of the present-day lunar surface.

Although anorthosite is the dominant rock type in the lunar
highlands, most of the rock samples collected there were not pure
anorthosite but rather **impact breccias.** These rocks are compos-
ites of different rock types that were broken apart, mixed, and
fused together by a series of meteoritic impacts (Figure 10-18).
The prevalence of breccias is evidence that the lunar highlands
have undergone eons of bombardment from space.

By carefully measuring the abundances of trace amounts of
radioactive elements in lunar samples and applying the principles of
radioactive dating (see Section 8-2), geologists have determined that
lunar rocks formed more than 3 billion years ago. Of these ancient
rocks, however, anorthosite is much older than the mare basalts.
Typical anorthositic specimens from the highlands are between
4.0 and 4.3 billion years old. All of these extremely ancient specimens are probably samples of the Moon's original crust. In sharp
contrast, all the mare basalts are between 3.1 and 3.8 billion years
old. With no plate tectonics to bring fresh material to the surface,
no new anorthosites or basalts have formed on the Moon for more
than 3 billion years.

The History of Lunar Cratering

By correlating the ages of Moon rocks with the density of craters
at the sites where the rocks were collected, geologists have found
that the rate of impacts on the Moon has changed over the ages.
The ancient, heavily cratered lunar highlands are evidence of an
intense bombardment that dominated the Moon's early history. For
around half a billion years, rocky debris left over from the formation of the planets rained down upon the Moon's young surface.

FIGURE 10-18   R I ☑ U X G

Impact Breccias These rocks are evidence of the Moon's long history of
bombardment from space. Parts of the Moon's original crust were shattered
and strewn across the surface by meteoritic impacts. Later impacts compressed
and heated this debris, welding it into the type of rock shown here. Such impact
breccias are rare on Earth but abundant on the Moon. (NASA)

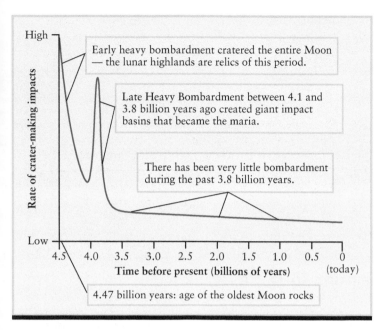

FIGURE 10-19

The Rate of Crater Formation on the Moon This graph shows the rate at which impact craters formed over the Moon's history. The impact rate today is only about 1/10,000 as great as during the most intense bombardment. (Adapted from T. Grotzinger, T. H. Jordan, F. Press, and R. Siever, *Understanding Earth*, 5th ed., W. H. Freeman, 2007)

As Figure 10-19 shows, this barrage extended from 4 billion years ago, when the Moon's surface solidified, until about 3.8 billion years ago. At its peak, this bombardment from space would have produced a new crater of about 1 km radius somewhere on the Moon about once per century. (If 100 years seems like a long time interval, remember that we are talking about a bombardment that lasted hundreds of millions of years and produced millions of craters over that time.)

The rate of impacts should have tapered off as meteoroids, asteroids, and the even larger planetesimals were swept up by the newly formed planets. In fact, radioactive dating indicates that the impact rate spiked upward between 4.1 and 3.8 billion years ago (see Figure 10-19), and this period is referred to as the Late Heavy Bombardment. This spike in impacts is thought to have occurred throughout the inner solar system as the Jovian planets scattered planetesimals—before the solar system settled into its present configuration (see Section 8-6).

Whatever the explanation for the epoch of heavy bombardment, major impacts during this period gouged out the mare basins, which later filled with lava to form the maria (see the *Cosmic Connections* figure). The relative absence of craters in the maria (see Figure 10-5) tells us that the impact rate has been quite low since the lava solidified. During these past 3.8 billion years, the Moon's surface has changed very little.

CONCEPT CHECK 10-4

Astronauts have left footprints on the Moon. With no wind or rain, do you expect that these features will last on the surface for a billion years?

Answer appears at the end of the chapter.

10-5 The Moon probably formed from debris cast into space when a huge protoplanet struck the young Earth

The lunar rocks collected by the *Apollo* and *Luna* missions have provided essential information about the history of the Moon. In particular, they have helped astronomers come to a consensus about one of the most important questions in lunar science: Where did the Moon come from?

The Receding Moon

Seventy years before the first spacecraft landed on the Moon, the British astronomer George Darwin (second son of the famous evolutionist Charles Darwin) deduced that the Moon must be slowly moving away from Earth. He began with the notion that the Moon's tidal forces elongate the oceans into a football shape (see Section 4-8, especially Figure 4-26). However, the long axis of this "football" does not point precisely at the Moon. Earth spins on its axis more rapidly than the Moon revolves around Earth, and this rapid rotation carries the tidal bulge about 10° ahead of a line connecting Earth and the Moon (Figure 10-20). This misaligned bulge produces a small but steady gravitational force that tugs the Moon forward, adding energy to the Moon's orbit. As we saw in Section 4-7, increasing the energy of an orbit increases its semimajor axis. Therefore, Earth's tidal bulge tugs the Moon into an ever larger orbit. In other words, Darwin predicted that the Moon must be spiraling away from Earth.

It became possible to test Darwin's predictions with the aid of a simple device that the *Apollo 11, 14,* and *15* and *Luna 21* missions left on the Moon—a set of reflectors, similar to the orange and red ones found in automobile taillights. Since 1969, scientists on Earth have been firing pulses of laser light at these reflectors and measuring very accurately the length of time it takes each

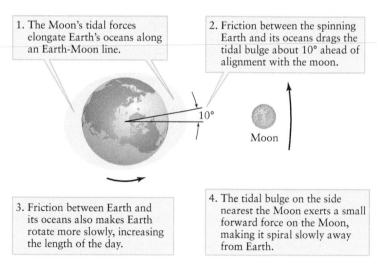

FIGURE 10-20

The Moon's Tidal Recession Earth's rapid rotation drags the tidal bulge of the oceans about 10° ahead of a direct alignment with the Moon. The bulge on the side nearest the Moon exerts more gravitational force than the other, more distant bulge. The net effect is a small forward force on the Moon that makes it spiral slowly away from Earth.

pulse to return to Earth. Knowing the speed of light, they can use these data to calculate the distance to the Moon to an accuracy of just 3 centimeters. They have found that the Moon is moving away from Earth at a very gradual rate of 3.8 cm per year—which is just what Darwin predicted!

As the Moon moves away from Earth, its sidereal period is getting longer in accordance with Kepler's third law (see Section 4-7). At the same time, friction between the oceans and the body of Earth is gradually slowing Earth's rotation. The length of Earth's day is therefore increasing, by approximately 0.002 seconds per century. In fact, 400 million years ago, the length of Earth's day was only 21 hours!

Theories of the Moon's Origins

These observations mean that in the distant past, Earth was spinning faster than it is now and the Moon was much closer. Darwin theorized that the early Earth may have been spinning so fast that a large fraction of its mass tore away, and this fraction coalesced to form the Moon. This is called the **fission theory** of the Moon's origin. Prior to the *Apollo* program, two other theories were in competition with the fission theory. The **capture theory** posits that the Moon was formed elsewhere in the solar system and then drawn into orbit about Earth by gravitational forces. By contrast, the **co-creation theory** proposes that Earth and the Moon were formed at the same time but separately.

One fact used to support the fission theory was that the Moon's average density (3344 kg/m^3) is comparable to that of Earth's outer layers, as would be expected if the Moon had been flung out of a rapidly rotating proto-Earth. However, the fission theory predicts that lunar and terrestrial rocks should be very similar. In fact, samples returned from the Moon by the *Apollo* and *Luna* missions show that the material forming lunar rocks has experienced higher temperatures than rocks on Earth. These higher temperatures are inferred by the presence of fewer **volatile elements** in lunar rock, which are elements that boil (and escape as gas) at temperatures under 900°C.

Arguing that the fission theory could not account for a temperature difference, some scientists maintained that the Moon formed elsewhere and was later captured by Earth. They also noted that the plane of the Moon's orbit is close to the plane of the ecliptic, as would be expected if the Moon had originally been in orbit about the Sun but was later captured by Earth.

There are, however, difficulties with the capture theory. If Earth did capture the Moon intact, then some necessary conditions must have been satisfied entirely by chance. It is easiest to understand the required coincidence in terms of orbital energy and speed. Again, from Section 4-7, each semimajor axis of an orbiting object has a certain energy, with larger energies corresponding to lower speeds and larger orbital distances. If an object were passing by too quickly, there would be nothing to slow it down into a captured orbit. If an object came in at a lower speed, nothing could speed it up into a steady orbit before it merged with Earth. Therefore, the only way to simply capture a passing Moon would be if it happened to pass by at just the right speed for its orbital distance from Earth; this seems very unlikely.

Proponents of the co-creation theory argued that it is easier for a planet to capture swarms of tiny rocks. In this theory the Moon formed from just such rocky debris. Great numbers of rock fragments orbited the newborn Sun in the plane of the ecliptic along with the protoplanets. Heat generated in collisions could have baked the volatile elements out of these smaller rock fragments. Then, just as planetesimals accreted to form the proto-Earth in orbit about the Sun, the fragments in orbit about Earth accreted to form the Moon.

The Collisional Ejection Theory

The present scientific consensus is that *none* of these three theories—fission, capture, and co-creation—is correct. Instead, the evidence points toward an idea proposed in 1975 by William Hartmann and Donald Davis and independently by Alastair Cameron and William Ward. In this **collisional ejection theory**, the proto-Earth was struck off-center by a Mars-sized object, and this collision ejected debris from which the Moon formed.

The collisional ejection theory agrees with what we know about the early history of the solar system. We saw in Section 8-5 that smaller objects collided and fused together to form the inner planets. Some of these collisions must have been quite spectacular, especially near the end of planet formation, when most of the mass of the inner solar system was contained in a few dozen large protoplanets. The collisional ejection theory proposes that one such object, a Mars-sized protoplanet, struck the proto-Earth about 4.5 billion years ago, about halfway between the age of the solar system (4.56 billion years) and the age of the oldest known Moon rocks (4.47 billion years). Figure 10-21 shows the results of a supercomputer simulation of this cataclysm. In the simulation, energy released during the collision produces a huge plume of vaporized rock that shoots out from the point of impact. As this ejected material cools, it coalesces to form the Moon. The tidal effects depicted in Figure 10-20 then cause the size of the Moon's orbit to gradually increase.

The collisional ejection theory also agrees with many properties of lunar rocks and of the Moon as a whole. For example, rock vaporized by the impact would have been depleted of volatile elements and water, leaving the Moon rocks we now know. If the off-center collision took place after chemical differentiation had occurred on Earth, so that our planet's iron had sunk to its center, then relatively little iron would have been ejected to become part of the Moon. The lack of iron on Earth's surface explains the Moon's low density and the small size of its iron core. It also explains the curiously low iron content found in the South Pole–Aitken Basin (see Figure 10-7), where an ancient impact excavated the surface down to a depth of 12 km. Data from the *Clementine* spacecraft show that iron concentration in this basin is only 10% (see Figure 10-11), which is far less than the 20–30% concentration at a corresponding depth below Earth's surface.

An Overview of the Moon's History

We can now summarize the entire geologic history of the Moon. The surface of the newborn Moon was probably molten for many years, both from heat produced during the formation of the Moon and from the decay of short-lived radioactive isotopes. As the Moon gradually cooled, low-density lava floating on the Moon's surface began to solidify into the anorthositic crust that exists today. The barrage of rock fragments that ended about 3.8 billion years ago produced the craters that cover the lunar highlands.

> The difference in chemical composition between lunar rocks and Earth rocks helps to constrain theories of the Moon's origin

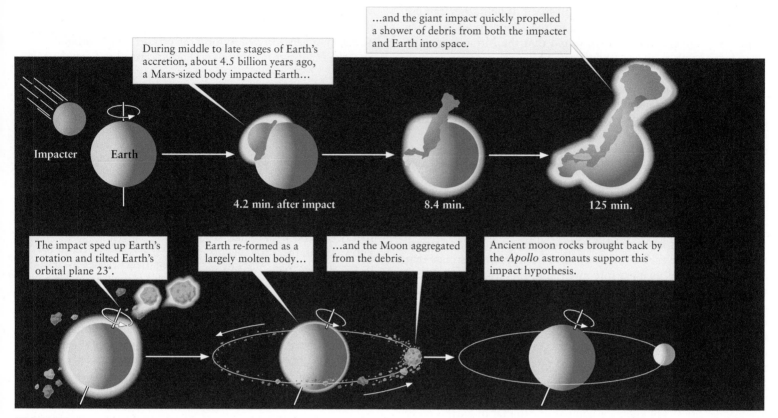

During middle to late stages of Earth's accretion, about 4.5 billion years ago, a Mars-sized body impacted Earth...

...and the giant impact quickly propelled a shower of debris from both the impacter and Earth into space.

Impacter   Earth

4.2 min. after impact

8.4 min.

125 min.

The impact sped up Earth's rotation and tilted Earth's orbital plane 23°.

Earth re-formed as a largely molten body...

...and the Moon aggregated from the debris.

Ancient moon rocks brought back by the *Apollo* astronauts support this impact hypothesis.

FIGURE 10-21

The Formation of the Moon This figure shows how the Moon could have formed in the aftermath of a collision between a Mars-sized protoplanet and the proto-Earth. (Adapted from T. Grotzinger, T. H. Jordan, F. Press, and R. Siever, *Understanding Earth, 5th ed.,* W. H. Freeman, 2007)

During the relatively brief period of the Late Heavy Bombardment between 4.1 and 3.8 billion years ago (see Figure 10-19), more than a dozen asteroid-sized objects, each measuring at least 100 km across, struck the Moon and blasted out the vast mare basins. Meanwhile, heat from the decay of long-lived radioactive elements like uranium and thorium began to melt the inside of the Moon. Great floods of molten rock gushed up from the lunar interior, filling the impact basins and creating the dark basaltic maria we see today.

Very little has happened on the Moon since those ancient times. The most recent geologic activity appears to be wrinklelike scarps forming as the cooling Moon shrinks in size, perhaps within the past few hundred million years. A few relatively fresh craters have been formed, but the astronauts visited a world that has remained largely unchanged for more than 3 billion years. During that same period on Earth, by contrast, the continents have been totally transformed time and time again (see Section 9-3).

Many questions and mysteries still remain. The six American and three Soviet lunar landings have brought back samples from only nine locations, barely scratching the lunar surface. We still know very little about the Moon's far side and poles. How many craters contain ice deposits at the poles? How much of the Moon's interior is molten? Just how large is its iron core? How old are the youngest lunar rocks? Did lava flows occur over western Oceanus Procellarum only 2 billion years ago, as crater densities there

suggest? Could there still be active volcanoes on the Moon? There is still much to be learned by exploring the Moon.

CONCEPTCHECK 10-5

In the collisional ejection theory, does it matter if Earth has undergone chemical differentiation before the impact?

Answer appears at the end of the chapter.

KEY WORDS

anorthosite, p. 268
capture theory (of Moon's formation), p. 271
center of mass, p. 256
co-creation theory (of Moon's formation), p. 271
collisional ejection theory (of Moon's formation), p. 271
crater, p. 258
far side (of the Moon), p. 256
fission theory (of Moon's formation), p. 271

impact breccia, p. 269
impact crater, p. 258
libration, p. 256
lunar highlands, p. 261
mare (*plural* maria), p. 258
mare basalt, p. 268
moonquake, p. 265
regolith, p. 268
synchronous rotation, p. 256
terminator, p. 256
terrae, p. 261
transient lunar phenomena, p. 268
volatile element, p. 271

Appearance of the Moon: The Earth-facing side of the Moon displays light-colored, heavily cratered highlands and dark-colored, smooth-surfaced maria. The Moon's far side has almost no maria.

• Virtually all lunar craters were caused by space debris striking the surface. There is no evidence of plate tectonic activity on the Moon.

Internal Structure of the Moon: Much of our knowledge about the Moon has come from human exploration in the 1960s and early 1970s and from more recent observations by unmanned spacecraft.

• Analysis of seismic waves and other data indicates that the Moon has a crust thicker than that of Earth (and thickest on the far side of the Moon), a thick mantle, and a small iron core.

• The Moon has a solid inner core surrounded by a liquid outer core.

• The Moon has no global magnetic field today, although it had a weak magnetic field billions of years ago.

Geologic History of the Moon: The anorthositic crust exposed in the highlands was formed between 4.3 and 4.0 billion years ago. The Late Heavy Bombardment formed the maria basins between 4.1 and 3.8 billion years ago, and the mare basalts solidified between 3.8 and 3.1 billion years ago.

• The Moon's surface has undergone very little change over the past 3 billion years.

• Meteoroid impacts and the solar wind have been the only significant "weathering" agents on the Moon. These weathering processes formed the Moon's regolith, or surface layer of powdered and fractured rock.

• All of the lunar rock samples are igneous rocks formed largely of the same minerals found in terrestrial rocks on Earth. However, material in lunar rocks appears to have been exposed to higher temperatures than terrestrial rocks.

Origin of the Moon: The collisional ejection theory of the Moon's origin holds that the proto-Earth was struck by a Mars-sized protoplanet and that debris from this collision coalesced to form the Moon. This theory successfully explains most properties of the Moon.

• The Moon was molten in its early stages, and the anorthositic crust solidified from low-density magma that floated to the lunar surface. The mare basins were created later by the impact of planetesimals and later filled with lava from the lunar interior.

• Tidal interactions between Earth and the Moon are slowing Earth's rotation and pushing the Moon away from Earth. This also causes the length of Earth's day to slowly increase.

Review Questions

1. Is it correct to say that the Moon orbits Earth? If not, what is a more correct description?

2. If the Moon always keeps the same face toward Earth, how is it possible for Earth observers to see more than half of the Moon's surface?

3. Why does the sky look black on the Moon even during daytime?

4. Why is it impossible for liquid water to exist on the surface of the Moon?

5. Describe two reasons why astronauts needed to wear spacesuits on the lunar surface.

6. Describe the kinds of features that can be seen on the Moon with a small telescope.

7. Are impact craters on the Moon the same size as the meteoroids that made the impact? Explain your answer.

8. Describe the differences between the maria and the lunar highlands. Which kind of terrain covers more of the Moon's surface? Which kind of terrain is more heavily cratered? Which kind of terrain was formed later in the Moon's history? How do we know?

9. Describe the differences between the near and far sides of the Moon. What is thought to be the likely explanation for these differences?

10. What does it mean to say the Moon is a "one-plate world"? What is the evidence for this statement?

11. Why was it necessary to send unmanned spacecraft to land on the Moon before sending humans there?

12. What is the evidence that water-ice exists at the lunar poles? Is this evidence definitive?

13. Why was it useful for the *Apollo* astronauts to bring magnetometers and seismometers to the Moon?

14. Could you use a magnetic compass to navigate on the Moon? Why or why not?

15. Does the absence of a global magnetic field around the Moon prove there is no liquid iron core?

16. Explain why moonquakes occur more frequently when the Moon is at perigee than at other locations along its orbit.

17. Why is Earth geologically active while the Moon is not?

18. What is the regolith? What causes its powdery character?

19. Why are there no sedimentary rocks on the Moon?

20. On the basis of Moon rocks brought back by the astronauts, explain why the maria are dark-colored but the lunar highlands are light-colored.

21. Briefly describe the main differences and similarities between Moon rocks and Earth rocks.

22. Rocks found on the Moon are between 3.1 and 4.47 billion years old. By contrast, the majority of Earth's surface is made of oceanic crust that is less than 200 million years old, and the very oldest Earth rocks are about 4 billion years old. If Earth and the Moon are essentially the same age, why is there such a disparity in the ages of rocks on the two worlds?

23. If Earth's tidal bulge pointed directly toward the Moon, would the Moon still be receding from Earth? Explain your answer.

24. Why do most scientists favor the collisional ejection theory of the Moon's formation?

25. Some people who supported the fission theory proposed that the Pacific Ocean basin is the scar left when the Moon pulled away from Earth. Explain why this idea is probably wrong.

Advanced Questions

Questions preceded by an asterisk (*) involve topics discussed in Box 10-1.

> **Problem-solving tips and tools**
>
> Recall that the average density of an object is its mass divided by its volume. The volume of a sphere is $4\pi r^3/3$, where r is the sphere's radius. The surface area of a sphere of radius r is $4\pi r^2$, while the surface area of a circle of radius r is πr^2. Recall also that the acceleration of gravity on Earth's surface is 9.8 m/s². You may find it useful to know that a 1-pound (1-lb) weight presses down on Earth's surface with a force of 4.448 newtons. You might want to review Newton's law of universal gravitation in Section 4-6. The time to travel a certain distance is equal to the distance traveled divided by the speed of motion. Consult Table 10-1 and the Appendices for any additional data.

26. Suppose two worlds (say, a planet and its satellite) have masses m_1 and m_2, and the center-to-center distance between the worlds is r. The distance d_{cm} from the center of world 1 to the center of mass of the system of two worlds is given by the formula

$$d_{cm} = \frac{m_2 r}{m_1 + m_2}$$

(a) Suppose world 1 is Earth and world 2 is the Moon. If Earth and the Moon are at their average center-to-center distance, find the distance from the center of Earth to the center of mass of the Earth-Moon system. (b) Is the Earth-Moon system's center of mass within Earth? How far below Earth's surface is it located? (c) Find the distance from the center of the Sun (mass 1.989×10^{30} kg) to the center of mass of the Sun-Earth system. How does this compare to the diameter of the Sun? Is it a good approximation to say that Earth orbits around the center of the Sun?

27. If you view the Moon through a telescope, you will find that details of its craters and mountains are more visible when the Moon is near first quarter phase or third quarter phase than when it is at full phase. Explain why.

28. In a whimsical moment during the *Apollo 14* mission, astronaut Alan Shepard hit two golf balls over the lunar surface. Give two reasons why they traveled much farther than golf balls do on Earth.

29. Temperature variations between day and night are much more severe on the Moon than on Earth. Explain why.

30. How much would an 80-kg person weigh on the Moon? How much does that person weigh on Earth?

31. Using the diameter and mass of the Moon given in Table 10-1, verify that the Moon's average density is about 3344 kg/m³. Explain why this average density implies that the Moon's interior contains much less iron than the interior of Earth.

*32. In Box 10-1 we calculated the tidal force that Earth exerts on two 1-kg rocks located on the near and far sides of the Moon. We assumed that the Earth-Moon distance was equal to its average value. Repeat this calculation (a) for the Moon at perigee and (b) for the Moon at apogee. (c) What is the ratio of the tidal force on the rocks at perigee to the tidal force at apogee?

33. The youngest lunar anorthosites are 4.0 billion years old, and the youngest mare basalts are 3.1 billion years old. Would you expect to find any impact breccias on the Moon that formed less than 3.1 billion years ago? Explain your answer.

34. In the maria, the lunar regolith is about 2 to 8 meters deep. In the lunar highlands, by contrast, it may be more than 15 meters deep. Explain how the different ages of the maria and highlands can account for these differences.

35. The mare basalts are volcanic rock. Is it likely that active volcanoes exist anywhere on the Moon today? Explain.

36. Calculate the round-trip travel time for a pulse of laser light that is fired from a point on Earth nearest the Moon, hits a reflector at the point on the Moon nearest Earth, and returns to its point of origin. Assume that Earth and the Moon are at their average separation from each other.

37. Before the *Apollo* missions to the Moon, there were two diametrically opposite schools of thought about the history of lunar geology. The "cold moon" theory held that all lunar surface features were the result of impacts. The most violent impacts melted the surface rock, which then solidified to form the maria. The opposite "hot moon" theory held that all lunar features, including maria, mountains, and craters, were the result of volcanic activity. Explain how lunar rock samples show that neither of these theories is entirely correct.

*38. When the Moon originally coalesced, it may have been only one-tenth as far from Earth as it is now. (a) When the Moon first coalesced, was Earth's tidal force strong enough to lift rocks off the lunar surface? Explain. (b) Compared with the net tidal force that Earth exerts on the Moon today, how many times larger was the net tidal force on the newly coalesced Moon? (This strong tidal force kept the one axis of the Moon oriented toward Earth, and the Moon kept that orientation after it solidified.)

Discussion Questions

39. Comment on the idea that without the presence of the Moon in our sky, astronomy would have developed far more slowly.

40. No *Apollo* mission landed on the far side of the Moon. Why do you suppose this was? What would have been the scientific benefits of a mission to the far side?

41. NASA is planning a new series of manned missions to the Moon. Compare the advantages and disadvantages of exploring the Moon with astronauts as opposed to using mobile, unmanned instrument packages.

42. Describe how you would empirically test the idea that human behavior is related to the phases of the Moon. What problems are inherent in such testing?

43. How would our theories of the Moon's history have been affected if astronauts had discovered sedimentary rock on the Moon?

44. Imagine that you are planning a lunar landing mission. What type of landing site would you select? Where might you land to search for evidence of recent volcanic activity?

Web/eBook Question

45. In 2005 the *SMART-1* spacecraft detected calcium on the lunar surface. Search the World Wide Web for information about the *SMART-1* mission and this discovery. How was the presence of calcium detected? What does this tell astronomers about the origin of the Moon?

ACTIVITIES

Observing Projects

> #### Observing tips and tools
>
> If you do not have access to a telescope, you can learn a lot by observing the Moon through binoculars. Note that the Moon will appear right side up through binoculars but inverted through a telescope; if you are using a map of the Moon to aid your observations, you will need to take this into account. Inexpensive maps of the Moon can be purchased from most good bookstores or educational supply stores. You can determine the phase of the Moon either by looking at a calendar (most of which tell you the dates of new moon, first quarter, full moon, and third quarter), by checking the weather page of your newspaper, by consulting the current issue of *Sky & Telescope* or *Astronomy* magazine, by using the *Starry Night*™ program, or by using the World Wide Web.

46. Observe the Moon through a telescope every few nights over a period of two weeks between new moon and full moon. Make sketches of various surface features, such as craters, mountain ranges, and maria. How does the appearance of these features change with the Moon's phase? Which features are most easily seen at a low angle of illumination? Which features show up best with the Sun nearly overhead?

47. Use *Starry Night*™ to examine the surface of the Moon. Select **Favourites > Explorations > Luna**. In order to see features on the surface of the Moon on the hemisphere facing away from the Sun, select **Options > Solar System >**

Planets-Moons... from the menu. In the **Planets-Moons Options** dialog box, ensure that the **Show dark side** checkbox is checked and then slide the control for this option all the way to the right, to **Brighter**. This will artificially brighten the dark side of the Moon. Use the **Zoom** controls and location scroller to examine the surface of the Moon. (a) Based on your observations, what evidence can you find that the Moon is geologically inactive? Explain. (b) From your observations, explain the hypothesis that the darker gray lunar *maria* are geologically younger than the brighter highland regions of the Moon. (c) Spreading outward from some of the largest craters on the Moon are straight lines of slightly lighter-colored material, called rays, which were caused by material ejected outward by the impact that caused the crater. Use the **Zoom** controls and the location scroller to get a better view of these lunar features. Follow these rays across the lunar surface. Is there any evidence that this ejected material has disturbed crater walls during the violent impacts that caused these craters? Place the cursor over several of these impact craters and open the contextual menu (**right-click** on a PC, **Ctrl-click** on a Mac). The name of the crater will be found in the command "Mark *crater name* on surface." (d) List at least three craters on the lunar surface that have rays.

48. Use the *Starry Night*™ program to view the Moon as it appears from Earth and compare this with the view of Earth as it would appear from the Moon. Open the view named **Favourites > Explorations > Moon From Earth**. With the **Time Flow Rate** set at **1 day**, **Step time forward** while observing the Moon in the view. Use the hand tool to adjust the screen as necessary to follow the Moon in the sky from day to day. When the Moon sets below the horizon, select **File > Revert** to return to the original view. **Right-click (Ctrl-click on a Mac)** over the image of the Moon and select **Magnify** from the contextual menu to center a magnified image of the Moon in the view. Next, type **Ctrl-H** (type **B** on a Mac) to remove the horizon from the view so that you may follow the appearance of the Moon over time without interference from the horizon. (a) **Step time forward** and observe the features on the Moon. Do you see a different view of the Moon from night to night or do you always see the same features on the Moon? Explain why the appearance of the Moon shows this behavior as time changes. Select **Favourites > Discovering the Universe > Tranquility Base** to change your viewing location to the *Apollo 11* landing site on the Moon, with Earth visible in the lunar sky. With the **Time Flow Rate** at **1 hours**, click the **Play** button in the toolbar. (b) Describe and explain your observations of the position of Earth in the lunar sky. Why does the lunar horizon not interfere with the visibility of Earth from this lunar viewing location?

49. Use *Starry Night*™ to observe the apparent size of the Moon as seen from Earth. Select **Favourites > Explorations > Moon Size**. The view shows the full moon as it would appear from the north pole of Earth. Surrounding the Moon is a yellow

circle that is precisely 30 arcminutes in diameter. As you can see, at the time and date of this view, the Moon does not quite fill this circle. To prevent the phases of the Moon from interfering with your observations of the size of the Moon in the sky, the dark hemisphere of the Moon is artificially brightened in this simulation. With the **Time Flow Rate** at **1 days, Step time forward** and observe the size of the Moon relative to the 30′ reference circle. (**a**) Does the apparent size of the Moon remain constant? If not, explain what this tells us about the shape of the Moon's orbit around Earth. (**b**) Approximately how many days does it take for the Moon to go through a complete cycle of variation of its apparent size? [*Hint:* Count the number of one-day time steps between two successive occasions when the Moon appears to fit precisely inside the 30′ reference circle. (**c**) As you make your observations, do the lunar surface features appear to remain stationary in the view? (They *would* remain stationary if the Moon's synchronous rotation were the *only* motion that the Moon had relative to Earth.) If not, explain. (*Hint:* A Web search on the word "libration" [not libation!] may help.)

Collaborative Exercise

50. The image of Crater Clavius in Figure 10-4 reveals numerous craters. Using the idea that the Moon's landscape can only be changed by impacts, make a rough sketch showing 10 of the largest craters and label them from oldest (those that showed up first) to youngest (the most recent ones). Explain your reasoning and any uncertainties.

ANSWERS

ConceptChecks

ConceptCheck 10-1: The darker color of the maria comes from solidified iron-rich lava that has partially filled large impact basins. These impact basins are deep (with the South Pole–Aitken Basin reaching depths of 12 km) and partial filling by lava still leaves the maria surfaces at lower elevations.

ConceptCheck 10-2: Figure 10-13 shows conclusive evidence of water on the Moon because the observed spectrum is such a good match to the known spectrum of water. An excess of hydrogen atoms is likely due to water, but is less certain because it could also arise from unusual minerals.

ConceptCheck 10-3: No. The Moon's liquid iron-rich core is too small for the convection that creates a magnetic dynamo. This is why the Moon today has no global magnetic field.

ConceptCheck 10-4: No. Impacting micrometeorites constantly "weather" the surface of the Moon. After a billion years, this weathering might even erase a 400-meter-wide crater.

ConceptCheck 10-5: Yes. The Moon has a lower average density than Earth, indicating a lower iron fraction. If Earth's iron sank before the impact, then outer material forming the Moon would be lower in iron.

| Mercury | Venus | Mars |
| R I V U X G | R I V U X G | R I V U X G |

(Mercury: NASA; Venus: NASA/JPL, MIT, and USGS; Mars: NASA, J. Bell/Cornell University, and M. Wolff/SSI)

Mercury, Venus, and Mars: Earthlike yet Unique

These images of Mercury, Venus, and Mars show the three planets that share the inner solar system with our Earth. All three have solid surfaces, and, in principle, a properly protected astronaut could stand on any of them. The differences between these three worlds, however, are pronounced.

Mercury is small, airless, and extensively cratered. It is also mysterious: Mercury has a perplexing and unexpected magnetic field. It also rotates in a manner unique in the solar system, spinning 3 times on its axis for every two orbits around the Sun.

Venus is nearly the same size as Earth, but it is shrouded by a perpetual cloud cover that hides its surface from view. Unmanned spacecraft have used radar to map Venus at close range and even land on its surface. The false-color topographic map shown in the image reveals that Venus has no true continents but merely highlands (shown in red) that rise gently above the planet's lower-lying areas (shown in blue). We have also learned that Venus's atmosphere is thick and searing hot, and that its clouds contain droplets of corrosive sulfuric acid.

Mars has captivated the popular imagination like no other planet. But rather than being the abode of warlike aliens, Mars proves to be an enigmatic world. Some parts of its surface are drier than any desert on Earth, while other locations show evidence of having been underwater for extended periods. Our challenge is to understand how these three terrestrial planets evolved to be so unique and so different than Earth.

11-1 Mercury, Venus, and Mars can all be seen with the naked eye

At various times Mercury, Venus, and Mars are among the brightest objects in the sky. At their greatest brilliance, each of these planets appears brighter than any star, which is why they have been known since ancient times and why their motions played a role in the religious beliefs of many ancient cultures.

Mercury and Venus are *inferior* planets, orbiting closer to the Sun than Earth. As a result, Mercury and Venus never appear very far from the Sun in our sky (Figure 11-1). Table 11-1 and Table 11-2 summarize some basic information about these planets and their orbits.

> Viewing Mercury is difficult because it is so close to the Sun; Venus is easier to see but covered with clouds; Mars is best seen at opposition

We get the best view of either Mercury or Venus when it appears as far from the Sun in the sky as it can be, at its greatest eastern or western elongation (Section 4-2). Of course, these planets cannot be viewed with the naked eye unless the Sun is below the horizon. On dates near its **greatest eastern elongation,** Mercury or Venus appears after sunset over the western horizon as an "evening star." On dates near its **greatest western elongation,** Mercury or Venus appears as a "morning star" that rises before the Sun in the eastern sky.

Unlike Mercury and Venus, Mars is a *superior* planet, which means that Mars has a semimajor axis larger than Earth's

orbit. As Table 11-3 shows, the Martian orbital semimajor axis is $a = 1.524$ AU. Hence, Mars and the Sun are sometimes on opposite sides of the sky as seen from Earth—that is when Mars is at opposition (see Figure 4-6)—and Mars appears high in our night sky.

Because Mercury's orbit is so close to the Sun, its maximum elongation is only 28° and it is quite difficult to observe. However, the orbit of Venus is about twice the size of Mercury's orbit. As a result, Venus can be seen about 47° away from the Sun and is easy to observe.

11-2 While Mars rotates much like Earth, Mercury's rotation is coupled to its orbital motion and Venus's rotation is slow and retrograde

Mars is relatively easy to observe with visible-light telescopes, and by tracking patches of color, its rotation has been measured. By contrast, only a few faint, hazy features can be seen on Mercury's surface, and the surface of Venus

> Radar technology revealed the curious rotation of Mercury and Venus

is perpetually hidden by clouds. These differences explain why the rotation of Mars has been well understood since the mid-seventeenth century, while it was not until the 1960s that astronomers accurately measured the rotation of Mercury and Venus. As we

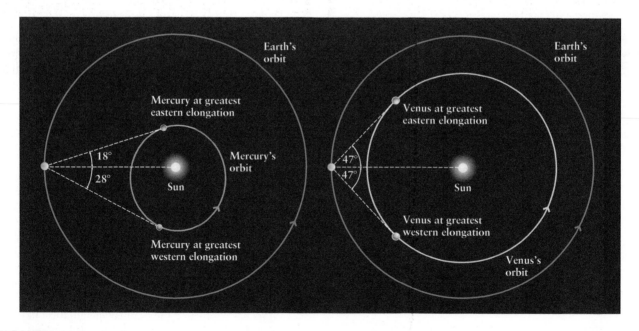

FIGURE 11-1

The Orbits of Mercury and Venus Mercury moves around the Sun every 88 days in a rather eccentric orbit. As seen from Earth, the angle between Mercury and the Sun at greatest eastern or western elongation can be as large as 28° (when Mercury is near aphelion) or as small as 18° (near perihelion). By contrast, Venus follows a larger, nearly circular orbit with a 224.7-day period. The angle between Venus and the Sun at eastern or western elongation is 47°.

TABLE 11-1 Mercury Data

Average distance from the Sun:	0.387 AU $= 5.79 \times 10^7$ km
Maximum distance from the Sun:	0.467 AU $= 6.98 \times 10^7$ km
Minimum distance from the Sun:	0.307 AU $= 4.60 \times 10^7$ km
Eccentricity of orbit:	0.206
Average orbital speed:	47.9 km/s
Orbital period:	87.969 days
Rotation period:	58.646 days
Inclination of equator to orbit:	0.5°
Inclination of orbit to ecliptic:	7° 00′ 16″
Diameter (equatorial):	4880 km = 0.383 Earth diameter
Mass:	3.302×10^{23} kg = 0.0553 Earth mass
Average density:	5430 kg/m^3
Escape speed:	4.3 km/s
Surface gravity (Earth = 1):	0.38
Albedo:	0.12
Average surface temperatures:	Day: 350°C = 662°F = 623 K Night: −170°C = −274°F = 103 K
Atmosphere:	**Essentially none**

R I **V** U X G

(NASA/Johns Hopkins U. Applied Physics Laboratory/Carnegie Institution of Washington)

will see, Mars rotates in a very Earthlike way, while Mercury and Venus rotate like no other objects in the solar system.

The Rotation of Mars

In 1659, the first reliable record of surface features on Mars was made by the Dutch scientist Christiaan Huygens, who observed a dark feature appear and disappear for several weeks. Huygens concluded that the rotation period of Mars—one day on Mars—is approximately 24 hours, surprisingly similar to a day on Earth.

Also like Earth, Mars's axis of rotation is not perpendicular to the plane of the planet's orbit, but is tilted by about 25° away from the perpendicular. This tilt is very close to Earth's 23½° tilt (see Figure 2-12). This striking coincidence means that Mars experiences Earthlike seasons, with opposite seasons in the northern and southern Martian hemispheres (see Section 2-5). Because Mars takes nearly 2 (Earth) years to orbit the Sun, the Martian seasons last nearly twice as long as on Earth.

The Challenge of Observing Mercury's Rotation

During the 1880s, the Italian astronomer Giovanni Schiaparelli attempted to make the first map of Mercury. Unfortunately, Schiaparelli's telescopic views of Mercury were so indistinct that he made a major error, which went uncorrected for more than half a century. He erroneously concluded that Mercury always keeps the same side facing the Sun. As we will see in Section 11-4, it was not Schiaparelli's last error.

Many objects in our solar system are in synchronous rotation, so that their rotation period equals their period of revolution—a situation also called **1-to-1 spin-orbit coupling.** We saw in Section 4-8 how Earth's tidal forces keep the Moon in synchronous rotation, so that it always keeps the same side toward Earth. Tidal forces also keep the two moons of Mars with the same side facing their parent planet, and likewise many of the satellites of Jupiter and Saturn.

It had been suggested as early as 1865 that the Sun's tidal forces would keep Mercury in synchronous rotation (resulting

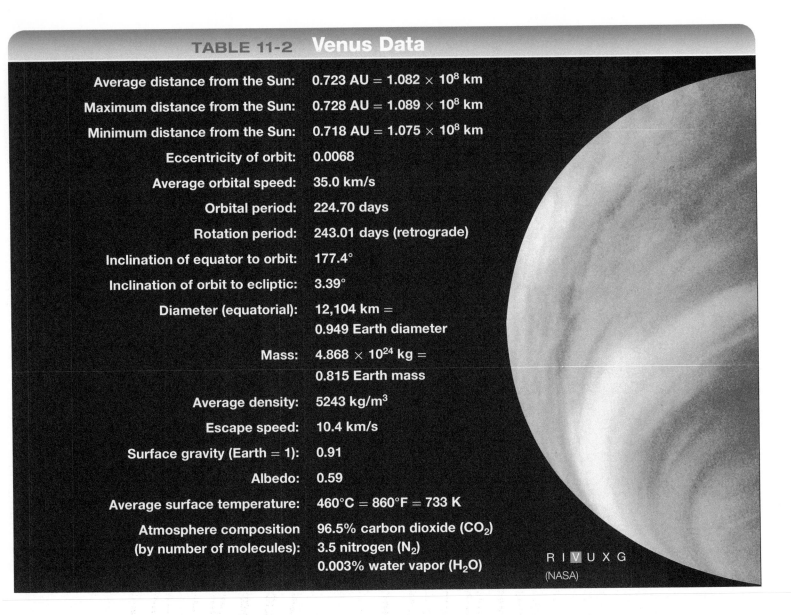

TABLE 11-2 Venus Data

Average distance from the Sun:	$0.723 \text{ AU} = 1.082 \times 10^8$ km
Maximum distance from the Sun:	$0.728 \text{ AU} = 1.089 \times 10^8$ km
Minimum distance from the Sun:	$0.718 \text{ AU} = 1.075 \times 10^8$ km
Eccentricity of orbit:	0.0068
Average orbital speed:	35.0 km/s
Orbital period:	224.70 days
Rotation period:	243.01 days (retrograde)
Inclination of equator to orbit:	177.4°
Inclination of orbit to ecliptic:	3.39°
Diameter (equatorial):	12,104 km = 0.949 Earth diameter
Mass:	4.868×10^{24} kg = 0.815 Earth mass
Average density:	5243 kg/m³
Escape speed:	10.4 km/s
Surface gravity (Earth = 1):	0.91
Albedo:	0.59
Average surface temperature:	460°C = 860°F = 733 K
Atmosphere composition (by number of molecules):	96.5% carbon dioxide (CO_2) 3.5 nitrogen (N_2) 0.003% water vapor (H_2O)

R I V U X G
(NASA)

in one side being perpetually illuminated by the Sun). But this is not how Mercury rotates. Because Mercury's features are undetectable in visible light, measuring the planet's rotation had to wait nearly a century for the development of advanced radar technology.

In 1965, Rolf B. Dyce and Gordon H. Pettengill used the giant 1000-ft radio telescope at the Arecibo Observatory in Puerto Rico to bounce powerful radar pulses off Mercury. The outgoing radiation consisted of microwaves of a very specific wavelength. In the reflected signal that echoed back from the planet, the wavelength had shifted as a result of the Doppler effect (see Section 5-9, especially Figure 5-26). As Mercury rotates, one side of the planet approaches Earth, while the other side recedes from Earth. Microwaves reflected from the planet's approaching side were shortened in wavelength, whereas those from its receding side were lengthened (Figure 11-2). By analyzing these variations in the Doppler-shifted microwaves, a rotation period of approximately 58.6 days was found for Mercury.

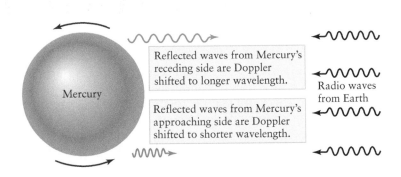

FIGURE 11-2

Measuring Mercury's Rotation Period As Mercury rotates, one side of the planet moves away from Earth and the other side toward Earth. If radio waves of a single wavelength are beamed toward Mercury, waves reflected from these two sides will be Doppler shifted to longer and shorter wavelengths, respectively. By measuring the wavelength shift of the reflected radiation, astronomers deduced how rapidly Mercury rotates and thus determined its rotation period.

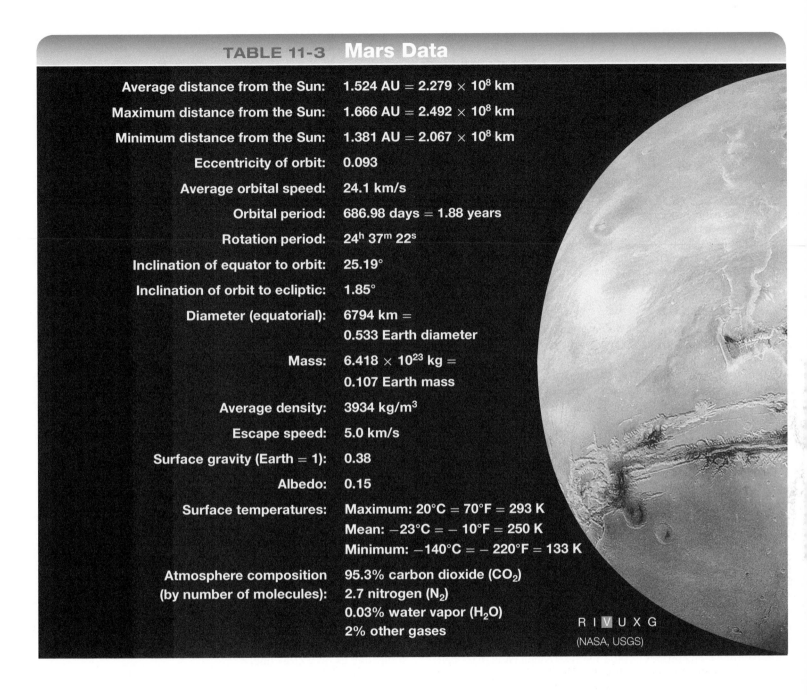

TABLE 11-3	Mars Data
Average distance from the Sun:	1.524 AU = 2.279×10^8 km
Maximum distance from the Sun:	1.666 AU = 2.492×10^8 km
Minimum distance from the Sun:	1.381 AU = 2.067×10^8 km
Eccentricity of orbit:	0.093
Average orbital speed:	24.1 km/s
Orbital period:	686.98 days = 1.88 years
Rotation period:	24^h 37^m 22^s
Inclination of equator to orbit:	25.19°
Inclination of orbit to ecliptic:	1.85°
Diameter (equatorial):	6794 km = 0.533 Earth diameter
Mass:	6.418×10^{23} kg = 0.107 Earth mass
Average density:	3934 kg/m³
Escape speed:	5.0 km/s
Surface gravity (Earth = 1):	0.38
Albedo:	0.15
Surface temperatures:	Maximum: 20°C = 70°F = 293 K Mean: −23°C = −10°F = 250 K Minimum: −140°C = −220°F = 133 K
Atmosphere composition (by number of molecules):	95.3% carbon dioxide (CO_2) 2.7 nitrogen (N_2) 0.03% water vapor (H_2O) 2% other gases

R I V U X G
(NASA, USGS)

Spin-Orbit Coupling and the Curious Rotation of Mercury

Giuseppe Colombo, an Italian physicist with a long-standing interest in Mercury, was intrigued by Dyce and Pettengill's results. Colombo noticed that their result of 58.6 days for the rotation period is very close to two-thirds of Mercury's accurately measured sidereal (orbital) period of 87.969 days:

$$\tfrac{2}{3} \, (87.969 \text{ days}) = 58.646 \text{ days}$$

Colombo therefore boldly speculated that Mercury's true rotation period is exactly 58.646 days, or 58 days and 15½ hours. He realized that this figure would mean that Mercury is locked into a **3-to-2 spin-orbit coupling:** The planet makes *three* complete rotations on its axis for every *two* complete orbits around the Sun.

No other planet or satellite in the solar system has this curious relationship between its rotation and its orbital motion.

Figure 11-3 shows how gravitational forces from the Sun cause Mercury's 3-to-2 spin-orbit coupling. The force of gravity decreases with increasing distance, which explains how the Moon is able to raise a tidal bulge in Earth's oceans (see Section 4-8). Mercury has no oceans, but it has a natural bulge of its own; thanks to the Sun's tidal forces, the planet is slightly elongated along one axis. The stronger gravitational force that the Sun exerts on the near side of the planet tends to twist the long axis to point toward the Sun, as Figure 11-3a shows. Indeed, Mercury's long axis would always point toward the Sun if its orbit were circular or nearly so. In this case the same side of Mercury would always face the Sun, and there would be synchronous rotation (Figure 11-3b). However, Mercury's orbit has a rather high eccentricity,

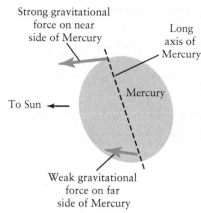

To Sun ←

Strong gravitational force on near side of Mercury

Long axis of Mercury

Mercury

Weak gravitational force on far side of Mercury

(a) Mercury is slightly elongated: The different gravitational forces on Mercury's two ends tend to twist the long axis to point toward the Sun.

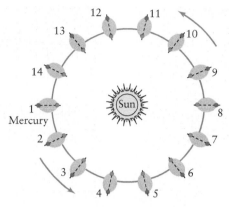

(b) If Mercury were in a circular orbit, its long axis would always point toward the Sun: Mercury would be in synchronous rotation (1-to-1 spin-orbit coupling).

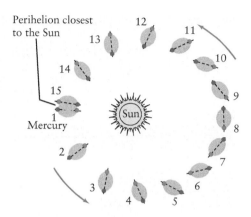

Perihelion closest to the Sun

(c) In fact Mercury is in an elliptical orbit, and its long axis only points toward the Sun at perihelion: Mercury spins on its axis 1½ times during each complete orbit (3-to-2 spin-orbit coupling).

FIGURE 11-3

Mercury's Spin-Orbit Coupling **(a)** The Sun's gravitational force on the near side of Mercury's long axis is greater than the force on the far side. This tends to twist the long axis to point toward the Sun. **(b)** If Mercury were in a circular orbit, the twisting effect shown in (a) would always keep the same side of Mercury (shown by a red dot) facing the Sun.

(c) Because Mercury's orbit is rather elongated, its rotation is more complex: The end of Mercury's long axis marked by a red dot faces the Sun at one perihelion (point 1), but at the next perihelion (point 15) the opposite end—marked by a blue dot—faces the Sun.

which leads to its 3-to-2 spin-orbit coupling. The high eccentricity leads to greater twisting forces at closer approaches, but the twisting effect on Mercury decreases rapidly as the planet moves away from the Sun. As a result, Mercury's long axis points toward the Sun only at perihelion (Figure 11-3c). From one perihelion to the next, alternate sides of Mercury face the Sun.

Because of the 3-to-2 spin-orbit coupling, the average time from sunrise to sunset on Mercury is just equal to its orbital period, or just under 88 days. This period helps explain the tremendous difference between daytime and nighttime temperatures on Mercury. Not only is sunlight about 7 times more intense on Mercury than on Earth (because of the smaller size of Mercury's orbit), but that sunlight has almost 3 Earth months to heat up the surface. As a result, daytime temperatures at the equator are high enough to *melt lead*—about 430°C (roughly 700 K, or 800°F). (In comparison, a typical kitchen oven reaches only about 232°C or 450°F.)

The time from sunset to the next sunrise is also 88 days, and the surface thus has almost 3 Earth months of darkness during which to cool down. Therefore, nighttime temperatures reach to below −170°C (about 100 K, or −270°F), which is cold enough to freeze carbon dioxide and methane. The surface of Mercury is truly inhospitable!

Observing Venus's Rotation: Penetrating the Clouds with Radio

Venus's perpetual cloud cover makes it impossible to measure the planet's rotation using visible-light telescopes. To penetrate the planet's perpetual cloud cover, astronomers applied to Venus the same technique of using radio waves and the Doppler effect that revealed the rotation of Mercury.

Clouds may contain gas, dust, haze, water droplets, or other small particles. In general, electromagnetic radiation can pass easily through such a cloud only if the wavelength is large compared to the size of the particles. We then say that the cloud is *transparent* to the radiation. For example, clouds in Earth's atmosphere are made of water droplets with an average diameter of about 20 μm (2×10^{-5} m, or 20,000 nm). Visible light has wavelengths between 400 to 700 nm, which is less than the droplet size. Hence, visible light cannot easily pass through such clouds, which is why cloudy days are darker than sunny days. But radio waves, with wavelengths of 0.1 m or more, and microwaves, with wavelengths from 10^{-3} m to 0.1 m, can pass through such a cloud with ease. So, your cell phone or radio works just as well on a cloudy or foggy day as on a clear one.

Like clouds in Earth's atmosphere, the clouds of Venus are transparent to radio waves and microwaves. Just as for Mercury (see Figure 11-2), the Doppler shift of waves reflected from the two sides of Venus reveal the speed and direction of the planet's rotation.

Venus: Where the Sun Rises in the West

These measurements show that Venus rotates very slowly; the sidereal rotation period of the planet is 243.01 days, even longer than the planet's 224.7 day orbital period. To an imaginary inhabitant of Venus who could somehow see through the cloud cover, stars would move across the sky at a rate of only 1½° per Earth day. As seen from Earth, by contrast, stars move 1½° across the sky in just 6 minutes.

Not only does Venus rotate slowly, it also rotates in an unusual direction. Most of the planets in the solar system spin on their axes

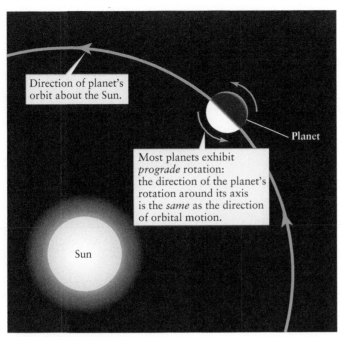

(a) Prograde rotation

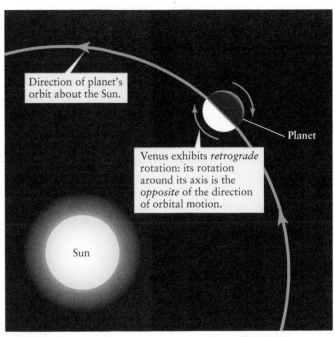

(b) Retrograde rotation

FIGURE 11-4

Prograde and Retrograde Rotation (a) If you could view the solar system from a point several astronomical units above Earth's north pole, you would see all of the planets orbiting the Sun in a counterclockwise direction. Most of the planets also rotate counterclockwise on their axes in prograde (forward) rotation. (b) Venus is an exception; from your perspective high above the plane of the solar system, you would see Venus rotating clockwise in retrograde (backward) rotation.

in **prograde rotation.** Prograde means "forward," and planets with prograde rotation spin on their axes in the same direction in which they orbit the Sun (**Figure 11-4a**). Venus is an exception to the rule: It spins in **retrograde rotation.** That is, the direction in which it spins on its axis is backward compared to the direction in which it orbits the Sun, as Figure 11-4b shows. (Recall that "retrograde" means "backward.")

CAUTION! Be careful not to confuse retrograde *rotation* with retrograde *motion*. When Venus is near inferior conjunction, so that it is passing Earth on its smaller, faster orbit around the Sun, it appears to us to move from east to west on the celestial sphere from one night to the next. This apparent motion is called retrograde motion, because it is backward to the apparent west-to-east motion that Venus displays most of the time. (You may want to review the discussion of retrograde motion in Sections 4-1 and 4-2.) By contrast, the retrograde rotation of Venus means that it spins on its axis from east to west, rather than west to east like Earth.

Venus's retrograde rotation adds to the puzzle of our solar system's origins. If you could view our solar system from a great distance above Earth's north pole, you would see all the planets orbiting the Sun counterclockwise. Closer examination would reveal that the Sun and most of the planets also *rotate* counterclockwise on their axes; the exceptions are Venus and Uranus. Most of the satellites of the planets also move counterclockwise along their orbits and rotate in the same direction. Thus, the rotation of Venus is not just opposite to its orbital motion; it is opposite to most of the orbital and rotational motions in the solar system!

As we saw in Section 8-4, the Sun and planets formed from a rotating solar nebula. Because the planets formed from the material of this rotating cloud, they tend to orbit and rotate in that same direction. It is difficult to imagine how Venus and Uranus could have bucked this trend. One theory suggests that a huge impact billions of years ago reversed Venus's direction of rotation.

CONCEPTCHECK 11-1

If we build bigger visible light telescopes, will astronomers finally be able to see the surface of Venus?

Answer appears at the end of the chapter.

11-3 Mercury is cratered like the Moon but has a surprising magnetic field

Imagine an alien astronomer observing Earth from Mercury, with the same telescope technology available to our own astronomers. Such an astronomer would be unable to resolve any features on Earth less than a few hundred kilometers across. It would be impossible for her to learn about Earth's

> Spacecraft observations at close range were necessary to reveal Mercury's unusual properties

mountain ranges or volcanoes. Earth-based astronomers face the same limitations in studying Mercury. To truly understand Mercury, it is essential to study this world at close range using spacecraft.

Mercury Spacecraft: *Mariner 10* and *MESSENGER*

 The first spacecraft to provide some knowledge about Mercury was *Mariner 10,* which flew close to Mercury on three occasions in 1974 and 1975. Two brief flybys in 2008 produced the first images of Mercury from the spacecraft *MESSENGER* (MErcury Surface, Space ENvironment, GEochemistry, and Ranging) before it settled into orbit in 2011. *MESSENGER* can distinguish features as small as 18 m—a vast improvement over *Mariner 10*'s 1-km resolution. Figure 11-5 shows a real-color image of Mercury from *MESSENGER.*

Mercury's Cratered Surface

As *Mariner 10* closed in on Mercury, scientists were struck by the Moonlike pictures appearing on their television screens. It was obvious that Mercury, like the Moon, is a barren, heavily cratered world, with no evidence for plate tectonics. These surface features are what we would expect based on Mercury's small size (see Table 7-1 and the figure that opens this chapter). As we saw in Section 7-6, a small world is expected to have little internal heat, and, hence, little or no geologic activity to erase an ancient,

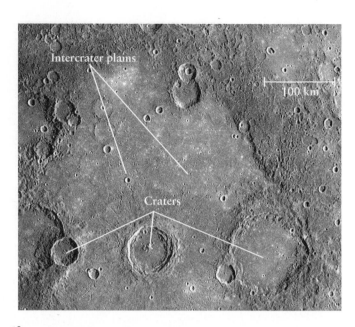

FIGURE 11-6　R I $\boxed{V}$ U X G
Mercurian Craters and Plains *MESSENGER* took this true-color image of a region near Mercury's equator at a range of 2800 km (1700 mi). Lava flows formed the plains inside and to the west of the crater on the right. (NASA/Johns Hopkins U. Applied Physics Laboratory/Carnegie Institution of Washington)

cratered surface (see Figure 7-11). But craters are not the only features of Mercury's surface; there are also gently rolling plains, long, meandering cliffs, and an unusual sort of jumbled terrain. As we will see, Mercury's distinctive surface shows that this planet is *not* merely a somewhat larger version of the Moon.

Figure 11-6 shows a typical close-up view of Mercury's surface. The consensus among astronomers is that most of the craters on both Mercury and the Moon were produced during the first 700 million years after the planets formed. Debris remaining after planet formation rained down on these young worlds, gouging out most of the craters we see today. We saw in Section 10-4 that the strongest evidence for this chronology comes from analysis and dating of Moon rocks. No spacecraft has landed on Mercury, so we are not able to make the same kind of direct analysis of rocks from the planet's surface.

Scarps: Evidence of a Shrunken Planet

Images also reveal numerous long cliffs, called **scarps,** meandering across Mercury's surface (Figure 11-7). Some scarps rise as much as 3 km (2 mi) above the surrounding plains and are 20 to 500 km long. These cliffs probably formed as Mercury cooled and contracted a few kilometers in size. Scarps are analogous to wrinkles on fruit—as the fruit dries out and shrinks in size, its surface is deformed. Just as we saw on the Moon (Figure 10-8), when a scarp cuts through a crater, it means that the scarp formed after the crater. Eventually, using age estimates related to craters, astronomers hope to date the scarps of Mercury. Some scarps on the Moon indicate very recent geologic activity: will Mercury

FIGURE 11-5　R I $\boxed{V}$ U X G
MESSENGER **Image of Mercury** This real-color image shows the natural gray color of Mercury's surface. Much like the Moon, Mercury's surface contains many craters, but none of the Moon's dark maria. The lack of darker maria is consistent with a low iron content measured on Mercury's surface. (NASA)

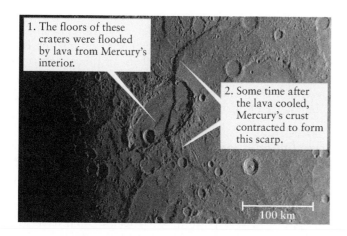

1. The floors of these craters were flooded by lava from Mercury's interior.

2. Some time after the lava cooled, Mercury's crust contracted to form this scarp.

100 km

FIGURE 11-7 R I ☑ U X G

ANIMATION 11-6

A Scarp A long cliff, or scarp, called Beagle Rupes runs across this *MESSENGER* image. This scrap is almost a kilometer high and extends for more than 600 km (370 mi) across a region near Mercury's equator. (NASA/ Johns Hopkins U. Applied Physics Laboratory/Carnegie Institution of Washington)

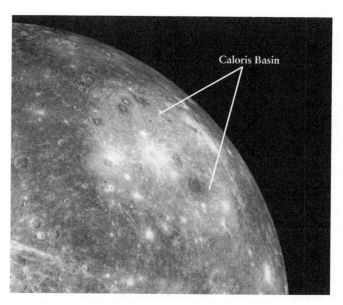

Caloris Basin

FIGURE 11-8 R I ☑ U X G

The Caloris Basin About 4 billion years ago an impact on Mercury created an immense basin, shown in orange in this enhanced-color *MESSENGER* image. The crater is about 1500 km (960 mi) in diameter. The impact fractured the surface extensively, forming several concentric chains of mountains. The mountains in the outermost ring are up to 2 km (6500 ft) high. Craters inside the basin (shown in blue) were formed later by smaller impacts. (NASA/Johns Hopkins U. Applied Physics Laboratory/Arizona State U./ Carnegie Institution of Washington)

reveal a similar surprise? Even with the scarps, there are no features on Mercury that resemble the boundaries of tectonic plates. Thus, we can regard Mercury's crust, like that of the Moon, as a single plate.

Fire and Ice: The Caloris Basin and Mercury's Poles

The most impressive feature discovered by *Mariner 10* was a huge impact scar called the Caloris Basin (*calor* is the Latin word for "heat"). The Sun is directly over the Caloris Basin during alternating perihelion passages, making it the hottest place on the planet once every 176 days. In Figure 11-8, enhanced-colors help to identify different chemical compositions present on the surface of the Caloris Basin. The human eye would see mostly gray, as seen in the other Mercury images. With enhanced coloring, bright orange spots near the rim of the basin are visible. These spots are thought to indicate volcanic material, while the blue areas suggest an above-average titanium oxide content. Material that appears blue is thought to have originated a few kilometers below the surface and was exposed through impact events.

The Caloris Basin, which measures 1500 km (960 mi) in diameter, is both filled with and surrounded by smooth lava plains. The basin was probably gouged out by the impact of a large meteoroid that penetrated the planet's crust, allowing lava to flood out onto the surface and fill the basin. Because fewer craters pockmark these lava flows than elsewhere on the surface, the Caloris impact must be relatively young. It must have occurred toward the end of the crater-making period—the Late Heavy Bombardment—that dominated the first 700 million years of our solar system's history.

The impact that created the huge Caloris Basin must have been so violent that it shook the entire planet. On the side of Mercury *opposite* the Caloris Basin, *Mariner 10* discovered a possible consequence of this impact—a jumbled, hilly region covering about 500,000 square kilometers, about twice the size of the state of Wyoming. Geologists theorize that seismic waves from the Caloris

impact became focused as they passed through Mercury. Perhaps as this concentrated seismic energy reached the opposite side of the planet, called the **antipode**, it deformed the surface and created jumbled terrain. This same process may also have taken place on the Moon: There is a region of chaotic hills on the side of the Moon directly opposite from the large impact basin called Mare Orientale. What about antipodal effects on Earth? Detailed calculations of Earth's dinosaur-killing asteroid impact (Chapter 15) indicate that the antipodal region, on the opposite side of Earth, would have been raised 3 to 5 meters!

Even though temperatures at Mercury's equator are high enough to melt lead, there is the possibility of water-ice at the poles. In 1991, radar waves that bounced from the surface showed an unexpectedly high reflectivity near the poles, and ice is highly reflective. Further analysis showed that the radar reflections came precisely from permanent shadows within craters near the pole, which are well below the freezing point of water. This idea gained even more credibility in 2010 with the discovery of water in a permanently shadowed crater on the Moon (Section 10-2).

Finally, in 2012, the *MESSENGER* spacecraft confirmed water-ice by detecting excess hydrogen in these shaded craters, where the hydrogen is part of water molecules. Analysis of hydrogen concentrations indicates a layer of almost pure water-ice more than 30 cm thick beneath a layer about 15 cm thick containing little water. Some astronomers speculate that the water was delivered by comets, but the source is unknown.

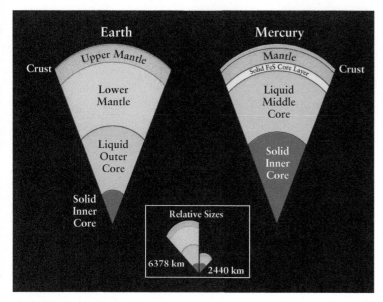

FIGURE 11-9

The Internal Structures of Mercury and Earth Compared to Earth's core, Mercury's iron-rich core takes up a much larger fraction of the planet's volume. Indeed, Mercury is the most iron-rich planet in the solar system. Mercury's iron core covers 83% of its radius and 68% of its volume, whereas Earth's core is only 55% of its radius and 17% of its volume. (NASA)

Mercury's Interior and Iron Core

By measuring how much *Mariner 10* was deflected by Mercury's gravity, scientists were able to make a precise determination of Mercury's mass. Given the mass and the diameter of Mercury, they could then calculate that the planet has an average density of 5430 kg/m³. This value is quite similar to Earth's average density of 5515 kg/m³. However, as shown in **Figure 11-9**, a much larger fraction of Mercury is composed of iron.

CAUTION! You might wonder how Mercury and Earth can have similar densities while Mercury's iron fraction is twice as large as Earth's. In Earth, which is 18 times more massive than Mercury, lighter material is compressed by gravity to create an overall high density. Mercury gets its own high density from its abundance of intrinsically heavy iron.

Several theories have been proposed to account for Mercury's high iron content. According to one theory, the inner regions of the primordial solar nebula were so warm that only those substances with high condensation temperatures—like iron-rich minerals—could have condensed into solids. Another theory suggests that a brief episode of very powerful solar winds could have stripped Mercury of its low-density mantle shortly after the Sun formed. A third possibility is that during the final stages of planet formation, Mercury was struck by a large planetesimal. Supercomputer simulations show that this cataclysmic collision would have ejected much of the lighter mantle, leaving a disproportionate amount of iron to reaccumulate to form the planet we see today (**Figure 11-10**).

Clues About the Core: Mercury's Magnetic Field

An important clue to the structure of Mercury's iron core came from the *Mariner 10* magnetometers, which discovered that Mercury has a magnetic field similar to that of Earth but only about 1% as strong. *MESSENGER* has confirmed the presence of a global magnetic field, which indicates that Mercury contains moving liquid material in its interior that conducts electricity. As in Earth, this moving material would act as a dynamo to generate the magnetic field. Hence, we conclude that at least part of Mercury's core must be in a liquid state.

Mercury's magnetic field comes as something of a surprise. The ancient, cratered surface of Mercury is evidence for a lack of geologic activity (that erases craters), which in turn shows that Mercury must have lost the internal heat that powers such activity. Again, we would expect that such a small planet would lose

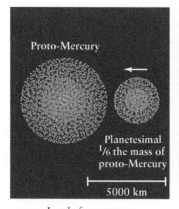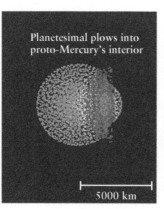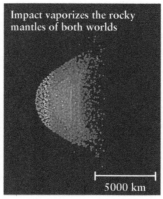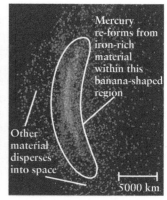

FIGURE 11-10

Stripping Mercury's Mantle by a Collision To account for Mercury's high iron content, one theory proposes that a collision with a planet-sized object stripped Mercury of most of its rocky mantle. These four images show a supercomputer simulation of a head-on collision between proto-Mercury and a planetesimal one-sixth its mass. (Courtesy of W. Benz, A. G. W. Cameron, and W. Slattery)

much of its internal heat (see Section 7-6). But the existence of a global magnetic field shows that Mercury is still partially molten inside, which is evidence that the planet has some internal heat. Thus, Mercury's small size and dense cratering are in contradiction to its global magnetic field.

Another mystery is that a planet must rotate fairly rapidly in order to stir its molten interior into the kind of motion that generates a magnetic field. But as we saw in Section 11-2, Mercury rotates very *slowly*. It takes 58.646 days to spin on its axis. We still have much to learn about the interior of Mercury.

Mercury's Unsolved Mysteries

A very unusual and unexplained property of Mercury's magnetic field is that it does not originate at the planet's center. Instead, *MESSENGER* found that Mercury's magnetic field originates about 488 km north of the planet's center—an offset equal to a whopping 20% of the planet's radius! No other planet or moon has shown such a large offset and this offset is currently one of Mercury's greatest mysteries. Yet, another related mystery is perhaps the most perplexing of all.

Mercury presumably formed close to the Sun under high temperatures. (The early migration expected in our solar system was mostly by the Jovian planets.) It is generally found that under higher temperatures, certain atomic elements—referred to as **volatile elements**—"boil off" out of planetary material. We saw in Section 10-5 that the Moon has fewer volatile elements than Earth, just as would be expected if the Moon was created from a fiery impact with Earth. However, *Mercury has an unexpectedly high abundance of volatile elements,* as if it formed at lower temperatures.

Not only is Mercury's high volatile content a problem in trying to explain its formation so close to the hot Sun, but it also poses a challenge to subsequent large impacts with Mercury. For example, we saw that one explanation for Mercury's large iron core is that a large impact blasted away some of the lighter outer material (Figure 11-10), but this hot impact event is inconsistent with a high volatile content; volatile elements would be expected to cook out of the remaining surface material, and then evaporate into space. It is also tempting to propose that a large impact created Mercury's off-center magnetic field, but this too runs against the observed high abundance of volatile elements on Mercury's surface.

Simply put, Mercury holds unsolved mysteries that are certain to keep astronomers busy. While additional data from *MESSENGER* will help to answer some questions, it will certainly open up new questions as well.

CONCEPT CHECK 11-2

Why is it thought that Caloris Basin would have formed near the end of the Late Heavy Bombardment?

CONCEPT CHECK 11-3

In Figure 11-10, rocky mantle is vaporized about 6 minutes after Mercury is hit by a large planetesimal. Would this be more consistent with a low or high abundance of volatile elements on Mercury?

Answers appear at the end of the chapter.

11-4 The first missions to Venus and Mars demolished decades of speculations about those planets

Prior to the 1960s, scientists developed a number of intriguing ideas about Venus and Mars. While these ideas were consistent with observations made using Earth-based telescopes, they needed to be radically revised with the advent of spacecraft data.

Speculations About Venus: A Lush Tropical Paradise?

Venus is bathed in more intense sunlight than Earth because it is closer to the Sun. If Venus had no atmosphere, and if its surface had an albedo like that of Mercury or the Moon, heating by the Sun would bring its surface temperature to around 45°C (113°F)—comparable to that found in the hottest regions on Earth. What was unclear until the 1960s was how much effect Venus's atmosphere has on the planet's surface temperature.

> Venus was once thought to be steamy and tropical, and Mars was thought to have vegetation that varied with the seasons

We saw in Section 9-1 that gases such as water vapor (H_2O) and carbon dioxide (CO_2) in Earth's atmosphere trap some of the infrared radiation that is emitted from our planet's surface. This trapped radiation elevates the temperatures of both the atmosphere and its surface, a phenomenon called the greenhouse effect. In 1932, Walter S. Adams and Theodore Dunham Jr. at the Mount Wilson Observatory found absorption lines of CO_2 in the spectrum of sunlight reflected from Venus, indicating that carbon dioxide is also present in the Venusian atmosphere. (Section 7-3 describes how astronomers use spectroscopy to determine the chemical compositions of atmospheres.) Thus, Venus's surface, like Earth's, should be warmed by the greenhouse effect.

However, Venus's perpetual cloud cover reflects back into space a substantial amount of the solar energy reaching the planet. This effect by itself acts to cool Venus's surface, just as clouds in our atmosphere can make overcast days cooler than sunny days. Some scientists suspected that this cooling kept the greenhouse effect in check, leaving Venus with surface temperatures below the boiling point of water. Venus might then have warm oceans and possibly even life in the form of tropical vegetation. But if the greenhouse effect were strong enough, surface temperatures could be so high that any liquid water would boil away. Then Venus would have a dry, desertlike surface with no oceans, lakes, or rivers.

Close-Up Observations of Venus: Revealing a Broiled Planet

This scientific controversy was resolved in 1962 when the unmanned U.S. spacecraft *Mariner 2* (Figure 11-11) made the first close flyby of Venus. As we learned in Sections 5-3 and 5-4, a dense object (like a planet's surface) emits radiation whose intensity and spectrum depend on the temperature of the object. *Mariner 2* carried instruments that measured radiation coming from the planet at two microwave wavelengths, 1.35 cm and 1.9 cm. Venus's atmosphere is transparent to both these wavelengths, so *Mariner 2* saw radiation that had been emitted by the

FIGURE 11-11

The *Mariner 2* Spacecraft *Mariner 2* was the first of all spacecraft from Earth to make a successful flyby of another planet. After coming within 34,773 km of Venus on December 14, 1962, it went into a permanent orbit around the Sun. (GSFC/NASA)

planet's surface and then passed through the atmosphere like visible light through glass.

From the amount of radiation that *Mariner 2* detected at these two wavelengths, astronomers concluded that the surface temperature of Venus was more than 400°C—well above the boiling point of water and even higher than the daytime temperatures on Mercury. Thus, there cannot be any liquid water on the planet's surface. Water vapor absorbs microwaves at 1.35 cm, so if there were substantial amounts of water vapor in Venus's atmosphere, it would have blocked this wavelength from reaching the detectors on board *Mariner 2*. In fact, the spacecraft did detect strong 1.35-cm emissions from Venus. Thus, the atmosphere, too, must be all but devoid of water.

This picture of Venus as a dry, hellishly hot world was confirmed in 1970 by the Soviet spacecraft *Venera 7*, which survived a descent through the Venusian atmosphere and managed to transmit data for a few seconds directly from the planet's surface. *Venera 7* and other successful Soviet landers during the early 1970s finally determined the surface temperature to be a nearly constant 460°C (860°F). Thus, Venus has an extraordinarily strong greenhouse effect, and the planet's surface is no one's idea of a tropical paradise.

Speculations About Mars: A World with Plant Life and Canals?

As we saw in Section 11-2, Mars's axis of rotation is tilted by almost the same angle as Earth's axis, and so Mars goes through seasons much like those on Earth. Perhaps biased by expectations of Earthlike conditions, in 1877 the Italian astronomer Giovanni

Schiaparelli said he was able to see 40 straight-line features crisscrossing the Martian surface (Figure 11-12). He called these dark linear features *canali,* an Italian word for "channels," which was soon mistranslated into English as "canals." The alleged discovery of canals implied that there were intelligent creatures on Mars capable of substantial engineering feats. This speculation helped motivate the American millionaire Percival Lowell to finance a major new observatory near Flagstaff, Arizona, primarily to study Mars. By the end of the nineteenth century, Lowell had reported observations of 160 Martian canals.

Although many astronomers observed Mars, not all saw the canals. In 1894, the American astronomer Edward Barnard, working at Lick Observatory in California, complained that "to save my soul I can't believe in the canals as Schiaparelli draws them." But objections by Barnard and others did little to sway the proponents of the canals.

As the nineteenth century drew to a close, speculation about Mars grew more and more fanciful. Perhaps the reddish-brown color of the planet meant that Mars was a desert world, and perhaps the Martian canals were an enormous planetwide irrigation network. From these ideas, it was a small leap to envision Mars as a dying world with canals carrying scarce water from melting polar

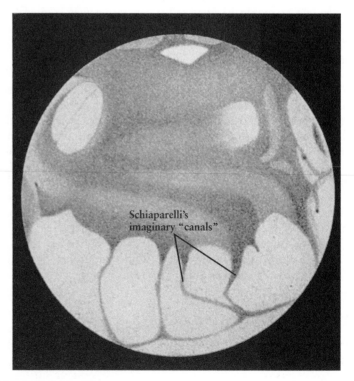

FIGURE 11-12

The Mirage of the Martian Canals Giovanni Schiaparelli studied the Martian surface using a 20-cm (8-in.) telescope, the same size used by many amateur astronomers today. He recorded his observations in drawings like this one, which shows a network of linear features that Percival Lowell and others interpreted as irrigation canals. Later observations with larger telescopes revealed that the "canals" were mere illusions. (© G.V. Schiaparelli)

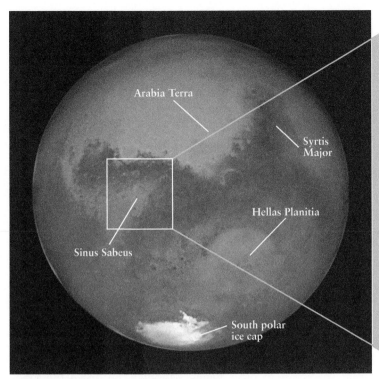

(a) Mars from the Hubble Space Telescope

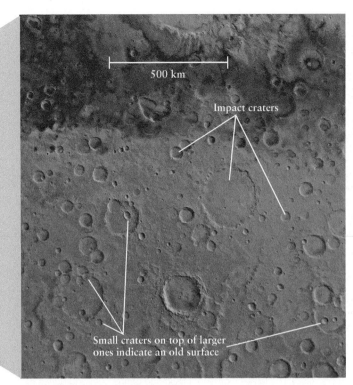

(b) Closeup of Sinus Sabeus region

FIGURE 11-13   R I ☒ **U X G**

Martian Craters (a) This image, made during the favorable opposition of 2003, shows cratered regions on the side of Mars opposite to that in the image that opens this chapter. Arabia Terra is dotted with numerous flat-bottom craters. Syrtis Major was first identified by Christiaan Huygens in 1659.

A single titanic impact carved out Hellas Planitia, which is 5 times the size of Texas. **(b)** This mosaic of images from the *Viking Orbiter 1* and *2* spacecraft shows an extensively cratered region located south of the Martian equator. (a: NASA; J. Bell, Cornell University; and M. Wolff, SSI; b: USGS)

caps to farmlands near the equator. And of special concern, some suggested that the Martians might be a warlike race who schemed to invade Earth for its abundant resources. (In Roman mythology, Mars was the god of war.) Stories of Martian invasions, from H. G. Wells's 1898 novel *The War of the Worlds* down to the present day, all owe their existence to the *canali* of Schiaparelli.

Observations of Mars: Revealing a Windblown, Cratered World

Another feature of Mars that gained attention is that its surface changes colors on a seasonal cycle. This observation lead to claims that Mars was covered in vegetation that varied season-ally, as happens on Earth. In the 1960s, three American spacecraft flew past Mars and sent back the first close-up pictures of the planet's surface. (Figure 11-13 shows even better images from a subsequent mission to Mars.) These images showed that the dark surface markings are just different-colored terrain.

The seasonal variations in color can be explained by winds in the Martian atmosphere that blow in different directions in different seasons. These seasonal winds blow fine dust across the Martian surface, producing planetwide dust storms that can be seen from Earth (Figure 11-14). The motion of dust covers some parts of the Martian terrain and exposes others. Thus, Earth

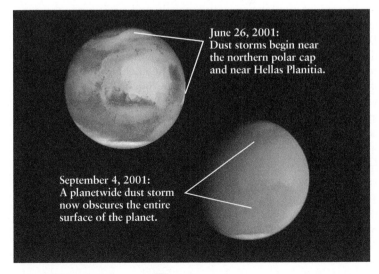

FIGURE 11-14   R I ☒ **U X G**

A Martian Dust Storm Dust storms on Mars can expand to cover the entire planet. These Hubble Space Telescope images show a particularly intense storm that took place in 2001, when it was spring in the Martian northern hemisphere and autumn in the southern hemisphere. (NASA; J. Bell, Cornell University; M. Wolff, SSI; and the Hubble Heritage Team, STScI/AURA)

FIGURE 11-15   R I <u>V</u> U X G

Victoria Crater This crater is about 800 meters wide (half a mile) and about 50 meters deep. The false-coloring of this image brings out sand dunes at the bottom. The robotic *Opportunity* rover just arrived when this image was taken, barely visible as a speck indicated by the arrow. This image was made by the HiRISE camera aboard the *Mars Reconnaissance Orbiter*. (NASA)

observers see large areas of the planet vary from dark to light with the passage of the seasons, mimicking the color changes that would be expected from vegetation.

Instead of showing irrigation canals and vegetation, the surface of Mars is dry and pockmarked with craters (Figure 11-13b). During the 1960s, scientists were learning that most of the craters on the Moon were formed by interplanetary debris that struck the lunar surface more than 3 billion years ago. The Martian craters, too, are the result of long-ago impacts from space. Since so many craters survive to the present day, at least part of the Martian surface must be extremely ancient (see Section 7-6). The spectacular Victoria Crater is shown in **Figure 11-15**, with sand dunes at its bottom.

CONCEPTCHECK 11-4

Microwaves are emitted from Venus's hot surface. What do 1.35-cm microwaves tell us about the presence of water on the surface of Venus and in its thick atmosphere?

Answer appears at the end of the chapter.

11-5 Both Venus and Mars have volcanoes— and Mars has signs of ancient plate tectonics

From a distance, Venus and Mars appear radically different: Venus is nearly the size of Earth and has a thick atmosphere, while Mars is much smaller and has only a thin atmosphere.

But spacecraft observations reveal that these two worlds have many surface features in common, including volcanoes and impact craters. Mars even shows some evidence for ancient plate tectonic activity. By comparing the surfaces of these two worlds to Earth we can gain insight into the similarities and differences between the three largest terrestrial planets.

> Venus has no plate tectonics because its crust is too thin, while on Mars the crust is too thick

Observing Venus and Mars from Orbit

To make a detailed study of a planet's surface, a spacecraft that simply flies by the planet will not suffice. Instead, it is necessary to place a spacecraft in orbit around the planet. Since the 1970s, a number of spacecraft have been placed in orbit around Venus and Mars.

In order to map the surface of Venus through the perpetual cloud layer, several of the Venus orbiters carried radar devices. A beam of microwave radiation from the orbiter easily penetrates Venus's clouds and reflects off the planet's surface; a receiver on the orbiter then detects the reflected beam. By measuring the time it takes for reflected waves to return to the orbiter, astronomers can determine the height and depth of Venus's terrain. The most recent spacecraft to orbit Venus, *Magellan*, produced the topographic map in **Figure 11-16**. The same radar method can be used on Mars, and the topographic map of Mars in **Figure 11-17** was produced by the *Mars Global Surveyor* spacecraft (which entered Mars's orbit in 1997).

The topographies of Mars and Venus differ in important ways from that of our Earth. Our planet has two broad classes of terrain: About 71% of Earth's surface is oceanic crust and about 27% is continental crust that rises above the ocean floors by about 4 to 6 km on average. On Venus, by contrast, about 60% of the terrain lies within 500 m of the average elevation, with only a few localized highlands (shown in yellow and red in Figure 11-16). Mars is different from both Earth and Venus: Rather than having elevated continents scattered among low-lying ocean floors, all of the high terrain on Mars (shown in red and orange in Figure 11-17) is in the southern hemisphere. Hence, planetary scientists refer to Mars as having **northern lowlands** and **southern highlands.** The implication is that the Martian crust is about 5 km thicker in the southern hemisphere than in the northern hemisphere, a situation called the **crustal dichotomy.** In this sense Mars resembles the Moon, which has a thicker crust on the far side than on the side that faces Earth (see Section 10-1).

CONCEPTCHECK 11-5

In Figure 11-17 the northern hemisphere is in the top half of the map, with the southern hemisphere in the bottom half. Which hemisphere has the youngest surface, and why?

Answer appears at the end of the chapter.

Tectonics on Venus: A Light, Flaky Crust

Before radar maps like Figure 11-16 were available, scientists wondered whether Venus had plate tectonics like those that have remolded the face of Earth. Venus is only slightly smaller than Earth and should have retained enough heat to sustain a molten

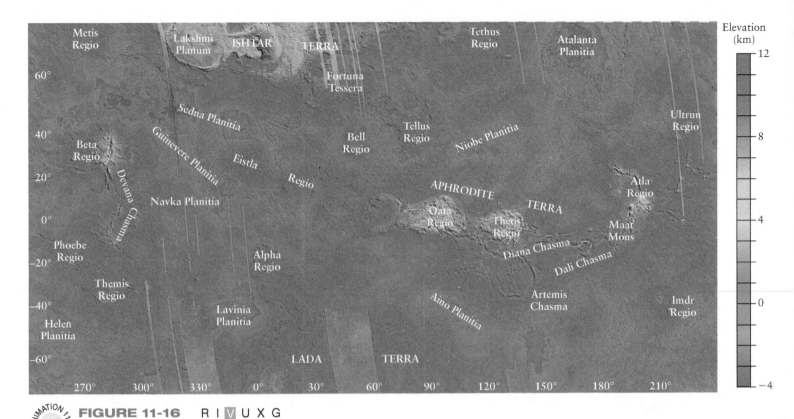

ANIMATION 11-7 **FIGURE 11-16** R I **V** U X G

A Topographic Map of Venus Radar altimeter measurements by *Magellan* were used to produce this topographic map of Venus. Color indicates elevations above (positive numbers) or below (negative numbers) the planet's average radius. (The blue areas are not oceans!) Gray areas were not mapped by *Magellan*. Flat plains of volcanic origin cover most of the planet's surface, with only a few continentlike highlands. (Peter Ford, MIT; NASA/JPL)

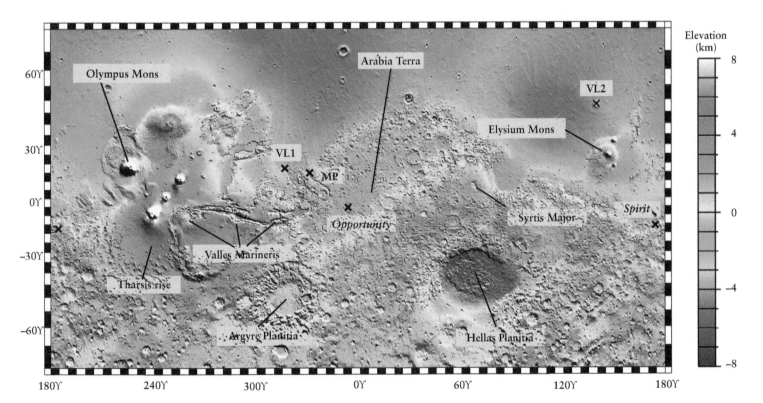

FIGURE 11-17 R I **V** U X G

A Topographic Map of Mars This map was generated from measurements made by the laser altimeter on board the *Mars Global Surveyor* spacecraft. As in Figure 11-15, color indicates elevations above or below the planet's average radius. Most of the southern hemisphere lies several kilometers above the northern hemisphere, with the exception of the immense impact feature called Hellas Planitia. The landing sites for *Viking Landers 1* and *2* (VL1 and VL2), *Mars Pathfinder* (MP), and the *Mars Exploration Rovers* (*Spirit* and *Opportunity*) are each marked with an X. (MOLA Science Team, NASA/GSFC)

interior and the convection currents that drive tectonic activity on Earth (see Section 9-3, especially Figure 9-15). If this were the case, then the same tectonic effects might also have shaped the surface of Venus. As we saw in Section 9-3, Earth's hard outer shell, or lithosphere, is broken into about a dozen large plates that move slowly across the globe.

Radar images from *Magellan* show *no* evidence of Earthlike plate tectonics on Venus. On Earth, long chains of volcanic mountains (like the Cascades in North America or the Andes in South America) form along plate boundaries where subduction is taking place. Mountainous features on Venus, by contrast, do not appear in chains. There are also no structures like Earth's Mid-Atlantic Ridge (Figure 9-13), which suggests that there is no seafloor spreading on Venus. With no subduction or seafloor spreading, there has been only limited horizontal displacement of Venus's lithosphere. Thus, like the Moon (see Section 10-1) and Mercury (see Section 11-3), Venus has a one-plate crust.

Unlike the Moon, however, Venus has had local, small-scale deformations and reshaping of the surface. One piece of evidence for this is that roughly a fifth of Venus's surface is covered by folded and faulted ridges. Further evidence comes from close-up *Magellan* images that show that Venus has about a thousand craters larger than a few kilometers in diameter, many more than have been found on Earth but only a small fraction of the number on the Moon or Mercury. We saw in Section 7-6 that the number of impact craters is a clue to the age of a planet's surface. Such impact craters formed at a rapid rate during the early history of the solar system, when considerable interplanetary debris still orbited the Sun, and have formed at a much slower rate since then. Consequently, the more craters a planet has, the older its surface. The number of craters on Venus indicates that the Venusian surface is roughly 500 million years old. This is about twice the age of Earth's surface but much younger than the surfaces of the Moon or Mercury, each of which is billions of years old. No doubt Venus was more heavily cratered in its youth, but localized activity in its crust has erased the older craters (Figure 11-18).

Surprisingly, Venus's craters are uniformly scattered across the planet's surface. We would expect that older regions on the surface—which have been exposed to bombardment for a longer time—would be more heavily cratered, while younger regions would be relatively free of craters. For example, the ancient highlands on the Moon are much more heavily cratered than the younger maria (see Section 10-4). Because such variations are not found on Venus, scientists conclude that the entire surface of the planet has essentially the *same* age. This is very different from Earth, where geological formations of widely different ages can be found.

One model that can explain these features suggests that the convection currents in Venus's interior are actually more vigorous than inside Earth, but that the Venusian crust is much thinner than the continental crust on Earth. Rather than sliding around like the plates of Earth's crust, the thin Venusian crust stays in roughly the same place but undergoes wrinkling and flaking (Figure 11-19). Hence, this model is called **flake tectonics**. Earth, too, may have displayed flake tectonics billions of years ago when its interior was hotter.

Although Venus almost certainly has molten material in its interior, it has no planetwide magnetic field. As we discussed in Section 7-7, Venus may have no magnetic field because it rotates too slowly to generate the kind of internal

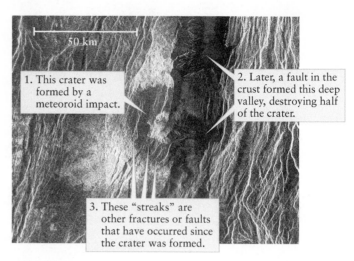

FIGURE 11-18   R I $\boxed{\text{V}}$ U X G

A Partially Obliterated Crater on Venus This *Magellan* image shows how the right half of an old impact crater 37 km (20 mi) in diameter was erased by a fault in the crust. This crater lies in Beta Regio (see the left side of Figure 11-16). Most features on Venus are named for women in history and legend; this crater commemorates Emily Greene Balch, an American economist and sociologist who won the 1946 Nobel Peace Prize. (NASA/JPL)

motions that would produce a magnetic field. With no such field, Venus has no magnetosphere. The planet is nonetheless shielded from the solar wind by ions in its upper atmosphere.

CONCEPTCHECK 11-6

How is it determined that Venus's surface is older than Earth's surface, yet younger than the surface of Mercury?

Answer appears at the end of the chapter.

Plate Tectonics on Mars

Like Venus, Mars lacks the global network of ridges and subduction zones like those produced by the seven major plates on Earth. However, evidence has been discovered for the existence of two plates on Mars associated with the 4000-km-long canyon named Valles Marineris (Figure 11-20). While this enormous chasm was first imaged in 1971, it was not until 2012 that astronomers realized the two sides of this canyon are actually plates that have slid past each other horizontally by 93 miles. So far, this is the only known plate boundary on Mars, giving Mars a total of two plates.

Another piece of evidence about ancient Mars comes from the presence of stripes, caused by crustal magnetism, that display alternating directions of the magnetic field. These magnetic stripes are shown as red and blue bands in the lower middle of Figure 7-15. It is thought that these stripes on Mars might be formed in a similar way that magnetic stripes are formed on Earth, as illustrated in Figure 9-22. On Earth, the stripes arise from an alternating global magnetic field that magnetizes lava as it cools at a rift. As the newly formed crust spreads away from the rift, stripes of magnetism with alternating directions spread across the surface. Two mechanisms are inherent in the forming of Earth's magnetic stripes that would also take place on Mars—plate tectonics and a

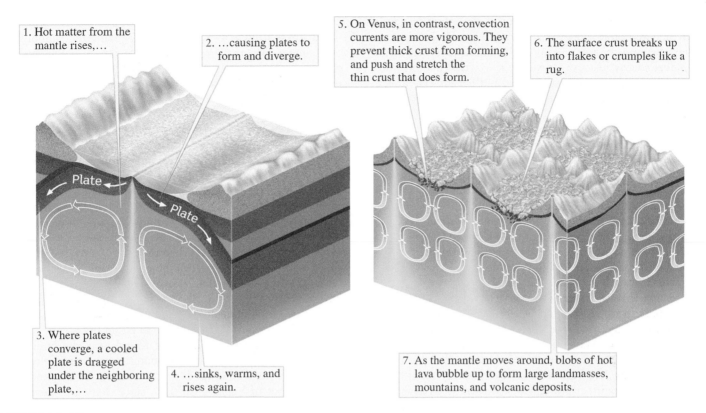

1. Hot matter from the mantle rises,...

2. ...causing plates to form and diverge.

Plate

Plate

3. Where plates converge, a cooled plate is dragged under the neighboring plate,...

4. ...sinks, warms, and rises again.

5. On Venus, in contrast, convection currents are more vigorous. They prevent thick crust from forming, and push and stretch the thin crust that does form.

6. The surface crust breaks up into flakes or crumples like a rug.

7. As the mantle moves around, blobs of hot lava bubble up to form large landmasses, mountains, and volcanic deposits.

FIGURE 11-19

Plate Tectonics Versus Flake Tectonics This illustration shows the difference between plate tectonics on Earth and the model of flake tectonics on Venus. (Courtesy of John Grotzinger)

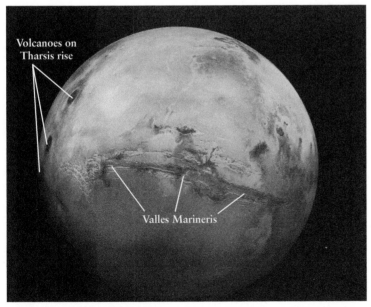

Volcanoes on Tharsis rise

Valles Marineris

(a) Mars and Valles Marineris

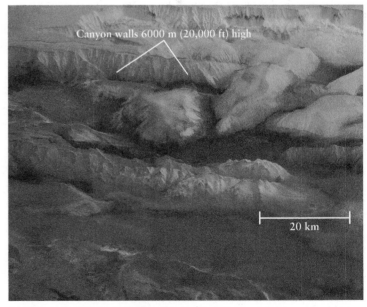

Canyon walls 6000 m (20,000 ft) high

20 km

(b) The central region of Valles Marineris

FIGURE 11-20   R I V U X G

Valles Marineris (a) This mosaic of *Viking Orbiter* images shows the huge rift valley of Valles Marineris, which extends from west to east for more than 4000 km (2500 mi), which is approximately the width of the United States, and is 600 km (400 mi) wide at its center. Its reaches depths of 8 km (5 mi) beneath the surrounding plateau, which is over 4 times as deep as the Grand Canyon, Arizona. At its western end is the Tharsis rise. (b) This perspective image from the *Mars Express* spacecraft shows what you would see from a point high above the central part of Valles Marineris. (a: USGS/NASA; b: ESA/DLR/FU Berlin, G. Neukum)

magnetic dynamo in a molten core with an alternating magnetic field. It is a remarkable achievement of geology that stripes of magnetism observed on Earth today might tell us so much about another planet's early history. However, if Martian plate tectonics helped to create these magnetic stripes, another mystery arises: Why can't we see more than the two tectonic plates associated with Valles Marineris?

We can also use magnetism detectable in craters to estimate when the magnetic dynamo of Mars shut down since craters that formed after the dynamo shut down would not show magnetism. Using age estimates of the craters without magnetism, we can estimate that the dynamo might have shut down around 500 million years after Mars formed.

Why didn't Mars experience more plate tectonic activity? Recall that smaller objects cool off more quickly than larger objects (Section 7-6). Because Mars is a much smaller world than Earth, the outer layers of the red planet have cooled more quickly than on Earth, which has led to a thicker crust. Thus, Mars lacks plate tectonics because its crust is too thick for one part of the crust to be subducted beneath another. We see that for a terrestrial planet to have extensive plate tectonics, the crust must not be too thin (like Venus) or too thick (like Mars), but just right (like Earth).

The idea that the Martian crust is too thick to allow for much plate tectonic activity has been verified by carefully monitoring the motion of a spacecraft orbiting Mars. If there is a concentration of mass (such as a thicker crust) in one region on the planet, gravitational attraction will make the spacecraft speed up as it approaches the concentration and slow down as it moves away. A team of scientists analyzed the orbit of *Mars Global Surveyor* in just this way. They found that unlike Earth's crust, which varies in thickness from 5 to 35 km, the Martian crust is about 40 km thick under the northern lowlands but about 70 km thick under the southern highlands. Both regions of the Martian crust are too thick to undergo subduction, making plate tectonics very difficult.

Volcanoes on Venus and Mars

Radar images of Venus and visible-light images of Mars show that both planets have a number of large volcanoes (Figure 11-21). *Magellan* observed more than 1600 major volcanoes and volcanic features on Venus, two of which are shown in Figure 11-21a. Both of these volcanoes have gently sloping sides. A volcano with this characteristic is called a **shield volcano,** because in profile it resembles an ancient Greek warrior's shield lying on the ground. Martian volcanoes are less numerous than those on Venus, but they are also shield volcanoes; the largest of these, Olympus Mons, is the largest volcano in the solar system (Figure 11-21b). Olympus Mons rises 24 km (15 mi) above the surrounding plains. By comparison, the highest volcano on Earth, Mauna Loa in the Hawaiian Islands, has a summit only 8 km (5 mi) above the ocean floor.

Most volcanoes on Earth are found near the boundaries of tectonic plates, where subducted material becomes molten magma and rises upward to erupt from the surface. This cannot explain the volcanoes of Venus and Mars, since there are no subduction zones on those planets. Instead, Venusian and Martian volcanoes probably formed by **hot-spot volcanism.** In this process, magma wells upward from a hot spot in a planet's mantle, elevating the overlying surface and producing a shield volcano.

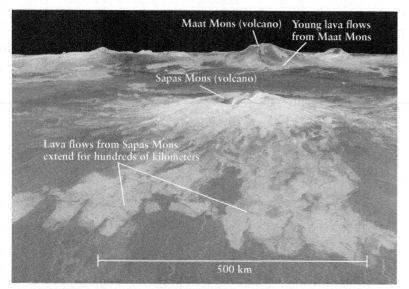

(a) Volcanoes and lava flows on Venus R I V U X G

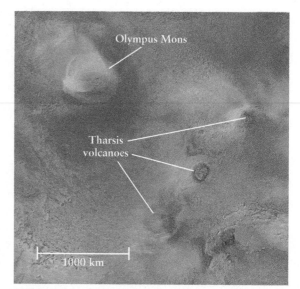

(b) Cloud-topped volcanoes on Mars R I V U X G

FIGURE 11-21

Volcanoes on Venus and Mars **(a)** The false color in this radar image approximates the real color of sunlight that penetrates Venus's thick clouds. The brighter color of the extensive lava flows indicates that they reflect radio waves more strongly. To emphasize the gently sloping volcanoes, the vertical scale has been exaggerated 10 times. **(b)** The volcanoes of Mars also have gently sloping sides. In this view looking down from Mars orbit you can see bluish clouds topping the summits of the volcanoes. These clouds, made of water ice crystals, form on most Martian afternoons. (a: NASA, JPL Multimission Image Processing Laboratory; b: NASA/JPL/Malin Space Science Systems)

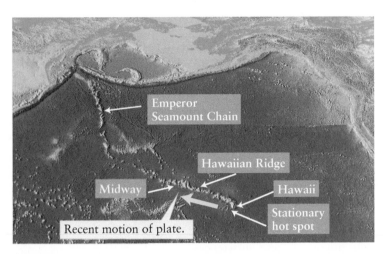

FIGURE 11-22

Hot-Spot Volcanoes on Earth A hot spot under the Pacific plate has remained essentially stationary for 70 million years while the plate has moved some 6000 km to the northwest. The upwelling magma has thus produced a long chain of volcanoes. The Hawaiian Islands are the newest of these; the oldest, the Emperor Seamount Chain, have eroded so much that they no longer protrude above the ocean surface. (World Ocean Floor, based on bathymetric studies by Bruce C. Heezen and Marie Tharp. Painting by Heinrich C. Berann. Copyright Marie Tharp, 1977)

On Earth, hot-spot volcanism is the origin of the Hawaiian Islands. These islands are part of a long chain of shield volcanoes that formed in the middle of the Pacific tectonic plate as that plate moved over a long-lived hot spot (Figure 11-22). On Venus and Mars, by contrast, the absence of significant plate tectonics means that the crust remains stationary over a hot spot. On Mars, a single hot spot under Olympus Mons probably pumped magma upward through the same vent for millions of years, producing one giant volcano rather than a long chain of smaller ones. The Tharsis rise and its volcanoes (see Figure 11-17, Figure 11-20a, and Figure 11-21b) may have formed from the same hot spot as gave rise to Olympus Mons; a different hot spot on the opposite side of Mars produced a smaller bulge centered on the volcano Elysium Mons, shown near the right-hand side in Figure 11-17. The same process of hot-spot volcanism presumably gave rise to large shield volcanoes on Venus like those shown in Figure 11-21a.

Volcanic Activity on Venus

About 80% of the surface of Venus is composed of flat plains of volcanic origin. In other words, essentially, the entire planet is covered with lava! This observation shows the tremendous importance of volcanic activity in Venusian geology.

To verify the volcanic nature of the Venusian surface, it is necessary to visit the surface and examine rock samples. **Figure 11-23**

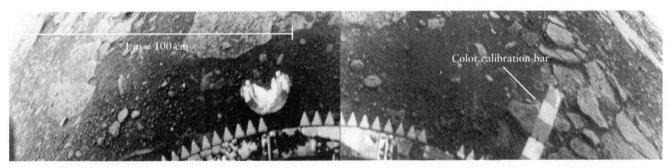

(a) Image from *Venera 13*

(b) Color-corrected image

FIGURE 11-23   R I V U X G

A Venusian Landscape **(a)** This wide-angle color photograph from *Venera 13* shows the rocky surface of Venus. The thick atmosphere absorbs the blue component of sunlight, giving the image an orange tint: The stripes on the spacecraft's color calibration bar that appear yellow are actually white in color.

(b) This color-corrected view shows that the rocks are actually gray in color. The rocky plates covering the ground may be fractured segments of a thin layer of lava. (a: NSSDC/NASA; b: GSFC/NASA)

is a panoramic view taken in 1981 by the Soviet spacecraft *Venera 13,* one of 10 unmanned spacecraft that the Soviet Union landed successfully on the surface of Venus. Russian scientists hypothesize that this region was covered with a thin layer of lava that fractured upon cooling to create the rounded, interlocking shapes seen in the photograph. This hypothesis agrees with information obtained from chemical analyses of surface material made by the spacecraft's instruments. These analyses indicate that the surface composition is similar to lava rocks called basalt, which are common on Earth (see Figure 9-18a) and in the maria of the Moon (see Figure 10-16). The results from *Venera 13* and other landers are consistent with the picture of Venus as a world whose surface and atmosphere have been shaped by volcanic activity.

Most of the volcanoes on Venus are probably inactive at present, just as is the case with most volcanoes on Earth. But in 2010, evidence was found for very recent volcanic activity on Venus. By analyzing infrared emission, the European Space Agency's *Venus Express* orbiter found material on volcanoes showing signs that it had not experienced significant weathering. However, due to weathering by Venus's hot and thick atmosphere, the surface of fresh volcanic rock should show weathering fairly quickly. This analysis points to very young lava flows, perhaps a few hundred years old, or as much as 2.5 million years old (which is still geologically young). The presence of such young lava flows suggests that Venus, like Earth, has some present-day volcanic activity.

Another piece of evidence for ongoing volcanic activity on Venus comes from the planet's atmosphere. An erupting volcano on Earth ejects substantial amounts of sulfur dioxide, sulfuric acid, and other sulfur compounds into the air. Many of these substances are highly reactive and short-lived, forming sulfate compounds that become part of the planet's surface rocks. For these substances to be relatively abundant in a planet's atmosphere, they must be constantly replenished by new eruptions. Sulfur compounds make up about 0.015% of the Venusian atmosphere, compared to less than 0.0001% of Earth's atmosphere. This evidence suggests that ongoing volcanic eruptions on Venus are ejecting sulfur compounds into the atmosphere to sustain the high sulfur content.

Volcanic Activity on Mars

Spectroscopic observations from Mars's orbit confirm that the planet's rocks and sands are made almost entirely of the three minerals feldspar, pyroxene, and olivine. These are the components of basalt, or solidified lava. Thus, Mars, like Venus, had a volcanic past.

Unlike lava flows on Venus, however, most of the lava flows on Mars have impact craters on them. These craters suggest that most Martian lava flows are very old and that most of the volcanoes on the red planet are no longer active. This is what we would expect from a small planet whose crust has cooled and solidified to a greater depth than on Earth, making it difficult for magma to travel from the Martian mantle to the surface.

However, a few Martian lava flows are crater-free, which suggests that they are only a few million years old. If volcanoes erupted on Mars within the past few million years, are they erupting now? Scientists have used infrared telescopes to search for telltale hot spots on the Martian surface, but have yet to discover any. Perhaps volcanism on Mars is rare but not yet wholly extinct.

Spacecraft that have landed on Mars have taught us a great deal about the geology and history of the red planet. Before we explore the Martian surface in detail, however, it is useful to examine the unique atmospheres of both Mars and Venus.

11-6 The dense atmosphere of Venus and the thin Martian atmosphere are dramatically different but have similar chemical compositions

We have seen that the three large terrestrial planets—Earth, Venus, and Mars—display very different styles of geology. As we will discover, these differences help to explain why each of these three worlds has a distinctively different atmosphere.

> On Venus, sulfuric acid rain evaporates before reaching the ground; on Mars, it snows frozen carbon dioxide

To explore a planet's atmosphere it is necessary to send a spacecraft to make measurements as it descends through that atmosphere. Both Soviet and U.S. spacecraft have done this for Venus (see Figure 11-23), and seven U.S. spacecraft have successfully landed on Mars. The results of these missions are summarized in **Figure 11-24**, which shows how pressure and temperature vary with altitude in the atmospheres of Earth, Venus, and Mars.

The Venusian Atmosphere: Dense, Hot, and Corrosive

Figure 11-24b depicts just how dense the Venusian atmosphere is. At the surface, the pressure is 90 atmospheres. This pressure is 90 times greater than the average air pressure at sea level on Earth and is about the same as the water pressure at a depth of about 1 kilometer below the surface of Earth's oceans. The density of the atmosphere at the surface of Venus is likewise high, more than 50 times greater than the sea-level density of our atmosphere. The atmosphere is so massive that once heated by the Sun, it retains its heat throughout the long Venusian night. As a result, temperatures on the day and night sides of Venus are almost identical.

The composition of atmosphere explains why Venus's surface temperatures are so high: 96.5% of the molecules in Venus's atmosphere are carbon dioxide, with nitrogen (N_2) making up most of the remaining 3.5% (see Table 9-4 for a comparison of the atmospheres of Venus, Earth, and Mars). Because Venus's atmosphere is so thick, and because most of the atmosphere is the greenhouse gas CO_2, the greenhouse effect (Section 9-1) has run wild. On Earth the greenhouse gases H_2O and CO_2 together make up only about 1% of our relatively sparse atmosphere, and the greenhouse effect has elevated the surface temperature by an additional 33°C (59°F). But on Venus, the dense shroud of CO_2 traps infrared radiation from the surface so effectively that the surface temperature is more than 400°C (720°F) greater than it would have been without the CO_2.

Soviet and U.S. spacecraft also discovered that Venus's clouds are primarily confined to three high-altitude layers. An upper cloud layer lies at altitudes between 68 and 58 km, a denser and more opaque cloud layer from 58 to 52 km, and an even more dense and opaque layer between 52 and 48 km. Above and below

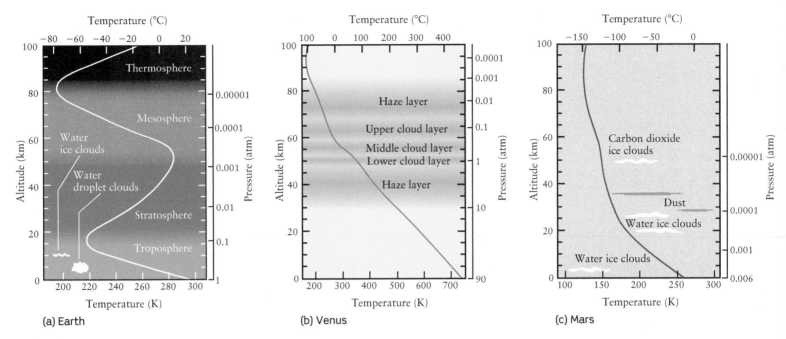

(a) Earth   (b) Venus   (c) Mars

FIGURE 11-24

Atmospheres of the Terrestrial Planets In each graph the curve shows how temperature varies with altitude from 0 to 100 km above the planet's surface. The scale on the right-hand side of each graph shows how pressure varies with altitude. (a) Clouds in Earth's atmosphere are seldom found above 12 km (40,000 ft). (b) Venus's perpetual cloud layers lie at much higher altitudes. The atmosphere is so dense that the pressure 50 km above the Venusian surface is 1 atm, the same as at sea level on Earth. (c) By contrast, the Martian atmosphere is so thin that the surface pressure is the same as the pressure at an altitude of 35 km on Earth. Wispy clouds can be found at extreme altitudes.

the clouds are 20-km-thick layers of haze. Below the lowest haze layer, the atmosphere is remarkably clear all the way down to the surface of Venus.

We saw in Section 11-5 that sulfur, which is not found in any appreciable amount in our atmosphere, plays an important role in the Venusian atmosphere. Sulfur combines with other elements to form gases such as sulfur dioxide (SO_2) and hydrogen sulfide (H_2S), along with sulfuric acid (H_2SO_4), the same acid used in automobile batteries. While Earth's clouds are composed of water droplets, Venusian clouds contain almost no water. Instead, they are composed of droplets of concentrated, corrosive sulfuric acid. Due to the high temperatures on Venus, these droplets never rain down on the planet's surface; they simply evaporate at high altitude. (You can see a similar effect on Earth. On a hot day in the desert of the U.S. Southwest, streamers of rain called *virga* appear out of the bottoms of clouds but evaporate before reaching the ground.)

Convection in the Venusian Atmosphere

We saw in Section 9-6 that Earth's atmosphere is in continuous motion: Hot gases rise and cooler gases sink, forming convection cells in the atmosphere. The same is true on Venus, but with some important differences. On Venus, gases in the equatorial regions are warmed by the Sun, then rise upward and travel in the upper cloud layer toward the cooler polar regions. At the polar latitudes, the cooled gases sink to the lower cloud layer, in which they are transported back toward the equator. The result is two huge convection cells, one in the northern hemisphere and another in

the southern hemisphere (Figure 11-25a). These cells are almost entirely contained within the main cloud layers shown in Figure 11-24b. The circulation is so effective at transporting heat around Venus's atmosphere that there is almost no temperature difference between the planet's equator and its poles. By contrast, Earth's atmosphere has a more complicated convection pattern (see Figure 11-25b and Figure 9-25) because Earth's rotation—which is far more rapid than Venus's—distorts the convection cells.

In addition to the north-south motions due to convection, high-altitude winds with speeds of 350 km/h (220 mi/h) blow from east to west, the same direction as Venus's retrograde rotation. Thus, the upper atmosphere rotates around the planet once every 4 days. The fast-moving winds stretch out the convection cells and produce V-shaped, chevronlike patterns in the clouds (Figure 11-26). In a similar way, Earth's rotation stretches out the convection cells in our atmosphere to produce the complicated circulation patterns shown in Figure 9-25.

On Earth, friction between the atmosphere and the ground causes wind speeds at the surface to be much lower than at high altitude. The same effect occurs on Venus: The greatest wind speed measured by spacecraft on Venus's surface is only about 5 km/h (3 mi/h). Thus, only slight breezes disturb the crushing pressures and infernal temperatures found on Venus's dry, lifeless surface.

The Martian Atmosphere: Not a Drop to Drink

As Figure 11-24c shows, the atmosphere of Mars is very cold and thin compared to the atmosphere of either Earth or Venus: The surface pressure is a mere 0.006 atmosphere. But its chemical

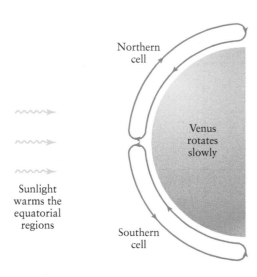

(a) Atmospheric circulation on Venus:
two convection cells

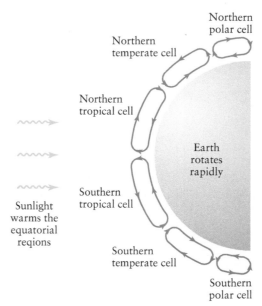

(b) Atmospheric circulation on Earth:
six convection cells

FIGURE 11-25

Atmospheric Circulation on Venus and Earth Solar heating causes convection in the atmospheres of both Venus and Earth, with warm air rising at the equator and cold air descending at the poles. (a) Because Venus rotates very slowly, it has little effect on the circulation. (b) Earth's rapid rotation distorts the atmospheric circulation into a more complex pattern (compare Figure 9-24).

composition is very close to that of Venus: The Martian atmosphere is 95.3% carbon dioxide (versus 96.5% for Venus) and 2.7% nitrogen (versus 3.5% for Venus). The predominance of carbon dioxide means that Mars, like Earth and Venus, is warmed by the greenhouse effect. However, the Martian atmosphere is so thin that the greenhouse effect is very weak and warms the Martian surface by only 5°C (versus 33°C on Earth). While the perpetual

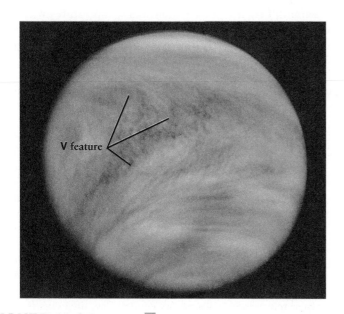

FIGURE 11-26   R I V U X G

Venus's Cloud Patterns This false-color image of Venus was made at ultraviolet wavelengths, at which the planet's atmospheric markings stand out best. The dark V feature is produced by the rapid motion of clouds around the planet's equator. The clouds move in the same retrograde direction as the rotation of the planet itself but at a much greater speed. (GSFC/NASA)

cloud cover on Venus maintains a steady surface temperature, the thin atmosphere on Mars provides very little thermal insulation. While daytime highs on Mars can be as high as 20°C (68°F), at night the temperature can plummet to −140°C (−220°F).

Water vapor makes up about 0.03% of the Martian atmosphere (versus about 1% for Earth), and this vapor can form clouds. Figure 11-21b shows clouds that form near the tops of Martian volcanoes when air moves up the volcanic slopes. The rising air cools until the water vapor condenses into ice crystals. The same process occurs on Earth, but the clouds that form in this way are made of droplets of liquid water rather than ice crystals. Why is there a difference?

The explanation is that liquid water cannot exist anywhere on the Martian surface or in the Martian atmosphere. Water is liquid over only a limited temperature range: If the temperature is too low, water becomes ice, and if the temperature is too high, it becomes water vapor. What determines this temperature range is the atmospheric pressure above a body of water or around a water drop. If the pressure is very low, molecules easily escape from the liquid's surface, causing the water to vaporize. Thus, at low pressures, water more easily becomes water vapor. The average surface temperature on Mars is only about 250 K (−23°C, or −10°F), and the average pressure is only 0.006 atmosphere. With this combination of temperature and pressure, water can exist as a solid (ice) and as a gas (water vapor) but not as a liquid. (You can see this same situation inside a freezer, where water vapor swirls around over ice cubes.) Hence, it never rains on Mars and there are no bodies of standing water anywhere on the planet.

In order to keep water on Mars in the liquid state, it would be necessary to increase both the temperature (to keep water from freezing) and the pressure (to keep the liquid water from evaporating). Later in this chapter we will see evidence that higher temperatures and pressures did exist on Mars billions of years ago, and that there was once liquid water on the surface.

Methane in the Martian Atmosphere

The discovery of large quantities of methane (CH_4) in the Martian atmosphere came as a big surprise in 2003. More specifically, while some methane is expected, the observed *variation* (over time, and for different regions) in the methane abundance is not understood. Because methane is destroyed in the Martian atmosphere after about 300 years, the discovery of large variations in methane abundance indicate an ongoing variable process for producing and removing methane. The source for more than 95% of Earth's methane is well known: *life!* Naturally, these observations raise some hopes for a microbial source of the Martian methane, possibly varying with the seasons. Another proposed source is geologic activity (such as volcanism), which would imply significantly more ongoing geologic activity than observed on Mars so far. Either way, solving the methane mystery seemed to promise breakthroughs in either Martian biology or geology.

However, in 2012, scientists found evidence for a third scenario involving meteorites to explain Martian methane. Small micrometeorites have showered Mars for billions of years, and, unlike on Earth, these meteorites do not burn up in the thin Martian atmosphere. Also, compared to Earth, ultraviolet (UV) light on the Martian surface is more intense without absorption in an atmospheric ozone layer. Taken together, a large quantity of meteoritic material on Mars is exposed to intense UV light.

By exposing UV light to actual meteorites on Earth with a similar carbon composition to meteoritic material on Mars, scientists found that the UV exposure produces methane. Furthermore, the amount of methane produced depends on temperature and would vary with the Martian seasons and latitude. And, by exposing new surfaces, even the movement of small meteorites by dust storms would produce significant variations in methane production. These results do not disprove that biology or geology are contributing to methane production, but it weakens the case for these causes. Fortunately, instruments on the Mars rover *Curiosity* can detect methane and should help to determine its origins.

The Martian Poles: Freezing the Atmosphere

Figure 11-24c shows that at very high altitudes above the Martian surface, atmospheric carbon dioxide can freeze into crystals and form clouds. (Frozen carbon dioxide can be made on Earth and is called "dry ice.") Such frozen carbon dioxide is also found in large quantities in ice caps at the north and south poles of Mars (**Figure 11-27**). Although no spacecraft has successfully landed on the ice caps, their composition can be measured from orbit by comparing the spectrum of sunlight reflected from the ice caps with the known reflection spectrum of frozen carbon dioxide. (Figure 7-4b shows another application of this idea.)

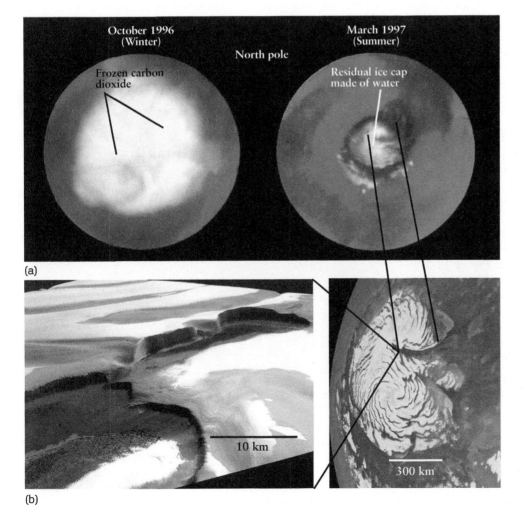

FIGURE 11-27 R I **V** U X G

Changing Seasons Reveal Water-Ice at the Martian North Pole **(a)** During the Martian winter, the temperature drops so low that carbon dioxide freezes out of the Martian atmosphere to make a large polar ice cap of carbon dioxide. During summer, the carbon dioxide returns to the atmosphere, revealing a residual ice cap made of water. **(b)** The dark canyon, Chasma Boreale, is more than 300 km long—similar in length and depth to the Grand Canyon, Arizona. Spectra reveal the residual ice is made of water and not CO_2. (a: S. Lee/J. Bell/M. Wolff/Space Science Institute/NASA; b: (left) NASA; (right) *Mars Reconnaissance Orbiter*, NASA/Caltech/JPL/E. DeJong/J. Craig/M. Stetson)

Group of rocks about 2 m
across, 8 m from the camera

Dunes

(a) A view from *Viking Lander 1*

(b) A wintertime view from *Viking Lander 2*

FIGURE 11-28 R I V U X G

Seasons on the Martian Surface **(a)** This springtime view from *Viking Lander 1* shows rocks that resemble volcanic rocks on Earth (see Figure 9-19a). They are thought to be part of an ancient lava flow that was broken apart by crater-forming asteroid impacts. Fine-grained debris has formed sand dunes. **(b)** This picture was taken during midwinter at the *Viking Lander 2* site.

Freezing carbon dioxide condenses onto water-ice crystals or dust grains in the atmosphere, causing them to fall to the ground and coat the surface with frost. This frost lasted for about a hundred days. (a: Dr. Edwin Bell II/NSSDC/GSFC/NASA; b: NASA)

As shown in Figure 11-27a, the size of the ice caps varies substantially with the seasons. When it is winter at one of the Martian poles, temperatures there are so low that atmospheric carbon dioxide freezes and the ice cap grows. (Think of it: It gets so cold at the Martian poles that the atmosphere freezes!) With the coming of warmer spring temperatures, much of the carbon dioxide returns to the gaseous state and the ice cap shrinks. With the arrival of summer, however, the rate of shrinkage slows abruptly and a **residual ice cap** remains solid throughout the summer. Scientists had long suspected that the north and south residual polar caps contain a large quantity of frozen water that is less easily evaporated (Figure 11-27b). The ESA's *Mars Express* spacecraft confirmed this idea by detecting the characteristic spectrum of water ice.

The wintertime freezing of atmospheric carbon dioxide was observed by the *Viking Lander 1* and *Viking Lander 2* spacecraft, which made the first successful landings on Mars in 1976. Both of these spacecraft touched down in the northern hemisphere of Mars (you can see the landing sites in Figure 11-17). Figure 11-28 compares the springtime and wintertime views seen from the *Viking Landers*. Closer to the north pole, seasonal warming of carbon dioxide has lead to a very strange landscape (Figure 11-29). The dark areas in this image are dust patches that only become exposed after carbon dioxide ice evaporates off the surface. On the steeper slopes, the dust cascades down, creating the dark treelike streaks.

Snowfall on Mars

Similar to the way that frost can form on a car's windshield on a cold morning, water and carbon dioxide in the Martian atmosphere can form frost as in Figure 11-28b. However, actual *snowfall* from clouds indicates a more complex climate system that also

occurs on Mars. In 2008, NASA's *Phoenix* lander detected water-snow from clouds about 4 km above the lander. As the Martian winter approached and temperatures dropped, reflections from a green laser shot into the clouds above the lander revealed large ice crystals, large enough to assume they fell to the ground. The snow

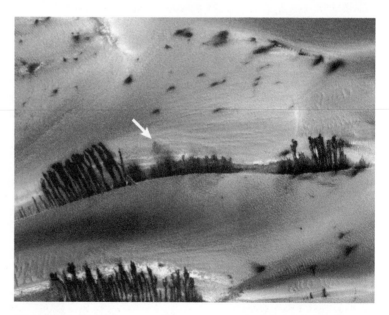

FIGURE 11-29 R I V U X G

Spring on Mars The dark areas in this enhanced-color image are dust patches uncovered by evaporating carbon dioxide ice near the north pole in springtime. On the steeper slopes, the dust cascades down, creating the dark treelike streaks. The yellow arrow points to a plume being kicked up by one of these dust-slides. (NASA/JPL/University of Arizona)

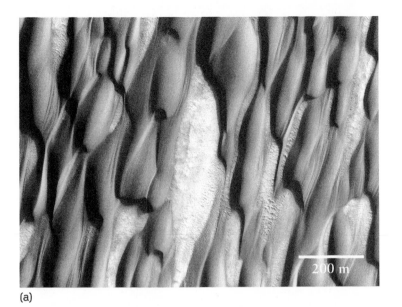

(a)

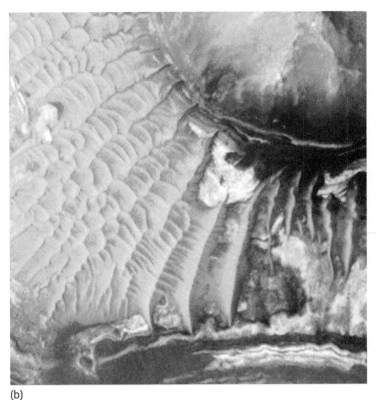

(b)

FIGURE 11-30   R I V **U X G**

Martian Sand Dunes **(a)** Enhanced-coloring brings out the composition of these dunes. Blue indicates sand that came from volcanic rock. The lighter colors on top are probably very fine-grained dust. **(b)** Sand from clay minerals form dunes that partially cover sedimentary rock. (NASA/JPL/University of Arizona)

was assumed to be water-snow because it was not yet cold enough for carbon dioxide to freeze.

Then, in 2012, astronomers reported that *Mars Reconnaissance Orbiter* found evidence of falling snow made from carbon dioxide (CO_2) during the deep cold of winter—the only known case of CO_2 snowfall anywhere in our solar system. The snow was detected above the Martian south pole through its visible and infrared spectrum, and was observed extending all the way to the ground.

Martian Dust, Dunes, and Devils

Figure 11-28a shows that the landscape around the *Viking Lander 1* is covered with a fine-grained dust. On much larger scales, wind can pile up the dust particles to form sand dunes (Figure 11-30). The dust particles are much smaller than ordinary household dust on Earth: They are only about 1μm (10^{-6} m) across, about the size of the particles in cigarette smoke.

Because Martian dust is so fine, it can be carried aloft and spread over the surface by Martian winds even though the atmosphere is so tenuous. (A wind of 100 km/h, or 60 mi/h, on Mars is only about as strong as an Earth breeze of 10 km/h or 6 mi/h.) Windblown dust can sometimes be seen from Earth as a yellow haze that obscures Martian surface features. Figure 11-14 shows a dust storm that covered the entire planet, triggered by the flow of carbon dioxide evaporating from the north polar ice cap with the coming of northern spring. On Earth, stormy weather means cloudy skies, wind, and falling rain; on Mars, where no raindrop falls, stormy weather means wind and dust.

Dust also plays a role in the ordinary daily weather on Mars. Each afternoon, columns of warm air rise from the heated surface

and form whirlwinds called **dust devils** (Figure 11-31). A similar phenomenon with the same name occurs in dry or desert terrain on Earth. Martian dust devils can reach an altitude of 6 km (20,000 ft) and help spread dust particles across the planet's surface. Martian landers have measured changes in air pressure

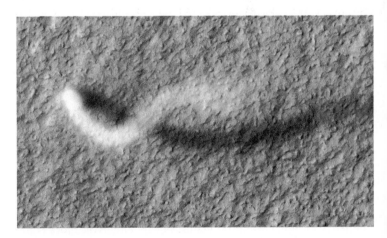

FIGURE 11-31   R I V **U X G**

A Martian Dust Devil This *Mars Global Surveyor* image shows a dust devil as seen from almost directly above. This tower of swirling air and dust casts a long shadow in the afternoon sun. Spotted in 2012, this dust devil is about 100 feet wide and half a mile in height. About a month later, another dust devil was seen reaching a height of 12 miles! (NASA/JPL-MRO/Caltech/University of Arizona)

as dust devils swept past and sometimes several of these would pass in a single afternoon. As the sun sets and afternoon turns to evening, the dust devils subside until the following day.

CONCEPTCHECK 11-7

How might Mars's environment become more hospitable to life if more CO_2 could be added to the atmosphere?

CONCEPTCHECK 11-8

What are three processes that could lead to variable amounts of methane on Mars?

Answers appear at the end of the chapter.

11-7 The atmospheres of Venus and Mars were very different billions of years ago

TUTORIAL 11-2 Why do Earth, Venus, and Mars have such dramatically different atmospheres? As we will see, the original atmospheres of all three worlds were essentially the same. However, each atmosphere evolved in a unique way determined by the planet's size and distance from the Sun. The *Cosmic Connections* figure summarizes the processes that led to the present-day atmospheres of the three large terrestrial planets.

> The early atmospheres of Earth, Venus, and Mars were predominantly water vapor and carbon dioxide

The Origin of Atmospheres on Earth, Venus, and Mars

The original atmospheres of Earth, Venus, and Mars all derived from gases that were emitted, or *outgassed,* from volcanoes (Figure 11-32). The gases released by present-day Earth volcanoes are predominantly water vapor (H_2O), followed by carbon dioxide (CO_2), sulfur dioxide (SO_2), and nitrogen (N_2). These gases should therefore have been important parts of the original atmospheres of Earth, Venus, and Mars. It is also possible—but far from certain—that icy comets from the outer solar system delivered significant quantities of water to these planets.

Due in part to intense volcanic activity that took place when the young planets had warmer interiors, water probably once dominated the atmospheres of Earth, Venus, and Mars. Studies of how stars evolve suggest that the early Sun was only about 70% as luminous as it is now, so the temperature in Venus's early atmosphere must have been quite a bit lower. Water vapor may actually have been able to liquefy and form oceans on Venus. Mars is smaller than either Earth or Venus, but its large volcanoes could also have emitted substantial amounts of water vapor. With a thicker atmosphere and hence greater pressure, liquid water could also have existed on Mars.

The Evolution of Earth's Atmosphere

On the present-day Earth most of the water is in the oceans. Nitrogen, which is not very reactive, is still in the atmosphere.

FIGURE 11-32 R I V U X G

A Volcanic Eruption on Earth The eruption of Mount St. Helens in Washington State on May 18, 1980, released a plume of ash and gas 40 km (25 mi) high. Eruptions of this kind on Earth, Venus, and Mars probably gave rise to those planets' original atmospheres. (USGS)

By contrast, carbon dioxide dissolves in water, which can fall as rain; as a result, rain removed most of the H_2O and CO_2 out of our planet's atmosphere long ago. The dissolved CO_2 reacted with rocks in rivers and streams, and the residue was ultimately deposited on the ocean floors, where it turned into carbonate rocks, such as limestone. Consequently, most of Earth's carbon dioxide is tied up in Earth's crust, and only about 1 in every 3000 molecules in our atmosphere is a CO_2 molecule. If Earth became as hot as Venus, much of its CO_2 would be boiled out of the oceans and baked out of the crust, and our planet would soon develop a thick, oppressive carbon dioxide atmosphere much like that of Venus.

The small amount of carbon dioxide in our atmosphere is sustained in part by plate tectonics. Tectonic activity causes carbonate rocks to cycle through volcanoes, where they are heated and forced to liberate their trapped CO_2. This liberated gas then rejoins the atmosphere. Without volcanoes, rainfall would remove all of the CO_2 from our atmosphere in only a few thousand years. This lack of CO_2 would dramatically reduce the greenhouse effect, our planet's surface temperature would drop precipitously, and the oceans would freeze. Plate tectonics is thus essential for maintaining our planet's livable temperatures.

The oxygen (O_2) in our atmosphere is the result of photosynthesis by plant life, which absorbs CO_2 and releases O_2. The net result is an atmosphere that is predominantly N_2 and O_2, with

Evolution of Terrestrial Atmospheres

Earth, Venus, and Mars formed with similar original atmospheres. However, these atmospheres changed dramatically over time due to factors such as planetary size and distance from the Sun.

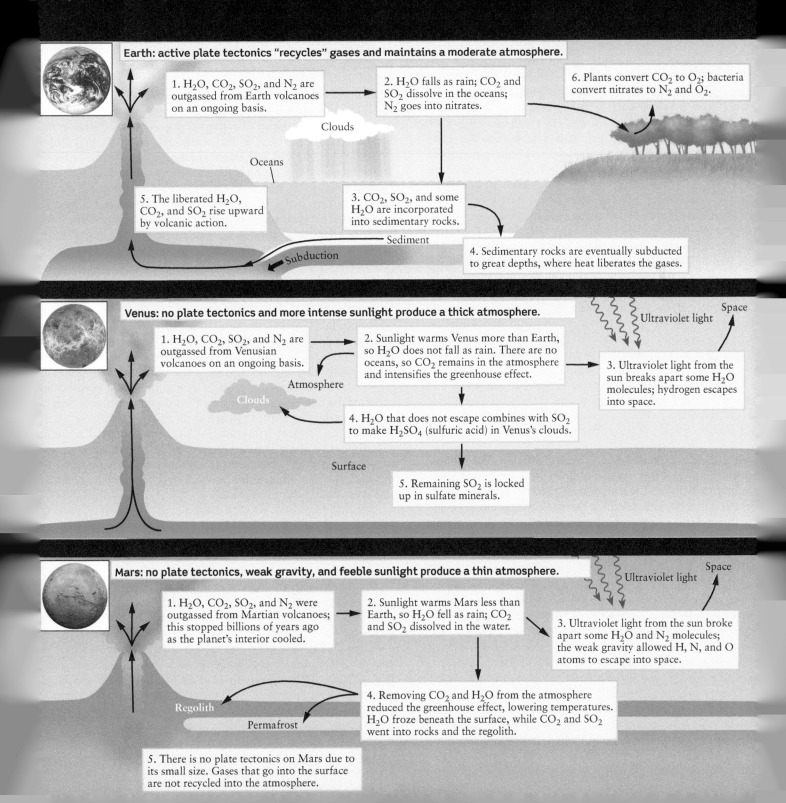

Earth: active plate tectonics "recycles" gases and maintains a moderate atmosphere.

1. H_2O, CO_2, SO_2, and N_2 are outgassed from Earth volcanoes on an ongoing basis.

2. H_2O falls as rain; CO_2 and SO_2 dissolve in the oceans; N_2 goes into nitrates.

6. Plants convert CO_2 to O_2; bacteria convert nitrates to N_2 and O_2.

Clouds

Oceans

5. The liberated H_2O, CO_2, and SO_2 rise upward by volcanic action.

3. CO_2, SO_2, and some H_2O are incorporated into sedimentary rocks.

Sediment

Subduction

4. Sedimentary rocks are eventually subducted to great depths, where heat liberates the gases.

Venus: no plate tectonics and more intense sunlight produce a thick atmosphere.

Space

Ultraviolet light

1. H_2O, CO_2, SO_2, and N_2 are outgassed from Venusian volcanoes on an ongoing basis.

2. Sunlight warms Venus more than Earth, so H_2O does not fall as rain. There are no oceans, so CO_2 remains in the atmosphere and intensifies the greenhouse effect.

3. Ultraviolet light from the sun breaks apart some H_2O molecules; hydrogen escapes into space.

Atmosphere

Clouds

4. H_2O that does not escape combines with SO_2 to make H_2SO_4 (sulfuric acid) in Venus's clouds.

Surface

5. Remaining SO_2 is locked up in sulfate minerals.

Mars: no plate tectonics, weak gravity, and feeble sunlight produce a thin atmosphere.

Space

Ultraviolet light

1. H_2O, CO_2, SO_2, and N_2 were outgassed from Martian volcanoes; this stopped billions of years ago as the planet's interior cooled.

2. Sunlight warms Mars less than Earth, so H_2O fell as rain; CO_2 and SO_2 dissolved in the water.

3. Ultraviolet light from the sun broke apart some H_2O and N_2 molecules; the weak gravity allowed H, N, and O atoms to escape into space.

Regolith

Permafrost

4. Removing CO_2 and H_2O from the atmosphere reduced the greenhouse effect, lowering temperatures. H_2O froze beneath the surface, while CO_2 and SO_2 went into rocks and the regolith.

5. There is no plate tectonics on Mars due to its small size. Gases that go into the surface are not recycled into the atmosphere.

enough pressure and a high enough temperature (thanks to the greenhouse effect) for water to remain a liquid—and hence for life, which requires liquid water, to exist.

The Evolution of the Venusian Atmosphere: A Runaway Greenhouse Effect

When Venus was young and had liquid water, the amount of atmospheric CO_2 was kept in check much as on Earth today: CO_2 was released by volcanoes, then dissolved in the oceans and bound up in carbonate rocks. Enough of the liquid water would have vaporized to create a thick cover of water vapor clouds. Since water vapor is a greenhouse gas, this humid atmosphere—perhaps denser than Earth's present-day atmosphere, but far less dense than the atmosphere that surrounds Venus today—would have efficiently trapped heat from the Sun. At first, this would have had little effect on the oceans of Venus. Although the temperature would have climbed above 100°C, the boiling point of water at sea level on Earth, the added atmospheric pressure from water vapor would have kept the water in Venus's oceans in the liquid state.

This hot and humid state of affairs may have persisted for several hundred million years. But as the Sun's energy output slowly increased over time, the temperature at the surface would eventually have risen above 374°C (647 K, or 705°F). Above this temperature, no matter what the atmospheric pressure, Venus's oceans would have begun to evaporate. The added water vapor in the atmosphere would have increased the greenhouse effect, making the temperature even higher and causing the oceans to evaporate faster. This temperature increase would have added even more water vapor to the atmosphere, further intensifying the greenhouse effect and making the temperature climb higher still. This is an example of a **runaway greenhouse effect,** in which an increase in temperature causes a further increase in temperature, and so on.

ANALOGY A runaway greenhouse effect is like a house in which the thermostat has accidentally been connected backward. Normally, when the temperature rises too high in a house, the heater shuts off. However, if a thermostat is connected backward, the hotter the house, the greater the heater output, raising the temperature even further.

Once Venus's oceans disappeared, so did the mechanism for removing carbon dioxide from the atmosphere. With no oceans to dissolve it, outgassed CO_2 began to accumulate in the atmosphere, making the greenhouse effect "run away" even faster. Temperatures eventually became high enough to "bake out" any CO_2 that was trapped in carbonate rocks. This liberated carbon dioxide formed the thick atmosphere of present-day Venus.

Sulfur dioxide (SO_2) is also a greenhouse gas released by volcanoes. Although present in far smaller amounts than carbon dioxide, it would have contributed to Venus's rising temperatures. Like CO_2, it dissolves in water, a process that helps moderate the amount of sulfur dioxide in our atmosphere. But when the oceans evaporated on Venus, the amount of atmospheric SO_2 would have increased.

The temperature in Venus's atmosphere would not have increased indefinitely. In time, solar ultraviolet radiation striking molecules of water vapor—the dominant cause of the runaway greenhouse effect—would have broken them into hydrogen and oxygen atoms. The lightweight hydrogen atoms would have then escaped into space (see Box 7-2 for a discussion of why lightweight atoms can escape a planet more easily than heavy atoms). The remaining atoms of oxygen, which is one of the most chemically active elements, would have readily combined with other substances in Venus's atmosphere. Eventually, almost all of the water vapor would have been irretrievably lost from Venus's atmosphere. With all the water vapor gone, and essentially all the carbon dioxide removed from surface rocks, the greenhouse effect would no longer have "run away," and the rising temperature would have leveled off.

Today, the infrared-absorbing properties of CO_2 have stabilized Venus's surface temperature at its present value of 460°C. Only minuscule amounts of water vapor—about 30 parts per million, or 0.003%—remain in the atmosphere. As on Earth, volcanic outgassing might still add small amounts of water vapor to the atmosphere, along with carbon dioxide and sulfur dioxide. But on Venus, these water molecules either combine with sulfur dioxide to form sulfuric acid clouds or break apart due to solar ultraviolet radiation. The great irony is that this state of affairs is the direct result of an earlier Venusian atmosphere that was predominantly water vapor!

The Evolution of the Martian Atmosphere: A Runaway Icehouse Effect

Mars probably had a thicker atmosphere 4 billion years ago. Thanks to Mars's greater distance from the Sun and hence less intense sunlight, temperatures would have been lower than on young Earth, and any water in the atmosphere would more easily have fallen as rain or snow. This precipitation would have washed much of the planet's carbon dioxide from its atmosphere, perhaps creating carbonate minerals in which the CO_2 is today chemically bound. Measurements from Mars orbit show only small amounts of carbonate materials on the surface, suggesting that the amount of atmospheric CO_2 that rained out was small. Hence, even the original Martian atmosphere was relatively thin, though thicker than the present-day atmosphere.

Because Mars is so small, it cooled early in its history and volcanic activity came to an end. Thus, any solid carbonates were not recycled through volcanoes as they are on Earth. The depletion of carbon dioxide from the Martian atmosphere into the surface would therefore have been permanent.

As the amount of atmospheric CO_2 declined, the greenhouse effect on Mars would have weakened and temperatures begun to fall. This temperature decrease would have caused more water vapor to condense into rain or snow and fall to the surface, taking even more CO_2 with them and further weakening the greenhouse effect. Thus, a decrease in temperature would have caused a further decrease in temperature—a phenomenon sometimes called a **runaway icehouse effect.** (This is the reverse of the runaway greenhouse effect that has taken place on Venus.) Ultimately, both water vapor and most of the carbon dioxide would have been removed from the Martian atmosphere. With only a very thin CO_2 atmosphere remaining, surface temperatures on Mars eventually stabilized at their present frigid values.

As both water vapor and carbon dioxide became depleted, ultraviolet light from the Sun could penetrate the thinning Martian atmosphere to strip it of nitrogen. Nitrogen molecules (N_2) normally do not have enough thermal energy to escape from Mars, but they can acquire that energy from ultraviolet photons, which break the molecules in two. Ultraviolet photons can also split carbon dioxide and water molecules, giving their atoms enough energy to escape. Indeed, in 1971 the Soviet *Mars 2* spacecraft found a stream of oxygen and hydrogen atoms (from the breakup of water molecules) escaping into space. Oxygen atoms that did not escape into space could have combined with iron-bearing minerals in the surface, forming rustlike compounds. Such compounds have a characteristic reddish-brown color and may be responsible for the overall color of the planet; if this is correct, Mars gained its color by losing its atmosphere!

A Magnetic Field Protects the Martian Atmosphere

Some scientists think a large asteroid impact triggered the loss of the Martian atmosphere. To understand the possible effect of an impact, we need to understand how the Martian atmosphere might have been protected. Ancient magnetic stripes on Mars's surface indicate that early on, Mars had a magnetic field arising from a magnetic dynamo in a molten core, similar to Earth's magnetic field (see Section 11-5). As with Earth's magnetic field, the Martian field would have protected the planet's atmosphere by deflecting the solar wind away from Mars (see Figure 9-19). The protection arises because if they are not deflected, high-speed particles in the solar wind can break apart atmospheric CO_2 and H_2O, allowing their lighter constituents to escape into space. Furthermore, the timeline of magnetic signatures in Martian craters suggest that this protective magnetic field persisted for about Mars's first 500 million years, which coincides with the Late Heavy Bombardment (see Section 8-6).

Putting all of these processes together, a new idea proposes that a very large asteroid impact on Mars disrupted the Martian magnetic dynamo, shut down its magnetic field, and led to the loss of the planet's atmosphere before its interior cooled down. Advocates of this scenario estimate that there are several large craters of the right age that indicate enough impact energy to significantly disrupt the core convection that produced a magnetic field. However, others are not convinced that such an impact could significantly alter the magnetic field.

Due to Mars's small size, it only contained enough heat for a short period of volcanism. Therefore, Mars may have been destined to lose its atmosphere anyway. Since planetary cooling shuts down the core convection that creates a global magnetic field (which protects CO_2 and H_2O), an atmosphere cannot outlast its planet's internal heat for very long. Whether or not a large impact brought an early end to the Martian atmosphere remains unknown.

Before it was lost, was the Martian atmosphere thick enough and warm enough for water to remain as a liquid on the planet's surface? As we will see in the next section, scientists have found convincing evidence that the answer is yes.

CONCEPTCHECK 11-9

What is "runaway" about Venus's runaway greenhouse?

CONCEPTCHECK 11-10

How would a global magnetic field on a young Mars help to keep the planet warm?

Answers appear at the end of the chapter.

11-8 Rovers have found evidence of ancient Martian water

Just as ancient fossils on Earth provide evidence of how life has evolved on our planet, ancient rocks and geological formations help us understand how our planet's surface and atmosphere have

> Parts of Mars are far drier than a bone, while others had standing water for extended periods

evolved. (For example, the oldest rocks on Earth have a chemical structure that shows the rocks formed during a time when there was little oxygen in the atmosphere and hence before the appearance of photosynthetic planet life.) Planetary scientists who want to learn about the history of Venus and Mars are therefore very interested in looking at ancient rocks on those worlds.

On Venus we cannot look very far into the past, since the planet's volcanic activity and flake tectonics have erased features more than about 500 million years old (see Section 11-5). On Mars, by contrast, we expect to find rocks that are billions of years old: The thin atmosphere causes very little erosion, there has been little recent volcanic activity to cover ancient rocks, and there is no subduction to drag old surface features back into the planet's interior. (The heavy cratering of the southern highlands bears testament to the great age of much of the Martian surface.) In a quest to find these ancient rocks and learn about the early history of Mars, scientists have sent spacecraft to land on the red planet and explore it at close range.

Landing on Mars

As of 2012, seven spacecraft have successfully landed on the Martian surface. *Viking Lander 1* and *Viking Lander 2*, which we first discussed in Section 11-6, touched down in 1976. They had no capability to move around on the surface, and so each *Viking Lander* could only study objects within reach of the scoop at the end of its mechanical arm (**Figure 11-33a**). Figure 11-28 shows the view from the two *Viking Lander* sites.

Three more recent unmanned explorers, *Mars Pathfinder* (landed 1997) and two *Mars Exploration Rovers* named *Spirit* and *Opportunity* (landed 2004), were equipped with electrically driven wheels that allow travel over the surface. Unlike the *Viking Landers,* these rovers did not have a heavy and expensive rocket engine to lower them gently to the Martian surface. Instead, the spacecraft was surrounded by airbags like those used in automobiles. Had there been anyone to watch *Mars Pathfinder* or one of the *Mars Exploration Rovers* land, they would have seen an oversized beach ball hit the surface at 50 km/h (30 mi/h), then bounce for a kilometer before finally rolling to a stop. Unharmed by its wild ride, the spacecraft then deflated its airbags and began to study the geology of Mars. In 2008, the *Phoenix Mars Lander* sampled a challenging north polar region over a period of a few months; it did not rove around.

Figure 11-17 shows the landing sites of the *Mars Pathfinder* rover (named *Sojourner*) and the two *Mars Exploration Rovers* (*Spirit* and *Opportunity*). Following directions from Earth, *Sojourner* spent three months exploring the rocks around its landing site. The *Mars Exploration Rovers* were also designed for a three-month lifetime, and both remained operational years after landing. In fact, as of this writing, *Opportunity* has traveled

Cameras   *Mars Exploration Rover*   *Mars Pathfinder* rover

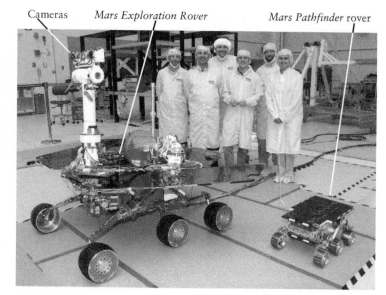

(a) Two generations of rovers

(b) *Curiosity* rover

FIGURE 11-33 R I V U X G

Martian Rovers **(a)** The technicians in this photograph pose with one of the *Mars Exploration Rovers* and with a replica of its predecessor, the *Mars Pathfinder* rover. **(b)** The rover *Curiosity* is much bigger than previous rovers; you can compare to the height of a person in parts (a) and (b). *Curiosity* will search for areas that may have been suitable for life. (a: JPL/NASA; b: NASA/JPL-Caltech)

more than 35 km over the Martian surface and continues to collect data. However, *Spirit* became stuck in soft sand in 2009 and communication was lost a year later. In addition to carrying cameras, most landers carried a device called an X-ray spectrometer to measure the chemical composition of rock samples. *Spirit* and *Opportunity* also carried a tool designed to grind holes into rocks, thus revealing parts of the rock's interior that have not been weathered by exposure to the Martian atmosphere.

The most recent Mars landing was by the rover *Curiosity* (Figure 11-33b). The stunning landing, engineered by NASA's Mars Science Laboratory mission, caught the public imagination in August 2012. *Curiosity* is too massive (900 kg) for the airbag landing scenario used to land earlier rovers. Instead, a parachute followed by rocket-thrusters provided the necessary speed reduction, but not all the way to the surface. To avoid stirring up too much dust onto the rover, a "sky crane" began lowering *Curiosity* down the last 20 meters. *Curiosity* is designed to analyze the habitability of Mars—the past or present conditions favorable for Martian microbial life. The rover is specially equipped to search for organic compounds thought to be necessary for life, as well as for chemical clues that life once existed. *Curiosity* is not equipped to study for life directly, and the promising ancient habitats it finds will need to be followed up by other missions in the search for life. As of early 2013, *Curiosity* was just beginning to analyze the surrounding terrain. For more about the *Curiosity* mission, see the *Scientific American* article at the end of this chapter.

The Martian Regolith

The *Viking Landers* were the first to reveal the nature of the Martian regolith, or surface material. The scoop on the extendable arm (Figure 11-33a) was used to dig into the regolith and retrieve samples for analysis. Bits of the regolith clung to a magnet mounted on the scoop, indicating that the regolith contains iron. The X-ray spectrometers showed that samples collected at both *Viking Lander* sites were rich in iron, silicon, and sulfur.

The *Viking Landers* each carried a miniature chemistry laboratory for analyzing the regolith. Curiously, when a small amount of water was added to a sample of regolith, the sample began to fizz and release oxygen. This experiment indicated that the regolith contains unstable chemicals called peroxides and superoxides, which break down in the presence of water to release oxygen gas.

The unstable chemistry of the Martian regolith probably comes from solar ultraviolet radiation that beats down on the planet's surface. (Unlike Earth, Mars has no ozone layer to screen out ultraviolet light.) Ultraviolet photons easily break apart molecules of carbon dioxide (CO_2) and water vapor (H_2O) by knocking off oxygen atoms, which then become loosely attached to chemicals in the regolith.

Among the most important scientific goals of the *Viking Landers* was to search for evidence of life on Mars. Although it was clear that Mars has neither civilizations nor vast fields of plants, it still seemed possible that simple microorganisms could have evolved on Mars during its early brief epoch of Earthlike climate. If so, perhaps some species had survived to the present day. To test this idea, the *Viking Landers* looked for biologically significant chemical reactions in samples of the Martian regolith, but they failed to detect any reactions of this type. Perhaps life never existed on Mars, or perhaps it existed once but has since vanished. (In Chapter 27, we discuss the *Viking Lander* experiments in detail.)

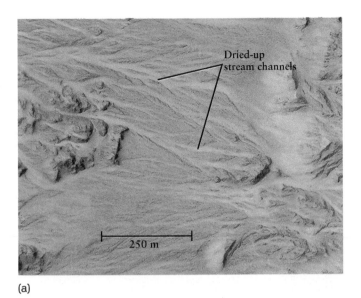

(a)

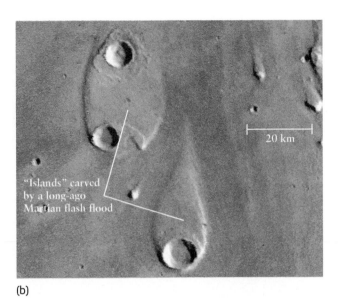

(b)

FIGURE 11-34   R I 🔽 U X G

Sign of Ancient Martian Water **(a)** This *Mars Reconnaissance Orbiter* image shows patterns from flowing water almost identical to those caused by rainfall in deserts on Earth. No rain falls in today's sparse Martian atmosphere, so these must date from an earlier era when the atmosphere was thicker.

(b) These features rise above the floor of Ares Valles. They were carved out by a torrent of water that flowed from the bottom of the image toward the top. Similar flood-carved islands are found on Earth in eastern Washington State. (a: NASA/JPL/U. of Arizona; b: NASA/USGS)

Exploring Martian Rocks

Images made from Mars orbit show a number of features that suggest water flowed there in the past (Figure 11-34). To investigate these, *Mars Pathfinder* landed in an area of the northern lowlands called Ares Valles, which appears to be an ancient flood plain. Figure 11-34b shows a portion of this plain. Flood waters can carry rocks of all kinds great distances from their original locations, so scientists expected that this landing site (Figure 11-35) would show great geologic diversity. They were not disappointed. Many of the rocks appeared to be igneous rock produced by volcanic action. Other rocks, which resemble the impact breccias found in the ancient highlands of the Moon (see Figure 10-18), appear to have come from areas that were struck by crater-forming meteoroids. Still others have a layered structure like sedimentary rocks on Earth (see Figure 9-18b). Such rocks form gradually at the bottom of bodies of water, suggesting that liquid water was stable in at least some regions of Mars for a substantial period of time.

Evidence for Ancient Martian Water

VIDEO 11-2   VIDEO 11-3   In the search for life beyond Earth, the search for signs of ancient water on Mars has been one of NASA's primary missions. If Mars ever supported life, biologists expect Martian life to have been dependent on water, just as it is for life on Earth. By finding where liquid water existed on Mars, rovers such as *Curiosity* can narrow the search further by looking for organic compounds, which are the building blocks of life (Chapter 27).

The two *Mars Exploration Rovers* landed in sites that were suspected to have been under water in the distant past. *Spirit* touched down in Gusev Crater, a large impact basin. A valley opens into this crater, resembling a river that runs into a lake. The destination for *Opportunity* was the flat plain of Sinus Meridianii, where orbiting spectrometers had spotted evidence of hematite—a type of iron oxide that, on Earth, tends to form in wet locations.

FIGURE 11-35   R I 🔽 U X G

Roving the Martian Surface This view from the stationary part of *Mars Pathfinder* shows the rover, called *Sojourner,* deploying its X-ray spectrometer against a rock named Moe. Scientists gave other rocks whimsical names such as Jiminy Cricket and Pooh Bear. (NASA/JPL)

(a) Volcanic rocks

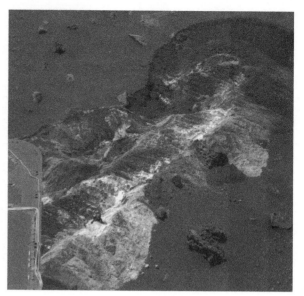

(b) Sulfur salts

FIGURE 11-36   R I V U X G

Dry Mars At the *Spirit* landing site, both **(a)** the dusty volcanic rocks that dominate the site and **(b)** these light-colored sulfur salts contain minerals that break apart upon contact with liquid water. Hence, this area must have been dry for billions of years. (a: NASA/JPL-Caltech/Cornell; b: NASA/JPL/Cornell)

Contrary to expectations, most of the rocks encountered by *Spirit* show *no* evidence that they have been exposed to water. The volcanic rocks found in and around Gusev Crater (Figure 11-36a) are composed of the minerals olivine and pyroxene, which break down if they are exposed to liquid water. In a few locations *Spirit* found deposits of sulfur salts, which can be formed by the action of water (Figure 11-36b). However, olivine and pyroxene were still found among the sulfur salts. While it remains a possibility that Gusev Crater was once a river-fed lake, the results from *Spirit* show that for the past billion years this region has been drier than any desert on Earth.

The results from *Opportunity* were entirely different. This rover discovered outcrops of sedimentary rocks on the plains of Sinus Meridianii (Figure 11-37). On these plains, *Opportunity* found millimeter-sized spheres of hematite that, like analogous spheres found on Earth, probably formed from iron-rich water circulating through sediment. These observations suggest Sinus Meridianii, which is among the flattest places known in the solar

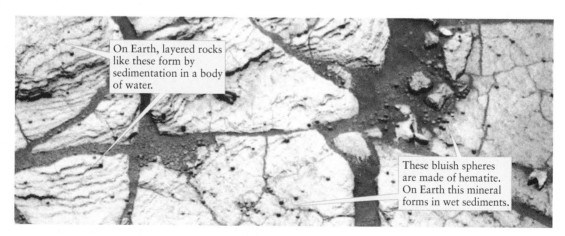

On Earth, layered rocks like these form by sedimentation in a body of water.

These bluish spheres are made of hematite. On Earth this mineral forms in wet sediments.

FIGURE 11-37   R I V U X G

Wet Mars This false-color image shows two signatures of water found by *Opportunity*. Both the sedimentary rock (exposed by erosion) and the hematite spheres appearing blue require standing water to form. So, this area must have been under water for an extended period. The small spheres are typically a few millimeters in size and in these false-color images are often referred to as "blueberries." (NASA/JPL/Cornell)

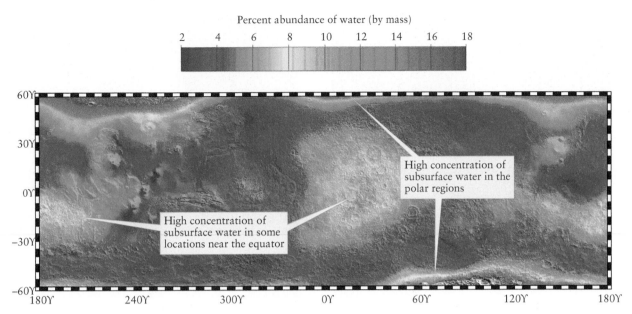

Percent abundance of water (by mass)

FIGURE 11-38

Water Beneath the Martian Surface This map of the abundance of water (probably in the form of ice) just below the surface of Mars is based on data collected from orbit by the *Mars Odyssey* spacecraft. These measurements can only reveal the presence of water or water-ice to a depth of about a meter; there may be much more water at greater depths. (LANL)

system, was once a lake bed. It is not known how long ago this area was under water, but the density of craters found there suggests that the surface formed 3 to 4 billion years ago.

When combined with observations from Mars orbit, the *Spirit* and *Opportunity* results show that ancient Mars was a very diverse planet. Most of the planet was probably very dry, like Gusev Crater, but there were isolated areas like Sinus Meridianii that had liquid water for some period of time. It remains a challenge to planetary scientists to explain why liquid water appeared on Mars only in certain places and at certain times.

Where Is the Water Now?

Whatever liquid water once coursed over parts of the Martian surface is presumably now frozen, either at the polar ice caps or beneath the planet's surface. To check this idea, scientists used a device on board the *Mars Odyssey* orbiter to measure the result of cosmic rays hitting Mars.

Cosmic rays are fast-moving subatomic particles that enter our solar system from interstellar space. When these particles strike and penetrate into the Martian soil, they collide with the nuclei of atoms in the soil and knock out neutrons. (Recall from Section 5-7 that atomic nuclei contain both positively charged protons and electrically neutral neutrons.) If the ejected neutrons originate within a meter or so of the surface, they can penetrate through the regolith and escape into space. However, if there are water molecules (H_2O) in the surface, the nuclei of the hydrogen atoms in these molecules tend to absorb the neutrons and prevent them from escaping. An instrument on board *Mars Odyssey* detects neutrons coming from Mars; where the instrument finds a *deficiency* of neutrons is where the soil contains hydrogen and, presumably, water.

The results from the *Mars Odyssey* neutron measurements show that there is abundant water beneath the surface at both Martian poles (Figure 11-38). But they also show a surprising amount of subsurface water in two regions near the equator, Arabia Terra (see Figure 11-13a and Figure 11-17) and another region on the opposite side of the planet. Recently, even small impacts have directly revealed ice just beneath the surface, but far away from the poles (Figure 11-39).

One idea explaining water-ice near the equator is that about a million years ago, gravitational forces from other planets caused a temporary change in the tilt of Mars's rotation axis. The ice caps were then exposed to more direct sunlight, causing some of the water to evaporate and eventually refreeze elsewhere on the planet. A big change in the tilt of Mars's axis would not be surprising for two reasons. First, Earth's large Moon stabilizes our rotation axis, but the Martian moons are too small to provide the planet with stabilizing forces. And second, Mars is close enough to Jupiter to be significantly influenced by Jupiter's strong gravity.

There are other surprising features on Mars related to the presence of water. A number of *Mars Global Surveyor* images show gullies apparently carved by water flowing down the walls of pits or craters (Figure 11-40a). These gullies could have formed when water trapped underground—where the pressure is greater and water can remain a liquid—seeped onto the surface. The gullies appear to be geologically young, which suggests that even today some liquid water survives below the Martian surface. Alternatively, these gullies may have formed during periods when the Martian climate was colder than usual and the slopes of craters were covered with a layer of dust and snow. Melting snow on the underside of this layer could have carved out gullies, which were later exposed when the climate

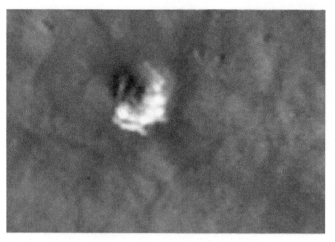

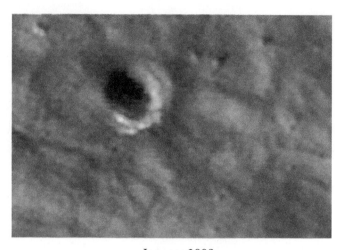

October 2008

January 2009

FIGURE 11-39   R I V U X G

Water Revealed by Recent Impacts A fresh impact observed in 2008 revealed water-ice. The small crater is only 6 meters wide and shows that ice is just beneath the Martian surface about halfway between the north pole and the equator. A spectrum taken of the white material shows that it is water, and not carbon dioxide. By January 2009, several months later, the water had evaporated. Quite surprisingly, the white material is 99% pure water; a mixture with about 50% dirt was expected, and this remains a mystery. (NASA/JPL-Caltech/University of Arizona)

warmed and the snow evaporated. Very recently, deposits have also appeared on crater walls (Figure 11-40b)—indicating ongoing erosion—but it is not clear if water is involved. At least in the past, there's evidence of liquid water resulting from a subsurface ice-melt. In Figure 11-41, flood channels emerge from a crater that would have been heated by the initial impact.

In 2012, *Curiosity* found strong evidence for the *steady flow* of water in an ancient stream or river, as opposed to a flash flood. Figure 11-42 shows rounded pebbles, just as we find in dry streambeds here on Earth. The pebbles are too big to have been moved by wind and indicate water-depths between ankle and knee high. Other nearby sites show similar rounding of pebbles,

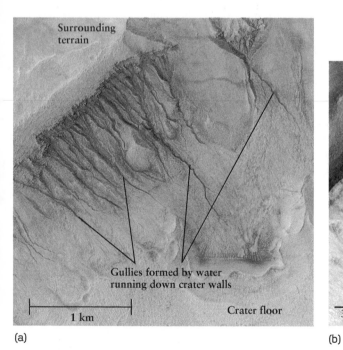

Surrounding terrain

Gullies formed by water running down crater walls

1 km

Crater floor

(a)

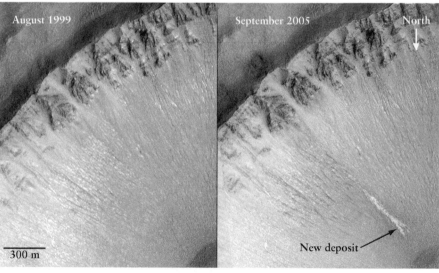

August 1999

September 2005

North

300 m

New deposit

(b)

FIGURE 11-40   R I V U X G

Martian Gullies (a) When the *Mars Global Surveyor* spacecraft looked straight down into this crater in the Martian southern hemisphere, it saw a series of gullies along the crater wall. These gullies may have formed by subsurface water seeping out to the surface, or by the melting of snow that fell on crater walls. (b) These two images of a southern hemisphere crater were taken six years apart. The new deposit seen in 2005 may have been caused by a subsurface melt, or might be a landslide unassociated with liquid water. (a and b: NASA/JPL/Malin Space Science Systems)

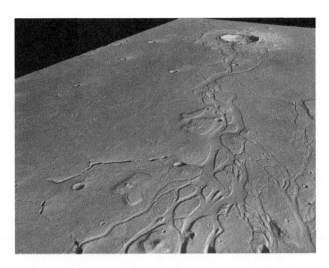

FIGURE 11-41 R I V U X G

Impact Melts Subsurface Ice The flood channels of Hephaestus Fossae run 600 km from a 20 km wide impact crater. Heat from the impact is thought to have caused frozen subsurface water-ice to melt. (NASA/DLR/FU Berlin [G. Neukum])

and based on images from orbit, the entire region has long been suspected as the outlet of an ancient river.

Just how much water is thought to have once been in liquid form on Mars? By examining flood channels at various locations around Mars, Michael Carr of the U.S. Geological Survey has estimated that there was enough water to cover the planet to a depth of 500 m (1600 ft). (By comparison, Earth has enough water to cover our planet to a depth of 2700 m.) Thus, while the total amount of water on Mars is less than on Earth, there may have

FIGURE 11-42 R I V U X G

Dry Streambed The rounding of the pebbles at this site indicate they were exposed to flowing water for a period lasting at least thousands of years. The pebbles, and certainly the largest rock (in the white circle), could not have been rounded or transported by winds in the thin Martian atmosphere. The jagged overhanging rock ledge (running down the middle of the image) appears to be the bank of the stream, undercut by erosion from the stream itself. Even the rocky ledge appears to be made of a sedimentary rock from prior exposure to water. (NASA)

been enough to form lakes or seas. At least some of this water is now frozen under the Martian surface and at its poles.

CONCEPTCHECK 11-11

What are two different lines of evidence that point to present-day subsurface ice on Mars?

Answer appears at the end of the chapter.

11-9 The two Martian moons resemble asteroids

Two moons move around Mars in orbits close to the planet's surface. These satellites are so tiny that they were not discovered until the favorable opposition of 1877. While Schiaparelli was seeing canals, the American astronomer Asaph Hall spotted the two moons, which orbit almost directly above its equator. He named them Phobos ("fear") and Deimos ("panic"), after the mythical horses that drew the chariot of the Greek god of war.

The Orbits of Phobos and Deimos

Phobos is the inner and larger of the Martian moons. It circles Mars in just 7 hours and 39 minutes at an average distance of only 9378 km (5828 mi) from

> As seen from Mars, Deimos rises in the east but Phobos rises in the west

the center of Mars. (Recall that our Moon orbits at an average distance of 384,400 km from Earth's center.) This orbital period is much shorter than the Martian day; hence, Phobos rises in the *west* and gallops across the sky in only 5½ hours, as viewed by an observer near the Martian equator. During this time, Phobos appears several times brighter in the Martian sky than Venus does from Earth.

Deimos, which is farther from Mars and somewhat smaller than Phobos, appears about as bright from Mars as Venus does from Earth. Deimos orbits at an average distance of 23,460 km (14,580 mi) from the center of Mars, slightly beyond the right distance for a synchronous orbit—an orbit in which a satellite seems to hover above a single location on the planet's equator. As seen from the Martian surface, Deimos rises in the east and takes about 3 full days to creep slowly from one horizon to the other. (Figure 11-43a).

Images from spacecraft in Mars's orbit reveal Phobos and Deimos to be jagged and heavily cratered (Figure 11-43b). Both moons are somewhat elongated from a spherical shape: Phobos is approximately 28 by 23 by 20 km in size, while Deimos is slightly smaller, roughly 16 by 12 by 10 km. Phobos and Deimos are both in synchronous rotation: The tidal forces that Mars exerts on these elongated moons cause them to always keep the same sides toward Mars, just as Earth's tidal forces ensure that the same side of the Moon always faces us (see Section 4-8).

We saw in Section 10-5 that the Moon raises a tidal bulge on Earth. Because Earth rotates on its axis more rapidly than the Moon orbits Earth, this bulge is dragged ahead of a line connecting Earth and the Moon (see Figure 10-20). The gravitational force of this bulge pulls the Moon into an ever-larger orbit. In the same way, and for the same reason, Deimos is slowly moving away from Mars.

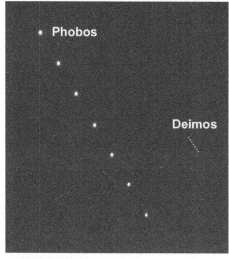

Phobos

Deimos

(a) As seen from Mars

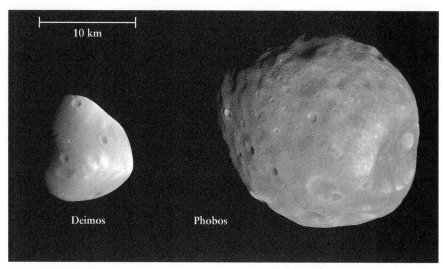

10 km

Deimos

Phobos

(b) As seen from Mars orbit

FIGURE 11-43   R I ☒ U X G

The Moons of Mars (a) The *Spirit* rover recorded seven images of the Martian night sky at 15-minute intervals to produce this composite view of the motions of Deimos and Phobos. **(b)** Two *Mars Reconnaissance Orbiter* images have been combined to show Deimos and Phobos and their heavily cratered surfaces to the same scale. The gravity on these moons is only about 1/1000 of Earth's, and not strong enough to shape them into spheres. (a: NASA/JPL/Cornell/Texas A&M; b: NASA/JPL Caltech/University of Arizona)

But for fast-moving Phobos, the situation is reversed: Mars rotates on its axis more slowly than Phobos moves around Mars. This slow rotation holds back the tidal bulge that Phobos raises in the solid body of Mars, and this bulge thus is *behind* a line connecting the centers of Mars and Phobos. (Imagine that the bulge in Figure 10-20 is tilted clockwise rather than counterclockwise.) This bulge exerts a gravitational force on Phobos that pulls it back, slowing the moon's motion and making it slowly spiral inward toward Mars. This inward spiral will cease in about 40 million years when Phobos smashes into the Martian surface!

The Nature of the Martian Moons

The origin of the Martian moons is unknown. One idea is that they are asteroids that Mars captured from the nearby asteroid belt (see Section 7-5). Like Phobos and Deimos, many asteroids reflect very little sunlight (typically less than 10%) because of their high carbon content. Phobos and Deimos also have very low densities (1900 kg/m³ and 1760 kg/m³, respectively), which are also characteristic of carbon-rich asteroids. Perhaps two of these asteroids wandered close enough to Mars to become permanently trapped by the planet's gravitational field. Alternatively, Phobos and Deimos may have formed in orbit around Mars out of debris left over from the formation of the planets. Some of this debris might have come from the asteroid belt, giving these two moons an asteroidlike character. Future measurements of the composition of the Martian moons—perhaps by landing spacecraft on their surfaces—may shed light on their origin.

CONCEPTCHECK **11-12**

What causes Phobos to slowly spiral in toward Mars?

Answer appears at the end of the chapter.

KEY WORDS

1-to-1 spin-orbit coupling, p. 279
3-to-2 spin-orbit coupling, p. 281
antipode, p. 285
cosmic rays, p. 309
crustal dichotomy, p. 290
dust devil, p. 301
flake tectonics, p. 292
greatest eastern elongation, p. 278
greatest western elongation, p. 278

hot-spot volcanism, p. 294
northern lowlands, p. 290
prograde rotation, p. 283
residual ice cap, p. 300
retrograde rotation, p. 283
runaway greenhouse effect, p. 304
runaway icehouse effect, p. 304
scarp, p. 284
shield volcano, p. 294
southern highlands, p. 290
volatile elements, p. 287

KEY IDEAS

Motions of Mercury, Venus, and Mars in Earth's Sky: Mercury and Venus can be seen in the morning or evening sky only, while it is possible to see Mars at any time of night depending on its position in its orbit.

• At their greatest eastern and western elongations, Mercury is only 28° from the Sun and Venus is only 47° from the Sun.

Rotation of Mercury, Venus, and Mars: Poor telescopic views of Mercury's surface led to the mistaken impression that the planet always keeps the same face toward the Sun (1-to-1 spin-orbit coupling).

• Radio and radar observations revealed that Mercury in fact has 3-to-2 spin-orbit coupling: The planet rotates on its axis three times every two orbits.

- Venus rotates very slowly in a retrograde direction. Its rotation period is longer than its orbital period.

- Mars rotates at almost the same rate as Earth, and its rotation axis is tilted by almost the same angle as Earth's axis.

Mercury's Surface, Interior, and Magnetic Field: Mercury's surface is pocked with craters, but there are extensive smooth plains between these craters.

- Long cliffs called scarps meander across the surface of Mercury. These probably formed as the planet's crust cooled, solidified, and shrank.

- Mercury has an iron core with a diameter equal to about 83% of the planet's diameter. By contrast, the diameter of Earth's core is only slightly more than one-half of Earth's diameter.

- Mercury has a weak magnetic field, which indicates that at least part of the iron core is liquid. However, the field is off-center by 20%, which remains unexplained.

- The high abundance of volatile elements that should have "boiled off" from Mercury's material during formation presents a mystery considering the planet's close proximity to the Sun.

Comparing Venus and Mars: Most of the surface of Venus is at about the same elevation, with just a few elevated regions. On Mars, the southern highlands rise several kilometers above the northern lowlands.

- Venus has a thick atmosphere and a volcanically active surface. Mars has a very thin atmosphere and little or no current volcanism.

- There is no evidence of plate tectonics on Venus, but Valles Marineris shows there are at least two plates on Mars. On Venus, there is vigorous convection in the planet's interior, but the crust is too thin to move around in plates; instead, it wrinkles and flakes. On Mars, the planet's smaller size means the crust cooled long ago and became too thick to allow widespread plate tectonic activity.

- Volcanoes on both Venus and Mars were produced by hot spots in the planet's interior.

- The entire Venusian surface is about 500 million years old and has relatively few craters. By contrast, most of the Martian surface is cratered and is probably billions of years old. The southern highlands on Mars are the most heavily cratered and hence the oldest part of the planet's surface.

The Atmospheres of Venus and Mars: Both planetary atmospheres are more than 95% carbon dioxide, with a few percent of nitrogen.

- The pressure at the surface of Venus is about 90 atmospheres. The greenhouse effect is very strong, which raises the surface temperature to 460°C. The pressure at the surface of Mars is only 0.006 atmosphere, and the greenhouse effect is very weak.

- The permanent high-altitude clouds on Venus are made primarily of sulfuric acid. By contrast, the few clouds in the Martian atmosphere are composed of water-ice and carbon dioxide ice.

- The circulation of the Venusian atmosphere is dominated by two huge convection currents in the cloud layers, one in the northern hemisphere and one in the southern hemisphere. The upper cloud layers of the Venusian atmosphere move rapidly around the planet in a retrograde direction, with a period of only about 4 Earth days.

- Weather on Mars is dominated by the north and south flow of carbon dioxide from pole to pole with the changing seasons. This can trigger planetwide dust storms.

Evolution of Atmospheres: Earth, Venus, and Mars all began with relatively thick atmospheres of carbon dioxide, water vapor, and sulfur dioxide.

- On Earth, most of the carbon dioxide went into carbonate rocks and most of the water into the oceans. Ongoing plate tectonics recycles atmospheric gases through the crust.

- On Venus, more intense sunlight and the absence of plate tectonics led to a thick carbon dioxide atmosphere and a runaway greenhouse effect.

- On Mars, a runaway icehouse effect resulted from weaker sunlight and a lack of strong plate tectonic activity.

Water on Mars: Liquid water cannot exist on present-day Mars because the atmosphere is too thin and cold. But there is evidence for frozen water at the polar ice caps and beneath the surface of the regolith.

- Geological evidence from unmanned rovers shows that much of the Martian surface has been dry for billions of years, but some regions had substantial amounts of liquid water in the past.

The Moons of Mars: Mars has two small, football-shaped satellites that move in orbits close to the surface of the planet. They may be captured asteroids or may have formed in orbit around Mars out of solar system debris.

QUESTIONS

Review Questions

1. Why is it impossible to see Mercury or Venus in the sky at midnight?

2. In his 1964 science fiction story "The Coldest Place," author Larry Niven described the "dark side" of Mercury as the coldest place in the solar system. What assumption did he make about the rotation of Mercury? Did this assumption turn out to be correct?

3. What is 3-to-2 spin-orbit coupling? How is the rotation period of an object exhibiting 3-to-2 spin-orbit coupling related to its orbital period? What aspects of Mercury's orbit cause it to exhibit 3-to-2 spin-orbit coupling? What telescopic observations proved this?

4. Why was it so difficult to determine the rate and direction of Venus's rotation? How were these finally determined? What is one proposed explanation for the slow, retrograde rotation of Venus?

5. Explain why Mercury does not have a substantial atmosphere.

6. What kind of surface features are found on Mercury? How do they compare to surface features on the Moon? Why are they probably much older than most surface features on Earth?

7. How do we know that the scarps on Mercury are younger than the lava flows? How can you tell that the scarp in

Figure 11-7 is younger than the vertically distorted crater at the center of the figure?

8. If Mercury is the closest planet to the Sun and has such a high average surface temperature, how is it possible that ice might exist on its surface?

9. Why do astronomers think that Mercury has a very large iron core?

10. Why is it surprising that Mercury has a global magnetic field? Why does the 58.646-day rotation period of Mercury imply that the planet can have only a weak magnetic field?

11. Is there a consistent explanation for why Mercury's magnetic field is off-center? If Earth's magnetic field was off-center by the same percentage, how many kilometers would that be?

12. Why are high concentrations of volatile elements a mystery on Mercury?

13. Why was it difficult to determine Venus's surface temperature from Earth? How was this finally determined?

14. The *Mariner 2* spacecraft did not enter Venus's atmosphere, but it was nonetheless able to determine that the atmosphere contains little water vapor. How was this done?

15. How did winds trick early observers of Mars into thinking that Mars had vegetation?

16. Do Venus and Mars have continents like those on Earth?

17. What is the Martian crustal dichotomy? What is the evidence that the southern highlands are older than the northern lowlands?

18. What is the evidence that the surface of Venus is only about 500 million years old?

19. *TUTORIAL 11-1* What is flake tectonics? Why does Venus exhibit flake tectonics rather than plate tectonics?

20. What geologic features (or lack thereof) on Mars have convinced scientists that extensive plate tectonics did not significantly shape the present Martian surface? Does Mars have any plates now?

21. What magnetic features indicate that Mars might have experienced plate tectonics in the past?

22. Compare the volcanoes of Venus, Earth, and Mars. Cite evidence that hot-spot volcanism is or was active on all three worlds.

23. Describe the evidence that there has been recent volcanic activity (a) on Venus and (b) on Mars.

24. Suppose all of Venus's volcanic activity suddenly stopped. (a) How would this affect Venus's clouds? (b) How would this affect the overall Venusian environment?

25. Why are the patterns of convection in the Venusian atmosphere so different from those in our atmosphere?

26. Why is it impossible for liquid water to exist on Mars today? If liquid water existed on Mars in the past, what must have been different then?

27. Why does the atmospheric pressure on Mars vary with the seasons? What is the relationship between this pressure variation and Martian dust storms?

28. *TUTORIAL 11-2* Why is it reasonable to assume that the primordial atmospheres of Earth, Venus, and Mars were roughly the same?

29. Carbon dioxide accounts for about 95% of the present-day atmospheres of both Mars and Venus. Why, then, is there a strong greenhouse effect on Venus but only a weak greenhouse effect on Mars?

30. (a) What is a runaway greenhouse effect? (b) What is a runaway icehouse effect?

31. Explain two ways that heat within Mars helped to maintain a greenhouse effect.

32. What is a dust devil? Why would you feel much less breeze from a Martian dust devil than from a dust devil on Earth?

33. (a) Why is Mars red? (b) Why is the Martian sky the color of butterscotch?

34. Were the *Viking Landers* able to determine whether life currently exists on Mars or whether it once existed there? Why or why not?

35. (a) The *Spirit* rover found the minerals olivine and pyroxene at its landing site on Mars. Explain how this shows that there has been no liquid water at that site for billions of years. (b) What evidence did the *Opportunity* rover find at its landing site to suggest that liquid water had once been present there?

36. How was the *Mars Odyssey* spacecraft able to detect water beneath the Martian surface without landing on the planet?

37. How do craters reveal evidence for water-ice beneath the Martian surface?

38. A full moon on Earth is bright enough to cast shadows. As seen from the Martian surface, would you expect a full Phobos or full Deimos to cast shadows? Why or why not?

Advanced Questions

> **Problem-solving tips and tools**
>
> We discussed the relationship between angular distance and linear distance in Box 1-1 and the idea of angular resolution in Section 6-3. You may need to refresh your memory about Kepler's third law, described in Section 4-4. You may also need to review the form of Kepler's third law that explicitly includes mass (see Section 4-7 and Box 4-4). Box 4-1 describes the relationship between a planet's sidereal orbital period and its synodic period (the time from one inferior conjunction to the next). You should recall that Wien's law (Section 5-4) relates the temperature of a blackbody to λ_{max}, its wavelength of maximum emission. Section 5-9 and Box 5-6 explain the Doppler effect and how to do calculations using it. The linear speed of a point on a planet's equator is the planet's circumference divided by its rotation period; recall that the circumference of a circle of radius r is $2\pi r$. The volume of a sphere of radius r is $4\pi r^3/3$. The speed of light is given in Appendix 7.

39. Figure 11-1 shows Mercury with a greatest eastern elongation of 18° and a greatest western elongation of 28°. On

November 25, 2006, Mercury was at a greatest western elongation of 20°. Was Mercury at perihelion, aphelion, or some other point on its orbit? Explain your answer.

40. Venus takes 440 days to move from greatest western elongation to greatest eastern elongation, but it needs only 144 days to go from greatest eastern elongation to greatest western elongation. With the aid of a diagram like Figure 11-1, explain why.

41. Venus's sidereal rotation period is 243.01 days and its orbital period is 224.70 days. Use these data to prove that a solar day on Venus lasts 116.8 days. (*Hint:* Develop a formula relating Venus's solar day to its sidereal rotation period and orbital period similar to the first formula in Box 4-1.)

42. In Section 11-2 we described the relationship between the length of Venus's synodic period and the length of an apparent solar day on Venus. Using this relationship and a diagram, explain why at each inferior conjunction the same side of Venus is turned toward Earth.

43. This time-lapse photograph was taken on May 7, 2003, during a *solar transit* of Mercury. Over a period of 5 hours and 19 minutes, Mercury appeared to move across the face of the Sun. Such solar transits of Mercury occur 13 or 14 times each century; they do *not* happen each time that Mercury is at inferior conjunction. Explain why not. (*Hint:* For a solar transit to occur, the Sun, Mercury, and Earth must be in a nearly perfect alignment. Does the orbit of Mercury lie in the plane of the ecliptic?)

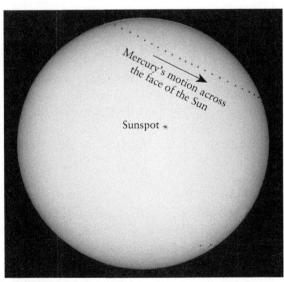

R I V U X G (Dominique Dierick)

44. Find the largest angular size that Mercury can have as seen from Earth. In order for Mercury to have this apparent size, at what point in its orbit must it be?

45. (a) Suppose you have a telescope with an angular resolution of 1 arcsec. What is the size (in kilometers) of the smallest feature you could have seen on the Martian surface during the opposition of 2005, when Mars was 0.464 AU from

Earth? (b) Suppose you had access to the Hubble Space Telescope (HST), which has an angular resolution of 0.1 arcsec. What is the size (in kilometers) of the smallest feature you could have seen on Mars with the HST during the 2005 opposition?

46. For a planet to appear to the naked eye as a disk rather than as a point of light, its angular size would have to be 1 arcmin, or 60 arcsec. (This is the same as the angular separation between lines in the bottom row of an optometrist's eye chart.) (a) How close would you have to be to Mars in order to see it as a disk with the naked eye? Does Mars ever get this close to Earth? (b) Would Earth ever be visible as a disk to an astronaut on Mars? Would she be able to see Earth and the Moon separately, or would they always appear as a single object? Explain your answers.

47. Imagine that you are part of the scientific team monitoring a spacecraft that has landed on Mars. At 5:00 P.M. in your control room on Earth, the spacecraft reports that the Sun is highest in the sky as seen from its location on Mars. When the Sun is next at its highest point as seen from the spacecraft, what time will it be in the control room on Earth?

48. (a) Mercury has a 58.646-day rotation period. What is the speed at which a point on the planet's equator moves due to this rotation? (*Hint:* Remember that speed is distance divided by time. What distance does a point on Mercury's equator travel as the planet makes one rotation?) (b) Use your answer to (a) to answer the following: As a result of rotation, what difference in wavelength is observed for a radio wave of wavelength 12.5 cm (such as is actually used in radar studies of Mercury) emitted from either the approaching or receding edge of the planet?

49. The orbital period of *Mariner 10* is twice that of Mercury. Use this fact to calculate the length of the semimajor axis of the spacecraft's orbit.

50. Consider the idea that Mercury has a solid iron-bearing mantle that is permanently magnetized like a giant bar magnet. Using the fact that iron demagnetizes at temperatures above 770°C, present an argument against this explanation of Mercury's magnetic field.

51. (a) At what wavelength does Venus's surface emit the most radiation? (b) Do astronomers have telescopes that can detect this radiation? (c) Why can't we use such telescopes to view the planet's surface?

52. The *Mariner 2* spacecraft detected more microwave radiation when its instruments looked at the center of Venus's disk than when it looked at the edge, or limb, of the planet. (This effect is called *limb darkening*.) Explain how these observations show that the microwaves are emitted by the planet's surface rather than its atmosphere.

53. In the classic Ray Bradbury science fiction story "All Summer in a Day," human colonists on Venus are subjected to continuous rainfall except for one day every few years when the clouds part and the Sun comes out for an hour or so. Discuss

how our understanding of Venus's atmosphere has evolved since this story was first published in 1954.

54. Suppose that Venus had no atmosphere at all. How would the albedo of Venus then compare with that of Mercury or the Moon? Explain your answer.

55. A hypothetical planet has an atmosphere that is opaque to visible light but transparent to infrared radiation. How would this affect the planet's surface temperature? Contrast and compare this hypothetical planet's atmosphere with the greenhouse effect in Venus's atmosphere.

56. Earth's northern hemisphere is 39% land and 61% water, while its southern hemisphere is only 19% land and 81% water. Thus, the southern hemisphere could also be called the "water hemisphere." The Moon also has two distinct hemispheres, the near side (which has a number of maria) and the far side (which has almost none). How are these hemispheric differences on Earth and on the Moon similar to the Martian crustal dichotomy? How are they different?

57. For a group of properly attired astronauts equipped with oxygen tanks, a climb to the summit of Olympus Mons would actually be a relatively easy (albeit long) hike rather than a true mountain climb. Give two reasons why.

58. On Mars, the difference in elevation between the highest point (the summit of Olympus Mons) and the lowest point (the bottom of the Hellas Planitia basin) is 30 km. On Earth, the corresponding elevation difference (from the peak of Mount Everest to the bottom of the deepest ocean) is only 20 km. Discuss why the maximum elevation difference is so much greater on Mars.

59. The *Mars Global Surveyor* (MGS) spacecraft was in a nearly circular orbit around Mars with an orbital period of 117 minutes. (a) Using the data in Table 11-3, find the radius of the orbit. (b) What is the average altitude of MGS above the Martian surface? (c) The orbit of MGS passed over the north and south poles of Mars. Explain how this makes it possible for the spacecraft to observe the entire surface of the planet.

60. The elevations in Figure 11-17 were measured using an instrument called MOLA (Mars Orbiter Laser Altimeter) on board *Mars Global Surveyor*. MOLA fired a laser beam downward, then measured how long it took for the beam to return to the spacecraft after reflecting off the surface. Suppose MOLA measured this round-trip time for the laser beam to reflect off the summit of Olympus Mons, as well as the round-trip time to reflect off the bottom of Hellas Planitia (see Question 58). Which round-trip time is longer? How much longer is it? Do you need to know the distance from *Mars Global Surveyor* to the surface to answer this question?

61. (a) The Grand Canyon in Arizona was formed over 15 to 20 million years by the flowing waters of the Colorado River, as well as by rain and wind. Contrast this formation scenario to that of Valles Marineris on Mars. (b) Valles Marineris is sometimes called the "Grand Canyon of Mars." Is this an appropriate description? Why or why not?

62. Water has a density of 1000 kg/m³, so a column of water n meters tall and 1 meter square at its base has a mass of $n \times$ 1000 kg. On either Earth or Venus, which have nearly the same surface gravity, a mass of 1 kg weighs about 9.8 newtons (2.2 lb). Calculate how deep you would have to descend into Earth's oceans for the pressure to equal the atmospheric pressure on Venus's surface, 90 atm or 9×10^6 newtons per square meter. Give your answer in meters.

63. Marine organisms produce sulfur-bearing compounds, some of which escape from the oceans into Earth's atmosphere. (These compounds are largely responsible for the characteristic smell of the sea.) Even more sulfurous gases are injected into our atmosphere by the burning of sulfur-rich fossil fuels, such as coal, in electric power plants. Both of these processes add more sulfur compounds to the atmosphere than do volcanic eruptions. On lifeless Venus, by contrast, volcanoes are the only source for sulfurous atmospheric gases. Why, then, are sulfur compounds so much rarer in our atmosphere than in the Venusian atmosphere?

64. The classic 1950 science fiction movie *Rocketship X-M* shows astronauts on the Martian surface with oxygen masks for breathing but wearing ordinary clothing. Would this be a sensible choice of apparel for a walk on Mars? Why or why not?

65. Suppose that the only information you had about Mars was the images of the surface in Figure 11-28. Describe at least two ways that you could tell from these images that Mars has an atmosphere.

66. Although the *Viking Lander 1* and *Viking Lander 2* landing sites are 6500 km apart and have different geologic histories, the chemical compositions of the dust at both sites are nearly identical. (a) What does this suggest about the ability of the Martian winds to transport dust particles? (b) Would you expect that larger particles such as pebbles would also have identical chemical compositions at the two *Viking Lander* sites? Why or why not?

67. Why do you suppose that Phobos and Deimos are not round like our Moon?

68. The orbit of Phobos has a semimajor axis of 9378 km. Use this information and the orbital period given in the text to calculate the mass of Mars. How does your answer compare with the mass of Mars given in Table 11-3?

69. Calculate the angular sizes of Phobos and Deimos as they pass overhead, as seen by an observer standing on the Martian equator. How do these sizes compare with that of the Moon seen from Earth's surface? Would Phobos and Deimos appear as impressive in the Martian sky as the Moon does in our sky?

70. You are to put a spacecraft into a synchronous circular orbit around the Martian equator, so that its orbital period is equal to the planet's rotation period. Such a spacecraft

would always be over the same part of the Martian surface. (**a**) Find the radius of the orbit and the altitude of the spacecraft above the Martian surface. (**b**) Suppose Mars had a third moon that was in a synchronous orbit. Would tidal forces make this moon tend to move toward Mars, away from Mars, or neither? Explain your answer.

Discussion Questions

71. Before about 350 B.C.E., the ancient Greeks did not realize that Mercury seen in the morning sky (which they called Apollo) and seen in the evening sky (which they called Hermes) were actually the same planet. Discuss why you think it took some time to realize this.

72. If you were planning a new mission to Mercury, what features and observations would be of particular interest to you?

73. What evidence do we have that the surface features on Mercury were not formed during recent geological history?

74. Describe the apparent motion of the Sun during a "day" on Venus relative to (**a**) the horizon and (**b**) the background stars. (Assume that you can see through the cloud cover.)

75. If you could examine rock samples from the surface of Venus, would you expect them to be the same as rock samples from Earth? Would you expect to find igneous, sedimentary, and metamorphic rocks like those found on Earth (see Section 9-3)? Explain your answers.

76. In 1978 the *Pioneer Venus Orbiter* spacecraft arrived at Venus. It carried an ultraviolet spectrometer to measure the chemical composition of the Venusian atmosphere. This instrument recorded unexpectedly high levels of sulfur dioxide and sulfuric acid, which steadily declined over the next several years. Discuss how this observation suggests that volcanic eruptions occurred on Venus not long before *Pioneer Venus Orbiter* arrived there.

Web/eBook Questions

77. **ANIMATION 11-1** **Elongations of Mercury.** Access the animation "Elongations of Mercury" in Chapter 11 of the *Universe* Web site or eBook. (**a**) View the animation and notice the dates of the greatest eastern and greatest western elongations. Which time interval is greater: from a greatest eastern elongation to a greatest western elongation, or vice versa? (**b**) Based on what you observe in the animation, draw a diagram to explain your answer to the question in (**a**).

78. Just as Mercury can pass in front of the Sun as seen from Earth (see Question 43), so can Venus. Transits of Venus are quite rare. The dates of the only transits in the twenty-first century are June 8, 2004, and June 6, 2012; the next ones will occur in 2117 and 2125. A number of European astronomers traveled to Asia and the Pacific islands to observe the transits of Venus in 1761 and 1769. Search the World Wide Web for information about these expeditions. Why were these events of such interest to astronomers? How definitive were the results of these observations?

79. Search the World Wide Web for information about possible manned missions to Mars. How long might such a mission take? How expensive would such a project be? What would be the advantages of a manned mission compared to an unmanned one?

80. **ANIMATION 11-2** **Conjunctions of Mars.** Access and view the animation "The Orbits of Earth and Mars" in Chapter 11 of the *Universe* Web site or eBook. (**a**) The animation highlights three dates when Mars is in opposition, so that Earth lies directly between Mars and the Sun. By using the "Stop" and "Play" buttons in the animation, find two times during the animation when Mars is in *conjunction*, so that the Sun lies directly between Mars and Earth (see Figure 4-6). For each conjunction, make a drawing showing the positions of the Sun, Earth, and Mars, and record the month and year when the conjunction occurs. (**b**) When Mars is in conjunction, at approximately what time of day does it rise as seen from Earth? At what time of day does it set? Is Mars suitably placed for telescopic observation when in conjunction?

ACTIVITIES

Observing Projects

> **Observing tips and tools**
>
> Mercury and Venus are visible in the morning sky when it is at or near greatest western elongation and in the evening sky when at or near greatest eastern elongation. These are also the times when Venus can be seen at the highest altitude above the horizon during the hours of darkness. Mars is most easily seen around an opposition. At other times Mars may be visible only in the early morning hours before sunrise or in the early evening just after sunset. Consult such magazines as *Sky & Telescope* and *Astronomy* or their Web sites for more detailed information about when and where to look for Mercury, Venus, and Mars during a given month. You can also use the *Starry Night*™ program on the CD-ROM that certain printed copies of accompanies this textbook.

81. Refer to the *Universe* Web site or eBook for a link to a Web site that calculates the dates of upcoming greatest elongations of Mercury. Consult such magazines as *Sky & Telescope* and *Astronomy,* or the Web sites for these magazines, to determine if any of these greatest elongations is going to be a favorable one. If so, make plans to be one of those rare individuals who has actually seen the innermost planet of the solar system. Set aside several evenings (or mornings) around the date of the favorable elongation to reduce the chances of being "clouded out." Select an observing site that has a clear, unobstructed view of the horizon where the Sun sets (or rises). If possible, make arrangements to have a telescope at your disposal. Search for the planet on the dates you have selected, and make a drawing of its appearance through your telescope.

82. *This observing project should be performed only under the direct supervision of an astronomer who knows how to point a telescope safely at Mercury.* Make arrangements to view Mercury during broad daylight. This is best done by visiting an observatory where the coordinates (right ascension and declination) of Mercury's position can be used to point the telescope. **DO NOT LOOK AT THE SUN! Looking directly at the Sun can cause blindness.**

83. Refer to the *Universe* Web site or eBook for a link to a Web site that calculates the dates of greatest elongations of Venus. If Venus is near a greatest elongation, view the planet through a telescope. Make a sketch of the planet's appearance. From your sketch, can you determine if Venus is closer to us or farther from us than the Sun?

84. Using a small telescope, observe Venus once a week for a month and make a sketch of the planet's appearance on each occasion. From your sketches, can you determine whether Venus is approaching us or moving away from us?

85. *This observing project should be performed only under the direct supervision of an astronomer who knows how to point a telescope safely at Venus.* Make arrangements to view Venus during broad daylight. This is best done by visiting an observatory where the coordinates (right ascension and declination) of Venus's position can be used to point the telescope. **DO NOT LOOK AT THE SUN! Looking directly at the Sun can cause blindness.**

86. If Mars is suitably placed for observation, arrange to view the planet through a telescope. Draw a picture of what you see. What magnifying power seems to give you the best image? Can you distinguish any surface features? Can you see a polar cap or dark markings? If not, can you offer an explanation for Mars's bland appearance?

87. Use *Starry Night*™ to examine Mercury. Select **Favourites > Explorations > Mercury** from the menu. Stop the advance of time and use the **Zoom** controls and the location scroller to examine the surface of the planet Mercury. Estimate the diameter of the largest craters on Mercury's surface by measuring their size on the screen with a ruler and comparing these diameters to the diameter of Mercury. (**a**) What are the diameters (in km) of the largest craters on Mercury? (**b**) Zoom in to examine the surface features of Mercury in more detail and compare these features with those of our Moon. Comment on the similarities and differences between these two planetary objects, neither of which has an atmosphere (e.g., presence of craters, light ray patterns, maria, peaks within craters, etc.).

88. Use the *Starry Night*™ program to observe the apparent motion of Mercury on the celestial sphere. Select **Favourites > Explorations > Mercury Motion.** This view, from the center of a transparent Earth, is locked and centered upon the Sun. The background stars have been removed but the planets are visible, including the inferior planets Mercury and Venus. With the **Time Flow Rate** set to **1 day**, click the **Play** button and observe the motions of these two inferior planets against the background stars as seen from Earth. (**a**) Use the **Time**

controls to find the first date after January 1, 2015, when Mercury is at its farthest point to the left of the Sun, and the first date after January 1, 2015, when Mercury is at its farthest point to the right of the Sun. What is your interpretation of these two dates and how would you label them? (**b**) If you wanted to make naked-eye observations of Mercury on these two dates, indicate the best time of day to make your observations for each date.

89. Use *Starry Night*™ to examine Mars. **Open Favourites > Explorations > Mars Surface.** Use the **Zoom** controls and the location scroller to explore this planet's surface features. You will notice that four volcanoes (Mons) and the Valles Marineris have been labeled. (**a**) Which of the volcanoes appears to be the largest? (**b**) Right-click on Mars (Ctrl-click on a Mac) and select **Markers and Outlines...** from the contextual menu. In the Mars **Markers and Outlines** dialog window, click the lower radio button to the left of the **List** label. Then in the dropdown box to the right of the **List** label, choose **Type.** In the rightmost dropdown box, select Crater as the type of feature and then click the **Check all Shown** button. Describe the distribution of craters on the Martian surface. (**c**) What does the distribution of craters in the region around the volcanoes suggest about the time at which these volcanoes formed on Mars?

90. Use *Starry Night*™ to observe solar transits of Venus. Select **Favourites > Explorations > Venus Transit** to open a view of the sky from Rome, Italy, at 5:30:00 A.M. local standard time on June 8, 2004. The view is centered upon the planet Venus. Click the **Zoom** panel in the toolbar and select 1° from the drop down menu to zoom in on Venus. Select **View > Hide Daylight** from the menu to improve the contrast in the view. Click the **Play** button and watch as Venus transits the disk of the Sun. (**a**) What is the name given to the planetary configuration of Venus relative to Earth during this transit? (**b**) Select **Edit > Undo Time Flow** to reset the time to 5:30 A.M. Select **View > Ecliptic Guides > The Ecliptic** from the menu. During the transit, is Venus precisely on the ecliptic? If not, approximately how far is it from the ecliptic? [*Hint:* You can use the angular separation tool to measure the perpendicular distance from Venus to the ecliptic.] (**c**) Use the **Zoom** and **Time** controls to find as accurately as possible the times at which the transit begins and ends. What is the duration of this transit? (**d**) **Stop** time flow and select **Options > Viewing Location...** from the menu. In the **Viewing Location** dialog window, click the **List** tab and then the **Show where** radio button. In the rightmost edit box, type **Fairbanks** and then highlight the entry for Fairbanks, Alaska, in the list and click the **Go To Location** button. Change the time and date to **12:00:00 P.M.** local standard time on **June 5, 2012.** The sky remains dark because the **Hide Daylight** option is still in effect. Open the **Find** pane and select **Magnify** from the dropdown menu for the Sun. Set the **Time Flow Rate** to **1 minute** and click **Play** to observe this transit of Venus. Select **Edit > Undo Time Flow** from the menu. Right-click (Ctrl-click on

a Mac) over Venus and select **Centre** from the contextual menu to center the view on Venus. Use the **Zoom** and **Time Flow** controls to determine the start time, end time, and duration of this transit as accurately as possible. What is the duration of this transit? (e) During the course of which of these two transits was Venus, on average, closer to the ecliptic? Explain your reasoning.

Collaborative Exercises

91. Figures 11-3b and 11-3c show a planet in synchronous rotation and Mercury with a 3-to-2 spin-orbit coupling, respectively. Stand up and demonstrate how planets move in each of these rotations by "orbiting" around a stationary classmate who represents our Sun. How would Mercury's motion be different if it had a 4-to-2 spin-orbit coupling instead?

92. In the nineteenth century, French mathematician and astronomer U. J. J. Le Verrier led a failed search for a hypothetical planet named Vulcan (after the mythical blacksmith of the gods) orbiting closer to our Sun than Mercury. If Vulcan had an orbit with the same eccentricity as Mercury's orbit but only one-half the size, what would have been its maximum eastern and western elongations?

93. Figure 11-4 shows prograde and retrograde rotation. Stand up and demonstrate how Venus rotates as it orbits our Sun, using a classmate as our stationary Sun. How is this different from how Earth rotates in its orbit?

94. The image of Mars that opens this chapter is from the Hubble Space Telescope. Draw a circle on your paper roughly 5 cm in diameter and, taking turns, have each person in your group sketch a different region of Mars. How is your collaborative sketch different than Schiaparelli's drawing shown in Figure 11-12?

ANSWERS

ConceptChecks

ConceptCheck 11-1: No. Venus is perpetually shrouded in a thick layer of clouds and because visible light does not penetrate the clouds, no light escapes for us to observe from Earth.

ConceptCheck 11-2: When Caloris Basin formed, all existing craters in that area would have been wiped out, and the craters we see today occurred afterward. The density of craters inside Caloris Basin is low compared to typical regions that still show high densities from the Late Heavy Bombardment. Therefore, it appears that Caloris Basin would have formed later.

ConceptCheck 11-3: The high temperatures of such an impact would "boil off" the volatile elements, and would be consistent with a low abundance on Mercury. However, we observe a high abundance of volatile elements, and this is a great mystery.

ConceptCheck 11-4: 1.35-cm microwaves are absorbed by water molecules. If the surface of Venus contained a significant amount of water, it would absorb these wavelengths. The same is true for the Venusian atmosphere. Since microwaves at 1.35 cm were readily detected, there can be little water on or around Venus.

ConceptCheck 11-5: The northern hemisphere has a younger surface. Compared to the southern hemisphere, the northern hemisphere is almost free of impact craters. This indicates that craters must have been erased on the northern hemisphere by a more recent process.

ConceptCheck 11-6: The age of a planetary surface is indicated by how many craters it has. Venus has more craters than Earth, yet fewer than Mercury. Detailed analysis estimates the age of Venus's surface at 500 million years old.

ConceptCheck 11-7: If CO_2 could be added to the atmosphere, the temperature on Mars might be increased through the greenhouse effect.

ConceptCheck 11-8: Methane-producing microbes, volcanic activity, and the interaction of UV sunlight with small meteorites could all produce variable amounts of methane on Mars.

ConceptCheck 11-9: The "runaway" term refers to the fact that an increase in temperature causes changes in the atmosphere, which causes the temperature to rise even higher and higher.

ConceptCheck 11-10: A global magnetic field can deflect the solar wind, protecting the Martian atmosphere from high-energy collisions. These collisions can break apart the greenhouse gases CO_2 and H_2O, allowing the constituent atoms to escape into space.

ConceptCheck 11-11: It appears that hydrogen, presumed to reside in frozen water, is absorbing neutrons on the surface of Mars (Figure 11-38). Also, impact craters have directly revealed water-ice (Figure 11-39).

ConceptCheck 11-12: Phobos raises a tidal bulge on Mars that lags behind Phobos due to Phobos orbiting faster than Mars rotates. Through gravity, this bulge pulls back on Phobos, slowing it down. As Phobos slowly loses orbital energy, it spirals in toward Mars.

SPACE

Reading the Red Planet

At 10:31 P.M. Pacific time on August 5, 2012, NASA's *Curiosity* rover began the first direct search for habitable environments on Mars

BY JOHN P. GROTZINGER AND ASHWIN VASAVADA

(from John P. Grotzinger and Ashwin Vasavada's "Reading the Red Planet," *Scientific American,* July 2012)

All science begins in a *Star Trek* mode: go where no one has gone before and discover new things without knowing in advance what they might be. As researchers complete their initial surveys and accumulate a long list of questions, they shift to a Sherlock Holmes mode: formulate specific hypotheses and develop ways to test them. The exploration of Mars is now about to make this transition. Orbiters have made global maps of geographic features and composition, and landers have pieced together the broad outlines of the planet's geologic history. It is time to get more sophisticated.

Our team has built the Mars Science Laboratory, also known as the *Curiosity* rover, on the hypothesis that Mars was once a habitable planet. The rover carries an analytic laboratory to test that hypothesis and find out what happened

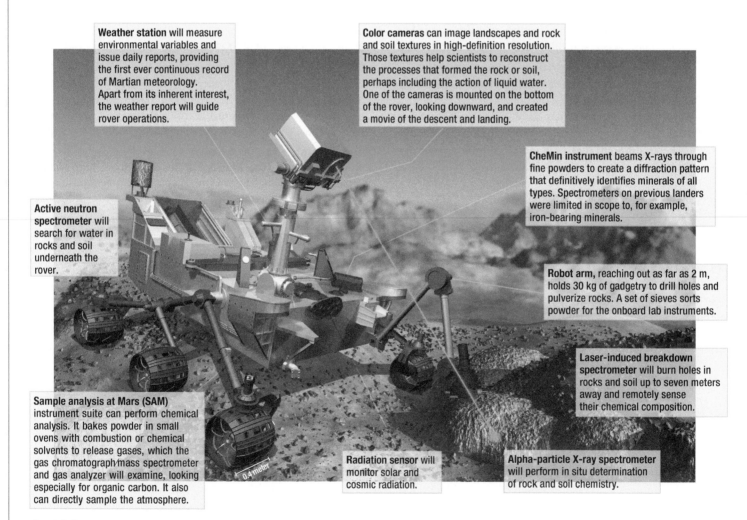

Weather station will measure environmental variables and issue daily reports, providing the first ever continuous record of Martian meteorology. Apart from its inherent interest, the weather report will guide rover operations.

Color cameras can image landscapes and rock and soil textures in high-definition resolution. Those textures help scientists to reconstruct the processes that formed the rock or soil, perhaps including the action of liquid water. One of the cameras is mounted on the bottom of the rover, looking downward, and created a movie of the descent and landing.

CheMin instrument beams X-rays through fine powders to create a diffraction pattern that definitively identifies minerals of all types. Spectrometers on previous landers were limited in scope to, for example, iron-bearing minerals.

Active neutron spectrometer will search for water in rocks and soil underneath the rover.

Robot arm, reaching out as far as 2 m, holds 30 kg of gadgetry to drill holes and pulverize rocks. A set of sieves sorts powder for the onboard lab instruments.

Laser-induced breakdown spectrometer will burn holes in rocks and soil up to seven meters away and remotely sense their chemical composition.

Sample analysis at Mars (SAM) instrument suite can perform chemical analysis. It bakes powder in small ovens with combustion or chemical solvents to release gases, which the gas chromatograph/mass spectrometer and gas analyzer will examine, looking especially for organic carbon. It also can directly sample the atmosphere.

0.4 meter

Radiation sensor will monitor solar and cosmic radiation.

Alpha-particle X-ray spectrometer will perform in situ determination of rock and soil chemistry.

(Don Foley)

to the early clement environment we believe the planet had. Loosely defined, a habitable environment has water, energy and carbon. Past missions have focused on the first requirement and confirmed that Mars had—and occasionally still has—liquid water [see "The Red Planet's Watery Past," by Jim Bell; *Scientific American,* December 2006]. Those missions have also seen hints of geochemical gradients that would provide energy for metabolism. But none has seen carbon in a form potentially suitable for life.

Like the twin *Viking* landers of the mid-1970s, *Curiosity* carries a gas chromatograph/mass spectrometer capable of sensing organic compounds, whether biological or abiological in origin. Unlike *Viking,* however, *Curiosity* is mobile and is touching down in a far more promising site. More important than finding carbon itself, the mission aims to discover how to conduct the search. Even on Earth, we are not entirely sure how to trawl the deep geologic record for preserved biosignatures. Paradoxically, the very characteristics that make so many environments habitable—water, oxidants, and chemical and temperature gradients—also tend to destroy organic compounds. Paleontologists have learned to seek the rare circumstances that facilitate preservation, such as geochemical conditions that favor very early mineralization. Silica, phosphate, clay, sulfate and, less commonly, carbonate are all known to entomb organics as they precipitate. Orbiters have made maps of some of these minerals at *Curiosity*'s landing site, which will guide its perambulations.

John P. Grotzinger, project scientist for the Mars Science Laboratory mission and a geologist at the California Institute of Technology, is interested in the evolution of surface environments on both Earth and Mars. He is a member of the National Academy of Sciences.

Ashwin Vasavada, deputy project scientist, is thrilled with the idea that someday people will be able to hike up Mount Sharp on Mars and retrace the path of Curiosity. *Based at the NASA Jet Propulsion Laboratory, he also participated in the* Galileo, Cassini, *and* Lunar Reconnaissance Orbiter *missions.*

R I V U X G   R I V U X G

Jupiter (left) and Saturn (right) to scale. (NASA/JPL/Space Science Institute)

Jupiter and Saturn: Lords of the Planets

LEARNING GOALS

By reading the sections of this chapter, you will learn

12-1 What observations from Earth reveal about Jupiter and Saturn

12-2 How Jupiter and Saturn rotate differently from terrestrial planets like Earth

12-3 The nature of the immense storms seen in the clouds of Jupiter and Saturn

12-4 How the internal heat of Jupiter and Saturn drives activity in their atmospheres

12-5 What the *Galileo* space probe revealed about Jupiter's atmosphere

12-6 How the shapes of Jupiter and Saturn indicate the sizes of their rocky cores

12-7 How Jupiter and Saturn's intense magnetic fields are produced by an exotic form of hydrogen

12-8 The overall structure and appearance of Saturn's ring system

12-9 What kinds of particles form the rings of Jupiter and Saturn

12-10 How spacecraft observations revealed the intricate structure of Saturn's rings

12-11 How Saturn's rings are affected by the presence of several small satellites

Among the most remarkable sights in our solar system are the colorful, turbulent atmosphere of Jupiter and the ethereal beauty of Saturn's rings. Both of these giant worlds dwarf our own planet: More than 1200 Earths would fit inside Jupiter's immense bulk, and more than 700 Earths inside Saturn. Unlike Earth, both Jupiter and Saturn are composed primarily of the lightweight elements hydrogen and helium, in abundances very similar to those in the Sun.

As these images show, the rapid rotations of both Jupiter and Saturn stretch their weather systems into colorful bands that extend completely around each planet. Both planets also have intense magnetic fields that are generated in their interiors, where hydrogen is so highly compressed that it becomes a metal.

Saturn, however, is not merely a miniature version of Jupiter. Its muted colors and more flattened shape are clues that Saturn's atmosphere and interior have important differences from those of Jupiter. The most striking difference is Saturn's elaborate system of rings, composed of countless numbers of icy fragments orbiting in the plane of the planet's equator. The rings display a complex and elegant structure, which is shaped by subtle gravitational influences from Saturn's retinue of moons. Jupiter, too, has rings, but they are made of dark, dustlike particles that reflect little light. These systems of rings, along with the distinctive properties of the planets themselves, make Jupiter and Saturn highlights of our tour of the solar system.

TABLE 12-1 Jupiter Data

Average distance from the Sun:	5.203 AU = 7.783 × 10⁸ km
Maximum distance from the Sun:	5.455 AU = 8.160 × 10⁸ km
Minimum distance from the Sun:	4.950 AU = 7.406 × 10⁸ km
Eccentricity of orbit:	0.048
Average orbital speed:	13.1 km/s
Orbital period:	11.86 years
Rotation period:	9ʰ 50ᵐ 28ˢ (equatorial)
	9ʰ 55ᵐ 29ˢ (internal)
Inclination of equator to orbit:	3.12°
Inclination of orbit to ecliptic:	1.30°
Diameter:	142,984 km = 11.209 Earth diameters (equatorial)
	133,708 km = 10.482 Earth diameters (polar)
Mass:	1.899×10^{27} kg = 317.8 Earth masses
Average density:	1326 kg/m³
Escape speed:	60.2 km/s
Surface gravity (Earth = 1):	2.36
Albedo:	0.44
Average temperature at cloudtops:	−108°C = −162°F = 165 K
Atmosphere composition: (by number of molecules)	86.2% hydrogen (H_2), 13.6% helium (He), 0.2% methane (CH_4), ammonia (NH_3), water vapor (H_2O), and other gases

R I **V** U X G
(Ocean/Corbis)

12-1 Jupiter and Saturn are the most massive planets in the solar system

Jupiter and Saturn are respectively the largest and second largest of the planets, and the largest objects in the solar system other than the Sun itself. Astronomers have known for centuries about the huge diameters and immense masses of these two planets. Given the distance to each planet and their angular sizes, they used the small-angle formula (Box 1-1) to calculate that Jupiter is about 11 times larger in diameter than Earth, while Saturn's diameter is about 9 times larger than Earth's (see Figure 7-2). By observing the orbits of Jupiter's four large moons and Saturn's large moon Titan (see Table 7-2) and applying Newton's form of Kepler's third law (see

Section 4-7 and Box 4-4), astronomers also determined that Jupiter and Saturn are, respectively, 318 times and 95 times more massive than Earth. In fact, Jupiter has 2½ times the combined mass of all the other planets, satellites, asteroids, meteoroids, and comets in the solar system. A visitor from interstellar space might well describe our solar system as the Sun, Jupiter, and some debris! Table 12-1 and Table 12-2 list some basic data about Jupiter and Saturn.

Observing Jupiter and Saturn

As for any planet whose orbit lies outside Earth's orbit, the best time to observe Jupiter or Saturn is when they are at opposition. At opposition, Jupiter can appear nearly 3 times brighter than Sirius, the brightest star in the sky. Only the Moon and Venus can

TABLE 12-2 Saturn Data

Average distance from the Sun:	9.572 AU = 1.432×10^9 km
Maximum distance from the Sun:	10.081 AU = 1.508×10^9 km
Minimum distance from the Sun:	9.063 AU = 1.356×10^9 km
Eccentricity of orbit:	0.053
Average orbital speed:	9.64 km/s
Orbital period:	29.37 years
Rotation period:	$10^h\ 13^m\ 59^s$ (equatorial)
	$10^h\ 39^m\ 25^s$ (internal)
Inclination of equator to orbit:	26.73°
Inclination of orbit to ecliptic:	2.48°
Diameter:	120,536 km = 9.449 Earth diameters (equatorial)
	108,728 km = 8.523 Earth diameters (polar)
Mass:	5.685×10^{26} kg = 95.16 Earth masses
Average density:	687 kg/m³
Escape speed:	35.5 km/s
Surface gravity (Earth = 1):	0.92
Albedo:	0.46
Average temperature at cloudtops:	−180°C = − 292°F = 93 K
Atmosphere composition (by number of molecules):	96.3% hydrogen (H_2), 3.3% helium (He), 0.4%, methane (CH_4), ammonia (NH_3), water vapor (H_2O), and other gases

R I V U X G
(STScI/Hubble Heritage Team)

outshine Jupiter at opposition. Saturn's brightness at opposition is only about 1/7 that of Jupiter, but it still outshines all of the stars except Sirius and Canopus.

Through a telescope, Jupiter at opposition presents a disk nearly 50 arcsec in diameter, approximately twice the angular diameter of Mars under the most favorable conditions (Figure 12-1a). Because Jupiter takes almost a dozen Earth years to orbit the Sun, it appears to meander slowly across the 12 constellations of the zodiac at the rate of approximately one constellation per year. Successive oppositions occur at intervals of about 13 months.

Saturn's orbital period is even longer, more than 29 Earth years, so it moves even more slowly across the zodiac. If Saturn did not move at all, successive oppositions would occur exactly one Earth year apart; thanks to its slow motion, Saturn's oppositions occur at intervals of about one year and two weeks. At opposition Saturn appears as a disk about 20 arcsec in diameter, with a dramatic difference from Jupiter: Saturn is surrounded by a magnificent system of rings (Figure 12-1b). Later in this chapter we will examine these rings in detail.

Dark Belts and Light Zones

As seen through an Earth-based telescope, both Jupiter and Saturn display colorful bands that extend around each planet parallel to its equator (Figure 12-1). More

Even a small telescope reveals colored bands on Jupiter and rings around Saturn

(a)

(b)

FIGURE 12-1 R I V U X G

Jupiter and Saturn as Viewed from Earth (a) The disk of Jupiter at opposition appears about 2½ times larger than (b) the disk of Saturn at opposition. Both planets display dark and light bands, though these are fainter on Saturn. (a: © Damian Peach; b: NASA)

detailed close-up views of Jupiter from a spacecraft (Figure 12-2a) show alternating dark and light bands parallel to Jupiter's equator in subtle tones of red, orange, brown, and yellow. The dark, reddish bands are called **belts,** and the light-colored bands are called **zones.** The image of Saturn in Figure 12-2b shows similar features, but with colors that are much less pronounced. (We will see in Section 12-4 how differences between the atmospheres of Jupiter and Saturn can explain Saturn's washed-out appearance.)

In addition to these conspicuous stripes, a huge, red-orange oval called the **Great Red Spot** is often visible in Jupiter's southern hemisphere. This remarkable feature was first seen by the English scientist Robert Hooke in 1664 but may be much older. It appears to be an extraordinarily long-lived storm in the planet's dynamic atmosphere. Saturn has no long-lived storm systems of this kind. However, many careful observers have reported smaller spots and blemishes in the atmospheres of both Jupiter and Saturn that last for only a few weeks or months.

(a) Jupiter

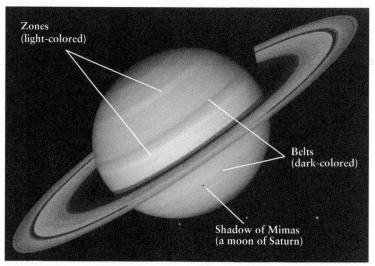

(b) Saturn

VIDEO 12-1 VIDEO 12-2 **FIGURE 12-2** R I V U X G

Jupiter and Saturn as Viewed from Space (a) This view of Jupiter is a composite of four images made by the *Cassini* spacecraft as it flew past Jupiter in 2000. (b) Images from the *Voyager 2* spacecraft as it flew past Saturn in 1981 were combined to show the planet in approximately natural color. (a: NASA/JPL/University of Arizona; b: USGS/NASA/JPL)

12-2 Unlike the terrestrial planets, Jupiter and Saturn exhibit differential rotation

Observations of features like the Great Red Spot and smaller storms allow astronomers to determine how rapidly Jupiter and Saturn rotate. At its equator, Jupiter completes a full rotation in only 9 hours, 50 minutes, and 28 seconds, making it not only the largest and most massive planet in the solar system but also the one with the fastest rotation. However, Jupiter rotates in a strikingly different way from Earth, the Moon, Mercury, Venus, or Mars.

Differential Rotation

If Jupiter were a solid body like a terrestrial planet (or, for that matter, a billiard ball), all parts of Jupiter's surface would rotate through one complete circle in this same amount of time (**Figure 12-3a**). But by watching features in Jupiter's cloud cover, Gian Domenico Cassini discovered in 1690 that the polar regions of the planet rotate a little more slowly than do the equatorial regions. (You may recall this Italian astronomer from Section 11-2 as the gifted observer who first determined Mars's rate of rotation.) Near the poles, the rotation period of Jupiter's atmosphere is about 5 minutes longer than at the equator (about 9 hours, 55 minutes, and 41 seconds). Saturn, too, has a longer rotation period near its poles, by about 25 minutes (10 hours, 39 minutes, and 24 seconds near the poles compared to 10 hours, 13 minutes, and 59 seconds at the equator).

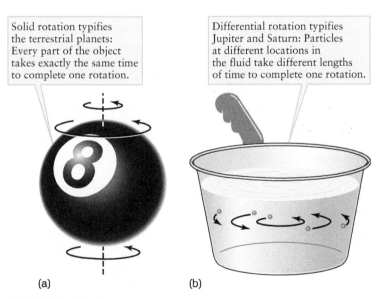

Solid rotation typifies the terrestrial planets: Every part of the object takes exactly the same time to complete one rotation.

Differential rotation typifies Jupiter and Saturn: Particles at different locations in the fluid take different lengths of time to complete one rotation.

(a)   (b)

FIGURE 12-3

Solid Rotation versus Differential Rotation **(a)** All parts of a solid object rotate together, but **(b)** a rotating fluid displays differential rotation. To see differential rotation, put some grains of sand, bread crumbs, or other small particles in a pot of water. Stir the water with a spoon to start it rotating, then take out the spoon. The particles near the center of the pot take less time to make a complete rotation than those away from the center.

ANALOGY You can see this kind of rotation, called **differential rotation**, in the kitchen. As you stir the water in a pot, different parts of the liquid take different amounts of time to make one "rotation" around the center of the pot (Figure 12-3b). Differential rotation shows that neither Jupiter nor Saturn can be solid throughout their volumes: They must be at least partially fluid, like water in a pot.

CONCEPTCHECK **12-1**

Will Jupiter's moons of rock and ice, such as Ganymede, exhibit differential rotation?

Answer appears at the end of the chapter.

The Compositions of Jupiter and Saturn

If Jupiter and Saturn have partially fluid interiors, they cannot be made of the rocky materials that constitute the terrestrial planets. An important clue to the compositions of Jupiter and Saturn are their average densities, which are only 1326 kg/m^3 for Jupiter and 687 kg/m^3 for Saturn. (By comparison, Earth's average density is 5515 kg/m^3.) To explain these low average densities, Rupert Wildt of the University of Göttingen in Germany suggested in the 1930s that Jupiter and Saturn are composed mostly of hydrogen and helium atoms—the two lightest elements in the universe—held together by their mutual gravitational attraction to form a planet. Wildt was motivated in part by his observations of prominent absorption lines of methane and ammonia in Jupiter's spectrum. (We saw in Section 7-3 how spectroscopy plays an important role in understanding the planets.) A molecule of methane (CH$_4$) contains four hydrogen atoms, and a molecule of ammonia (NH$_3$) contains three. The presence of these hydrogen-rich molecules was strong, but indirect, evidence of abundant hydrogen in Jupiter's atmosphere.

Direct evidence for hydrogen and helium in Jupiter's atmosphere, however, was slow in coming. The problem was that neither gas produces prominent | **Spacecraft observations were needed to determine the compositions of Jupiter and Saturn** spectral lines in the visible sunlight reflected from the planet. To show the presence of these elements conclusively, astronomers had to look for spectral lines in the ultraviolet part of the spectrum. These lines are very difficult to measure from Earth, because almost no ultraviolet light penetrates our atmosphere (see Figure 6-25). Astronomers first detected the weak spectral lines of hydrogen molecules in Jupiter's spectrum in 1960. The presence of helium on Jupiter and Saturn was finally confirmed in the 1970s and 1980s, when spacecraft first flew past these planets and measured their hydrogen spectra in detail.

Today we know that the chemical composition of Jupiter's atmosphere is 86.2% hydrogen molecules (H$_2$), 13.6% helium atoms, and 0.2% methane, ammonia, water vapor, and other gases. The percentages in terms of *mass* are somewhat different because a helium atom is twice as massive as a hydrogen molecule. Hence, by mass, Jupiter's atmosphere is approximately 75% hydrogen, 24% helium, and 1% other substances, quite similar to that of the Sun. We will see evidence in Section 12-6 that Jupiter has a large rocky core made of heavier elements. It is estimated that the breakdown by mass of the planet as a whole (atmosphere plus interior) is approximately 71% hydrogen, 24% helium, and 5% all heavier elements.

Saturn's Missing Helium

Like Jupiter, Saturn is thought to have a large rocky core. But unlike Jupiter, data from Earth-based telescopes and spacecraft show that the atmosphere of Saturn has a serious helium deficiency: Its chemical composition is 96.3% hydrogen molecules, 3.3% helium, and 0.4% other substances (by mass, 92% hydrogen, 6% helium, and 2% other substances). Saturn's atmosphere is a puzzle because Jupiter and Saturn are thought to have formed in similar ways from the gases of the solar nebula (see Section 8-4), and so both planets (and the Sun) should have essentially the same abundances of hydrogen and helium. So where did Saturn's helium go?

The explanation may be simply that Saturn is smaller than Jupiter, and as a result Saturn probably cooled more rapidly. (We saw in Section 7-6 why a small world cools down faster than a large one.) This cooling would have triggered a process analogous to the way rain develops here on Earth. When the air is cool enough, humidity in Earth's atmosphere condenses into raindrops that fall to the ground. On Saturn, however, it is droplets of liquid helium that condense within the planet's cold, hydrogen-rich outer layers. In this scenario, helium is deficient in Saturn's upper atmosphere simply because it has fallen deeper into the planet. By contrast, Jupiter's helium has not rained out because its upper atmosphere is warmer and the helium does not form droplets.

ANALOGY An analogy to helium "rainfall" within Saturn is what happens when you try to sweeten tea by adding sugar. If the tea is cold, the sugar does not dissolve well and tends to sink to the bottom of the glass even if you stir the tea with a spoon. But if the tea is hot, the sugar dissolves with only a little stirring. In the same way, it is thought that the descending helium droplets once again dissolve in hydrogen once they reach the warmer depths of Saturn's interior.

In this scenario, Jupiter and Saturn both have about the same overall chemical composition. But Saturn's smaller mass, less than a third that of Jupiter, results in less gravitational force tending to compress its hydrogen and helium. This lack of compression explains why Saturn's density is only about half that of Jupiter, and is in fact the lowest of any planet in the solar system.

CAUTION! Because Jupiter and Saturn are almost entirely hydrogen and helium, it would be impossible to land a spacecraft on either planet. An astronaut foolish enough to try would notice the hydrogen and helium around the spacecraft becoming denser, the temperature rising, and the pressure increasing as the spacecraft descended. But the hydrogen and helium would never solidify into a surface on which the spacecraft could touch down. Long before reaching the planet's rocky core, the pressure of the hydrogen and helium would reach such unimaginably high levels that any spacecraft, even one made of the strongest known materials, would be crushed.

CONCEPTCHECK **12-2**

If Jupiter and Saturn formed with nearly the same chemical composition, why might Jupiter be observed to have more helium than Saturn?

Answer appears at the end of the chapter.

12-3 Spacecraft images show remarkable activity in the clouds of Jupiter and Saturn

Most of our detailed understanding of Jupiter and Saturn comes from a series of robotic spacecraft that have examined these remarkable planets at close range. They found striking evidence of stable, large-scale weather patterns in both planets' atmospheres, as well as evidence of dynamic changes on smaller scales.

Spacecraft to Jupiter and Saturn

The first several spacecraft to visit Jupiter and Saturn each made a single flyby of the planet. *Pioneer 10* flew past Jupiter in December 1973; it was followed a year later by the nearly identical *Pioneer 11*, which went on to make the first-ever flyby of Saturn in 1979. Also in 1979, another pair of spacecraft, *Voyager 1* and *Voyager 2*, sailed past Jupiter. These spacecraft sent back spectacular close-up color pictures of Jupiter's dynamic atmosphere. Both *Voyagers* subsequently flew past Saturn.

The first spacecraft to go into Jupiter orbit was *Galileo*, which carried out an extensive program of observations from 1995 to 2003. The *Cassini* spacecraft went into orbit around Saturn in 2004. A cooperative project of NASA, the European Space Agency, and the Italian Space Agency, this spacecraft is named for astronomer Gian Domenico Cassini. On the way to its destination it viewed Jupiter at close range, recording detailed images such as Figure 12-2a. The latest spacecraft, *Juno*, was launched in 2011, and will reach Jupiter in 2016. By measuring Jupiter's composition, structure, and magnetic field, *Juno* should shed light on planet formation, the early solar system, and magnetic dynamos.

Observing Jupiter's Dynamic Atmosphere

While the general pattern of Jupiter's atmosphere stayed the same during the four years between the *Pioneer* and *Voyager* flybys, there were some remarkable changes in the area surrounding the centuries-old Great Red Spot. During the *Pioneer* flybys, the Great Red Spot was embedded in a broad white zone that dominated the planet's southern hemisphere (Figure 12-4a). By the time of the *Voyager* missions, a dark belt had broadened and encroached on the Great Red Spot from the north (Figure 12-4b). In the same way that colliding weather systems in our atmosphere can produce strong winds and turbulent air, the interaction in Jupiter's atmosphere between the belt and the Great Red Spot embroiled the entire region in turbulence (Figure 12-5). By 1995, the Great Red Spot was once again centered within a white zone (Figure 12-4c); when *Cassini* flew past Jupiter in 2000, the dark

> Immense storms on Jupiter and Saturn can last for months or years

The Great Red Spot lies within a white zone.

(a) *Pioneer 11*, December 1974

A dark belt has encroached on the Great Red Spot.

(b) *Voyager 2*, July 1979

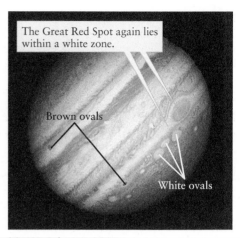

The Great Red Spot again lies within a white zone.

Brown ovals

White ovals

(c) HST, February 1995

FIGURE 12-4   R I V U X G

Jupiter's Changing Appearance These images from **(a)**, **(b)** spacecraft and **(c)** the Hubble Space Telescope show major changes in the planet's upper atmosphere over a 20-year period. (a and b: NASA/JPL; c: Reta Beebe and Amy Simon, New Mexico State University, and NASA)

belt to the north of the Great Red Spot embroiled the entire region in turbulence.

Over the past three centuries, Earth-based observers have reported many long-term variations in the Great Red Spot's size and color. At its largest, it measured 40,000 by 14,000 km—so large that three Earths could fit side by side across it! At other times (as in 1976 and 1977), the spot almost faded from view. During the *Voyager* flybys of 1979, the Great Red Spot was comparable in size to Earth (see Figure 12-5).

Clouds at different heights within Jupiter's atmosphere reflect different wavelengths of infrared light. Using this effect, astronomers used an infrared telescope on board the *Galileo* spacecraft to help clarify the vertical structure of the Great Red Spot. Most of the Great Red Spot is made of clouds at relatively high altitudes, surrounded by a collar of very low-level clouds about 50 km (30 miles, or 160,000 feet) below the high clouds at the center of the spot. This same kind of structure is seen in high-pressure areas in Earth's atmosphere, although on a much smaller scale.

Cloud motions in and around the Great Red Spot reveal that the spot rotates counterclockwise and takes about 6 days to complete a full rotation. Furthermore, winds to the north of the spot blow to the west, and winds south of the spot move toward the east. The circulation around the Great Red Spot is thus like a wheel spinning between two oppositely moving surfaces (see Figure 12-5). Weather patterns on Earth tend to change character and eventually dissipate when they move between plains and mountains or between land and sea. But because no solid surface or ocean underlies Jupiter's clouds, no such changes can occur for the Great Red Spot—which may explain how this wind pattern has survived for at least three centuries.

Other persistent features in Jupiter's atmosphere are the **white ovals.** Several white ovals are visible in Figure 12-4c. As in the Great Red Spot, the wind flow in white ovals is counterclockwise. White ovals are also apparently long-lived; Earth-based observers have reported seeing them in the same location since 1938.

Most of the white ovals are observed in Jupiter's southern hemisphere, whereas **brown ovals** are more common in Jupiter's northern hemisphere. Brown ovals appear dark in a visible-light image like Figure 12-4c, but they appear bright in an infrared image. For

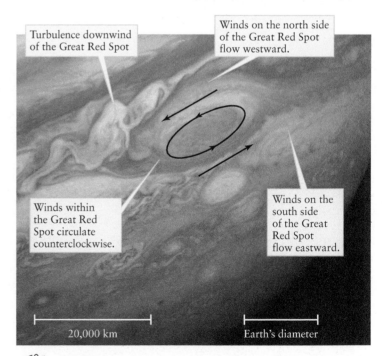

Turbulence downwind of the Great Red Spot

Winds on the north side of the Great Red Spot flow westward.

Winds within the Great Red Spot circulate counterclockwise.

Winds on the south side of the Great Red Spot flow eastward.

20,000 km   Earth's diameter

VIDEO 12.3   **FIGURE 12-5**   R I V U X G

The Great Red Spot This image of the Great Red Spot shows the counterclockwise circulation of gas in the Great Red Spot that takes about 6 days to make one rotation. The clouds that encounter the spot are forced to pass around it, and when other oval features are near it, the entire system becomes particularly turbulent, like batter in a two-bladed blender. (NASA/JPL)

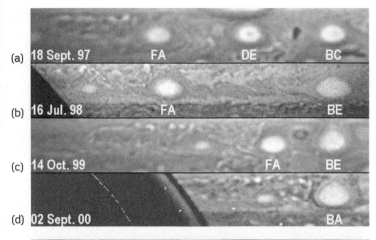

FIGURE 12-6 R I V U X G

Creating Red Spot Jr. **(a–d)** For 60 years prior to 1998, the three white ovals labeled FA, DE, and BC traveled together at the same latitude on Jupiter. Between 1998 and 2000, they combined into one white oval, labeled BA, which **(e)** became a red spot, named Red Spot Jr., in 2006. (a–d: NASA/JPL/WFPC2; e: NASA, ESA, A. Simon-Miller [NASA/GSFC], and I. de Pater [University of California Berkeley])

this reason, brown ovals are understood to be holes in Jupiter's cloud cover. They permit us to see into the depths of the Jovian atmosphere, where the temperature is higher and the atmosphere emits infrared light more strongly. White ovals, by contrast, have relatively low temperatures. They are areas with cold, high-altitude clouds that block our view of the lower levels of the atmosphere.

Between 1998 and 2000 three white ovals merged in Jupiter's southern hemisphere. The merger of these ovals, each of which had been observed for 60 years, led to the formation of a massive storm about half the size of the Great Red Spot (Figure 12-6). Time will tell whether this new feature—called Red Spot Jr.—is as long-lived as its larger cousin. Even Jupiter's large belts can disappear. In 2009, Jupiter's dark southern belt began to fade and was completely gone by June 2010 (Figure 12-7). This disappearing act has occurred about 14 times in the last century.

FIGURE 12-7 R I V U X G

Dark Belt Fades on Jupiter The south equatorial belt completely faded from view between mid-2009 and mid-2010. White-colored ammonia ice crystals appeared at higher elevations, blocking the darker belt from view, but the full mechanism is not understood. By late 2010, the belt began to reappear. (NASA, ESA, UC Berkeley, STScI, Jupiter Impact Science Team)

CONCEPTCHECK 12-3

Besides color, what are some differences between Jupiter's white ovals and brown ovals?

Answer appears at the end of the chapter.

Storms and Patterns on Saturn

While Saturn has belts and zones like Jupiter, it has no storm systems as long-lived as Jupiter's Great Red Spot. But about every 30 years—roughly the orbital period of Saturn—Earth-based observers have reported seeing storms in Saturn's clouds that last for several days or even months. Figure 12-8 shows one such storm that appeared in Saturn's southern hemisphere in 2004.

VIDEO 12-4 Saturn's storms are thought to form when warm atmospheric gases rise upward and cool, causing gaseous ammonia to solidify into crystals and form white clouds. On Earth, similar rapid lifting of air occurs within thunderstorms and can cause water droplets to solidify into hailstones. Earth thunderstorms produce electrical discharges (lightning) that generate radio waves, which you hear as "static" on an AM radio. Radio receivers on board *Cassini* have recorded similar "static" emitted by storms on Saturn, which strongly reinforces the idea that these storms are like thunderstorms on Earth.

Around Saturn's north pole, an enigmatic hexagonal pattern was glimpsed by *Voyager* in 1980. *Cassini* captured this pattern in greater detail, establishing that it persists for more than 26 years (Figure 12-9a). Amazingly, experiments on Earth that simulate Saturn's atmosphere have recreated this pattern (Figure 12-9b).

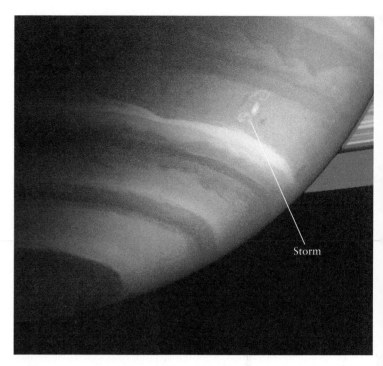

VIDEO 12-5 **FIGURE 12-8** R I V U X G

A New Storm on Saturn This infrared image from the *Cassini* spacecraft shows a storm the size of Earth that appeared in Saturn's southern hemisphere in 2004. *Cassini* has observed several storms in this same region of Saturn. (NASA/JPL/Space Science Institute)

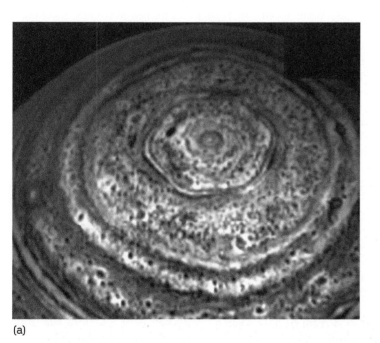

(a)

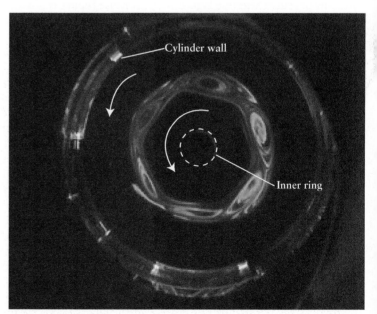

(b)

FIGURE 12-9 R I V U X G

Hexagonal Pattern Around Saturn's North Pole (a) An infrared image made by *Cassini*. Each side of the hexagonal pattern is about the diameter of Earth. Inside the hexagon is a rapidly circulating jet stream. (b) Experiments with fluids have recreated the green hexagonal pattern in a cylinder. The cylinder, filled with fluid, spins slowly to simulate Saturn's overall rotation, while an inner ring is spun faster than the fluid to simulate the jet stream observed around Saturn's pole. Simulating Saturn's gaseous atmosphere can be performed with fluids because under the right conditions, fluids and gas behave similarly. (a: Cassini/NASA; b: Ana Claudia Barbosa Aguiar/Science Magazine)

12-4 The internal heat of Jupiter and Saturn has a major effect on the planets' atmospheres

TUTORIAL 12-1 Weather patterns on Earth are the result of the motions of air masses. Winter storms in the midwestern United States occur when cold, moist air moves southward from Canada, and tropical storms in the southwest United States are caused by the northeastward motion of hot, moist air from Hawaii. All such motions, as well as the overall global circulation of our atmosphere, are powered by the Sun. Sunlight is absorbed by Earth's surface, and the heated surface in turn warms the atmosphere and stirs it into motion (see Sections 9-1 and 9-5). The energy of sunlight also powers the motions of the atmospheres of Venus and Mars (see Section 11-6). On both Jupiter and Saturn, however, atmospheric motions are powered both by solar energy and by the internal energy of the planet.

Jupiter and Saturn: Radiating Energy into Space

In the late 1960s, astronomers using Earth-based telescopes made the remarkable discovery that Jupiter emits more energy in the form of infrared radiation than it absorbs from sunlight—in fact, about twice as much. (By comparison, the internal heat that Earth radiates into space is only about 0.005% of what it absorbs from the Sun.) About 4.56 billion years ago when Jupiter formed, it would have been about twice as large. Since then, it has contracted under its own gravity, converting gravitational energy into heat (see Kelvin-Helmholtz contractions, Section 8-4). More specifically, Jupiter shrinks as its outer layers are pulled in closer, but as this gas picks up speed, collisions between gas molecules randomize their motion and this produces heat (see Figure 5-11). Because Jupiter is so large, it has retained substantial heat, or thermal energy, even after billions of years of cooling, and so still has plenty of energy to emit as infrared radiation. Even today, the contraction continues so that Jupiter shrinks by about 2 centimeters per year.

Saturn, too, radiates into space more energy than it receives as sunlight. Because Saturn is smaller than Jupiter, it should have begun with less internal heat trapped inside, and it should have radiated that heat away more rapidly. Hence, we would expect Saturn to be radiating very little energy today. In fact, when we take Saturn's smaller mass into account, we find that Saturn releases about 25% *more* energy from its interior on a per-kilogram basis than does Jupiter. The explanation for this seeming paradox may be that helium is raining out of Saturn's upper atmosphere, as we described in Section 12-2. The helium condenses into droplets at cold upper altitudes, but generates heat at lower altitudes due to friction between the falling droplets and the surrounding gases. This heat eventually escapes from the planet's surface as infrared radiation. This "raining out" of helium from Saturn's upper layers is calculated to have begun 2 billion years ago. The amount of thermal energy released by this process adequately accounts for the extra heat radiated by Saturn since that time.

Heat naturally flows from a hot place to a colder place, never the other way around. Hence, in order for heat to flow upward through the atmospheres of Jupiter and Saturn and radiate out into space, the temperature must be warmer deep inside the atmosphere than it is at the cloudtops. Infrared measurements have confirmed that the temperature within the atmospheres of both planets does indeed increase with increasing depth (Figure 12-10).

ANALOGY You can understand the statement that "heat naturally flows from a hot place to a colder place" by visualizing an ice cube placed on a hot frying pan. Experience tells you that the hot frying pan cools down a little while the ice warms up and melts. If heat were to flow the other way, the ice would get colder and the frying pan would get hotter—which never happens in the real world.

A Convection Paradox

When a fluid is warm at the bottom and cool at the top, like water being warmed in a pot, one effective way for energy to be distributed through the fluid is by the up-and-down motion called *convection*. (We introduced the idea of convection in Section 9-1; see, in particular, Figure 9-5.) For many years it was thought that Jupiter's light-colored zones are regions where warm gas from low levels is rising and cooling, forming high-altitude clouds, while the dark-colored belts are regions where cool high-altitude gas is descending and being heated (Figure 12-11). However, this picture has been called into question with images made by *Cassini* during its 2000–2001 flyby of Jupiter, which show evidence for rising material in the dark belts.

The Fastest Winds in the Solar System

VIDEO 12-6 In addition to the vertical motion of gases within the belts and zones, the very rapid rotation of both Jupiter and Saturn creates a global pattern of eastward and westward **zonal winds.** Wind speeds on Jupiter can exceed 500 km/h (300 mi/h). Earth's atmospheric circulation has a similar pattern of eastward and westward flow (see Figure 9-25), but with slower wind speeds. The faster winds on Jupiter presumably result from the planet's more rapid rotation as well as the substantial flow of heat from the planet's interior.

The zonal winds on Jupiter are generally strongest at the boundaries between belts and zones. Within a given belt or zone, the wind reverses direction between the northern and southern boundaries. As Figure 12-5 shows, such reversals in wind direction are associated with circulating storms such as the Great Red Spot.

Saturn rotates at about the same rate as Jupiter, but its interior releases less heat and it receives only a quarter as much energy in sunlight as does Jupiter. With less energy available to power the motions of Saturn's atmosphere, we would expect its zonal winds to be slower than Jupiter's. Surprisingly, Saturn's winds are *faster!* In 1980 the *Voyager* spacecraft measured wind speeds near Saturn's equator that approach 1800 km/h (1100 mi/h), approximately two-thirds the speed of sound in Saturn's atmosphere and faster than winds on any other planet. Since then, the equatorial winds seem to have abated somewhat. Between 1996 and 2004, astronomers using the Hubble Space Telescope showed that the wind speed had slowed to about 1000 km/h, and in 2005 *Cassini* measured wind speeds of about 1350 km/h at the equator. (These changes may be due to seasonal variations or changes in Saturn's clouds.) Why Saturn's winds should be even faster than Jupiter's is not yet completely understood.

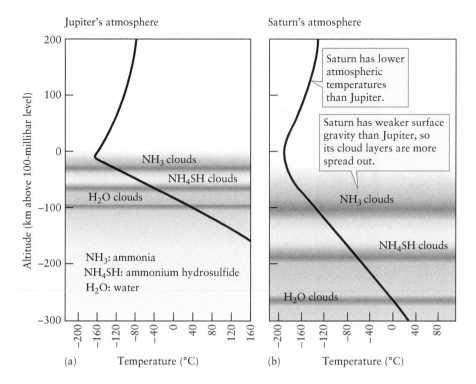

FIGURE 12-10

The Upper Atmospheres of Jupiter and Saturn The black curves in these graphs show temperature versus altitude in each atmosphere, as well as the probable arrangements of the cloud layers. Zero altitude in each atmosphere is chosen to be the point where the pressure is 100 millibars, or one-tenth of Earth's atmospheric pressure. Beneath both planets' cloud layers, the atmosphere is composed almost entirely of hydrogen and helium. (Adapted from A. P. Ingersoll)

Jupiter and Saturn's Cloud Layers

From spectroscopic observations and calculations of atmospheric temperature and pressure, scientists conclude that both Jupiter and Saturn have three main cloud layers of differing chemical composition. The uppermost cloud layer is composed of crystals of frozen ammonia. Deeper in the atmosphere, ammonia (NH_3) and hydrogen sulfide (H_2S)—a compound of hydrogen and sulfur—combine to produce ammonium hydrosulfide (NH_4SH) crystals. At even greater depths, the clouds are composed of crystals of frozen water.

Jupiter's strong surface gravity compresses these cloud layers into a region just 75 km deep in the planet's upper atmosphere (Figure 12-10a). But Saturn has a smaller mass and hence weaker surface gravity, so the atmosphere is less compressed and the same three cloud layers are spread out over a range of nearly 300 km (Figure 12-10b). The colors of Saturn's clouds are less dramatic than Jupiter's (see Figure 12-2) because deeper cloud layers are partially obscured by the hazy atmosphere above them.

The colors of clouds on Jupiter and Saturn depend on the temperatures of the clouds and, therefore, on the depth of the clouds within the atmosphere. Brown clouds are the warmest and are thus the deepest layers that we can see. Whitish clouds form the next layer up, followed by red clouds in the highest layer. The whitish zones on each planet are therefore somewhat higher than the brownish belts, while the red clouds in Jupiter's Great Red Spot are among the highest found anywhere on that planet.

Despite all the data from various spacecraft, we still do not know what gives the clouds of Jupiter and Saturn their colors. The crystals of NH_3, NH_4SH, and water-ice in the three main cloud layers are all white. Thus, other chemicals must cause the browns, reds, and oranges. Certain molecules closely related to ammonium hydrosulfide, which can form into long chains that have a yellow-brown color, may play a role. Compounds of sulfur or phosphorus, which can assume many different colors, depending on their temperature, might also be involved. The Sun's ultraviolet radiation may induce the chemical reactions that produce these colorful compounds.

CONCEPTCHECK 12-4

Why are Jupiter's cloud layers more compressed than Saturn's?
Answer appears at the end of the chapter.

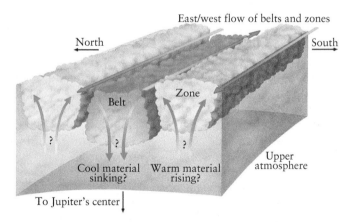

FIGURE 12-11

Original Model of Jupiter's Belts and Zones The light-colored zones and dark-colored belts in Jupiter's atmosphere were believed until recently to be regions of rising and descending gases, respectively. In the zones, gases warmed by heat from Jupiter's interior were thought to rise upward and cool, forming high-altitude clouds. In the belts, cooled gases were thought to descend and undergo an increase in temperature; the cloud layers seen there are at lower altitudes than in the zones. Observations by the *Cassini* spacecraft on its way to Saturn suggest that just the opposite might be correct, with the darker belts rising (these new doubts are indicated by the question marks in the figure above). *Cassini* was able to spot numerous small clouds—too small to be identified from Earth—rising within the dark belts, suggesting that the belts themselves might be rising.

12-5 The *Galileo* space probe explored Jupiter's deep atmosphere

By observing Jupiter's and Saturn's cloud layers from a distance, we have been able to learn a great deal about the atmospheres beneath the clouds. But there is no substitute for making measurements on site, and for many years scientists planned to send a spacecraft to explore deep into Jupiter's atmosphere. While still 81.52 million kilometers (50.66 million miles) from Jupiter, the *Galileo* spacecraft released the *Galileo Probe*, a cone-shaped body about the size of an office desk. While *Galileo* itself later fired its rockets to place itself into an orbit around Jupiter, the *Galileo Probe* continued on a course that led it on December 7, 1995, to a point in Jupiter's clouds just north of the planet's equator.

Exploring Jupiter's Atmosphere

A heat shield protected the *Galileo Probe* as air friction slowed its descent speed from 171,000 km/h (106,000 mi/h) to 40 km/h (25 mi/h) in just 3 minutes. The spacecraft then deployed a parachute and floated down through the atmosphere (Figure 12-12). For the next hour, the probe observed its surroundings and radioed its findings back to the main *Galileo* spacecraft, which in turn radioed them to Earth. The mission ceased at a point some 200 km (120 mi) below Jupiter's upper cloud layer, where the tremendous pressure (24 atmospheres) and high temperature (152°C = 305°F) finally overwhelmed the probe's electronics.

Although the *Galileo Probe* did not have a camera, it did carry a variety of instruments that made several new discoveries. A radio-emissions detector found evidence for lightning discharges that, while less frequent than on Earth, are individually much stronger than lightning bolts in our atmosphere. Other measurements showed that Jupiter's winds, which are brisk in the atmosphere above the clouds, are even stronger beneath the clouds. The *Galileo Probe* measured a nearly constant wind speed of 650 km/h (400 mi/h) throughout its descent. This shows convincingly that the energy source for the winds is Jupiter's interior heat. If the winds were driven primarily by solar heating, as is the case on Earth, the wind speed would have decreased with increasing depth.

A central purpose of the *Galileo Probe* mission was to test the three-layer cloud model depicted in Figure 12-10a by making direct measurements of Jupiter's clouds. But the probe saw only traces of the NH_3 and NH_4SH cloud layers and found no sign at all of the low-lying water clouds. (One explanation of this surprising result is that the probe by sheer chance entered an unusually warm and cloud-free part of Jupiter's atmosphere.)

The *Galileo Probe* also made measurements of the chemical composition of Jupiter's atmosphere. These data revealed that the relative proportions of hydrogen and helium are almost exactly the same as in the Sun, a finding in line with the accepted picture of how Jupiter formed from the solar nebula (see Section 8-4). But a number of heavy elements, including carbon, nitrogen, and sulfur, were found to be significantly more abundant on Jupiter than in the Sun, which was an expected result. Over the past several billion years, Jupiter's strong gravity should have pulled in many pieces of interplanetary debris, thus building up a small but appreciable abundance of heavy elements.

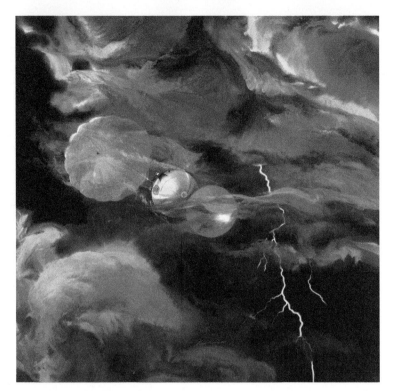

FIGURE 12-12

The *Galileo Probe* Enters Jupiter's Atmosphere This artist's impression shows the *Galileo Probe* descending under a parachute through Jupiter's clouds. The jettisoned heat shield (shown to the right of the probe) protected the probe during its initial high-speed entry into the atmosphere. The *Galileo Probe* returned data for 58 minutes before it was crushed and melted by the pressure and temperature of the atmosphere. (NASA)

Noble Gases and the Origin of Jupiter

Another surprising and potentially important result from the *Galileo Probe* concerns three elements, argon (Ar), krypton (Kr), and xenon (Xe). These elements are three of the six so-called **noble gases,** which do not combine with other atoms to form molecules. (The noble gases, which also include helium, appear in the rightmost column of the periodic table in Box 5-5.) All three of these elements appear in tiny (or trace) amounts in Earth's atmosphere and in the Sun, and so must also have been present in the solar nebula. But the *Galileo Probe* found that argon, krypton, and xenon are about 3 times as abundant in Jupiter's atmosphere as in the Sun's atmosphere.

If these elements were incorporated into Jupiter directly from the gases of the solar nebula, they should be equally as abundant in Jupiter as in the Sun (which formed from the gas at the center

> Measuring trace amounts of rare gases on Jupiter provides clues about the origins of the planets

of the solar nebula; see Section 8-3). So, the excess amounts of argon, krypton, and xenon must have entered Jupiter in the form of solid planetesimals, just as did carbon, nitrogen, and sulfur. But at Jupiter's distance from the Sun, the presumed temperature of the solar nebula was too high to permit argon, krypton, or xenon to solidify. Left in their warmer gaseous form, it is unclear how Jupiter could have accumulated so much of these gases. How, then, did the excess amounts of these noble gases become part of Jupiter?

One hypothesis proposes that Jupiter originally formed much farther from the Sun. Out in the cold, remote reaches of the solar nebula, argon, krypton, and xenon could have solidified into icy particles and fallen into the young Jupiter, thus enriching it in those elements. Interactions between Jupiter and the material of the solar nebula could then have forced the planet to spiral inward, eventually reaching its present distance from the Sun. Although we saw in Section 8-7 that some migration of planets is expected, current models do not support Jupiter forming so far away. When the *Juno* spacecraft arrives at Jupiter in 2016, it can help to solve the mystery of Jupiter's noble gases.

CONCEPTCHECK 12-5

Why is it difficult to understand that Jupiter has greater abundances of noble gases than the Sun?

Answer appears at the end of the chapter.

12-6 The oblateness of Jupiter and Saturn reveals their rocky cores

The *Galileo Probe* penetrated only a few hundred kilometers into Jupiter's atmosphere before it was crushed. To learn about Jupiter's structure at greater depths, as well as that of Saturn, astronomers must use indirect clues. The shapes of Jupiter and Saturn are important clues, because they indicate the size of the rocky cores at the centers of the planets. When the *Juno* spacecraft reaches Jupiter, we will learn even more about the planet's internal structure.

Oblateness and Rotation

Even a casual glance through a small telescope shows that Jupiter and Saturn are not spherical but slightly flattened or **oblate.** (You can see this in Figures 12-1 and 12-2.) The diameter across Jupiter's equator (142,980 km) is 6.5% larger than its diameter from pole to pole (133,700 km). Thus, Jupiter is said to have an **oblateness** of 6.5%, or 0.065. Saturn has an even larger oblateness of 9.8%, or 0.098, making it the most oblate of all the planets. By comparison, the oblateness of Earth is just 0.34%, or 0.0034.

If Jupiter and Saturn did not rotate, both would be perfect spheres. A massive, nonrotating object naturally settles into a spherical shape so that every atom on its surface experiences the same gravitational pull aimed directly at the object's center. Because Jupiter and Saturn do rotate, however, the body of each planet tends to fly outward and away from the axis of rotation.

ANALOGY You can demonstrate this effect for yourself. Stand with your arms hanging limp at your sides, then spin yourself around. Your arms will naturally tend to fly outward, away from the vertical axis of rotation of your body. Likewise, if you drive your car around a circular road, you feel thrown toward the outside of the circle; the car as a whole is rotating around the center of the circle, and you tend to move away from the rotation axis. (We discussed the physical principles behind this in Box 4-3.) For Jupiter and Saturn, this same effect deforms the planets into their nonspherical, oblate shapes.

Modeling Jupiter and Saturn's Interiors

The oblateness of a planet depends both on its rotation rate and on how the planet's mass is distributed throughout its volume. In general, the more a planet's mass is concentrated in a solid core, the less oblate (or more spherical) the planet's shape. The challenge to planetary scientists is to create a model of a planet's mass distribution that is consistent with the observed oblateness and rotation rate. One such model for Jupiter suggests that 2.6% of its mass is concentrated in a dense, rocky core (Figure 12-13a). Although this core has about 8 times the mass of Earth, the crushing weight of the remaining bulk of Jupiter compresses it to a diameter of just 11,000 km, slightly *smaller* than Earth's diameter of 12,800 km. The pressure at the center of the core is estimated to be about 70 million atmospheres, and the central temperature is probably about 22,000 K. By contrast, the temperature at Jupiter's cloudtops is only 165 K.

If Jupiter formed by accretion of gases onto a rocky "seed," the present-day core is presumably that very seed. Additional rocky

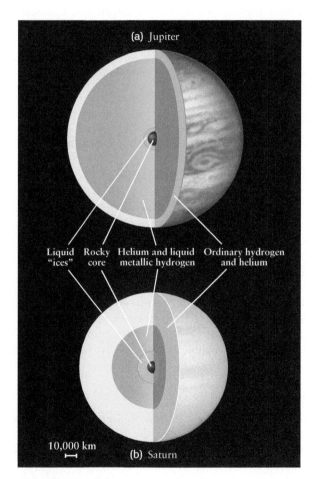

FIGURE 12-13

The Internal Structures of Jupiter and Saturn These diagrams of the interiors of Jupiter and Saturn are drawn to the same scale. Each planet's rocky core is surrounded by an outer core of liquid "ices," a layer of helium and liquid metallic hydrogen, and a layer of helium and ordinary molecular hydrogen (H_2). Saturn's rocky core contains a larger fraction of the planet's mass than does Jupiter's, while Saturn has a smaller volume of liquid metallic hydrogen than does Jupiter.

material was presumably added later by meteoritic material that fell into the planet and sank to the center. If, however, Jupiter formed directly from the gases of the solar nebula, the meteoritic material that later fell in makes up the present-day core. Material from icy planetesimals, too, would have sunk deep within Jupiter and added to the core. In the model we have been describing, these "ices"—principally water (H_2O), methane (CH_4), ammonia (NH_3), and other molecules made by chemical reactions among these substances—form a layer some 3000 km thick that surrounds the rocky core. (Because these substances are less dense than rock, they "float" on top of the rocky core.) Despite the high pressures in this layer, the temperature is so high that the "ices" are probably in a liquid state.

Saturn rotates at about the same rate as Jupiter but has less mass, and therefore less gravity to pull its material inward. Hence, Saturn's rotation should cause material at its equator to bulge outward more than on Jupiter, giving Saturn a greater oblateness. In fact, if Saturn and Jupiter had the same internal structure, we would actually expect Saturn to be even more oblate than it really is. We can account for this discrepancy if Saturn has a different mass distribution than Jupiter. Detailed calculations suggest that about 10% of Saturn's mass is contained in its rocky core, compared to 2.6% for Jupiter's rocky core (see Figure 12-13b). Like Jupiter, the rocky core of Saturn is probably surrounded by an outer core of liquid "ices."

CONCEPTCHECK 12-6

If Jupiter's rotation rate was made large enough, could it be more oblate than Saturn?

Answer appears at the end of the chapter.

12-7 Metallic hydrogen inside Jupiter and Saturn endows the planets with strong magnetic fields

The oblateness of Jupiter and Saturn gives information about the size of the planets' rocky cores. However, oblateness does not provide much insight into the liquid surrounding the planets' cores. Just as with Earth, a window into a planet's inner workings can come from a global magnetic field.

Jupiter's Magnetic Field

Measurements by the *Pioneer* and *Voyager* spacecraft detected a magnetic field and indicate that the field at Jupiter's equator is 14 times stronger than at Earth's equator. Like Earth's magnetic field, Jupiter's field is thought to be generated by motions of an electrically conducting fluid in the planet's interior. But wholly unlike Earth, the moving fluid within Jupiter is an exotic form of hydrogen called **liquid metallic hydrogen.**

As we discussed in Section 7-7, hydrogen becomes a liquid metal only when the pressure exceeds about 1.4 million atmospheres. This pressure occurs at depths more than about 7000 km below Jupiter's cloudtops. Thus, the internal structure of Jupiter consists of four distinct regions, as shown in Figure 12-13a: a rocky core, a roughly 3000-km thick layer of liquid "ices," a layer of helium and liquid

metallic hydrogen about 56,000 km thick, and a layer of helium and ordinary hydrogen about 7000 km thick. The colorful cloud patterns that we can see through telescopes are located in the outermost 100 km of the outer layer.

Figure 12-13a shows that much of the enormous bulk of Jupiter is composed of liquid metallic hydrogen. Jupiter's rapid rotation sets the liquid metallic hydrogen into motion, giving rise to a powerful dynamo 20,000 times stronger than Earth's and generating an intense magnetic field.

Saturn's Magnetic Field

Saturn also has a substantial magnetic field, but it pales by Jupiter standards: Saturn's internal dynamo is only 3% as strong as Jupiter's. (To be fair, it is still 600 times stronger than Earth's dynamo!) This difference between the planets suggests that even though Saturn rotates nearly as rapidly as Jupiter, there must be far less liquid metallic hydrogen within Saturn to be stirred up by the rotation and thereby generate a magnetic field. This conclusion makes sense: Compared to Jupiter, Saturn has less mass, less gravity, and less internal pressure to compress ordinary hydrogen into liquid metallic hydrogen. As a consequence, as Figure 12-13 shows, liquid metallic hydrogen probably makes up only a relatively small fraction of Saturn's bulk.

Jupiter's Magnetosphere

Jupiter's strong magnetic field surrounds the planet with a magnetosphere so large that it envelops the orbits of many of its moons (Figure 12-14). The *Pioneer* and *Voyager* spacecraft that carried instruments to detect charged particles and magnetic fields, revealed the awesome dimensions of the Jovian magnetosphere. If you could see Jupiter's magnetosphere from Earth, it would cover an area in the sky 16 times larger than the Moon!

> Both Jupiter and Saturn have magnetospheres that are millions of kilometers across

As charged particles in the solar wind collide with the Jovian magnetosphere, they slow down, creating a region called a **bow shock.** The bow shock marks the transition between the magnetosphere and the surrounding space. Deeper within Jupiter's magnetosphere, charged particles—also referred to as ions—are trapped in belts that are analogous to Earth's Van Allen belts (recall Figure 9-19). An astronaut venturing into these belts would quickly receive a lethal dose of this charged-particle radiation.

A major source for charged particles in Jupiter's magnetosphere is the moon Io (see Table 7-2). Io's active volcanoes eject about a ton of material into Jupiter's magnetosphere each second. As shown in Figure 12-14, these particles form a torus.

Saturn's Magnetosphere

Saturn's magnetosphere is similar to Jupiter's, but is only about 10 to 20% as large and contains far fewer charged particles than Jupiter's. There are two reasons for this deficiency. First, Saturn lacks a highly volcanic moon like Io to inject particles into the magnetosphere. Second, and more important, many charged particles are absorbed by the icy particles that make up Saturn's rings (see Section 12-10). As with Jupiter, the charged particles that do exist in Saturn's magnetosphere are concentrated in radiation belts similar to the Van Allen belts in Earth's magnetosphere.

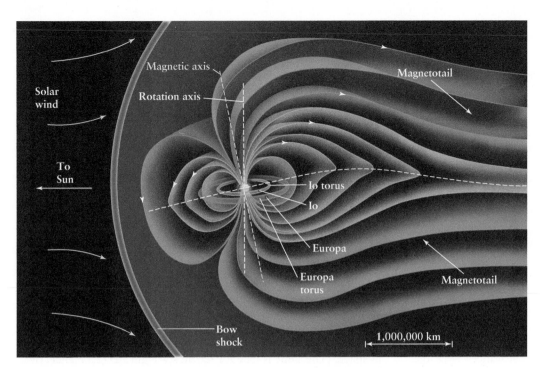

FIGURE 12-14

Jupiter's Magnetosphere Created by the planet's rotation, the ion-trapping regions of Jupiter's magnetosphere (in orange, analogous to the Van Allen belts) extend into the realm of the Galilean moons. Gases originating from Jupiter's moons Io and Europa form tori (doughnut-shaped regions) in the magnetosphere. Some of Io's particles are pulled by the field onto Jupiter. Swept back by the solar wind, the magnetosphere has a "magnetotail" pointing away from the Sun. The magnetotail is often over 500 million km long and sometimes it reaches all the way to Saturn. (J. Clarke, University of Michigan, and NASA)

Aurorae on Jupiter and Saturn

On Earth, charged particles trapped in our planet's magnetic field stream onto the north and south magnetic poles, creating the aurora (see Section 9-4). The same effect takes place on both Jupiter and Saturn. **Figure 12-15** shows images from the Hubble Space Telescope that reveal glowing auroral rings centered on the magnetic poles of Jupiter and of Saturn. As on Earth, the emission comes from a region high in each planet's atmosphere, hundreds of kilometers above the cloudtops.

CONCEPTCHECK 12-7

Consider Figures 12-14 and 12-15a. Could charged particles from volcanoes on the moon Io lead to brighter aurorae on Jupiter?

Answer appears at the end of the chapter.

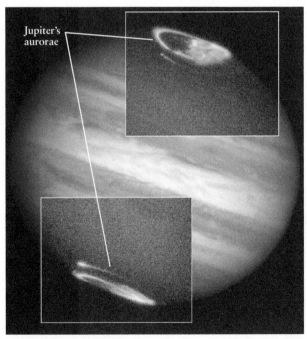

(a) Jupiter   R I V U X G

(b) Saturn   R I V U X G

FIGURE 12-15

Aurorae on Jupiter and Saturn These images combine visible wavelengths of the planets, and ultraviolet light from the aurorae. On both Jupiter and Saturn, charged particles from the magnetosphere are funneled onto the planet's magnetic poles. When the particles collide with and excite molecules in the upper atmosphere, the molecules emit ultraviolet light. In **(b)**, a large disturbance in the solar wind approaching Saturn coincided with the brightening of the aurora in the middle image of the three. (a: J. Clarke, University of Michigan, and NASA; b: NASA/Hubble/Z. Levay and J. Clark)

12-8 Earth-based observations reveal three broad rings encircling Saturn

Compared to Saturn, Jupiter is unquestionably larger, more massive, more colorful, and more dynamic in both its atmosphere and magnetosphere. But as seen through even a small telescope, Saturn stands out as perhaps the most beautiful of all the worlds of the solar system, thanks to its majestic rings (Figure 12-16). As we will see in Section 12-9, Jupiter also has rings, but they are so faint that they were not discovered until spacecraft could observe Jupiter at close range.

Discovering Saturn's Rings

In 1610, Galileo became the first person to see Saturn through a telescope. He saw few details, but he did notice two puzzling lumps protruding from opposite edges of the planet's disk. Curiously, these lumps disappeared in 1612, only to reappear in 1613. Other observers saw similar appearances and disappearances over the next several decades.

In 1655, the Dutch astronomer Christiaan Huygens began to observe Saturn with a better telescope than was available to any of his predecessors. (We discussed in Section 11-2 how Huygens used this same telescope to observe Mars.) On the basis of his observations, Huygens suggested that Saturn was surrounded by a thin, flattened ring. At times this ring was edge-on as viewed from Earth, making it almost impossible to see. At other times Earth observers viewed Saturn from an angle either above or below the plane of the ring, and the ring was visible, as in Figure 12-1. (The lumps that Galileo saw were the parts of the ring to either side of Saturn, blurred by the poor resolution of his rather small telescope.) Astronomers confirmed this brilliant deduction over the next several years as they watched the ring's appearance change just as Huygens had predicted.

As the quality of telescopes improved, astronomers realized that Saturn's "ring" is actually a *system* of rings, as Figure 12-16 shows. In 1675, the Italian astronomer Cassini discovered a dark division in the ring. The **Cassini division** is a gap about 4800 kilometers wide that separates the outer **A ring** from the brighter **B ring** closer

to the planet. In the mid-1800s, astronomers managed to identify the faint **C ring**, or *crepe* ring, that lies just inside the B ring. To appreciate the enormous scale of the ring system, consider that Mercury could fit snugly within the Cassini division!

Saturn and its rings are best seen when the planet is at or near opposition. A modest telescope gives a good view of the A and B rings, but a large telescope and excellent observing conditions are needed to see the C ring.

How the Rings Appear from Earth

Earth-based views of the Saturnian ring system change as Saturn slowly orbits the Sun. Huygens was the first to understand that this change occurs because the

> As seen from Earth, Saturn's rings change their orientation as the planet orbits the Sun

rings lie in the plane of Saturn's equator, and this plane is tilted 27° from the plane of Saturn's orbit. As Saturn orbits the Sun, its rotation axis and the plane of its equator maintain the same orientation in space. (The same is true for Earth; see Figure 2-12.) Hence, over the course of a Saturnian year, an Earth-based observer views the rings from various angles (Figure 12-17). At certain times Saturn's north pole is tilted toward Earth and the observer looks "down" on the "top side" of the rings. Half a Saturnian year later, Saturn's south pole is tilted toward us and the "underside" of the rings is exposed to our Earth-based view.

When our line of sight to Saturn is in the plane of the rings, the rings are viewed edge-on and seem to disappear entirely (see the images at the upper left and lower right corners of Figure 12-17). This disappearance indicates that the rings are very thin. In fact, they are thought to be only about 10 meters thick (about 30 feet). In proportion to their diameter, Saturn's rings are thousands of times thinner than the sheets of paper used to print this book.

The last edge-on presentation of Saturn's rings of the twentieth century was in 1995–1996; the first two of the twenty-first century are in 2008–2009 and in 2025. From 1996 to 2008, we Earth observers see the "underside" of the rings (as in the image at the lower left corner of Figure 12-17); from 2009 until 2025, we see the "top" of the rings.

A ring

B ring

C ring

Cassini division

FIGURE 12-16   R I ☑ U X G

Saturn's System of Rings This *Cassini* image shows many details of Saturn's rings. The C ring is so faint that it is almost invisible in this view. At 274,000 km, the outer diameter of the A ring is almost twice the diameter of Jupiter. Saturn is visible through the rings (look near the very bottom of the image), which shows that the rings are not solid. The ring system is so large that Mercury could fit inside the Cassini division. (NASA/cassini3d.com/images)

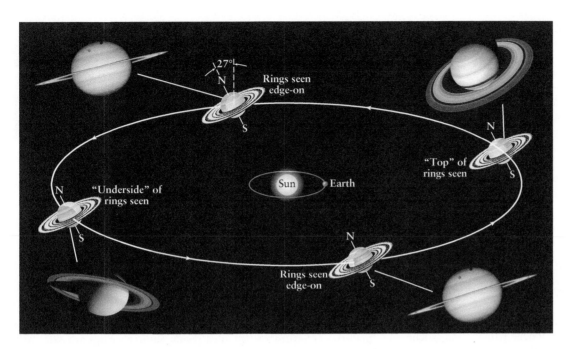

FIGURE 12-17

R I **V** U X G

The Changing Appearance of Saturn's Rings Saturn's rings are aligned with its equator, which is tilted 27° from Saturn's orbital plane. As Saturn moves around its orbit, the rings maintain the same orientation in space, so observers on Earth see the rings at various angles. The accompanying images of Saturn are made from the "underside" of the rings, the "top" side, and nearly edge-on. (NASA/JPL/Space Science Institute)

CONCEPTCHECK 12-8

What causes Saturn's brilliant rings to be sometimes nearly invisible and at other times easily observable with the smallest of telescopes?

Answer appears at the end of the chapter.

12-9 Saturn's rings are composed of numerous icy fragments, while Jupiter's rings are made of small rocky particles

Astronomers have long known that Saturn's rings could not possibly be solid sheets of matter. In 1857, the Scottish physicist James Clerk Maxwell (whom we last encountered in Section 5-2) proved theoretically that if the rings were solid, differences in Saturn's gravitational pull on different parts of the rings would cause the rings to tear apart. He concluded that Saturn's rings are made of "an indefinite number of unconnected particles."

Ring Particles

In 1895, James Keeler at the Allegheny Observatory in Pittsburgh became the first to confirm by observation that the rings are not rigid. He made this conclusion by photographing the spectrum of sunlight reflected from Saturn's rings. As the rings orbit Saturn, the spectral lines from the side approaching us are blueshifted by the Doppler effect (recall Section 5-9, especially Figure 5-26). At the same time, spectral lines from the receding side of the rings are redshifted. Keeler noted that the size of the wavelength shift increased inward across the rings—the closer to the planet, the greater the shift. This variation in Doppler shift proved that the inner portions of Saturn's rings are moving around the planet more rapidly than the outer portions. Indeed, the orbital speeds

across the rings are in complete agreement with Kepler's third law: The square of the orbital period about Saturn at any place in the rings is proportional to the cube of the distance from Saturn's center (see Section 4-7 and Box 4-4). This result is exactly what would be expected if the rings consisted of numerous tiny "moonlets," or **ring particles,** each individually circling Saturn.

Saturn's rings are quite bright; they reflect 80% of the sunlight that falls on them. (By comparison, Saturn itself reflects 46% of incoming sunlight.) Astronomers therefore long suspected that the ring particles are made of ice and ice-coated rock. This hunch was confirmed in the 1970s, when the American astronomers Gerard P. Kuiper and Carl Pilcher identified absorption features of frozen water in the rings' near-infrared spectrum. The *Voyager* and *Cassini* spacecraft have made even more detailed infrared measurements that indicate the temperature of the rings ranges from −180°C (−290°F) in the sunshine to less than −200°C (−330°F) in Saturn's shadow. Water-ice is in no danger of melting or evaporating at these temperatures.

To determine the sizes of the particles that make up Saturn's rings, astronomers analyzed the radio signals received from a spacecraft as it passed behind the rings. How easily radio waves can travel through the rings depends on the relationship between the wavelength and the particle size. The results show that **most of the particles range in size from pebble-sized fragments about 1 cm in diameter to chunks about 5 m across,** the size of large boulders. Most abundant are snowball-sized particles about 10 cm in diameter. Recall that Mercury could fit inside the Cassini division, so seeing even the largest boulder-sized ring particles is not possible in images such as Figure 12-16.

The Roche Limit

All of Saturn's material may be ancient debris that failed to accrete (fall together) into satellites. The total amount of material in the rings is quite small. If Saturn's entire ring system were compressed together to make a satellite, it would be no more than 100 km (60 mi) in diameter. But, in fact, the ring particles are so close to Saturn that they will never be able to form moons.

To see why, imagine a collection of small particles orbiting a planet. Gravitational attraction between neighboring particles tends to pull the particles together. However, because the various particles are at differing distances from the parent planet, they also experience different amounts of gravitational pull from the planet. This difference in gravitational pull is a **tidal force** that tends to keep the particles separated. (We discussed tidal forces in detail in Section 4-8. You may want to review that section, and in particular Figure 4-26.)

> Tidal forces prevent the material of Saturn's rings from coalescing into a moon

The closer a pair of particles is to the planet, the greater the tidal force that tries to pull the pair apart. At a certain distance from the planet's center, called the **Roche limit**, the disruptive tidal force is just as strong as the gravitational force between the particles. (The concept of this limit was developed in the mid-1800s by the French mathematician Edouard Roche.) Inside the Roche limit, the tidal force overwhelms the gravitational pull between neighboring particles, and these particles cannot accrete to form a larger body. Instead, they tend to spread out into a ring around the planet. (The *Cosmic Connections* figure depicts this process.) Indeed, most of Saturn's system of rings visible in Figure 12-16 lies within the planet's Roche limit.

All large planetary satellites are found outside their planet's Roche limit. If any large satellite were to come inside its planet's Roche limit, the planet's tidal forces would cause the satellite to break up into fragments. We will see in Chapter 14 that such a catastrophic tidal disruption may be the eventual fate of Neptune's large satellite, Triton.

CAUTION! It may seem that it would be impossible for any object to hold together inside a planet's Roche limit. But the ring particles inside Saturn's Roche limit survive and do not break apart. The reason is that the Roche limit applies only to objects held together by the *gravitational* attraction of each part of the object for the other parts. By contrast, the forces that hold a rock or a chunk of ice together are *chemical* bonds between the object's atoms and molecules. These chemical forces are much stronger than the disruptive tidal force of a nearby planet, so a rock or chunk of ice does not break apart. In the same way, people walking around on Earth's surface (which is inside Earth's Roche limit) are in no danger of coming apart, because we are held together by comparatively strong chemical forces rather than gravity.

Jupiter's Ring and the Roche Limit

When the *Voyager 1* spacecraft flew past Jupiter in 1979, it trained its cameras not just on the planet but also on the space around the planet's equator. It discovered that Jupiter, too, has a system of rings that lies within its Roche limit (**Figure 12-18**). These rings differ from Saturn's in two important ways. First, Jupiter's rings are composed of tiny particles of rock with an average size of only about 1 μm (= 0.001 m = 10^{-6} m) and that reflect less than 5% of

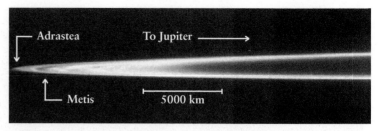

VIDEO 12-7 **FIGURE 12-18** R I **V** U X G
Jupiter's Main Ring This *Galileo* image shows Jupiter's main ring almost edge-on. The ring's outer radius, about 129,000 km from the center of Jupiter, is close to the orbit of Adrastea, the second closest (after Metis) of Jupiter's moons. Not shown is an even larger and even more tenuous pair of "gossamer rings," one of which extends out to 181,000 km and the other out to 222,000 km. (NASA/JPL; Cornell University)

the sunlight that falls on them. Second, there is very little material in the rings of Jupiter, less than 1/100,000 (10^{-5}) the amount of material in Saturn's rings. As a result, Jupiter's rings are extremely faint, which explains why their presence was first revealed by a spacecraft rather than an Earth-based telescope. The ring particles are thought to originate from meteorite impacts on Jupiter's four small, inner satellites, two of which are visible in Figure 12-18.

We will see in Chapter 14 that the rings of Uranus and Neptune are also made of many individual particles orbiting inside each planet's Roche limit. Like the rings of Jupiter, these rings are quite dim and difficult to see from Earth.

CONCEPTCHECK 12-9

If an asteroid entered a low-Earth orbit inside Earth's Roche limit, what would become of it?

Answer appears at the end of the chapter.

12-10 Saturn's rings consist of thousands of narrow, closely spaced ringlets

The photograph in Figure 12-1b, which was made in 1974, indicates what astronomers understood about the structure of Saturn's rings in the mid-1970s. Each of the A, B, and C rings appeared to be rather uniform, with little or no evidence of any internal structure. Within a few years, however, close-up observations from spacecraft revealed the true complexity of the rings, as well as their chemical composition.

Spacecraft Views of Saturn's Rings

Pioneer 11, Voyager 1, and *Voyager 2* recorded images of Saturn's rings during their flybys in 1979, 1980, and 1981, respectively. *Pioneer 11* had only a relatively limited capability to make images, but cameras on board the two *Voyager* spacecraft sent back a number of pictures showing the detailed structure of Saturn's rings. Since entering Saturn orbit in 2004, the *Cassini* spacecraft has provided scientists with even higher-resolution images of the rings.

Planetary Rings and the Roche Limit

If a small moon wanders too close to a planet, tidal forces tear the moon into smaller particles. These particles form a ring around the planet. The critical distance from the planet at which this happens is called the Roche limit. This helps us understand why the rings of Jupiter and Saturn are made of small particles, as are the rings of Uranus and Neptune (discussed in Chapter 14).

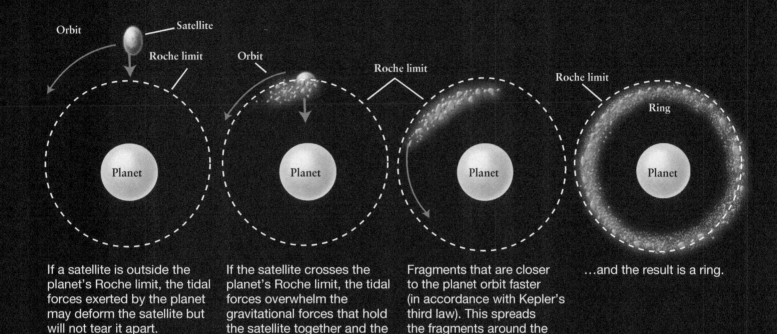

If a satellite is outside the planet's Roche limit, the tidal forces exerted by the planet may deform the satellite but will not tear it apart.

If the satellite crosses the planet's Roche limit, the tidal forces overwhelm the gravitational forces that hold the satellite together and the satellite fragments.

Fragments that are closer to the planet orbit faster (in accordance with Kepler's third law). This spreads the fragments around the planet…

…and the result is a ring.

Jupiter, Saturn, Uranus, and Neptune all have systems of rings that lie mostly within the Roche limit. This diagram shows each of the four ring systems scaled to the radius of the planet.

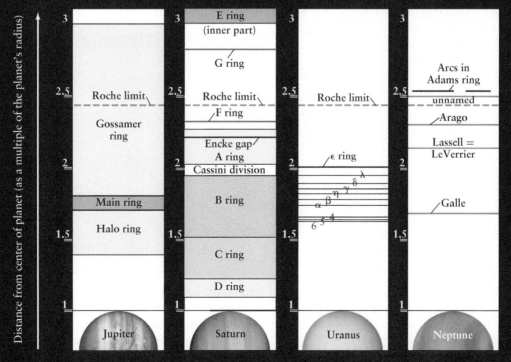

A planet's Roche limit is about 2.4 times the radius of the planet, provided the material orbiting the planet has the same density as the planet itself. For denser material the Roche limit is closer to the planet.

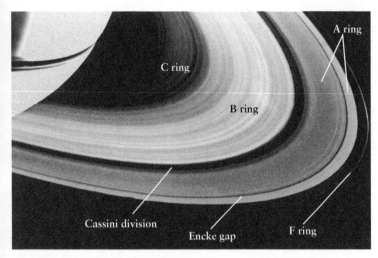

FIGURE 12-19 R I V U X G

Details of Saturn's Rings The colors in this *Voyager 1* image have been enhanced to emphasize small differences between different portions of the rings. The broad Cassini division is clearly visible, as is the narrow Encke gap in the outer A ring. The very thin F ring lies just beyond the outer edge of the A ring. (NASA/JPL)

Figure 12-19 shows a *Voyager 1* image. Some of the features seen in this image were expected, including the **Encke gap**, a 270-km-wide division in the outer A ring observed by the German astronomer Johann Franz Encke in 1838. But to the amazement of scientists, images like the one in Figure 12-19 revealed that the broad A, B, and C rings are not uniform at all but instead consist of hundreds upon hundreds of closely spaced bands or **ringlets**. These ringlets are arrangements of ring particles that have evolved from the combined gravitational forces of neighboring particles, of Saturn's moons, and of the planet itself.

Pioneer 11 first detected the narrow **F ring**, which you can see in Figure 12-19. It is only about 100 km wide and lies 4000 km beyond the outer edge of the A ring. Close-up views from the *Voyagers* and *Cassini* show that the F ring is made up of several narrow ringlets. Small moons that orbit close to the F ring exert gravitational forces on the ring particles, thus warping and deforming these ringlets (Figure 12-20).

Exploring Saturn's Ring Particles

Spacecraft have done more than show Saturn's rings in greater detail; they have also viewed the rings from perspectives not possible from Earth. Through Earth-based telescopes, we can see only the sunlit side of the rings. From this perspective, the B ring appears very bright, the A ring moderately bright, the C ring dim, and the Cassini division dark (see Figure 12-19). The fraction of sunlight reflected back toward the Sun (the *albedo*) is directly related to the concentration and size of the particles in the ring. The B ring is bright because it has a high concentration of relatively large, icy, reflective particles, whereas the darker Cassini division has a lower concentration of such particles.

The *Voyagers* and *Cassini* have expanded our knowledge of the ring particles by imaging the shaded side of the rings, which cannot be seen from Earth. Figure 12-21 shows such an image. Because the spacecraft was looking back toward the Sun, this picture shows sunlight that has passed through the rings. The B ring looks darkest in this image because little sunlight gets through its dense collection of particles. If the Cassini division were completely empty, it would look black, because we would then see through it to the blackness of empty space. But the Cassini division looks *bright* in

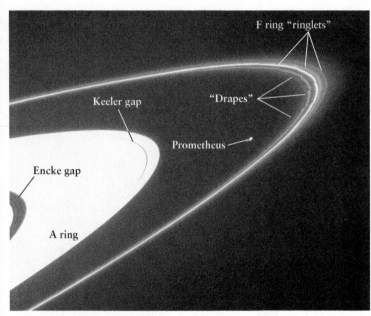

 VIDEO 12-8 **FIGURE 12-20** R I V U X G

Details of Saturn's F Ring This *Cassini* image shows Saturn's F ring (compare Figure 12-19). The bright central ringlet is about 50 km wide. As the small satellite Prometheus orbits just inside the F ring, its gravitational pull causes disturbances called "drapes" in the ringlets. The narrowness of the F ring is the result of gravitational forces exerted by Prometheus and another small satellite called Pandora (see Section 12-11). This image also shows two divisions in the A ring, the Encke gap and the Keeler gap. (NASA/JPL/Space Science Institute)

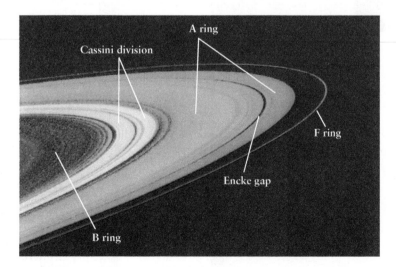

FIGURE 12-21 R I V U X G

The View from the Far Side of Saturn's Rings This false-color view of the side of the rings away from the Sun was taken by *Voyager 1*. The Cassini division appears white, not black like the empty gap between the A and F rings. Hence, the Cassini division cannot be empty, but must contain a number of relatively small particles that scatter sunlight like dust motes. (NASA/JPL)

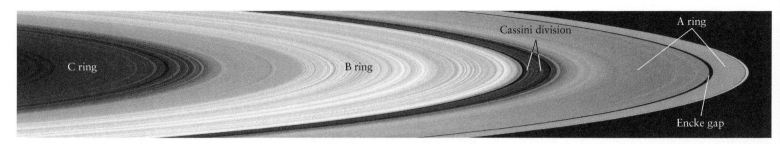

FIGURE 12-22   R I V U X G

Color Variations in Saturn's Rings This mosaic of *Cassini* images shows the rings in natural color. The color variations are indicative of slight differences in chemical composition among particles in different parts of the rings. (NASA/JPL/Space Science Institute)

Figure 12-21. This shows that the Cassini division is not empty but contains a relatively small number of particles. (In the same way, dust particles in the air in your room become visible when a shaft of sunlight passes through them.)

The process by which light bounces off particles in its path is called **light scattering.** The proportion of light scattered in various directions depends upon both the size of the particles and the wavelength of the light. (See Box 5-4 for examples of light scattering on Earth.) Because visible light has a much smaller wavelength than radio waves, light scattering allows scientists to measure the sizes of particles too small to detect using the radio technique described in Section 12-10. As an example, by measuring the amount of light scattered from the rings at various wavelengths and from different angles as the two *Voyager* spacecraft sped past Saturn, scientists showed that the F ring contains a substantial number of tiny particles about 1 μm ($= 10^{-6}$ m) in diameter. This size is about the same as the size of the particles found in smoke.

There are also subtle differences in color from one ring to the next, as the computer-enhanced image in **Figure 12-22** shows. Although the main chemical constituent of the ring particles is frozen water, trace amounts of other chemicals—perhaps coating the surfaces of the ice particles—are probably responsible for the different colors. These trace chemicals have not yet been identified.

The color variations in Figure 12-22 suggest that the icy particles do not migrate substantially from one ringlet to another. Had such migration taken place, the color

> Different colors in the rings reveal variations in chemical composition

differences would have been smeared out over time. The color differences may also indicate that different sorts of material were added to the rings at different times. In this scenario, the rings did not all form at the same time as Saturn but were added to over an extended period. New ring material could have come from small satellites that shattered after being hit by a stray asteroid or comet.

Discovering New Ring Systems

In addition to revealing new details about the A, B, C, and F rings, the *Voyager* cameras also discovered three new ring systems: the D, E, and G rings. The drawing in **Figure 12-23** shows the layout of all

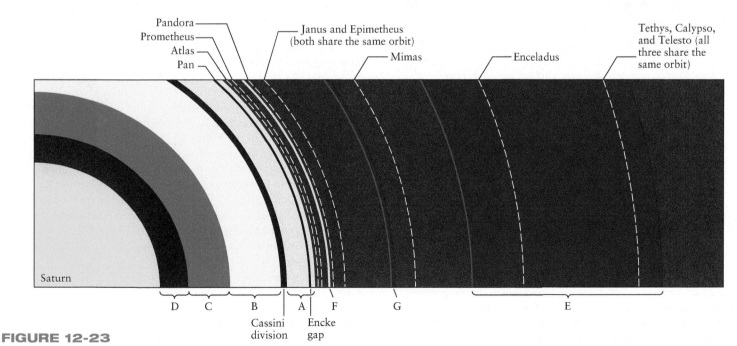

FIGURE 12-23

The Arrangement of Saturn's Rings This scale drawing shows the location of Saturn's rings along with the orbits of the 11 inner satellites. Only the A, B, and C rings can readily be seen from Earth. The remaining rings are very faint, and their existence was only confirmed by spacecraft flybys.

Cassini division   G ring   E ring

FIGURE 12-24   R I V U X G

Backlit Saturn With Saturn eclipsing the Sun, *Cassini* took 165 images to form this breathtaking color-enhanced portrait. The contrast was also increased to reveal fainter rings, and the outermost E ring is visible. The E ring contains water-ice particles *as well as* the moon Enceladus that ejected this water through ice-volcanoes. The ring labels point to the prominent feature of each ring, although the rings are noticeably wide. (NASA/JPL/Space Science Institute)

of Saturn's known rings along with the orbits of some of Saturn's satellites. The **D ring** is Saturn's innermost ring system. It consists of a series of extremely faint ringlets located between the inner edge of the C ring and the Saturnian cloudtops. The **E ring** and the **G ring** both lie far from the planet, well beyond the narrow F ring. Both of these outer ring systems are extremely faint and fuzzy, but can be seen in Figure 12-24. Each lacks the ringlet structure so prominent in the main ring systems. The E ring encloses the orbit of Enceladus, one of Saturn's icy moons. Some scientists suspected that water ejected from Enceladus is the source of ice particles in the E ring, much as Io's volcanoes supply material to Jupiter's magnetosphere (see Section 12-8). Then, in 2010, the *Cassini* spacecraft observed directly that water geysers on Enceladus feed the E ring (Chapter 13).

CONCEPTCHECK 12-10

How would Figure 12-22 look different if ring particles migrated quickly through Saturn's rings?

Answer appears at the end of the chapter.

12-11 Saturn's inner satellites affect the appearance and structure of its rings

Why do Saturn's rings have such a complex structure? The answer is that, like any object in the universe, the particles that make up the rings are affected by the force of gravity. Saturn's gravitational pull keeps the ring particles in orbit around the planet. But the satellites of Saturn also exert gravitational forces on the rings. These forces can shape the orbits of the ring particles and help give rise to the rings' structure.

Resonance and Shepherd Satellites

An arrangement in which one orbiting object has twice the orbital period of another is called an **orbital resonance.** (More generally, an orbital resonance occurs whenever the orbital periods of two objects are related by a ratio of small integers.) In an orbital resonance, gravitational interactions occur with just the right rhythm to transfer the greatest amount of energy. As an analogy, you naturally push someone on a swing in resonance with the swing's

motion; in this way a little energy is added with each push, and the swing gets very high with repeated pushes. Now let's see how the Cassini division results from an orbital resonance.

Astronomers have long known that one of Saturn's moons, Mimas, has a gravitational effect on Saturn's ring system. Mimas is a moderate-sized satellite that orbits Saturn every 22.6 hours. According to Kepler's third law, ring particles in the Cassini division—if they existed—would orbit Saturn approximately every 11.3 hours. Consequently, a given set of ring particles in the Cassini division would find itself between Saturn and Mimas on every second orbit (that is, every 22.6 hours). In other words, particles in the Cassini division are in a 2-to-1 orbital resonance with Mimas. Because of these repeated alignments, the combined gravitational forces of Saturn and Mimas cause small particles to deviate from their original orbits, sweeping them out of the Cassini division. This helps explain why the Cassini division is relatively empty (although not completely so, as Figure 12-21 shows).

While Mimas tends to spread ring particles apart, other satellites exert gravitational forces that herd particles together. Cassini observed two such satellites orbiting on either side of the narrow F ring (Figure 12-25). Pandora, the outer of the two satellites, moves around Saturn at a slightly slower speed than the particles in the ring. As ring particles pass Pandora, they experience a tiny gravitational tug that slows them down slightly. Hence, these particles lose a little energy and fall into orbits a bit closer to Saturn. At the same time, Prometheus, the inner satellite, orbits the planet somewhat faster than the F ring particles. The gravitational pull of Prometheus makes the ring particles speed up a little, giving them some extra energy and nudging them into slightly higher orbits. The competition between the pulls exerted by these two satellites focuses the F ring particles into a well-defined, narrow band.

Because of their confining influence, Prometheus and Pandora are called **shepherd satellites.** We will see in Chapter 14 that Uranus is encircled by narrow rings that are likewise kept confined by shepherd satellites.

An effect similar to shepherding explains two divisions in the A ring, the Encke gap and the even narrower Keeler gap (Figure 12-20). In 2005, *Cassini* images revealed a miniature moon, Daphnis, which actually orbits inside the Keeler gap (Figure 12-26). In a fashion similar to the F ring's shepherds, Daphnis's gravitational forces cause ring particles outside its orbit to move farther out and cause those inside its orbit to move farther inward toward

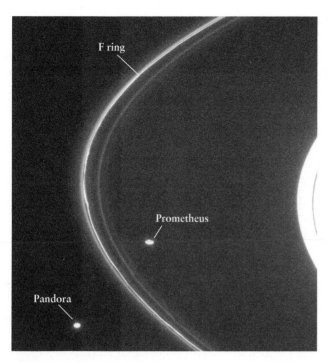

FIGURE 12-25 R I V U X G
Saturn's F Ring and Its Two Shepherds This *Cassini* image shows the tiny satellites Prometheus (102 km across) and Pandora (84 km across). (Compare with Figure 12-20.) These orbit Saturn on either side of its F ring, shepherding the ring particles into a narrow band. The satellites lap each other every 25 days. (NASA/JPL)

Saturn. In this way, Daphnis helps to maintain a gap in the rings that is 42 km (26 mi) wide, many times wider than Daphnis's 7-km (4-mi) diameter. In a similar way, a 20-km wide moon called Pan orbits within and maintains the 270-km wide Encke gap. Both of these moons cause wavelike disturbances in the ring material on either side of the gap through which they travel (see Figure 12-25).

> Gravitational effects carve gaps in the rings and focus ring particles into narrow ringlets

In addition to their dynamic ring systems, Jupiter and Saturn together have an exotic collection of moons. These include Saturn's moon Titan, with a thick atmosphere rich in hydrocarbons; Jupiter's satellite Io, the most volcanic world in the solar system;

and Saturn's satellite Enceladus, which has exotic geysers that spew a mixture of water vapor and ice. We will examine these alien worlds in Chapter 13.

CONCEPTCHECK 12-11

If Saturn's moon Mimas orbited somewhat farther than it currently does, how would this affect the Cassini division?
Answer appears at the end of the chapter.

KEY WORDS

A ring, p. 338	light scattering, p. 343
B ring, p. 338	liquid metallic hydrogen, p. 336
belts, p. 326	
bow shock, p. 336	noble gases, p. 334
brown oval, p. 330	oblate, oblateness, p. 335
C ring, p. 338	orbital resonance, p. 344
Cassini division, p. 338	ring particles, p. 339
D ring, p. 344	ringlets, p. 342
differential rotation, p. 327	Roche limit, p. 340
E ring, p. 344	shepherd satellite, p. 344
Encke gap, p. 342	tidal force, p. 340
F ring, p. 342	white oval, p. 329
G ring, p. 344	zonal winds, p. 332
Great Red Spot, p. 326	zones, p. 326

KEY IDEAS

Composition and Structure: Jupiter and Saturn are both much larger than Earth. Each is composed of 71% hydrogen, 24% helium, and 5% all other elements by mass. Both planets have a higher percentage of heavy elements than does the Sun.

• Jupiter probably has a rocky core several times more massive than Earth. The core is surrounded by a layer of liquid "ices" (water, ammonia, methane, and associated compounds). On top of this ice layer is a layer of helium and liquid metallic hydrogen and an outermost layer composed primarily of ordinary hydrogen and helium. All of Jupiter's visible features are near the top of this outermost layer.

• Saturn's internal structure is similar to that of Jupiter, but its core makes up a larger fraction of its volume and its liquid metallic hydrogen mantle is shallower than that of Jupiter.

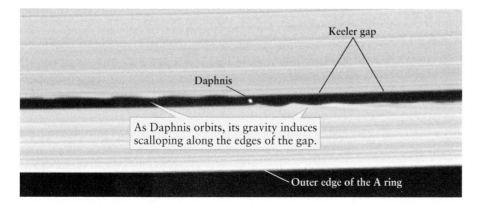

VIDEO 12-9 **FIGURE 12-26** R I V U X G
Daphnis and the Keeler Gap This *Cassini* spacecraft shows Saturn's tiny moon Daphnis orbiting within the Keeler gap (see also Figure 12-20). Although it is just 7 km across, Daphnis exerts enough gravitational force to keep the 42-km wide gap open and induce wavelike disturbances in the two edges of the gap. (NASA/JPL/Space Science Institute)

- Jupiter and Saturn both rotate so rapidly that these planets are noticeably flattened, or oblate

- Both Jupiter and Saturn emit more energy than they receive from the Sun. Presumably both planets are still cooling.

Atmospheres: The visible "surfaces" of Jupiter and Saturn are actually the tops of their clouds. The rapid rotation of the planets twists the clouds into dark belts and light zones that run parallel to the equator. Strong zonal winds run along the belts and zones.

- The outer layers of both planets' atmospheres show differential rotation: The equatorial regions rotate slightly faster than the polar regions.

- The colored ovals visible in the Jovian atmosphere represent gigantic storms. Some, such as the Great Red Spot, are quite stable and persist for many years. Storms in Saturn's atmosphere seem to be shorter-lived.

- There are presumed to be three cloud layers in the atmospheres of Jupiter and Saturn. The reasons for the distinctive colors of these different layers are not yet known. The cloud layers in Saturn's atmosphere are spread out over a greater range of altitude than those of Jupiter, giving Saturn a more washed-out appearance.

- Saturn's atmosphere contains less helium than Jupiter's atmosphere. This lower abundance may be the result of helium raining downward into the planet.

Magnetic Fields and Magnetospheres: Jupiter has a strong magnetic field created by currents in a metallic hydrogen layer. With less liquid metallic hydrogen, Saturn's magnetic field is much weaker than Jupiter's.

Rings: Saturn is circled by a system of thin, broad rings lying in the plane of the planet's equator. This system is tilted away from the plane of Saturn's orbit, which causes the rings to be seen at various angles by an Earth-based observer over the course of a Saturnian year.

Structure of the Rings: Saturn has three major, broad rings (A, B, and C) that can be seen from Earth. Other, fainter rings were found by the *Voyager* spacecraft.

- The principal rings of Saturn are composed of numerous particles of ice and ice-coated rock ranging in size from a few micrometers to about 10 m. Most of the rings exist inside the Roche limit of Saturn, where disruptive tidal forces are stronger than the gravitational forces attracting the ring particles to each other.

- Each of Saturn's major rings is composed of a great many narrow ringlets. The faint F ring, which is just outside the A ring, is kept narrow by the gravitational pull of shepherd satellites.

- Jupiter's faint rings are composed of a relatively small amount of small, dark, rocky particles that reflect very little light.

QUESTIONS

Review Questions

1. Mars passes closer to Earth than Jupiter does, but with an Earth-based telescope it is easier to see details on Jupiter than on Mars. Why is this?

2. Saturn is the most distant of the planets visible without a telescope. Is there any way we could infer this from naked-eye observations? Explain your answer. (*Hint:* Think about how Saturn's position on the celestial sphere must change over the course of weeks or months.)

3. As seen from Earth, does Jupiter or Saturn undergo retrograde motion more frequently? Explain your answer.

4. *TUTORIAL 12-1* In what ways are the motions of Jupiter's atmosphere like the motion of water stirred in a pot (see Figure 12-3b)? In what ways are they different?

5. Is the chemical composition of Jupiter as a whole the same as that of its atmosphere? Explain any differences.

6. Astronomers can detect the presence of hydrogen in stars by looking for the characteristic absorption lines of hydrogen in the star's visible spectrum (Figure 5-21). They can also detect hydrogen in glowing gas clouds by looking for hydrogen's characteristic emission lines (Figure 5-18). Explain why neither of these techniques helped Earth-based astronomers to detect hydrogen in Jupiter's atmosphere.

7. On a warm, humid day, water vapor remains in the atmosphere. But if the temperature drops suddenly, the water vapor forms droplets, clouds appear, and it begins to rain. Relate this observation to why there is relatively little helium in Saturn's atmosphere compared to the atmosphere of Jupiter.

8. What would happen if you tried to land a spacecraft on the surface of Jupiter?

9. What are the belts and zones in the atmospheres of Jupiter and Saturn? Is the Great Red Spot more like a belt or a zone? Explain your answer.

10. Give one possible explanation why weather systems on Jupiter are longer-lived than weather systems on Earth.

11. What are white ovals and brown ovals? What can we infer about them from infrared observations?

12. Compare and contrast the source of energy for motions in Earth's atmosphere with the energy source for motions in the atmospheres of Jupiter and Saturn.

13. Both Jupiter and Saturn emit more energy than they receive from the Sun in the form of sunlight. Compare the internal energy sources of the two planets that produce this emission.

14. What observations from the *Cassini* spacecraft contradict the accepted picture of zone and belt convection in Jupiter's atmosphere?

15. Compare the atmospheres of Jupiter and Saturn. Why does Saturn's atmosphere look "washed out" in comparison to that of Jupiter?

16. Which data from the *Galileo Probe* were in agreement with astronomers' predictions? Which data were surprising?

17. Fewer than one in every 10^5 atoms in Jupiter's atmosphere is an argon atom, and fewer than one in 10^8 is an atom of krypton or xenon. If these atoms are so rare, why are scientists concerned about them? How do the abundances of these

elements in Jupiter's atmosphere compare to the abundances in the Sun? What hypothesis has been offered to explain these observations?

18. Why is Jupiter oblate? What do astronomers learn from the value of Jupiter's oblateness?

19. What is liquid metallic hydrogen? What is its significance for Jupiter?

20. Describe the internal structures of Jupiter and Saturn, and compare them with the internal structure of Earth.

21. Explain why Saturn is more oblate than Jupiter, even though Saturn rotates more slowly.

22. Compare and contrast Jupiter's magnetosphere with the magnetosphere of a terrestrial planet like Earth.

23. Why is Saturn's magnetosphere less extensive than Jupiter's?

24. *TUTORIAL 12-2* What observations of Saturn's rings proved that they are not solid?

25. If Saturn's rings are not solid, why do they look solid when viewed through a telescope?

26. Although the *Voyager* and *Cassini* spacecraft did not collect any samples of Saturn's ring particles, measurements from these spacecraft allowed scientists to determine the sizes of the particles. Explain how this was done.

27. The Space Shuttle and other spacecraft orbit Earth well within Earth's Roche limit. Explain why these spacecraft are not torn apart by tidal forces.

28. How do Jupiter's rings differ from those of Saturn?

29. Describe the structure of Saturn's rings. What evidence is there that ring particles do not migrate significantly between ringlets?

30. During the planning stages for the *Pioneer 11* mission, when relatively little was known about Saturn's rings, it was proposed to have the spacecraft fly through the Cassini division. Why would this have been a bad idea?

31. What is the relationship between Saturn's satellite Mimas and the Cassini division?

32. Why is the term "shepherd satellite" appropriate for the objects so named? Explain how a shepherd satellite operates.

Advanced Questions

The question preceded by an asterisk (*) involves topics discussed in Box 7-2.

> ### Problem-solving tips and tools
>
> Box 4-4 describes how to use Newton's form of Kepler's third law. Newton's law of universal gravitation, discussed in Section 4-6, is the basic equation from which you can calculate a planet's surface gravity. Box 7-2 discusses escape speed, and Sections 5-3 and 5-4 discuss the properties of thermal radiation (including the Stefan-Boltzmann law, which relates the temperature of a body to the amount of thermal radiation that it emits).

33. The angular diameter of Jupiter at opposition varies little from one opposition to the next. By contrast, the angular diameter of Mars at opposition is quite variable. Explain why there is a difference between these two planets.

34. Jupiter was at opposition on June 5, 2007. On that date Jupiter appeared to be in the constellation Ophiuchus. Approximately when will Jupiter next be at opposition in this same region of the celestial sphere? Explain your answer.

35. Jupiter's equatorial diameter and the rotation period at Jupiter's equator are both given in Table 12-1. Use these data to calculate the speed at which an object at the cloudtops along Jupiter's equator moves around the center of the planet.

36. Using orbital data for a Jovian satellite of your choice (see Appendix 3), calculate the mass of Jupiter. How does your answer compare with the mass quoted in Table 12-1?

37. The density of water is 1,000 kg/m^3. It has been claimed that Saturn would float if one had a large enough bathtub. Using the mass and size of Saturn given in Table 12-2, confirm that the planet's average density is about 690 kg/m^3, and comment on this somewhat fanciful claim.

38. Roughly speaking, Jupiter's composition (by mass) is three-quarters hydrogen and one-quarter helium. The mass of a single hydrogen atom is given in Appendix 7; the mass of a single helium atom is about 4 times greater. Use these numbers to calculate how many hydrogen atoms and how many helium atoms there are in Jupiter.

39. Use the information given in Section 12-3 to estimate the wind velocities in the Great Red Spot, which rotates with a period of about 6 days.

40. An astronaut floating above Saturn's cloudtops would see a blue sky, even though Saturn's atmosphere has a very different chemical composition than Earth's. Explain why. (*Hint:* See Box 5-4.)

41. If Jupiter emitted just as much energy per second (as infrared radiation) as it receives from the Sun, the average temperature of the planet's cloudtops would be about 107 K. Given that Jupiter actually emits twice this much energy per second, calculate what the average temperature must actually be.

42. (a) From Figure 12-10a, by how much does the temperature increase as you descend from the 100-millibar level in Jupiter's atmosphere to an altitude 100 km below that level? (b) Use Figure 12-10b to answer the same question for Saturn's atmosphere. (c) In Earth's troposphere (see Section 9-5), the air temperature increases by 6.4°C for each kilometer that you descend. In which planet's atmosphere—Earth, Jupiter, or Saturn—does the temperature increase most rapidly with decreasing altitude?

*43. Consider a hypothetical future spacecraft that would float, suspended from a balloon, for extended periods in Jupiter's upper atmosphere. If we want this spacecraft to return to Earth after completing its mission, calculate the speed at which the spacecraft's rocket motor would have to accelerate it in order to escape Jupiter's gravitational pull. Compare with the escape speed from Earth, equal to 11.2 km/s.

44. The *Galileo Probe* had a mass of 339 kg. On Earth, its weight (the gravitational force exerted on it by Earth) was 3320 newtons, or 747 lb. What was the gravitational force that Jupiter exerted on the *Galileo Probe* when it entered Jupiter's clouds?

45. From the information given in Section 12-6, calculate the average density of Jupiter's rocky core. How does this compare with the average density of Earth? With the average density of Earth's solid inner core? (See Table 9-1 and Table 9-2 for data about Earth.)

46. In the outermost part of Jupiter's outer layer (shown in yellow in the upper part of Figure 12-13), hydrogen is principally in the form of molecules (H_2). Deep within the liquid metallic hydrogen layer (shown in orange in Figure 12-13), hydrogen is in the form of single atoms. Recent laboratory experiments suggest that there is a gradual transition between these two states, and that the transition layer overlaps the boundary between the ordinary hydrogen and liquid metallic hydrogen layers. Use this information to redraw the upper part of Figure 12-12 and to label the following regions in Jupiter's interior: (i) ordinary (nonmetallic) hydrogen molecules; (ii) nonmetallic hydrogen with a mixture of atoms and molecules; (iii) liquid metallic hydrogen with a mixture of atoms and molecules; (iv) liquid metallic hydrogen atoms.

47. When Saturn is at different points in its orbit, we see different aspects of its rings because the planet has a 27° tilt. If the tilt angle were different, would it be possible to see the upper and lower sides of the rings at all points in Saturn's orbit? If so, what would the tilt angle have to be? Explain your answers.

48. As seen from Earth, the intervals between successive edge-on presentations of Saturn's rings alternate between about 13 years, 9 months, and about 15 years, 9 months. Why do you think these two intervals are not equal?

49. (a) Use Newton's form of Kepler's third law to calculate the orbital periods of particles at the outer edge of Saturn's A ring and at the inner edge of the B ring. (b) Saturn's rings orbit in the same direction as Saturn's rotation. If you were floating along with the cloudtops at Saturn's equator, would the outer edge of the A ring and the inner edge of the B ring appear to move in the same or opposite directions? Explain.

50. VIDEO 12-10 This *Cassini* close-up image of Saturn's rings shows a number of bright, straight features elongated outward called *spokes*. As these features orbit around Saturn, they tend to retain their shape like the rigid spokes on a rotating bicycle wheel. The spokes rotate at the same rate as Saturn's magnetic field and are thought to be clouds of tiny, electrically charged particles kept in orbit by magnetic forces. Explain why the spokes could *not* maintain their shape if they were kept in orbit by gravitational forces alone.

R I V U X G   (NASA/JPL/Space Science Institute)

51. The Cassini division involves a 2-to-1 resonance with Mimas. Does the location of the Encke gap—133,500 km from Saturn's center—correspond to a resonance with one of the other satellites? If so, which one? (See Appendix 3.)

Discussion Questions

52. Describe some of the semipermanent features in Jupiter's atmosphere. Compare and contrast these long-lived features with some of the transient phenomena seen in Jupiter's clouds.

53. Suppose you were asked to design a mission to Jupiter involving an unmanned airplanelike vehicle that would spend many days (months?) flying through the Jovian clouds. What observations, measurements, and analyses should this aircraft be prepared to make? What dangers might the aircraft encounter, and what design problems would you have to overcome?

54. What sort of experiment or space mission would you design in order to establish definitively whether Jupiter has a rocky core?

55. The classic science fiction films *2001: A Space Odyssey* and *2010: The Year We Make Contact* both involve manned spacecraft in orbit around Jupiter. What kinds of observations could humans make on such a mission that cannot be made by robotic spacecraft? What would be the risks associated with such a mission? Do you think that a manned Jupiter mission would be as worthwhile as a manned mission to Mars? Explain your answers.

56. Suppose that Saturn were somehow moved to an orbit around the Sun with semimajor axis 1 AU, the same as Earth's. Discuss what long-term effects this would have on the planet and its rings.

Web/eBook Question

57. On Jupiter, the noble gases argon, krypton, and xenon provide important clues about Jupiter's past. On Earth, xenon is used in electronic strobe lamps because it emits a very white

light when excited by an electric current. Argon, by contrast, is one of the gases used to fill ordinary incandescent lightbulbs. Search the World Wide Web for information about why argon is used in this way. Why do premium, long-life lightbulbs use krypton rather than argon?

58. Search the World Wide Web, especially the Web sites for NASA's Jet Propulsion Laboratory and the European Space Agency, for information about the current status of the *Cassini* mission. What recent discoveries has *Cassini* made?

59. The two *Voyager* spacecraft were launched from Earth along a trajectory that took them directly to Jupiter. The force of Jupiter's gravity then gave the two spacecraft a "kick" that helped push them onward to Saturn. The much larger *Cassini* spacecraft, by contrast, was first launched on a trajectory that took it past Venus. Search the World Wide Web, especially the Web sites for NASA's Jet Propulsion Laboratory and the European Space Agency, for information about the trajectory that *Cassini* took through the solar system to reach Saturn. Explain why this trajectory was so different from that of the *Voyagers*.

60. **The Rotation Rate of Saturn.** Access and view the video "Saturn from the Hubble Space Telescope" in Chapter 12 of the *Universe* Web site or eBook. The total amount of time that actually elapses in this video is 42.6 hours. Using this information, identify and follow an atmospheric feature and determine the rotation period of Saturn. How does your answer compare with the value given in Table 12-2?

ACTIVITIES

Observing Projects

Observing tips and tools

Like Mars, Jupiter and Saturn are most easily seen around opposition. (The dates of Jupiter's and Saturn's oppositions are available on the World Wide Web.) Each planet is visible in the night sky for several months before or after its opposition. At other times, Jupiter and Saturn may be visible in either the predawn morning sky or the early evening sky. Consult such magazines as *Sky & Telescope* and *Astronomy* or their Web sites for more detailed information about when and where to look for Jupiter during a given month. You can also use the *Starry Night*™ program on the CD-ROM that accompanies selected copies of this textbook. A relatively small telescope with an objective diameter of 15 to 20 cm (6 to 8 inches), used with a medium-power eyepiece to give a magnification of $25\times$ or so, should enable you to see some of the dark belts on Jupiter and the rings of Saturn.

61. If Jupiter is visible in the night sky, make arrangements to view the planet through a telescope. What magnifying power seems to give you the best view? Draw a picture of what you see. Can you see any belts and zones? How many? Can you see the Great Red Spot?

62. Make arrangements to view Jupiter's Great Red Spot through a telescope. Consult the *Sky & Telescope* Web site, which lists the times when the center of the Great Red Spot passes across Jupiter's central meridian. The Great Red Spot is well placed for viewing for 50 minutes before and after this happens. You will need a refractor with an objective of at least 15 cm (6 in.) diameter or a reflector with an objective of at least 20 cm (8 in.) diameter. Using a pale blue or green filter can increase the color contrast and make the spot more visible. For other useful hints, see the article "Tracking Jupiter's Great Red Spot" by Alan MacRobert (*Sky & Telescope*, September 1997).

63. View Saturn through a small telescope. Make a sketch of what you see. Estimate the angle at which the rings are tilted to your line of sight. Can you see the Cassini division? Can you see any belts or zones in Saturn's clouds? Is there a faint, starlike object near Saturn that might be Saturn's satellite Titan? What observations could you perform to test whether the starlike object is a Saturnian satellite?

64. Use *Starry Night*™ to observe the appearance of Jupiter. Open **Favourites > Explorations > Jupiter** and use the **Time Flow** controls to determine the rotation period of the planet. (*Hint:* You may want to track the motion of an easily recognizable feature, such as the Great Red Spot.) (**a**) What is the approximate rotation period of Jupiter? (**b**) Measure Jupiter's equatorial radius and its polar radius. You can use the angular separation tool to make these measurements. (*Hint:* Click on the cursor selection tool at the left side of the toolbar and select the angular measurement tool from the dropdown menu.) Measure from the center of the planet to its east or west limb (edge) along the equator of the planet to determine its equatorial radius and then measure from the center of the planet to its north or south pole to measure the polar radius. Are these two radii the same? If not, which is larger? Offer an explanation for this observation.

65. Use *Starry Night*™ to examine Jupiter. Open **Favourites > Explorations > Jupiter**. Select **Options > Solar System > Planets-Moons…** from the menu. In the **Planets-Moons Options** dialog window, click the checkbox for **Surface guides** and click **OK** to remove the pole sticks and equator from the image. Click the **Stop** button and then use the **Zoom** controls and location scroller to examine Jupiter's atmosphere. (**a**) Describe the appearance of the atmosphere of the planet. (**b**) Compare the number of white ovals and brown ovals in Jupiter's southern hemisphere (the hemisphere in which the Great Red Spot is located) with the number of white ovals and brown ovals visible in the northern hemisphere. What general rule can you state about the abundance of these storms in the two hemispheres.

66. Use *Starry Night*™ to observe the changing appearance of Saturn as seen from Earth. Select **Favourites > Explorations > Atlas** from the menu and click the **Now** button in the toolbar. Use the **Find** pane to locate and lock the view on Saturn (double-click the entry for **Saturn**). **Zoom** in until Saturn and its rings are clearly visible. Set the **Time Flow Rate** to

1 year. Now use the single-step forward button to observe the changing orientation of the rings as time advances in 1-year steps. (**a**) Describe qualitatively how the rings change orientation, as seen from Earth. (**b**) Use the **Time** controls to determine approximately how long it takes for Saturn's rings, as seen from Earth, to go through a complete cycle from edge-on to fully open to edge-on to fully open and then finally to edge-on again. (**c**) Click the **Now** button to return to the present time. Use the **Time** controls to find during which of the next 30 years Saturn's rings will appear edge-on as seen from Earth. (**d**) Why do we see the orientation of Saturn's rings change in the way that you found in part (b)?

67. Use *Starry Night*™ to examine the rings of Saturn. Open **Favourites > Explorations > Saturn** and use the location scroller to view Saturn so that you are looking straight down on the plane of the rings. (**a**) Draw a copy of what you see and label the different rings and divisions. (**b**) Using the location scroller, adjust the view so that Saturn's rings appear edge-on and then rotate the image until the Sun comes into view. Which of the following is correct?

    (i) The Sun is in the same plane as the rings of Saturn.

    (ii) The rings are in Saturn's equatorial plane.

    (iii) Neither (i) nor (ii) is correct.

    (iv) Both (i) and (ii) are correct.

Collaborative Exercises

68. Using a ruler with millimeter markings on five various images of Jupiter in the text, Figures 12-1a, 12-2a, and 12-4, determine the ratio of the longest width of the Great Red Spot to the full diameter of Jupiter. Each group member should measure a different image and all values should be averaged.

69. The text provides different years that spacecraft have flown by Jupiter and Saturn. List these dates and create a time line by listing one important event that was occurring on Earth during each of those years.

70. If the largest circle you can draw on a piece of paper represents the largest diameter of Saturn's rings, about how large would Saturn be if scaled appropriately? Which item in a group member's backpack is closest to this size?

ConceptChecks

ConceptCheck 12-1: No. Solid objects cannot undergo differential rotation because all parts of a solid object rotate together.

ConceptCheck 12-2: If the planets formed with the same chemical composition, then helium in Saturn's colder atmosphere could have condensed and moved deeper inside, resulting in Saturn appearing to be mostly hydrogen.

ConceptCheck 12-3: White ovals are long-lasting, low-temperature, high-altitude features that obscure views of what lies underneath. Brown ovals are openings in Jupiter's cloud cover that emit more infrared energy from higher-temperature material at lower altitudes.

ConceptCheck 12-4: Jupiter has significantly more mass, and as a result the cloud layers are gravitationally compressed much more on Jupiter than on less massive Saturn.

ConceptCheck 12-5: We expect the Sun and Jupiter to have formed from the same material—the solar nebula—each with the same abundances of noble gases. Because the noble gases do not undergo chemical reactions with other atoms and molecules, their observed abundance today is thought to be representative of the abundance when the Sun and Jupiter formed.

ConceptCheck 12-6: Yes. All else being equal, increasing the rotation rate increases the oblateness.

ConceptCheck 12-7: Yes. Some charged particles ejected by Io's volcanism can join the ion-trapping belts around Jupiter. Just as on Earth, when these ions stream along the magnetic field, they can cause aurorae at Jupiter's magnetic poles.

ConceptCheck 12-8: As Saturn travels around the Sun, the tilted and quite thin rings are sometimes seen somewhat "face-on" and are easily observed, whereas at other times the view from Earth makes the rings "edge-on" and nearly invisible.

ConceptCheck 12-9: An asteroid orbiting Earth within Earth's Roche limit for enough time would be ripped apart by Earth's gravitational attraction differing on the near side and the far side of the asteroid (which is a tidal force), resulting in a tiny ring of debris orbiting Earth.

ConceptCheck 12-10: Slight color variations indicate differences in material making the rings. If this material migrated quickly, color variations would be smeared out; instead, there are some very narrow ring features with different colors in Figure 12-22.

ConceptCheck 12-11: The Cassini division is created by a destabilizing 2-to-1 orbital resonance with Saturn's moon Mimas. Due to this resonance, repeated gravitational tugs clear out any particles at this location. If Mimas orbited farther away, its period would increase, and the Cassini division would appear farther out as well (where its dislodged particles also had a larger period).

This is Jupiter's moon Io, the most volcanically active body in the solar system. The faint plume on the left is an active volcano. This image was taken by the *Galileo* spacecraft. (*Galileo* Project, JPL, NASA) R I V U X G

Jupiter and Saturn's Satellites of Fire and Ice

LEARNING GOALS

By reading the sections of this chapter, you will learn

13-1 What the Galilean satellites are and how they orbit Jupiter

13-2 The similarities and differences among the Galilean satellites

13-3 How the Galilean satellites formed

13-4 Why Io is the most volcanically active world in the solar system

13-5 How Io interacts with Jupiter's magnetic field

13-6 The evidence that Europa may have an ocean beneath its surface

13-7 The kinds of geologic activity found on Ganymede and Callisto

13-8 The nature of Titan's thick, hydrocarbon-rich atmosphere

13-9 Why most of Jupiter's moons orbit in the "wrong" direction

13-10 What powers the volcanoes on Saturn's icy moon Enceladus

To many astronomers, the moons of Jupiter and Saturn are so unique and complex that they are best thought of as *worlds*. Io, shown in the above image, is the most geologically active world in the solar system. Its surface is covered with colorful sulfur compounds deposited by dozens of active volcanoes. Europa, another of Jupiter's large satellites, has no volcanoes but does have a weirdly cracked surface of ice. Beneath Europa's ice there is probably a worldwide ocean, within which there could possibly be small, single-celled organisms.

No less exotic is Saturn's large satellite Titan, which is enveloped by a thick, nitrogen-rich atmosphere from which liquid hydrocarbons fall as rain and form massive lakes. Another moon of Saturn, Enceladus, shocks us with its ice-volcanoes, which *Cassini* has caught in the act of erupting. In this chapter we will explore these and other intriguing aspects of Jupiter and Saturn's extensive collection of satellites.

13-1 Jupiter's Galilean satellites are easily seen with Earth-based telescopes

TUTORIAL 13-1   When Galileo trained his telescope on Jupiter in January 1610, he became the first person to observe satellites, or moons, orbiting another planet (see Section 4-5). He called them the Medicean stars, in order to curry the favor of a wealthy Florentine

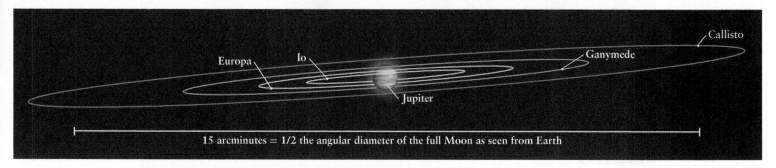

FIGURE 13-1

The Orbits of the Galilean Satellites This illustration shows the orbits of the four Galilean satellites as seen from Earth. All four orbits lie in nearly the same plane as Jupiter's equator. The apparent angular size and orientation of the orbits depends on the relative positions of Earth and Jupiter.

patron of the arts and sciences. We now call them the **Galilean satellites.** Individually, these moons are named Io, Europa, Ganymede, and Callisto, after four mythical lovers and companions of the god whom the Greeks called Zeus and the Romans called Jupiter.

When viewed through an Earth-based telescope, the Galilean satellites look like pinpoints of light (see Figure 4-16). All four satellites orbit Jupiter in nearly the same plane as the planet's equator (Figure 13-1). From Earth, we always see that plane nearly edge-on, so the satellites appear to move back and forth relative to Jupiter (see Figure 4-17). The orbital periods are fairly short, ranging from 1.8 (Earth) days for Io to 16.7 days for Callisto, so we can easily see the satellites move from one night to the next and even during a single night. These motions led Galileo to realize that he was seeing objects orbiting around Jupiter, in much the same way that Copernicus said that planets move around the Sun (see Section 4-2).

Each of the four Galilean satellites is bright enough to be visible to the naked eye. Why, then, did Galileo need a telescope to discover them? The reason is that as seen from Earth, the angular separation between Jupiter and the satellites is quite small—never more than 10 arcminutes for Callisto and even less for the other three Galilean satellites. To the naked eye, these satellites are lost in the overwhelming glare of Jupiter. But a small telescope or even binoculars increases the apparent angular separation by enough to make the Galilean satellites visible.

Synchronous Rotation and Orbital Resonance

The brightness of each satellite varies slightly as it orbits Jupiter, because the satellites also spin on their axes as they orbit Jupiter, and dark and light areas on their surfaces are alternately exposed to and hidden from our view. Remarkably, each Galilean satellite goes through one complete cycle of brightness during one orbital period. This observation tells us that each satellite is in *synchronous rotation,* so that it rotates exactly once on its axis during each trip around its orbit (see Figure 3-4b). As we saw in Section 4-8, our Moon's synchronous rotation is the result of gravitational forces exerted on the Moon by Earth. Likewise, Jupiter's gravitational forces keep the Galilean satellites in synchronous rotation.

Much more important, however, are the effects of an **orbital resonance.** Recall that an orbital resonance occurs when bodies exert a regular, repeating gravitational influence on each other. As we will see in this chapter, heat generated by this effect has a profound influence on Io and Europa.

ANIMATION 13-1 Synchronous rotation means that the *rotation* period and orbital period are in a 1-to-1 ratio for each Galilean satellite. Remarkably, there is also a simple ratio of the different *orbital* periods of the three inner Galilean satellites, Io, Europa, and Ganymede, which leads to an orbital resonance. Io orbits Jupiter in 1.77 days, while Europa has twice the orbital period of Io, and Ganymede has 4 times the orbital period of Io. Thus, the orbital periods of Io, Europa, and Ganymede are in the ratio 1:2:4, which you can verify from the data in Table 13-1. This is an example of a stabilizing orbital resonance. (In Section 12-11, we saw that a destabilizing resonance cleared out the Cassini division.)

This orbital resonance—a special relationship among the satellites' orbits—is maintained by the gravitational forces that they exert on one another. Those forces can be quite strong, because the three

> Gravitational interactions between Io, Europa, and Ganymede induce a rhythmic relationship among their orbits

inner Galilean satellites pass relatively close to one another; at their closest approach, Io and Europa are separated by only two-thirds the distance from Earth to the Moon. Such a close approach occurs once for every two of Io's orbits, so Europa's gravitational pull acts on Io in a rhythmic way. Indeed, Io, Europa, and Ganymede all exert rhythmic gravitational tugs on one another. Just as a drummer's rhythm keeps musicians on the same beat, this gravitational rhythm maintains the simple ratio of orbital periods among the satellites. As we will see, this "rhythmic drumming" provides a significant source of energy that heats the interiors of Io and Europa.

By contrast, Callisto orbits at a relatively large distance from the other three large satellites. Hence, the gravitational forces on Callisto from the other satellites are relatively weak, and there is no simple relationship between Callisto's orbital period and the period of Io, Europa, or Ganymede.

Transits, Eclipses, and Occultations

Because the orbital planes of the Galilean satellites are nearly edge-on to our line of sight, we see these satellites undergoing transits, eclipses, and occultations. In a *transit,* a satellite passes between us and Jupiter, and we see the satellite's shadow as a black dot against the planet's colorful cloudtops. (Figure 12-2a shows Europa's shadow on Jupiter.) In an *eclipse,* one of the satellites disappears and then reappears as it passes into and out of Jupiter's enormous

TABLE 13-1 Jupiter's Galilean Satellites Compared with the Moon, Mercury, and Mars

	Average distance from Jupiter (km)	Orbital period (days)	Diameter (km)	Mass (kg)	Mass (Moon = 1)	Average density (kg/m³)	Albedo
Io	421,600	1.769	3642	8.932×10^{22}	1.22	3529	0.63
Europa	670,900	3.551	3120	4.791×10^{22}	0.65	3018	0.64
Ganymede	1,070,000	7.155	5268	1.482×10^{23}	2.02	1936	0.43
Callisto	1,883,000	16.689	4800	1.077×10^{23}	1.47	1851	0.17
Moon	—	—	3476	7.349×10^{22}	1.00	3344	0.11
Mercury	—	—	4880	3.302×10^{23}	4.49	5430	0.12
Mars	—	—	6794	6.419×10^{23}	8.73	3934	0.15

5000 km

Io   Europa   Ganymede   Callisto

Jupiter   421,600 km   670,900 km   1,070,000 km   1,883,800 km

Average distance from Jupiter

Note: Jupiter is shown to the same scale as the distances of the satellites from Jupiter. Compared to this scale, the images of the satellites themselves have been enlarged 74×.   R I ⊽ U X G   (NASA/JPL)

shadow. In an **occultation,** a satellite passes completely behind Jupiter, and the satellite is temporarily blocked from our Earth-based view. (The word *occultation* comes from a Latin verb meaning "to cover.")

Before spacecraft first ventured to Jupiter, astronomers used eclipses in a very clever way to estimate the diameters of the Galilean satellites. When a satellite emerges from Jupiter's shadow, it does not blink on instantly. Instead, there is a brief interval during which the satellite gets progressively brighter as more of its surface is exposed to sunlight. The satellite's diameter is calculated from the duration of this interval and the satellite's orbital speed (which is known from Kepler's laws).

As Table 13-1 shows, the smallest Galilean satellite (Europa) is slightly smaller than our Moon; the largest satellite (Ganymede) is larger than Mercury and more than three-quarters the size of Mars. Thus, the Galilean satellites truly are worlds in their own right.

CONCEPTCHECK **13-1**

If you know the speed of a moon or satellite (from Kepler's laws), and you measure how long it takes to disappear during orbit behind its planet (which is an occultation), what can you learn about the satellite?

Answer appears at the end of the chapter.

13-2 Data from spacecraft reveal the unique properties of the Galilean satellites

Even the finest images made with Earth-based telescopes have revealed relatively few details about the Galilean satellites. Almost everything we know about these satellites has come from observations made at close range by spacecraft.

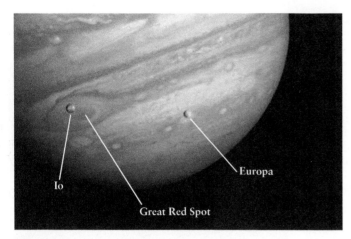

Io

Great Red Spot

Europa

FIGURE 13-2 R I $\boxed{\text{V}}$ U X G

Io and Europa from *Voyager 1* *Voyager 1* recorded this image of Jupiter and the Galilean satellites Io and Europa in February 1979, when it was 20 million kilometers (12.4 million miles) from the giant planet. The satellites are actually smaller than some of Jupiter's cloud features, such as the Great Red Spot (see Section 12-3). *Voyager 1* and *Voyager 2* also recorded many images of the Galilean satellites at closer range. (NASA/JPL)

The first close-range observations of the Galilean satellites were made by the *Pioneer 10* and *Pioneer 11* spacecraft as they flew past Jupiter in 1973 and 1974. The images from these missions were of relatively low resolution, however, and much more information has come from three subsequent missions. *Voyager 1* and *Voyager 2* flew past Jupiter in 1979 and recorded tens of thousands of images, one of which is shown in Figure 13-2. The *Galileo* spacecraft, which made 35 orbits around Jupiter between 1995 and 2003, carried out an even more extensive investigation of Jupiter's satellites. The *Cassini* and *New Horizons* spacecraft also imaged Jupiter on their way to Saturn and Pluto, respectively.

Measuring Satellite Densities

One key goal of these missions was to determine the densities of the Galilean satellites to very high accuracy. Given the density of a planet or satellite, astronomers can draw conclusions about its chemical composition and internal structure. To find the densities, scientists first determined the satellite masses. They did this by measuring how the gravity of Io, Europa, Ganymede, and Callisto deflected the trajectories of the *Voyager* and *Galileo* spacecraft. Given the masses and diameters of the satellites, they then calculated each satellite's average density (mass divided by volume). Table 13-1 lists our current knowledge about the sizes and masses of the Galilean satellites.

Of the four Galilean satellites, Europa proves to be the least massive: Its mass is only two-thirds that of our Moon. Ganymede is by far the most massive of the four, with more than double the mass of our Moon. In fact, Ganymede is the most massive satellite anywhere in the solar system. In second place is Callisto, with about 1½ times the Moon's mass.

However, Ganymede and Callisto are *not* merely larger versions of our Moon. The reason is that both Ganymede and Callisto have very low average densities of less than 2000 kg/m³. By comparison, typical rocks in Earth's crust have densities around 3000 kg/m³,

and the average density of our Moon is 3344 kg/m³. The low densities of Ganymede and Callisto mean that these satellites cannot be composed primarily of rock. Instead, they are probably roughly equal parts of rock and water-ice. Water-ice is substantially less dense than rock, but it becomes as rigid as rock under high pressure (such as is found in the interiors of the Galilean satellites). Furthermore, water molecules are relatively common in the solar system. Enough water would have been available in the early solar system to make up a substantial portion of a large satellite such as Ganymede or Callisto.

Water-ice, however, cannot be a major constituent of the two inner Galilean satellites. The innermost satellite, Io, has the highest average density, 3529 kg/m³, slightly greater than the density of our Moon. The next satellite out, Europa, has an average density of 3018 kg/m³. Both of these values are close to the densities of typical rocks in Earth's crust. Hence, it is reasonable to suppose that both Io and Europa are made primarily of rocky material.

> Three of the Galilean satellites are made of a mixture of ice and rock; only Io is ice-free

The most definitive evidence for water-ice on three of Jupiter's Galilean satellites has come from spectroscopic observations. The spectra of sunlight reflected from Europa, Ganymede, and Callisto show absorption at the infrared wavelengths characteristic of water-ice molecules. You can see this in Figure 7-4, which compares the spectrum of Europa to the spectrum of water-ice. Europa's density shows that it is composed mostly of rock, so its ice must be limited to the satellite's outer regions. Among the Galilean satellites, only Io shows no trace of water-ice on its surface or in its interior.

CONCEPTCHECK 13-2

By using Kepler's laws on any of the moons orbiting Jupiter, the mass of Jupiter is easily determined from Earth-based observations. Can the masses of Jupiter's moons be determined from Earth as well?

Answer appears at the end of the chapter.

13-3 The Galilean satellites formed like a solar system in miniature

As Table 13-1 shows, the more distant a Galilean satellite is from Jupiter, the greater the proportion of ice within the satellite and the lower its average density. But why should this be? An important clue is that the average densities of the planets show a similar trend: Moving outward from the Sun, the average density of the inner six planets steadily decreases, from more than 5000 kg/m³ for Mercury to less than 1000 kg/m³ for Saturn (recall Table 7-1). The best explanation for this similarity is that the Galilean satellites formed around Jupiter in much the same way that the planets formed around the Sun, although on a much smaller scale.

> The differences in composition among the Galilean satellites mimic those among planets orbiting the Sun

We saw in Section 8-5 that dust grains in the solar nebula played an important role in planet formation. In the inner parts of the nebula, close to the hot protosun, only dense rock and metal grains were able to remain solid. These grains accumulated over time into the dense, rocky inner planets. But in the cold outer reaches of the nebula (beyond the "snow line"), dust grains were able to retain icy coatings of low-density material like water and ammonia. Hence, when the Jovian planets formed in the outer solar nebula and incorporated these ice-coated grains, the planets ended up with both rock and ices in their cores as well as substantial amounts of ammonia and water throughout their interiors and atmospheres.

As Jupiter coalesced, the gas accumulating around it formed a rotating "Jovian nebula." The central part of this nebula became the huge envelope of hydrogen and helium that makes up most of Jupiter's bulk. But in the outer parts of this nebula, dust grains could have accreted to form small solid bodies. These grew to become the Galilean satellites.

Although the "Jovian nebula" was far from the protosun, not all of it was cold. Jupiter, like the protosun, must have emitted substantial amounts of radiation due to Kelvin-Helmholtz contraction, which we discussed in Section 8-4. (We saw in Section 12-4 that even today, Jupiter emits more energy due to its internal heat than it absorbs from sunlight.) Hence, temperatures very close to Jupiter must have been substantially higher than at locations farther from the planet. Calculations show that only rocky material would condense at the orbital distances of Io and Europa, but frozen water could be retained and incorporated into satellites at the distances of Ganymede and Callisto. In this way Jupiter ended up with two classes of Galilean satellites: primarily rocky satellites versus satellites made of mixture of rock and water (Figure 13-3).

The analogy between the solar nebula and the "Jovian nebula" is not exact. Unlike the Sun, Jupiter can be thought of as a "failed star." Its internal temperatures and pressures never became high enough to ignite nuclear reactions that convert hydrogen into helium. (Jupiter's mass would have had to be about 80 times larger for these reactions to have begun.) Furthermore, the icy worlds

that formed in the outer region of the "Jovian nebula"—namely, Ganymede and Callisto—were of relatively small mass. Hence, they were unable to attract gas from the nebula and become Jovian planets in their own right. Nonetheless, it is remarkable how the main processes of planet formation seem to have occurred twice in our solar system—once in the solar nebula around the Sun, and once in microcosm in the gas cloud around Jupiter.

The diverse densities of the Galilean satellites suggest that these four worlds may be equally diverse in their geology. This conclusion turns out to be entirely correct. Because each of the Galilean satellites is unique, we devote the next several sections to each satellite in turn.

CONCEPTCHECK 13-3

Why do the average densities of the Galilean moons suggest a Jovian nebula developing like the formation of a miniature solar system?

Answer appears at the end of the chapter.

13-4 Io is covered with colorful sulfur compounds ejected from active volcanoes

Before the two *Voyager* spacecraft flew past Jupiter, the nature of Io was largely unknown. On the basis of Io's size and density, however, it was expected that Io would be a world much like our own Moon—geologically dead, with little internal heat available to power tectonic or volcanic activity (see Section 7-6). Therefore, it was thought that Io's surface would be extensively cratered, because there would have been little geologic activity to erase those craters over the satellite's history.

Io proved to be utterly different from these naive predictions, as you can see in Figure 13-4 and the opening figure of this chapter. For our first views, *Voyager 1* came within 21,000 kilometers (13,000 miles) of Io and began sending back a series of bizarre and unexpected pictures of the satellite. These images showed that Io has *no* impact craters at all! Instead, the surface is pockmarked by irregularly shaped pits and is blotched with vibrant colors.

The scientists were seeing confirmation of a prediction published in the journal *Science* just three days before *Voyager 1* flew past Io in March 1979. In this article, Stanton Peale of the University of California, Santa Barbara, and Patrick Cassen and Ray Reynolds of NASA's Ames Research Center reported their conclusion that Io's interior must be kept hot by Jupiter's tidal forces, and that this should lead to intense volcanic activity.

Tidal Heating on Io

As we saw in Section 4-8, tidal forces are differences in the gravitational pull on different parts of a planet or satellite. These forces tend to deform the shape of the planet or satellite (see Figure 4-28). Io, for example, is somewhat distorted from a spherical shape by tidal forces from Jupiter. If Io were in a perfectly circular orbit so that it traveled around Jupiter at a steady speed, Io's rotation would always be in lockstep with its orbital motion. In this case, Io would be in perfect synchronous rotation, with its long axis always pointed directly toward Jupiter. But as Io moves around

> The theory of tidal forces helped predict Io's volcanic nature

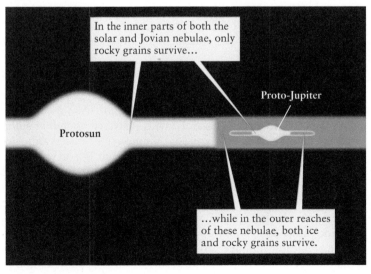

FIGURE 13-3

Formation of the Galilean Satellites Heat from the protosun made it impossible for icy grains to survive within the innermost 2.5 to 4 AU of the solar nebula. In the same way, Jupiter's heat evaporated any icy grains that were too close to the center of the "Jovian nebula." Hence, the two inner Galilean satellites were formed primarily from rock, while the outer two incorporated both rock and ice.

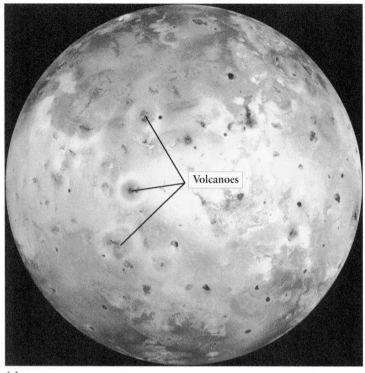

(a)

(b)

VIDEO 13-1 **FIGURE 13-4**   R I V U X G

The Colors of Io **(a)** This true-color image of Io shows a striking yellow surface. Numerous volcanoes dot Io's surface, spewing out sulfur compounds. Sulfur (S) appears yellow, sulfur dioxide (SO_2) appears white, and S_3 and S_4 appear red. **(b)** Here on Earth, we find yellow and white sulfur formations around volcanic vents. (a: JPL/ NASA; b: Cubo Images/Superstock)

its orbit, Europa and Ganymede exert gravitational tugs on it in a regular, rhythmic fashion, thanks to the 1:2:4 ratio among the orbital periods of these three satellites. These tugs distort Io's orbit into an ellipse, and so its speed varies as it moves around its orbit. Consequently, Io's long axis "nods" back and forth by about ½° as seen from Jupiter. Due to this "nodding," the tidal stresses that Jupiter exerts on Io vary rhythmically as Io moves around its orbit. These varying tidal stresses alternately squeeze and flex Io. Just as a ball of clay or bread dough gets warm as you knead it between your fingers, this squeezing and flexing causes **tidal heating** of Io's interior. (Earth's Moon also "nods" as it orbits our planet, but experiences almost no tidal heating because Earth's tidal forces are very weak compared to those of massive Jupiter.)

Tidal heating adds energy to Io at a rate of about 10^{14} watts, equivalent to 24 tons of TNT exploding every second. As this energy makes its way to the satellite's surface, it provides about 2.5 watts of power to each square meter of Io's surface. By comparison, the average global heat flow through Earth's crust is 0.06 watts per square meter. Only in volcanically active areas on Earth do we find heat flows that even come close to Io's average. Thus, Peale, Cassen, and Reynolds predicted "widespread and recurrent surface volcanism" on Io.

Io's Active Volcanoes

No one expected that *Voyager 1* would obtain images of erupting volcanoes on Io. After all, a spacecraft making a single trip past Earth would be highly unlikely to catch a large volcano actually erupting. But, in fact, *Voyager 1* images of Io revealed *eight* different giant plumes of gas from volcanic eruptions. (Figure 13-5a shows two of these.) More recently, the *Galileo* and *New Horizons* spacecraft returned detailed images of several such

plumes (Figure 13-5b, c). These observations resoundingly confirm the Peale-Cassen-Reynolds prediction: Io is by far the most volcanic world in the solar system.

Io's volcanic plumes rise to astonishing heights of 70 to 290 km above Io's surface. To reach such altitudes, the material must emerge from volcanic vents with speeds between 300 and 1000 m/s (1100 to 3600 km/h, or 700 to 2200 mi/h). Even the most violent volcanoes on Earth have eruption speeds of only around 100 m/s (360 km/h, or 220 mi/h). Scientists thus began to suspect that Io's volcanic activity must be fundamentally different from volcanism here on Earth.

The Nature of Io's Volcanic Eruptions

An important clue about volcanism on Io came from the infrared spectrometers aboard *Voyager 1*, which detected abundant sulfur and sulfur dioxide in Io's volcanic plumes. This observation led to the idea that the plumes are actually more like geysers than volcanic eruptions. In a geyser on Earth, water seeps down to volcanically heated rocks, where it changes to steam and erupts explosively through a vent. Planetary geologists Susan Kieffer, Eugene Shoemaker, and Bradford Smith suggested that sulfur dioxide rather than water could be the principal propulsive agent driving volcanic plumes on Io. Sulfur dioxide is a solid at the frigid temperatures found on most of Io's surface, but it should be molten at depths of only a few kilometers. Just as the explosive conversion of water into steam produces a geyser on Earth, the conversion of liquid sulfur dioxide into a high-pressure gas could result in eruption velocities of up to 1000 m/s.

VIDEO 13-2 Io's dramatic coloration (see Figure 13-4) is probably due to sulfur and sulfur dioxide, which are ejected in volcanic plumes and later fall back to the surface. Sulfur is normally bright

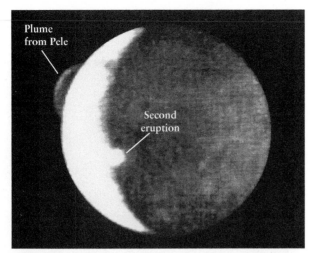

Plume from Pele

Second eruption

(a) *Voyager 1*, March 1979

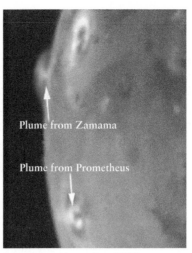

Plume from Zamama

Plume from Prometheus

(b) *Galileo*, November 1997

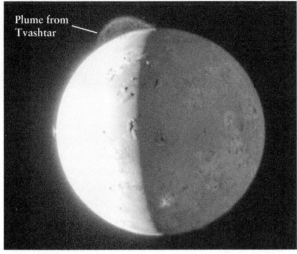

Plume from Tvashtar

(c) *New Horizons*, 2007

FIGURE 13-5   R I V U X G

Volcanic Plumes on Io **(a)** Io's volcanic eruptions were first discovered on this *Voyager 1* image. The plume at upper left, which extends 260 km (150 mi) above Io's surface, came from the volcano Pele. **(b)** This *Galileo* image shows plumes from the volcanoes Prometheus and Zamama. The plumes rise to about 100 km (60 mi) above Io's surface and are about 250 km (150 mi) wide. **(c)** At the top of this *New Horizons* image, the plume from the volcano Tvashtar reaches 290 km high. Another plume is visible at the far left edge. The mountains near the center, just to the left of the terminator (the line between day and night) are about as tall as Mount Everest. (a: NASA; b: Planetary Image Research Laboratory/University of Arizona/JPL/NASA; c: NASA/Johns Hopkins University Applied Physics Laboratory/Southwest Research Institute)

yellow, which explains the dominant color of Io's surface. But if sulfur is heated and suddenly cooled, as would happen if it were ejected from a volcanic vent and allowed to fall to the surface, it can assume a range of colors from orange and red to black. Indeed, these colors are commonly found around active volcanic vents

(Figure 13-6). Whitish surface deposits (examine Figure 13-4 and Figure 13-6a), by contrast, are probably due to sulfur dioxide (SO_2). Volcanic vents on Earth commonly discharge SO_2 in the form of an acrid gas. But on Io, when hot SO_2 gas is released by an eruption into the cold vacuum of space, it crystallizes into white

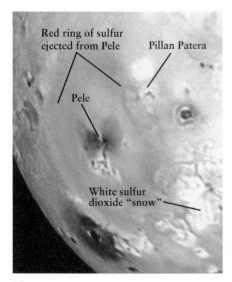

Red ring of sulfur ejected from Pele

Pillan Patera

Pele

White sulfur dioxide "snow"

(a)

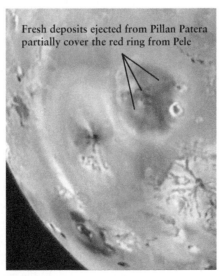

Fresh deposits ejected from Pillan Patera partially cover the red ring from Pele

(b)

FIGURE 13-6   R I V U X G

Rapid Changes on Io **(a)** This *Galileo* image shows a portion of Io's southern hemisphere. Sulfur ejected from the active volcanic vent Pele produced a red ring that resembles paint sprayed from a can. **(b)** A few months later, a volcanic plume erupted from a vent called Pillan Patera, spraying dark material over an area some 400 km (250 mi) in diameter (about the size of Arizona). Other than Earth, Io is the only world in the solar system that shows such noticeable changes over short time spans. (NASA/JPL)

snowflakes. This sulfur dioxide "snow" then falls back down onto Io's surface.

The plumes are not the whole story of volcanism on Io, however. The sources of the plumes appear in *Voyager* images as black spots 10 to 50 km in diameter, which are actually volcanic vents (see Figure 13-4a). Many of these black spots, which cover 5 percent of Io's surface, are surrounded by dark lava flows. Figure 13-7 is a *Galileo* image of one of Io's most active volcanic regions, where in 1999 lava was found spouting to altitudes of thousands of meters along a fissure 40 km (25 mi) in length. The glow from this lava fountain was so intense that it was visible with Earth-based telescopes. Thus, Io has two distinct styles of volcanic activity: unique geyserlike plumes, and lava flows that are more dramatic versions of volcanic flows on Earth.

While sulfur and sulfur compounds are the key constituents of Io's volcanic plumes, the lava flows must be made of something else. Infrared measurements from the *Galileo* spacecraft (Figure 13-8) show that lava flows on Io are at temperatures of 1700 to 2000 K (about 1450 to 1750°C, or 2600 to 3150°F). At these temperatures sulfur could not remain molten but would evaporate almost instantly. Io's lavas are also unlikely to have the same chemical composition as typical lavas on Earth, which have temperatures of only 1300 to 1450 K. Instead, lavas on Io are probably enriched in magnesium and iron, which give the lava a higher melting temperature.

Similar lavas enriched in magnesium and iron are found on Earth, but primarily in solidified lava beds that formed billions of years ago when Earth's interior was much hotter than today. The presence of this enriched lava on Io suggests that its interior, too, is substantially hotter than that of the present-day Earth. One model proposes that Io has a 100-km (60-mi) thick crust floating atop a worldwide ocean of liquid magma 800 km (500 mi) deep. (As we saw in Section 9-2, the mantle that lies underneath Earth's crust is a plastic solid rather than a true liquid.) The global extent of this

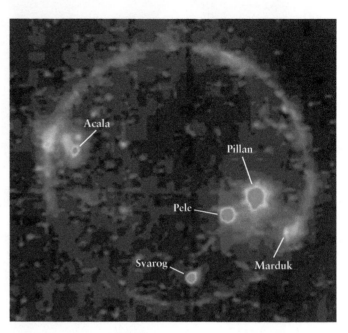

FIGURE 13-8   R I $\boxed{V}$ U X G

Io's Glowing Volcanoes This false-color *Galileo* image (made while Io was in the darkness of Jupiter's shadow) shows the infrared and visible glow from several volcanic regions, including Pele and Pillan (see Figure 13-6). These regions are substantially hotter than typical lava flows on Earth. (Planetary Image Research Laboratory/University of Arizona/JPL/NASA)

"magma ocean" would explain why volcanic activity is found at all points on Io's surface, rather than being concentrated in pockets as on Earth (see Section 9-3). It could also explain the origin of Io's mountains, which reach heights up to 10 km (30,000 ft). In this model, these mountains are blocks of Io's crust that have tilted before sinking into the depths of the magma ocean.

Io's worldwide volcanic activity is remarkably persistent. When *Voyager 2* flew through the Jovian system in July 1979, four months after *Voyager 1*, almost all of the volcanic plumes seen by *Voyager 1* were still active. And when *Galileo* first viewed Io in 1996, about half of the volcanoes seen by the *Voyager* spacecraft were still ejecting material, while other, new volcanoes had become active.

Io may have as many as 425 active volcanoes, each of which ejects an estimated 10,000 tons of material per second. Altogether, volcanism on Io may eject as much as 10^{13} tons of matter each year. This amount of material is sufficient to cover the satellite's entire surface to a depth of 1 meter in a century, or to cover an area of 1000 square kilometers in a few weeks (see Figure 13-6). Thanks to this continual "repaving" of the surface, there are probably no long-lived features on Io, and any impact craters are quickly covered up.

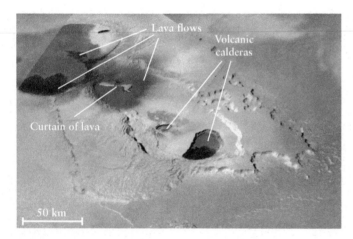

FIGURE 13-7   R I $\boxed{V}$ U X G

Io's Lava Flows and a Curtain of Fire This mosaic of *Galileo* spacecraft images shows a nested chain of volcanic calderas (pits) with black lava flows. The red streak at upper left is a fissure in the surface about 40 km (25 mi) long, through which lava is erupting upward in a vertical sheet or curtain. The erupting lava reaches the amazing height of 1500 m (5000 ft). (University of Arizona/JPL/NASA)

CONCEPTCHECK 13-4

The strongest tidal forces on Io are exerted by Jupiter. Then why are the orbital properties of Europa and Ganymede important for generating the heat that powers Io's volcanism?

Answer appears at the end of the chapter.

13-5 Jupiter's magnetic field makes electric currents flow through Io

Most of the material ejected from Io's volcanoes falls back onto the satellite's surface. But some material goes on a remarkable journey into space, by virtue of Io's location deep within Jupiter's magnetosphere.

The Io Torus

The Jovian magnetosphere contains charged particles, or ions, some of which collide with Io and its volcanic plumes. The impact of these collisions knocks more ions out of the plumes and off the surface, and these additional charged particles become part of Jupiter's magnetosphere. Indeed, material from Io is the main source of material in Jupiter's magnetosphere as a whole. Some of the ejected material forms the **Io torus,** a huge doughnut-shaped ring of ionized gas that circles Jupiter at the distance of Io's orbit. A gas that is hot enough for electrons to dissociate from atoms forms a gas of charged particles called a **plasma.** The plasma's glow can be detected from Earth (Figure 13-9a).

Jupiter's magnetic field has other remarkable effects on Io. As Jupiter rotates, its magnetic field rotates with it. This field sweeps over Io at high speed and generates a voltage

> Jupiter and Io together act like an immense electric generator

of 400,000 volts across the satellite, causing 5 million amperes of electric current to flow through Io.

ANALOGY Whenever you go shopping with a credit card, you use the same physical principle that generates an electric current within Io. The credit card's number is imprinted as a magnetic code in the dark stripe on the back of the card. The salesperson who swipes your card through the card reader is actually moving the card's magnetic field past a coil of wire within the reader. This action generates within the coil an electric current that carries the same coded information about your credit card number. That information is transmitted to the credit card company, which (it is hoped) approves your purchase. An electric generator at a power plant works in the same way. A coil of wire is moved through a strong magnetic field, which makes current flow in the coil. This current is delivered to transmission lines and eventually to your home.

In fact, a current flows not only through Io but also through the sea of charged particles in Jupiter's magnetosphere and through Jupiter's atmosphere. This forms a gigantic, oval-shaped electric circuit that connects Io and Jupiter in the same way that wires connect the battery and the lightbulb in a flashlight (Figure 13-9b). Jupiter's aurora, which we discussed in Section 12-7, is particularly strong at the locations where this current strikes the upper atmosphere of Jupiter.

This important phenomenon occurs on all the Galilean moons, leading to profound insights and discoveries, so let us summarize what we have learned for further use in the chapter. As Jupiter's magnetic field sweeps over an electrically conductive medium (in this case, a plasma concentrated around Io), there is an induced flow of electricity in the conductive medium called an **induced electric current.** In addition to inducing the flow of electricity, all electricity produces its own magnetic field, which, in this scenario, is called an **induced magnetic field.**

Probing Io's Interior

The electric current within and near Io also creates a weak magnetic field, which was first detected by *Voyager 1*. But what about

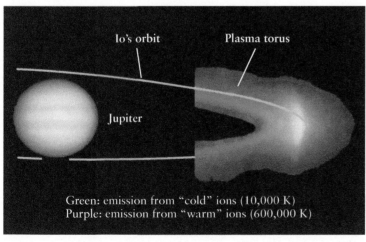

(a)

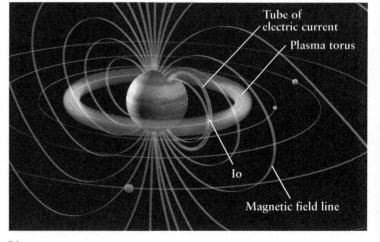

(b)

VIDEO 13-3

FIGURE 13-9   R I V U X G

The Io Torus **(a)** An Earth-based telescope recorded this false-color infrared image of Io's plasma torus of charged particles. Because of Jupiter's glare, only the outer edge of the torus could be photographed; an artist added the visible-light picture of Jupiter and the yellow line indicating

the rest of the torus. **(b)** A tube of electric current carries a charged-particle plasma between Jupiter and Io, forming an immense electric circuit. (a: Courtesy of J. Trauger; b: Alfred T. Kamajian and Torrence V. Johnson, "The Galileo Mission to Jupiter and Its Moons," *Scientific American,* February 2000, p. 44)

an intrinsic magnetic field generated by Io's interior? *Galileo* data determined that Io does not generate a magnetic field. This means that Io does not contain a convecting iron core.

If Io has undergone enough tidal heating to melt its interior, chemical differentiation must have taken place and the satellite should have a dense core. (We described chemical differentiation in Box 7-1 and Section 8-5.) To test for a dense core, scientists measured how Io's gravity deflected the trajectory of *Galileo* as the spacecraft flew past. From these measurements, they could determine not only Io's mass but also the satellite's oblateness (how much it deviates from a spherical shape because of its rotation). The amount of oblateness indicates the size of the core; the greater the fraction of the satellite's mass contained in its core, the less oblate the satellite will be. The *Galileo* observations suggest that Io has a dense core composed of iron and iron sulfide (FeS), with a radius of about 900 km (about half the satellite's overall radius). Surrounding the core is a mantle of partially molten rock, on top of which is Io's visible crust.

CONCEPTCHECK 13-5

Is Jupiter's magnetic field generated by the induced electric currents flowing between Jupiter and Io?

Answer appears at the end of the chapter.

13-6 Europa is covered with a smooth layer of ice that may cover a worldwide ocean

Europa, the second of the Galilean satellites, is the smoothest body in the solar system. There are no mountains and no surface features greater than a few hundred meters high (Figure 13-10). There are almost no craters, indicating a young surface that has been reprocessed by geologic activity. The dominant surface feature is a worldwide network of stripes and cracks (Figure 13-11). Like Io, Europa is an exception to the general rule that a small world should be cratered and geologically dead (see Section 7-6).

Europa's Surface: Icy but Active

Neither *Voyager 1* nor *Voyager 2* flew very near Europa, so most of our knowledge of this satellite comes from the close passes made by *Galileo*. (Both Figure 13-10 and Figure 13-11 are *Galileo* images.) But even before these spacecraft visited Jupiter, spectroscopic observations from Earth indicated that Europa's surface is almost pure frozen water (see Figure 7-4). This observation was confirmed by instruments on board *Galileo*, which showed that Europa's infrared spectrum is a close match to that of a thin layer of fine-grained water-ice frost on top of a surface of pure water-ice. (The brown areas in Figure 13-10 show where the icy surface contains deposits of rocky material from meteoritic impacts, from Europa's interior, or from a combination of these sources.)

> Geology on Europa is based on solid ice and liquid water, not rock and molten magma

The purity of Europa's ice suggests that water is somehow brought upward from the moon's interior to the surface, where it solidifies to make a fresh, smooth layer of ice. Indeed, some *Galileo* images show what appear to be lava flows on Europa's

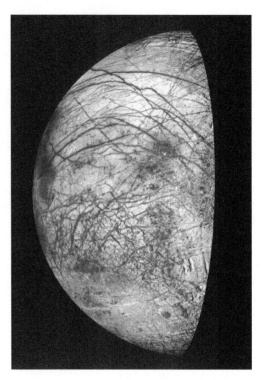

FIGURE 13-10 R I V U X G

Europa Dark lines crisscross Europa's smooth, icy surface in this false-color composite of visible and infrared images from *Galileo*. These lines are fractures in Europa's crust that can be as much as 20 to 40 km (12 to 25 mi) wide. Only a few impact craters are visible on Europa, which indicates that this satellite has a very young surface on which all older craters have been erased. (NASA/JPL/University of Arizona)

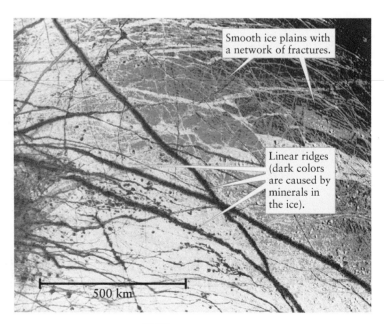

Smooth ice plains with a network of fractures.

Linear ridges (dark colors are caused by minerals in the ice).

500 km

FIGURE 13-11 R I V U X G

Europa's Fractured Crust False colors in this *Galileo* composite image emphasize the difference between the linear ridges and the surrounding plains. The smooth ice plains (shown in blue) are the basic terrain found on Io. (NASA/JPL)

surface, although the "lava" in this case is mostly ice. This idea helps to explain why Europa has very few craters (any old ones have simply been covered up) and why its surface is so smooth. Europa's surface may thus represent a water-and-ice version of plate tectonics.

Although Europa's surface is almost pure water-ice, keep in mind that Europa is *not* merely a giant ice ball. The satellite's density shows that rocky material makes up about 85 to 90 percent of Europa's mass. Hence, only a small fraction of the mass, about 10 to 15 percent, is water-ice. Because the surface is icy, we can conclude that the rocky material is found within Europa's interior.

Europa is too small to have retained much of the internal heat that it had when it first formed. But there must be internal heat nonetheless to power the geologic processes that erase craters and bring fresh water to Europa's surface. What keeps Europa's interior warm? The most likely answer, just as for Io, is heating by Jupiter's tidal forces. The rhythmic gravitational tugs exerted by the orbital resonance of Io, Ganymede, and Europa deform Europa's orbit, causing variations in the tidal stresses from Jupiter that make Europa flex. But because Europa is farther from Jupiter, tidal effects on Europa are only about one-fourth as strong as those on Io, which may explain why no ongoing volcanic activity has yet been seen on Europa.

Some features on Europa's surface, such as the fracture patterns shown in Figure 13-11, may be the direct result of the crust being stretched and compressed by tidal flexing. Figure 13-12 shows other features, such as networks of ridges and a young, very smooth circular area, that were probably caused by the internal heat that tidal flexing generates. The rich variety of terrain depicted in Figure 13-12, with stress ridges going in every direction, shows that Europa has a complex geologic history.

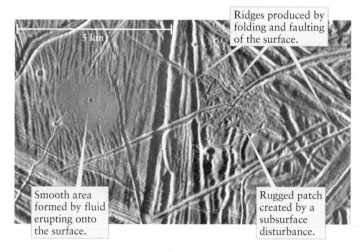

FIGURE 13-12   R I **V** U X G

Europa in Close-up This high-resolution *Galileo* image shows a network of overlapping ridges, part of which has been erased to leave a smooth area and part of which has been jumbled into a rugged patch of terrain. Europa's interior must be warm enough to power this complex geologic activity. (NASA/JPL)

Among the unique structures found on Europa's surface are **ice rafts**. The area shown in Figure 13-13a was apparently subjected to folding, producing the same kind of linear features as those in Figure 13-12. But a later tectonic disturbance broke the surface into small chunks of crust a few kilometers across, which then "rafted" into new positions. A similar sort of rafting happens in Earth's Arctic Ocean every spring, when the winter's

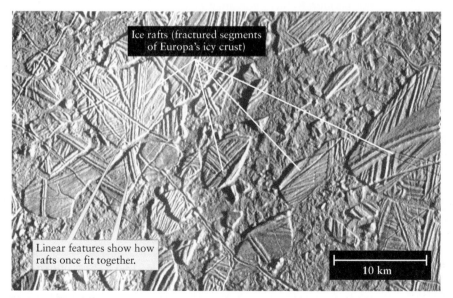

(a) Ice rafts on Europa

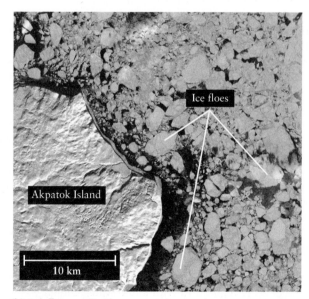

(b) Ice floes on Earth

FIGURE 13-13   R I **V** U X G

Moving Ice on Europa and Earth (a) Some time after a series of ridges formed in this region of Europa's surface, the icy crust broke into "rafts" that were moved around by an underlying liquid or plastic layer. The colors in this *Galileo* image may be due to minerals that were released from beneath the surface after the crust broke apart. (b) Europa's ice rafts are analogous to ice floes created when pack ice breaks up, as in this spacecraft view of part of the Canadian arctic. (a: NASA/JPL; b: USGS and NASA)

accumulation of surface ice breaks up into drifting ice floes (Figure 13-13b). The existence of such structures on Europa suggests that there is a subsurface layer of liquid water or soft ice over which the ice rafts can slide with little resistance.

But questions remain: Is Europa's heat mostly gone, leaving behind a frozen interior today? Is there a large subsurface ocean, or just thin pockets of liquid water? To answer these questions, we return to measurements of magnetic fields, which act like faithful messengers from a planet's interior.

An Underground Ocean?

Magnetic field measurements made by the *Galileo* spacecraft have provided key evidence for a large saltwater ocean within Europa. Unlike Earth or Jupiter, Europa does not seem to create a steady magnetic field of its own. But as Europa moves through Jupiter's intense magnetic field, electric currents are induced within the satellite's interior, just as they are around Io (see Section 13-5), and these currents generate a weak but measurable induced magnetic field. (The strength and direction of this induced magnetic field varies, and can even reverse within hours, as Europa moves through different parts of the Jovian magnetosphere; these quick reversals could not occur if Europa generated its field by itself.)

To explain these observations, there must be an electrically conducting fluid beneath Europa's crust—a perfect description of an underground ocean of water with dissolved minerals. (Pure water is a very poor conductor of electricity, so electrically charged particles, such as dissolved minerals, would make the ocean water conducting.) The dissolved minerals might be the familiar salt found in Earth's saltwater oceans, but the actual minerals are not yet known.

If some of this water should penetrate upward to Europa's surface through cracks in the crust, it would vaporize and leave the dissolved minerals on the surface. In fact, these minerals might explain the dark linear features in Figure 13-11.

By combining measurements of Europa's induced magnetic field, gravitational pull, and oblateness, scientists estimate that Europa has a frozen ice crust about 10 km thick on top of a liquid water ocean about 100 km deep (**Figure 13-14**). Beneath the ocean, models suggest a rocky mantle surrounding a metallic core some 600 km (400 mi) in radius. Since water is known to partially dissolve some of the minerals that make rock, contact with this mantle would be the source of the ocean's electrical conductivity.

Interestingly, Europa also has an extremely thin atmosphere of oxygen (producing an atmospheric pressure only a trillionth of Earth's). We saw in Section 9-5 that oxygen in Earth's atmosphere is produced by plants through photosynthesis. But Europa's oxygen atmosphere is probably the result of ions from the solar wind and Jupiter's magnetosphere striking the satellite's icy surface. These collisions break apart water molecules, liberating atoms of hydrogen (which escape into space) and oxygen.

The existence of a warm, subsurface ocean on Europa, if proved, would make Europa one of the few worlds in the solar system other than Earth with liquid water. Liquid water on Europa would have dramatic implications. On Earth, water and warmth are essentials for the existence of life. Perhaps single-celled organisms have evolved in the water beneath Europa's crust, where they would use dissolved minerals and organic compounds as food sources. In light of this possibility, NASA has taken steps to prevent biological contamination of Europa. At the end of the *Galileo*

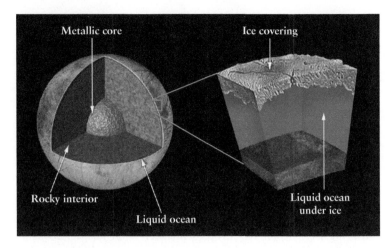

FIGURE 13-14 R I V U X G

Europa's Ocean While small compared to Europa's overall size, its ocean is still enormous. Estimated at about 100 km deep, it would be over 10 times deeper than Earth's oceans and contain more than twice the volume of water. (NASA/JPL)

mission in 2003, the spacecraft (which may have carried traces of organisms from Earth) was sent to burn up in Jupiter's atmosphere, rather than remaining in orbit where it might someday crash into Europa. An appropriately sterilized spacecraft may one day visit Europa and search for evidence of life within this exotic moon.

CONCEPTCHECK 13-6

If the orbital resonance of Io, Europa, and Ganymede contributes to the heating of their interiors, why isn't Europa hot enough to have volcanism like Io?

CONCEPTCHECK 13-7

Why does an induced magnetic field coming from Europa indicate a large subsurface ocean?

Answers appear at the end of the chapter.

13-7 Liquid water may also lie beneath the cratered surfaces of Ganymede and Callisto

Unlike Io and Europa, the two outer Galilean satellites have cratered surfaces. In this respect, Ganymede and Callisto bear a superficial resemblance to our

> Both Ganymede and Callisto show evidence of past geologic activity

Moon (see the illustration with Table 13-1). But unlike the Moon's craters, the craters on both Ganymede and Callisto are made primarily of ice rather than rock, and Ganymede has a number of surface features that indicate a geologically active past.

The Two Faces of Ganymede's Surface

Ganymede is the largest satellite in our solar system and is even larger than the planet Mercury. Like our Moon, Ganymede has

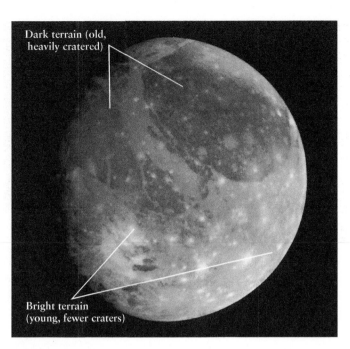

Dark terrain (old, heavily cratered)

Bright terrain (young, fewer craters)

FIGURE 13-15   R I V U X G

Ganymede Two distinct types of terrain—one dark, heavily cratered, and hence old, the other bright, less cratered, and hence younger—are visible in this *Galileo* image of the hemisphere of Ganymede that always faces away from Jupiter. Craters in general appear bright, suggesting that the impacts that formed the craters excavated the surface to reveal ice underneath. (NASA/JPL)

two distinct kinds of terrain, called **dark terrain** and **bright terrain** (Figure 13-15). On the Moon, the dark maria are younger than the light-colored lunar highlands (see Section 10-1). On Ganymede, by

contrast, the dark terrain is older, as indicated by its higher density of craters. The bright terrain is much less cratered and therefore younger. Because ice is more reflective than rock, even Ganymede's dark terrain is substantially brighter than the lunar surface.

Voyager and *Galileo* images show noticeable differences between young and old impact craters on Ganymede. The youngest craters are surrounded by bright rays of freshly exposed ice that has been excavated by the impact (see the lower left of Figure 13-15). Older craters are darker, perhaps because of chemical processes triggered by exposure to sunlight.

The close-up *Galileo* image in Figure 13-16 shows the contrast between dark and bright terrain. The area on the left-hand side of this image is strewn with large and small craters, reinforcing the idea that the dark terrain is billions of years old. The dark terrain is also marked by long, deep furrows, which are deformations of Ganymede's crust that have partially erased some of the oldest craters. Since other craters lie on top of the furrows, the stresses that created the furrows must have occurred long ago. Since geologic activity is a result of a planet or satellite's internal heat, this shows that Ganymede must have had a warm interior in the distant past.

Linear features are also seen in the bright terrain on the right in Figure 13-16. The *Voyager* missions discovered that the bright terrain is marked by long grooves, some of which extend for hundreds of kilometers and are as much as a kilometer deep. Images from *Galileo* show that some adjacent grooves run in different directions, indicating that the bright terrain has been subjected to a variety of tectonic stresses. *Galileo* also found a number of small craters overlying the grooves, some of which are visible in Figure 13-16. The density of craters shows that the bright terrain, while younger than the dark terrain, is at least a billion years old.

After the *Voyager* missions, it was thought that the bright terrain represented fresh ice that had flooded through cracks in Ganymede's crust, analogous to the maria on our Moon (see

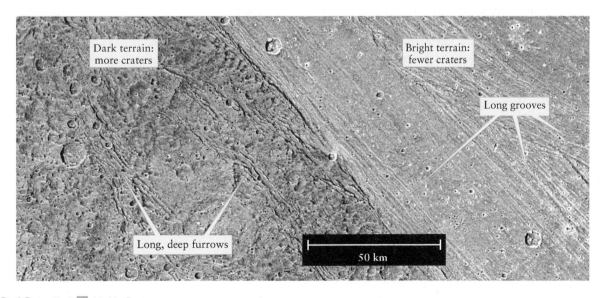

Dark terrain: more craters

Bright terrain: fewer craters

Long grooves

Long, deep furrows

50 km

FIGURE 13-16   R I V U X G

Evidence for Geologic Activity on Ganymede Stresses in Ganymede's crust have produced linear features that extend for hundreds of kilometers across both the dark and bright terrain. These features must be quite ancient, since numerous impact craters have formed on top of them. (NASA)

Section 10-1). However, high-resolution *Galileo* images do not show the kinds of surface features that would be expected if this terrain were produced by ice-volcanoes. Yet there are several low-lying areas within the bright terrain that appear to have been flooded by a watery fluid that subsequently froze. These low-lying areas may be *rift valleys* where the crust has been pulled apart, like Valles Marineris on Mars (see Section 11-5). In order to flood the floors of these valleys, there must have been liquid water beneath Ganymede's surface when the valleys formed. This is additional evidence that Ganymede must have had a warm interior in the past, and that this internal warmth persisted at least until the era when the bright terrain formed a billion years ago. The persistence of internal warmth is rather surprising, because Ganymede is too small to have retained much of the heat left over from its formation or from radioactive elements, and it orbits too far from Jupiter to have any substantial tidal heating.

Ganymede's Interior and Bizarre Magnetic Field

An equally great surprise from *Galileo* observations was the discovery that Ganymede has its own magnetic field, and that this field is twice as strong as that of Mercury. This field is sufficiently strong to trap charged particles, giving Ganymede its own "mini-magnetosphere" within Jupiter's much larger magnetosphere. The presence of a magnetic field shows that electrically conducting material must be in motion within Ganymede to act as a dynamo, which means that the satellite must have substantial internal heat even today. A warm interior causes differentiation, and *Galileo* measurements indeed showed that Ganymede is highly differentiated. It has a metallic core some 500 km in radius, surrounded by a rocky mantle and by an outer shell of ice 800 km thick.

An even more surprising result is that Ganymede's magnetic field varies somewhat as it orbits Jupiter, and that these variations are correlated with Ganymede's location within Jupiter's magnetic field. One explanation for this variation is that Ganymede has a layer of salty, electrically conducting liquid water some 170 km beneath its surface. As the satellite moves through Jupiter's magnetic field, electric currents are induced in this layer just as happens within Europa (see Section 13-6), and these currents generate a weak field that adds to the stronger field produced by Ganymede's dynamo.

A possible explanation for Ganymede's internal heat is that gravitational forces from the other Galilean satellites may have affected its orbit. While Ganymede's present-day orbit around Jupiter is quite circular, calculations show that the orbit could have been more elliptical in the past. Just as for Io and Europa, such an elliptical orbit would have given rise to substantial tidal heating of Ganymede's interior. It may have retained a substantial amount of that heat down to the present day.

Callisto's Perplexing Surface

In marked contrast to Io, Europa, and Ganymede is Callisto, Jupiter's outermost Galilean satellite. Images from *Voyager* showed that Callisto has numerous impact craters scattered over an icy crust (**Figure 13-17**). Callisto's ice is not as reflective as that on Europa or Ganymede, however; the surface appears to be covered with some sort of dark mineral deposit.

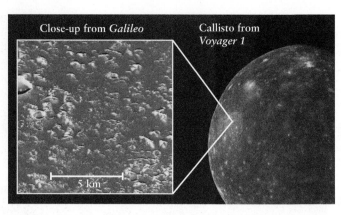

FIGURE 13-17   R I **V** U X G

Callisto Numerous craters pockmark Callisto's icy surface. The inset shows a portion of a huge impact basin called Valhalla. Most of the very smallest craters in this close-up view have been completely obliterated, and the terrain between the craters is blanketed by dark, dusty material. (Left: Arizona State University/JPL/NASA; right: JPL/NASA)

CAUTION! Figure 13-17, as well as the images that accompany Table 13-1, shows that Callisto is the darkest of the Galilean satellites. Nonetheless, its surface is actually more than twice as reflective as that of our own Moon. Callisto's surface may be made of "dirty" ice, but even dirty ice reflects more light than the gray rock found on the Moon. If you could somehow replace our Moon with Callisto, a full moon as seen from Earth would be 4 times brighter than it is now (thanks also in part to Callisto's larger size).

The *Voyager* images led scientists to conclude that Callisto is a dead world, with an ancient surface that has never been reshaped by geologic processes. There is no simple relationship between Callisto's orbital period and those of the other Galilean satellites, so it would seem that Callisto never experienced any tidal heating and hence was unable to power any geologic processes at its surface. But once the *Galileo* spacecraft went into orbit around Jupiter and began returning high-resolution images, scientists realized that Callisto's nature is not so simple. In fact, Callisto proves to be the most perplexing of the Galilean satellites.

One curious discovery from *Galileo* is that while Callisto has numerous large craters, there are very few with diameters less than 1 km. Such small craters are abundant on Ganymede, which has probably had the same impact history as Callisto. The lack of small craters on Callisto implies that most of these craters have been eroded away. How this erosion could have happened is not known. Another surprising result is that Callisto's surface is covered by a blanket of dark, dusty material (see the inset in Figure 13-17). Where this material came from, and how it came to be distributed across the satellite's surface, is not understood.

Callisto's Interior: Hot or Cold?

Callisto's most unexpected feature is that, like Europa and Ganymede, it appears to have an induced magnetic field that varies as the satellite moves through different parts of Jupiter's

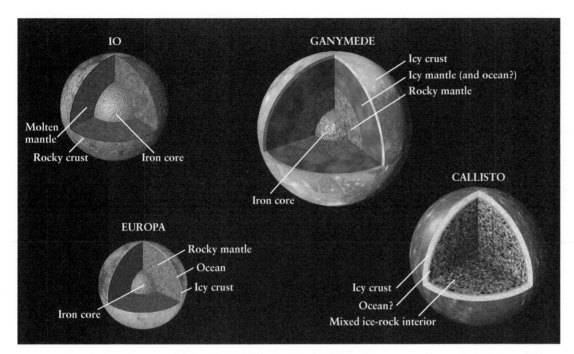

FIGURE 13-18
Interiors of the Galilean
Satellites These cross-sectional
diagrams show the probable internal
structures of the four Galilean
satellites of Jupiter, based on
information from the *Galileo* mission.
(NASA/JPL)

magnetosphere. Callisto must have a layer of electrically conducting material in order to generate such a field. One model suggests that this material is in the form of a subsurface ocean like Europa's but only about 10 km (6 mi) deep. However, without tidal heating, Callisto's interior is probably so cold that a layer of liquid water would quickly freeze. One proposed explanation is that Callisto's ocean is a mixture of water and a form of "antifreeze." In cold climates on Earth, antifreeze is added to the coolant in automobile radiators to lower the liquid's freezing temperature and prevent it from solidifying. Ammonia (NH_3) in Callisto's proposed subsurface ocean could play the same role: There was plenty of ammonia available at these far reaches of the solar system, and the ability for ammonia to act like an antifreeze is well-established.

The presence of a liquid layer beneath Callisto's surface is difficult to understand. Even with ammonia acting as an antifreeze, this layer cannot be too cold or it would solidify. This argues for a relatively *warm* interior. But *Galileo* measurements suggest that Callisto's interior is not highly differentiated. The lack of differentiation implies that the interior is *cold,* and may never have been heated enough to undergo complete chemical differentiation. The full story of Callisto's interior, it would seem, has yet to be told.

Although Ganymede and Callisto have many differences, one feature they share is an extremely thin and tenuous atmosphere. Like Europa, Ganymede has a thin oxygen (O_2) atmosphere thought to be created when radiation splits water-ice (H_2O) on the surface into oxygen and hydrogen, with the lighter H_2 escaping Ganymede's weak gravity. On Callisto, the atmosphere is known to contain carbon dioxide (CO_2). This gas would escape in several days and is thought to be replenished by evaporation of CO_2 ice from Callisto's cold polar regions.

The diagrams in Figure 13-18 summarize the probable interior structures of the Galilean satellites. They demonstrate the remarkable variety of these terrestrial worlds.

CONCEPTCHECK **13-8**

Why is it thought that Callisto contains ammonia?

Answer appears at the end of the chapter.

13-8 Titan has a thick atmosphere and hydrocarbon lakes

Unlike Jupiter, Saturn has only one large satellite that is comparable in size to our Moon. This satellite, Titan, was discovered by Christiaan Huygens in 1665. Titan's diameter of 5150 km makes it second in size among satellites to Ganymede (Section 13-7). Like Ganymede, Titan has a low density that suggests it is made of a mixture of ice and rock. (See Table 7-2 for a comparison between Titan, the Moon, and the Galilean satellites.) But unlike Ganymede or any other satellite in the solar system, Titan has a thick atmosphere with a unique chemical composition.

Titan's Atmosphere

By the early 1900s, scientists had begun to suspect that Titan (Figure 13-19) might have an atmosphere, because it is cool enough and massive enough to retain heavy gases. (Box 7-2 explains the criteria for a planet or satellite to retain an atmosphere.) These suspicions were confirmed in 1944, when Gerard Kuiper discovered the spectral lines of methane in sunlight reflected from Titan. (We discussed Titan's spectrum and its interpretation in Section 7-3.)

Because of its atmosphere, Titan was a primary target for the *Voyager* missions. To the chagrin of mission scientists and engineers, however, *Voyager 1* spent hour after precious hour of its limited, preprogrammed observation time sending back featureless images like the one at the center of Figure 13-19. Titan's atmosphere is so

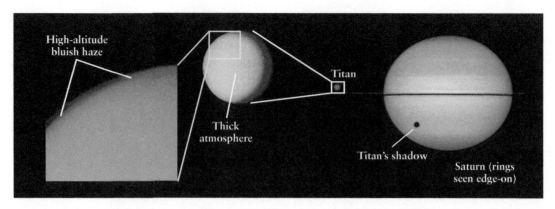

FIGURE 13-19 R I V U X G

Titan The Hubble Space Telescope image at far right shows Titan and the shadow it casts on Saturn's clouds. The *Voyager* image in the middle shows how Titan's atmosphere gives it a nearly featureless appearance. At far left is a color-enhanced *Voyager* image that shows a haze extending well above the satellite's visible edge, or limb. (Left: JPL/NASA; center: NASA; right: Erich Karkoschka, LPL/STScI/NASA)

thick—about 200 km (120 mi) deep, or about 10 times the depth of Earth's troposphere—that it blocked *Voyager's* view of the surface. The haze surrounding Titan is so dense that little sunlight penetrates to the ground; noon on Titan is only about 1/1000 as bright as noon on Earth, though still 350 times brighter than night on Earth under a full moon.

By examining how the radio signal from *Voyager 1* was affected by passing through Titan's atmosphere, scientists inferred that the atmospheric pressure at Titan's surface is 50 percent greater than the atmospheric pressure at sea level on Earth. Titan's surface gravity is weaker than that of Earth, so to produce this pressure Titan must have considerably more gas in its atmosphere than Earth. Indeed, calculations show that about 7 times more gas (by mass) lies above a square meter of Titan's surface than above a square meter of Earth. With Titan's combination of low gravity and thick atmosphere, estimates even suggest that humans could literally fly through the atmosphere by flapping special "wings" attached to their arms!

Voyager data show that more than 95 percent of Titan's atmosphere is nitrogen gas, although it is not clear where this nitrogen came from. One idea suggests that most of this nitrogen came from ammonia (NH_3), a compound of nitrogen and hydrogen that is quite common in the outer solar system. In this scenario, the ammonia is released from Titan's interior, and then is easily broken into hydrogen and nitrogen by the Sun's ultraviolet radiation. Titan's gravity is too weak to retain hydrogen atoms (see Box 7-2), so these atoms escape into space and leave the more massive nitrogen gas molecules behind. However, not all of the details for this scenario fit well with recent *Cassini* data, prompting another model.

A 2011 experiment at the University of Tokyo indicates that *comets smashing into Titan's surface* might have created its nitrogen atmosphere. The experiment had two steps. First, a laser was shot into water-ice that was rich in ammonia. The energy of the laser blast simulates a cometary impact, and the ammonia-rich ice approximates Titan's crust. Second, measurements were made of the amount of nitrogen gas liberated by the simulated impact.

Finally, by assuming the frequency of impacts expected during the Late Heavy Bombardment about 3.9 billion years ago (see Section 8-6), researchers could account for the nitrogen seen in Titan's present-day atmosphere.

The remaining 5 percent of Titan's atmosphere is predominantly methane (CH_4), which is the principal component of the "natural gas" used on Earth as fuel. However, in the absence of oxygen, Titan's atmosphere is in no danger of catching on fire. Methane is an example of a **hydrocarbon,** which is any compound made only from hydrogen and carbon atoms. The interaction of methane in Titan's atmosphere with ultraviolet light from the Sun produces other hydrocarbons, including ethane (C_2H_6). As we will see, methane and ethane play a very significant role on Titan beyond its atmosphere.

What gives Titan its reddish-brown color (seen in Figure 13-19)? Nitrogen in Titan's atmosphere can combine with airborne hydrocarbons to produce other compounds, such as hydrogen cyanide (HCN). Hydrogen cyanide, along with other molecules, can join together in long, repeating molecular chains. Some of these long molecular chains remain suspended in Titan's atmosphere and might be responsible for its unique color.

Titan's Methane Cycle

On Earth, water can be a gas (like water vapor in the atmosphere), a liquid (as in the oceans), or a solid (as in snow and ice). On Titan, the atmospheric pressure is so high and the surface temperature is so low—about 95 K (−178°C, or −288°F)—that any water is frozen solid. But temperatures and pressures on Titan are just right for methane to exist as a gas, liquid, or solid. This single revelation by the *Voyager* missions—that Titan could support three phases of matter for methane—led to the bold prediction that vapor clouds of methane might rain down onto Titan's surface to form the only lakes known outside of Earth (Figure 13-20). Are these predictions correct? Does Titan have something analogous to Earth's

> A faint mist of methane rain falls continually on Titan

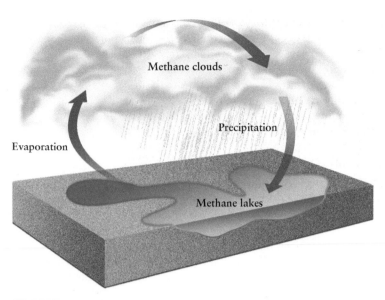

FIGURE 13-20

Methane Cycle on Titan On Titan, small variations in temperature and pressure can change methane between its gas and liquid states, just as water can change states to form the hydrological cycle on Earth. Even though Titan receives only about 1 percent as much sunlight as Earth does, it is enough to drive a methane cycle.

hydrologic cycle, with methane playing the role of water? As we will see, the answer is definitely yes!

There is one difference though. While Titan's surface consists mostly of water-ice, methane-ice is only expected at high altitudes in Titan's atmosphere, where this frozen methane-snow melts to rain before hitting the surface. Even though methane-ice appears to play little role in shaping Titan, it can shape plans for future spacecraft orbits that might want to avoid this methane-snow.

Exploring Titan with *Cassini* and *Huygens*

The designers of the *Cassini* mission to Saturn—which continues until 2017—made the exploration of Titan a central goal. Carried with *Cassini* was the *Huygens* probe, which landed successfully on Titan; *Huygens* was the first landing on any object in the outer solar system. By penetrating Titan's atmosphere, *Cassini* and *Huygens* lifted a veil on the largest unexplored surface in the solar system.

Titan's atmosphere is more transparent to infrared wavelengths than to visible light, so *Cassini* has used its infrared telescope to obtain the first detailed images of Titan's surface (Figure 13-21a). Most of the surface is light-colored, but a swath of dark terrain extends around most of Titan's equator.

Cassini has used its atmosphere-penetrating radar to map the dark terrain. (Like infrared light, radio waves can pass through

(a) R I V U X G

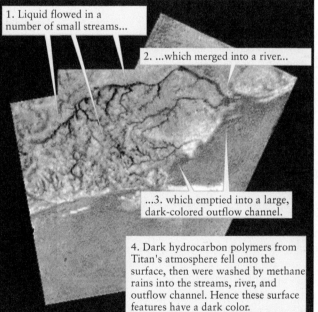

1. Liquid flowed in a number of small streams...

2. ...which merged into a river...

...3. which emptied into a large, dark-colored outflow channel.

4. Dark hydrocarbon polymers from Titan's atmosphere fell onto the surface, then were washed by methane rains into the streams, river, and outflow channel. Hence these surface features have a dark color.

(b) R I V U X G

15 cm

(c) R I V U X G

VIDEO 13-4 VIDEO 13-5 **FIGURE 13-21**

Beneath Titan's Clouds (a) This mosaic of Titan was constructed from *Cassini* images made at visible and infrared wavelengths during a close flyby in December 2005. (b) From 6.5 kilometers high, the descending *Huygens* lander recorded this view of features on Titan. Carved by flowing liquid, rivers and streams seem to have flowed between ridges about 200 meters high. It is too cold on Titan for water to be a liquid, so these streams and rivers must have carried liquid methane or ethane. (c) This nearly true-color *Huygens* image shows the view from Titan's surface. The 15-cm (6-in.) wide "rock" is about 85 cm (33 in.) from the camera. Unlike Earth rocks, Titan "rocks" are chunks of water-ice. (NASA/JPL/ESA/University of Arizona)

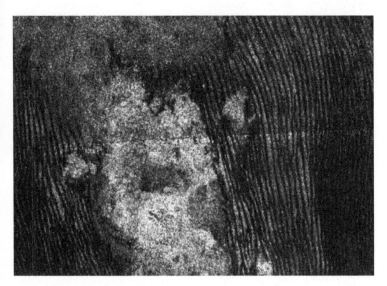

FIGURE 13-22 Ⓡ I V U X G

Sand Dunes on Titan Unlike sand on Earth, which is made of small particles of silicate rock, sand on Titan is probably made from a mixture of water-ice and organic compounds. The organic compounds give the dark sand their color, which might look like coffee grounds. A mountainous region in the center of this image prevents formation of dunes. Concentrated around the equator, at least 13 percent of Titan's total surface area is covered in dunes. (NASA/JPL/Caltech/ASI/ESA and USGS/ESA)

Titan's atmosphere.) These images reveal a series of long, parallel lines of sand dunes about 1 to 2 km apart (Figure 13-22). These are aligned west to east, in the same direction that winds blow at Titan's equator, and so presumably formed by wind action. The dunes can be 100 meters high and stretch for hundreds of kilometers.

The most remarkable images of Titan's surface have come from *Huygens,* the small landing craft that *Cassini* carried on its journey from Earth. Named for the discoverer of Titan, *Huygens* entered Titan's atmosphere on January 14, 2005. The lander took 2½ hours to descend to the surface under a parachute, during which time it made detailed images of the terrain (see Figure 13-21b). Chunks of water-ice appear on the surface like rocks (see Figure 13-21c). These are thought to come from the breakup of Titan's mostly water-ice crust. After touchdown, *Huygens* continued to return data for another 70 minutes before its batteries succumbed to Titan's low temperatures.

Huygens images such as Figure 13-21b show that liquids have indeed flowed on Titan like water on Earth. Yet none of the images made by the lander during its descent or after touchdown showed any evidence of standing liquid. However, dampness in the ground detected by *Huygens* indicates recent methane rainfall or flooding at the landing site.

What would a rainstorm be like on Titan? First, consider that the strength of a downpour on Earth depends on how much water vapor builds up in the atmosphere before larger droplets form and fall as rain. Compared to Earth, measurements indicate that Titan's atmosphere holds much more moisture before droplets of methane would precipitate out as rain. However, the evaporation rate of methane from Titan's surface is slower than the evaporation of water into Earth's atmosphere. As a result, it can take

100 to 1000 years before enough methane builds up to rain down from Titan's skies. But, when it does rain, so much methane has built up that flash floods are the likely result. The methane raindrops are also expected to be several times larger than the water drops on Earth, and to fall more slowly in Titan's lower gravity and thicker atmosphere. Grape-sized, slow-falling raindrops are nice to imagine, but keep in mind that they are similar in composition to liquefied natural gas.

Cassini Discovers Vast Lakes

Huygens' view of Titan was limited to the equatorial region where it landed. While evidence near the equator pointed to the presence of liquid in the past, it took *Cassini's* radar maps of the poles to find present-day liquid on Titan. The radar maps (Figure 13-23) have revealed the *first standing bodies of liquid* in the solar system beyond Earth. The terrain includes vast lakes that are most likely methane and ethane. If correct, evaporation from these large reservoirs could help explain how Titan's atmosphere contains so much methane even though ultraviolet sunlight breaks it apart.

How Titan's lakes were discovered is more subtle than the images imply. To analyze Titan's terrain through its atmosphere, radar waves are reflected like echoes off its surface. This works for imaging typical surfaces with mountains and canyons, but very flat surfaces reflect radar at only one angle—like a ball bouncing off a wall—and no radar echo is returned (except when *Cassini* is directly overhead). Therefore, Titan's lakes are inferred by their *flatness:* They are surface regions where very little radar echo is detected. In fact, the measurements are so precise that the lakes present another mystery: *They are too flat.*

Most large lakes have wind-driven waves, and with Titan's thick atmosphere and low gravity, wind is expected even at the poles. Indeed, we mentioned that near the equator, windblown sand dunes can reach 100 meters in height, and stretch for hundreds of kilometers (see Figure 13-22). Thus, the flatness of the polar lakes remains a mystery.

Overall, there are many more lakes near Titan's north pole than its south pole. Why? Titan's northern hemisphere is currently in its winter season. Since lower temperatures lead to more condensation, more lakes are expected. Over the course of a Titan year (which lasts 29.37 Earth years) the supply of liquid will migrate with the seasons from pole to pole, much as frozen carbon dioxide migrates seasonally on Mars (see Section 11-6).

More than the seasons however, it is expected that an even larger effect accounts for a long-term predominance of lakes in the north. Due to the shape of Saturn's orbit around the Sun, during the Titan summer the entire Saturn-Titan system is 12 percent farther from the Sun than during the winter, making Titan's north cooler, on average, than the south. Over long timescales, when you consider the combined orbital properties of Saturn and Titan, lakes are expected to predominate at one pole or another for about 32,000 years before reversing, and we just happen to be in a northern part of that cycle.

Cryovolcanoes on Titan?

Combining infrared and radar images, *Cassini* also detected surface features that indicate a **cryovolcano** (Figure 13-24). Instead of erupting molten rock, a cryovolcano (or ice volcano) on Titan

(a)

(b)

FIGURE 13-23 R I V U X G

Hydrocarbon Seas on Titan *Cassini* used its onboard radar to record these false-color views of the region near Titan's north pole. **(a)** The dark regions show where the surface does not reflect radio waves at all, which leads to the interpretation that these are flat lakes (presumably filled with liquid

methane and ethane). **(b)** Lake Ligeia Mare is one of the larger lakes and is about the size of Lake Superior (about 350 miles wide). In this high resolution image, many tributaries can be seen around the lake's perimeter. (NASA/JPL/ USGS)

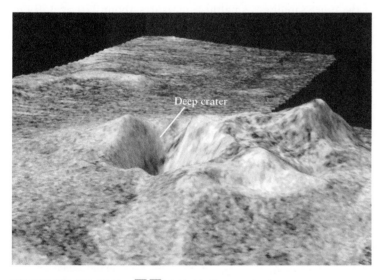

FIGURE 13-24 R I V U X G

Cryovolcanism The Sorta Facula region contains Titan's best candidate for a cryovolcano made of water-ice. Some of the taller peaks are more than 1,000 meters high, next to a deep 1,500-meter volcanic crater. There are also examples on Earth where a crater is deeper than the height of the volcano. At the moment, the cryovolcano does not appear to be active. The blue areas in this false-colored radar map are thought to be exposed water-ice, with other colors possibly due to a thin layer of organic compounds covering water-ice. (NASA/JPL-Caltech/USGS/University of Arizona)

is expected to erupt or ooze icy mixtures of water, ammonia, and methane. Acting as an "antifreeze," the ammonia (NH_3) is key, and it allows the mixture to remain a liquid within Titan's interior. This picture is quite consistent, as the ammonia (broken down by ultraviolet sunlight) was already expected as the source of Titan's atmospheric nitrogen.

In addition to evaporation from lakes, the presence of ice volcanoes on Titan can explain why there is so much methane in the satellite's atmosphere. Methane molecules are broken apart by ultraviolet photons from the Sun, and the hydrogen atoms escape into space. This process would destroy all of Titan's methane within 10 million years unless a fresh supply is added to the atmosphere, perhaps aided by cryovolcanoes. While entirely fascinating, Titan's cryovolcanism isn't entirely unique: We'll see in Section 13-10 that another moon of Saturn (Enceladus) shows ongoing cryovolcanism.

CONCEPTCHECK 13-9

Explain two roles that ammonia (NH_3) seems to play on Titan; one role on or above the surface, and one below.

Answer appears at the end of the chapter.

Titan's Geology

There are some indications that Titan might have a subsurface ocean of liquid water. Electrical measurements made by *Huygens* are consistent with the top of such an ocean about 45 kilometers beneath the surface. Furthermore, models with a subsurface ocean might provide a better fit, compared to a solid body, for Titan's carefully measured shape and gravitational field. However, due to Titan's atmosphere, *Cassini* cannot orbit closely enough to make the additional magnetic field measurements that would more clearly reveal a subsurface ocean on Titan. Why is water so interesting? As we will see in Chapter 27 on the search for extraterrestrial life, liquid water either in a subsurface ocean or associated with cryovolcanoes means that Titan could be habitable to simple forms of life.

The predicted interior of present-day Titan is shown in Figure 13-25, but what was Titan like in the past? We have seen that Titan was probably volcanically active to provide its atmosphere with methane, but what has powered the volcanic activity? Titan is too small to have retained the internal heat from its formation. Furthermore, although Titan's rotation is synchronous like that of Io, its orbit around Saturn is so circular that it cannot be heated by tidal flexing in the way that Io is (see Section 13-4).

One model to power cryovolcanism proposes that about 500 million years ago, the last of Titan's internal heat from radioactivity caused convection to take place in the satellite's icy crust. The resulting ice volcanoes would repave the entire surface, explaining Titan's unexpectedly low crater density. The amount of methane released during this proposed episode is enough to sustain the gas in the atmosphere, but—very importantly—not necessarily enough to create the extensive hydrocarbon lakes. Thus, we have much to learn about Titan's present condition, as well as its past.

What already seems clear from the data is that the early predictions were correct: Titan's methane does indeed take liquid and gas forms, producing a cycle similar to water's hydrological cycle on Earth. With its storms, lakes, and volcanoes, Titan is a truly fascinating world.

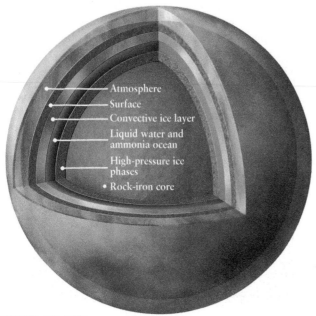

FIGURE 13-25

Titan's Interior This cross-sectional diagram is based on models of Titan's interior. As shown, there could be an outer layer of convective ice, and a liquid water subsurface ocean could extend hundreds of kilometers deep. Based on Titan's density, about half the moon is made of iron, with the other half water in various states. (Illustration by Kevin Hand for *Scientific American*)

Labels in figure:
- Atmosphere
- Surface
- Convective ice layer
- Liquid water and ammonia ocean
- High-pressure ice phases
- Rock-iron core

CONCEPTCHECK 13-10

If lakes were not found in images made by the *Huygens* probe, and *Cassini* cannot see through Titan's thick haze, how were lakes discovered?

CONCEPTCHECK 13-11

What insight from *Voyager* data led scientists to predict that Titan's methane might be involved in a cycle similar to Earth's hydrological cycle?

Answers appear at the end of the chapter.

13-9 Jupiter has dozens of small satellites that have different origins

Besides the four Galilean satellites, Jupiter has at least 63 other confirmed satellites (as of 2012). Each is named for a mythological character associated with Jupiter. Appendix 3 gives details of some of the satellites' orbits. It is helpful to sort Jupiter's satellites into three groups according to their sizes and orbits.

Four planet-sized satellites, Io, Europa, Ganymede, and Callisto, can be thought of as four of the terrestrial worlds of the solar system. All are comparable in size to the planet Mercury (see Table 13-1). The semimajor axes of their orbits range from 5.9 to 26 times the radius of Jupiter. (By comparison, the distance from Earth to the Moon is about 60 times Earth's radius.)

Four small inner satellites, Metis, Adrastea, Amalthea, and Thebe, have orbital semimajor axes between 1.8 and 3.1 times the radius of Jupiter. As Figure 13-26 shows, these satellites have irregular shapes, much like a potato (or an asteroid). The largest of the inner satellites, Amalthea, measures about 270 by 150 km (170 by 95 mi)—about 10 times larger than the moons of Mars, but only about one-tenth the size of the Galilean satellites. At these small sizes, gravity is too weak to pull the moons into a spherical shape. Discovered in 1892, Amalthea has a distinct reddish color due to sulfur that Jupiter's magnetosphere removed from Io's volcanic plumes, then deposited onto Amalthea's surface. The three other inner satellites are even smaller than Amalthea. They were first discovered in images from *Voyager 1* and *Voyager 2.*

Fifty-nine small outer satellites have diameters from 184 km for the largest to about 1 km for the smallest. All orbit far beyond Callisto, with semimajor axes that range from about 100 to 360 times the radius of Jupiter. (Twenty-three of these satellites were discovered in 2003 alone, and even more may be discovered as telescope technology continues to evolve.)

The Galilean satellites and the four inner satellites all orbit Jupiter in the plane of the planet's equator. (The same is true of Jupiter's rings,

> Only a few of Jupiter's satellites orbit the planet in the same direction as Jupiter rotates

discussed in Section 12-9.) Furthermore, these orbits are all **prograde orbits,** which means that these objects orbit Jupiter in the same direction as Jupiter's rotation. We would expect orbits in the same direction if these satellites formed from the same primordial, rotating cloud as Jupiter itself. (In a similar way, the planets all orbit the Sun in nearly the same plane and in the same direction as the Sun's rotation. As we saw in Section 8-4, this reinforces the idea that the Sun and planets formed from the same solar nebula.)

FIGURE 13-26   R I V U X G

Jupiter's Four Inner Satellites First observed in 1892, Amalthea was the last satellite of any planet to be discovered without the aid of photography. Metis, Adrastea, and Thebe, by contrast, are so small that they were not discovered until 1979. These images from the *Galileo* spacecraft are the first to show these three satellites as anything more than points of light. (NASA/JPL; Cornell University)

In contrast, the 59 outer satellites all circle Jupiter along orbits that are inclined at steep angles to the planet's equatorial plane. These satellites are thought not to have formed along with Jupiter; rather, they are probably wayward asteroids captured by Jupiter's powerful gravitational field. One piece of evidence for this is that most of the outer satellites are in **retrograde orbits**: They orbit Jupiter in a direction opposite the planet's rotation. Calculations show that it is easier for Jupiter to capture a satellite into a retrograde orbit than a prograde one.

CONCEPTCHECK **13-12**

Why does a retrograde orbit suggest that a moon is a captured asteroid?

Answer appears at the end of the chapter.

13-10 The icy surfaces of Saturn's six moderate-sized moons provide clues to their histories

Saturn, too, has an extensive collection of satellites: 62 are known as of 2012. As we do for Jupiter's satellites in Section 13-9, we can subdivide the satellites of Saturn into three categories based on their sizes and their orbits. (Appendix 3 has information about the orbits of all of Saturn's satellites.) We will see that while Saturn does not have a set of large moons like the Galilean satellites of Jupiter, it does have a remarkably diverse collection of moderate-sized satellites.

Saturn's Retinue of Moons

One planet-sized satellite, Titan, is actually one of the terrestrial worlds of our solar system. With its diameter of 5150 km, Titan is intermediate in size between Mercury and Mars. Its prograde orbit has a semimajor axis of just over 20 times the radius of Saturn and lies in nearly the same plane as Saturn's equator and rings.

VIDEO 13-6 *Six moderate-sized satellites* were all discovered before 1800. These six satellites have properties and diameters that lead us to group them in pairs: Mimas and Enceladus (400 to 500 km in diameter), Tethys and Dione (roughly 1000 km in diameter), and Rhea and Iapetus (about 1500 km in diameter). The semimajor axes of their orbits range from 3.1 to 59 times the radius of Saturn. Like Titan, all of their orbits are prograde and close to the plane of Saturn's equator.

Fifty-five small satellites have sizes from 3 to 266 km. Some are in prograde orbits as small as 2.2 times the radius of Saturn, and lie within or close to the rings (see Figure 12-26). These satellites may be jagged fragments of ice and rock produced by impacts and collisions. The shepherd satellites that shape Saturn's F ring (see Figure 12-25) may be such fragments. Others are in retrograde orbits with semimajor axes as large as 380 times Saturn's radius and may be captured asteroids. A little farther out than Titan, *Cassini* made a close flyby of one strange moon named Hyperion (**Figure 13-27**).

CONCEPTCHECK **13-13**

In Figure 13-27, Hyperion appears almost like a sponge on the outside. Why would we, in fact, expect a lot of empty space throughout Hyperion?

Answer appears at the end of the chapter.

Six Strange Satellites

The most curious and mysterious of these worlds may be Saturn's six moderate-sized satellites; four of these are shown

FIGURE 13-27   R I V U X G

Hyperion This heavily pockmarked moon is about 410 kilometers across its elongated axis. Its density is about *half* that of water, so it is probably very porous. Rather than rotating around a single axis, the gravitational tugs on its elongated body by Titan and Saturn lead to a non-repeating chaotic tumble. (NASA/JPL/Space Science Institution)

(a) Mimas
(diameter 392 km)

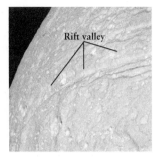

(b) Tethys
(diameter 1060 km)

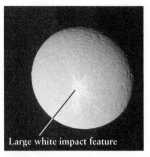

(c) Dione
(diameter 1120 km)

(d) Rhea
(diameter 1530 km)

FIGURE 13-28   R I V U X G

Saturn's Mid-Sized Satellites **(a–d)** These images from *Cassini* show the variety of surface features found on four of Saturn's six moderate-sized satellites. The satellites depicted are not shown to scale. (NASA/JPL/Space Science Institute)

in Figure 13-28. The six satellites have very low average densities of less than 1400 kg/m³. This implies that unlike the satellites of the inner solar system or the Galilean satellites of Jupiter, they are composed primarily of ices (mostly frozen water and ammonia) with very little rock. The motions of these mid-sized satellites are not surprising: All six orbit the planet in the plane of Saturn's equator and in the direction of the planet's rotation, and all six exhibit synchronous rotation. What is surprising is the curious variety of surface features that the *Voyager* and *Cassini* spacecraft discovered on these satellites.

Mimas, the innermost of the six middle-sized moons, has the heavily cratered surface we would expect on such a small world. Its most remarkable feature is a crater so large (Figure 13-28a) that the impact that formed it must have come close to shattering Mimas into fragments.

Most of Tethys, like Mimas, is heavily cratered. But Tethys also has an immense rift valley called Ithaca Chasma that extends for 2000 km, or about ¾ of the way around Tethys (Figure 13-28b). This rift may have formed when the water inside Tethys cooled and solidified, or it may be the result of a shock wave that traveled through Tethys after an immense asteroid impact. (A large impact crater lies at the antipode, which is on the side of Tethys opposite Ithaca Chasma.)

Dione, slightly larger than Tethys, has a strange surface dichotomy: The surfaces of its leading hemisphere (that is, the hemisphere facing toward its direction of orbital motion) and trailing hemisphere are quite different. Dione's leading hemisphere is heavily cratered, but its trailing hemisphere is covered by a network of bright stress cracks caused by some sort of tectonic activity (Figure 13-28c). Like Dione, the moon Rhea also has more craters on its leading hemisphere (Figure 13-28d).

Enceladus has become one of the most interesting moons in the entire solar system (Figure 13-29). *Cassini* discovered that Enceladus is volcanically active, though with ice taking the place of lava. While the moon Titan shows evidence of ice-volcano formations (Figure 13-24), *Cassini* has caught the cryovolcanoes on Enceladus in the act of erupting! Figure 13-29b shows several distinct plumes ejecting water vapor

> Although just 500 km across, Enceladus has powerful ice-volcanoes at its south pole

and ice from the features referred to as "tiger stripes" near the satellite's south pole. By measuring the distribution of speeds for water-ice particles ejected from Enceladus, high-speed ejecta were found that could escape Enceladus's gravity to leave behind a ring in its orbit (Figure 13-29c). This is Saturn's E ring (Figure 12-24) and solves a mystery about how this ring formed.

At first there was a major debate as to whether the ejected water came from a small patch of melted ice localized under the tiger stripes, or whether the water came from a global subsurface ocean. *Cassini* has shed light on this question when passing directly through the plumes and analyzing their content. By detecting dissolved minerals (or "salts") within the water, *Cassini* data indicate that the water steadily dissolves a rocky mineral interior or mantle (just as Earth's oceans do), which strongly supports the model of a large subsurface ocean.

With ocean water conveniently shot into space by Enceladus for our sampling and analysis, some scientists argue that Enceladus is the solar system's best place to search for microbial life beyond Earth (in a future mission). Recall that Europa's oceans are probably buried under 10 km of ice, making the liquid water very difficult to sample. We'll consider this further in Chapter 27 (The Search for Extraterrestrial Life).

What could be powering the geologic activity on Enceladus? One guess is tidal heating, which provides the internal heat for the inner Galilean satellites of Jupiter (see Section 13-4). The satellite Dione orbits Saturn in 65.7 hours, almost exactly twice Enceladus's orbital period of 32.9 hours, and the 1:2 ratio of orbital periods should set Enceladus up to be caught in a tidal tug-of-war between Saturn and Dione. Calculations suggest that the amount of tidal heating produced in this way is modest, but should be enough to melt the ices of Enceladus at least part of the time.

Another mysterious moon of Saturn, Iapetus, is also associated with a newly discovered ring of Saturn (Figure 13-30). Iapetus has been known to be unusual ever since its discovery in 1671: Unlike the other satellites, its brightness varies greatly as it orbits Saturn. Images from *Cassini* show extreme differences between the leading and trailing hemispheres of Iapetus (Figure 13-30a); the leading side faces in the direction of Iapetus's orbit. Most of the leading hemisphere is as black as asphalt, but the trailing hemisphere is highly reflective, like the surfaces of the other moderate-sized moons.

(a)

(b)

Tiger stripes

Water-ice plumes from cryovolcanoes

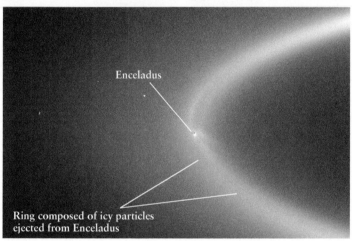

Enceladus

Ring composed of icy particles ejected from Enceladus

(c)

FIGURE 13-29   R I [V] **U X G**

Cryovolcanoes on Enceladus **(a)** This *Cassini* image reveals an icy Enceladus. The lack of significant cratering suggests that some geologic process resurfaced the moon within the past 100 million years. **(b)** Plumes of water vapor and ice are erupting from cryovolcanoes associated with the "tiger stripes" near the south pole of Enceladus. The faint plumes are best seen when they are backlit by the Sun, so Enceladus appears as a crescent. **(c)** A backlit view from *Cassini* shows how particles ejected from Enceladus become part of Saturn's E ring, within which Enceladus orbits (see Figure 12-24). (NASA/JPL/Space Science Institute)

What could account for the two-sided nature of Iapetus? Models for Iapetus suggest that a small initial difference in coloration between the two sides would be amplified. A slightly darker surface (less reflective) would absorb more sunlight, warm up more, and evaporate more ice. The residue left behind would leave the surface even darker and therefore warmer, amplifying the effect even further. In 2009, this idea gained support when a new ring of Saturn was discovered near to Iapetus that could provide the initial debris to start the runaway process of darkening (Figure 13-30b).

CONCEPTCHECK **13-14**

Some early models to explain water ejected from Enceladus avoided reliance on a global subsurface ocean (although a large ocean is certainly the more exciting case), and instead considered smaller pockets of melted water. What subsequent data shifted models to a large ocean, and why?

Answer appears at the end of the chapter.

(a)   **R** [I] [V] **U X G**

(b)   **R** [I] **V U X G**

FIGURE 13-30

The Two Faces of Iapetus **(a)** This color-enhanced *Cassini* image shows nonreflective material on leading face of Iapetus. The contrast is not very strong in true color, but the contrast is very strong when comparing reflectivity: The lighter side reflects like somewhat dirty ice, whereas the darker side reflects like black charcoal. **(b)** The leading model to explain the two faces requires debris to create

some initial color differences, which is thought to come from a very large ring observed in the infrared by the Spitzer Space Telescope. Discovered in 2009, this ring has a diameter about 300 times larger than Saturn. This is an artist's illustration of the ring, with an inset of Saturn in the infrared. (a: NASA/JPL/Space Science Institute; b: NASA/JPL-Caltech/R. Hurt (SSC)

KEY WORDS

bright terrain (Ganymede),
  p. 363
cryovolcano, p. 368
dark terrain (Ganymede),
  p. 363
Galilean satellites, p. 352
hydrocarbon, p. 366
ice rafts (Europa), p. 361
induced electric current, p. 359

induced magnetic field, p. 359
Io torus, p. 359
occultation, p. 353
orbital resonance, p. 352
plasma, p. 359
prograde orbit, p. 370
retrograde orbit, p. 371
tidal heating, p. 356

KEY IDEAS

Nature of the Galilean Satellites: The four Galilean satellites orbit Jupiter in the plane of its equator. All are in synchronous rotation.

• The orbital periods of the three innermost Galilean satellites, Io, Europa, and Ganymede, are in the ratio 1:2:4. This forms an orbital resonance.

• The two innermost Galilean satellites, Io and Europa, have roughly the same size and density as our Moon. They are composed principally of rocky material. The two outermost Galilean satellites, Ganymede and Callisto, are roughly the size of Mercury. Lower in density than either the Moon or Mercury, they are made of roughly equal parts ice and rock.

• The Galilean satellites probably formed in a similar fashion to planets in our solar system but on a smaller scale.

Io: Io is covered with a colorful layer of sulfur compounds deposited by frequent explosive eruptions from volcanic vents. These eruptions resemble terrestrial geysers.

• The energy to heat Io's interior and produce the satellite's volcanic activity comes from tidal forces that flex the satellite. This tidal flexing is aided by an orbital resonance: the 1:2:4 ratio of orbital periods among the inner three Galilean satellites.

• The Io torus is a ring of electrically charged particles circling Jupiter at the distance of Io's orbit.

• When Jupiter's magnetic field sweeps over the Galilean satellites, it induces electric currents, which in turn, create an induced magnetic field.

Europa: While composed primarily of rock, Europa is covered with a smooth layer of water-ice.

• The surface has hardly any craters, indicating a geologically active history. Other indications are a worldwide network of long cracks and ice rafts that indicate a subsurface layer of liquid water or soft ice. As with Io, tidal heating and an orbital resonance is responsible for Europa's internal heat.

• There is probably an ocean beneath Europa's frozen surface. Minerals dissolved in this ocean may explain Europa's induced magnetic field.

Ganymede: Two types of terrain are found on the icy surface of Ganymede: areas of dark, ancient, heavily cratered surface and regions of heavily grooved, lighter-colored, younger terrain.

• Ganymede is highly differentiated and probably has a metallic core. It has a surprisingly strong magnetic field and a magnetosphere of its own.

• While there is at present little tidal heating of Ganymede, it may have been heated in this fashion in the past. An induced magnetic field suggests that it, too, has a layer of liquid water beneath the surface.

Callisto: Callisto has a heavily cratered crust of water-ice. The surface shows little sign of geologic activity, because there was never any significant tidal heating of Callisto. However, some unknown processes have erased the smallest craters and blanketed the surface with a dark, dusty substance.

• Magnetic field data seem to suggest that Callisto has a shallow subsurface ocean.

Titan: The largest Saturnian satellite, Titan, is a terrestrial world with a dense nitrogen atmosphere and an important 5 percent methane. Temperatures and pressures on Titan are near the point where methane can be found in a solid, liquid, or gas phase.

• Titan shows evidence for flooding expected from methane rain, and many lakes near its north pole. Methane appears to work in a cycle similar to water in Earth's hydrological cycle.

Other Satellites: As of 2012, Jupiter has a total of 67 confirmed satellites and Saturn has a total of 62.

• Beyond the Galilean satellites, Jupiter has many small satellites that move in much larger orbits that are noticeably inclined to the plane of Jupiter's equator. Many of these orbit in the direction opposite to Jupiter's rotation. These do not appear to have formed with Jupiter, and they are thought to be captured asteroids.

• In addition to Titan, six moderate-sized moons circle Saturn in regular orbits: Mimas, Enceladus, Tethys, Dione, Rhea, and Iapetus. They are probably composed largely of ice, but their surface features and histories vary significantly. The other, smaller moons include shepherd satellites that control the shapes of Saturn's rings and captured asteroids in large retrograde orbits.

• Enceladus ejects plumes of water from a region near its south pole. Minerals detected in these plumes indicate a subsurface ocean in contact with a rocky mantle.

QUESTIONS

Review Questions

1. *TUTORIAL 13-1* If you observed the Galilean satellites through a telescope for a single night, could you notice their motions around Jupiter?

2. Why can't the Galilean satellites be seen with the naked eye?

3. During the time it takes Ganymede to complete one orbit, how many orbits do Io and Europa complete?

4. In what ways does the system of Galilean satellites resemble our solar system? In what ways is it different?

5. No spacecraft from Earth has ever landed on any of the Galilean satellites. How, then, can we know if they are made of rock, water, or a combination of the two?

6. In what ways did the formation of the Galilean satellites mimic the formation of the planets? In what ways were the two formation processes different?

7. In the classic science-fiction film *2010: The Year We Make Contact,* an alien intelligence causes Jupiter to contract so much that nuclear reactions begin at its center. As a result, Jupiter becomes a star like the Sun. Is this possible in principle? Explain your answer.

8. All of the Galilean satellites orbit Jupiter in the same direction. Furthermore, the planes of their orbits all lie within 0.5° of Jupiter's equatorial plane. Explain why these observations are consistent with the idea that the Galilean satellites formed from a "Jovian nebula."

9. Most of Jupiter's smaller moons orbit in a direction opposite to Jupiter's rotation. What does this tell us about those moons?

10. What is the source of energy that powers Io's volcanoes? How is it related to the orbits of Io and the other Galilean satellites?

11. Io has no impact craters on its surface, while our Moon is covered with craters. What is the explanation for this difference?

12. Despite all the gases released from its interior by volcanic activity, Io does not possess a thick atmosphere. Explain why not.

13. Long before the *Voyager* flybys, Earth-based astronomers reported that Io appeared brighter than usual for the few hours after it emerged from Jupiter's shadow. From what we know about the material ejected from Io's volcanoes, suggest an explanation for this brief brightening of Io.

14. How do lavas on Io differ from typical lavas found on Earth? What does this difference tell us about Io's interior?

15. What accounts for the different colors found on Io's surface?

16. What is the Io torus? What is its source?

17. What is the origin of the electric current that flows through Io?

18. How was the *Galileo* spacecraft used to determine the internal structure of Io and the other Galilean satellites?

19. What surface features on Europa provide evidence for geologic activity?

20. What is the evidence for an ocean of liquid water beneath Europa's icy surface? What is the evidence that substances other than water are dissolved in this ocean?

21. What aspects of Europa lead scientists to speculate that life may exist there? What precautions have been taken to prevent contaminating Europa with Earth organisms?

22. Why is ice an important constituent of Ganymede and Callisto, but not of Earth's Moon?

23. How do scientists know that the dark terrain on Ganymede is older than the bright terrain?

24. In what ways is Ganymede like our own Moon? In what ways is it different? What are the reasons for the differences?

25. Why were scientists surprised to learn that Ganymede has a magnetic field? What does this field tell us about Ganymede's history?

26. What leads scientists to suspect that there is liquid water beneath Ganymede's surface?

27. Why are numerous impact craters found on Ganymede and Callisto but not on Io or Europa?

28. Describe the surprising aspects of Callisto's surface and interior that were revealed by the *Galileo* spacecraft. Why did these come as a surprise?

29. Compare and contrast the surface features of the four Galilean satellites. Discuss their relative geological activity and the evolution of these four satellites.

30. The larger the orbit of a Galilean satellite, the less geologic activity that satellite has. Explain why.

31. Explain how orbital resonance from the 1:2:4 ratio of the orbital periods of Io, Europa, and Ganymede is related to the geologic activity on these satellites.

32. *TUTORIAL 13-2* Describe Titan's atmosphere. Explain how the Sun's ultraviolet radiation might have altered Titan's original atmosphere?

33. In what ways do hydrocarbons like methane on Titan behave like water on Earth?

34. Why does the presence of methane in Titan's atmosphere imply that Titan has had recent volcanic activity, or has some other source?

35. What is a cryovolcano? How can water be anything but frozen solid in Titan's low temperatures?

36. What is the evidence that there were liquid hydrocarbons on Titan in the past? That there are liquid hydrocarbons now?

37. Is the bulk of Titan's surface made of water-ice or methane-ice?

38. How would you account for the existence of the satellites of Jupiter other than the Galilean ones?

39. Which of Saturn's moderate-sized satellites show evidence of geologic activity? What might be the energy source for this activity?

40. Explain the present-day difference in the reflectivity on parts of Iapetus. Explain what might cause this if Saturn's largest ring—the newly discovered ring around Saturn in Figure 13-30—produces an initial difference on Iapetus's surface.

41. Saturn's equator is tilted by 27° from the ecliptic, while Jupiter's equator is tilted by only 3°. Use these data to explain why we see fewer transits, eclipses, and occultations of Saturn's satellites than of the Galilean satellites.

Advanced Questions

Questions preceded by an asterisk (*) involve topics discussed in Box 1-1 or Box 7-2.

> **Problem-solving tips and tools**
>
> Newton's form of Kepler's third law (Box 4-4) relates the masses of two objects in orbit around each other to the period and size of the orbit. The small-angle formula is discussed in Box 1-1. The best seeing conditions on Earth give a limiting angular resolution of ¼ arcsec. Because the orbits of the Galilean satellites are almost perfect circles, you can easily

(continued on page 376)

(continued from page 375)
calculate the orbital speeds of these satellites from the data listed in Table 13-1. Data about Jupiter itself are given in Table 12-1 and data about all of the satellites of Jupiter and Saturn can be found in Appendix 3. For a discussion of escape speed and how planets retain their atmospheres, see Box 7-2.

42. Using the orbital data in Table 13-1, demonstrate that the Galilean satellites obey Kepler's third law.

*43. What is the size of the smallest feature you should be able to see on a Galilean satellite through a large telescope under conditions of excellent seeing when Jupiter is near opposition? How does this compare with the best *Galileo* images, which have resolutions around 25 meters?

44. Using the diameter of Io (3642 km) as a scale, estimate the height to which the plume of Pele rises above the surface of Io in Figure 13-5a. (You will need to make measurements on this figure using a ruler.) Compare your answer to the value given in the figure caption.

45. Explain why the volcanic plumes in Figure 13-5b have a bluish color. (*Hint:* See Box 5-4.)

46. Jupiter, its magnetic field, and the charged particles that are trapped in the magnetosphere all rotate together once every 10 hours. Io takes 1.77 days to complete one orbit. Using a diagram, explain why particles from Jupiter's magnetosphere hit Io primarily from behind (that is, on the side of Io that trails as it orbits the planet).

47. Assuming material is ejected from Io into Jupiter's magnetosphere at the rate of 1 ton per second (1000 kg/s), how long will it be before Io loses 10 percent of its mass? How does your answer compare with the age of the solar system?

48. Volcanic eruptions from Io add about 1 ton of material to Jupiter's magnetosphere every second. Based on the information given in Section 13-4, what fraction of the material ejected from Io's volcanoes goes into Jupiter's magnetosphere?

49. How long does it take for Ganymede to entirely enter or entirely leave Jupiter's shadow? Assume that the shadow has a sharp edge.

50. Figure 13-16 shows that Ganymede's dark terrain has both more craters and larger craters than the bright terrain. Explain what this tells you about the sizes of meteoroids present in the solar system in the distant past and in the more recent past.

51. Put a few cubes of ice in a glass, then fill the glass with warm water to make the ice cubes crack. The fracture lines in each cube are clearly visible because they reflect light. Use this observation to suggest why Ganymede's icy bright terrain, which may have been badly fractured by the stresses that produced the grooves (Figure 13-16), is 25 percent more reflective than the dark terrain.

*52. Find the escape speed on Titan. What is the limiting molecular weight of gases that could be retained by Titan's gravity?

(*Hint:* Use the ideas presented in Box 7-2 and assume an average atmospheric temperature of 95 K.)

53. Many of the gases in the atmosphere of Titan, such as methane, ethane, and acetylene, are highly flammable. Why, then, doesn't Titan's atmosphere catch fire? (*Hint:* What gas in our atmosphere is needed to make wood, coal, or gasoline burn?)

54. Suppose Earth's Moon were removed and replaced in its orbit by Titan. What changes would you expect to occur in Titan's atmosphere? Would solar eclipses be more or less common as seen from Earth? Explain your answers.

55. At infrared wavelengths, the Hubble Space Telescope can see details on Titan's surface as small as 580 km (360 mi) across. Determine the angular resolution of the Hubble Space Telescope using infrared light. If visible light is used, is the angular resolution better, worse, or the same? Explain your answer.

56. (a) To an observer on Enceladus, what is the time interval between successive oppositions of Dione? Explain your answer. (b) As seen from Enceladus, what is the angular diameter of Dione at opposition? How does this compare to the angular diameter of the Moon as seen from Earth (about ½°)?

Discussion Questions

57. If you could replace our Moon with Io, and if Io could maintain its present amount of volcanic activity, what changes would this cause in our nighttime sky? Do you think that Io could in fact remain volcanically active in this case? Why or why not?

58. Speculate on the possibility that Europa, Ganymede, or Callisto might harbor some sort of life. Explain your reasoning.

59. Comment on the suggestion that Titan may harbor life-forms.

60. Jupiter's satellite Io and Saturn's satellite Enceladus are both geologically active, and both are in 2-to-1 resonances with other satellites. However, the amount of geologic activity of Enceladus is far less than on Io. Discuss some possible reasons for this difference.

61. Imagine that you are in charge of planning a successor to the *Cassini* spacecraft to further explore the Saturnian system. In your opinion, which satellites in the system should be examined more closely? What data should be collected? What kinds of questions should the new mission attempt to answer?

Web/eBook Questions

62. Various spacecraft missions have been proposed to explore Europa in greater detail. Search the World Wide Web for information about these. How would these missions test for the presence of an ocean beneath Europa's surface?

63. Two new satellites of Jupiter were discovered in 2011. Search the World Wide Web for information about these satellites. What do the orbital characteristics of these moons tell you about their origins?

ACTIVITIES

Observing Projects

> **Observing tips and tools**
>
> You can easily find the apparent positions of the Galilean satellites for any date and time using the *Starry Night*™ software if you have access. For even more detailed information about satellite positions, consult the "Satellites of Jupiter" section in the *Astronomical Almanac* for the current year. If your goal is to view Saturn's satellites, consult the section entitled "Satellites of Saturn" in the *Astronomical Almanac* for the current year. This includes a diagram showing the orbits of Mimas, Enceladus, Tethys, Dione, Rhea, Titan, and Hyperion. Plan your observing session by looking up the dates and times of the most recent greatest eastern elongations of the various satellites. You will have to convert from universal time (UT), also known as Greenwich Mean Time, to your local time zone. Then, using the tick marks along the orbits in the diagram, estimate the positions of the satellites relative to Saturn at the time you will be at the telescope. Another useful resource is the "Celestial Calendar" section of *Sky & Telescope.* During months when Saturn is visible in the night sky, this section of the magazine includes a chart of Saturn's satellites.

64. Observe Jupiter through a pair of binoculars. Can you see all four Galilean satellites? Make a drawing of what you observe. If you look again after an hour or two, can you see any changes?

65. Observe Jupiter through a small telescope on three or four consecutive nights. Make a drawing each night showing the positions of the Galilean satellites relative to Jupiter. Record the time and date of each observation. Consult the sources listed above in the "Observing tips and tools" to see if you can identify the satellites by name.

66. If you have access to a moderately large telescope, make arrangements to observe several of Saturn's satellites. At the telescope, you should have no trouble identifying Titan. Tethys, Dione, and Rhea are about one-sixth as bright as Titan and should be the next easiest satellites to find. Can you confidently identify any of the other satellites?

67. Use *Starry Night*™ to observe the moons of Jupiter. Select **Favourites > Explorations > Galilean Moons** from the menu. (a) With the **Time Flow Rate** set to **2 hours**, use the **Step time forward** button (just to the right of the **Play** button) to observe and draw the positions of the moons relative to Jupiter at 2-hour intervals. From your drawings, which moon orbits closest to Jupiter and which orbits farthest away? Explain your reasoning. (b) Determine the periods of orbits of these moons (change the **Time Flow Rate** if necessary). (c) Are there times when one or more of the satellites are not visible? What happens to the moons at those times?

68. Use *Starry Night*™ to examine some of the satellites of Saturn. Open **Favourites > Explorations > Saturn's Moons**. The view shows Saturn from a position in space hovering above the planet's equator in the plane of the rings. Many of Saturn's moons are labeled and their orbits are also shown. You can use the elevation controls in the toolbar (the buttons to the left of the **Home** button) to move closer to and further away from Saturn. This allows you to identify the satellites of Saturn, since they will appear to move whereas the distant stars will remain stationary as you move with respect to the planet. Alternately, you can reduce the confusion in identifying these moons by clicking on **View > Stars > Stars** to remove the stars from the view. You can also use the location scroller to obtain a different perspective and help you to identify more of Saturn's moons. Finally, you can position the cursor near to the planet precisely in the plane of the rings and when one of the inner moons, which are dimmer and more difficult to see, passes the cursor, the **HUD** will briefly display its name. You should be able to find at least eight satellites. (a) Which satellites are these? (b) Saturn's rings are in the same plane as Saturn's equator. Which satellite's orbit appears to be the farthest from this plane?

69. Use the *Starry Night*™ program to view Saturn from the surface of its satellite Enceladus. Open **Favourites > Explorations > Enceladus** from the menu. The view shows Saturn as it might appear if you were observing it from the crater named Ahmad on the surface of Enceladus. With the **Time Flow Rate** at **1 minute**, click the **Play** button and observe the position of Saturn in the sky as seen from Enceladus. (a) How do the stars appear to move? (b) How does Saturn appear to move in the sky? (c) What do these observations suggest about the orbital and rotational periods of Enceladus?

70. Use *Starry Night*™ to explore the orbits of some of Saturn's smaller outer satellites. Open the view **Favourites > Explorations > Small Outer Saturn Moons** from the menu. This view from a location in space hovering above Saturn in the plane of the rings shows the orbits of a number of the smaller, outer moons of the planet. For comparison, the orbit of Titan is also shown in the view. (a) What are some of the striking differences between the orbits of these smaller outer moons and that of Titan? (b) What hypothesis might account for these differences?

Collaborative Exercises

71. Imagine that scientists are proposing to send a robotic lander to Jupiter's satellite Callisto. Create a 100-word written proposal describing a robotic lander mission to another of the Galilean satellites, explaining why your group found it to be the most interesting and why the government should allocate the money for your alternative project. In your proposal, be sure to demonstrate your knowledge of Callisto and the other satellite.

72. From the data and the accompanying images for Table 13-1: Jupiter's Galilean Satellites Compared with the Moon, Mercury, and Mars, use someone's shoe to represent the 150,000-km diameter of Jupiter and determine about how many "shoes" away would each of the Galilean satellites be from Jupiter.

ANSWERS

ConceptChecks

ConceptCheck 13-1: You can determine the diameter of the satellite, even though you cannot see the physical extent of the satellite. As the satellite is increasingly blocked or occulted by the planet (from our view), we receive less and less of the satellite's reflected light. The time it takes for the satellite's light to completely fade away tells you the transit time, and, combined with the satellite's speed, the diameter can be calculated.

ConceptCheck 13-2: No. In order to measure the masses of Jupiter's moons, the gravitational deflection of a spacecraft by these moons had to be measured.

ConceptCheck 13-3: Because the average density of each moon decreases with distance from Jupiter (Table 13-1). This implies a warmer interior in an early Jovian nebula, where rock and metal grains would form the bulk of solid moon-forming material. Farther away from Jupiter, water would freeze with solid grains and help build moons with a lower average density. This is analogous to the formation of the solar system discussed in Section 8-5.

ConceptCheck 13-4: It is not enough for Jupiter to exert a constant tidal force on Io; that would be like squeezing a ball into one deformed shape and keeping it there. A varying tidal force is necessary, and this arises due to Europa and Ganymede rhythmically tugging the long axis of Io's orbit back and forth by about ½°.

ConceptCheck 13-5: No. Jupiter has its own magnetic field that comes from electric currents within its interior. As Jupiter's field sweeps over Io, electric currents and a magnetic field are induced. This induced magnetic field is a separate and much weaker field than Jupiter's.

ConceptCheck 13-6: The orbital resonance of Io, Europa, and Ganymede causes variations in the tidal forces exerted onto these moons by Jupiter. These tidal forces generate heat, but are stronger on Io because it is closer to Jupiter than Europa.

ConceptCheck 13-7: The induced magnetic field comes from the induced electric current in Europa created by Jupiter. For this current to flow through Europa's water-interior, the water must be liquid and therefore not solid ice. When water comes in contact with rock (probably Europa's rocky mantle), the water dissolves some of the rock's minerals, making the water electrically conductive. Modeling of Europa's induced magnetic field points to a large subsurface ocean.

ConceptCheck 13-8: Callisto has an unexpected induced magnetic field: unexpected because this indicates a liquid ocean, yet at Callisto's temperature any interior water should be frozen. However, ammonia allows water to remain a liquid at temperatures where pure water would freeze.

ConceptCheck 13-9: Ninety-five percent of Titan's atmosphere is nitrogen, which is thought to come from either ultraviolet sunlight breaking down ammonia or comet impacts generating nitrogen from ammonia. Ammonia in water also lowers the temperature at which that water will freeze, and this allows water below the surface to act like lava in cryovolcanoes.

ConceptCheck 13-10: *Cassini* discovered lakes using radar. By reflecting radio waves, radar can be used to map the height of surface features. The lakes show up as very flat regions, and tributary channels support this interpretation.

ConceptCheck 13-11: *Voyager* found that temperatures and pressures on Titan are just right for methane to exist as a gas, liquid, or solid. Titan's methane cycle depends on these changes between liquid and gas, just as Earth's hydrologic cycle depends on changes between liquid and gas phases of water.

ConceptCheck 13-12: If Jupiter formed like a mini solar system as expected, any moon that formed with Jupiter would have a prograde orbit (in the same directions as Jupiter's rotation), and would also orbit approximately in Jupiter's equatorial plane. The retrograde moons (orbiting in the direction opposite to Jupiter's rotation at high inclinations to Jupiter's equator) must have a different origin, and the capture of asteroids is most likely.

ConceptCheck 13-13: Hyperion's average density is less than half that of water. The only obvious way to account for Hyperion's size then, is to assume it is very "porous" and has a lot of empty space inside.

ConceptCheck 13-14: *Cassini* passed through the plumes and detected dissolved minerals (or salts). The simplest interpretation is that a large subsurface ocean partially dissolves minerals from a rocky mantle.

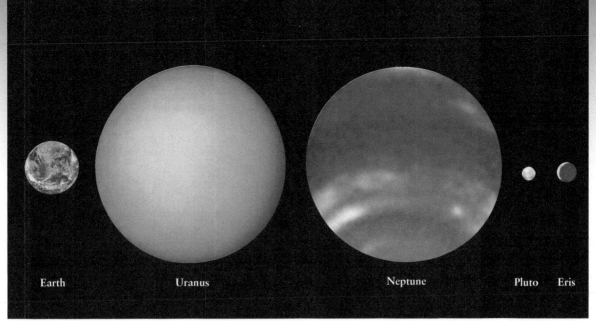

Earth, Uranus, Neptune, Pluto, and Eris to scale. (Alan Stern, Southwest Research Institute; Marc Buie, Lowell Observatory; NASA; and ESA)

R I V U X G

Uranus, Neptune, Pluto, and the Kuiper Belt: Remote Worlds

LEARNING GOALS

By reading the sections of this chapter, you will learn

14-1   How Uranus and Neptune were discovered

14-2   The unusual properties of the orbit and atmosphere of Uranus

14-3   What gives Neptune's clouds and atmosphere their distinctive appearance

14-4   The internal structures of Uranus and Neptune

14-5   The unique orientations of the magnetic fields of Uranus and Neptune

14-6   Why the rings of Uranus and Neptune are hard to see

14-7   What Uranus's larger moons tell us about the planet's history

14-8   Why Neptune's satellite Triton is destined to be torn apart

14-9   How Pluto came to be discovered

14-10   Why Pluto is no longer classified as a planet

Beyond Saturn, in the cold, dark recesses of the solar system, orbit three worlds that have long been shrouded in mystery. These worlds—Uranus, Neptune, and Pluto (all shown here to the same scale as Earth)—are so distant, so dimly lit by the Sun, and so slow in their motion against the stars that they were unknown to ancient astronomers and were discovered only after the invention of the telescope. Even then, little was known about Uranus and Neptune until *Voyager 2* flew past these planets during the 1980s.

Surrounded by a system of small moons and thin, dark rings, Uranus is tipped on its side so that its axis of rotation lies nearly in its orbital plane. This remarkable orientation suggests that Uranus may have been the victim of a staggering impact by a massive planetesimal. Neptune is, at first glance, a denser, more massive version of Uranus, but it is a far more active world. It has an internal energy source that Uranus seemingly lacks, as well as atmospheric bands and storm activity resembling those on Jupiter.

Neptune also has dark rings, small, icy moons, and an intriguing large satellite, Triton, which is nearly devoid of impact craters and has geysers that squirt nitrogen-rich vapors. Triton's retrograde orbit suggests that this strange world may have been gravitationally captured by Neptune.

Close cousins of Triton are Pluto and thousands of other trans-Neptunian objects, among the most remote members of the solar system. These objects harbor many mysteries, in part because they have not yet been visited by spacecraft.

TUTORIAL 14-1

14-1 Uranus was discovered by chance, but Neptune's existence was predicted by applying Newtonian mechanics

Before the eighteenth century, only six planets were known to orbit the Sun: Mercury, Venus, Earth, Mars, Jupiter, and Saturn. Another planet was discovered by William Herschel, a German-born musician who moved to England in 1757 and became fascinated by astronomy.

Discovering the "Georgian Star"

Using a telescope that he built himself, Herschel was systematically surveying the sky on March 13, 1781, when he noticed a faint, fuzzy object that he first thought to be a distant comet. By the end of 1781, however, his observations had revealed that the object's orbit was relatively circular and was larger than Saturn's orbit. Comets, by contrast, can normally be seen only when they follow elliptical orbits that bring them much closer to the Sun.

Herschel had discovered the seventh planet from the Sun. In doing so, he had doubled the radius of the known solar system from 9.5 AU (the semimajor axis of Saturn's orbit) to 19.2 AU (the distance from the Sun to Uranus). Herschel originally named his discovery Georgium Sidus (Latin for "Georgian star") in honor of the reigning monarch, George III. The name Uranus—in Greek mythology, the personification of Heaven—came into currency only some decades later.

Although Herschel received the credit for discovering Uranus, he was by no means the first person to have seen it. At opposition, Uranus is just barely bright enough to be seen with the naked eye under good observing conditions, so it was probably seen by the ancients. Many other astronomers with telescopes had sighted

TABLE 14-1	Uranus Data
Average distance from the Sun:	19.194 AU = 2.871×10^9 km
Maximum distance from the Sun:	20.017 AU = 2.995×10^9 km
Minimum distance from the Sun:	18.371 AU = 2.748×10^9 km
Eccentricity of orbit:	0.0429
Average orbital speed:	6.83 km/s
Orbital period:	84.099 years
Rotation period (internal):	17.24 hours
Inclination of equator to orbit:	97.86°
Inclination of orbit to ecliptic:	0.77°
Diameter:	51,118 km = 4.007 Earth diameters (equatorial)
Mass:	8.682×10^{25} kg = 14.53 Earth masses
Average density:	1318 kg/m³
Escape speed:	21.3 km/s
Surface gravity (Earth = 1):	0.90
Albedo:	0.56
Average temperature at cloudtops:	−218°C = −360°F = 55 K
Atmosphere composition (by number of molecules):	82.5% hydrogen (H_2), 15.2% helium (He), 2.3% methane (CH_4)

R I V U X G

(NASA/JPL)

this planet before Herschel; it is plotted on at least 20 star charts drawn between 1690 and 1781. But all these other observers mistook Uranus for a dim star. Herschel was the first to track its motion relative to the stars and recognize it as a planet. Tracking Uranus was no small task, because Uranus moves very slowly on the celestial sphere, just over 4° in the space of a year (compared to about 12° for Saturn and about 35° for Jupiter).

The Discovery of Neptune

It was by carefully tracking Uranus's slow motions that astronomers were led to discover Neptune. By the beginning of the nineteenth century, it had become painfully clear to astronomers that they could not accurately predict the orbit of Uranus using Newtonian mechanics. By 1830 the discrepancy between the planet's predicted and observed positions had become large enough (2 arcmin) that some scientists suspected that Newton's law of gravitation might not be accurate at great distances from the Sun.

By the mid-1840s, two scientists working independently—the French astronomer Urbain Jean Joseph Le Verrier and the English mathematician John Couch Adams—were exploring an earlier yet sounder suggestion. Perhaps the gravitational pull of an as yet undiscovered planet was causing Uranus to deviate slightly from its predicted orbit. Calculations by both scientists concluded that Uranus had indeed caught up with and had passed a more distant planet. Uranus had accelerated slightly as it approached the unknown planet, then decelerated slightly as it receded from the planet.

Inspired by Le Verrier's results, astronomers at Cambridge University Observatory undertook a six-week search for the proposed new planet in the summer of 1846. They were unsuccessful, in part because they lacked accurate star maps of the part of the sky being searched.

Meanwhile, Le Verrier wrote to Johann Gottfried Galle at the Berlin Observatory with detailed predictions of where to search for the new planet. Galle received the letter on September 23, 1846. That very night, aided by more complete star maps and after just a half hour of searching, Galle and Heinrich d'Arrest located an uncharted star with the expected brightness in the predicted location. Subsequent observations confirmed that this new "star" showed a planetlike motion with respect to other stars.

Le Verrier proposed that the planet be called Neptune. After years of debate between English and French astronomers, the credit for its discovery came to be divided between Adams and Le Verrier. Neptune is the only planet in our solar system whose existence was revealed by calculation rather than chance discovery.

At opposition, Neptune can be bright enough to be visible through small telescopes. The first person thus to have seen it was probably Galileo. His drawings from January 1613, when he was using his telescope to observe Jupiter and its four large satellites, show a "star" less than 1 arcmin from Neptune's location. Galileo even noted in his observation log that on one night this star seemed to have moved in relation to the other stars. But Galileo would have been hard pressed to identify Neptune as a planet, because its motion against the background stars is so slow (just over 2° per year).

Observing Uranus and Neptune

Through a large, modern telescope, both Uranus and Neptune are dim, uninspiring sights. Each planet appears as a hazy, featureless disk with a faint greenish-blue tinge. Although Uranus and

> Uranus and Neptune are in such large orbits that they move very slowly on the celestial sphere

Neptune are both about 4 times larger in diameter than Earth, they are so distant that their angular diameters as seen from Earth are tiny—no more than 4 arcsec for Uranus and just over 2 arcsec for Neptune. To an Earth-based observer, Uranus is roughly the size of a golf ball seen at a distance of 1 kilometer. Table 14-1 and Table 14-2 give basic data about Uranus and Neptune, respectively.

CONCEPTCHECK 14-1

How was Neptune predicted to exist at the right location before it was ever observed?

Answer appears at the end of the chapter.

14-2 Uranus is nearly featureless and has an unusually tilted axis of rotation

TUTORIAL 14-2 Scientists had hoped that *Voyager 2* would reveal cloud patterns in Uranus's atmosphere when it flew past the planet in January 1986. But even images recorded at close range showed Uranus to be remarkably featureless (Figure 14-1). Faint

South pole

FIGURE 14-1   R I V U X G

Uranus from *Voyager 2* This image looks nearly straight down onto Uranus's south pole, which was pointed almost directly at the Sun when *Voyager 2* flew past in 1986. None of the *Voyager 2* images of Uranus shows any pronounced cloud patterns. The color is due to methane in the planet's atmosphere, which absorbs red light but reflects green and blue. (JPL/NASA)

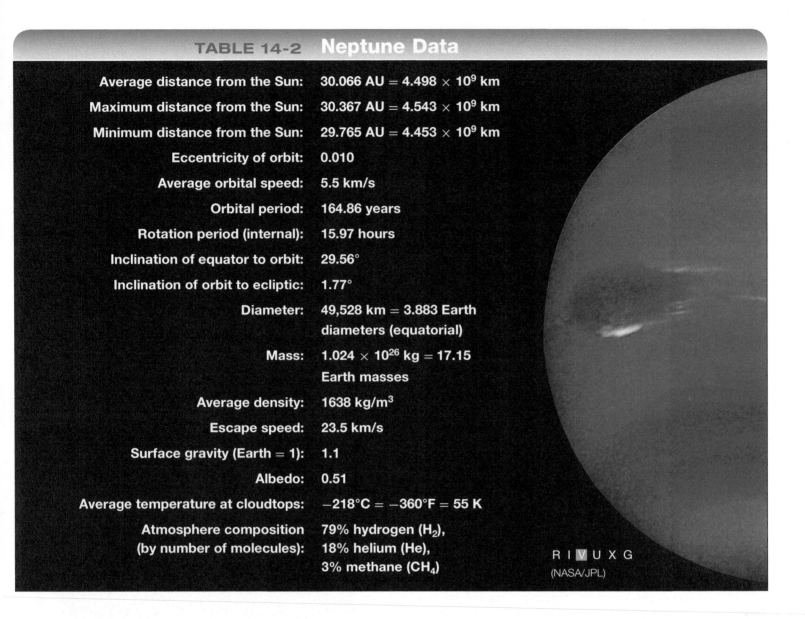

TABLE 14-2	Neptune Data
Average distance from the Sun:	30.066 AU = 4.498×10^9 km
Maximum distance from the Sun:	30.367 AU = 4.543×10^9 km
Minimum distance from the Sun:	29.765 AU = 4.453×10^9 km
Eccentricity of orbit:	0.010
Average orbital speed:	5.5 km/s
Orbital period:	164.86 years
Rotation period (internal):	15.97 hours
Inclination of equator to orbit:	29.56°
Inclination of orbit to ecliptic:	1.77°
Diameter:	49,528 km = 3.883 Earth diameters (equatorial)
Mass:	1.024×10^{26} kg = 17.15 Earth masses
Average density:	1638 kg/m^3
Escape speed:	23.5 km/s
Surface gravity (Earth = 1):	1.1
Albedo:	0.51
Average temperature at cloudtops:	−218°C = −360°F = 55 K
Atmosphere composition (by number of molecules):	79% hydrogen (H$_2$), 18% helium (He), 3% methane (CH$_4$)

R I V U X G
(NASA/JPL)

cloud markings became visible in images of Uranus only after extreme computer enhancement (**Figure 14-2**).

Uranus's Atmosphere

Voyager 2 data confirmed that the Uranian atmosphere is dominated by hydrogen (82.5%) and helium (15.2%), similar to the atmospheres of Jupiter and Saturn. Uranus differs, however, in that 2.3% of its atmosphere is methane (CH$_4$), which is 5 to 10 times the percentage found on Jupiter and Saturn. In fact, Uranus has a higher percentage of all heavy elements—including carbon atoms, which are found in molecules of methane—than Jupiter and Saturn. (In Section 14-4 we will investigate how this could have come about.)

Methane preferentially absorbs the longer wavelengths of visible light, so sunlight reflected from Uranus's upper atmosphere is depleted of its reds and yellows. Fewer reds and yellows in the reflected light gives the planet its distinct greenish-blue appearance. As on Saturn's moon Titan (see Section 13-8), ultraviolet light from the Sun turns some of the methane gas into a hydrocarbon haze, making it difficult to see the lower levels of the atmosphere.

Ammonia (NH$_3$), which constitutes 0.01 to 0.03% of the atmospheres of Jupiter and Saturn, is almost completely absent from the Uranian atmosphere. The reason for this absence is that Uranus is colder than Jupiter or Saturn: The temperature in its upper atmosphere is only 55 K (−218°C = −360°F). Ammonia freezes at these very low temperatures, so any ammonia has long since precipitated out of the atmosphere and into the planet's interior. For the same reason, Uranus's atmosphere is also lacking in water. Hence, the substances that make up the clouds on Jupiter and Saturn—ammonia, ammonium hydrosulfide (NH$_4$SH), and water—are not available in Uranus's atmosphere. This helps to explain the bland, uniform appearance of the planet shown in Figure 14-1.

The few clouds found on Uranus are made primarily of methane, which condenses into droplets only if the pressure is sufficiently high. Hence, methane clouds form only at lower levels within the atmosphere, where they are difficult to see.

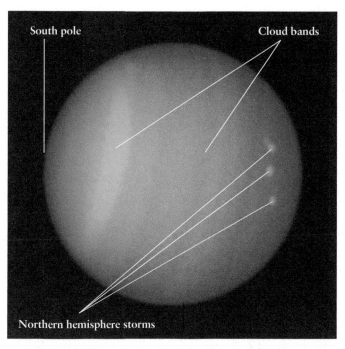

South pole

Cloud bands

Northern hemisphere storms

FIGURE 14-2   R I V U X G

Uranus from the Hubble Space Telescope Images made at
infrared, visible, and ultraviolet wavelengths were combined and enhanced
to give this false-color view of cloud features on Uranus. These images were
captured in August 2004 (18 years after the image in Figure 14-1), during
springtime in Uranus's northern hemisphere. (NASA and Erich Karkoschka,
University of Arizona)

An Oddly Tilted Planet

The rotation period of Uranus's atmosphere is about 16 hours.
Like Jupiter and Saturn, Uranus rotates differentially, so this period
depends on the latitude and can be measured by tracking the
motions of clouds. To determine the rotation period for the underlying body of the planet, scientists looked to Uranus's magnetic field,
which is presumably anchored in the planet's interior, or at least in
the deeper and denser layers of its atmosphere. Data from *Voyager
2* indicate that Uranus's internal period of rotation is 17.24 hours.

Voyager 2 also confirmed that Uranus's rotation axis is tilted
in a unique and bizarre way. Herschel found the first evidence of
this in 1787, when he discovered two moons orbiting Uranus in a
plane that is almost perpendicular to the plane of the planet's orbit
around the Sun. Thus, compared to the other planets, Uranus is
nearly tipped on its side (Figure 14-3).

Careful measurement shows that Uranus's axis of rotation is
tilted by 98°, as compared to 23½° for Earth (compare Figure 14-3
with Figure 2-12). A tilt angle greater than 90° means that Uranus
exhibits retrograde (backward) rotation like that of Venus, shown
in Figure 11-4b. Astronomers suspect that Uranus might have
acquired its large tilt angle billions of years ago, when one or more
massive bodies collided with Uranus while the planet was still
forming. It turns out that the moons of Uranus place constraints on
such axis-tilting impacts, and we will discuss this scenario further
in Section 14-7.

The radical tilt of its axis means that as Uranus moves along its
84-year orbit, its north and south poles alternately point toward or
away from the Sun. This tilt produces highly exaggerated seasonal
changes. For example, when *Voyager 2* flew by in 1986, Uranus's

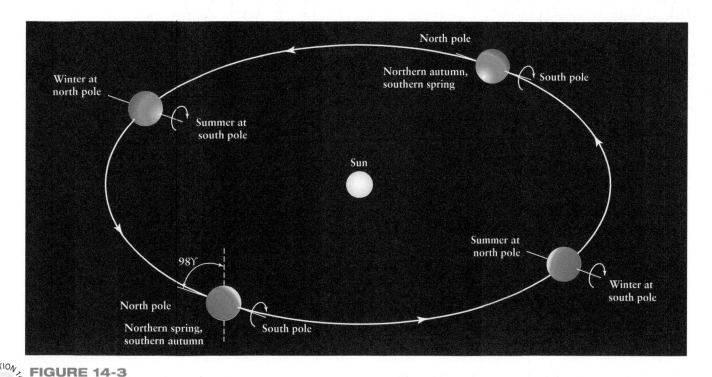

North pole

Northern autumn,
southern spring

South pole

Winter at
north pole

Summer at
south pole

Sun

98γ

North pole

Northern spring,
southern autumn

South pole

Summer at
north pole

Winter at
south pole

ANIMATION 14-1   **FIGURE 14-3**

Exaggerated Seasons on Uranus For most planets, the
rotation axis is roughly perpendicular to the plane of the planet's orbit
around the Sun. But for Uranus the rotation axis is tilted by 98° from the
perpendicular. This tilt causes severely exaggerated seasons. For example,

during midsummer at Uranus's south pole, the Sun appears nearly overhead
for many Earth years, while the planet's northern regions are in continuous
darkness. Half an orbit later, the seasons are reversed.

south pole was pointed toward the Sun. Most of the planet's southern hemisphere was bathed in continuous sunlight, while most of the northern hemisphere was subjected to a continuous, frigid winter night. But over the following quarter of a Uranian year later (21 of our years), sunlight has gradually been returning to the northern hemisphere, triggering immense storms there. Figure 14-2 shows some of these storms, which are much more visible at infrared wavelengths than with visible light.

Atmospheric Motions on Uranus

By following the motions of clouds and storm systems on Uranus, scientists find that the planet's winds flow to the east—that is, in the same direction as the planet's rotation—at northern and southern latitudes, but to the west near the equator. This wind pattern is quite unlike the situation on Jupiter and Saturn, where the zonal winds alternate direction many times between the north and south poles (see Section 12-4). The fastest Uranian winds (about 700 km/h, or 440 mi/h) are found at the equator.

We saw in Section 12-4 that the internal heat of Jupiter and Saturn plays a major role in driving atmospheric activity on those worlds. Uranus is different: It appears to have little or no internal source of thermal energy. Measurements to date show that unlike Jupiter or Saturn, Uranus radiates into

> Unlike the other Jovian planets, Uranus's interior heat has no effect on its atmosphere's motions

space only as much energy as it receives from the Sun. With only feeble sunlight to provide energy to its atmosphere, and little internal thermal energy, Uranus lacks the dramatic wind and cloud dynamics found on Jupiter and Saturn. It is not known why Uranus lacks a significant source of internal thermal energy.

Although Uranus's equatorial region was receiving little sunlight at the time of the *Voyager 2* flyby, the atmospheric temperature there (about 55 K = −218°C = −359°F) was not too different from that at the sunlit pole. Heat must therefore be efficiently transported from the poles to the equator. This north-south heat transport, which is perpendicular to the wind flow, may have mixed and homogenized the atmosphere to make Uranus nearly featureless.

CONCEPTCHECK 14-2

If methane were absent from Uranus's atmosphere, what color would it appear?

Answer appears at the end of the chapter.

14-3 Neptune is a cold, bluish world with Jupiterlike atmospheric features

VIDEO 14-1

At first glance, Neptune appears to be the twin of Uranus. (See the image that opens this chapter, and compare Tables 14-1 and 14-2.) But these two planets are by no means identical. While Neptune and Uranus have almost the same diameter, Neptune is 18% more massive. Neptune's axis of rotation also has a more modest 29½° tilt. When *Voyager 2* flew past Neptune in August 1989, it revealed that the planet has a more active and dynamic atmosphere than Uranus. This activity suggests that Neptune, unlike Uranus, has a powerful source of energy in its interior.

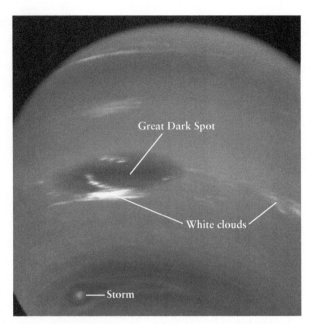

VIDEO 14-2 **FIGURE 14-4** R I **V** U X G
Neptune from *Voyager 2* When this picture of Neptune's southern hemisphere was taken in 1989, the Great Dark Spot measured about 12,000 by 8000 km, comparable in size to Earth. (A smaller storm appears at the lower left.) The white clouds are thought to be composed of crystals of methane-ice. The color contrast in this image has been exaggerated to emphasize the differences between dark and light areas. (NASA/JPL)

Neptune's Atmosphere

The *Voyager 2* data showed that Neptune has essentially the same atmospheric composition as Uranus: 79% hydrogen, 18% helium, 3% methane, and almost no ammonia or water vapor. As for Uranus, the presence of methane gives Neptune a characteristic bluish-green color. The temperature in the upper atmosphere is also the same as on Uranus, about 55 K. That these temperatures should be so similar, even though Neptune is much farther from the Sun, is further evidence that Neptune has a strong internal heat source.

Unlike Uranus, however, Neptune has clearly visible cloud patterns in its atmosphere. At the time that *Voyager 2* flew past Neptune, the most prominent feature in the

> Unlike its close relative Uranus, Neptune has truly immense storms

planet's atmosphere was a giant storm called the Great Dark Spot (Figure 14-4). The Great Dark Spot had a number of similarities to Jupiter's Great Red Spot (see Sections 12-1 and 12-3). The storms on both planets were comparable in size to Earth's diameter, both appeared at about the same latitude in the southern hemisphere, and the winds in both storms circulated in a counterclockwise direction (see Figure 12-5). But Neptune's Great Dark Spot appears not to have been as long-lived as the Great Red Spot on Jupiter. When the Hubble Space Telescope first viewed Neptune in 1994, the Great Dark Spot had disappeared.

Voyager 2 also saw a few conspicuous whitish clouds on Neptune. These clouds are thought to be produced when winds carry methane gas into the cool, upper atmosphere, where it condenses into crystals of methane-ice. *Voyager 2* images show these high-altitude clouds casting shadows onto lower levels of Neptune's atmosphere (Figure 14-5). Images from the Hubble Space Telescope also show the presence of high-altitude clouds (Figure 14-6).

Neptune's Internal Heat

Due to its greater distance from the Sun, Neptune receives less than half as much energy from the Sun as Uranus. But with less solar energy available to power atmospheric motions, why are there high-altitude clouds and huge, dark storms on Neptune but not on Uranus? At least part of the answer is probably that Neptune is still slowly contracting, thus converting gravitational energy into thermal energy that heats the planet's core. (The same is true for Jupiter and Saturn, as we saw in Section 12-4.) The evidence for this contraction is that Neptune, like Jupiter and Saturn but unlike Uranus, emits more energy than it receives from the Sun. The combination of a warm interior and a cold outer atmosphere can cause convection in Neptune's atmosphere, producing the up-and-down motion of gases that generates clouds and storms. Neptune also resembles Jupiter in having faint belts and zones parallel to the planet's equator (see Figure 14-6).

Like those on Uranus, most of Neptune's clouds are probably made of droplets of liquid methane. Because these droplets form fairly deep within the atmosphere, the clouds are more difficult to see than the ones on Jupiter. Hence, Neptune's belts and zones are less pronounced than Jupiter's, although more so than those on Uranus (thanks to the extra cloud-building energy from Neptune's

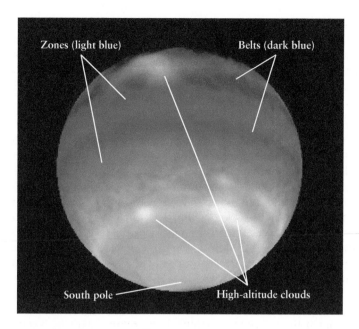

FIGURE 14-6   R I V̅ U X G

Neptune's Banded Structure This enhanced-color Hubble Space Telescope image shows Neptune's belts and zones. White areas denote high-altitude clouds; the very highest clouds (near the top of the image) are shown in yellow-red. The green belt near the south pole is a region where the atmosphere absorbs blue light, perhaps indicating a different chemical composition there. (Lawrence Sromovsky, University of Wisconsin-Madison; STScI/NASA)

Long, streaky high-altitude clouds composed of crystals of methane ice.

High-altitude clouds cast dark shadows on the cloud deck below.

FIGURE 14-5   R I V̅ U X G

Cirrus Clouds over Neptune *Voyager 2* recorded this image of clouds near Neptune's terminator (the border between day and night on the planet). Like wispy, high-altitude cirrus clouds in Earth's atmosphere, these clouds are thought to be made of ice crystals. The difference is that Neptune's cirrus clouds are probably methane-ice, not water-ice as on Earth. (NASA/JPL)

interior). As described above, Neptune's high-altitude clouds (see Figure 14-5) are probably made of *frozen* methane.

Although Neptune displays more evidence of up-and-down motion in its atmosphere than Uranus, the global pattern of east and west winds is almost identical on the two planets. This similarity is rather strange. The two planets are heated very differently by the Sun, thanks to their different distances from the Sun and the different tilts of their axes of rotation, so we might have expected the wind patterns on Uranus and Neptune also to be different. Perhaps the explanation of these wind patterns will involve understanding how heat is transported not only within the atmospheres of Uranus and Neptune but within their interiors as well.

CONCEPTCHECK 14-3

With Neptune even farther from the Sun than Uranus, where does Neptune get the energy to power more atmospheric activity than Uranus?

Answer appears at the end of the chapter.

14-4 Uranus and Neptune contain a higher proportion of heavy elements than Jupiter and Saturn

At first glance, Uranus and Neptune might seem to be simply smaller and less massive versions of Jupiter or Saturn (see Table 7-1). But like many first impressions, this one is misleading because it fails to take into account what lies beneath the surface. We saw in Section 14-2 that the interiors of both Jupiter and Saturn are composed primarily of hydrogen and helium, in nearly the same abundance as the Sun. But Uranus and Neptune must have a different composition, because their average densities are too high.

Uranus and Neptune: Curiously Dense

If Uranus and Neptune formed from gas with similar abundances of elements as Jupiter and Saturn, the smaller masses of Uranus and Neptune would produce less gravitational compression, and Uranus and Neptune would both have lower average densities than Jupiter or Saturn. In fact, however, Uranus and Neptune have average densities (1320 kg/m^3 and 1640 kg/m^3, respectively) that are comparable to or greater than those of Jupiter (1330 kg/m^3) or Saturn (690 kg/m^3). Therefore, we can conclude that Uranus and Neptune contain greater proportions of the heavier elements than do Jupiter or Saturn.

This picture is not what we might expect. According to our discussion in Section 8-5 of how the solar system formed, hydrogen and helium should be relatively more abundant as distance from the vaporizing heat of the Sun increases. But Uranus and Neptune contain a greater percentage of heavy elements and, therefore, a *smaller* percentage of hydrogen and helium, than Jupiter and Saturn.

A related problem is how to explain the masses of Uranus and Neptune. At the locations of Uranus (19 AU from the Sun) and Neptune (30 AU from the Sun) the solar nebula was probably so sparse that these planets would have taken thousands of millions of years to grow to their present sizes around a core of icy planetesimals. But protoplanetary disks observed around other stars (see Section 8-4) do not appear to survive for that long before they

dissipate. In other words, the problem is not how to explain why Uranus and Neptune are smaller than Jupiter and Saturn; it is how to explain why Uranus and Neptune should exist at all!

Models of How Uranus and Neptune Formed

In one model for the origin of Uranus and Neptune, the planets formed in denser regions of the solar nebula closer to the Sun (see the Nice model, Section 8-6). In this region they could have grown rapidly to their present sizes. Had the planets remained at these locations, they could eventually have accumulated enough hydrogen and helium to become as large as Jupiter or Saturn. But before that could happen, gravitational interactions with Jupiter, Saturn, and the disk of planetesimals would have deflected Uranus and Neptune outward into their present large orbits. Because the solar nebula was very sparse at those greater distances, Uranus and Neptune stopped growing and remained at the sizes we see today.

> Explaining the origins of Uranus and Neptune poses a challenge to planetary scientists

Figure 14-7 shows one model for the present-day internal structures of Uranus and Neptune. (Compare this to the internal structures of Jupiter and Saturn, shown in Figure 12-13.) In this model each planet has a rocky core roughly the size of Earth, although more massive. Each planet's core is surrounded by a mantle of liquid water and ammonia. (This mixture means that the mantle is chemically similar to household window cleaning fluid.) Around the mantle is a layer of liquid molecular hydrogen and liquid helium, with a small percentage of liquid methane. This layer is relatively shallow compared to those on Jupiter and Saturn, and the pressure is not high enough to convert the liquid hydrogen into liquid metallic hydrogen.

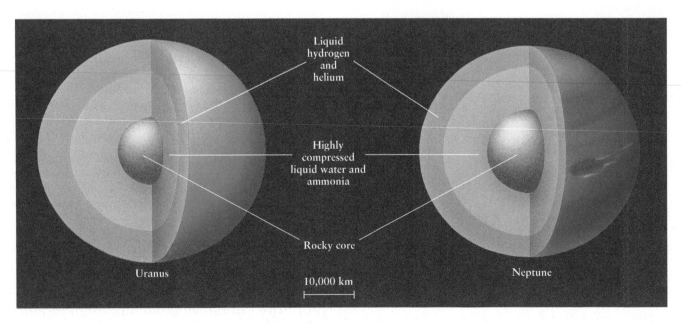

FIGURE 14-7

The Internal Structures of Uranus and Neptune In the model shown here, both Uranus and Neptune have a rocky core (resembling a terrestrial planet), a mantle of liquid water with ammonia dissolved in it, and an outer layer of liquid molecular hydrogen and liquid helium. The atmospheres are very thin shells on top of the hydrogen-helium layer. Uranus and Neptune have about the same diameter, but since Neptune is more massive it may have a somewhat larger core.

Why is it thought that Neptune had to form closer to the Sun than its present distance?

Answer appears at the end of the chapter.

14-5 The magnetic fields of both Uranus and Neptune are oriented at unusual angles

Astronomers were quite surprised by the data sent back from *Voyager 2*'s magnetometer as the spacecraft sped past Uranus and Neptune. These data, as well as radio emissions from charged particles in their magnetospheres, showed that the magnetic fields of both Uranus and Neptune are tilted at steep angles to their axes of rotation (**Figure 14-8**). Uranus's **magnetic axis**, the line connecting its north and south magnetic poles, is inclined by 59° from its axis of rotation; Neptune's magnetic axis is tilted by 47°. By contrast, the magnetic axes of Earth, Jupiter, and Saturn are all nearly aligned with their rotation axes; the angle between their magnetic and rotation axes is 12° or less. Scientists were also surprised to find that the magnetic fields of Uranus and Neptune are offset from the centers of the planets.

CAUTION! The drawing of Earth's magnetic field at the far left of Figure 14-8 may seem to be mislabeled, because it shows the *south* pole of a magnet at Earth's *north* pole. But, in fact, this is correct, as you can understand by thinking about how magnets work. On a magnet that is free to swivel, like the magnetized needle in a compass, the "north pole" is called that because it tends to point north on Earth. Likewise, a compass needle's south pole points toward the south on Earth. Furthermore, opposite magnetic poles attract. If you take two magnets and put them next to each other, they try to align themselves so that one magnet's north pole is next to the other magnet's south pole. Now, if you think of a compass needle as one magnet and the entire Earth as the other magnet, it makes sense that the compass needle's north pole is being drawn toward a magnetic *south* pole—which happens to be located near Earth's geographic north pole. Earth's magnetic pole nearest its geographic North Pole is called the "magnetic north pole." Note that the magnets drawn inside Jupiter, Saturn, and Neptune are oriented opposite to Earth's. On any of these planets, the north pole of a compass needle would point south, not north!

Explaining Misaligned Magnetism

Why are the magnetic axes and axes of rotation of Uranus and Neptune so badly misaligned? And why are the magnetic fields offset from the centers of the planets? One possibility is that their magnetic fields might be undergoing a reversal; geological data show that Earth's magnetic field has switched north to south and back again many times in the past. Another possibility is that the misalignments resulted from catastrophic collisions with planet-sized bodies. As we discussed in Section 14-2, the tilt of Uranus's rotation axis and its system of moons suggest that such collisions occurred long ago. As we will see in Section 14-7, Neptune may have gravitationally captured its largest moon, Triton, but no one knows if that incident was responsible for offsetting Neptune's magnetic axis.

Because neither Uranus nor Neptune is massive enough to compress hydrogen to a metallic state, their magnetic fields cannot be generated in the same way as those of Jupiter and Saturn. Instead, under the high pressures found in the watery mantles of Uranus and Neptune, dissolved molecules such as ammonia lose one or more electrons and become electrically charged (that is, they become *ionized*; see Section 5-8). Water is a good conductor of electricity when it has such electrically charged molecules dissolved in it, and electric currents in this fluid are probably the source of the magnetic fields of Uranus and Neptune.

> Unlike any other planets, Uranus and Neptune have magnetic fields caused by ionized ammonia

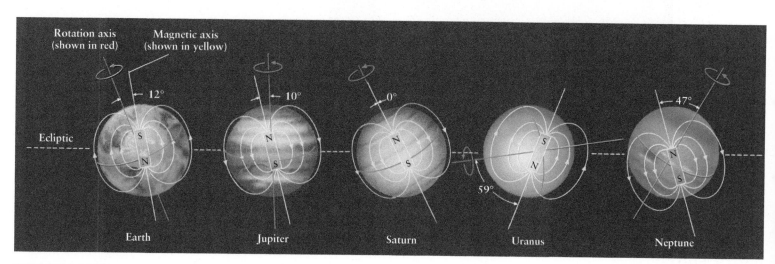

FIGURE 14-8

The Magnetic Fields of Five Planets This illustration shows how the magnetic fields of Earth, Jupiter, Saturn, Uranus, and Neptune are tilted relative to their rotation axes (shown in red). For both Uranus and Neptune, the magnetic fields are offset from the planet's center and steeply inclined to the rotation axis.

The *Cosmic Connections* figure summarizes the key properties of Uranus and Neptune and how they compare with those of Jupiter and Saturn.

CONCEPTCHECK 14-5

In order for a magnetic dynamo to create a global magnetic field around Uranus and Neptune, these planets must have an electrically conducting fluid. What is that conducting fluid thought to be?

Answer appears at the end of the chapter.

14-6 Uranus and Neptune each has a system of thin, dark rings

On March 10, 1977, Uranus was scheduled to move in front of a faint star, as seen from the Indian Ocean. A team of astronomers headed by James L. Elliot of Cornell University observed this event, called an **occultation**, from a NASA airplane equipped with a telescope. They hoped that by measuring how long the star

was hidden, they could accurately measure Uranus's size. In addition, by carefully measuring how the starlight faded when Uranus passed in front of the star, they planned to deduce important properties of Uranus's upper atmosphere.

The Surprising Rings of Uranus

To everyone's surprise, the background star briefly blinked on and off several times just before the star passed behind Uranus and again immediately after (Figure 14-9). The

> Like the planet itself, the rings of Uranus were discovered by accident

astronomers concluded that Uranus must be surrounded by a series of nine narrow rings. In addition to these nine rings, *Voyager 2* discovered two others that lie even closer to the cloudtops of Uranus. Two other extremely faint rings, much larger in diameter than any of the others, were found in 2005 using the Hubble Space Telescope (Figure 14-10).

Unlike Saturn's rings, the rings of Uranus are dark and narrow: Most are less than 10 km wide. Typical particles in Saturn's rings are chunks of reflective water-ice the size of snowballs (a few centimeters across), but typical particles in Uranus's rings—thought to be mostly water-ice—are 0.1 to 10 meters in size and are no more

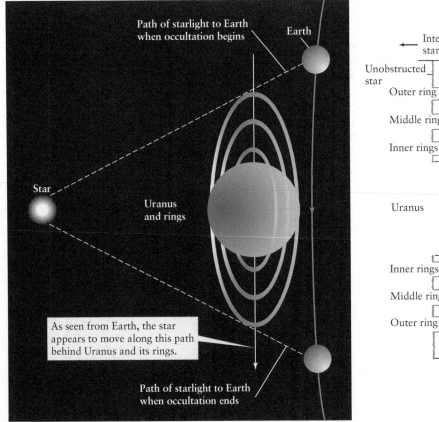

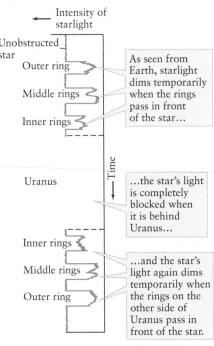

FIGURE 14-9

How the Rings of Uranus Were Discovered As seen from Earth, Uranus occasionally appears to move in front of a distant star. Such an event is called an occultation. The star's light is completely blocked when it is

behind the planet. But before and after the occultation by Uranus, the starlight dims temporarily as the rings pass in front of the star.

The Outer Planets: A Comparison

Uranus and Neptune are not simply smaller versions of Jupiter and Saturn. This table summarizes the key differences among the four Jovian planets.

	Interior	Surface	Rings	Atmosphere	Magnetic Field
Jupiter	Terrestrial core, thick liquid metallic hydrogen layer, molecular hydrogen	No solid surface, atmosphere gradually thickens to liquid state, belt and zone structure, hurricanelike features	Yes	Primarily H, He	19,000× Earth's total field; at its cloud layer, 14× stronger than Earth's surface field
Saturn	Similar to Jupiter, with bigger terrestrial core and less metallic hydrogen	No solid surface, less distinct belt and zone structure than Jupiter	Yes	Primarily H, He	570× Earth's total field; at its cloud layer, $^2/_3×$ Earth's surface field
Uranus	Terrestrial core, liquid water shell, liquid hydrogen and helium mantle	No solid surface, weak belt and zone system, hurricanelike features, color from methane absorption of red, orange, yellow	Yes	Primarily H, He, some CH_4	50× Earth's total field, at its cloud layer, 0.7× Earth's surface field
Neptune	Similar to Uranus	Similar to Uranus	Yes	Primarily H, He, some CH_4	35× Earth's total field, at its cloud layer, 0.4× Earth's surface field

For detailed comparisons between planets, see Appendices 1 and 2.

*To see the orientations of these magnetic fields relative to the rotation axes of the planets, see Figure 14-8.

Earth   Jupiter   Saturn   Uranus   Neptune

1AU
|←4.2AU→|←4.3AU→|←——9.7AU——→|←——10.9AU——→|
Sun   Earth   Jupiter   Saturn   Uranus   Neptune

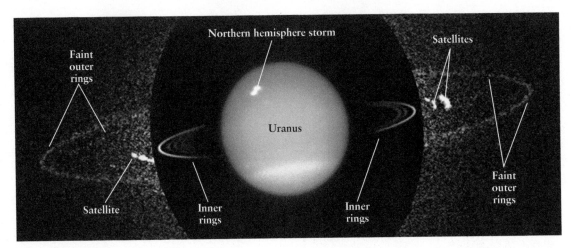

FIGURE 14-10   R I V U X G

Uranus's Rings from the Hubble Space Telescope Several Hubble Space Telescope images were assembled into this view of Uranus's rings. Relatively short exposure times reveal the planet and the inner rings, while much longer exposure times were needed to reveal the outer rings. During these long exposures several of Uranus's satellites moved noticeably, leaving bright trails on the image. (NASA; ESA; and M. Showalter, SETI Institute)

reflective than lumps of coal. It is not surprising that these narrow, dark rings escaped detection for so long. **Figure 14-11** is a *Voyager 2* image of the rings from the side of the planet away from the Sun, where light scattering from exceptionally small ring particles makes them more visible. (We discussed light scattering in Section 5-6 and Box 5-4.) Figure 14-10 shows the Sun-facing side of the rings.

The Uranian rings are so dark because sunlight at Uranus is only one-quarter as intense as at Saturn. As a result, Uranus's ring particles are so cold that they can retain some methane-ice. Scientists speculate that eons of impacts by electrons trapped in the magnetosphere of Uranus have converted this methane-ice into dark carbon compounds. This **radiation darkening** can account for the low reflectivity of the rings.

Uranus's major rings are located less than 2 Uranian radii from the planet's center, well within the planet's Roche limit (see Section 12-9). Some sort of mechanism, possibly one involving shepherd satellites, efficiently confines particles to their narrow orbits. (In Section 12-11 we discussed how shepherd satellites help keep planetary rings narrow.) *Voyager 2* searched for shepherd satellites but found only two. The others may be so small and dark that they have simply escaped detection.

The two faint rings discovered in 2005 (see Figure 14-10) are different: They lie well *outside* Uranus's Roche limit. The outer of these two rings owes its existence to a miniature satellite called Mab (named for a diminutive fairy in English folklore) that orbits within this ring. Meteorites colliding with Mab knock dust off the satellite's surface. Since Mab is so small (just 36 km in diameter) and hence has little gravity, the ejected dust escapes from the satellite and goes into orbit around Uranus, forming a ring. The inner of the two faint rings may be formed in the same way; however, no small satellite has yet been found within this ring.

The Decaying Rings of Neptune

Like Uranus, Neptune is surrounded by a system of thin, dark rings that were first detected in stellar occultations. **Figure 14-12** is a *Voyager 2* image of Neptune's rings. The ring particles vary in size from a few micrometers (1 μm = 10^{-6} m) to about 10 meters. As for Uranus's rings, the particles that make up Neptune's rings reflect very little light because they have undergone radiation darkening.

In 2002 and 2003, a team led by Imke de Pater of the University of California, Berkeley, used the 10-meter Keck telescope in Hawaii to observe in detail the rings of Neptune. Remarkably, they found that all of the rings had become fainter since the *Voyager 2* flyby in 1989. Apparently, the rings are losing particles faster then new ones are being added. If the rings continue to decay at the same rate, one of them may vanish completely within a century. More research into the nature of Neptune's rings is needed to explain this curious and rapid decay.

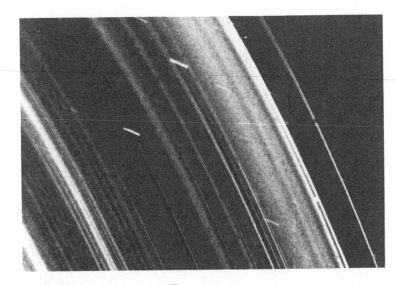

FIGURE 14-11   R I V U X G

The Shaded Side of Uranus's Rings This view from *Voyager 2,* taken when the spacecraft was in Uranus's shadow, looks back toward the Sun. Numerous fine-grained dust particles gleam in the spaces between the main rings. This dust is probably debris from collisions between larger particles in the main rings. The short horizontal streaks are star images blurred by the spacecraft's motion during the exposure. (NASA/JPL)

VIDEO 14.3 **FIGURE 14-12**   R I V U X G

Neptune's Rings The two main rings of Neptune and a faint, inner ring can easily be seen in this composite of two *Voyager 2* images. There is also a faint sheet of particles whose outer edge is located between the two main rings and extends inward toward the planet. The overexposed image of Neptune itself has been blocked out. (NASA)

CONCEPTCHECK 14-6

Why are Uranus's rings so much more difficult to see than Saturn's rings?

Answer appears at the end of the chapter.

14-7 Uranus's larger and smaller satellites

Before the *Voyager 2* flyby of Uranus, five moderate-sized satellites—Titania, Oberon, Ariel, Umbriel, and Miranda—were known

to orbit the planet (Figure 14-13). These are Uranus's largest moons. Most are named after sprites and spirits in Shakespeare's plays. They range in diameter from about 1600 km (1000 mi) for Titania and Oberon to less than 500 km (300 mi) for Miranda. These moons have average densities around 1500 kg/m³, which is consistent with a mixture of half water-ice and half rock (although Miranda is mostly ice).

Voyager 2 discovered 11 other small Uranian satellites, most of which have diameters of less than 100 km (60 mi); 11 more were found using ground-based telescopes and the Hubble Space Telescope between 1997 and 2005. Only Jupiter and Saturn have more known satellites than Uranus. In the infrared, Uranus, its rings, and some of its moons are easily visible (Figure 14-14).

The moons of Uranus are generally divided into three groups: the larger moons, the smaller inner moons orbiting near the rings, and the smaller outer moons that are probably captured asteroids.

Uranus's Larger Satellites

Umbriel and Oberon both appear to be geologically dead worlds, with surfaces dominated by impact craters. By contrast, Ariel's surface appears to have been cracked at some time in the past, allowing some sort of ice lava to flood low-lying areas. A similar process appears to have taken place on Titania. This geologic activity may be due to a combination of the satellites' internal heat and tidal heating like that which powers volcanism on Jupiter's satellite Io (see Section 13-4). Io's tidal heating is only possible because of the 1:2:4 ratio of the orbital periods of Io, Europa, and Ganymede. While there are no such simple ratios between the present-day orbital periods of Uranus's satellites, there may have been in the past. If so, tidal heating could have helped reshape the surfaces of some of the satellites.

Unique among Uranus's satellites is Miranda, which has a landscape unlike that of any other world in the solar system. Much of the surface is heavily cratered,

> Uranus's satellite Miranda has cliffs twice as high as Mount Everest

as we would expect for a satellite only 470 km in diameter, but several regions have unusual and dramatic topography

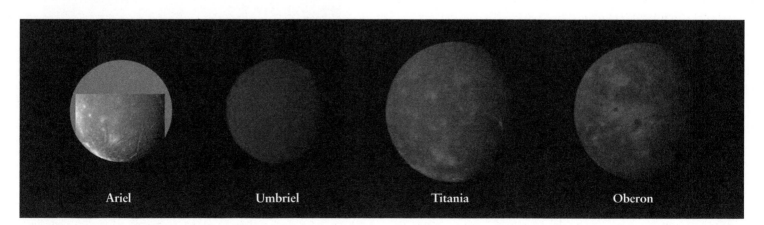

FIGURE 14-13   R I V U X G

Uranus's Principal Satellites This "family portrait" (a montage of five *Voyager 2* images) shows Uranus's five largest moons to the same scale and correctly displays their respective reflectivities. (The darkest satellite, Umbriel, is actually more reflective than Earth's Moon.) All five satellites have grayish surfaces, with only slight variations in color. (NASA/JPL-Caltech/R. Hurt [SSC])

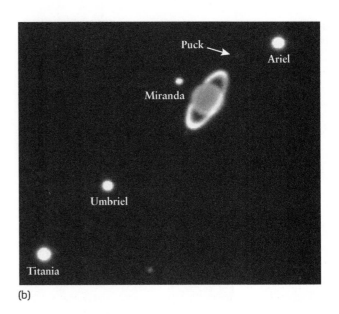

(a)

(b)

FIGURE 14-14 R I V U X G

Uranus's Rings and Moons **(a)** This false-color infrared image from the Hubble Space Telescope shows eight of Uranus's inner moons, all of which were discovered by *Voyager 2* when it flew past Uranus in 1986. They all lie within 86,000 km of the planet's center (only about one-fifth of the distance from Earth to our Moon). The arcs show how far each moon moves around its orbit in 90 minutes. Note the pole and equator; unlike the other planets, Uranus nearly lies on its side with its rotational axis close to its orbital plane

around the Sun. **(b)** This infrared image from the Very Large Telescope shows five of Uranus's larger moons. All the moons visible in this image are in prograde orbit, orbiting in the same direction as Uranus rotates. The rings in this image appear even brighter than Uranus: Gaseous methane in Uranus's atmosphere absorbs infrared light, while ice in the rings reflects this wavelength. (a: Erich Karkoschka, University of Arizona; and NASA, b: European Southern Observatory)

(Figure 14-15). Detailed analysis of Miranda's geology suggests that this satellite's orbital period was once in a whole-number ratio with that of more massive Umbriel or Ariel. The resulting tidal heating melted Miranda's interior, causing dense rocks in some locations on the surface to settle toward the satellite's center as blocks of less dense ice were forced upward toward the surface, thus creating Miranda's resurfaced terrain. Tidal heating must have ceased before this process could run its full course, which explains why some ancient, heavily cratered regions remain on Miranda's surface.

Uranus's Larger Satellites Constrain Models of an Axis-Tilting Impact

The five large moons of Uranus help us understand the collision that is proposed to have tilted the rotation axis of Uranus. A collision large enough to turn a planet on its side would be quite a violent event. Such an impactor is estimated to be several times the mass of Earth—quite large!—though there might instead have been several smaller impacts. While several impacts might seem less likely to occur, modeling indicates that a single large impact would have resulted in retrograde orbits for Uranus's large moons. On the other hand, two or more smaller impactors, each the size of Earth, would better explain the observed prograde orbits of the large Uranian moons.

Another constraint on the model of impacts with Uranus is that if these impacts occurred after the large moons formed, the moons would have remained in their original orbital plane, which is assumed to be in Uranus's original equatorial plane. Instead, what we see today is both Uranus *and* the orbital plane of its moons

tilted together, so that the larger moons are still in the equatorial plane (see Figure 14-14b). In order to tilt Uranus and its moons together, models suggest that impacts might have rotated a disk of

VIDEO 14-4 **FIGURE 14-15** R I V U X G

Miranda This composite of *Voyager 2* images shows that part of Miranda's surface is ancient and heavily cratered, while other parts are dominated by parallel networks of valleys and ridges. At the very bottom of the image—where a "bite" seems to have been taken out of Miranda—is a range of enormous cliffs that jut upward to an elevation of 20 km, twice as high as Mount Everest. (NASA/JPL)

protoplanetary material from which both Uranus and its moons formed. While frequent impacts are expected during the early solar system, impacts of this magnitude are not expected, and the origin of Uranus's tilt is still an open question.

Uranus's Perplexing Small Satellites

While Uranus's small satellites are unlikely to show the kind of geology found in Miranda, some of them move in curious orbits. Eight of the nine smaller outer satellites in large orbits beyond Oberon are in retrograde orbits, moving around Uranus opposite to the direction in which Uranus rotates. Many of the small outer satellites of Jupiter and Saturn have retrograde orbits and are probably captured asteroids (see Section 13-9 and Section 13-10); the same is probably true of the outer satellites of Uranus.

The 13 inner satellites that orbit closest to Uranus (inside the orbit of Miranda) are all in prograde orbits, so they travel around Uranus in the same direction as the planet rotates (Figure 14-14a). Just like the rings of Uranus, these inner moons are very dark; this is probably due to the same process of radiation darkening. In fact, it is thought that the Uranian rings were probably created by the fragmentation of one or more inner moons.

Will the inner moons orbit Uranus indefinitely? Probably not. When Mark Showalter and Jack Lissauer compared 11 of these satellites' orbits in 2005 with their orbits in 1994, they found surprisingly large differences. (The other two inner satellites were only discovered in 2003, so it is not known whether their orbits underwent similar changes.) These satellites can pass within a few thousand to a few hundred kilometers from each other, so they can exert strong gravitational forces on each other. Over time these forces can modify their orbits. Computer simulations of these gravitational interactions indicate that the inner satellites might begin colliding with each other within a hundred million years. If these inner satellites are in such unstable orbits, have they really been in orbit around Uranus since it formed more than 4.56 billion years ago? Or did Uranus somehow acquire these satellites in the relatively recent past?

CONCEPTCHECK **14-7**

Why is there doubt about most of Uranus's inner moons being roughly as old as Uranus?

Answer appears at the end of the chapter.

14-8 Neptune's satellite Triton is an icy world with a young surface and a tenuous atmosphere

Neptune has 13 known satellites, listed in Appendix 3. They are named for mythological beings related to bodies of water. (Neptune itself is named for the Roman god of the sea.) Most of these worlds are small, icy bodies, probably similar to the smaller satellites of Uranus. The one striking exception is Triton, Neptune's largest satellite (see Table 7-2). In many ways Triton is quite unlike any other world in the solar system.

Triton's Backward Orbit and Young Surface

Like many of the small satellites of Jupiter, Saturn, Uranus, and Neptune, Triton is in a retrograde orbit, so that it moves around Neptune opposite to the direction of Neptune's rotation. Furthermore, the plane of Triton's orbit is inclined by 23° from the plane of Neptune's equator. It is difficult to imagine how a satellite might form out of the same cloud of material as a planet and end up orbiting in a direction opposite the planet's rotation and in such a tilted plane. Hence, Triton probably formed elsewhere in the solar system, collided long ago with a now-vanished satellite of Neptune, and was captured by Neptune's gravity. With a diameter of 2706 km, a bit smaller than our Moon but much larger than any other satellite in a retrograde orbit, Triton is certainly the largest captured satellite.

Figure 14-16 shows the icy, reflective surface of Triton as imaged by *Voyager 2*. Most of Triton's ice is water, with some additional nitrogen and methane-ice. There is a conspicuous absence of large craters, which immediately tells us that Triton has a young surface on which the scars of ancient impacts have largely been erased by tectonic activity. There are areas that resemble frozen lakes and may be the calderas of extinct ice volcanoes. Some of Triton's surface features resemble the long cracks seen on Europa (Section 13-6) and Ganymede (Section 13-7). Still other features are unique to Triton. For example, in the upper portion of Figure 14-16 you can see dimpled, wrinkled terrain that resembles the skin of a cantaloupe.

Triton's tectonically active history is probably related to its having been captured into orbit around Neptune. After its capture, Triton most likely started off in a highly elliptical orbit, but today the satellite's orbit is quite circular. The satellite's original elliptical orbit would have been made circular by tidal forces exerted on Triton by Neptune's gravity. These tidal forces would also have stretched and flexed Triton, causing enough tidal

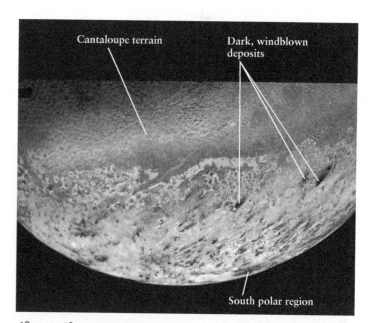

VIDEO 14-5 VIDEO 14-6 **FIGURE 14-16** R I **V** U X G

Triton Several high-resolution *Voyager 2* images were combined to create this mosaic. The pinkish material surrounding Triton's south polar region is probably nitrogen frost. Some of this presumably evaporates when summer comes to the south pole; the northward flow of the evaporated gas may cause the dark surface markings. Farther north is a brown area of "cantaloupe terrain" that resembles the skin of a melon. (NASA/JPL)

heating to melt much of the interior. The resulting volcanic activity (with lavas made of ice rather than molten rock) would have obliterated Triton's original surface features, including craters.

There is still enough warmth in Triton's interior to keep it active with cryovolcanoes (where the lava consists of water that is prevented from freezing due to dissolved ammonia). *Voyager 2* observed plumes of dark material being ejected from the surface to a height of 8 km (5 mi). The plumes contain invisible nitrogen gas, along with small and dark dust particles. These plumes may have been generated from a hot spot far below Triton's surface, similar to geysers on Earth. Alternatively, the energy source for the plumes may be sunlight that warms the surface, producing subsurface pockets of gas and creating fissures in the icy surface through which the gas can escape.

Triton's surface temperature is only 38 K (−235°C = −391°F), the lowest of any world yet visited by spacecraft. This temperature is low enough to solidify nitrogen, and indeed the spectrum of sunlight reflected from Triton's surface shows absorption lines due to nitrogen-ice as well as methane-ice. But Triton is also warm enough to allow some nitrogen to evaporate from the surface, like the steam that rises from ice cubes when you first take them out of the freezer. *Voyager 2* confirmed that Triton has a very thin nitrogen atmosphere with a surface pressure of only 1.6×10^{-5} atmosphere, about the same as at an altitude of 100 km above Earth's surface. Despite its thinness, Triton's atmosphere was revealed by its winds. *Voyager 2* saw areas on Triton's surface where dark dusty material has been blown downwind by a steady breeze (see Figure 14-16). Dark dusty material ejected from the geyserlike plumes was carried as far as 150 km by high-altitude winds.

Tidal Forces and Triton's Doom

Just as tidal forces presumably played a large role in Triton's history, they also determine its future. Triton raises a tidal bulge on Neptune, just as our Moon distorts Earth (recall Figure 10-20). In the case of the Earth-Moon system, the gravitational pull of Earth's tidal bulge causes the Moon to spiral away

> Within 100 million years, Triton will be torn apart by Neptune's tidal forces

from Earth. But because Triton's orbit is retrograde, the tidal bulge on Neptune exerts a force on Triton that makes the satellite slow down rather than speed up. (In Figure 10-20, imagine that the moon is orbiting toward the bottom of the figure rather than toward the top.) This is causing Triton to spiral gradually in toward Neptune. In approximately 100 million years, Triton will be inside Neptune's Roche limit, and the satellite will eventually be torn to pieces by tidal forces. When this happens, the planet will develop a spectacular ring system—overshadowing by far its present-day set of narrow rings—as rock fragments gradually spread out along Triton's former orbit.

Prior to *Voyager 2*, only one other satellite was known to orbit Neptune. Nereid, which was first sighted in 1949, is in a prograde orbit. Hence, it orbits Neptune in the direction opposite to Triton. Nereid also has the most eccentric orbit of any satellite in the solar system; its distance from Neptune varies from 1.4 million to 9.7 million kilometers. One possible explanation is that when Triton was captured by Neptune's gravity, the interplay of gravitational forces exerted on Nereid by both Neptune and Triton moved Nereid from a relatively circular orbit (like those of Neptune's other, smaller moons) into its present elliptical one.

CONCEPTCHECK 14-8

How was an atmosphere detected on Triton?

Answer appears at the end of the chapter.

14-9 Pluto is smaller than any planet

Speculations about worlds beyond Neptune date back to the late 1800s, when a few astronomers suggested that Neptune's orbit was being perturbed by an unknown object. Encouraged by the fame of Adams and Le Verrier, several people set out to discover "Planet X." Two prosperous Boston gentlemen, William Pickering and Percival Lowell, were prominent in this effort. (Lowell also enthusiastically promoted the idea of canals on Mars; see Section 11-4.) Modern calculations show that there are, in fact, no

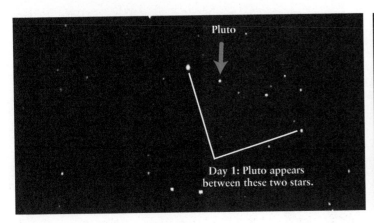

Day 1: Pluto appears between these two stars.

Day 2: Pluto has moved relative to the stars.

FIGURE 14-17   R I ◼V◼ U X G

Pluto's Motion across the Sky Pluto was discovered in 1930 by searching for a dim, starlike object that moves slowly in relation to the background stars. These two photographs were taken three days apart.

Even when its apparent motion on the celestial sphere is fastest, Pluto moves only about 1 arcmin per day relative to the stars. (John Chumack/Science Source)

unaccounted perturbations of Neptune's orbit. It is thus not surprising that no orbiting body was found at the positions predicted by Pickering, Lowell, and others. Yet the search continued.

The Discovery of Pluto

Before he died in 1916, Lowell urged that a special camera with a wide field of view be constructed to help search for Planet X. After many delays, the camera was finished in 1929 and installed at the Lowell Observatory in Flagstaff, Arizona, where a young astronomer, Clyde W. Tombaugh, had recently joined the project. On February 18, 1930, Tombaugh finally discovered the long-sought object. It was disappointingly faint—a thousand times dimmer than the dimmest object visible with the naked eye and 250 times dimmer than Neptune at opposition—and presented no discernible disk. The new world, originally labeled a planet, was named by the eleven-year-old student Venetia Burney for Pluto, the mythological god of the underworld. In 2006, Pluto was reclassified as a **dwarf planet,** which is an object orbiting the Sun (but not a satellite) with enough mass to gravitationally pull itself into a spherical shape, yet not enough gravity to clear out planetesimals from its surroundings. (We will further discuss Pluto's reclassification in the next section.)

Pluto's orbit (see **Figure 14-17**) around the Sun is more elliptical (eccentricity 0.2501) and more steeply inclined to the plane of the ecliptic (17.15°) than the orbit of any of the planets. In fact, Pluto's orbit is so eccentric that it is sometimes closer to the Sun than Neptune is. This was the case from 1979 until 1999; indeed, when Pluto was at perihelion in 1989, it was more than 10^8 km closer to the Sun than was Neptune. Happily, the orbits of Neptune and Pluto are such that the two worlds will never collide.

Pluto's rotation axis is tilted by more than 90° and has retrograde rotation like Uranus. However, seeing the surface of this interesting dwarf planet has been very difficult: Pluto is so far away that its diameter subtends only a very small angle of 0.15 arcsec. Hence, it is extraordinarily difficult to resolve any of its surface features. But a real-color map of Pluto's surface has been made by carefully combining multiple observations of Pluto from the Hubble Space Telescope (**Figure 14-18**). The darker colors in Figure 14-18 are thought to come from solar ultraviolet light interacting with methane on the surface to leave behind a carbon-rich residue. The bright spot in the center image is thought to be carbon-monoxide frost. Not only were the colors unexpected—with their dark orange and charcoal black patches—but a big surprise came when comparison with data from a decade earlier showed the north pole region getting brighter and the south pole getting darker.

Pluto's Satellites

In 1978, while examining some photographs of Pluto, James W. Christy of the U.S. Naval Observatory noticed that the image of Pluto on a photographic plate appeared slightly elongated, as though Pluto had a lump on one side. He promptly examined a number of other photographs of Pluto, and found a series that showed the lump moving clockwise around Pluto with a period of about 6 days.

Christy concluded that the lump was actually a satellite of Pluto. He proposed that the newly discovered moon be christened Charon (pronounced KAR-en), after the mythical boatman who ferried souls across the River Styx to Hades, the domain ruled by Pluto. (Christy also chose the name because of its similarity to Charlene, his wife's name.) The average distance between Charon and Pluto is a scant 19,640 km, less than 5% of the distance between Earth and our Moon. In 2005 a team of astronomers using the Hubble Space Telescope discovered two additional, smaller satellites that orbit Pluto at 2 to 3 times the distance of Charon. Then, in 2011 and 2012, Hubble discovered two more moons (**Figure 14-19**).

Observations show that Charon's orbital period of 6.3872 days is the same as the rotational period of Pluto *and* the rotational period of Charon. In other words, both Pluto and

> Pluto and its satellite Charon both rotate in lockstep, always keeping the same faces toward each other

Charon rotate synchronously with their orbital motion, and so both always keep the same face toward each other. As seen from the Charon-facing side of Pluto, the satellite remains perpetually suspended above the horizon. Likewise, Pluto neither rises nor sets as seen from Charon.

Soon after the discovery of Charon, astronomers witnessed an alignment that occurs only once every 124 years. From 1985 through 1990, Charon's orbital plane appeared nearly edge-on as seen from Earth, allowing astronomers to view mutual eclipses of Pluto and its moon. As the bodies passed in front of each other, their combined brightness diminished in ways that revealed their sizes and surface characteristics. Data from the eclipses combined

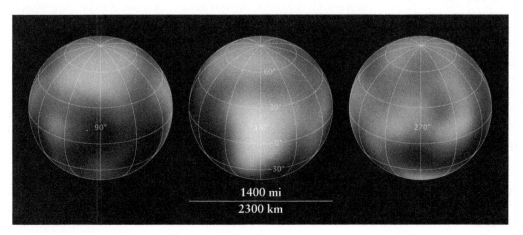

1400 mi
2300 km

FIGURE 14-18   R I **V** U X G

Three Faces of Pluto These real-color maps of Pluto were created using a technique called *dithering*. By comparing images taken at slightly different orientations, small changes in brightness can be traced back to spots on the surface that are too small to be directly resolved. To carry out this special analysis, it took 20 computers running continuously for four years to create these maps! The surface changes are thought to result from Pluto's seasonal variation in sunlight exposure: As one region receives more light, ice evaporates and then refreezes onto colder regions. (NASA, ESA, and Z. Levay [STScI], and M. Buie [Southwest Research Institute])

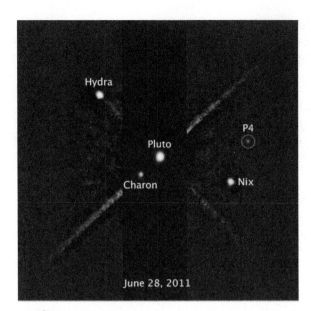

June 28, 2011

VIDEO 14-7 VIDEO 14-8 **FIGURE 14-19**   R I �V U X G

Pluto and Its Satellites This Hubble Space Telescope image shows Pluto, its large satellite Charon, and the two small satellites Nix and Hydra. The fourth and fifth moons of Pluto (referred to as P4 and P5) do not have official names as of this writing. Estimates put the sizes of P4 and P5 between 10 and 34 km across. Nix is named for the mythological goddess of the night who was the mother of Charon; Hydra in mythology was a nine-headed, poisonous serpent who guarded the entrance to Hades. (*Note:* The streaks are image artifacts.) (NASA/ESA/Mark Showalter [SETI Institute])

with subsequent measurements give Pluto's diameter as about 2274 km and Charon's as about 1190 km. For comparison, our Moon's diameter (3476 km) nearly equals the diameters of Pluto and Charon added together. The brightness variations during eclipses suggest that Charon has a bright southern polar cap.

The average densities of Pluto and Charon, at about 2000 kg/m^3, are essentially the same as that of Triton (2070 kg/m^3). All three worlds are therefore probably composed of a mixture of rock and ice, as we might expect for small bodies that formed in the cold outer reaches of the solar nebula.

Pluto's spectrum shows absorption lines of various solid ices that cover its surface, including nitrogen (N_2), methane (CH_4), and carbon monoxide (CO). Stellar occultation measurements have shown that Pluto, like Triton, has a very thin atmosphere. Exposed to a daytime temperature of around 40 K, N_2 and CO ices turn to gas more easily than frozen CH_4. For this reason, most of Pluto's tenuous atmosphere probably consists of N_2 and CO. Charon's weaker gravity has apparently allowed most of the N_2, CH_4, and CO to escape into space; only water-ice is found on its surface.

The Origin of Pluto's Satellites

Pluto and Charon are similar to each other in mass, size, and density. Throughout the solar system, planets are typically many times larger and more massive than any of their satellites. The similarities between Pluto (a dwarf planet) and Charon suggest to some astronomers that this binary system probably formed when Pluto collided with a similar body. Perhaps chunks of matter were stripped from the second body, leaving behind a mass, now called

Charon, that was captured into orbit by Pluto's gravity. This same collision probably also left behind Nix, Hydra, and the moons P4 and P5 shown in Figure 14-19.

Supporting this collision hypothesis is the observation that all satellites have almost the same color and reflection spectrum, suggesting a common origin. Pluto, by contrast, has a redder color than its satellites, which is consistent with the idea that Pluto and the satellites formed in different ways.

For this scenario to be feasible, there must have been many Plutolike objects in the outer regions of the solar system. Astronomers estimate that in order for a collision or close encounter between two of them to have occurred at least once since the solar system formed some 4.56 billion years ago, there must have been at least a thousand Plutos. As we will see, astronomers have begun to discover this population of Plutolike objects (called plutoids) in the dark recesses of the solar system beyond Neptune.

CONCEPTCHECK **14-9**

What is the difference between a planet and a dwarf planet like Pluto?

Answer appears at the end of the chapter.

14-10 Trans-Neptunian Objects and Pluto's Reclassification

For many years astronomers attempted to find other worlds beyond Neptune using the same technique used to discover Pluto (see Figure 14-17). The first to succeed were David Jewitt and Jane Luu, who in 1992 found an object with a semimajor axis of 42 AU and a diameter estimated to be only 240 km. This object, named 1992 QB1, has a reddish color like Pluto, possibly because frozen methane has been degraded by eons of radiation exposure. As of 2013, more than 1200 **trans-Neptunian objects**—icy worlds whose orbits have semimajor axes larger than that of Neptune—have been discovered. Most of these are relatively small, like 1992 QB1, but a number are larger than Charon and one is larger than Pluto itself (Figure 14-20). In light of these recent discoveries, Pluto is best understood as a particularly large (but not the largest) trans-Neptunian object, rather than a planet.

Reclassifying Pluto as both a dwarf planet and a trans-Neptunian object (both new terms) instead of a planet, was controversial. However, Pluto was already known to be somewhat of an oddball planet, as it is much smaller than the true planets and has an orbit inclined from the ecliptic plane significantly more than the planets. Furthermore, compared to the planets, Pluto also has a significantly lower density, suggesting it has a much larger proportion of water-ice. However, once large trans-Neptunian objects were discovered—one even larger than Pluto—a new class of objects came into view, of which Pluto was only a member. The main decision then facing astronomers was whether to *add* several more oddball objects like Pluto to the list of planets, or to consider Pluto and these new objects members of a different category. The solution adopted by the International Astronomical Union in 2006 was to create a category called trans-Neptunian objects, and in addition, to determine that a larger object like Pluto anywhere in the solar system could also be called a dwarf planet if it met certain requirements (see Section 14-9).

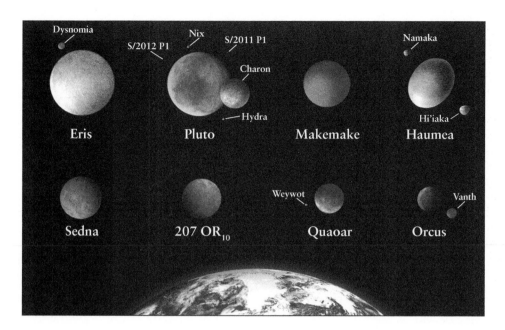

FIGURE 14-20

The Largest Trans-Neptunian Objects This artist's impression depicts Earth and the largest objects known beyond Neptune (as of 2012) to the same scale. The largest of these, Eris, has a diameter between 2300 km and 2400 km. There is less uncertainty about Pluto's diameter, which is close to 2300 km. Note the differences in color among these objects and that five of them have satellites of their own. Table 7-4 in Section 7-5 lists the properties of seven of these objects. (NASA; ESA; and A. Field [STScI])

With improvements in the sensitivity of telescopes, new objects beyond Neptune are being discovered at a rapid pace. Based on these observations, it is thought that there could be 35,000 or more objects beyond Neptune with diameters greater than 100 km. One of the larger objects could have collided with Pluto in the past, giving rise to Pluto's retinue of satellites (see Section 14-9). At least 40 trans-Neptunian objects have satellites, which suggests that such collisions have taken place many times.

The Kuiper Belt and Beyond

Most of the trans-Neptunian objects lie within the **Kuiper belt,** which extends from about 30 to 50 AU from the Sun and is relatively close to the ecliptic. When the solar system was young, a large number of icy planetesimals formed in the region beyond Jupiter. Over time, the gravitational forces of the massive Jovian planets deflected most of these planetesimals beyond Neptune's orbit, concentrating them into a belt centered on the plane of the ecliptic. Most of the trans-Neptunian objects within the belt are in orbits that are only slightly inclined to the ecliptic; these objects are thought to have formed beyond Neptune and to be in roughly their original orbits. Other objects such as Makemake and Haumea (see Figure 14-20 and Table 7-4) are in orbits that are inclined by about 30° from the ecliptic. These objects are thought to have been pushed into their steeply inclined orbits by gravitational interactions with Neptune.

The processes that gave rise to the Kuiper belt in our solar system also appear to have taken place around other stars. Figure 14-21 shows a disk of material surrounding the young star HD 139664. This disk has dimensions comparable to our Kuiper belt. A number of other young stars have been found with disks of this same type.

There are relatively few members of the Kuiper belt between the orbits of Neptune and Pluto. Remarkably, there are about 100 objects that have nearly the same semimajor axis as Pluto. These so-called **plutinos,** which

> Neptune's gravity shapes the orbits of many icy worlds, including Pluto

include Pluto itself, have the property that they complete nearly two orbits around the Sun in the same time that Neptune completes three orbits. The plutinos thus experience rhythmic gravitational pulls from Neptune, and these pulls keep them in their orbits. (In Section 13-1 we saw a similar relationship among the orbital periods of Jupiter's satellites Io, Europa, and Ganymede, though the ratio of their orbital periods is 1:2:4 rather than the 2:3 ratio

FIGURE 14-21   R I V U X G

A "Kuiper Belt" Around Another Star The star HD 139664 is only slightly more massive than the Sun but is thought to be just 300 million years old. This Hubble Space Telescope image shows a ring of material around HD 139664 that is similar in size to the Kuiper belt in our solar system. HD 139664 is 57 light-years (17 parsecs) from Earth in the constellation Lupus (the Wolf). (NASA; ESA; and P. Kalas [University of California, Berkeley])

for Neptune and the plutinos.) Most Kuiper belt objects orbit at distances beyond the plutinos but within about 50 AU from the Sun, at which distance an object completes one orbit for every two orbits of Neptune. At this distance the rhythmic gravitational forces of Neptune appear to pull small objects inward, giving the Kuiper belt a relatively sharp outer edge.

Two examples of trans-Neptunian objects that are not members of the Kuiper belt are shown in Figure 14-20. Eris, the largest known trans-Neptunian object, has a semimajor axis of more than 68 AU and an orbital eccentricity of 0.442. This orbit takes Eris from inside the orbit of Pluto to far beyond the Kuiper belt. Eris is also the most reflective body in the solar system—more reflective than ice or snow, and even more reflective than a mirror! Eris reflects about 96% of the light striking its surface, whereas a mirror reflects about 90–95%. It is thought that this unusual reflectivity comes from a very thin layer (less than a millimeter) of nitrogen-rich ice mixed with methane frost. Due to Eris's highly elongated orbit, as it moves farther from the Sun, its atmosphere can freeze onto the surface to produce this reflective layer.

Figure 14-20 also shows an even more extreme case with Sedna: Its orbital semimajor axis is 518 AU, giving it an orbital period of more than 10,000 years, and the orbital eccentricity has the remarkably high value of 0.85. It is not well understood how Sedna could have been moved into such an immense and elongated orbit.

New Horizons

Excitement about the worlds beyond Neptune has motivated the development of a spacecraft called *New Horizons*. Launched in 2006, *New Horizons* swung by Jupiter in 2007 and will make the first-ever flyby of Pluto and Charon in 2015. To make the journey in nine years, *New Horizons* was launched fast enough to pass our Moon nine hours after launch, while it took more than three days for the *Apollo 11* mission; *New Horizons* had the greatest launch speed of any man-made object.

One of many surveys *New Horizons* will attempt is to observe methane frost gathering on the nighttime side of Pluto. But isn't the nighttime side dark? No. Just as moonlight can illuminate portions of Earth during nighttime, Pluto's moon Charon can do the same for Pluto. Even though sunlight is weaker at Pluto compared to Earth, Charon is so close to Pluto that Charon-moonlight is *twice as bright* as the light we experience from our Moon. Thus, an astronaut could read a book by the bright Charon moonlight!

Ironically, the nine years it takes to get to Pluto might save the *New Horizons* mission by allowing time to solve a potential problem. Some recent estimates suggest that small but frequent impacts in the Kuiper belt could endow Pluto with a large cloud of dangerous dust; even colliding with dust a few millimeters in size could destroy the spacecraft. We are searching for this dangerous dust cloud from Earth, and one "thread the needle" idea is to send *New Horizons* through such a cloud on a path that has already been cleared of dust by Charon. The alternative is to veer farther away from Pluto, but that would lower the resolution of features imaged on Pluto's surface.

After surveying Pluto and its moons, *New Horizons* will go on to explore other bodies in the Kuiper belt. The high-resolution images and other data to be returned by *New Horizons* promise to revolutionize our understanding of these most remote members of the solar system.

CONCEPTCHECK 14-10

How do Pluto's properties differ from the planets?
Answer appears at the end of the chapter.

KEY WORDS

dwarf planet, p. 395
Kuiper belt, p. 398
magnetic axis, p. 387
occultation, p. 388

plutino, p. 398
radiation darkening, p. 390
trans-Neptunian object,
   p. 397

KEY IDEAS

Discovery of the Outer Planets: Uranus was discovered by chance, while Neptune was discovered at a location predicted by applying Newtonian mechanics.

Atmospheres of Uranus and Neptune: Both Uranus and Neptune have atmospheres composed primarily of hydrogen, helium, and a few percent methane.

• Methane absorbs red light, giving Uranus and Neptune their greenish-blue color.

• No white ammonia clouds are seen on Uranus or Neptune. Presumably the low temperatures have caused almost all the ammonia to precipitate into the interiors of the planets. All of these planets' clouds are composed of methane.

• Much more cloud activity is seen on Neptune than on Uranus. This is because Uranus lacks a substantial internal heat source.

Interiors and Magnetic Fields of Uranus and Neptune: Both Uranus and Neptune may have a rocky core surrounded by a mantle of water and ammonia. Electric currents in these mantles may generate the magnetic fields of the planets.

• The magnetic axes of both Uranus and Neptune are steeply inclined from their axes of rotation. The magnetic and rotational axes of all the other planets are more nearly parallel. The magnetic fields of Uranus and Neptune are also offset from the centers of the planets.

Uranus's Unusual Rotation: Uranus's axis of rotation lies nearly in the plane of its orbit, producing greatly exaggerated seasonal changes on the planet.

• This unusual orientation may be the result of one or more collisions with planetlike objects early in the history of our solar system. Such collisions could have knocked Uranus on its side.

Ring Systems of Uranus and Neptune: Uranus and Neptune are both surrounded by systems of thin, dark rings. The low reflectivity of the ring particles may be due to radiation-darkened methane-ice.

Satellites of Uranus and Neptune: Uranus has five satellites similar to the moderate-sized moons of Saturn, plus at least 22 more small satellites. Neptune has 13 satellites, one of which (Triton) is comparable in size to our Moon or the Galilean satellites of Jupiter.

• Triton has a young, icy surface indicative of tectonic activity. The energy for this activity may have been provided by tidal heating that occurred when Triton was captured by Neptune's gravity into a retrograde orbit.

• Triton has a tenuous nitrogen atmosphere.

Dwarf Planet: An object orbiting the Sun (but is not a moon) with enough mass to gravitationally pull itself into a spherical shape, yet not enough gravity to clear out planetesimals from its surroundings.

• This term was introduced in 2006 to help reclassify Pluto as a dwarf planet.

Trans-Neptunian Object: An object that orbits the Sun (other than planets and comets) and on average orbits at a distance greater than Neptune.

• More than a thousand icy worlds have been discovered beyond Neptune. Pluto and Charon are part of this population.

• Most trans-Neptunian objects lie in a band called the Kuiper belt that extends from 30 to 50 AU from the Sun. Neptune's gravity shapes the orbits of objects within the Kuiper belt.

QUESTIONS

Review Questions

1. *TUTORIAL 14-1* Could astronomers in antiquity have seen Uranus? If so, why was it not recognized as a planet?

2. Why do you suppose that the discovery of Neptune is rated as one of the great triumphs of science, whereas the discoveries of Uranus and Pluto are not?

3. Why is it so difficult to see features in the atmosphere of Uranus?

4. (a) Draw a figure like Figure 14-3, and indicate on it where Uranus was in 1986 and 2004. Explain your reasoning. (*Hint:* See Figure 14-1 and Figure 14-2.) (b) In approximately what year will the Sun next be highest in the sky as seen from Uranus's south pole? Explain your reasoning.

5. Why do you suppose the tilt of Uranus's rotation axis was deduced from the orbits of its satellites and not by observing the rotation of the planet itself?

6. Describe the seasons on Uranus. In what ways are the Uranian seasons different from those on Earth?

7. *TUTORIAL 14-2* Explain the statement "Methane is to Uranus's atmosphere as water is to Earth's atmosphere."

8. A number of storms in the Uranian atmosphere can be seen in Figure 14-2, but none are visible in Figure 14-1. How can you account for the difference?

9. Why are Uranus and Neptune distinctly blue-green in color, while Jupiter and Saturn are not?

10. How does the energy source for Uranus's atmospheric motions differ from those from Jupiter, Saturn, and Neptune?

11. Why are fewer white clouds seen on Uranus and Neptune than on Jupiter and Saturn?

12. Why do Uranus and Neptune have higher densities than Jupiter and Saturn?

13. Discuss the main reason why Uranus and Neptune are substantially smaller than Jupiter and Saturn.

14. How do the orientations of Uranus's and Neptune's magnetic axes differ from those of other planets?

15. Briefly describe why it is thought that Uranus was struck by at least one large planetlike object several billion years ago.

16. Compare the rings that surround Jupiter, Saturn, Uranus, and Neptune. Briefly discuss their similarities and differences.

17. The 1977 occultation that led to the discovery of Uranus's rings was visible from the Indian Ocean. Explain why it could not be seen from other parts of Earth's night side.

18. As *Voyager 2* flew past Uranus, it produced images only of the southern hemispheres of the planet's satellites. Why do you suppose this was?

19. Why do astronomers think that the energy needed to resurface parts of Miranda came from tidal heating rather than the satellite's own internal heat?

20. Using the data in Appendix 3, explain why Uranus's satellites Caliban and Stephano were probably captured from space rather than having formed at the same time as the planet itself.

21. Briefly describe the evidence supporting the idea that Triton was captured by Neptune.

22. If you were floating in a balloon in Neptune's upper atmosphere, in what part of the sky would you see Triton rise? Explain your reasoning.

23. Why is it reasonable to suppose that Neptune will someday be surrounded by a broad system of rings, perhaps similar to those that surround Saturn?

24. How can astronomers distinguish a faint solar system object like Pluto from background stars within the same field of view?

25. What is the evidence that the five moons of Pluto might have a common origin?

26. How does the presence of Pluto's moons suggest that there must be other worlds beyond Neptune?

27. Why are there a large number of objects with the same semimajor axis as Pluto?

28. Are there any trans-Neptunian objects that are not members of the Kuiper belt? Are there any members of the Kuiper belt that are not trans-Neptunian objects? Explain your answers.

Advanced Questions

Questions preceded by an asterisk (*) involve topics discussed in Box 1-1.

> **Problem-solving tips and tools**
>
> See Box 1-1 for the small-angle formula. Recall that the volume of a sphere of radius r is $4\pi r^3/3$. Section 4-7 discusses the gravitational force between two objects and Newton's form of Kepler's third law. You will find discussions of the original form of Kepler's third law in Section 4-4, tidal forces in Section 4-8, Wien's law for blackbody radiation in Section 5-4, and the transparency of Earth's atmosphere to various wavelengths of light in Section 6-7.

29. For which configuration of the Sun, Uranus, and Neptune is the gravitational force of Neptune on Uranus at a maximum?

For this configuration, calculate the gravitational force exerted by the Sun on Uranus and by Neptune on Uranus. Then calculate the fraction by which the sunward gravitational pull on Uranus is reduced by Neptune at that configuration. Based on your calculations, do you expect that Neptune has a relatively large or relatively small effect on Uranus's orbit?

30. At certain points in its orbit, a stellar occultation by Uranus would *not* reveal the existence of the rings. What points are those? How often does this circumstance arise? Explain using a diagram.

31. According to one model for the internal structure of Uranus, the rocky core and the surrounding shell of water-ices and methane-ices together make up 80% of the planet's mass. This interior region extends from the center of Uranus to about 70% of the planet's radius. (a) Find the average density of this interior region. (b) How does your answer to (a) compare with the average density of Uranus as a whole? Is this what you would expect? Why?

32. Uranus's epsilon (ε) ring has a radius of 51,150 km. (a) How long does it take a particle in the ε ring to make one complete orbit of Uranus? (b) If you were riding on one of the particles in the ε ring and watching a cloud near Uranus's equator, would the cloud appear to move eastward or westward as Uranus rotates? Explain your answer.

33. Suppose you were standing on Pluto. Describe the motions of Charon relative to the Sun, the stars, and your own horizon. Would you ever be able to see a total eclipse of the Sun? (*Hint:* You will need to calculate the angles subtended by Charon and by the Sun as seen by an observer on Pluto.) In what circumstances would you *never* see Charon?

34. It is thought that Pluto's tenuous atmosphere may become even thinner as the planet moves toward aphelion (which it will reach in 2113), then regain its present density as it again moves toward perihelion. Why should this be?

35. The brightness of sunlight is inversely proportional to the square of the distance from the Sun. For example, at a distance of 4 AU from the Sun, sunlight is only $(\frac{1}{4})^2 = \frac{1}{16} = 0.0625$ as bright as at 1 AU. Compared with the brightness of sunlight on Earth, what is its brightness (a) on Pluto at perihelion (29.649 AU from the Sun) and (b) on Pluto at aphelion (49.425 AU from the Sun)? (c) How much brighter is it on Pluto at perihelion compared with aphelion? (Even this brightness is quite low. Noon on Pluto is about as dim as it is on Earth a half hour after sunset on a moonless night.)

*36. If Earth-based telescopes can resolve angles down to 0.25 arcsec, how large could a trans-Neptunian object be at Pluto's average distance from the Sun and still not present a resolvable disk?

*37. Calculate the maximum angular separation between Pluto and Charon as seen from Earth. Assume that Pluto is at perihelion (29.649 AU from the Sun) and that Pluto is at opposition as seen from Earth.

38. Presumably Pluto and Charon raise tidal bulges on each other. Explain why the average distance between Pluto and Charon is probably constant, rather than increasing like the Earth-Moon distance or decreasing like the Neptune-Triton distance. Include a diagram like Figure 10-20 as part of your answer.

39. (a) Find the semimajor axis of the orbit of an object whose period is 3/2 of the orbital period of Neptune. How does your result compare to the semimajor axis of Pluto's orbit? (b) A number of Kuiper belt objects called plutinos have been discovered with the same orbital period and hence the same semimajor axis as Pluto. Explain how these objects can avoid colliding with Pluto.

40. Find the semimajor axis of the orbit of an object whose period is twice the orbital period of Neptune. How does your result compare to the outer limit of the Kuiper belt?

41. Suppose you wanted to search for trans-Neptunian objects. Why might it be advantageous to do your observations at infrared rather than visible wavelengths? (*Hint:* At visible wavelengths, the light we see from planets is reflected sunlight. At what wavelengths would you expect distant planets to *emit* their own light most strongly? Use Wien's law to calculate the wavelength range best suited for your search.) Could such observations be done at an observatory on Earth's surface? Explain your answer.

42. The *New Horizons* spacecraft swung by Jupiter to get a boost from that planet's gravity, enabling it to reach Pluto relatively quickly. To see what would happen if this technique were not used, consider a spacecraft trajectory that is an elliptical orbit around the Sun. The perihelion of this orbit is at 1 AU from the Sun (at Earth) and the aphelion is at 30 AU (at Pluto's position). Calculate how long it would take a spacecraft in this orbit to make the one-way trip from Earth to Pluto. Based on the information in Section 14-10, how much time is saved by making a swing by Jupiter instead?

Discussion Questions

43. Discuss the evidence presented by the outer planets that suggests that catastrophic impacts of planetlike objects occurred during the early history of our solar system.

44. Some scientists are discussing the possibility of placing spacecraft in orbit about Uranus and Neptune. What kinds of data should be collected, and what questions would you like to see answered by these missions?

45. If Triton had been formed along with Neptune rather than having been captured, would you expect it to be in a prograde or retrograde orbit? Would you expect the satellite to show signs of tectonic activity? Explain your answers.

46. Would you expect the surfaces of Pluto and Charon to be heavily cratered? Explain why or why not.

47. Imagine that you are in charge of planning the *New Horizons* flyby of Pluto and Charon. In your opinion, what data should

be collected and what kinds of questions should the mission attempt to answer?

48. In 2006 the International Astronomical Union changed Pluto's designation from *planet* to *dwarf planet*. One criterion that Pluto failed to meet was that a planet must have "cleared the neighborhood" around its orbit. In what sense has Pluto not done so? In what sense have the eight planets (Mercury through Uranus) cleared their neighborhoods? Do you agree with this criterion?

Web/eBook Questions

49. **Miranda.** Access and view the video "Uranus's Moon Miranda" in Chapter 14 of the *Universe* Web site or eBook. Discuss some of the challenges that would be involved in launching a spacecraft from Earth to land on the surface of Miranda.

50. Charon was discovered by an astronomer at the U.S. Naval Observatory. Why do you suppose the U.S. Navy carries out work in astronomy? Search the World Wide Web for the answer.

51. Search the World Wide Web for a list of trans-Neptunian objects. What are the largest and smallest objects of this sort that have so far been found, and how large are they? Have any objects larger than Eris been found?

ACTIVITIES

Observing Projects

Observing tips and tools

You can find Uranus and Neptune with binoculars if you know where to look (a good star chart is essential), but Pluto is so dim that it can be a challenge to spot even with a 25-cm (10-inch) telescope. Each year, star charts that enable you to find these planets are printed in the issue of *Sky & Telescope* for the month in which each planet is first visible in the nighttime sky. You can also locate the outer planets using the *Starry Night*™ program if you have access.

52. Make arrangements to view Uranus through a telescope. The planet is best seen at or near opposition. Use a star chart at the telescope to find the planet. Are you certain that you have found Uranus? Can you see a disk? What is its color?

53. If you have access to a large telescope, make arrangements to view Neptune. Like Uranus, Neptune is best seen at or near opposition and can most easily be found using a star chart. Can you see a disk? What is its color?

54. If you have access to a large telescope (at least 25 cm in diameter), make arrangements to view Pluto. Using the star chart from *Sky & Telescope* referred to above, view the part of the sky where Pluto is expected to be seen and make a careful

sketch of all of the stars that you see. Repeat this process on a later night. Can you identify the "star" that has moved?

55. Use the *Starry Night*™ program to observe Uranus. Open **Favourites > Explorations > Uranus Satellites** to see an image of Uranus and its five largest moons as they would have appeared through a telescope situated at the south pole of Earth in February 1986. Select **Labels > Planets-Moons** to turn labels to these moons on or off. With the **Time Flow Rate** at 5 minutes, click **Play** and observe the motion of the moons around the planet. (a) Describe how the satellites move, and relate your observations to Kepler's third law. (b) Change the year in the toolbar to **2007** and observe the motion of the moons as time runs forward. How do the orbits look different than in (a)? Explain any differences.

56. Use the *Starry Night*™ program to examine the satellites of Uranus. Open **Favourites > Explorations > Uranus** and select **View > Stars > Stars** from the menu to remove the background stars from the view. Use the **Elevation** buttons in the **Viewing Location** section of the toolbar to change the distance from the planet to about **0.01 AU**. You should now be able to see at least five satellites of the planet Uranus. Open the **Find** pane and expand the layer for **Uranus**. Click the checkboxes to the left of each of the listed moons to display their names and click the checkboxes to the right of the listed moons to display their orbits in the view. Use the location scroller to adjust the view so that the orbits of the moons appear edge-on. (a) Do all of the satellites appear to lie in this same plane? (*Hint:* **Zoom** in on Uranus if necessary to check your conclusion.) (b) How do you imagine that this plane relates to the plane of Uranus's equator?

57. Use the *Starry Night*™ program to observe Neptune and two of its moons. Select **Favourites > Explorations > Triton and Nereid**. The view shows Neptune and its moons, Triton and Nereid, from a location in space hovering 0.01 AU above Neptune's equator. The stars have been removed from the view. With the **Time Flow Rate** at 1 hour, click the **Play** button and observe the motion of these satellites around Neptune. You can use the location scroller to change your viewpoint and you can use the **Elevation** buttons to move closer to or farther from Neptune. (a) Does Triton orbit Neptune in the same direction as Neptune's direction of rotation or in the opposite direction? (b) How will Triton's orbital direction relative to the rotational direction of Neptune affect this moon's ultimate fate? (c) Nereid is in a rather elongated orbit that takes it far from Neptune. Does Nereid orbit Neptune in the same direction as Neptune's direction of rotation or in the opposite direction? In which direction (clockwise or counterclockwise) does Nereid orbit Neptune when viewed from above the north pole of the planet? (d) When it is closest to Neptune, does Nereid approach closer to Neptune than does Triton? If you decrease the elevation of the observing location above Neptune to about 500,000 km, you will see guides on Neptune indicating the planet's poles, equator, and a meridian.

Use the **Time Flow** controls and the position of the meridian on the planet to determine the approximate rotational period of Neptune. (**e**) What is the planet's rotation period?

58. Use the *Starry Night*™ program to observe Pluto and Charon. Select **Favourites > Explorations > Pluto**. The view is centered on Pluto from about 42,000 km above its surface. One of Pluto's moons, Charon, is also shown, along with its orbit. Note that the south pole stick on Charon lines up with the star Atria and that the red meridian and pole sticks on Pluto are aligned. Make a note of the time shown in the toolbar. With the **Time Flow Rate** at 30,000×, click the **Play** button and observe the motion of Charon around Pluto and the rotation of Pluto. Use the time controls to stop time once Charon has completed a full orbit with its south pole stick once again aligned with the star Atria. (**a**) What is the period of Charon's orbit? (**b**) What is Pluto's rotation period? (**c**) How do these two periods compare? Explain this result. (**d**) If Earth and the Moon were to have the same orbital and rotational relationship between them that Pluto and Charon have, which of the following statements would NOT be true?

  (i) The Moon would always appear above one particular location on Earth.

  (ii) Everyone on Earth would see the Moon pass through their sky every month.

  (iii) Earth would rotate with a period of 27.3 days.

Collaborative Exercise

59. Sir William Herschel, a British astronomer, discovered Uranus in 1781 and named it Georgium Sidus, after the reigning monarch, George III. What name might Uranus have been given in 1781 if an astronomer in your country had discovered it? Why? What if it had been discovered in your country in 1881? In 1991?

ANSWERS

ConceptChecks

ConceptCheck 14-1: The trajectory of Uranus did not show slow changes in speed predicted by its orbit around the Sun. Instead, Uranus deviated by slightly speeding up and slowing down, and this could be attributed to a new planet at just the right location.

ConceptCheck 14-2: Methane in Uranus's atmosphere absorbs the Sun's longer wavelengths of light, so if it were absent, Uranus would have more reds and, as a result, appear more yellow-white, like Saturn.

ConceptCheck 14-3: It is thought that Neptune is still contracting, which converts gravitational energy into thermal energy to power its atmosphere.

ConceptCheck 14-4: At Neptune's present distance, there would not have been enough matter to form such a large planet. Furthermore, even if there were more matter, Neptune has a higher proportion of heavy elements than expected in the outer reaches of the solar system.

ConceptCheck 14-5: Ammonia dissolved in water under high pressure loses electrons, which become free to move, and this makes the fluid electrically conductive.

ConceptCheck 14-6: Due to radiation darkening, Uranian rings are black as coal, unlike Saturn's icy light-reflecting rings. Uranus's rings are also very narrow, at about 10 km wide, compared to rings on Saturn that can be over 1000 times wider.

ConceptCheck 14-7: The orbits of some moons were seen to change between observations made in 1994 and 2005. Based on the degree of observed change, it might only have taken a few hundred million years before these moons would have collided, so they might have been acquired somewhat recently.

ConceptCheck 14-8: Geyserlike plumes observed by *Voyager 2* showed streaking, indicating that they were carried downwind by a thin atmosphere.

ConceptCheck 14-9: Pluto's gravity is too weak to clear its neighborhood of nearby planetesimals in the Kuiper belt. According to new classifications adopted in 2006—that a planet must clear its neighborhood—this makes Pluto a dwarf planet instead of a planet.

ConceptCheck 14-10: Pluto is much smaller than any planet, and has a much higher fraction of water-ice.

Comet Hale-Bopp, the great comet of 1997. (John Chumack/Science Photo Library/Science Source) R I V U X G

Vagabonds of the Solar System

On the Tuesday night after Easter in the year 1066, a strange new star appeared in the European sky. Seemingly trailing fire, this "star" hung in the night sky for weeks. Some, including the English king Harold, may have taken it as an ill omen; others, such as Harold's rival William, Duke of Normandy, may have regarded it as a sign of good fortune. Perhaps they were both right, because by the end of the year Harold was dead and William was installed on the English throne.

Neither man ever knew that he was actually seeing the trail of a city-sized ball of ice and dust, part of which evaporated as it rounded the Sun to produce a long tail. They were seeing a comet—remarkably, the same Comet Halley that last appeared in 1986 and will next be seen in 2061. Although we now understand that comets are natural phenomena rather than supernatural omens, they still have the power to awe and inspire us, as did Comet Hale-Bopp (shown here) in 1997.

While comets are made of ice and dust, asteroids are rocky bits of the solar nebula that never formed into a full-sized planet. Some asteroids break into fragments, as do some comets, and some of these fragments fall to Earth as meteorites. On extraordinarily rare occasions, an entire asteroid strikes Earth. Such an asteroid impact may have led to the extinction of the dinosaurs some 65 million years ago. Thus, these minor members of our solar system can have major consequences for our planet.

15-1 A search for a planet between Mars and Jupiter led to the discovery of asteroids

TUTORIAL 15-1 After William Herschel's discovery of Uranus in 1781 (which we described in Section 14-1), many astronomers began to wonder if there were other, as yet undiscovered, planets. If these planets were too dim to be seen by the naked eye, they might still be visible through telescopes. These planets would presumably be found close to the plane of the ecliptic, because all the other planets orbit the Sun in or near that plane (see Section 7-1). But how far from the Sun might these additional planets be found?

The Hunt for the "Missing Planet"

LOOKING DEEPER 15-1 Astronomers in the late eighteenth century had a simple rule of thumb, called the *Titius-Bode law,* which relates the sizes of planetary orbits. The law states that from one planet to the next, the semimajor axis of the orbit increases by a factor between approximately 1.4 and 2. For example, the semimajor axis of Mercury's orbit is 0.39 AU. Venus's orbit has a semimajor axis of 0.72 AU, which is larger by a factor of $(0.72)/(0.39) = 1.85$. (Unlike Newton's laws, the Titius-Bode "law" is not a fundamental law of nature. It is probably just a reflection of how our solar system happened to form.)

The glaring exception is Jupiter (semimajor axis 5.20 AU), which is more than *3 times* farther from the Sun than Mars (semimajor axis 1.52 AU). Table 7-1 depicts this large space between the orbits of Mars and Jupiter. If the rule of thumb is correct, astronomers reasoned, there should be a "missing planet" with an orbit about 1.4 to 2 times larger than that of Mars. This planet should therefore have a semimajor axis between 2 and 3 AU.

Six German astronomers, who jokingly called themselves the "Celestial Police," organized an international group to begin a careful search for this missing planet. Before their search began, however, surprising news reached them from Giuseppe Piazzi, a Sicilian astronomer. Piazzi had been carefully mapping faint stars in the constellation of Taurus when on January 1, 1801, the first night of the nineteenth century, he noticed a dim, previously uncharted star. This star's position shifted slightly over the next several nights. Suspecting that he might have found the "missing planet," Piazzi excitedly wrote to Johann Bode, the director of the Berlin Observatory and a member of the Celestial Police.

Unfortunately, Piazzi's letter did not reach Bode until late March. By that time, Earth had moved around its orbit so that Piazzi's object appeared too near the Sun to be visible in the night sky. With no way of knowing where to look after it emerged from the Sun's glare, astronomers feared that Piazzi's object might have been lost.

Upon hearing of this dilemma, the brilliant young German mathematician Karl Friedrich Gauss took up the challenge. He developed a general method of computing an object's orbit from only three separate measurements of its position on the celestial sphere. (With slight modifications, this same method is used by astronomers today.) In November 1801, Gauss predicted that Piazzi's object would be found in the predawn sky in the constellation of Virgo. And indeed Piazzi's object was sighted again on December 31, 1801, only a short distance from the position Gauss had calculated. Piazzi named the object Ceres (pronounced SEE-reez), after the patron goddess of Sicily in Roman mythology.

FIGURE 15-1 R I V U X G
Ceres Compared with Earth and Moon A drawing of Ceres, the largest of the asteroids and the first one discovered, is shown here to the same scale as Earth and the Moon. Ceres is too small to be considered a planet, and it cannot be regarded as a moon because it does not orbit any other body. To denote their status, asteroids are also called minor planets. (NASA)

Ceres orbits the Sun once every 4.6 years at an average distance of 2.77 AU, just where astronomers had expected to find the missing planet. But Ceres is very small; its equatorial diameter is a mere 974.6 km (Figure 15-1). (The pole-to-pole diameter is even smaller, just 909.4 km.) Hence, Ceres reflects only a little sunlight, which is why it cannot be seen with the naked eye even at opposition; it can be seen with binoculars, but it looks like just another faint star.

On March 28, 1802, Heinrich Olbers discovered another faint, starlike object that moved against the background stars. He called it Pallas, after the Greek goddess of wisdom. Like Ceres, Pallas orbits the Sun every 4.6 years at an average distance of 2.77 AU, but its orbit is more steeply inclined from the plane of the ecliptic and is somewhat more eccentric. Pallas is even smaller and dimmer than Ceres, with an estimated diameter of only 522 km. Obviously, Pallas was also not the missing planet. Ceres and Pallas are **asteroids.** Asteroids are small objects with various mixtures of rock, ice, and metal that orbit the Sun between Mars and Jupiter. (Like all asteroids, Ceres and Pallas are classified as *small solar system bodies,* or **minor planets**; these are all objects other than planets and comets that are in direct orbit around the Sun.)

A new question then faced the astronomers of the early nineteenth century: Did a missing planet even exist? Some speculated that perhaps there had once been such a full-sized planet, but it had somehow broken apart or exploded to produce the asteroids. The search was on to discover this population. Two more such asteroids were discovered in the next few years, Juno in 1804 and Vesta in 1807. As telescope technology improved, hundreds of more asteroids were discovered. Going beyond telescopes in 2007—two hundred years after Vesta's discovery—NASA's spacecraft *Dawn* was launched on a mission to visit Vesta and Ceres (more in Section 15-3).

An Abundance of Asteroids

The next major breakthrough came in 1891, when the German astronomer Max Wolf began using photographic techniques to search for asteroids. Before this, asteroids had to be painstakingly discovered by scrutinizing the skies for faint, uncharted, starlike objects whose positions move slightly from one night to the next. With photography, an astronomer simply aims a camera-equipped telescope at the stars and takes a long exposure. If an asteroid happens to be in the field of view, its orbital motion during the long exposure leaves a distinctive trail on the photographic plate (Figure 15-2). Using this technique, Wolf alone discovered 228 asteroids.

Today, more than a million asteroids have been discovered, of which more than 100,000 have been studied well enough so that their orbits are known. About 5000 more are discovered every month, some by amateur astronomers. After the discoverer reports a new find to the Minor Planet Center of the Smithsonian Astrophysical Observatory, the new asteroid is given a provisional designation. For example, asteroid 1980 JE was the fifth ("E") to be discovered during the second half of May (the tenth half-month of the year, "J") 1980. The same scheme for provisional designations is also used for small objects found in the outer solar system beyond Neptune, or trans-Neptunian objects (see Section 14-10).

If the asteroid is located again on at least four succeeding oppositions—a process that can take decades—the asteroid is assigned an official sequential number (Ceres is 1, Pallas is 2, and so forth), and the discoverer is given the privilege of suggesting a name for the asteroid. The suggested name must then be approved by the International Astronomical Union. For example, the asteroid 1980 JE was officially named 3834 Zappafrank (after the American musician Frank Zappa) in 1994.

Ceres (formally called "1 Ceres") is unquestionably the largest asteroid at 975 km; it has one-third the mass of all the asteroids combined (including Ceres). In recognition of its size, Ceres is also called a dwarf planet—large enough to pull itself into a sphere, but unable to gravitationally clear out its surroundings (see Section 14–9).

Besides Ceres, only two asteroids—Pallas and Vesta—have diameters greater than 500 km. A general feature of asteroids is that smaller ones are more numerous. Thus, we find other asteroids have diameters between 200 and 300 km, and 200 more are bigger than 100 km across. Infrared observations indicate that

there are between 700,000 and 1.7 million asteroids greater than 1 km in size. And while a million asteroids might sound like a lot, the vast majority of asteroids are smaller than 1 km across.

CAUTION! Be careful not to confuse asteroids with the trans-Neptunian objects that we discussed in Section 14-10. Asteroids are found in the inner solar system, are made primarily of rocky materials, and are generally quite small. By contrast, trans-Neptunian objects are found primarily in the Kuiper belt beyond Neptune, are a mixture of ices and rock, and range in size from a few kilometers across to 2900 km in diameter—more than 3 times the diameter of Ceres. The combined mass of all trans-Neptunian objects is not known, but is probably thousands of times greater than the combined mass of all asteroids. Asteroids make up only a tiny fraction of the total mass of the solar system!

Interestingly, the dwarf planets Ceres and Pluto have a common story line. Like Pluto, Ceres was considered a planet at first and retained that designation for about 50 years. Just as Pluto recently went from being the smallest planet to being one of the largest trans-Neptunian objects, Ceres had been reclassified and went from being the smallest planet to the largest asteroid. Ceres and Pluto were simply the first objects discovered in these new categories.

The Asteroid Belt

Like Ceres, Pallas, Vesta, and Juno, most asteroids orbit the Sun at distances between 2 and 3.5 AU. This region of our solar system between the orbits of Mars and Jupiter is called the **asteroid belt** (Figure 15-3).

CAUTION! You may wonder how spacecraft such as *Voyager 1, Voyager 2, Galileo,* and *Cassini* were able to cross the asteroid belt to reach Jupiter or Saturn. Aren't asteroids hazards to space navigation, as they are sometimes depicted in science-fiction movies? Wouldn't these spacecraft have been likely to run into an asteroid? Happily, the answer to both questions is no! While it is true that there are more than 10^5 asteroids, they are spread over a belt with a total area (as seen from above the plane of the ecliptic) of about 10^{17} square km—about 10^8 times greater than Earth's entire surface area. Hence, the average distance between asteroids in the ecliptic plane is about 10^6 km, about twice the distance between Earth and the Moon. Furthermore, many asteroids have orbits that are tilted out of the ecliptic. The *Galileo* spacecraft did pass within a few thousand kilometers of two asteroids, but these passes were intentional and required that the spacecraft be carefully aimed.

> Science-fiction movies to the contrary, asteroids are not hazards to space navigation

What of the nineteenth-century notion of a missing planet? This idea has long since been discarded, because the combined matter of all the asteroids (including an estimate for those not yet officially known) would produce an object barely 1500 km in diameter. This combined matter is considerably smaller than anything that could be considered a missing planet. We now understand that asteroids are actually debris left over from the formation of the solar system.

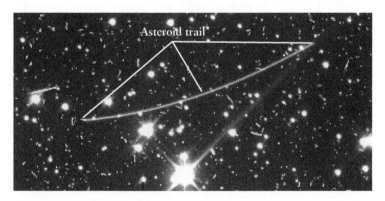

FIGURE 15-2   R I ⃒V⃒ U X G

The Trail of an Asteroid Telescopes used to photograph the stars are motorized to follow the apparent motion of the celestial sphere. Because asteroids orbit the Sun, their positions change with respect to the stars, and they leave blurred trails on time exposure images of the stars. This bluish asteroid trail was recorded by accident by the Hubble Space Telescope. The shorter streaks (mostly yellow) are imaging artifacts. (R. Evans and K. Stapelfeldt, Jet Propulsion Laboratory; and NASA)

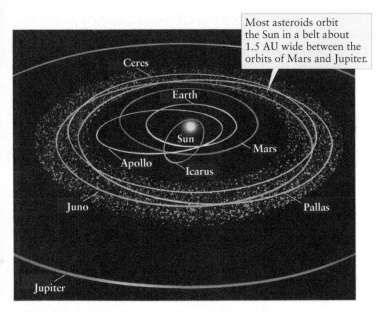

Most asteroids orbit the Sun in a belt about 1.5 AU wide between the orbits of Mars and Jupiter.

FIGURE 15-3

The Asteroid Belt Most asteroids, including the large asteroids Ceres, Pallas, and Juno, have roughly circular orbits that lie within the asteroid belt. By contrast, some asteroids, such as Apollo and Icarus, have eccentric orbits that carry them inside the orbit of Earth.

CONCEPTCHECK 15-1

Why was a "missing planet" expected around the orbital distance of the asteroid belt?

CONCEPTCHECK 15-2

Is Ceres an asteroid or dwarf planet?

Answers appear at the end of the chapter.

15-2 Jupiter's gravity helped shape the asteroid belt

Where did the asteroid belt come from? Why did asteroids, rather than a planet, form in the solar nebula (see Section 8-5) in the region between Mars and Jupiter? Important insights into these questions come from supercomputer simulations of the formation of the terrestrial planets, like the one depicted in Figure 8-11 but more elaborate.

Jupiter and the Formation of the Asteroid Belt

A typical simulation of the inner solar system starts off with a billion (10^9) or so planetesimals, each with a mass of 10^{15} kg or more, so that their combined mass equals that of the terrestrial planets. The computer then follows these planetesimals as they collide and accrete to form the planets. By selecting different speeds and positions for the planetesimals at the start of the simulation, a scientist can study alternative scenarios of the development of the inner solar system.

If the effects of Jupiter's gravity are not included in a simulation, an Earth-sized planet usually forms in the asteroid belt, giving us five terrestrial planets instead of four. But when Jupiter's gravity is added, this fifth terrestrial planet is less likely to form. Jupiter's strong pull "clears out" the asteroid belt by disrupting the orbits of planetesimals in the belt, ejecting most of them from the solar system altogether. As a result, the asteroid belt becomes depleted of planetesimals within about 1 million years, before a planet has a chance to form. The few planetesimals that remain in the simulation—only about 0.1% of those that originally orbited between Mars and Jupiter—become the asteroids that we see today.

The asteroids were never part of a larger planet

Kirkwood Gaps

Jupiter's gravity continues to influence the asteroid belt down to the present day. The American astronomer Daniel Kirkwood found the first evidence for this in 1867, when he discovered gaps in the asteroid belt. These features, today called **Kirkwood gaps,** can best be seen in a histogram like Figure 15-4, which shows asteroid orbital periods. The gaps in this histogram show regions where there are relatively few asteroids. Curiously, these gaps occur for asteroid orbits whose periods are simple fractions (such as 1/3, 2/5, 3/7, and 1/2) of Jupiter's orbital period.

To understand why the Kirkwood gaps exist, imagine an asteroid within the belt that circles the Sun once every 5.93 years, exactly half of Jupiter's orbital period. On every *second* trip around the Sun, the asteroid finds itself lined up between Jupiter and the Sun again, always at the same location and with the same orientation. Because of these repeated alignments, called a *2-to-1* **orbital resonance,** Jupiter's gravity deflects the asteroid from its

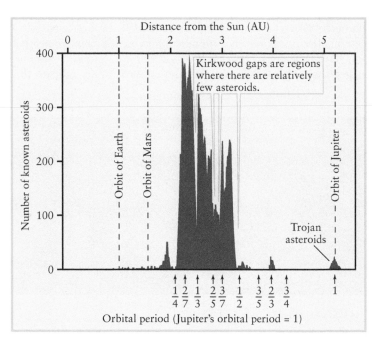

FIGURE 15-4

The Kirkwood Gaps This graph displays the numbers of asteroids at various distances from the Sun. Very few asteroids are found in orbits whose orbital periods are equal to the simple fractions 1/3, 2/5, 3/7, and 1/2 of Jupiter's orbital period. The Trojan asteroids (at far right) follow the same orbit as Jupiter and thus have the same orbital period. We describe these in Section 15-4.

original 5.93-year orbit, ultimately ejecting it from the asteroid belt (this is the gap labeled as 1/2 in Figure 15-4). By contrast, for an object not in resonance—whose orbital period does not result in repeated Sun-asteroid-Jupiter alignments—Jupiter's repeated tugs on the object occur at different locations around the Sun and do not add up in a manner to eventually pull it out of orbit.

Another Kirkwood gap corresponds to an orbital period of 1/3 Jupiter's period, or 3.95 years (a *3-to-1 orbital resonance*). Although not necessarily as strong, additional gaps exist for other simple ratios between the periods of asteroids and Jupiter. Elsewhere in the solar system, similar orbital resonances help clear out the Cassini division in Saturn's rings (see Section 12-11), and give an outer edge to the Kuiper belt beyond Neptune (see Section 14-10). These are examples of *unstable* resonances, but under special conditions a resonance can actually be a *stabilizing* influence (see the clump of asteroids with a period of 2/3 in Figure 15-4).

CONCEPTCHECK 15-3

What prevented a planet from forming where the asteroid belt currently exists, and did this effect eject mass from the region or just keep the existing mass from forming a planet?

Answer appears at the end of the chapter.

15-3 Astronomers use a variety of techniques to study asteroids

Although the asteroid belt is mostly empty space, collisions between asteroids should occur from time to time. Asteroids move in orbits with a variety of different eccentricities and inclinations (see Figure 15-3), and some of these orbits intersect each other. A collision between asteroids must be an awe-inspiring event. Typical collision speeds are estimated to be from 1 to 5 km/s (3600 to 18,000 km/h, or 2000 to 11,000 mi/h), which is more than sufficient to shatter rock. Recent observations show that these titanic impacts play an important role in determining the nature of asteroids.

> Many asteroids have been battered by collisions with other asteroids

Observing Asteroids with Telescopes, Radar, and Spacecraft

State-of-the-art optical telescopes barely have the resolution to reveal surface details on large asteroids such as Ceres, Pallas, and Vesta (Figure 15-5). These asteroids resemble miniature terrestrial planets. They have enough mass, and hence enough gravity, to pull themselves into roughly spherical shapes. These large asteroids are likely to have extensively cratered surfaces due to collisions with other asteroids in the asteroid belt.

Optical telescopes reveal much less detail about the smaller asteroids. But by studying how the brightness of an asteroid changes as it rotates, astronomers can infer the asteroid's shape. An elongated asteroid appears dimmest when its long axis points toward the Sun so that it presents less of its surface to reflect sunlight. By contrast, a rotating spherical asteroid will have a more constant brightness. Such observations show that most smaller asteroids are not spherical at all. Instead, they retain the odd shapes produced by previous collisions with other asteroids.

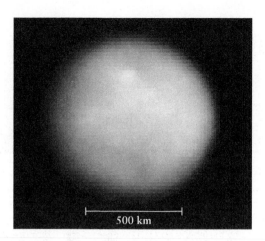

FIGURE 15-5 R I V U X G
A Hubble Space Telescope View of a Large Asteroid With an equatorial diameter of 974.6 km (605.6 mi), Ceres is the largest of all the asteroids. Its very nearly spherical shape suggests that it was once molten throughout its volume, in which case it probably underwent chemical differentiation like the terrestrial planets (see Box 7-1 and Section 8-5). (NASA; ESA; J. Parker, SwRI; P. Thomas, Cornell U.; L. McFadden, U. of Maryland; and M. Mutchler and Z. Levay, STScI)

An important recent innovation in studying asteroids has been the use of radar. An intense radio beam is sent toward an asteroid, and the reflected signal is detected and analyzed to determine the asteroid's shape (Figure 15-6). The amount of the radio signal that is reflected also gives clues about the texture of the surface.

The most powerful technique for studying asteroids is to send a spacecraft to make observations at close range. The Jupiter-bound *Galileo* spacecraft gave us our first close-up view of an asteroid in 1991, when it flew past 951 Gaspra. Two years later *Galileo* made a close pass by 243 Ida. The *NEAR Shoemaker* spacecraft made a close approach to 253 Mathilde in

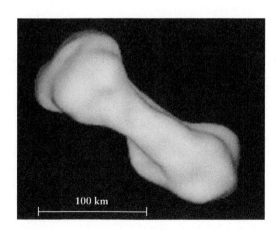

FIGURE 15-6 R I V U X G
A Radar View of a Medium-Sized Asteroid This radar image reveals the curious dog-bone shape of the asteroid 216 Kleopatra. The 300-m Arecibo radio telescope in Puerto Rico was used both to send radio waves toward Kleopatra and to detect the waves that were reflected back. Scientists then created this image by computer analysis of the reflected waves. (Arecibo Observatory, JPL/NASA)

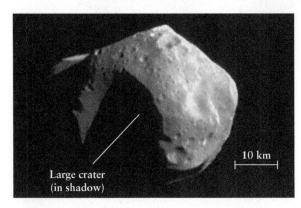

FIGURE 15-7 R I V U X G

A Spacecraft View of a Small Asteroid The asteroid 253 Mathilde has an albedo of only 0.04, making it half as reflective as a charcoal briquette. It has an irregular shape, with dimensions 66 × 48 × 46 km. The partially shadowed crater indicated by the label is about 20 km across and 10 km deep. The *NEAR Shoemaker* spacecraft imaged Mathilde at close range during its close flyby in June 1997. (Science Source)

1997 (Figure 15-7). It later went into orbit around 433 Eros (see Figure 7-7) and touched down on Eros in 2001. It was the first spacecraft to orbit and land on an asteroid. A number of other spacecraft have since made close flybys of asteroids.

One of the most ambitious asteroid missions to date is *Hayabusa* ("Falcon"), a project of the Japan Aerospace Exploration Agency. In 2005 it touched down briefly on the surface of 25143 Itokawa, where it gathered small samples of dust from the asteroid's surface (Figure 15-8). In 2010 the dust samples were safely returned to Earth, completing the first-ever asteroid sample return. Analysis indicates that the dust is probably fragments of a larger asteroid, and that the dust has been exposed to space for about 8 million years. As expected, the composition of the asteroid dust also matches some of the meteoritic material that makes its way to Earth (more on meteorites in Section 15-5).

Hayabusa also contained a mini-lander (about half a pound and 12 cm wide) designed to hop around in Itokawa's low gravity. Unfortunately, the mini-lander did not deploy correctly and has floated off into space.

Dawn Visits Vesta

In 2011, NASA's *Dawn* spacecraft went into orbit around Vesta, the second largest asteroid after Ceres (Figure 15-9). Vesta is a remarkable asteroid that some scientists think of as a transitional object between an asteroid and a planet. Vesta has an average diameter of 525 km and is thought to be a remnant protoplanet that never merged with similar objects to build a planet during formation of the solar system. Vesta is not quite massive enough to gravitationally pull itself into a sphere, so it is not a dwarf planet like Ceres. However, Vesta is large enough that its own radioactivity is thought to have melted the asteroid early on, allowing iron to sink to the core (we will learn more about *differentiated asteroids* in Section 15-5).

Estimates suggest that Vesta lost about 1% of its mass in a huge collision about one or two billion years ago. Amazingly, debris from this impact with Vesta has actually made it to Earth as meteorites. In fact, about 6% of meteorites landing on Earth originated from Vesta. The only other known bodies in the solar system whose rocky debris has been identified on Earth are the Moon and Mars.

FIGURE 15-8 R I V U X G

Sample Returned from Itokawa The surface of Itokawa is rough and studded with boulders. It has been described as a "rubble pile." Compared to Mathilde and other asteroids, Itokawa has a surprising lack of craters. The asteroid may even be the result of two asteroids merging together. The sample returned to Earth contained about 1500 small grains, and is certain to shed light on Itokawa and asteroids in general. (Atlas Photo Bank/Science Source)

The *Dawn* spacecraft has actually imaged the crater produced by the impact that ejected these rocks from Vesta (Figure 15-10). By analyzing infrared spectra, *Dawn* has confirmed that the chemical composition around Vesta's south pole crater matches the meteorites long-suspected to come from this asteroid. The collision at Vesta's south pole was so energetic that grooves formed around the equator as the entire asteroid vibrated like a ringing bell.

The *Dawn* mission does not end with Vesta, which it left in September 2012. If all goes according to plan, *Dawn* will go into orbit around Ceres in 2015.

Collisions and the Nature of Asteroids

What have spacecraft and Earth-based radar told us about the asteroids? To appreciate the answer, we should first consider what scientists expected to find with these tools. Until recently, the consensus among planetary scientists was that the smaller

FIGURE 15-9 R I V U X G

Vesta This view of Vesta is a mosaic composed of the best views from the *Dawn* spacecraft. Many craters pockmark the surface. Vesta's average diameter is 525 km. (NASA/JPL-Caltech/UCLA/MPS/DLR/IDA)

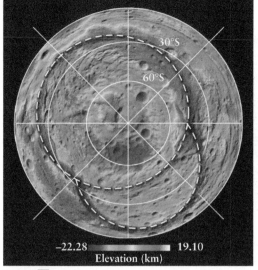

(a) R I V̲ U X G

FIGURE 15-10

Impact on Vesta (a) The impact at Vesta's south pole appears to have created its unique equatorial grooves. (b) This topographic map looking directly over the south pole shows the outlines of a massive crater named Rheasilvia. The crater is so wide that it covers most of the southern hemisphere and is about 90% the width of the entire asteroid. The central

(b) R I̲ V U X G

peak (shown in red, about 14 miles high) is similar in size to the largest mountain in the solar system—Olympus Mons on Mars. There is speculation about another large crater underneath Rheasilvia (jutting out in the lower right), but this is uncertain. (a, b: NASA/JPL-Caltech/UCLA/MPS/DLR/IDA)

asteroids were solid, rocky objects. The surfaces were presumed to be cratered by impacts, but only relatively small craters were expected. The reason is that an impact forceful enough to make a very large crater would probably fracture a rocky asteroid into two or more smaller asteroids.

The truth proves to be rather different. It is now suspected that most asteroids larger than about a kilometer are not entirely solid, but are actually many smaller solid pieces held together by gravity. An extreme example is Mathilde (see Figure 15-7). This asteroid has such a low density (1300 kg/m^3) that it cannot be made of solid rock, which typically has densities of 2500 kg/m^3 or higher. Itokawa is also well below the density of solid rock. Therefore, a good description of an asteroid like this is a "rubble pile" of small fragments that fit together loosely. Most likely, billions of years of impacts have totally shattered this asteroid. But the impacts were sufficiently gentle that the resulting fragments drifted away relatively slowly, only to be pulled back onto the asteroid by gravitational attraction.

Another surprising discovery is that some asteroids have extremely large craters, comparable in size to the asteroid itself. In addition to Vesta's massive south pole crater (Figure 15-10b), an example is the immense crater on Mathilde, shown in Figure 15-7. The object that formed this crater probably collided with Mathilde at a speed of 1 to 5 km/s (2000 to 11,000 mi/h), which could have easily shattered solid rock. Why didn't this collision break the asteroid completely apart, sending the pieces flying in all directions? The explanation may be that the asteroid had already been broken into a loose collection of fragments before this major impact took place. Such a fragmented asteroid can more easily absorb the energy of a collision than can a solid, rocky one.

ANALOGY If you fire a rifle bullet at a wine glass, the glass will shatter into tiny pieces. But if you fire the same bullet into a bag full of sand (which has much the same chemical composition as glass), the main damage will be a hole in the bag. In a similar way, an asteroid that is a collection of small pieces can survive a major impact more easily than a solid asteroid.

In some cases a collision that produces a rubble-pile asteroid can eject a piece that goes into orbit around the asteroid. This may explain why a handful of asteroids are known to have satellites. There is even one known case of an asteroid with two satellites, like a miniature planetary system (Figure 15-11). If someone were lucky

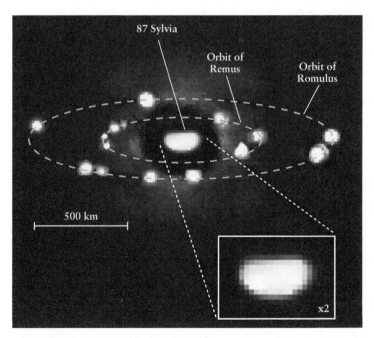

FIGURE 15-11   R I̲ V U X G

An Asteroid with Two Satellites Although it is only about 280 km across, the potato-shaped asteroid 87 Sylvia has two satellites of its own. The outer satellite, Romulus, is 18 km across; the inner satellite, Remus, is just 7 km across. This image is actually a composite of nine observations made with the adaptive optics system at the Yepun telescope of the European Southern Observatory (see Figure 6-17). (F. Marchis, U. of California; P. Descamps, D. Hestroffer, and J. Berthier, Observatoire de Paris; and ESO)

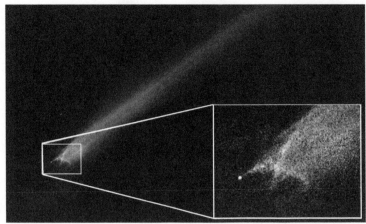

(a) R I [V] U X G

FIGURE 15-12

Asteroid Collisions **(a)** This X-shaped profile of debris is suspected to result from the collision of two asteroids. The debris tails are made of dust and gravel, and this object (named P/2010 A2) is seen orbiting within the asteroid belt. **(b)** The asteroid Scheila unexpectedly doubled in brightness, which prompted a closer look at UV wavelengths. While UV images reveal dust plumes kicked

(b) R I [V] [U] X G

up by a collision, they also indicate Scheila is not a comet (due to a lack of any water vapor emission lines in the plumes). Weeks after the collision's plume was imaged, it faded from view. Within the white circle, images are combined at UV and visible wavelengths, and other than Scheila, the bright spots are stars.

(a: NASA/ESA/D. Jewitt [UCLA]; b: NASA)

enough to observe an asteroid soon after an impact, the short-lived debris from head-on collisions could be observed, as in **Figure 15-12**.

If a collision is sufficiently energetic, it may shatter an asteroid permanently. Evidence for this was first pointed out in 1918 by the Japanese astronomer Kiyotsugu Hirayama. He drew attention to families of asteroids that share nearly identical orbits. Each of these groupings, now called **Hirayama families,** presumably resulted from a parent asteroid that was broken into fragments by a high-speed, high-energy collision with another asteroid.

CONCEPTCHECK 15-4

How do we know that some meteoritic dust comes from asteroids?

CONCEPTCHECK 15-5

Without a returned sample, how did the *Dawn* spacecraft verify that some meteors on Earth came from the asteroid Vesta?

Answers appear at the end of the chapter.

15-4 Asteroids are found outside the asteroid belt—and have struck Earth

While the majority of asteroids lie within the asteroid belt, there are some notable exceptions. A few actually share the same orbit as Jupiter. Other asteroids venture well inside the orbit of Mars, and can pose a serious threat to life on Earth.

The Lagrange Points and Trojan Asteroids

We saw in Section 15-2 that Jupiter's gravitational pull by itself depletes certain orbits in the asteroid belt, forming the Kirkwood gaps. But the gravitational forces of the Sun and Jupiter work

together to *capture* asteroids at two locations outside the asteroid belt called the **stable Lagrange points.** One of these points is one-sixth of the way around Jupiter's orbit ahead of the planet, while the other point is the same distance behind the planet (**Figure 15-13a**). The French mathematician Joseph Louis Lagrange predicted the existence of these points in 1772; asteroids were first discovered there in 1906.

The asteroids trapped at Jupiter's Lagrange points are called **Trojan asteroids,** with some named individually after heroes of the Trojan War. More than 5000 Trojan asteroids have been catalogued so far (see Figure 15-13b). (Nine objects have also been found at the Lagrange points of Neptune. These are icy trans-Neptunian objects rather than rocky asteroids.)

Near-Earth Objects and Impacts from Space

Some asteroids have highly elliptical orbits that bring them into the inner regions of the solar system. Asteroids that *cross* Mars's orbit, or whose orbits lie completely within that of Mars, are called **near-Earth objects** or **NEOs.** These are shown in red in Figure 15-13b.

Occasionally, a near-Earth object may pass relatively close to Earth. For example, in 2009 the 35-m-wide asteroid 2009 DD45 came within 72,000 km of Earth, which is much closer than the Moon and only twice as far as some communications satellites. Near misses by objects this size are expected every few years, and indeed, a similar asteroid came twice as close in 2004. Passing by even closer on February 15, 2013, the 30-m-wide asteroid DA14 missed hitting us by only 27,000 km. This is about half the orbital distance of some communications satellites and is the closest approach on record for an asteroid of this size (more on that historic day later in this section). More than 9000 near-Earth objects are known, and there may be thousands more that we have not yet detected. Fortunately, space is a big place and Earth is a rather small target, so large asteroids rarely strike Earth.

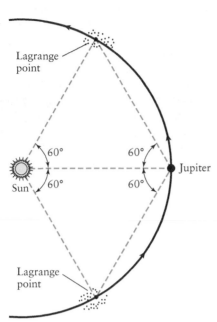

Lagrange
point

60° 60°

Sun 60° 60° Jupiter

Lagrange
point

(a) A clump of asteroids is found at each
of Jupiter's Lagrange points

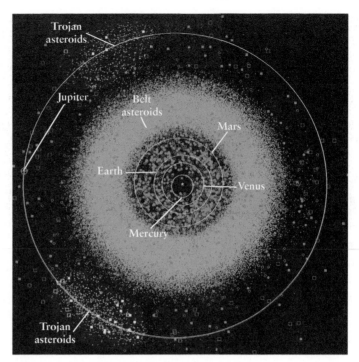

Trojan
asteroids

Jupiter

Belt
asteroids

Mars

Earth

Venus

Mercury

Trojan
asteroids

(b) A map of all asteroids within Jupiter's orbit

FIGURE 15-13

Asteroids Outside the Belt **(a)** The combined gravitational forces of
Jupiter and the Sun tend to trap asteroids near the two stable Lagrange
points along Jupiter's orbit. Note that the Sun, Jupiter, and either one of
these points lie at the vertices of an equilateral triangle. **(b)** This plot shows
the actual positions of all known asteroids at Jupiter's orbit or closer. Green

dots denote belt asteroids, deep blue squares denote Trojan asteroids, and
red circles denote asteroids that come within 1.3 AU of the Sun. Most of the
red dots are called near-Earth objects (NEOs). Comets are filled and unfilled
light blue squares. (b: Gareth Williams/Minor Planet Center)

However, collisions between asteroids produce numerous
smaller chunks of rock, and many of these do eventually rain down
on the terrestrial planets. Fortunately for us, the majority of these
fragments are quite small. But on rare occasions a large fragment
collides with our planet. When such a collision takes place, the
result is an impact crater whose diameter depends on the mass and
speed of the impinging object.

A relatively young, pristine terrestrial impact crater is the
Barringer Crater near Winslow, Arizona (**Figure 15-14**). This crater
was formed 50,000 years ago when an iron-rich asteroid approximately 50 m across struck the ground at a speed of more than
11 km/s (40,000 km/h, or 25,000 mi/h). The resulting blast was
equivalent to the detonation of a 20-megaton hydrogen bomb and
left a crater 1.2 km (3/4 mi) wide and 200 m (650 ft) deep at its
center. This energy was also large enough to vaporize the asteroid
and no fragments of the impacting object have been found. (Figure
7-10b shows a much larger but much older crater in Canada.)

A rough rule of thumb for smaller asteroids that strike Earth
is that asteroids initially larger than about 10 m in size can make
it through the atmosphere, with about half of the asteroid burning up or breaking up in the process. Then, the resulting crater is
roughly 20 times larger than the main fragment impacting Earth's
surface. However, the asteroid's composition and impact speed
also influence how much of the asteroid burns up in the atmosphere, and as in the Barringer event (Figure 15-14), craters can
form without large fragments hitting Earth's surface.

Iridium and the Demise of the Dinosaurs

The element iridium plays a special role in deciphering the history
of asteroid impacts on Earth. Iridium is common in asteroids, but

FIGURE 15-14   R I $\boxed{\text{V}}$ U X G

The Barringer Crater This 1.2 km (3/4 mi) wide crater was formed by an
asteroid impact about 50,000 years ago. Although not as well preserved as
the young Barringer crater, almost 200 other impact craters—typically tens of
kilometers across—have been identified on Earth. Few of these are more than
500 million years old, because the reshaping of Earth's surface by erosion and
by tectonic processes tends to erase craters over time. (Meteor Crater Enterprises)

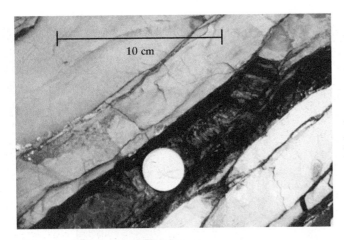

FIGURE 15-15 R I V U X G

Iridium-Rich Clay This photograph of strata in Italy's Apennine Mountains shows a dark-colored layer of iridium-rich clay. This clay is sandwiched between older layers of white limestone (lower right) and younger layers of grayish limestone (upper left). The iridium-rich layer is thought to be the result of an asteroid impact 65 million years ago. The coin is the size of a U.S. quarter. (Professor Walter Alvarez/Science Source)

rare on Earth, so iridium's presence is a reliable indication of a material's asteroid origins. Measurements of iridium in Earth's crust can therefore tell us the rate at which meteoritic material has been deposited on Earth over the ages. The geologist Walter Alvarez and his physicist father, Luis Alvarez, from the University of California, Berkeley, made such measurements in the late 1970s.

Working at a site of exposed marine limestone in the Apennine Mountains in Italy, the Alvarez team discovered an *exceptionally high abundance of iridium* in a dark-colored layer of clay between limestone strata (Figure 15-15). Since this discovery in 1979, a comparable layer of iridium-rich material has been uncovered at numerous sites around the world. In every case, geological dating reveals that this apparently worldwide layer of iridium-rich clay was deposited about 65 million years ago.

> An asteroid impact probably ended the age of dinosaurs 65 million years ago

Paleontologists were quick to realize the significance of this date, because it was 65 million years ago that the dinosaurs rather suddenly became extinct. In fact, more than 85% of all the species on Earth disappeared within a relatively brief span of time.

The Alvarez discovery suggests a startling explanation for this dramatic mass extinction: Perhaps an asteroid hit Earth at that time. An asteroid 10 km in diameter slamming into our planet would have produced a fiery plume that ignited tremendous wildfires, killing untold numbers of animals and consuming much of Earth's vegetation. The energy of this impact would have been a *billion* times stronger than the nuclear bomb dropped on Hiroshima, Japan, during World War II. In this scenario, the dust thrown up from the impact and soot from wildfires hung in the atmosphere, blackening the sky for about a decade after the impact. The dust eventually drifted to the ground, forming a worldwide layer of iridium-rich clay. (Evidence for this picture is that a layer of soot 65 million years old is also found at various sites around the world. From the thickness of this layer, scientists calculate that the

wildfires produced nearly 70 *billion* tons of soot.) Temperatures would have plummeted due to the blocking out of sunlight by dust and soot, and the loss of many plants would have killed larger animals higher up in the food chain. The *Cosmic Connections* figure depicts the sequence of events that may have followed the asteroid impact.

Tiny rodentlike creatures capable of ferreting out seeds and nuts were among the animals that managed to survive this catastrophe, setting the stage for the rise of mammals in the eons that followed. As the dominant mammals on the planet, we may owe our very existence to an ancient asteroid.

In 1992 a team of geologists suggested that this asteroid crashed into a site in Mexico. They based this conclusion on glassy debris and violently shocked grains of rock ejected from the 180-km-diameter Chicxulub Crater on the Yucatán Peninsula (Figure 15-16). Using radioactive dating techniques (see Section 8-3 and Box 8-1), the scientists found that the asteroid struck 64.98 million years ago, in remarkable agreement with the age of the iridium-rich layer of clay.

In 2010, a special international panel of 41 scientists, spanning fields ranging from astronomy to geology to paleontology, reviewed 20 years of data and several different hypotheses for extinction of the dinosaurs. It was their strong conclusion that the current evidence points to an asteroid that triggered the mass extinction 65 million years ago.

The Tunguska Event

On June 30, 1908, a spectacular explosion occurred over the Tunguska region of Siberia that released about 10^{15} joules of energy, equivalent to a nuclear detonation of several hundred kilotons. The blast knocked a man off his porch some 60 km away and could be heard more than 1000 km away. Millions of tons of dust were injected into the atmosphere, darkening the sky as far away as California.

Preoccupied with political upheaval and World War I, a successful expedition from Russia did not reach the site until 1927. Researchers found that trees had been seared and felled radially outward in an area about 50 km in diameter (Figure 15-17). There was no clear evidence of a crater. In fact, the trees at "ground zero" were left standing upright, although they were completely stripped of branches and leaves.

Several teams of modern astronomers have argued that the Tunguska explosion was caused by an asteroid or comet traveling at supersonic speed. They arrived at this conclusion after assessing the effects of various impactor sizes, speeds, and compositions. Even data from above-ground nuclear detonations of the 1940s and 1950s were worked into the calculations. The Tunguska event is well-matched by an asteroid or comet about 80 m (260 ft) in diameter entering Earth's atmosphere at 22 km/s (50,000 mph) and exploding about 7 km high in the atmosphere.

The Historic Asteroid Impact of 2013

On February 15, 2013, astronomers were eagerly awaiting the flyby of asteroid DA14. Discovered a year earlier, its predictable and historic near miss—passing closer than some satellites—was sure to make headlines on astronomy news sites. Instead, by sheer coincidence, an unexpected asteroid with nearly the opposite

iller Asteroid

is compelling evidence that 65 million years ago Earth was struck by an asteroid about ... in diameter. This impact triggered a biological cataclysm that wiped out 85% of all plant and ...al species on our planet. [From "The Day the World Burned" by D. A. Kring and O. O. Durda, *Scientific American*, ...ber 2003: art by Chris Butler]

The Day Before

...ate Cretaceous swamps and rivers in North America had a mix of coniferous, broad-...eaved evergreen, and deciduous trees. They ...ormed canopied forests and open woodlands ...with understories of ferns, aquatic plants and ...lowering shrubs.

Impact

The Chicxulub Impact occurred in a shallow sea and immediately lofted rocky, molten, and vapor... debris into the atmosphere. The bulk of the debr... rained down on nearby continental regions, but much of it rose all the way into space.

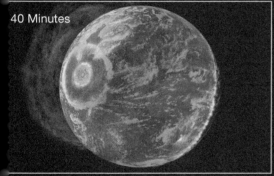

40 Minutes

...he vapor-rich plume of material expanded to ...envelop Earth. As material in that plume fell back ...o the ground, it streaked through the atmosphere ...ike trillions of meteors, heating it in some places ...y hundreds of degrees.

5 Years

After fires had ravaged the landscape, only a few stark trunks and skeletons remained. Soot from the fires and dust from the impact slowly settled to the ground. Sunlight was dramatically, if not totally attenuated for years.

10 Years

...he postimpact environment was less diverse. ...erns and algae were the first to recover. Plant ...pecies in swamps and swamp margins generally ...urvived better than species in other types of ...cosystems. Conifers fared particularly badly.

50 Years

Shrubs took advantage of the vacant landscape and began to cover it. Species pollinated by the wind did better than those that relied on insects. Trees began to grow, but it took years for forest canopies to rebuild. The recovery time is

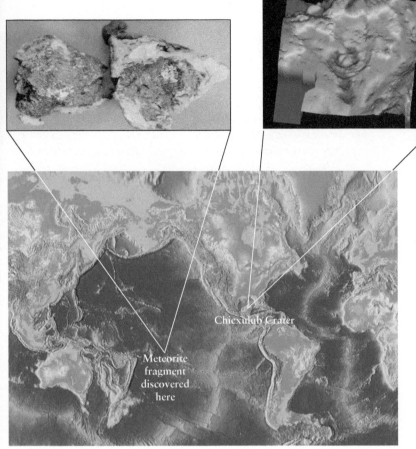

FIGURE 15-16

Confirming an Extinction-Level Impact Site Measuring the gravitational force from slight variations in density (resulting from the impact) can reveal a buried crater. Concentric rings of the underground Chicxulub Crater (right inset) lie under a portion of the Yucatán Peninsula. This crater has been dated to 65 million years ago and is believed to be the site of the impact that led to the extinction of the dinosaurs. A piece of 65-million-year-old meteorite discovered in the middle of the Pacific Ocean in 1998 is believed to be a fragment of that meteorite. The fragment, about 0.3 cm (0.1 in.) long, was cut into two pieces for study (left inset). (left inset: Frank T. Kyte, UCLA; right inset: Mark Pilkington/Geological Survey of Canada/Science Source; map: Peter W. Sloss, NOAA-NESDIS-NGDCD)

trajectory struck Earth first, and it made headlines around the world (Figure 15-18).

Named after a lake in Russia where a fragment appears to have landed, the Chebarkul asteroid was about 15 m wide and weighed about 10,000 tons. The asteroid slammed into Earth's atmosphere at about 60 times the speed of sound; at that tremendous speed, an object could orbit Earth in less than 40 minutes. Most of the asteroid vaporized in a bright fireball about 20 km above Earth's surface. However, the fireball produced a damaging blast wave when it released around as much energy as 20 nuclear bombs of the size detonated in World War II.

The Chebarkul asteroid impact is the largest since the Tunguska event in 1908, and impacts of this size are expected about once every hundred years. Unlike any other impact event, it caused over a thousand human injuries and damaged thousands of buildings. Drawn to their windows to view a fireball that outshone the Sun, over a thousand people were injured by flying shards of glass when the slower-travelling shock wave hit their windows nearly three minutes later.

The Dangers of Asteroid Impacts

Large objects, such as that which formed the Chicxulub Crater and may have led to the demise of the dinosaurs, occasionally strike Earth with the destructive force of as much as a million megatons of TNT. An asteroid 1 or 2 km in diameter, striking Earth at 30 km/s, would devastate an area the size of California. Such an impact would throw more than 10^{13} kg of microscopic

FIGURE 15-17 R I V U X G

Aftermath of the Tunguska Event In 1908 a stony asteroid with a mass of about 10^8 kilograms entered Earth's atmosphere over the Tunguska region of Siberia. Its passage through the atmosphere made a blazing trail in the sky some 800 km long. The asteroid apparently exploded before reaching the surface, blowing down trees for hundreds of kilometers around "ground zero." (Sovfoto)

(a)

(b)

(c)

FIGURE 15-18 R I V̄ U X G

Russian Asteroid Impact of 2013 **(a)** The fireball lasted about 30 seconds, leaving a smoke trail about 400 km across the sky over the Russian town of Chelyabinsk. **(b)** Impact with the atmosphere produced a powerful shockwave that shattered over a million square feet of glass, and damaged many buildings such as this zinc plant. With winter temperatures around −15° C (5° F), broken windows posed an ongoing threat. **(c)** This 6-m wide hole in Lake Chebarkul is a likely impact site for a larger piece of meteoritic debris, although divers could not find the suspected piece in the first few weeks after the impact. Many smaller meteorites were found around this hole and, to a lesser degree, throughout the city. The collected meteorites are mostly rock, with about 10% iron. (a: Xinhua/RIA/Xinhua Press/Corbis; b: Jiang Kehong/Xinhua Press/Corbis; c: HANDOUT/Reuters/Corbis)

dust particles into the atmosphere, which would decrease the amount of sunlight reaching Earth's surface by enough to threaten the health of the world's agriculture. The loss of a year's crops could lead to the demise of a quarter of the world's population and would place our civilization—although perhaps not the survival of our species—in grave jeopardy.

The good news is that dangerously large objects strike Earth *very* infrequently. On average, an asteroid about 5 to 10 m wide harmlessly strikes Earth once a year, releasing as much energy in the upper atmosphere as the atomic bomb on Hiroshima. A more serious impact from an object greater than 50 m, like the one that caused the Tunguska event, is expected every thousand years or so. Getting hit by a 1-km asteroid would certainly be dangerous, but is expected only about once every 500,000 years. For asteroids 10 km or larger, impacts are so rare that the last known case is the dinosaur-killer 65 million years ago.

Smaller objects, on the other hand, strike Earth more frequently. Basketball-size objects hit us about once a day. Even refrigerator-size objects hit us about once a month. These smaller objects mostly burn up in the atmosphere, but next we consider the debris that occasionally reaches Earth's surface from objects large and small.

CONCEPTCHECK **15-6**

What are two pieces of evidence for a massive asteroid impact 65 million years ago?

CALCULATIONCHECK **15-1**

If an impact crater on Earth is 2000 m wide, approximately how big would the asteroid fragment have been that made it through Earth's atmosphere?

Answers appear at the end of the chapter.

15-5 Meteorites are classified as stones, stony irons, or irons, depending on their composition

Asteroids can only be studied at close range using spacecraft. But from time to time pieces of asteroids fall to Earth where they can be examined by scientists. These asteroid fragments are called *meteorites*.

Meteoroids, Meteors, and Meteorites

A **meteoroid**, like an asteroid, is a chunk of rock or metal in space. There is no official dividing line between meteoroids and asteroids, but as of 2010 some astronomers define meteoroids as objects measuring less than 1 m across.

A **meteor** is the brief flash of light (sometimes called a *shooting star*) that is visible at night when a meteoroid enters Earth's atmosphere (**Figure 15-19**). The shooting star begins with an incoming rock about 50 km high in Earth's atmosphere. As the fast-moving meteoroid rams into the air in front of it, the compressed air heats up to about 3000 K and produces the bright trail of light we see. The width of the bright streak might only be a meter or so wide, and cools in less than a second. However, due to the rock's high speed, the bright trail can be several kilometers long.

FIGURE 15-19 R I V U X G

A Meteor A meteor is produced when a meteoroid—a piece of interplanetary rock or dust—strikes Earth's atmosphere at high speed (typically 5 to 30 km/s). The light comes from heat as the air in front of the meteor is highly compressed. This heat also vaporizes rock and most meteoroids burn up completely at altitudes of 100 km (60 mi) or so. This long exposure (notice the star trails) shows an exceptionally bright meteor called a fireball. (SPL/Science Source)

While an impressive sight, the size of a typical meteoroid entering Earth's atmosphere is between a sand grain and a pebble. For incoming rocks less than 10 m wide, the same heat that produces its bright trail vaporizes most of the object. Therefore, the typical meteoroid is completely burned up in the atmosphere. On the other hand, small particles of cosmic dust (see Figure 8-9) slow down quickly, produce little heat, and continuously drift down to Earth's surface without being noticed.

If a piece of rock or metal survives its fiery descent through the atmosphere and reaches the ground, it is called a **meteorite.** As incredible as it may seem, an estimated total of 80 tons of extraterrestrial matter falls on Earth *each day.* However, most of this matter is cosmic dust rather than meteorites. It is fortunate that large meteorite falls are rather rare events—direct hits can cause damage (**Figure 15-20**). Because they are so rare, meteorites are prized by scientists and collectors alike. Almost all meteorites are asteroid fragments and thus provide information about the chemical composition of asteroids. A very few meteorites have been identified as pieces of the Moon, Mars, or the asteroid Vesta, that were blasted off the surfaces of those worlds by violent impacts.

> Studying meteorites reveals the properties of the asteroids from which they came

People have been finding specimens of meteorites for thousands of years, and descriptions of them appear in ancient Chinese, Greek, and Roman literature. Many civilizations have regarded meteorites as objects of veneration. The sacred Black Stone, enshrined at the Ka'aba in the Great Mosque at Mecca, may be a relic of a meteorite impact. More recently, selling meteorites has become profitable enough that when scientists show up to investigate an impact, oftentimes the fragments have already been taken.

The extraterrestrial origin of meteorites was hotly debated until as late as the eighteenth century. Upon hearing a lecture by two Yale professors, President Thomas Jefferson is said to have remarked, "I could more easily believe that two Yankee professors could lie than that stones could fall from Heaven." Although several falling meteorites had been widely witnessed and specimens from them had been collected, many scientists were reluctant to accept the idea that rocks could fall to Earth from outer space. Conclusive evidence for the extraterrestrial origin of meteorites came on April 26, 1803, when fragments pelted the French town of L'Aigle. This event was analyzed by the noted physicist Jean-Baptiste Biot, and his conclusions helped to finally convince scientists that meteorites were indeed extraterrestrial.

Types of Meteorites

Meteorites are classified into three broad categories: stones, stony irons, and irons. As their name suggests, **stony meteorites,** or **stones,** look like ordinary rocks at first glance, but they are sometimes covered with a **fusion crust** (Figure 15-21a). This crust is produced by the melting of the meteorite's outer layers during its fiery descent through the atmosphere. When a stony meteorite is cut in two and polished, tiny flecks of iron can sometimes be found in the rock (Figure 15-21b). Meteorites collected from the 2013 Chebarkul impact event are of this type and contain about 10% iron.

Although stony meteorites account for about 95% of all the meteoritic material that falls to Earth, they are the most difficult meteorite specimens to find. If they lie undiscovered and become exposed to the weather for a few years, they look almost indistinguishable from common terrestrial rocks. (The asteroid that caused the Tunguska explosion, described in Section 15-4, was probably of stony composition, which would account for the lack of obvious meteoritic debris on the ground.) Meteorites with high

FIGURE 15-20 R I V U X G

A Meteorite "Fender-Bender" On the evening of October 9, 1992, Michelle Knapp of Peekskill, New York, heard a noise from outside that sounded like a car crash. She discovered that the trunk of her car had been smashed by a 12-kilogram (27-pound) meteorite, which was lying beside the car and was still warm to the touch. Meteorite damage is not covered by automobile insurance, but Ms. Knapp was offered several tens of thousands of dollars for the meteorite (and the car). (Ingo Wagner/epa/Corbis)

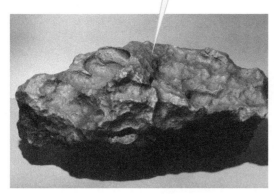

Many stony meteorites are coated with dark fusion crusts...

(a)

...but when cut and polished they reveal tiny specks of iron in the rock.

(b)

FIGURE 15-21 R I V U X G

Stony Meteorites Of all meteorites that fall on Earth, 95% are stones. **(a)** Many freshly discovered specimens, like the one shown here, are coated with dark fusion crusts. This particular stone fell in Texas. **(b)** Most stony meteorites contain small grains of nickel-rich iron distributed throughout the rock. Chondrules are also evident in this specimen from Australia. (a: Tom McHugh/Science Source; b: Daniel Ball/School of Earth and Space Exploration/ASU/ Courtesy of the Center for Meteorite Studies)

iron content can be found much more easily, because they can be located with a metal detector. Consequently, iron and stony iron meteorites dominate most museum collections.

Stony iron meteorites consist of roughly equal amounts of rock and iron. Figure 15-22, for example, shows the mineral olivine suspended in a matrix of iron. Only about 1% of the meteorites that fall to Earth are stony irons.

Iron meteorites (Figure 15-23), or **irons**, account for about 4% of the material that falls on Earth. Iron meteorites usually contain no stone, but many contain from 10% to 20% nickel. Before humans began extracting iron from ores around 2000 B.C.E., the only source of the metal was from iron meteorites. Because these are so rare, iron was regarded as a precious metal like gold or silver.

In 1808, Count Alois von Widmanstätten, director of the Imperial Porcelain Works in Vienna, discovered a conclusive test for the most common type of iron meteorite. About 75% of all iron meteorites have a unique crystalline structure. These **Widmanstätten** patterns (pronounced VIT-mahn-shtetten) become visible when the meteorite is cut, polished, and briefly dipped into a dilute solution of acid (Figure 15-23b). Nickel-iron crystals of this type can form only if the molten metal cools slowly over many millions of years. Therefore, Widmanstätten patterns are never found in counterfeit meteorites, or "meteorwrongs."

Meteorites and the Early History of Asteroids

Widmanstätten patterns suggest that some asteroids were originally at least partly molten inside and remained partly molten long after they were formed. As soon as an asteroid accreted from planetesimals some 4.56 billion years ago, rapid decay of short-lived radioactive isotopes could have heated the asteroid's interior to temperatures above the melting point of rock. If the asteroid was large enough—at least 200 to 400 km in diameter—to insulate the interior and prevent it from cooling, the interior would have remained molten over the next few million years and chemical differentiation would have occurred (see Box 7-1 and Section 8-5). Iron and nickel would have sunk toward the asteroid's center, forcing less dense rock upward toward the asteroid's surface.

Like a terrestrial planet, such a **differentiated asteroid** would have a distinct core and crust. After a differentiated asteroid cooled and its core solidified, collisions with other asteroids would have fragmented the parent body into meteoroids. Iron meteorites are therefore specimens from a differentiated asteroid's core, while stony irons come from regions between a differentiated asteroid's core and its crust.

Unlike irons and stony irons, stony meteorites may or may not come from differentiated asteroids. Smaller asteroids would not have retained their internal heat long enough for chemical differentiation to occur. Some stony meteorites are fragments of such **undifferentiated asteroids** (also called *primitive* asteroids), while others are relics of the crust of differentiated asteroids.

FIGURE 15-22 R I V U X G

A Stony Iron Meteorite Stony irons account for about 1% of all the meteorites that fall to Earth. This particular specimen, a variety of stony iron called a pallasite, fell in Chile. (Maurice Savage/Alamy Images)

CALCULATIONCHECK 15-2

If a meteor passes through Earth's atmosphere at about 5 km/s, how long is the tail if the atmosphere heated by the meteor only remains hot enough to be visible for about 1 second?

Iron meteorites are composed of nickel-iron minerals and are characterized by a surface covered with depressions...

(a)

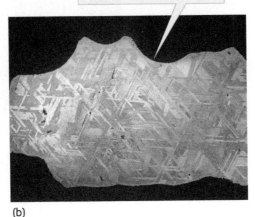

...and when cut and polished, by interlocking crystals in a Widmanstätten pattern.

(b)

FIGURE 15-23   R I V U X G

Iron Meteorites **(a)** The surface of a typical iron is covered with small depressions caused by ablation (removal by melting) during its high-speed descent through the atmosphere. Note the one-cent coin for scale. **(b)** When cut, polished, and etched with a weak acid solution, most iron meteorites exhibit interlocking crystals in designs called Widmanstätten patterns.

(a: Detlev van Ravenswaay/Science Source; b: The Natural History Museum/Alamy)

CONCEPTCHECK 15-7

Why does a Widmanstätten pattern ensure that a sample is truly a meteorite?

Answers appear at the end of the chapter.

15-6 Some meteorites retain traces of the early solar system

A class of rare stony meteorites called **carbonaceous chondrites** shows no evidence of ever having been melted while in space. As suggested by their name, carbonaceous chondrites contain substantial amounts of carbon and carbon compounds, including complex organic molecules and as much as 20% water bound into the minerals. These compounds would have been broken down and the water driven out if these meteorites had been subjected to heating and melting. Carbonaceous chondrites may therefore be samples of *the original material from which our solar system was created.* The asteroid 253 Mathilde, shown in Figure 15-7, has a very dark gray color and the same sort of spectrum as a carbonaceous chondrite. It, too, is likely composed of material that predates the formation of the solar system.

Amino acids, the building blocks of proteins upon which terrestrial life is based, are among the organic compounds occasionally found inside carbonaceous chondrites. Interstellar organic material has thus probably been falling on our planet since its formation. Some scientists suspect that carbonaceous chondrites may have provided some of the original building blocks in the development of life on Earth.

Shortly after midnight on February 8, 1969, the night sky around Chihuahua, Mexico, was illuminated by a brilliant blue-white light moving across the heavens. As the light crossed the sky, it disintegrated in a spectacular, noisy explosion that dropped thousands of rocks and pebbles over terrified onlookers. Within hours, teams of scientists were on their way to collect specimens of a carbonaceous chondrite, collectively named the Allende meteorite after the locality (**Figure 15-24**).

Specimens were scattered in an elongated ellipse approximately 50 km long by 10 km wide. Most fragments were coated with a fusion crust, but minerals immediately beneath the crust showed no signs of damage. Surface material is peeled away as it becomes heated during flight through the atmosphere; this process forms the fusion crust. Because the heat has little time to penetrate the meteorite's interior, compounds there are left intact.

One of the most striking discoveries to come from the Allende meteorite was evidence suggesting that a nearby star exploded into a **supernova** about 4.6 billion years ago. In a supernova, a massive star reaches the end of its life cycle and blows itself apart in a cataclysm that hurls matter outward at tremendous speeds (see Figure 1-8). During this detonation, violent collisions between nuclei produce a host of radioactive elements, including ^{26}Al, a radioactive isotope of aluminum.

Researchers found clear evidence for the former presence of ^{26}Al in the Allende meteorite: Chemical analyses revealed a high abundance of a stable isotope of magnesium (^{26}Mg), which is produced by the radioactive decay of ^{26}Al. Some astronomers interpret this as evidence for a supernova in our vicinity at about the time the Sun was born. Indeed, by compressing interstellar gas and dust, the supernova's shock wave may have triggered the birth of our solar system.

FIGURE 15-24   R I V U X G

A Piece of the Allende Meteorite This carbonaceous chondrite fell near Chihuahua, Mexico, in February 1969. The meteorite's dark color is due to its high abundance of carbon. Radioactive age-dating indicates that this meteorite is 4.56 billion years old, suggesting that this meteorite is a specimen of primitive planetary material that predates the formation of the planets. The ruler is 15 cm (6 in.) long. (Courtesy of J. A. Wood)

Why are carbonaceous chondrites thought to preserve some of the primal material that formed our solar system?

Answer appears at the end of the chapter.

15-7 A comet is a chunk of ice and dust that partially vaporizes as it passes near the Sun

TUTORIAL 15-3 Just as heat from the protosun produced two classes of planets, the terrestrial and the Jovian, two main types of small bodies formed in the solar system. Near the Sun, interplanetary debris consists of the rocky objects called asteroids or meteoroids. Far from the Sun, where temperatures in the early solar system were low enough to permit ices of water, methane, ammonia, and carbon dioxide to form, interplanetary debris took the form of loose collections of ices and small rocky particles. These objects are like icy versions of "rubble pile" asteroids such as 253 Mathilde, which we described in Section 15-3. The Harvard astronomer Fred Whipple, who devised this model in 1950, called them "icy conglomerates" or "dirty snowballs." Some of these "snowballs" are trans-Neptunian objects that orbit forever in the outer regions of the solar system. Some of them, however, are in noncircular orbits that take them inward toward the Sun, where the ice begins to evaporate. These objects are known as **comets.**

The Structure of a Comet

Kuiper belt objects and asteroids travel around the Sun along roughly circular orbits that lie close to the plane of the ecliptic (see Figure 2-14). In sharp contrast, comets travel around the Sun along highly elliptical orbits inclined at random angles to the ecliptic. Comets are just a few kilometers across, so when they are far from the Sun they are difficult to spot with even a large telescope. As a comet approaches the Sun, however, solar heat begins to vaporize the comet's ices, liberating gases as well as dust particles. The liberated gases begin to glow, producing a fuzzy, luminous ball called a **coma** that is typically 1 million km in diameter. Some of these luminous gases stream outward into a long, flowing **tail**. This tail, which can be more than 100 million km in length—comparable to the distance from Earth to the Sun—is one of the most awesome sights that can be seen in the night sky (see Figure 15-25 and the image that opens this chapter).

LOOKING DEEPER 15-2 Many comets are discovered when they are still far from the Sun, long before their tails grow to full size. A number of these discoveries are made by amateur astronomers, many of whom use only binoculars or small telescopes to aid in their search.

Figure 15-26 depicts the structure of a comet. The solid part of the comet, from which the coma and tail emanate, is called the **nucleus.** It is a mixture of ice and dust that typically measures a few kilometers across. Before 1986, no one had seen the nucleus of a comet. It is small, dim, and buried in the glare of the bright coma. In that year, the first close-up pictures of a comet's nucleus were obtained by a spacecraft that flew past Comet Halley (Figure 15-27a). Comet Halley's potato-shaped nucleus is darker than coal, reflecting only about 4% of the light that strikes it. This dark color is probably caused by a layer of carbon-rich compounds and dust that is left behind as the comet's ice evaporates.

FIGURE 15-25   R I V U X G

Comet Hyakutake Using binoculars, Japanese amateur astronomer Yuji Hyakutake first noticed this comet on the morning of January 30, 1996. Two months later, Comet Hyakutake came within 0.1 AU (15 million km, or 9 million mi) of Earth and became one of the brightest comets of the twentieth century. This image was made in April 1996, when the comet's tail extended more than 30° across the sky (more than 60 times the diameter of the full moon). The stars appear as short streaks because the camera followed the comet's motion during the exposure. (Galaxy Picture Library/Alamy)

Several 15-km-long jets of dust particles were seen emanating from bright areas on Comet Halley's nucleus. These jets seem to be active only when exposed to the Sun. The bright areas, which probably cover about 10% of the surface of the nucleus, are presumably places where the dark layer covering the nucleus is particularly thin. When exposed to the Sun, these areas evaporate

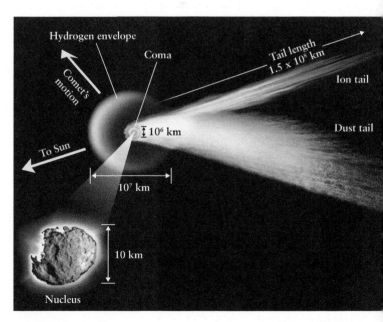

FIGURE 15-26

The Structure of a Comet The solid part of a comet is the nucleus. It is typically about 10 km in diameter. The coma typically measures 1 million km in diameter, and the hydrogen envelope usually spans 10 million km. A comet's tail can be as long as 1 AU. Comet Wild 2 (inset) is examined further in Figure 15-28. (This drawing is not to scale.) (Inset: NASA/JPL)

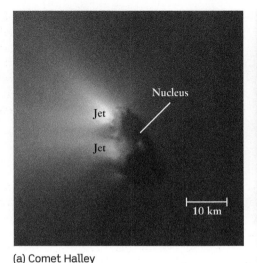

(a) Comet Halley

(b) Comet Hartley 2

FIGURE 15-27   R I V U X G

Comet Nuclei and Jets When sunlight reaches exposed ice on the surface of the nuclei, evaporation produces jets. **(a)** This image from the European Space Agency spacecraft *Giotto* shows the dark, potato-shaped nucleus of Comet Halley in March 1986. Sunlight from the left illuminates half of the nucleus, which is about 15 km long and 8 km wide. Two bright jets of dust extend from the nucleus toward the Sun. **(b)** Comet Hartley 2 imaged by NASA's EPOXI spacecraft. This small comet is only about 1.2 miles wide from end to end. A spectrum of the jets indicates that they emit about 200 kg/sec of water, and even more carbon dioxide. (a: Max Planck Institute for Solar System Research; b: NASA/JPL-Caltech/UMD)

rapidly, producing the jets. The jets build the coma, which feeds the comet's tails. When the nucleus's rotation brings the bright areas into darkness, away from the Sun, the jets shut off.

In 2010, NASA's EPOXI mission flew by Comet Hartley 2 (Figure 15-27b) and measured the wavelength spectrum of light coming from the jets that stream off its nucleus. Earlier data indicated that within the jets, water evaporates at a rate of about 200 kg/sec (440 lbs/sec), and water's evaporation was thought to power the jets. However, the spectrum measured by EPOXI reveals that there is a lot of carbon dioxide (CO_2) in the jets as well. Carbon dioxide undergoes the same transformation as water, changing from solid form (familiar as "dry ice") directly to vapor, when exposed to sunlight. Surprisingly, it appears that the evaporation of carbon dioxide powers the jets even more than water.

After imaging comets and analyzing their composition with spectra, the next steps in exploration are to sweep up material from their jets, smash projectiles into their surfaces, and even land on their surfaces for detailed analysis.

Sweep, Smash, and Land

In 2006, NASA's *Stardust* probe crashed to Earth. Far from being an accident, *Stardust*'s landing was the final phase of a huge success after having swept up dust surrounding the nucleus of Comet Wild 2 (Figure 15-28). Some of the most interesting materials collected from the comet are amino acids. We already saw in Section 15-6 that amino acids are present in some meteorites as well; both comets *and* asteroids contain these basic building blocks of life.

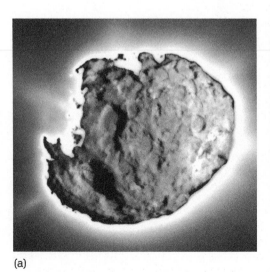

(a)

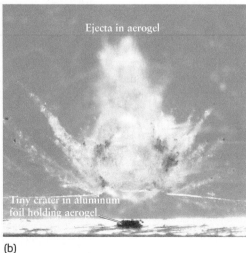

(b)

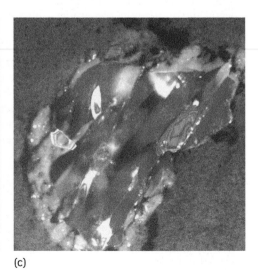

(c)

FIGURE 15-28   R I V U X G

Comet Wild 2 **(a)** This picture shows two images combined. The inner portion is a high-resolution photograph showing the surprisingly heavily cratered comet. The bright outer portion shows gas and dust jetting away from the comet. Its tails are millions of kilometers long. **(b)** A substance called aerogel was used to capture particles from Comet Wild 2's dust tail. A piece of space debris pierced the aluminum foil holding the aerogel and embedded in it, along with pieces of the foil. **(c)** A 2-μm piece of comet dust, composed of a mineral called forsterite. On Earth this mineral is used to make gems called peridot. (a: NASA/JPL; b: NASA/JPL-Caltech/UMD; c: NASA/JPL-Caltech/University of Washington.)

The *Stardust* mission also undermines the long-held notion that comets never get warm enough to contain liquid water. In 2011, analysis of samples revealed minerals that could only form in liquid water, indicating that the full story on comets is still far from clear.

VIDEO 15-3 To assess the structural properties of a comet, a mission was designed to smash a projectile into a comet's surface and analyze the debris. To carry out this bold mission, the *Deep Impact* spacecraft flew to within 500 km of Comet Tempel 1 in 2005 (**Figure 15-29**). Then, *Deep Impact* launched a 372-kg copper projectile that slammed into the surface of the nucleus at 37,000 km/h (23,000 mi/h). The collision excavated about 11,000 tons of material from the nucleus for analysis. The character of the ejected material confirmed that the nucleus of Tempel 1 is not held together by chemical forces (the way a solid block of ice is held together). Instead, the nucleus is like a "rubble pile" and is held together by gravitational forces between numerous parts.

The *Deep Impact* mission also measured the mass and hence the density of the comet nucleus. The very low density of 600 kg/m³ (60% of the density of water) shows that *the nucleus cannot be solid*. It is best regarded as a porous jumble of rock, fine dust, and ice.

An even more ambitious mission to explore a comet is the European Space Agency's *Rosetta*. This spacecraft was launched in 2004 and is scheduled to intercept Comet 67P/Churyumov-Gerasimenko in 2014. If all goes well, *Rosetta* will go into orbit around the comet nucleus and release a small probe that will actually land on the nucleus. Due to the low gravity around a comet,

tethered harpoons will be fired into the surface to prevent the lander from bouncing off into space. In addition to studying the comet's overall characteristics, the lander is designed to detect organic molecules (including amino acids) and some of their subtle properties. By comparing the results to organic molecules on Earth, *Rosetta* will help us assess whether or not some of the organic molecules found in living organisms on Earth could have actually come from comets.

Comet Envelopes and Comet Tails

A part of a comet that is not visible to the human eye is the **hydrogen envelope,** a huge sphere of tenuous gas surrounding the nucleus (Figure 15-26). This hydrogen comes from water molecules (H_2O) that escape from the comet's evaporating ice and then break apart when they absorb ultraviolet photons from the Sun. The hydrogen atoms also absorb solar ultraviolet photons, which excite the atoms from the ground state into an excited state (see Section 5-8). When the atoms return to the ground state, they emit more ultraviolet photons.

Unfortunately for astronomers, Earth-based telescopes cannot detect the emission from a comet's hydrogen envelope because our atmosphere absorbs ultraviolet light (see Section 6-7). Instead, cameras above Earth's atmosphere must be used. **Figure 15-30** shows two views of the same comet, one as it appeared in visible light to Earth-based observers and one as photographed by an ultraviolet camera aboard a rocket. The ultraviolet image shows the enormous extent of the hydrogen envelope, which can span 10 million km.

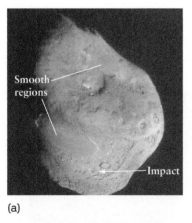

(a)

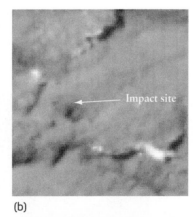

(b)

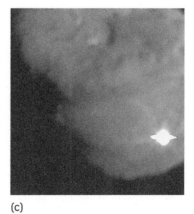

(c)

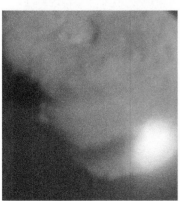

(d)

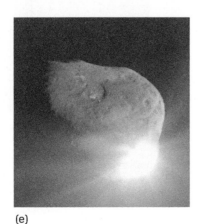

(e)

FIGURE 15-29 R I V U X G

Comet Tempel 1 **(a)** This composite image of Comet Tempel 1 has higher resolution at the bottom, as the projectile from *Deep Impact* headed in that direction. The smooth regions on the comet have yet to be explained. **(b)** Thirty seconds before the projectile struck the comet. **(c)** Seconds after impact, hot debris explodes away from the comet nucleus. The horizontal white streak through the center is where the camera's CCD became overloaded with light from the event. **(d)** Moments later, the gases and dust were expanding outward. **(e)** An image taken 67 s after impact. Within minutes, the cloud of debris became much larger than the entire nucleus. (a: NASA/JPL/UMD; b–e: NASA/JPL-Caltech/UMD)

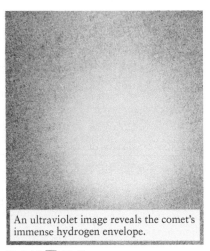

A visible-light image shows the comet's tail.

An ultraviolet image reveals the comet's immense hydrogen envelope.

R I V U X G   R I V U X G

FIGURE 15-30

A Comet and Its Hydrogen Envelope These visible-light and ultraviolet images of Comet Kohoutek, which made an appearance in 1974, are reproduced to the same scale. The hydrogen envelope is only visible in the ultraviolet image. (Johns Hopkins University, Naval Research Laboratory)

As the diagram in Figure 15-26 suggests, comet tails always point away from the Sun. This observation is true regardless of the direction of the comet's motion (Figure 15-31). The implication that something from the Sun was "blowing" the comet's gases radially outward led Ludwig Biermann to predict the existence of a *solar wind*, a stream of particles rushing away from the Sun (see Section 8-5). A decade later, in 1962, Biermann's prediction was confirmed when the solar wind was detected for the first time by instruments on the *Mariner 2* spacecraft (see Figure 11-11).

> A comet's tail does not stream behind it; instead, it points generally away from the Sun

In fact, the Sun usually produces *two* comet tails—an **ion tail** and a **dust tail** (Figure 15-32). Ionized atoms and molecules—that is, atoms and molecules missing one or more electrons—are swept directly away from the Sun by the solar wind to form the relatively straight ion tail. The distinct blue color of the ion tail is caused by emissions from ionized carbon-bearing molecules such as CN and C_2. The dust tail is formed when photons of light strike dust particles freed from the evaporating nucleus. (Figure 8-9 shows a dust particle of this type.) Light exerts a pressure on any object that absorbs or reflects it. This pressure, called **radiation pressure**, is quite weak but is strong enough to make fine-grained dust particles in a comet's coma drift away from the comet, thus producing a dust tail. The solar wind has less of an effect on dust particles than on (much smaller) ions, so the dust tail ends up being curved rather than straight. On rare occasions a comet is oriented in such a way that its dust tail appears to stick out in front of the comet (Figure 15-33).

After the comet passes perihelion, it recedes from the Sun back into the cold regions of the outer solar system. The ices stop vaporizing, the coma and tail disappear, and the comet goes back into an inert state—until the next time its orbit takes it toward the inner solar system. An object in an elliptical orbit spends most of its time far from the Sun, so it is only during a relatively brief period before and after perihelion that a comet can have a

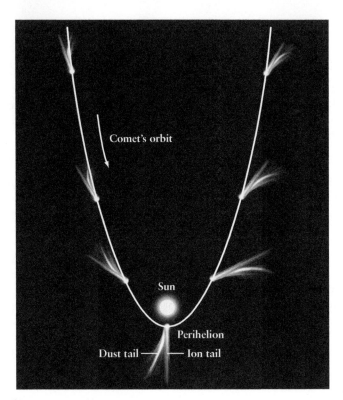

Comet's orbit

Sun

Perihelion

Dust tail —   — Ion tail

ANIMATION 15-1 **FIGURE 15-31**

The Orbit and Tail of a Comet The solar wind and the pressure of sunlight blow a comet's dust particles and ionized atoms away from the Sun. Consequently, a comet's tail points generally away from the Sun. In particular, the tail does not always stream behind the nucleus. At the upper right of this figure, the comet is moving upward along its orbit and is literally chasing its own tail.

prominent tail. An example is Comet Halley, which orbits the Sun along a highly elliptical path that stretches from just inside Earth's orbit to slightly beyond the orbit of Neptune (Figure 15-34). The comet's tail is visible to the naked eye or with binoculars only

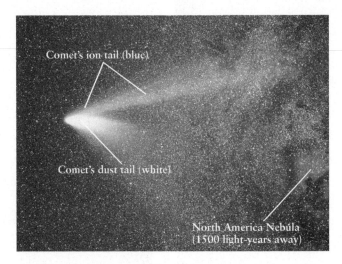

Comet's ion tail (blue)

Comet's dust tail (white)

North America Nebula (1500 light-years away)

FIGURE 15-32   R I V U X G

The Two Tails of Comet Hale-Bopp A comet's dust tail is white because it reflects sunlight, while the molecules in the ion tail emit their own light with a characteristic blue color. When this picture was taken on March 8, 1997, the ion tail extended more than 10° across the sky. The red object to the right is the North America Nebula, a star-forming region some 1500 light-years beyond the solar system. (Tony and Daphne Hallas/Science Source)

FIGURE 15-33 R I V U X G

The Antitail of Comet Hale-Bopp In January 1998, nine months after passing perihelion, Comet Hale-Bopp was oriented in such a way that the end of its arched dust tail looked like a spike sticking out of the comet's head. This spike is called an "antitail" because it appears to protrude in front of the comet. (European Southern Observatory)

during a few months around perihelion, which last occurred in 1986 and will happen again in 2061.

Other comets have orbits that are larger and more elongated, with even longer periods. The orbit of Comet Hyakutake (see Figure 15-25) takes it to a distance of about 2000 AU from the Sun, 50 times the size of Pluto's orbit, with an orbital period of around 108,304 years.

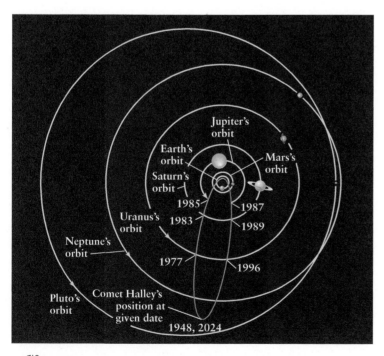

ANIMATION 15-2 **FIGURE 15-34**

Comet Halley's Eccentric Orbit Like most comets, Comet Halley has an elongated orbit and spends most of its time far from the Sun. Figure 4-25 shows the comet's tail, which appears only when the comet is close to perihelion. Comet Halley has been observed at intervals of about 76 years—the period of its orbit—since 88 B.C.E. (see Section 4-7).

15-8 Comets originate either from the Kuiper belt or from the Oort cloud

Both comets and Kuiper belt objects are mixtures of ice and rock that must have formed in the outer solar system. (Had they formed in the warm inner solar system, they would not have included ices.) But most Kuiper belt objects have relatively circular orbits that lie close to the plane of the ecliptic, while comets have very elongated orbits that are often steeply inclined to the ecliptic. What gives comets their distinctive orbits? As we will see, the answer to this question depends on whether the comet's orbital period is relatively short or relatively long.

Jupiter-Family Comets and the Kuiper Belt

As we saw in Section 14-10, the gravitational effects of Neptune shape the orbits of objects in the **Kuiper belt.** Most of these objects are found between the orbit of Pluto (where an object makes two orbits for every three orbits of Neptune) and about 50 AU from the Sun (where an object makes one orbit for every two orbits of Neptune).

However, collisions between Kuiper belt objects can break off relatively small chunks. Gravitational perturbations from Neptune can occasionally launch one of these chunks into a highly elliptical orbit that takes it close to the Sun. When one of these chunks approaches the Sun, it develops a visible tail and appears as a comet. These are called **Jupiter-family comets** because their orbits tend to be influenced strongly by Jupiter's gravitational pull. Jupiter-family comets orbit the Sun in fewer than 20 years. One example is Comet Tempel 1, which has an orbital period of 5.5 years.

There may be tens of thousands of objects in the Kuiper belt that could eventually be perturbed into an elongated orbit and become comets. It is also thought that a few of these refugees from the Kuiper belt have been captured by the gravitational pull of Saturn, becoming that planet's small outer satellites (see Section 13-10).

Short-Period Comets, Long-Period Comets, and the Oort Cloud

The majority of comets are **short-period comets,** with orbital periods between 20 and 200 years, and **long-period comets,** which take more than 200 years to complete one orbit around the Sun. Comet Halley, shown in Figure 15-27a, has an orbital period of 76 years and is a short-period comet; Comet Hyakutake (see Figure 15-25) is a long-period comet that takes about 70,000 years to complete one orbit.

The short-period comets are thought to come from the Kuiper belt (Figure 15-35). Due to their longer periods, the long-period comets are thought to come from much farther out—a reservoir of icy objects that extends from the Kuiper belt to some 50,000 AU from the Sun, or about one-fifth of the way to the nearest star. This reservoir, first hypothesized by Dutch astronomer Jan Oort in 1950 and now called the **Oort cloud,** is thought to contain two

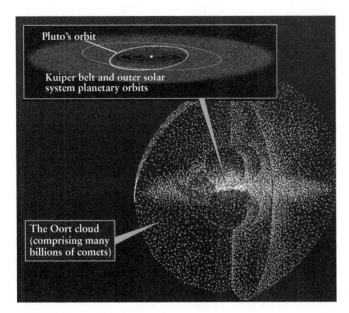

FIGURE 15-35

The Kuiper Belt and Oort Cloud Short-period comets are thought to originate in the Kuiper belt, which begins just beyond the orbit of Neptune at 30 AU, and extends to about 50 AU from the Sun. Most of the Kuiper belt objects lie in a disk close to the plane of the solar system. The Oort cloud, by contrast, contains a doughnut-shaped distribution in the plane of the ecliptic, but also a spherical distribution of icy objects whose orbits are randomly oriented relative to the ecliptic. The inner edge of the doughnut-shaped distribution is about 2000 AU, and the Kuiper belt is so small compared to the Oort cloud that it would appear as a dot on the scale used in this figure. Long-period comets are thought to originate in the Oort cloud. (NASA and A. Field/Space Telescope Science Institute)

vast arrangements of comets around the Sun—an inner doughnut-shaped distribution and an outer spherical distribution. The outer spherical distribution was proposed in order to account for the observation that long-period comets can have large orbital inclinations relative to the ecliptic (the orbital plane of the planets).

Because astronomers discover long-period comets at the rate of about one per month, it is reasonable to suppose that there is an enormous population of comets in the Oort cloud. The number of Oort cloud objects greater than 1 km in size range is as high as 5 trillion (5×10^{12}). Only such a large reservoir of comet nuclei would explain why we see so many long-period comets, even though each one takes tens of thousands of years to travel once around its orbit. Even with trillions of "dirty snowballs" in the Oort cloud, their average separation from each other is about 10 million km. Taken together, the total mass in the Oort cloud is equal to about 5 Earth masses.

Because the Oort cloud is so distant, direct detection of objects in the Oort cloud is a daunting challenge. The best candidate Oort cloud object is the trans-Neptunian object Sedna (Figure 14-20), with an aphelion of 960 AU.

The Oort cloud was probably created 4.56 billion years ago from numerous icy planetesimals that orbited the Sun in the vicinity of the newly formed Jovian planets. When the outward migration of Saturn, Uranus, and Neptune deflected these icy objects

inward toward Jupiter, Jupiter's large gravity catapulted most of these objects clear out of the solar system. However, a small portion of these icy objects failed to leave the solar system entirely and formed the Oort cloud (Section 8-6). Gravitational perturbations from nearby stars could have tilted the planes of the orbits in all directions, giving the Oort cloud its spherical shape. Computer simulations even suggest the possibility that the Sun could have captured a substantial fraction of the existing Oort cloud objects from the protoplanetary disks of *other* stars.

Changing a Comet's Orbit

The distinction between long-period, short-period, and Jupiter-family comets can be blurred by the effects of gravitational perturbations. During a return trip toward the Sun, an encounter with a Jovian planet may force a comet into a much larger orbit. Alternatively, such an encounter can move a long-period comet into a smaller orbit (Figure 15-36). Several comets have been perturbed in this way into orbits that always remain within the inner solar system. One example is Comet Tempel, which now orbits between Mars and Jupiter.

Comet Breakup and Meteor Showers

Because they evaporate away part of their mass each time they pass near the Sun, comets cannot last forever. A typical comet may lose about 0.5% to 1% of its ice each time it

> Comets eventually break apart, and their fragments give rise to meteor showers

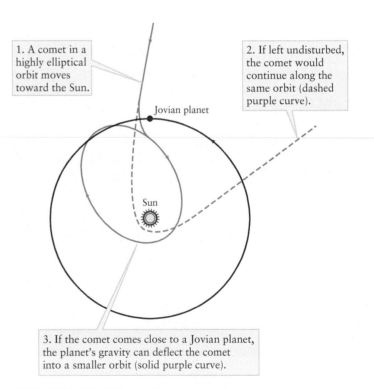

1. A comet in a highly elliptical orbit moves toward the Sun.

2. If left undisturbed, the comet would continue along the same orbit (dashed purple curve).

Jovian planet

Sun

3. If the comet comes close to a Jovian planet, the planet's gravity can deflect the comet into a smaller orbit (solid purple curve).

FIGURE 15-36

Transforming a Comet's Orbit The gravitational force of a planet can deflect a comet from a highly elliptical orbit into a less elliptical one. In some cases a comet ends up with an orbit that keeps it within the inner solar system.

FIGURE 15-37   R I V U X G

The Fragmentation of Comets **(a)** Comet LINEAR was discovered by the Lincoln Near Earth Asteroid Research (LINEAR) project in 1999. As it passed the perihelion of its orbit in July 2000, Comet LINEAR broke into more than a dozen small fragments. These continue to orbit the Sun along the same trajectory that the comet followed prior to its breakup. This image was made using the Very Large Telescope (see Figure 6-18). **(b)** This comet, with a 5.4-year orbit, has been coming apart for decades. In 2006, it further fragmented after passing perihelion. One piece, Fragment B, shed at least 30 smaller pieces, shown here. (a: European Southern Observatory; b: NASA; ESA; H. Weaver [APL/JHU]; M. Mutchler and Z. Levay [STScI])

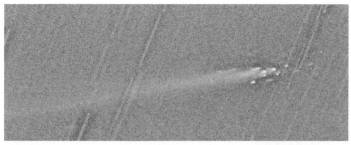

(a)

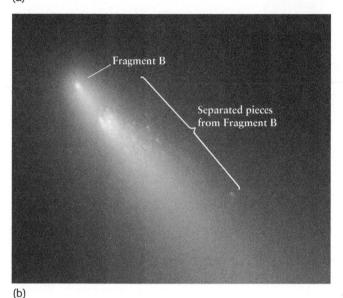

Fragment B

Separated pieces from Fragment B

(b)

passes near the Sun. Hence, the ice completely vaporizes after about 100 or 200 perihelion passages, leaving only a swarm of dust and pebbles. Astronomers have observed some comet nuclei in the process of fragmenting (Figure 15-37).

One remarkable comet, called Shoemaker-Levy 9, broke apart for a different reason. As the comet swung by Jupiter in July 1992, the giant planet's tidal forces tore the nucleus into more than 20 fragments. Jupiter's gravity then deflected the trajectories of these fragments so that they plummeted into the planet's atmosphere (Figure 15-38).

As a comet's nucleus evaporates, residual dust and rock fragments form a **meteoritic swarm,** a loose collection of debris that continues to circle the Sun along the comet's orbit (Figure 15-39). If Earth's orbit happens to pass through this swarm, a **meteor shower** is seen as the dust particles strike Earth's upper

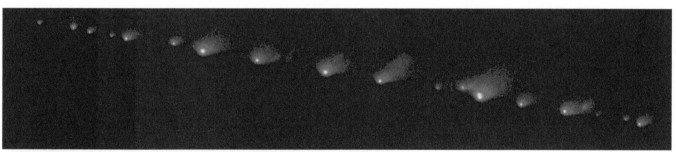

(a)   R I V U X G

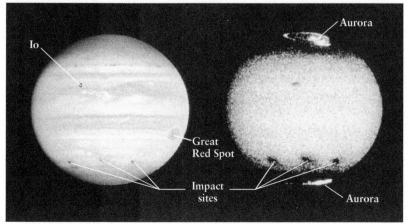

Io

Great Red Spot

Impact sites

Aurora

Aurora

(b)   R I V U X G   R I V U X G

FIGURE 15-38

Comet Shoemaker-Levy 9 and Its Encounter with Jupiter **(a)** This comet, originally orbiting Jupiter, was torn apart by the planet's gravitational force on July 7, 1992, fracturing into at least 21 pieces. Its returning debris, shown here in May 1994, struck the planet between July 16 and July 22, 1994. **(b)** Shown here are visible (left) and ultraviolet (right) images of Jupiter taken by the Hubble Space Telescope after three pieces of Comet Shoemaker-Levy 9 struck the planet. Astronomers had expected white remnants (the color of condensing ammonia or water vapor); the darkness of the impact sites may have come from carbon compounds in the comet debris. Note the aurorae in the ultraviolet image. (a: H. A. Weaver, T. E. Smith, STScI, and NASA; b: NASA)

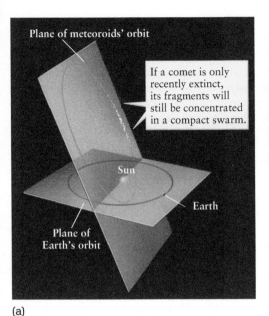

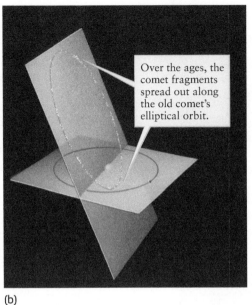

(a)

(b)

FIGURE 15-39

Meteoritic Swarms Rock fragments and dust from "burned out" comets continue to circle the Sun. **(a)** The most spectacular meteor showers occur when Earth passes through compact swarms of debris from recently extinct comets. **(b)** Meteor showers associated with older comets are more predictable, because Earth passes through the evenly distributed swarm on each trip around the Sun.

Labels in figure (a): Plane of meteoroids' orbit; If a comet is only recently extinct, its fragments will still be concentrated in a compact swarm.; Sun; Earth; Plane of Earth's orbit

Labels in figure (b): Over the ages, the comet fragments spread out along the old comet's elliptical orbit.

atmosphere. Nearly a dozen meteor showers can be seen each year (Table 15-1). Two of the bigger meteor showers—Perseids in summer and Geminids in winter—have almost a meteor per minute, which makes for great viewing.

CONCEPTCHECK 15-10

What motivated the Oort cloud hypothesis of a distant collection of icy objects?

Answer appears at the end of the chapter.

KEY WORDS

amino acids, p. 418
asteroid, p. 404
asteroid belt, p. 405
carbonaceous chondrite, p. 418
coma (of a comet), p. 419
comet, p. 419
differentiated asteroid, p. 417
dust tail, p. 422

fusion crust, p. 416
Hirayama family, p. 410
hydrogen envelope, p. 421
iron meteorite (iron), p. 417
ion tail, p. 422
Jupiter-family comet, p. 423
Kirkwood gaps, p. 406
Kuiper belt, p. 423

TABLE 15-1	**Prominent Yearly Meteor Showers**			
Shower name	Date of maximum intensity*	Typical hourly rate	Average speed (km/s)	Radiant constellation
Quadrantids	January 3	40	40	Boötes
Lyrids	April 22	15	50	Lyra
Eta Aquarids	May 4	20	64	Aquarius
Delta Aquarids	July 30	20	40	Aquarius
Perseids	August 12	50	60	Perseus
Orionids	October 21	20	66	Orion
Taurids	November 4	15	30	Taurus
Leonids	November 16	15	70	Leo
Geminids	December 13	50	35	Gemini
Ursids	December 22	15	35	Ursa Minor

* The date of maximum intensity is the best time to observe a particular shower, although good displays can often be seen a day or two before or after the maximum. The typical hourly rate is given for an observer under optimum viewing conditions. The average speed refers to how fast the meteoroids are moving when they strike the atmosphere. The radiant indicates which constellation, and thus the direction, the meteors appear to come from.

KEY IDEAS

Discovery of the Asteroids: Astronomers first discovered the asteroids while searching for a "missing planet."

• Thousands of asteroids with diameters ranging from a few kilometers up to about 1000 km orbit within the asteroid belt between the orbits of Mars and Jupiter.

Origin of the Asteroids: The asteroids are the relics of planetesimals that failed to accrete into a full-sized planet, thanks to the effects of Jupiter and other Mars-sized objects.

• Even today, gravitational perturbations by Jupiter deplete certain orbits within the asteroid belt. The resulting gaps, called Kirkwood gaps, occur at simple fractions of Jupiter's orbital period.

• Jupiter's gravity also captures asteroids in two locations, called Lagrangian points, along Jupiter's orbit.

Asteroid Collisions: Asteroids undergo collisions with one another, causing them to break up into smaller fragments.

• Some asteroids, called near-Earth objects, move in elliptical orbits that cross the orbits of Mars and Earth. If such an asteroid strikes Earth, it forms an impact crater whose diameter depends on both the mass and the speed of the asteroid.

• An asteroid struck Earth 65 million years ago, probably causing the extinction of the dinosaurs and many other species.

Meteoroids, Meteors, and Meteorites: Small rocks in space less than 1 m in size are called meteoroids. If a meteoroid enters Earth's atmosphere, it produces a bright trail called a meteor (or shooting star). A rock that does not burn up in the atmosphere and reaches Earth's surface is called a meteorite; most of the meteoroids are too small to make it to Earth before being completely vaporized.

• Meteorites are grouped into three major classes, according to composition: iron, stony iron, and stony meteorites. Irons and stony irons are fragments of the core of an asteroid that was large enough and hot enough to have undergone chemical differentiation, just like a terrestrial planet. Some stony meteorites come from the crust of such differentiated meteorites, while others are fragments of small asteroids that never underwent differentiation.

• Rare stony meteorites called carbonaceous chondrites may be relatively unmodified material from the solar nebula. These meteorites often contain organic material and may have provided building blocks for the origin of life on Earth.

• Analysis of isotopes in certain meteorites suggests that a nearby supernova may have triggered the formation of the solar system 4.56 billion years ago.

Comets: A comet is a chunk of ice with imbedded rock fragments that generally moves in a highly elliptical orbit about the Sun.

• As a comet approaches the Sun, its icy nucleus develops a luminous coma, surrounded by a vast hydrogen envelope. An ion tail and a dust tail extend from the comet, pushed away from the Sun by the solar wind and radiation pressure, respectively.

Origin and Fate of Comets: Comets are thought to originate from two regions, the Kuiper belt and the Oort cloud.

• The Kuiper belt lies in the plane of the ecliptic at distances between 30 and 50 AU from the Sun. Its icy objects are thought to be the source of short-period comets, whose periods are less than 200 years.

• The Oort cloud is thought to contain billions of icy objects in a spherical distribution that extends out to 50,000 AU from the Sun. The long-period comets, with periods greater than 200 years, are thought to originate in the Oort cloud.

• Fragments of "burned out" comets produce meteoritic swarms. A meteor shower is seen when Earth passes through a meteoritic swarm.

QUESTIONS

Review Questions

1. What led astronomers to suspect that there were members of the solar system that orbit between Mars and Jupiter?

2. How did the first asteroid come to be discovered? What role did theoretical calculations play in confirming the discovery? How did this discovery differ from what astronomers had expected to find?

3. How do modern astronomers discover new asteroids?

4. What are the differences between asteroids and trans-Neptunian objects?

5. Describe the asteroid belt. Does it lie completely within the plane of the ecliptic? What are its inner and outer radii?

6. If Jupiter was not present in our solar system, would the asteroid belt exist? Why or why not?

7. What are Kirkwood gaps? What causes them?

8. Consider the early solar system when Jupiter's semimajor axis changed as the planet migrated. Would the effect of the unstable orbital resonances (called Kirkwood gaps in the present-day asteroid belt) lead to more disruption of the original belt objects, or less?

9. Compare the explanation of the Kirkwood gaps in the asteroid belt to the way in which Saturn's moons help produce divisions in that planet's rings (see Section 12-11).

10. TUTORIAL 15-1 How is it possible to tell that some asteroids are nonspherical even though we do not have images of those asteroids?

11. What is the evidence that some asteroids are made of a loose conglomeration of smaller pieces?

12. What evidence links some meteorites on Earth to the asteroid Vesta?

13. The asteroid 243 Ida, which was viewed by the *Galileo* spacecraft, is a member of a Hirayama family. Discuss what this tells us about the history of this asteroid. Where might you look to find other members of the same family?

14. What are the Trojan asteroids, and where are they located? What holds them in this location?

15. What are near-Earth objects? What is the evidence that Earth has been struck by these objects?

16. What is the evidence that an asteroid impact could have contributed to the demise of the dinosaurs?

17. **TUTORIAL 15-2** What is the difference between a meteoroid, a meteor, and a meteorite?

18. Is there anywhere on Earth where you might find large numbers of stony meteorites that are not significantly weathered? If so, where? If not, why not?

19. Scientists can tell that certain meteorites came from the interior of an asteroid rather than from its outer layers. Explain how this is done.

20. Why are some asteroids differentiated while others are not?

21. Suppose you found a rock you suspect might be a meteorite. Describe some of the things you could do to determine whether it was a meteorite or a "meteorwrong."

22. What is the evidence that carbonaceous chondrites are essentially unaltered relics of the early solar system? What do they suggest about how the solar system may have formed?

23. **TUTORIAL 15-3** With the aid of a drawing, describe the structure of a comet.

24. Why is the phrase "dirty snowball" an appropriate characterization of a comet's nucleus?

25. What process produces jets that stream off a comet's nucleus?

26. What did we learn from the *Stardust* mission about the possible connection between comets and the origins of life on Earth?

27. What did scientists learn about the structure of comet nuclei from the *Deep Impact* mission?

28. Why do the ion tail and dust tail of a comet point in different directions?

29. Why do comets have prominent tails for only a short time during each orbit?

30. Why is it that Jupiter and Saturn can be seen in the night sky every year, while seeing specific comets such as Halley and Hyakutake is a once-in-a-lifetime event?

31. What is the relationship between the Kuiper belt and comets?

32. What is the Oort cloud? How might it be related to planetesimals left over from the formation of the solar system?

33. Why are comets more likely to break apart at perihelion than at aphelion?

34. Why do astronomers think that meteorites come from asteroids, while meteor showers are related to comets?

35. Why are asteroids, meteorites, and comets all of special interest to astronomers who want to understand the early history and subsequent evolution of the solar system?

Advanced Questions

Questions preceded by an asterisk (*) involve topics discussed in Box 1-1 or 7-2.

> **Problem-solving tips and tools**
>
> The small-angle formula is described in Box 1-1. We discussed retrograde motion in Section 4-1 and described its causes in Section 4-2. You will need to use Kepler's third law, described in Section 4-4 and Box 4-2, in some of the problems below. Box 7-2 discusses the concept of escape speed. A spherical object of radius r intercepts an amount of sunlight proportional to its cross-sectional area, equal to pr^2. The volume of a sphere of radius r is $4pr^3/3$.

36. When Olbers discovered Pallas in March 1802, the asteroid was moving from east to west relative to the stars. At what time of night was Pallas highest in the sky over Olbers's observatory? Explain your reasoning.

37. Consider the Kirkwood gap whose orbital period is two-fifths of Jupiter's 11.86-year period. Calculate the distance from the Sun to this gap. Does your answer agree with Figure 15-4?

*38. When the image in Figure 15-5 was made, the asteroid Ceres was 1.63 AU, or 2.44×10^8 km, from the Hubble Space Telescope. (a) What was the angular size of the asteroid as a whole? (b) You can see individual pixels in the image shown in Figure 15-5. Using a ruler and the scale bar in the figure, determine how many kilometers on the surface of Ceres are contained in the width of one pixel. (c) What is the angular width of each pixel in arcseconds?

39. Suppose that a binary asteroid (two asteroids orbiting each other) is observed in which one member is 16 times brighter than the other. Suppose that both members have the same albedo and that the larger of the two is 120 km in diameter. What is the diameter of the other member?

40. The accompanying image from the *Galileo* spacecraft shows the asteroid 243 Ida, which has dimensions 56 × 24 × 21 km. *Galileo* discovered a tiny moon called Dactyl, just 1.6 × 1.2 km in size, which orbits Ida at a distance of about 100 km. (In Greek mythology, the Dactyli were beings who lived on the slopes of Mount Ida.) Describe a scenario that could explain how Ida came to have a moon.

R I **V** U X G

(JPL/NASA)

*41. Assume that Ida's tiny moon Dactyl (see Question 40) has a density of 2500 kg/m³. (**a**) Calculate the mass of Dactyl in kilograms. For simplicity, assume that Dactyl is a sphere 1.4 km in diameter. (**b**) Calculate the escape speed from the surface of Dactyl. If you were an astronaut standing on Dactyl's surface, could you throw a baseball straight up so that it would never come down? Professional baseball pitchers can throw at speeds around 40 m/s (140 km/h, or 90 mi/h); your throwing speed is probably a bit less.

42. Imagine that you are an astronaut standing on the surface of a Trojan asteroid. How will you see the phase of Jupiter change with the passage of time? How will you see Jupiter move relative to the distant stars? Explain your answers.

43. Use the percentages of stones, irons, and stony iron meteorites that fall to Earth to estimate what fraction of their parent asteroids' interior volume consisted of an iron core. Assume that the percentages of stones and irons that fall to Earth indicate the fractions of a parent asteroid's interior volume occupied by rock and iron, respectively. How valid do you think this assumption is?

44. On March 8, 1997, Comet Hale-Bopp was 1.39 AU from Earth and 1.00 AU from the Sun. Use this information and that given in the caption to Figure 15-32 to estimate the length of the comet's ion tail on that date. Give your answer in kilometers and astronomical units.

45. *Sun-grazing comets* come so close to the Sun that their perihelion distances are essentially zero. Find the orbital periods of Sun-grazing comets whose aphelion distances are (**a**) 100 AU, (**b**) 1000 AU, (**c**) 10,000 AU, and (**d**) 100,000 AU. Assuming that these comets can survive only a hundred perihelion passages, calculate their lifetimes. (*Hint:* Remember that the semimajor axis of an orbit is one-half the length of the orbit's long axis.)

46. Comets are generally brighter a few weeks after passing perihelion than a few weeks before passing perihelion. Explain why might this be. (*Hint:* Water, including water-ice, does an excellent job of retaining heat.)

47. The hydrogen envelopes of comets are especially bright at an ultraviolet wavelength of 122 nm. Use Figure 5-24 to explain why.

48. A *very* crude model of a typical comet nucleus is a cube of ice (density 1000 kg/m³) 10 km on a side. (**a**) What is the mass of this nucleus? (**b**) Suppose 1% of the mass of the nucleus evaporates away to form the comet's tail. Suppose further that the tail is 100 million (10^8) km long and 1 million (10^6) km wide. Estimate the average density of the tail (in kg/m³). For comparison, the density of the air you breathe is about 1.2 kg/m³. (**c**) In 1910 Earth actually passed through the tail of Comet Halley. At the time there was some concern among the general public that this could have deleterious effects on human health. Was this concern justified? Why or why not?

49. For many years it was thought that the Tunguska event was caused by a comet striking Earth. This idea was rejected because a small comet would have broken up too high in the atmosphere to cause significant damage on the ground. Explain why, using your knowledge of a comet's structure.

Discussion Questions

50. Discuss the idea that Ceres should be regarded as the smallest dwarf planet rather than the largest asteroid. What are the advantages of this scheme? What are the disadvantages?

51. From the abundance of craters on the Moon and Mercury, we know that numerous asteroids and meteoroids struck the inner planets early in the history of our solar system. Is it reasonable to suppose that numerous comets also pelted the planets 3.5 to 4.5 billion years ago? Speculate about the effects of such a cometary bombardment, especially with regard to the evolution of the primordial atmospheres on the terrestrial planets.

52. In the 1998 movie *Armageddon,* an asteroid "the size of Texas" is on a collision course with Earth. The asteroid is first discovered by astronomers just 18 days prior to impact. To avert disaster, a team of astronauts blasts the asteroid into two pieces just 4 hours before impact. Discuss the plausibility of this scenario. (*Hint:* On average, the state of Texas extends for about 750 km from north to south and from east to west. How does this compare with the size of the largest known asteroids?)

53. Suppose astronomers discover that a near-Earth object the size of 1994 XM1 is on a collision course with Earth. Describe what humanity could do within the framework of present technology to counter such a catastrophe.

Web/eBook Questions

54. Search the World Wide Web for information about the present status of NASA's *Dawn* mission. Describe one result from Vesta that is not described in this textbook.

55. Search the World Wide Web to find out why some scientists disagree with the idea that a tremendous impact led to the demise of the dinosaurs. (They do not dispute that the impact took place, only what its consequences were.) What are their arguments? From what you learn, what is your opinion?

56. Several scientific research programs are dedicated to the search for near-Earth objects (NEOs), especially those that might someday strike our planet. Search the World Wide Web for information about at least one of these programs. How does this program search for NEOs? How many NEOs has this program discovered? Will any of these pose a threat in the near future?

57. ⟨VIDEO 15.4⟩ **Estimating the Speed of a Comet.** Access and view the video "Two Comets and an Active Sun" in Chapter 15 of the *Universe* Web site or eBook. (**a**) Why don't the comets reappear after passing the Sun? (**b**) The white circle shows the size of the Sun, which has the diameter 1.39×10^6 km. Using this to set the scale, step through the video and measure how the position of one of the comets changes. Use your measurements and the times displayed in the video to estimate the comet's speed in km/h and km/s. (Assume that the comet moved in the same plane as that shown in the video.) As part of your answer, explain the technique and calculations you used. (**c**) How does your answer in (b) compare with the orbital speed of Mercury, the innermost and fastest-moving planet (see Table 11-1)? Why is there a difference? (**d**) If the comet's motion was not in the same plane as that shown in the video, was its actual speed more or less than your estimate in (b)? Explain.

ACTIVITIES

Observing Project

> **Observing tips and tools**
>
> *Meteors:* Informative details concerning upcoming meteor showers appear on the Web sites for *Sky & Telescope* and *Astronomy* magazines.
>
> *Comets:* Because astronomers discover dozens of comets each year, there is usually a comet visible somewhere in the sky. Unfortunately, they are often quite dim, and you will need to have access to a moderately large telescope (at least 35 cm, or 14 in.). You can get up-to-date information from the Minor Planet Center Web site. Also, if there is an especially bright comet in the sky, useful information about it might be found at the Web sites for *Sky & Telescope* and *Astronomy* magazines.
>
> *Asteroids:* At opposition, some of the largest asteroids are bright enough to be seen through a modest telescope. Check the Minor Planet Center Web site to see if any bright asteroids are near opposition. If so, check the current issue as well as the most recent January issue of *Sky & Telescope* magazine for a star chart showing the asteroid's path among the constellations. You will need such a chart to distinguish the asteroid from background stars. Also, you can locate Ceres, Pallas, Juno, and Vesta using the *Starry Night*™ program if you have access.

58. Make arrangements to view a meteor shower. Table 15-1 lists the dates of major meteor showers. Choose a shower that will occur near the time of a new moon. Set your alarm clock for the early morning hours (1 to 3 A.M.). Get comfortable in a reclining chair or lie on your back so that you can view a large portion of the sky. Record how long you observe, how many meteors you see, and what location in the sky they seem to come from. How well does your observed hourly rate agree with published estimates, such as those in Table 15-1? Is the radiant of the meteor shower apparent from your observations?

59. If a comet is visible with a telescope at your disposal, make arrangements to view it. Can you distinguish the comet from background stars? Can you see its coma? Can you see a tail?

60. Make arrangements to view an asteroid. Observe the asteroid on at least two occasions, separated by a few days. On each night, draw a star chart of the objects in your telescope's field of view. Has the position of one starlike object shifted between observing sessions? Does the position of the moving object agree with the path plotted on published star charts? Do you feel confident that you have in fact seen the asteroid?

61. Use *Starry Night*™ to explore some of the dwarf planets of the solar system. Select **Favourites > Explorations > Dwarf Planets** from the menu. This view, from a position in space about 97 AU from the Sun shows Neptune's orbit as well as the orbits of several dwarf planets. Right-click (Ctrl-click on a Mac) on the Sun and select **Centre** from the contextual menu. (**a**) Do the dwarf planets revolve about the Sun in the same or opposite direction as the planets? (**b**) Use the location scroller to adjust the view so that the plane of Neptune's orbit appears edge-on. How do the orbital planes of the dwarf planets compare to those of the planets?

62. Use *Starry Night*™ to locate the largest asteroid Ceres, now designated a dwarf planet, and to examine the asteroid belt. Select **Favourites > Explorations > Asteroids.** The view shows the orbits of the major planets and of the dwarf planet Ceres, centered on the Sun from a distance just beyond the orbit of Jupiter, at about 5.3 AU. Numerous asteroids are represented by the bright green dots. Use the location scroller to examine the extent of the asteroid belt and observe the orbits of some of the asteroids from different perspectives, particularly along the plane of the ecliptic and perpendicular to the plane of the ecliptic. (**a**) In which region of the solar system is the major concentration of asteroids? (**b**) How would you describe the plane of Ceres's orbit relative to the plane of the ecliptic? (**c**) Look for asteroids whose motion brings them within Earth's orbit. If you find such an asteroid, click the **Stop** button and use the **HUD** to identify it. You can display an asteroid's orbit by right-clicking on the asteroid (Ctrl-click on a Mac) to open the contextual menu and selecting **Orbit.** Use the location scroller to study the asteroid's orbit. Are the orbits of these asteroids generally in the plane of the ecliptic?

63. Use *Starry Night*™ to observe the motions of two smaller objects in the solar system, the dwarf planet Ceres and the minor planet Pallas, both members of the asteroid belt, as seen from Earth. Open **Favourites > Explorations > Atlas** to view the sky from the center of a transparent Earth. Select **View > Ecliptic Guides > The Ecliptic** from the menu and set the **Time Flow Rate** to **1 sidereal day.** Open the **Find** pane, type the name **Ceres** in the search box, and press the **Enter** key to center the view on this dwarf planet. Click the **Play** button and observe the motion of Ceres over the course of at least 2 years of simulated time. (**a**) Describe how Ceres moves. (**b**) How can you tell that Ceres orbits the Sun in the same direction as the planets? (**c**) Return to the **Find** pane, type the name **Pallas** in the search box and press the **Enter** key to center the view on this minor planet. Watch the motion of Pallas in the sky for at least 2 years of simulated time. (**d**) How does the motion of Pallas compare with that of Ceres? (**e**) Which object's orbit is more steeply inclined to the plane of Earth's orbit? How can you tell? (**f**) One distinguishing feature that identifies a dwarf planet from a minor planet is its shape. A dwarf planet is approximately spherical in shape whereas a minor planet is more irregular. Find and zoom in on each of the above objects in turn. Which object has the more irregular shape? (**g**) Find the objects Juno and Iris. On the basis

of the above criterion, what is the classification of each of these objects, dwarf planet or minor planet?

64. Use *Starry Night™* to observe a comet. (**a**) Select **Favourites > Explorations > Hale-Bopp**. From the appearance of this comet, predict the direction of the Sun relative to the comet and explain how you made your prediction. Use the hand tool to find the Sun in the view to verify your hypothesis. (**b**) Select **File > Revert** from the menu. From the appearance of the comet, can you predict what its motion will be in the future against the background stars? Click the **Play** button to see if your prediction was correct.

65. Use *Starry Night™* to study the motion of two comets. (**a**) Open **Favourites > Explorations > Hyakutake** to observe Comet Hyakutake from space as it approaches the inner solar system. Click the **Play** button and observe the comet as time progresses. Allow time to continue and observe the approach of a second comet, Hale-Bopp. To repeat the animation, select **File > Revert** from the menu and, if you wish, decrease the **Time Flow Rate** to observe the motion of the comet more closely. Use the location scroller to gain different perspectives on the motion of both comets. Describe what you see. Is each comet's orbit in about the same plane as the orbits of the inner planets, or is it steeply inclined to that plane? (*Hint:* You can stop **Time Flow** and right-click on each comet in turn and click on **Orbit** in the dropdown menu to display each comet's orbit.) How does the comet's speed vary as it moves along its orbit? During which part of the orbit is the tail visible? In what direction does the tail point? (**b**) Select **Favourites > Explorations > Atlas** from the menu. Set the date in the toolbar to **January 1, 1995**. Open the **Find** pane and enter **Hyakutake** in the search box to center the view on the comet. Change the **Time Flow Rate** to **1 sidereal day** and click the **Play** button. Observe the comet's motion for at least 2 years of simulated time. Describe the motion, and explain why it is more complicated than the motion you observed in part (a). (**c**) Click the **Now** button in the toolbar. Set the time step to **1 lunar m**, and click the **Play** button. Comet Hyakutake is currently moving almost directly away from the Sun and so, as seen from the Sun, its position on the celestial sphere should not change. Is this what you see in *Starry Night™*? Explain your observation. (*Hint:* You are observing from Earth, not the Sun.)

66. Use *Starry Night™* to examine the regions of the solar system where comets originate. Open **Favourites > Explorations > Oort cloud**. The view is centered on the Sun and shows the orbits of the major planets as well as the inner and outer boundaries of the region in the solar system in which the Kuiper belt is found. (**a**) What type of comets originates from this belt? (**b**) Use the location scroller to view the boundaries of the Kuiper belt region. The inner boundary corresponds approximately to the orbit of which major planet? (**c**) Beyond the Kuiper belt lies the Oort cloud. Use the **Increase current elevation** button to move farther from the Sun until you encounter the boundary of the Oort cloud. What type of comet arises from this region of the solar system?

ConceptChecks

ConceptCheck 15-1: The solar system is filled with planets that have semimajor axes increasing from one to the next by about a factor between 1.4 and 2. The exception to this pattern is Jupiter, but the pattern would hold up if there was a planet where the asteroid belt currently exists.

ConceptCheck 15-2: Ceres is an asteroid and also a dwarf planet—large enough to pull itself into a sphere but unable to gravitationally clear out its surroundings.

ConceptCheck 15-3: Jupiter's gravity prevented a planet from forming where the asteroid belt currently exists. The planet would have probably been about the size of Earth, but mass was ejected and all the mass in the asteroid belt today could only form a very small object 1500 km in diameter (less than the distance from Los Angeles to Chicago).

ConceptCheck 15-4: The *Hayabusa* mission returned samples of the asteroid Itokawa, where analysis on Earth showed a match to meteoritic dust.

ConceptCheck 15-5: The *Dawn* spacecraft was able to assess Vesta's composition by measuring the infrared spectra of the asteroid's surface. Analysis finds a match with about 6% of the meteorites found on Earth.

ConceptCheck 15-6: Iridium is common in asteroids, but rare in Earth's rocks. A thick layer of iridium is found around the world dating back 65 million years, implying a massive impact at that time. More evidence comes from a massive impact event dated at 65 million years old near the Yucatán Peninsula.

ConceptCheck 15-7: If a nickel-iron sample cools too quickly, it will not contain a Widmanstätten pattern, which takes millions of years to form. This is not a feature that can be artificially produced on Earth.

ConceptCheck 15-8: The buildup of larger objects in the formation of the solar system involved heating (due to radioactivity and collisions) leading to alteration of their chemical composition. However, the carbonaceous chondrites do not show chemical alteration that would have occurred if they were heated. Therefore, this class of meteorites appears to contain some original unaltered material present in the early solar nebula.

ConceptCheck 15-9: The *Deep Impact* mission measured the density of Comet Tempel 1 and found it be significantly less dense than water—only 60%. Because the comet is mostly water (with smaller amounts of other material), the comet must be porous with empty spaces to have a density lower than water.

ConceptCheck 15-10: From Kepler's third law, the longer the orbital period, the larger the semimajor axis of the orbit. The short-period comets already orbit beyond Neptune in the Kuiper belt, and the long-period comets require a collection of objects much farther out; this is called the Oort cloud.

CalculationChecks

CalculationCheck 15-1: 100 m. Since the crater is about 20 times larger than the fragment hitting Earth, the fragment is about (2000 m)/20 = 100 m across.

CalculationCheck 15-2: 5 km. (5 km/s) × (1 s) = 5 km.

Pluto and the Kuiper Belt by Scott Sheppard

The "what is a planet" controversy began with the discovery of Ceres and several other objects between Mars and Jupiter in the early 1800s. At first, Ceres and then Pallas, Juno, and Vesta were considered planets, even though they were orders of magnitude less massive than the other known planets and all had similar orbits. The definition of a planet at this time was simple. Any object in orbit about the Sun that did not show cometary effects was a planet.

By the mid-1800s, the discovery of more and more objects between Mars and Jupiter saw the words "asteroid" and "minor planet" start to be used in order to signify these objects as an ensemble of bodies. That is, astronomers felt a major planet should be a rare and unique thing. The major planet club thus was left with eight members: Mercury, Venus, Earth, Mars, Jupiter, Saturn, Uranus, and Neptune.

The planet controversy arose once again in 1930 with the discovery of Pluto, which was quickly regarded as the ninth planet. During the ensuing years, new observations downgraded Pluto's size from being near that of Earth to being smaller than our Moon. Its orbit also proved to be peculiar because it was highly inclined and eccentric, having it even crossing Neptune's orbit. Its uniqueness kept Pluto in the planet club with relatively few detractors until 1992, when the first Kuiper belt object was discovered. As it became apparent that hundreds of thousands of Kuiper belt objects existed on stable orbits just beyond Neptune, Pluto's planet status was becoming less and less certain. With the discovery of Eris (2003 UB313) in 2003, an object larger than Pluto, Pluto's last unique claim of being the largest Kuiper belt object was lost. Thus, the planet controversy had finally reached a point where a major planet's minimum size must be precisely defined.

No matter how you count them, there are no longer nine major planets in our solar system. The same situation of the main belt asteroids in the mid-1800s is now apparent with Pluto, Eris, and the rest of the Kuiper belt objects. Although the largest Kuiper belt objects are significantly bigger than the largest main belt asteroids, they are still an ensemble of bodies with a continuous size distribution and similar formation and evolution histories. The reason Pluto's status as a planet is a controversy now is simply because it took 62 years between the discovery of Pluto and the next Kuiper belt object in 1992. Within that time span Pluto became known as the ninth planet through several generations and was ingrained in society as such.

Classification is not that important to scientists but it is the understanding of how the object formed and its history. Even so, any classification should have a scientific base and not be based on tradition, as seems to be the case for Pluto in the recent past. Eris as well as Pluto and the rest of the Kuiper belt objects had very similar histories and formation scenarios and thus they should be considered as an ensemble of objects just like the main belt asteroids between Mars and Jupiter. As technology advances, our knowledge and thus understanding advances and so should our classification.

Several minimum-size planet definitions were debated recently. The first is that anything larger than a certain minimum size is a planet. The chosen size would likely be arbitrary and may be a nice round number like 1000 km or the size of Pluto. This stems from the traditional belief that Pluto has been considered a planet in the past and should continue as such. This would give us 10 current planets including Eris. A second possible definition is that anything that is spherical should be a planet. This definition is based on physics in which the gravitational force of an object is large enough to overcome any other forces in determining the overall shape of the object. Determining if an object meets this criteria or not is complicated. This definition would also significantly change the status quo by increasing the number of known planets by several factors overnight. A third definition is that any object with a unique orbit and gravitationally dominates its local orbital environment should be a planet. This definition is the most scientifically based, in that objects with similar formation and evolutionary histories would be grouped together. Using this definition would give us eight known major planets, dropping Pluto from the list.

A final proposal as to what is a planet takes into account that the current known major planets themselves are remarkably different and can be split into individual categories themselves. Mercury, Venus, Earth, and Mars are terrestrial planets whose compositions are dominated by rock. Jupiter and Saturn are gas giant planets dominated by their hydrogen and helium envelopes. Uranus and Neptune are ice giant planets dominated by gases other than hydrogen and helium. The trans-Neptunian objects or "ice dwarf planets," as some are calling them, are probably composed of large amounts of volatiles such as methane-ice and water-ice.

The International Astronomical Union (IAU) recently chose to use a combination of the second and third definitions above to define a planet. That is, a planet is an object that is both spherical and has a unique orbit in which it is gravitationally dominant. An object that only satisfies definition two above and not three is now being called a dwarf planet, of which Ceres, Pluto, and Eris qualify, as well as several more objects in the Kuiper belt. In reality, it is the public that must accept this definition, and we will only know if that is the case a few generations from now.

Defining the word "planet" is like defining what an ocean is. At first it appears to be a simple term, but is very hard to precisely define. Further planet discoveries will be made and there are sure to be borderline cases. This is just the way nature is; it does not have little bins to nicely classify objects, but there is usually a continuous array of objects. The important thing to take away from all of this is to understand this array of objects by using our rapidly expanding scientific knowledge and understanding our place in the solar system.

Scott S. Sheppard is an astronomer at the Carnegie Institution of Washington in the Department of Terrestrial Magnetism located in Washington, DC. His main research currently involves discovering small bodies in our solar system to understand their dynamical and physical properties that help constrain solar system and planet formation. He has discovered several small moons around all the giant planets, been a pioneer in the discovery of Neptune Trojans, as well as found tens of Kuiper Belt objects.

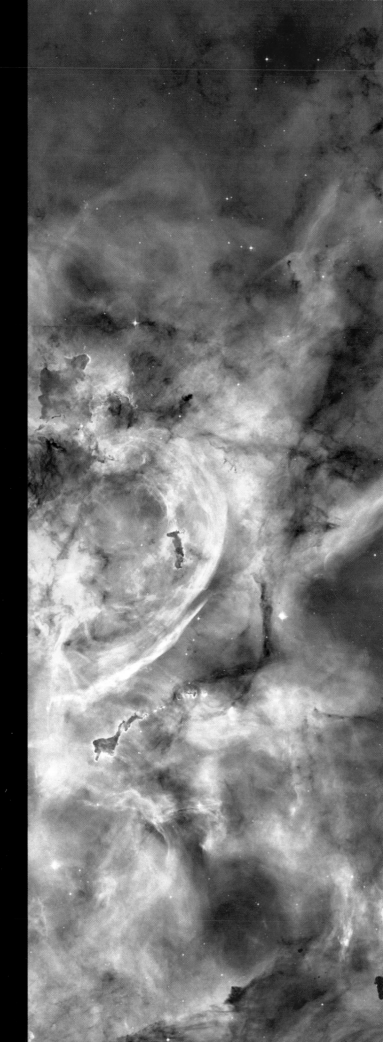

III Stars and Stellar Evolution

Spanning more than 300 light-years across, the Carina Nebula shown here is one of the largest and brightest nebulae in our Milky Way Galaxy. In fact, if aliens were to gaze upon us from another galaxy, this nebula would be one of the prominent features they would see. Within this nebula shines one of our Galaxy's brightest stars, Eta Carinae, at more than 5 million times the luminosity of the Sun. Ultraviolet light from the brightest stars excite the nebula's gas and the enhanced coloring in this image tracks emission from hydrogen, oxygen, nitrogen, and sulfur. In the darkest regions, dense gas is forming a new generation of stars. (NASA)

R I V U X G

Zooming in to a small portion of the Carina
Nebula, this feature 3 light-years long is
called Mystic Mountain. The dense hydrogen
body of this feature is slowly being eroded by
ultraviolet light from surrounding stars. In the
coolest and densest regions of this feature,
new stars are forming. New stars can form
a pair of oppositely directed jets, which are
visible at the peak of this mystic mountain.
On the flat-topped structure to the left of
the peak, another jet can be seen, while its
other half is presumably obscured by the
nebula. (NASA/ESA/M. Livio and the Hubble 20th
Anniversary Team [STScI])
R I **V** U X G

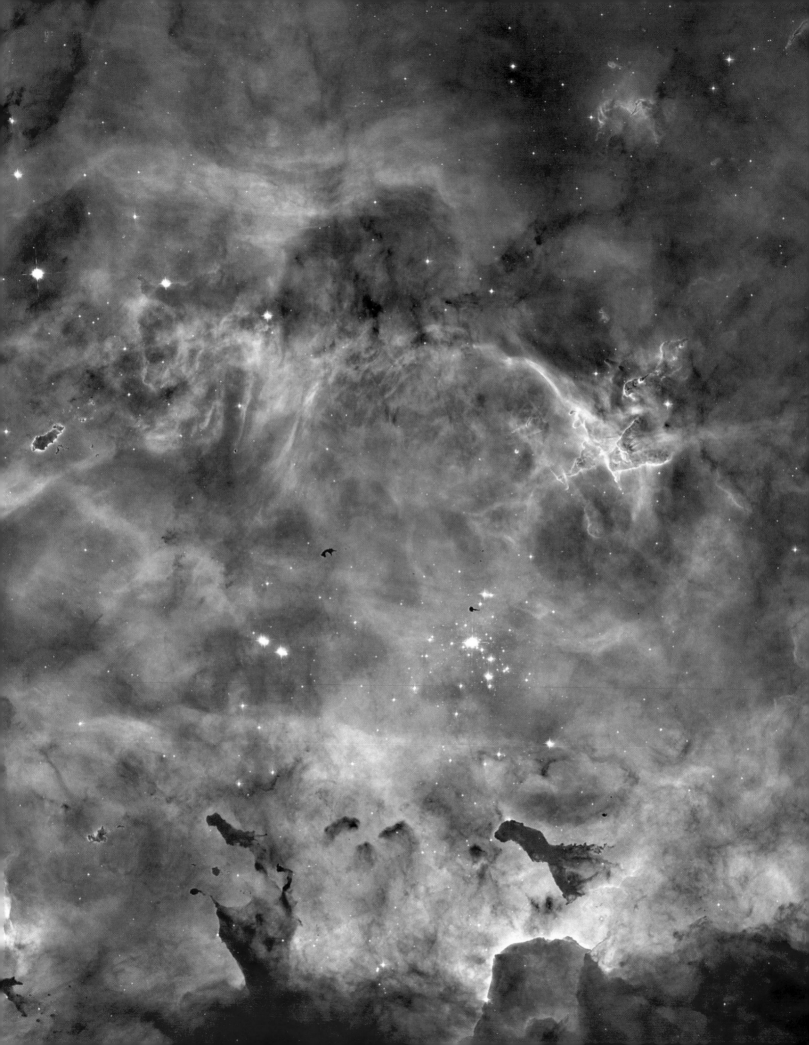

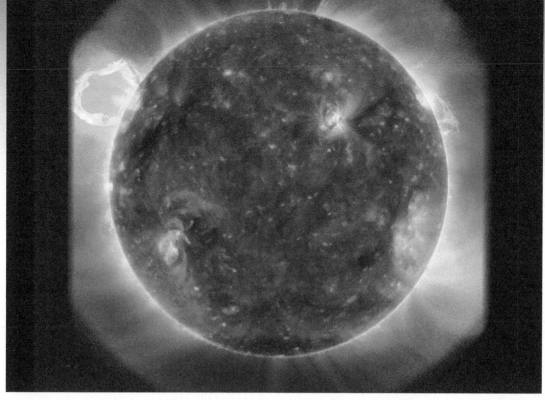

This ultraviolet image of the Sun, taken in 2010 by the Solar Dynamics Observatory, shows a prominence (upper left) in which magnetic fields carry gases above the Sun's surface in a loop that can extend upward hundreds of thousands of kilometers. (NASA/GSFC/AIA)

R I V U X G

Our Star, the Sun

The Sun is by far the brightest object in the sky. By earthly standards, the temperature of its glowing surface is remarkably high, about 5800 K, but there are regions of the Sun that reach far higher temperatures of tens of thousands or even millions of kelvins. Gases at such temperatures emit ultraviolet light or X-rays, which makes the Sun's outer regions prominent in the accompanying image from an ultraviolet telescope in space. Far from being smooth and placid, the Sun is a dynamic object with many features. Some of the hottest and most energetic regions on the Sun spawn immense disturbances and can propel solar material across space to reach Earth and other planets.

In recent decades, we have learned that the Sun shines because at its core hundreds of millions of tons of hydrogen are converted to helium every second. We have confirmed this picture by detecting the by-products of this transmutation—strange, ethereal particles called neutrinos—streaming outward from the Sun into space. We have discovered that the Sun has a surprisingly violent atmosphere, with a host of features such as sunspots whose numbers rise and fall on a predictable 11-year cycle. By studying the Sun's vibrations, we have begun to probe beneath its surface into hitherto unexplored realms. And we have just begun to investigate how changes in the Sun's activity can affect Earth's environment as well as our technological society.

16-1 The Sun's energy is generated by thermonuclear reactions in its core

The Sun is the largest member of the solar system. It has almost a thousand times more mass than all the planets, moons, asteroids, comets, and meteoroids put together. But the Sun is also a star. In fact, it is a remarkably typical star, with a mass, size, surface temperature, and chemical composition that are roughly midway between the extremes exhibited by the myriad other stars in the heavens. Table 16-1 lists essential data about the Sun.

Solar Energy

For most people, what matters most about the Sun is the energy that it radiates into space. Without the Sun's warming rays, our oceans would freeze. Furthermore, since most life is part of a food chain dependent on sunlight for energy, an Earth without sunlight would be nearly dead. To understand the nature of life on Earth, we must understand the Sun.

> The Sun's surface is hot enough to emit visible light

Why is the Sun such a strong source of energy? Recall that the hotter an object, the more light it emits (see Figure 5-9). The Sun's spectrum is close to that of an idealized blackbody with a temperature of 5800 K (see Sections 5-3 and 5-4, especially Figure 5-13). Thanks to this high temperature, each square meter of the Sun's surface emits a tremendous amount of radiation, principally at visible wavelengths. Indeed, the Sun is the only object in the solar system that *emits* substantial amounts of visible light. The surface temperatures of the Moon and planets are very low compared to the Sun; the visible light that we see from bodies in the solar system is actually sunlight that struck those worlds and was *reflected* toward Earth.

The bright Sun emits a tremendous amount of energy. The total amount of energy emitted by the Sun each second, called its **luminosity**, is about 3.9×10^{26} watts (3.9×10^{26} joules of energy emitted per second). (We discussed the relationship between the Sun's surface temperature, radius, and luminosity in Box 5-2.) Astronomers denote the Sun's luminosity by the symbol $L_\odot$. A circle with a dot in the center is the astronomical symbol for the Sun and was also used by ancient astrologers.

Does the Sun emit so much energy because it is hot or because it is big? To put the Sun's temperature and size into perspective,

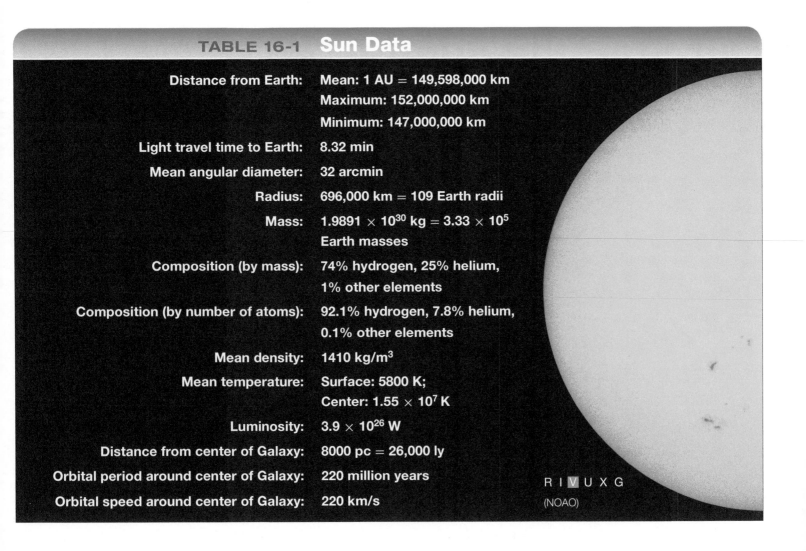

TABLE 16-1	**Sun Data**
Distance from Earth:	Mean: 1 AU = 149,598,000 km
	Maximum: 152,000,000 km
	Minimum: 147,000,000 km
Light travel time to Earth:	8.32 min
Mean angular diameter:	32 arcmin
Radius:	696,000 km = 109 Earth radii
Mass:	1.9891×10^{30} kg = 3.33×10^5 Earth masses
Composition (by mass):	74% hydrogen, 25% helium, 1% other elements
Composition (by number of atoms):	92.1% hydrogen, 7.8% helium, 0.1% other elements
Mean density:	1410 kg/m³
Mean temperature:	Surface: 5800 K; Center: 1.55×10^7 K
Luminosity:	3.9×10^{26} W
Distance from center of Galaxy:	8000 pc = 26,000 ly
Orbital period around center of Galaxy:	220 million years
Orbital speed around center of Galaxy:	220 km/s

R I V U X G
(NOAO)

we can make a comparison with ovens and lightbulbs. The solar surface is about 10 times hotter than the maximum temperature of a household oven. A factor of 10 might not sound like a great amount between the Sun and an oven, but we saw in Section 5-4, for the Stefan-Boltzmann law, F (energy flux) $= \sigma T^4$, that raising an object's temperature 10 times will increase the energy it emits per second by a factor of 10,000! But the Sun is also much bigger than an oven, and to appreciate the effect of size, let us consider lightbulbs.

Suppose you stood in a room that contained an object at the same temperature as the Sun but with only a surface area of 1 square millimeter. Would it burn you up? No. Such an object would emit around 100 watts, similar to a household lightbulb. Just like a lightbulb, you would have to put your hand quite close to feel any warmth. What helps make the Sun so strong is that each square meter patch of the solar surface emits about as much as a million 100-watt lightbulbs, and the Sun's surface area has more than a million trillion of these 1-square-meter patches. Taken together, it is the high temperature *and* large size of the Sun that make it so bright.

CONCEPTCHECK 16-1

The Sun emits light at nearly all possible wavelengths. At which range of wavelengths is most of the Sun's energy emitted?

Answer appears at the end of the chapter.

The Source of the Sun's Energy: Early Ideas

The bright Sun leads us to a more fundamental question: What keeps the Sun's visible surface so hot? Or, put another way, what is the fundamental source of the tremendous energy that the Sun radiates into space? For centuries, this question was one of the greatest mysteries in science. The mystery deepened in the nineteenth century, when geologists and biologists found convincing evidence that life has existed on Earth for at least several hundred million years. Since life as we know it depends crucially on sunlight, the conclusion was that the Sun was at least several hundred million years old. (We now know that Earth is 4.54 billion years old and that life has existed on it for most of its history.) The source of the Sun's energy posed a severe problem for physicists. What source of energy could have kept the Sun shining for so long (**Figure 16-1**)?

One attempt to explain solar energy was made in the mid-1800s by the English physicist Lord Kelvin (for whom the temperature scale is named) and the German scientist Hermann von Helmholtz. They argued that the tremendous weight of the Sun's outer layers should cause the Sun to contract gradually, compressing its interior gases. Whenever a gas is compressed, its temperature rises. (You can demonstrate this with a bicycle pump: As you pump air into a tire, the temperature of the air increases and the pump becomes warm to the touch.) Kelvin and Helmholtz thus suggested that gravitational contraction could cause the Sun's gases to become hot enough to radiate energy out into space.

This process, called *Kelvin-Helmholtz contraction,* actually does occur during the earliest stages of the birth of a star like the Sun (see Section 8-4). But Kelvin-Helmholtz contraction cannot be the major source of the Sun's energy today. If it were, the Sun would have had to be much larger in the relatively recent past.

FIGURE 16-1   R I [V] **U X G**

The Sun The Sun's visible surface has a temperature of about 5800 K. At this temperature, all solids and liquids vaporize to form gases. It was only in the twentieth century that scientists discovered what has kept the Sun so hot for billions of years—the thermonuclear fusion of hydrogen nuclei in the Sun's core. (Jeremy Woodhouse/PhotoDisc)

Helmholtz's own calculations showed that the Sun could have started its initial collapse from the solar nebula no more than about 25 million years ago. But the geological and fossil record shows that Earth is far older than that, and so the Sun must be as well. Hence, this model of a Sun that shines because it shrinks cannot be correct.

On Earth, a common way to produce heat and light is by burning fuel, such as a log in a fireplace or charcoal in a barbeque grill. Is it possible that a similar process explains the energy released by the Sun? The answer is no, because this process could not continue for a long enough time to explain the age of Earth. It might seem difficult to estimate how long the Sun could last if its energy came from burning, but the calculation only requires a few steps. To begin with, the chemical reactions involved in burning release roughly 10^{-19} joule of energy per atom. Therefore, the number of atoms that would have to undergo chemical reactions each second to generate the Sun's luminosity of 3.9×10^{26} joules per second is approximately

$$\frac{3.9 \times 10^{26} \text{ joules per second}}{10^{-19} \text{ joule per atom}} = 3.9 \times 10^{45} \text{ atoms per second}$$

From its mass and chemical composition, we know that the Sun contains about 10^{57} atoms. Thus, the length of time that would be required to consume the entire Sun by burning is

$$\frac{10^{57} \text{ atoms}}{3.9 \times 10^{45} \text{ atoms per second}} = 3 \times 10^{11} \text{ seconds}$$

There are about 3×10^7 seconds in a year. Hence, in this model, the Sun would burn itself out in a mere 10,000 (10^4) years! This period of time is far shorter than the known age of Earth, so chemical reactions cannot explain how the Sun shines. This is a very powerful result because even though there are numerous types of chemical reactions that release energy (anything from fire to dynamite), the maximum possible chemical energy in the Sun's total mass could not keep it burning for very long.

The Source of the Sun's Energy: Discovering Thermonuclear Fusion

The source of the Sun's luminosity could be explained if there were a process that was like burning but released much more energy per atom. Then the rate at which atoms would have to be consumed would be far less, and the lifetime of the Sun could be long enough to be consistent with the known age of Earth. Albert Einstein discovered the key to such a process in 1905. According to his *special theory of relativity,* a quantity m of mass can in principle be converted into an amount of energy E according to a now-famous equation:

Einstein's mass-energy equation

$$E = mc^2$$

m = quantity of mass, in kg

c = speed of light = 3×10^8 m/s

E = amount of energy into which the mass can be converted, in joules

The speed of light c is a large number, so c^2 is huge. Therefore, a small amount of matter can release an awesome amount of energy. For example, if the mass in just one penny (about 2.5 grams) were completely converted into energy, it would release about 5 times the energy of the nuclear bomb detonated over Hiroshima, Japan, during World War II. (While the energy in a penny's worth of mass is impressive, the largest nuclear bombs release the energy contained in the mass of 10 rolls of quarters; see Figure 1-6.)

Inspired by Einstein's ideas, astronomers began to wonder if the Sun's energy output might come from the conversion of matter into energy. The Sun's low density of 1410 kg/m^3 indicates that it must be made of the very lightest atoms, primarily hydrogen and helium. In the 1920s, the British astronomer Arthur Eddington showed that temperatures near the center of the Sun must be so high that atoms become completely ionized. Hence, at the Sun's center we expect to find hydrogen nuclei and electrons flying around independent of each other.

Another British astronomer, Robert Atkinson, suggested that under these conditions hydrogen nuclei could fuse together to produce helium nuclei in a *nuclear reaction* that transforms a tiny amount of mass into a large amount of energy. Experiments in the laboratory using individual nuclei show that such reactions can indeed take place. The process of converting hydrogen into helium is called **hydrogen fusion.** (It is also sometimes called *hydrogen burning,* even though nothing is actually burned in the conventional sense. Ordinary burning involves chemical reactions that rearrange the outer electrons of atoms but have no effect on the atoms' nuclei.)

> Ideas from relativity and nuclear physics led to an understanding of how the Sun shines

The fusing together of nuclei is also called **thermonuclear fusion,** because it can take place only at extremely high temperatures. This is because all nuclei have a positive electric charge and so tend to repel one another. But in the extreme heat and pressure at the Sun's center, hydrogen nuclei (protons) are moving so fast that they can overcome their electric repulsion and actually touch one another. When that happens, thermonuclear fusion can take place.

ANALOGY You can think of protons as tiny electrically charged spheres that are coated with a very powerful glue. If the spheres are not touching, the repulsion between their charges pushes them apart. But if the spheres are forced into contact, the strength of the glue "fuses" them together. In this fusion, a little bit of mass is lost, which gets converted into energy.

CAUTION! Be careful not to confuse thermonuclear fusion with the similar-sounding process of *nuclear fission.* In nuclear fusion, energy is released by joining or fusing together nuclei of lightweight atoms such as hydrogen. In nuclear fission, by contrast, the nuclei of very massive atoms such as uranium or plutonium release energy by fragmenting into smaller nuclei. Nuclear power plants produce energy using fission, not fusion. (Generating power using fusion has been a goal of researchers for decades, but no one has yet devised a commercially viable way to do this.)

CONCEPTCHECK 16-2

If hydrogen nuclei are positively charged, under what conditions can two hydrogen nuclei overcome the repulsive force of their electric charges and fuse together, thus releasing energy according to Einstein's equation, $E = mc^2$?

CALCULATIONCHECK 16-1

How much energy is released when just 5 kg of mass is converted into energy?

Answers appear at the end of the chapter.

Converting Hydrogen to Helium

We learned in Section 5-8 that the nucleus of a hydrogen atom (H) consists of a single proton. The nucleus of a helium atom (He) consists of two protons and two neutrons. In the nuclear process that Atkinson described, four hydrogen nuclei combine to form one helium nucleus, with a concurrent release of energy:

$$4 \text{ H} \rightarrow \text{He} + \text{energy}$$

In several separate reactions, two of the four protons are changed into neutrons, and eventually combine with the remaining protons to produce a helium nucleus. This sequence of reactions is called the **proton-proton chain.** The *Cosmic Connections* figure depicts the proton-proton chain in detail.

Each time this process takes place, a small fraction (0.7%) of the initial combined mass of the hydrogen nuclei does not show up in the final mass of the helium nucleus. This "lost" mass is converted into energy. Box 16-1 describes how to use Einstein's mass-energy equation to calculate the amount of energy released.

COSMIC CONNECTIONS
The Proton–Proton Chain

The most common form of hydrogen fusion in the Sun involves three steps, each of which releases energy.

Hydrogen fusion in the Sun usually takes place in a sequence of steps called the proton-proton chain. Each of these steps releases energy that heats the Sun and gives it its luminosity.

STEP 1

(a) Two protons (hydrogen nuclei, ^{1}H) collide.

(b) One of the protons changes into a neutron (shown in blue). The proton and neutron form a hydrogen isotope (^{2}H).

(c) One byproduct of converting a proton to a neutron is a neutral, nearly massless neutrino (ν), which escapes from the Sun.

(d) The other byproduct of converting a proton to a neutron is a positively charged electron, or positron (e^+). This encounters an ordinary electron (e^-), annihilating both particles and converting them into gamma-ray photons (γ). The energy of these photons goes into sustaining the Sun's internal heat.

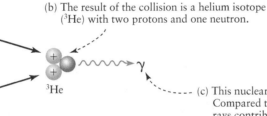

STEP 2

(a) The ^{2}H nucleus produced in Step 1 collides with a third proton (^{1}H).

(b) The result of the collision is a helium isotope (^{3}He) with two protons and one neutron.

(c) This nuclear reaction releases another gamma-ray photon (γ). Compared to other sources in these three steps, these gamma rays contribute the most energy to the Sun's internal heat.

STEP 3

(a) The ^{3}He nucleus produced in Step 2 collides with another ^{3}He nucleus produced from three other protons.

(b) Two protons and two neutrons from the two ^{3}He nuclei rearrange themselves into a different helium isotope (^{4}He).

(c) The two remaining protons are released. The energy of their motion contributes to the Sun's internal heat.

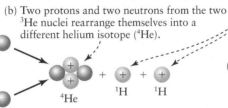

(d) Six ^{1}H nuclei went into producing the two ^{3}He nuclei, which combine to make one ^{4}He nucleus. Since two of the original ^{1}H nuclei are returned to their original state, we can summarize the three steps as:

$$4\ ^1\text{H} \longrightarrow\ ^4\text{He} + \text{energy}$$

ANIMATION 16-1

Hydrogen fusion also takes place in all of the stars visible to the naked eye. (Fusion follows a different sequence of steps in the most massive stars, but the net result is the same.)

BOX 16-1 TOOLS OF THE ASTRONOMER'S TRADE

Converting Mass into Energy

The *Cosmic Connections: The Proton-Proton Chain* figure shows the steps involved in the thermonuclear fusion of hydrogen at the Sun's center. In these steps, four protons are converted into a single nucleus of ^{4}He, the most common isotope of helium with two protons and two neutrons. (As we saw in Box 5-5, different isotopes of the same element have the same number of protons but different numbers of neutrons.) The reaction depicted in the *Cosmic Connections* Step 1 also produces a neutral, nearly massless particle called the *neutrino*. Neutrinos respond hardly at all to ordinary matter, so they travel almost unimpeded through the Sun's massive bulk. Hence, the energy that neutrinos carry is quickly lost into space. This loss is not great, however, because the neutrinos carry relatively little energy. (See Section 16-4 for more about these curious particles.)

Most of the energy released by thermonuclear fusion appears in the form of gamma-ray photons. The energy of these photons remains trapped within the Sun for a long time, thus maintaining the Sun's intense internal heat. Some gamma-ray photons are produced by the reaction shown as Step 2 in the *Cosmic Connections* figure. Others appear when an electron in the Sun's interior annihilates a positively charged electron, or **positron,** which is a by-product of the reaction shown in Step 1 in the *Cosmic Connections* figure. An electron and a positron are respectively matter and antimatter, and they convert their mass entirely into energy when they meet. (You may have thought that "antimatter" was pure science fiction. In fact, tremendous amounts of antimatter are being created and annihilated in the Sun as you read these words.)

We can summarize the thermonuclear fusion of hydrogen as follows:

$$4\ ^1\text{H} \rightarrow\ ^4\text{He} + \text{neutrinos} + \text{gamma-ray photons}$$

To calculate how much energy is released in this process, we use Einstein's mass-energy formula: The energy released is equal to the amount of mass consumed multiplied by c^2, where c is the speed of light. To see how much mass is consumed, we compare the combined mass of four hydrogen atoms (the ingredients) to the mass of one helium atom (the product):

$$
\begin{aligned}
4 \text{ hydrogen atoms} &= 6.693 \times 10^{-27} \text{ kg}\\
-1 \text{ helium atom} &= 6.645 \times 10^{-27} \text{ kg}\\
\hline
\text{Mass lost} &= 0.048 \times 10^{-27} \text{ kg}
\end{aligned}
$$

Thus, a small fraction (0.7%) of the mass of the hydrogen going into the nuclear reaction does not show up in the mass of the helium. This lost mass is converted into an amount of energy $E = mc^2$:

$$
\begin{aligned}
E = mc^2 &= (0.048 \times 10^{-27} \text{ kg})(3 \times 10^8 \text{ m/s})^2\\
&= 4.3 \times 10^{-12} \text{ joule}
\end{aligned}
$$

This amount of energy is released by the formation of a single helium atom. It would light a 10-watt lightbulb for almost one-half of a trillionth of a second.

EXAMPLE: How much energy is released when 1 kg of hydrogen is converted to helium?

Situation: We are given the initial mass of hydrogen. We know that a fraction of the mass is lost when the hydrogen undergoes fusion to make helium; our goal is to find the quantity of energy into which this lost mass is transformed.

Tools: We use the equation $E = mc^2$ and the result that 0.7% of the mass is lost when hydrogen is converted into helium.

Answer: When 1 kg of hydrogen is converted to helium, the amount of mass lost is 0.7% of 1 kg, or 0.007 kg. (This means that 0.993 kg of helium is produced.) Using Einstein's equation, we find that this missing 0.007 kg of matter is transformed into an amount of energy equal to

$$E = mc^2 = (0.007 \text{ kg})(3 \times 10^8 \text{ m/s})^2 = 6.3 \times 10^{14} \text{ joules}$$

Review: The energy released by the fusion of 1 kg of hydrogen is the same as that released by burning *20,000 metric tons* (2×10^7 kg) of coal! Hydrogen fusion is a *much* more efficient energy source than ordinary burning.

The Sun's luminosity is 3.9×10^{26} joules per second. To generate this much power, hydrogen must be consumed at a rate of

$$\frac{3.9 \times 10^{26} \text{ joules per second}}{6.3 \times 10^{14} \text{ joules per kilogram}}$$

$$= 6 \times 10^{11} \text{ kilograms per second}$$

That is, the Sun converts 600 million metric tons of hydrogen into helium every second.

CAUTION! You may have heard the idea that mass is always conserved (that is, it is neither created nor destroyed), or that energy is always conserved in a reaction. Einstein's ideas show that neither of these statements is quite correct, because mass can be converted into energy and vice versa. A more accurate statement is that the total amount of mass *plus* energy is conserved. Hence, the destruction of mass in the Sun does not violate any laws of nature.

For every four hydrogen nuclei converted into a helium nucleus, 4.3×10^{-12} joule of energy is released—only enough energy to power a small Christmas light for a trillionth of a second. This amount of energy may seem tiny, but it is about 10^7 times larger than the amount of energy released in a typical chemical reaction, such as occurs in ordinary burning. Thus, thermonuclear fusion explains how the Sun could have been shining for billions of years.

To produce the Sun's luminosity of 3.9×10^{26} joules per second, 6×10^{11} kg (600 million metric tons) of hydrogen must be converted into helium each second—this is about 100 times the mass of Egypt's largest pyramid. This rate is prodigious, but there is, literally, an astronomical amount of hydrogen in the Sun. In particular, the Sun's core contains enough hydrogen to have been giving off energy at the present rate for as long as the solar system has existed, about 4.56 billion years, and to continue doing so for more than 6 billion years into the future.

The proton-proton chain is also the energy source for many of the stars in the sky. In stars with central temperatures that are much hotter than that of the Sun, however, hydrogen fusion proceeds according to a different set of nuclear reactions, called the **CNO cycle**, in which carbon, nitrogen, and oxygen nuclei absorb protons to produce helium nuclei. Still other thermonuclear reactions, such as helium fusion, carbon fusion, and oxygen fusion, occur late in the lives of many stars.

CONCEPTCHECK 16-3

If 1 kg of hydrogen combines to form helium in the proton-proton chain, why is only 0.007 kg (0.7%) available to be converted into energy?

CONCEPTCHECK 16-4

How do astronomers estimate that our Sun can continue to shine for more than 6 billion years?

Answers appear at the end of the chapter.

16-2 A theoretical model of the Sun shows how energy gets from its center to its surface

While thermonuclear fusion is the source of the Sun's energy, this process cannot take place everywhere within the Sun. As we have seen, extremely high temperatures—in excess of 10^7 K—are required for atomic nuclei to fuse together to form larger nuclei. The temperature of the Sun's visible surface, about 5800 K, is too low for these reactions to occur there. Hence, thermonuclear fusion can be taking place only within the Sun's interior. But precisely where does it take place? And how does the energy produced by fusion make its way to the surface, where it is emitted into space in the form of photons?

To answer these questions, we must understand conditions in the Sun's interior. Ideally, we would send an exploratory spacecraft to probe deep into the Sun; in practice, the Sun's intense heat would vaporize even the sturdiest spacecraft. Instead, astronomers use the laws of physics to construct a theoretical model of the Sun, and then compare predictions of the model to a variety of observations. (We discussed the use of models in science in Section 1-1.) Let's see what ingredients go into building a model of this kind.

Hydrostatic Equilibrium

Note first that the Sun is not undergoing any dramatic changes. The Sun is not exploding or collapsing, nor is it significantly heating or cooling. The Sun is thus said to be in both *hydrostatic equilibrium* and *thermal equilibrium*.

To understand what is meant by **hydrostatic equilibrium**, imagine a slab of material in the solar interior (Figure 16-2a). In equilibrium, the slab on average will move neither up nor down. (In fact, there are upward and downward motions of material inside the Sun, but these motions average out.) Equilibrium is maintained by a balance among three forces that act on this slab:

1. The downward pressure of the layers of solar material above the slab.

2. The upward pressure of the hot gases beneath the slab.

3. The slab's weight—that is, the downward gravitational pull it feels from the rest of the Sun.

The pressure from below must balance both the slab's weight and the pressure from above. Hence, the pressure below the slab

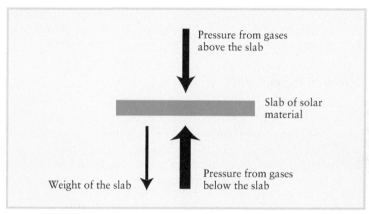

(a) Material inside the sun is in hydrostatic equilibrium, so forces balance

(b) A fish floating in water is in hydrostatic equilibrium, so forces balance

FIGURE 16-2

Hydrostatic Equilibrium **(a)** Material in the Sun's interior tends to move neither up nor down. The upward forces on a slab of solar material (due to pressure of gases below the slab) must balance the downward forces (due to the slab's weight and the pressure of gases above the slab). Hence, the pressure must increase with increasing depth. **(b)** The same principle applies to a fish floating in water. In equilibrium, the forces balance and the fish neither rises nor sinks. (Ken Usami/PhotoDisc)

must be greater than that above the slab. In other words, pressure has to increase with increasing depth. For the same reason, pressure increases as you dive deeper into the ocean (Figure 16-2b) or as you move toward lower altitudes in our atmosphere.

ANALOGY If you had to join a vertical stack of 10 students piled on top of each other, you would naturally prefer to be on the top. Each student lower in the stack must bear the weight of all the students above them, so the pressure on each body increases with depth.

Hydrostatic equilibrium also tells us about the density of the slab of solar material. If the slab is too dense, its weight will be too large and it will sink; if the density is too low, the slab will rise. To prevent this, the density of solar material must have a certain value at each depth within the solar interior. (The same principle applies to objects that float beneath the surface of the ocean. Scuba divers wear weight belts and inflatable vests to adjust their average density so that they will neither rise nor sink but will stay submerged at the same level.)

Thermal Equilibrium

Another consideration is that the Sun's interior is so hot that it is completely gaseous. Gases compress and become more dense when you apply greater pressure to them, so density must increase along with pressure as you go to greater depths within the Sun. Furthermore, when you compress a gas, its temperature tends to increase, so the temperature must also increase as you move toward the Sun's center.

While the temperature in the solar interior is different at different depths, the temperature at each depth remains constant in time. This principle is called **thermal equilibrium.** For the Sun to be in thermal equilibrium, all the energy generated by thermonuclear reactions in the Sun's core must be transported to the Sun's glowing surface, where it can be radiated into space. If too much energy flowed from the core to the surface to be radiated away, the Sun's interior would cool down; alternatively, the Sun's interior would heat up if too little energy flowed to the surface.

CONCEPTCHECK 16-5

If our Sun were much less massive, how would the pressure at the Sun's center be different from what it actually is?

Answer appears at the end of the chapter.

Transporting Energy Outward from the Sun's Core

But exactly how is energy transported from the Sun's center to its surface? There are three methods of energy transport: *conduction, convection,* and *radiative diffusion.* Only the last two are important inside the Sun.

If you heat one end of a metal bar with a blowtorch, energy flows to the other end of the bar so that it too becomes warm. The efficiency of this method of energy transport, called **conduction,** varies significantly from one substance to another. For example, metal is a good conductor of heat, but plastic is not (which is why metal cooking pots often have plastic handles). Conduction is *not* an efficient means of energy transport in substances with low average densities, including the gases inside stars like the Sun.

Inside stars like our Sun, energy moves from center to surface by two other means: convection and radiative diffusion. **Convection** is the circulation of fluids—gases or liquids—between hot and cool regions. Hot gases (with lower density) rise toward a star's surface, while cool gases (with higher density) sink back down toward the star's center. This physical movement of gases transports heat energy outward in a star, just as the physical movement of water boiling in a pot transports energy from the bottom of the pot (where the heat is applied) to the cooler water at the surface.

In **radiative diffusion,** photons created in the thermonuclear inferno at a star's center diffuse outward toward the star's surface. Individual photons are absorbed and reemitted by atoms and electrons inside the star. The overall result is an outward migration from the hot core, where photons are constantly created, toward the cooler surface, where they escape into space.

CONCEPTCHECK 16-6

Why is the energy transport process of conduction relatively unimportant when studying how energy moves toward the Sun's surface?

Answer appears at the end of the chapter.

Modeling the Sun

To construct a model of a star like the Sun, astrophysicists express the ideas of hydrostatic equilibrium, thermal equilibrium, and energy transport as a set of equations. To ensure that the model applies to the particular star under study, they also make use of astronomical observations of the star's surface. (For example, to construct a model of the Sun, they use the data that the Sun's surface temperature is 5800 K, its luminosity is 3.9×10^{26} watts, and the gas pressure and density at the surface are almost zero.) The astrophysicists then use a computer to solve their set of equations and calculate conditions layer by layer in toward the star's center. The result is a model of how temperature, pressure, and density increase with increasing depth below the star's surface (see Figure 9-5).

Table 16-2 and Figure 16-3 show a theoretical model of the Sun that was calculated in just this way. Different models of the Sun use slightly different assumptions, but all models give essentially the same results as those shown here. From such computer models we have learned that at the Sun's center the density is 160,000 kg/m^3 (14 times the density of lead!), the temperature is 1.55×10^7 K, and the pressure is 3.4×10^{11} atm. (One atmosphere, or 1 atm, is the average atmospheric pressure at sea level on Earth.)

Table 16-2 and Figure 16-3 show that the solar luminosity rises to 100% at about one-quarter of the way from the Sun's center to its surface. In other words, the Sun's energy production occurs within a volume that extends out only to 0.25 R$_\odot$. (The symbol R$_\odot$ denotes the solar radius, or radius of the Sun as a whole, equal to 696,000 km.) Outside 0.25 R$_\odot$, the density and temperature are too low for thermonuclear reactions to take place. Also note that 94% of the total mass of the Sun is found within the inner 0.5 R$_\odot$. Hence, the outer 0.5 R$_\odot$ contains only a relatively small amount of material.

How energy flows from the Sun's center toward its surface depends on how easily photons move through the gas. If the solar gases are relatively transparent, photons can travel moderate distances before being scattered or absorbed, and energy is thus transported by radiative diffusion. You can think of radiative

TABLE 16-2   A Theoretical Model of the Sun

Distance from the Sun's center (solar radii)	Fraction of luminosity	Fraction of of mass	Temperature ($\times 10^6$ K)	Density (kg/m^3)	Pressure relative to pressure at center
0.0	0.00	0.00	15.5	160,000	1.00
0.1	0.42	0.07	13.0	90,000	0.46
0.2	0.94	0.35	9.5	40,000	0.15
0.3	1.00	0.64	6.7	13,000	0.04
0.4	1.00	0.85	4.8	4,000	0.007
0.5	1.00	0.94	3.4	1,000	0.001
0.6	1.00	0.98	2.2	400	0.0003
0.7	1.00	0.99	1.2	80	4×10^{-5}
0.8	1.00	1.00	0.7	20	5×10^{-6}
0.9	1.00	1.00	0.3	2	3×10^{-7}
1.0	1.00	1.00	0.006	0.00030	4×10^{-13}

Note: The distance from the Sun's center is expressed as a fraction of the Sun's radius ($R_\odot$). Thus, 0.0 is at the center of the Sun and 1.0 is at the surface. The fraction of luminosity is that portion of the Sun's total luminosity produced within each distance from the center; this is equal to 1.00 for distances of 0.25 $R_\odot$ or more, which means that all of the Sun's nuclear reactions occur within 0.25 solar radius from the Sun's center. The fraction of mass is that portion of the Sun's total mass lying within each distance from the Sun's center. The pressure is expressed as a fraction of the pressure at the center of the Sun.

diffusion as the "free flight" of photons, and the energy is transported in these photons. If the gases are relatively *opaque*, photons are frequently scattered or absorbed and cannot easily get through the gas. In an opaque gas, heat builds up and convection then becomes the most efficient means of energy transport. In convection, the energy is in the form of heat; the movement of hot gas transports the energy. The gases start to churn, with hot lower-density gas moving outward and cooler gas sinking inward.

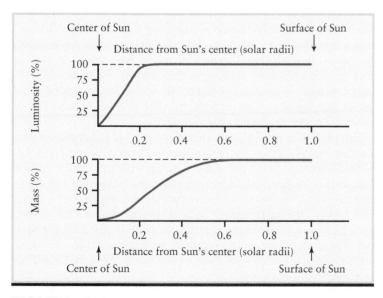

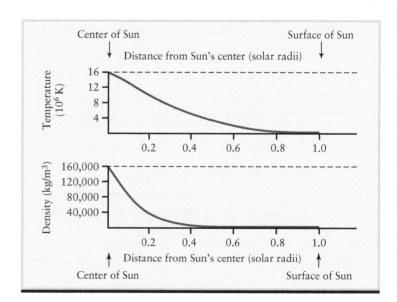

FIGURE 16-3

A Theoretical Model of the Sun's Interior These graphs depict what percentage of the Sun's total luminosity is produced within each distance from the center (upper left), what percentage of the total mass lies within each distance from the center (lower left), the temperature at each distance (upper right), and the density at each distance (lower right). (See Table 16-2 for a numerical version of this model.)

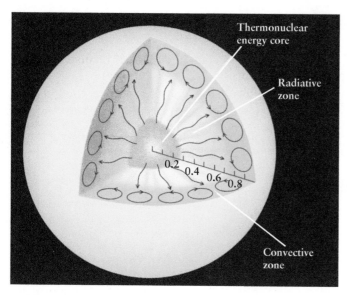

FIGURE 16-4

The Sun's Internal Structure Thermonuclear reactions occur in the Sun's core, which extends out to a distance of 0.25 $R_\odot$ from the center. Energy is transported outward, via radiative diffusion, to a distance of about 0.71 $R_\odot$. In the outer layers between 0.71 $R_\odot$ and 1.00 $R_\odot$, energy flows outward by convection.

From the center of the Sun out to about 0.71 $R_\odot$, energy is transported by radiative diffusion. Hence, this region is called the **radiative zone.** Beyond about 0.71 $R_\odot$, the temperature is low enough (a mere 2×10^6 K or so) for electrons and hydrogen nuclei to join into hydrogen atoms. These atoms are very effective at absorbing photons, much more so than free electrons or nuclei, and this absorption chokes off the outward flow of photons. Therefore, beyond about 0.71 $R_\odot$, radiative diffusion is not an effective way to transport energy. Instead, convection dominates the energy flow in this outer region, which is why it is called the **convective zone.** Figure 16-4 shows these aspects of the Sun's internal structure.

Although energy travels through the radiative zone in the form of photons, the photons have a difficult time of it. Table 16-2 shows that the material in this zone is extremely dense, so photons from the Sun's core take a long time to diffuse through the radiative zone. As a result, it takes approximately 170,000 years for energy created at the Sun's center to travel 696,000 km to the solar surface and finally escape as sunlight. The energy flows outward at an average rate of 50 cm per hour, or about 20 times slower than a snail's pace.

Once the energy works its way out of the Sun, it travels much faster—at the speed of light! Thus, solar energy that reaches you today took only 8 minutes to travel the 150 million km from the Sun's surface to Earth. But this energy was actually produced by thermonuclear reactions that took place in the Sun's core hundreds of thousands of years ago.

> The sunlight that reaches Earth today results from thermonuclear reactions that took place about 170,000 years ago

CONCEPTCHECK 16-7

Which of the following decreases when we move from the Sun's central core toward its surface: temperature or luminosity?

CALCULATIONCHECK 16-2

By how much (in percent) does the Sun's temperature drop from its temperature at the central core to its temperature at 0.5 $R_\odot$?

Answers appear at the end of the chapter.

16-3 Astronomers probe the solar interior using the Sun's own vibration

We have described how astrophysicists construct models of the Sun. But since we cannot see into the Sun's opaque interior, how can we check these models to see if they are accurate? What is needed is a technique for probing the Sun's interior. A very powerful technique of just this kind involves measuring vibrations of the Sun as a whole. This field of solar research is called **helioseismology.**

Vibrations are a useful tool for examining the hidden interiors of all kinds of objects. Food shoppers test whether melons are ripe by tapping on them and listening to the sound made by the resulting vibrations. Since the vibrations of a melon depend on its density profile, the sounds from tapping provide clues about the melon's interior. Similarly, geologists can determine the structure of Earth's interior by using seismographs to record vibrations during earthquakes.

Although there are no true "sunquakes," the Sun does vibrate at a variety of frequencies, somewhat like a ringing bell. These vibrations were first noticed in 1960 by Robert Leighton of the California Institute of Technology, who made high-precision Doppler shift observations of the solar surface. These measurements revealed that parts of the Sun's surface move up and down about 10 m every 5 minutes. Since the mid-1970s, several astronomers have reported slower vibrations, having periods ranging from 20 to 160 minutes. The detection of extremely slow vibrations has inspired astronomers to organize networks of telescopes around and in orbit above Earth to monitor the Sun's vibrations on a continuous basis.

The vibrations of the Sun's surface can be compared with sound waves. If you could somehow survive within the Sun's outermost layers, you would first notice a deafening roar, somewhat like a jet engine, produced by turbulence in the Sun's gases. But superimposed on this noise would be a variety of nearly pure tones. You would need greatly enhanced hearing to detect these tones, however; the strongest has a frequency of just 0.003 hertz, 13 octaves below the lowest frequency audible to humans. (Recall from Section 5-2 that one hertz is one oscillation per second.)

In 1970, Roger Ulrich, at UCLA, pointed out that sound waves moving upward from the solar interior would be reflected back inward after reaching the solar surface. However, as a reflected sound wave descends back into the Sun, the increasing density and pressure bend the wave severely, turning it around and aiming it back toward the solar surface. In other words, sound waves bounce back and forth between the solar surface and layers deep within the Sun. These sound waves can reinforce each other if their wavelength is the right size, just as sound waves of a particular wavelength resonate inside an organ pipe.

The Sun oscillates in millions of ways as a result of waves resonating in its interior. Figure 16-5 is a computer-generated illustration of one such mode of vibration.

> The Sun's oscillations resemble those inside an organ pipe but are much more complex

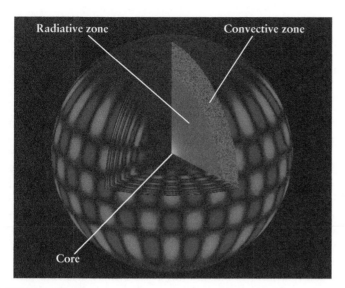

FIGURE 16-5

A Sound Wave Resonating in the Sun This computer-generated image shows one of the millions of ways in which the Sun's interior vibrates. The regions that are moving outward are colored blue, those moving inward, red. As the cutaway shows, these oscillations are thought to extend into the Sun's radiative zone (compare Figure 16-4). (National Solar Observatory)

Helioseismologists can deduce information about the solar interior from measurements of these oscillations. In addition to providing information about interior temperatures and densities, helioseismology sets limits on the amount of helium in the Sun's core and convective zone and determines the thickness of the transition region between the radiative zone and convective zone.

CONCEPTCHECK 16-8

Do solar sound waves travel exclusively on the surface of the Sun—like ripples across a pond—or do they travel through the Sun's interior as well?

CONCEPTCHECK 16-9

What can be determined from carefully monitoring the Sun's vibrations?

Answers appear at the end of the chapter.

16-4 Neutrinos reveal information about the Sun's core—and have surprises of their own

We have seen circumstantial evidence that thermonuclear fusion is the source of the Sun's power. To be certain, however, we need more definitive evidence. How can we show that thermonuclear fusion really is taking place in the Sun's core?

What Sunlight Cannot Tell Us

Although the light energy that we receive from the Sun originates in the core, it provides few clues about conditions there. The problem is that this energy has changed form repeatedly during its passage from the core: It appeared first as photons diffusing through the radiative zone, then as heat transported through the outer layers by convection, and then again as photons emitted from the Sun's glowing surface. As a result of these transformations, much of the information that the Sun's radiated energy once carried about conditions in the core has been lost.

ANALOGY If you make a photocopy of a photocopy of a photocopy of an original document, the final result may be so blurred as to be unreadable. In an analogous way, because solar energy is transformed many times while en route through the Sun, the story it could tell us about the Sun's core is hopelessly blurred.

Solar Neutrinos

Happily, there is a way for scientists to learn about conditions in the Sun's core and to get direct evidence that thermonuclear fusion really does happen there. The trick is to detect the subatomic by-products of thermonuclear fusion reactions.

> Scientists use the most ethereal of subatomic particles to learn about the Sun, and vice versa

As part of the process of hydrogen fusion, protons change into neutrons and release **neutrinos** (see the *Cosmic Connections* figure in Section 16-1 as well as Box 16-1). Like photons, neutrinos are particles that have no electric charge. Unlike photons, however, neutrinos interact only very weakly with matter. Even the vast bulk of the Sun offers little impediment to their passage, so neutrinos must be streaming out of the core and into space. Indeed, the conversion of hydrogen into helium at the Sun's center produces 10^{38} neutrinos each second. Some of these neutrinos hit Earth, and about 10^{14} neutrinos from the Sun—that is, **solar neutrinos**—must pass through each square meter of Earth every second. Perhaps the exceedingly weak interaction between neutrinos and matter is a good thing, since about a trillion neutrinos pass through our heads each second.

If it were possible to detect these solar neutrinos, we would have direct evidence that thermonuclear reactions really do take place in the Sun's core. Beginning in the 1960s, scientists began to build neutrino detectors for precisely this purpose.

The challenge is that neutrinos are exceedingly difficult to detect. Just as neutrinos pass unimpeded through the Sun, they also pass through Earth almost as if it were not there. We stress the word "almost," because neutrinos can and do interact with matter, albeit infrequently.

On rare occasions a neutrino will strike a neutron and convert it into a proton. This effect was the basis of the original solar neutrino detector, designed and built by Raymond Davis of the Brookhaven National Laboratory in the 1960s. This device used 100,000 gallons of perchloroethylene (C_2Cl_4), a fluid used in dry cleaning, in a huge tank buried deep underground. Most of the solar neutrinos that entered Davis's tank passed right through it with no effect whatsoever. But occasionally a neutrino struck the nucleus of one of the chlorine atoms (^{37}Cl) in the cleaning fluid and converted one of its neutrons into a proton, creating a radioactive atom of argon (^{37}Ar). Fortunately for experimenters, what individual neutrinos lack in barely interacting with matter, the Sun partially compensates for by producing neutrinos in extraordinarily large numbers.

The rate at which argon is produced is related to the neutrino flux—that is, the number of neutrinos from the Sun arriving at Earth per square meter per second. By counting the number of newly created argon atoms, Davis was able to determine the neutrino flux from the Sun. (Other subatomic particles besides neutrinos can also induce reactions that create radioactive atoms. By placing the experiment deep underground, however, the overlying Earth absorbs essentially all such particles—with the exception of neutrinos.)

The Solar Neutrino Problem

Davis and his collaborators found that solar neutrinos created one radioactive argon atom in the tank every three days. But this rate corresponded to only one-third of the neutrino flux predicted from standard models of the Sun. This troubling discrepancy between theory and observation, called the **solar neutrino problem,** motivated scientists around the world to conduct further experiments to measure solar neutrinos.

One key question was whether the neutrinos that Davis had detected had really come from the Sun. (The Davis experiment had no way to determine the direction from which neutrinos had entered the tank of cleaning fluid.) The direction was resolved by an experiment in Japan called Kamiokande, which was designed by the physicist Masatoshi Koshiba. A large underground tank containing 3000 tons of water was surrounded by 1100 light detectors. From time to time, a high-energy solar neutrino struck an electron in one of the water molecules, dislodging it and sending it flying like a pin hit by a bowling ball. The recoiling electron produced a streaking flash of light, which was sensed by the detectors. By analyzing the flashes, scientists could tell the direction from which the neutrinos were coming and confirmed that they emanated from the Sun. These results in the late 1980s gave direct evidence that thermonuclear fusion is indeed occurring in the Sun's core. (Davis and Koshiba both received the 2002 Nobel Prize in Physics for their pioneering research on solar neutrinos.)

Like Davis's experiment, however, Koshiba and his colleagues at Kamiokande detected only a fraction of the expected flux of neutrinos. Where, then, were the missing solar neutrinos?

One proposed solution had to do with the energy of the detected neutrinos. The vast majority of neutrinos from the Sun are created during the first step in the proton-proton chain, in which two protons combine to form a heavy isotope of hydrogen (see the *Cosmic Connections* figure in Section 16-1). But these neutrinos have too little energy to convert chlorine into argon. Both Davis's and Koshiba's experiments responded only to high-energy neutrinos produced by reactions that occur only part of the time near the end of the proton-proton chain. (The *Cosmic Connections* figure does not show these reactions.) Could it be that the discrepancy between theory and observation would go away if the flux of low-energy neutrinos could be measured?

To test this idea, two teams of physicists constructed neutrino detectors that used several tons of gallium (a liquid metal) rather than cleaning fluid. Low-energy neutrinos convert gallium (^{71}Ga) into a radioactive isotope of germanium (^{71}Ge). By chemically separating the germanium from the gallium and counting the radioactive atoms, the physicists were able to measure the flux of low-energy solar neutrinos. These experiments—GALLEX in Italy and SAGE (Soviet-American Gallium Experiment) in Russia—detected only 50% to 60% of the expected neutrino flux. Hence, the solar neutrino problem was a discrepancy between theory and observation for neutrinos of all energies.

Another proposed solution to the neutrino problem was that the Sun's core is cooler than predicted by solar models. If the Sun's central temperature were only 10% less than the current estimate,

fewer neutrinos would be produced and the neutrino flux would agree with experiments. However, a lower central temperature would cause other obvious features, such as the Sun's size and surface temperature, to be different from what we observe.

Finding the Missing Neutrinos

Only very recently has the solution to the neutrino problem been found. The answer lies not in how neutrinos are produced, but rather in what happens to them between the Sun's core and detectors on Earth. Motivated by the lack of observed solar neutrinos, physicists have discovered that there are actually three types of neutrinos. Only one of these types is produced in the Sun, and it is only this type that can be detected by the experiments we have described. But if some of the solar neutrinos change *in flight* into a different type of neutrino, the detectors in these experiments would record only a fraction of the total neutrino flux. This effect, where one type of neutrino can spontaneously transform into another type, is called **neutrino oscillation.** In June 1998, scientists at the Super-Kamiokande neutrino observatory (a larger and more sensitive device than Kamiokande) revealed evidence that neutrino oscillation does indeed take place.

The best confirmation of this idea has come from the Sudbury Neutrino Observatory (SNO) in Canada. Like Kamiokande and Super-Kamiokande, SNO uses a large tank of water placed deep underground (**Figure 16-6**). But unlike those earlier experiments, SNO can detect all three types of neutrinos. It detects neutrinos by using *heavy water.* In ordinary, or "light," water, each hydrogen atom in the H_2O molecule has a solitary proton as its nucleus. In heavy water, by contrast, each of the hydrogen nuclei has a nucleus made up of a proton and a neutron. (This isotope ^2H is shown in the *Cosmic Connections* figure in Section 16-1.) If a high-energy solar neutrino of any type passes through SNO's tank of heavy water, it can knock the neutron out of one of the ^2H nuclei. The ejected neutron can then be captured by another nucleus, and this capture releases energy that manifests itself as a tiny burst of light. As in Kamiokande, detectors around SNO's water tank record these light flashes.

Scientists using SNO have found that the combined flux of all three types of neutrinos coming from the Sun is *equal* to the theoretical prediction. Together with the results from earlier neutrino experiments, this result strongly suggests that the Sun is indeed producing neutrinos at the predicted rate as a by-product of thermonuclear reactions. But before these neutrinos can reach Earth, about two-thirds of them undergo spontaneous oscillation and change their type. Thus, there is really no solar neutrino problem (no lack of neutrinos)—scientists merely needed the right kind of detectors to observe all the neutrinos, including the ones that transformed in flight.

The story of the solar neutrino problem illustrates how two different branches of science—in this case, studies of the solar interior and investigations of subatomic particles—can sometimes interact, to the mutual benefit of both. While there is still much we do not understand about the Sun and about neutrinos, a new generation of neutrino detectors in Japan, Canada, and elsewhere promises to further our knowledge of these exotic realms of astronomy and physics.

CONCEPTCHECK 16-10

Why did the SNO experiment find three times as many neutrinos as the Kamiokande experiment?

Answer appears at the end of the chapter.

FIGURE 16-6 R I V U X G

A Solar Neutrino Experiment Located 2073 m (6800 ft) underground in the Creighton nickel mine in Sudbury, Canada, the Sudbury Neutrino Observatory is centered around a tank that contains 1000 tons of water. Occasionally, a neutrino entering the tank interacts with one or another of the particles already there. Such interactions create flashes of light, called Cherenkov radiation. Some 9600 light detectors sense this light. The numerous silver protrusions are the back sides of the light detectors prior to their being wired and connected to electronics in the lab (seen at the bottom of the photograph). (Science Source)

16-5 The photosphere is the lowest of three main layers in the Sun's atmosphere

Although the Sun's core is hidden from our direct view, we can easily see sunlight coming from the high-temperature gases that make up the Sun's atmosphere. These outermost layers of the Sun prove to be the sites of truly dramatic activity, much of which has a direct impact on our planet. By studying these layers, we gain further insight into the character of the Sun as a whole.

Observing the Photosphere

A visible-light photograph like Figure 16-7 makes it appear that the Sun has a definite surface. This surface is actually an illusion; the Sun is gaseous throughout its volume because of its high internal temperature, and the gases simply become less and less dense as you move farther away from the Sun's center.

Why, then, does the Sun appear to have a sharp, well-defined surface? The reason is that essentially all of the Sun's visible light emanates from a single, thin layer of gas called the **photosphere** ("sphere of light"). Just as you can see only a certain distance through Earth's atmosphere before objects vanish in the haze, we can see only about 400 km into the photosphere. This distance is so small compared with the Sun's radius of 696,000 km that the photosphere appears to be a definite surface.

The photosphere is actually the lowest of the three layers that together constitute the solar atmosphere. Above it are the *chromosphere* and the *corona,* both of which are transparent to visible light. We can see them only using special techniques, which we discuss later in this chapter. Everything below the photosphere is called the *solar interior.*

The Lesson of Limb Darkening

The photosphere is heated from below by energy streaming outward from the solar interior. Hence, temperature should decrease as you go upward in the photosphere, just as in the solar interior (see Table 16-2 and Figure 16-3). We know this is the case because the photosphere appears darker around the edge, or *limb,* of the Sun than it does toward the center of the solar disk, an effect called **limb darkening** (examine Figure 16-7). This happens because when we look near the Sun's limb, we do not see as deeply into the photosphere as we do when we look near the center of the disk (Figure 16-8). The high-altitude gas we observe at the limb is not as hot and thus does not glow as brightly as the deeper, hotter gas seen near the disk center.

The spectrum of the Sun's photosphere confirms how its temperature varies with altitude. As we saw in Section 16-1, the photosphere shines like a nearly perfect blackbody with an average temperature of about 5800 K. However, superimposed on this spectrum are many dark absorption lines (see Figure 5-12). As discussed in Section 5-6, we see an *absorption line spectrum* of this sort whenever we view a hot, glowing object through a relatively cool gas. In this case, the hot object is the lower part of the photosphere; the cooler gas is in the upper part of the photosphere, where the temperature declines to about 4400 K. All the absorption lines in the Sun's spectrum are produced in this relatively cool layer, as atoms selectively absorb photons of various wavelengths streaming outward from the hotter layers below.

CAUTION! You may find it hard to think of 4400 K as "cool." But keep in mind that the ratio of 4400 K to 5800 K, the temperature in the lower photosphere, is the same as the ratio of the temperature on a Siberian winter night to that of a typical day in Hawaii.

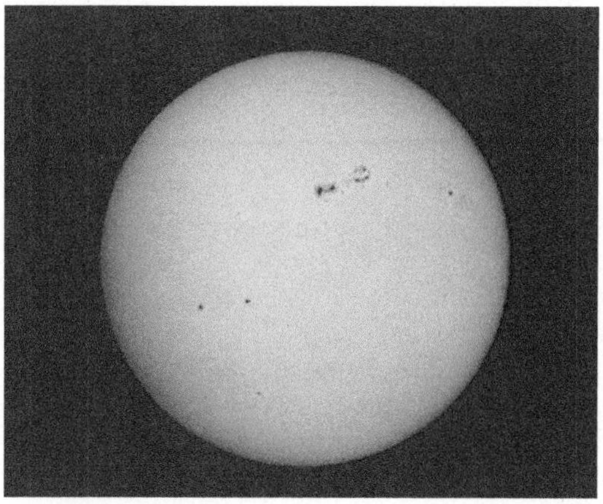

FIGURE 16-7 R I V U X G

The Photosphere The photosphere is the layer in the solar atmosphere from which the Sun's visible light is emitted. Note that the Sun appears darker around its limb, or edge; here we are seeing the upper photosphere, which is relatively cool and thus glows less brightly. (The dark sunspots, which we discuss in Section 16-8, are also relatively cool regions.) (Celestron International)

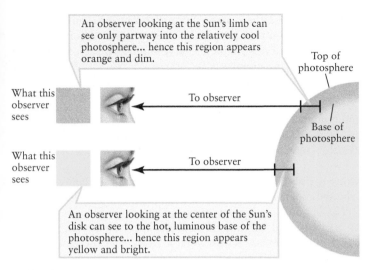

An observer looking at the Sun's limb can see only partway into the relatively cool photosphere... hence this region appears orange and dim.

Top of photosphere

To observer

Base of photosphere

What this observer sees

What this observer sees

To observer

An observer looking at the center of the Sun's disk can see to the hot, luminous base of the photosphere... hence this region appears yellow and bright.

FIGURE 16-8

The Origin of Limb Darkening Light from the Sun's limb and light from the center of its disk both travel about the same straight-line distance through the photosphere to reach us. Because of the Sun's curved shape, light from the limb comes from a greater height within the photosphere, where the temperature is lower and the gases glow less brightly. Hence, the limb appears darker and more orange.

Granules and Supergranules in the Photosphere

We can learn still more about the photosphere by examining it with a telescope—but only when using special dark filters to prevent eye damage. *Looking directly at the Sun without the correct filter, whether with the naked eye or with a telescope, can cause permanent blindness!* Under good observing conditions, astronomers using such filter-equipped telescopes can often see a blotchy pattern in the photosphere, called **granulation** (**Figure 16-9**). Each light-colored **granule** measures about 1000 km (600 mi) across—equal in size to the areas of Texas and Oklahoma combined—and is surrounded by a darkish boundary. The difference in brightness between the center and the edge of a granule corresponds to a temperature drop of about 300 K.

Granulation is caused by convection of the gas in the photosphere. The inset in Figure 16-9 shows how hot gas from lower levels rises upward in granules, cools off, spills over the edges of the granules, and then plunges back down into the Sun. Convection can occur only if the gas is heated from below, like a pot of water being heated on a stove (see Section 16-2). Along with limb darkening and the Sun's absorption line spectrum, granulation shows that the upper part of the photosphere must be cooler than the lower part.

> Like water boiling on a stove, the photosphere bubbles with convection cells

Time-lapse photography reveals more of the photosphere's dynamic activity. Granules form, disappear, and reform in cycles lasting only a few minutes. At any one time, about 4 million granules cover the solar surface.

Superimposed on the pattern of granulation are even larger convection cells called **supergranules** (**Figure 16-10**). As in granules, gases rise upward in the middle of a supergranule, move horizontally outward toward its edge, and descend back into the Sun. The difference is that a typical supergranule is about 35,000 km in diameter, large enough to enclose several hundred granules. This large-scale convection moves at only about 0.4 km/s (1400 km/h, or 900 mi/h), about one-tenth the speed of gases churning in a regular-sized granule. A given supergranule lasts about a day.

ANALOGY Similar patterns of large-scale and small-scale convection can be found in Earth's atmosphere. On the large scale, air rises gradually at a low-pressure area, then sinks gradually at a high-pressure area, which might be hundreds of kilometers away. This pattern is analogous to the flow in a supergranule. Thunderstorms in our atmosphere are small but intense convection cells within which air moves rapidly up and down. Like granules, they last only a relatively short time before they dissipate.

The Photosphere: Hot, Thin, and Opaque

Although the photosphere is a very active place, it actually contains relatively little material. Careful examination of the spectrum shows that it has a density of only about 10^{-24} kg/m^3, roughly 0.01% the density of Earth's atmosphere at sea level. The photosphere is made primarily of hydrogen and helium, the most abundant elements in the solar system (see Figure 8-4).

Despite being such a thin gas, the photosphere is surprisingly opaque to visible light. (This means that visible light is scattered numerous times as it passes through the photosphere.) If it were

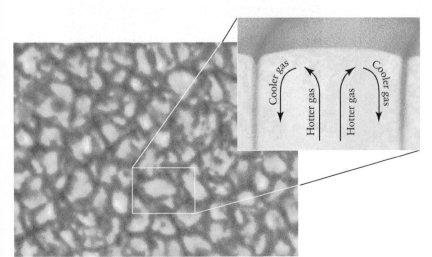

Cooler gas   Hotter gas   Hotter gas   Cooler gas

VIDEO 16-1   VIDEO 16-2   **FIGURE 16-9**   R I **V** U X G

Solar Granulation High-resolution photographs of the Sun's surface reveal a blotchy pattern called granulation. Granules are convection cells about 1000 km (600 mi) wide in the Sun's photosphere. **Inset:** Rising hot gas produces bright granules. Cooler gas sinks downward along the boundaries between granules; this gas glows less brightly, giving the boundaries their dark appearance. This convective motion transports heat from the Sun's interior outward to the solar atmosphere. (MSFC/NASA; inset: Goran Scharmer, Lund Observatory)

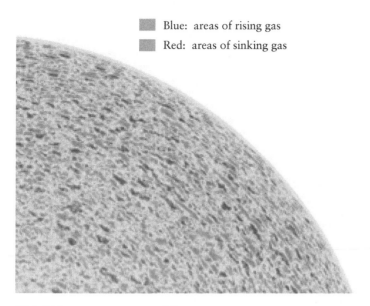

Blue: areas of rising gas
Red: areas of sinking gas

FIGURE 16-10   R I $\boxed{V}$ U X G

Supergranules and Large-Scale Convection Supergranules display relatively little contrast between their center and edges, so they are hard to observe in ordinary images. But they can be seen in a false-color Doppler image like this one. Light from gas that is approaching us (that is, rising) is shifted toward shorter "bluer" wavelengths, while light from receding gas (that is, descending) is shifted toward longer "redder" wavelengths (see Section 5-9). (David Hathaway, MSFC/NASA)

not so opaque, we could see into the Sun's interior to a depth of hundreds of thousands of kilometers, instead of the mere 400 km that we can see down into the photosphere. What makes the photosphere so opaque is that its hydrogen atoms sometimes acquire an extra electron, becoming **negative hydrogen ions.** The scattering of photons can be pictured in two steps. In the first step, the extra electron of a negative hydrogen ion is only loosely attached and can be dislodged if it absorbs a photon of any visible wavelength. In the second step, a similar photon can be emitted when the dislodged electron once again combines with hydrogen to form another negative hydrogen ion, but the emitted photon *heads in a random direction each time.* Hence, negative hydrogen ions absorb light very efficiently, and there are enough of these light-absorbing ions in the photosphere to make it quite opaque. Because it is so opaque, the photosphere's spectrum is close to that of an ideal blackbody (blackbodies are discussed in Section 5-3).

CONCEPTCHECK **16-11**

What causes the photosphere to bubble like water boiling on the stove?

Answer appears at the end of the chapter.

16-6 Spikes of rising gas extend through the Sun's chromosphere

An ordinary visible-light image such as Figure 16-7 gives the impression that the Sun ends at the top of the photosphere. But during a total solar eclipse, the Moon blocks the photosphere from our view, revealing a glowing, pinkish layer of gas above the

photosphere (Figure 16-11). This layer is the tenuous **chromosphere** ("sphere of color"), the second of the three major levels in the Sun's atmosphere. The chromosphere is only about one ten-thousandth (10^{-4}) as dense as the photosphere, or about 10^{-8} as dense as our own atmosphere. No wonder it is normally invisible!

Comparing the Chromosphere and Photosphere

Unlike the photosphere, which has an absorption line spectrum, the chromosphere has a spectrum dominated by emission lines. An emission line spectrum is produced by the atoms of a hot, thin gas (see Section 5-6 and Section 5-8). As their electrons fall from higher to lower energy levels, the atoms emit photons.

One of the strongest emission lines in the chromosphere's spectrum is the H_α line at 656.3 nm, which is emitted by a hydrogen atom when its single electron falls from the $n = 3$ level to the $n = 2$ level (recall Figure 5-23b). This wavelength is in the red part of the spectrum, which gives the chromosphere its characteristic pinkish color. The spectrum also contains emission lines of singly ionized calcium, as well as lines due to ionized helium and ionized metals. In fact, helium—named after Helios, a Greek reference to the Sun—was originally discovered through detection in the chromospheric spectrum in 1868, almost 30 years before helium was discovered on Earth.

Analysis of the chromospheric spectrum shows that temperature *increases* with increasing height in the chromosphere. This trend is just the opposite of the situation in the photosphere, where

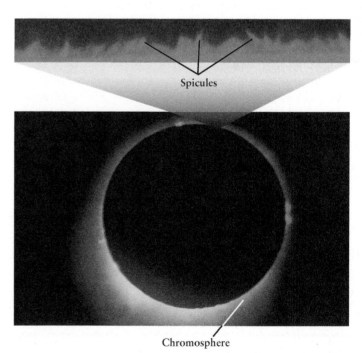

Spicules

Chromosphere

FIGURE 16-11   R I $\boxed{V}$ U X G

The Chromosphere During a total solar eclipse, the Sun's glowing chromosphere can be seen around the edge of the Moon. It appears pinkish because its hot gases emit light at only certain discrete wavelengths, principally the H_α emission of hydrogen at a red wavelength of 656.3 nm. The expanded area above shows spicules, jets of chromospheric gas that surge upward into the Sun's outer atmosphere. (NOAO)

temperature decreases with increasing height. The temperature is about 4400 K at the top of the photosphere; 2000 km higher, at the top of the chromosphere, the temperature is nearly 25,000 K. This temperature difference is very surprising, since temperatures should decrease as you move away from the Sun's interior. In Section 16-10 we will see how solar scientists explain this seeming paradox.

The photosphere's spectrum is dominated by absorption lines at certain wavelengths, while the spectrum of the chromosphere has emission lines at these same wavelengths. In other words, the photosphere appears dark at the specific wavelengths at which the chromosphere emits most strongly, such as the H_α wavelength of 656.3 nm. By viewing the Sun through a special filter that is transparent to light only at the wavelength of H_α, astronomers can screen out light from the photosphere and make the chromosphere visible. Filters make it possible to see the chromosphere at any time, not just during a solar eclipse.

Spicules

The top photograph in Figure 16-11 is a high-resolution image of the Sun's chromosphere taken through an H_α filter. This image

> Jets of gas thousands of kilometers in height rise through the chromosphere

shows numerous vertical spikes, which are actually jets of rising gas called **spicules**. A typical spicule lasts just 15 minutes or so: It rises at the rate of about 20 km/s (72,000 km/h, or 45,000 mi/h), can reach a height of several thousand kilometers, and then collapses and fades away (Figure 16-12). Approximately 300,000 spicules exist at any one time, covering about 1% of the Sun's surface. When viewed

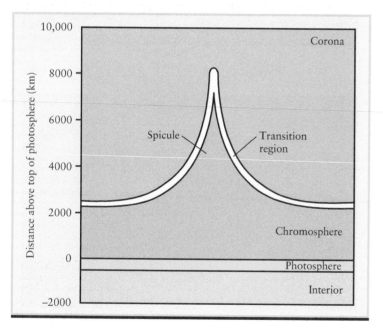

FIGURE 16-12

The Solar Atmosphere This schematic diagram shows the three layers of the solar atmosphere. The lowest, the photosphere, is about 400 km thick. The chromosphere extends about 2000 km higher, with spicules jutting up to nearly 10,000 km above the photosphere. Above a transition region is the Sun's outermost layer, the corona, which we discuss in Section 16-7. It extends many millions of kilometers out into space. (Adapted from J. A. Eddy)

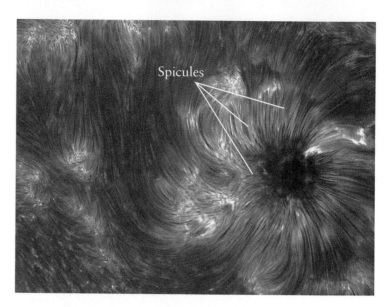

FIGURE 16-13   R I $\boxed{\text{V}}$ U X G

Spicules from Above Surrounding this sunspot are numerous spicules, also called "fibrils" because of their long and thin fiberlike shape. These spicules are about as long as Earth's diameter and contain very hot gas confined by magnetic fields. This is one of the highest-resolution images of the Sun ever made. (K. Reardon [Osservatorio Astrofisico di Arcetri, INAF], IBIS, DST, NSO)

from above, spicules can surround a sunspot, as in Figure 16-13 (we will learn more about sunspots in Section 16-8). While the Sun's magnetic field appears to be involved in the production of spicules, the details are not understood.

CONCEPTCHECK 16-12

Why do the spicules in Figure 16-11 appear red? Do spicules emit light at other wavelengths?

Answer appears at the end of the chapter.

16-7 The corona ejects mass into space to form the solar wind

The **corona**, or outermost region of the Sun's atmosphere, begins at the top of the chromosphere. It extends out to a distance of several million kilometers. Despite its tremendous extent, the corona is only about one-millionth (10^{-6}) as bright as the photosphere—no brighter than the full moon. Hence, the corona can be viewed only when the light from the photosphere is blocked out, either by use of a specially designed telescope or during a total solar eclipse.

Figure 16-14 is an exceptionally detailed photograph of the Sun's corona taken during a solar eclipse. It shows that the corona is not merely a spherical shell of gas surrounding the Sun. Rather, numerous streamers extend in different directions far above the solar surface. The shapes of these streamers vary on timescales of days or weeks. (For another view of the corona during a solar eclipse, see Figure 3-10b.)

Comparing the Corona, Chromosphere, and Photosphere

Like the chromosphere that lies below it, the corona has an emission line spectrum characteristic of a hot, thin gas. When the

FIGURE 16-14   R I V̄ U X G

The Solar Corona This striking photograph of the corona was taken during the total solar eclipse of July 11, 1991. Numerous streamers extend for millions of kilometers above the solar surface. The unearthly light of the corona is one of the most extraordinary aspects of experiencing a solar eclipse. (Courtesy of R. Christen and M. Christen, Astro-Physics, Inc.)

spectrum of the corona was first measured in the nineteenth century, astronomers found a number of emission lines at wavelengths that had never been seen in the laboratory. Their explanation was

that the corona contained elements that had not yet been detected on Earth. However, laboratory experiments in the 1930s revealed that these unusual emission lines were in fact caused by the same atoms found elsewhere in the universe—but in highly ionized states. For example, a prominent green line at 530.3 nm is caused by highly ionized iron atoms, each of which has been stripped of 13 of its 26 electrons. In order to strip that many electrons from atoms, temperatures in the corona must reach 2 million kelvins (2×10^6 K) or even higher—far greater than the temperatures in the chromosphere. Figure 16-15 shows how temperature varies in the chromosphere and corona.

CAUTION! The corona is actually not very "hot"—that is, it contains very little thermal energy. The reason is that the corona is nearly a vacuum. In the corona there are only about 10^{11} atoms per cubic meter, compared with about 10^{23} atoms per cubic meter in the Sun's photosphere and about 10^{25} atoms per cubic meter in the air that we breathe. Because of the corona's high temperature, the atoms there are moving at very high speeds. But because there are so few atoms in the corona, the total amount of energy in these moving atoms (a measure of how "hot" the gas is) is rather low. If you flew a spaceship into the corona, you would have to worry about becoming overheated by the intense light coming from the photosphere, but you would notice hardly any heating from the corona's ultrathin gas.

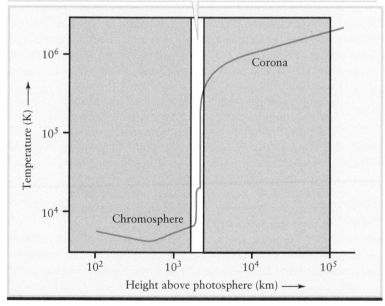

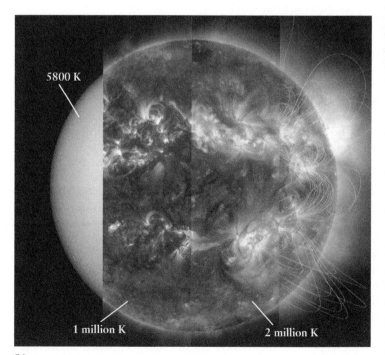

(a)   (b)

FIGURE 16-15   R I V̄ U X G

Temperatures in the Sun's Upper Atmosphere (a) This graph shows how temperature varies with altitude in the Sun's chromosphere and corona and in the narrow transition region between them. In order to show a large range of values, both the vertical and horizontal scales are nonlinear. **(b)** This composite portrait shows the different features that appear as wavelengths

corresponding to higher temperatures are imaged. At visible wavelengths, the 5800 K photosphere is visible, while ultraviolet light images features at millions of degrees. On the far right, modeling has been used to draw lines of the Sun's magnetic field in the corona. (a: Adapted from A. Gabriel; b: NASA)

ANALOGY The situation in the corona is similar to that inside a conventional oven that is being used for baking. Both the walls of the oven and the air inside the oven are at the same high temperature, but the air contains very few atoms and thus carries little energy. If you put your hand in the oven momentarily, the lion's share of the heat you feel is radiation from the oven walls.

The low density of the corona explains why it is so dim compared with the photosphere. In general, the higher the temperature of a gas, the brighter it glows. But because there are so few atoms in the corona, the net amount of light that it emits is very feeble compared with the light from the much cooler, but also much denser, photosphere.

CONCEPTCHECK 16-13

Why is the corona so difficult to see if it is so much hotter than the photosphere?

Answer appears at the end of the chapter.

The Solar Wind and Coronal Holes

Earth's gravity keeps our atmosphere from escaping into space. In the same way, the Sun's powerful gravitational attraction keeps most of the gases of the photosphere, chromosphere, and corona from escaping. But the corona's high temperature means that its atoms and ions are moving at very high speeds, around a million km per hour. As a result, some of the coronal gas can and does escape. This outflow of gas, which we first encountered in Section 8-5, is called the **solar wind.**

Each second the Sun ejects about a million tons (10^9 kg) of material into the solar wind. But the Sun is so massive that, even over its entire lifetime, it will eject only a few tenths of a percent of its total mass. The solar wind is composed almost entirely of electrons and nuclei of hydrogen and helium. About 0.1% of the solar wind is made up of ions of more massive atoms, such as silicon, sulfur, calcium, chromium, nickel, iron, and argon. The aurorae seen at far northern or southern latitudes on Earth are produced when electrons and ions from the solar wind enter our upper atmosphere.

Special telescopes enable astronomers to see the origin of the solar wind. To appreciate what sort of telescopes are needed, note that because the temperature of the coronal gas is so high, ions in the corona are moving very fast (see Box 7-2). When ions collide, the energy of the impact is so great that the ion's electrons are boosted to very high energy levels. As the electrons fall back to lower levels, they emit high-energy photons in the ultraviolet and X-ray portions of the spectrum—wavelengths at which the photosphere and chromosphere are relatively dim. Hence, telescopes sensitive to these short wavelengths are ideal for studying the corona and the flow of the solar wind.

Earth's atmosphere absorbs most ultraviolet light and X-rays, so telescopes for these wavelengths must be placed above the atmosphere on spacecraft (see Section 6-7, especially Figure 6-25). Figure 16-16 shows an ultraviolet view of the corona from the *SOHO* spacecraft (*Solar and Heliospheric Observatory*), a joint project of the European Space Agency (ESA) and NASA.

Figure 16-16 reveals that the corona is not uniform in temperature or density. The densest, highest-temperature regions appear bright, while the thinner, lower-temperature regions are dark. Note the large dark area, called a **coronal hole** because it is

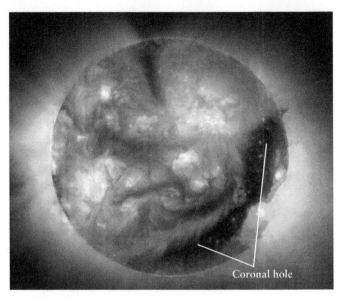

Coronal hole

VIDEO 16-3 **FIGURE 16-16** R I V U X G

The Ultraviolet Corona The *SOHO* spacecraft recorded this false-color ultraviolet view of the solar corona. The dark feature running across the Sun's disk from the top is a coronal hole, a region where the coronal gases are thinner than elsewhere. Such holes are often the source of strong gusts in the solar wind. (SOHO/EIT/ESA/NASA)

almost devoid of luminous gas. Particles streaming away from the Sun can most easily flow outward through these particularly thin regions. Therefore, it

> Unlike the lower levels of the Sun's atmosphere, the corona has immense holes that shift and reshape

is thought that coronal holes are the main corridors through which particles of the solar wind escape from the Sun.

Evidence in favor of this picture has come from the *Ulysses* spacecraft, another joint ESA/NASA mission. In 1994 and 1995, *Ulysses* became the first spacecraft to fly over the Sun's north and south poles, where there are apparently permanent coronal holes. The spacecraft indeed measured a stronger solar wind emanating from these holes.

The temperatures in the corona and the chromosphere are not at all what we would expect. Just as you feel warm if you stand close to a campfire but cold if you move away, we would expect that the temperature in the corona and chromosphere would *decrease* with increasing altitude and, hence, increasing distance from the warmth of the Sun's photosphere. Why, then, does the temperature in these regions *increase* with increasing altitude? This question has been one of the major unsolved mysteries in astronomy for the past half-century.

In 2011, the Solar Dynamics Observatory found evidence that spicules (Section 16-6) play a role in heating the corona. Some spicules can get quite large—the width of a typical state in the United States and as tall as Earth—and shoot gas at about 150,000 miles/hour. Observations show that some of the gas ejected by spicules is heated to millions of degrees as it is shot into the corona. With so many spicules present at any given time (about 300,000), estimates indicate this ejected gas makes a significant contribution to coronal heating.

However, spicules are only one contribution to coronal heating, and as astronomers have tried to resolve this dilemma, they have found important clues in one of the Sun's most familiar features—sunspots.

CONCEPTCHECK **16-14**

What is the connection between the solar wind and coronal holes?

Answer appears at the end of the chapter.

16-8 Sunspots are low-temperature regions in the photosphere

Granules, supergranules, spicules, and the solar wind occur continuously. These features are said to be aspects of the *quiet* Sun. But other, more dramatic features appear periodically, including massive eruptions and regions of concentrated magnetic fields. When these features are present, astronomers refer to the *active* Sun. The features of the active Sun that can most easily be seen with even a small telescope are sunspots (although only with a safety filter attached).

Observing Sunspots

Sunspots are irregularly shaped dark regions in the photosphere. Sometimes sunspots appear in isolation (Figure 16-17a), but frequently they are found in sunspot groups (Figure 16-17b; see also Figure 16-7). Although sunspots vary greatly in size, typical ones measure a few tens of thousands of kilometers across—comparable to the diameter of Earth. Sunspots are not permanent features of the photosphere but last between a few hours and a few months.

Each sunspot has a dark central core, called the *umbra,* and a brighter border called the *penumbra.* We used these same terms in Section 3-4 to refer to different parts of Earth's shadow or the Moon's shadow. But a sunspot is not a shadow: It is a region in the photosphere where the temperature is relatively low, which makes it appear darker than its surroundings. As we saw in Section 5-4, Wien's law relates the color of a blackbody, such as the photosphere, to the blackbody's temperature; the cooler the blackbody, the longer the wavelength at which it emits the most light. If the surrounding hot photosphere is blocked from view, a sunspot's cooler umbra appears red and the warmer penumbra appears orange. The colors of a sunspot indicate that the temperature of the umbra is typically 4300 K and that of the penumbra is typically 5000 K. While high by earthly standards, these temperatures are quite a bit lower than the average photospheric temperature of 5800 K.

The Stefan-Boltzmann law (see Section 5-4) tells us that the energy flux from a blackbody is proportional to the fourth power of its temperature. This law lets us compare the amounts of light energy emitted by a square meter of a sunspot's umbra and by a square meter of undisturbed photosphere. The ratio is

$$\frac{\text{flux from umbra}}{\text{flux from photosphere}} = \left(\frac{4300 \text{ K}}{5800 \text{ K}}\right)^4 = 0.30$$

That is, the umbra emits only 30% as much light as an equally large patch of undisturbed photosphere, which is why sunspots appear so dark.

Sunspots and the Sun's Rotation

Occasionally, a sunspot group is large enough to be seen without a telescope. Chinese astronomers recorded such sightings 2000 years ago, and huge sunspot groups visible to the naked eye (with an appropriate filter) were seen in 1989 and 2003. But it was not until Galileo introduced the telescope into astronomy (see Section 4-5) that anyone was able to examine sunspots in detail.

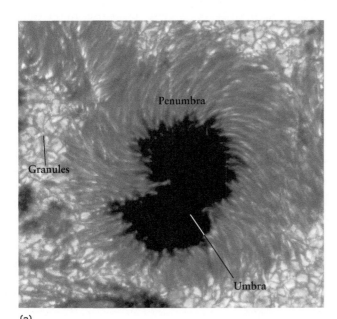

(a)

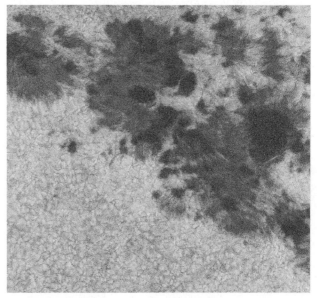

(b)

VIDEO 16-4 **FIGURE 16-17** R I **V** U X G

Sunspots **(a)** This high-resolution photograph of the photosphere shows a mature sunspot. The dark center of the spot is called the umbra and has the lowest temperature. It is bordered by the penumbra, which is less dark, is hotter than the umbra, and has a featherlike appearance. (Even the umbra emits visible light, but not enough to capture in an image against the glare of the Sun.) **(b)** In this view of a typical sunspot group, several sunspots are close enough to overlap. In both images you can see granulation in the surrounding, undisturbed photosphere.

(a: Scharmer et al., Royal Swedish Academy of Sciences/Science Source; b: NOAO)

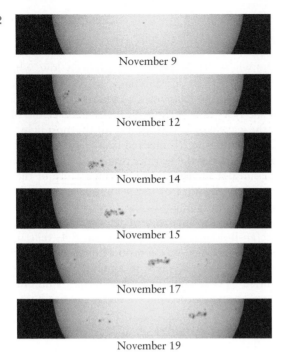

November 9

November 12

November 14

November 15

November 17

November 19

VIDEO 16-5 **FIGURE 16-18** R I **V** U X G

Tracking the Sun's Rotation with Sunspots This series of photographs taken in 1999 shows the rotation of the Sun. By observing the same group of sunspots from one day to the next, Galileo found that the Sun rotates once in about four weeks. (The equatorial regions of the Sun actually rotate somewhat faster than the polar regions.) Notice how the sunspot group shown here changed its shape. (The Carnegie Observatories)

Galileo discovered that he could determine the Sun's rotation rate by tracking sunspots as they moved across the solar disk (Figure 16-18). He found that the Sun rotates once in about four weeks. A typical sunspot group lasts about two months, so a specific one can be followed for two solar rotations.

Further observations by the British astronomer Richard Carrington in 1859 demonstrated that the Sun does not rotate as a rigid body. Instead, the equatorial regions rotate more rapidly than the polar regions. This phenomenon is known as **differential rotation.** Thus, while a sunspot near the solar equator takes only 25 days to go once around the Sun, a sunspot at 30° north or south of the equator takes 27½ days. The rotation period at 75° north or south is about 33 days, while near the poles it may be as long as 35 days.

CONCEPTCHECK **16-15**

If the center of a sunspot has a temperature of about 4300 K, why does it appear dark?

Answer appears at the end of the chapter.

The Sunspot Cycle

The average number of sunspots on the Sun is not constant, but varies in a predictable **sunspot cycle** (Figure 16-19a). This phenomenon was first reported by the German astronomer Heinrich Schwabe in 1843 after many years of observing. As Figure 16-19a shows, the average number of sunspots varies with a period of about 11 years. A period of exceptionally many sunspots is a **sunspot maximum** (Figure 16-19b), as occurred in 1979, 1989, and 2000. Conversely, the Sun is almost devoid of sunspots at a **sunspot minimum** (Figure 16-19c), as occurred in 1976, 1986, 1996, and 2008. During the 2008 minimum in sunspot activity, there were fewer sunspots observed than in any other year since 1913.

> The number of sunspots increases and decreases on an 11-year cycle

The locations of sunspots also vary with the same 11-year sunspot cycle. At the beginning of a cycle, just after a sunspot minimum, sunspots first appear at latitudes around 30° north and south of the solar equator (Figure 16-20). Over the succeeding years, the sunspots occur closer and closer to the equator.

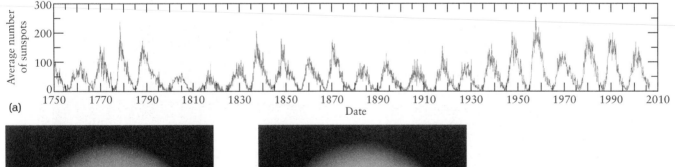

(a)

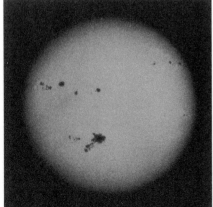

(b) Near sunspot maximum

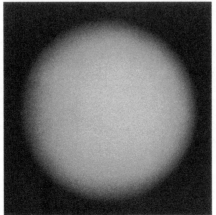

(c) Near sunspot minimum

FIGURE 16-19 R I **V** U X G

The Sunspot Cycle **(a)** The number of sunspots on the Sun varies with a period of about 11 years. The most recent sunspot maximum occurred in 2000. **(b)** This photograph, taken near sunspot maximum in 1989, shows a number of sunspots and large sunspot groups. The sunspot group visible near the bottom of the Sun's disk has about the same diameter as the planet Jupiter. **(c)** Near sunspot minimum, as in this 1986 photograph, essentially no sunspots are visible. (NOAO)

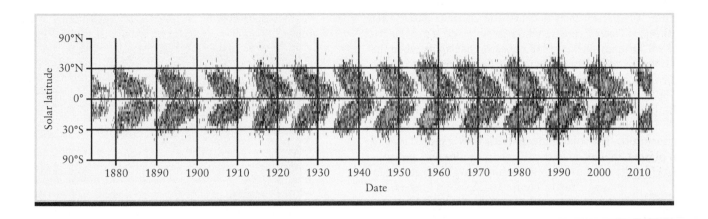

FIGURE 16-20

Variations in the Average Latitude of Sunspots The dots in this graph (sometimes called a "butterfly diagram") record how far north or south of the Sun's equator sunspots were observed. At the beginning of each sunspot cycle, most sunspots are found near latitudes 30° north or south. As the cycle goes on, sunspots typically form closer to the equator. (NASA Marshall Space Flight Center)

16-9 Sunspots are produced by a 22-year cycle in the Sun's magnetic field

TUTORIAL 16-3 Why should the number of sunspots vary with an 11-year cycle? Why should their average latitude vary over the course of a cycle? And why should sunspots exist at all? The first step toward answering these questions came in 1908, when the American astronomer George Ellery Hale discovered that sunspots are associated with intense magnetic fields on the Sun.

Probing Solar Magnetism

LOOKING DEEPER 16-2 When Hale focused a spectroscope on sunlight coming from a sunspot, he found that many spectral lines appear to be split into several closely spaced lines (Figure 16-21).

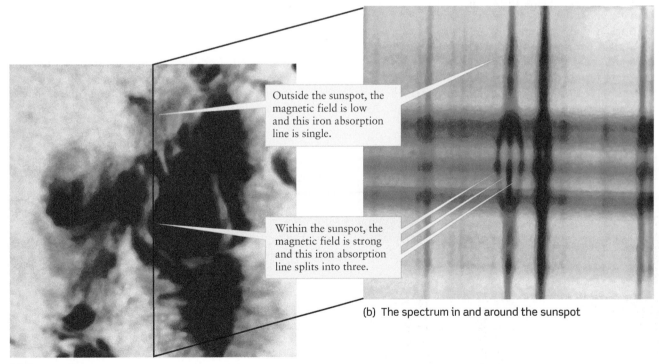

Outside the sunspot, the magnetic field is low and this iron absorption line is single.

Within the sunspot, the magnetic field is strong and this iron absorption line splits into three.

(a) A sunspot

(b) The spectrum in and around the sunspot

FIGURE 16-21   R I U X G

Sunspots Have Strong Magnetic Fields (a) A black line in this image of a sunspot shows where the slit of a spectrograph was aimed. (b) This is a portion of the resulting spectrum, including a dark absorption line caused by iron atoms in the photosphere. The splitting of this line by the sunspot's magnetic field can be used to calculate the field strength. Typical sunspot magnetic fields are over 5000 times stronger than Earth's field at its north and south poles. (NOAO)

This "splitting" of spectral lines is called the **Zeeman effect,** after the Dutch physicist Pieter Zeeman, who first observed it in his laboratory in 1896. Zeeman showed that a spectral line splits when the atoms are subjected to an intense magnetic field. The more intense the magnetic field, the wider the separation of the split lines.

Hale's discovery showed that sunspots are places where the hot gases of the photosphere are bathed in a concentrated magnetic field. Many of the atoms of the Sun's atmosphere are ionized due to the high temperature. The solar atmosphere is thus a special type of gas called a **plasma,** in which electrically charged ions and electrons can move freely. Like any moving, electrically charged objects, they can be deflected by magnetic fields. Figure 16-22 shows how a magnetic field in the laboratory bends a beam of fast-moving electrons into a curved trajectory. Similarly, the paths of moving ions and electrons in the photosphere are deflected by the Sun's magnetic field. In particular, magnetic forces act on the hot plasma that rises from the Sun's interior due to convection. Where the magnetic field is particularly strong, these forces push the hot plasma away. The result is a localized region where the gas is relatively cool and thus glows less brightly—in other words, a sunspot.

To get a fuller picture of the Sun's magnetic fields, astronomers take images of the Sun at two wavelengths, one just less than and one just greater than the wavelength of a magnetically split spectral line. From the difference between these two images, they can construct a picture called a **magnetogram,** which displays the magnetic fields in the solar atmosphere. Figure 16-23a is an ordinary white-light photograph of the Sun taken at the same time as the magnetogram in Figure 16-23b. In the magnetogram, dark blue indicates areas of the photosphere with one magnetic polarity (north), and yellow indicates areas with the opposite (south) magnetic polarity. This image shows that many sunspot groups have roughly comparable areas covered by north and south magnetic polarities (see also Figure 16-23c). Thus, a sunspot group resembles a giant bar magnet, with a north magnetic pole at one end and a south magnetic pole at the other.

If different sunspot groups were unrelated to one another, their magnetic poles would be randomly oriented, like a bunch of compass needles all pointing in random directions. As Hale

FIGURE 16-22   R I V U X G

Magnetic Fields Deflect Moving, Electrically Charged Objects
In this laboratory experiment, a beam of negatively charged electrons (shown by a blue arc) is aimed straight upward from the center of the apparatus. The entire apparatus is inside a large magnet, and the magnetic field deflects the beam into a curved path. (Andrew Lambert Photography/Science Source)

discovered, however, there is a striking regularity in the magnetization of sunspot groups. As a given sunspot group moves with the Sun's rotation, the sunspots in front are called the "preceding members" of the group. The spots that follow behind are referred to as the "following members." Hale compared the sunspot groups in the two solar hemispheres, north or south of the Sun's equator. He found that the preceding members in one solar hemisphere all have the same magnetic polarity, while the preceding members in the other hemisphere have the opposite polarity. Furthermore, in the hemisphere where the Sun has its

(a) Visible-light image

(b) Magnetogram

(c) Magnetogram of a sunspot group

FIGURE 16-23   R I V U X G

Mapping the Sun's Magnetic Field **(a)** This visible-light image and **(b)** this false-color magnetogram were recorded at the same time. Dark blue and yellow areas in the magnetogram have north and south magnetic polarity, respectively; blue-green regions have weak magnetic fields. The

highly magnetized regions in (b) correlate with the sunspots in (a). **(c)** The two ends of this large sunspot group have opposite magnetic polarities (colored blue and yellow), like the ends of a giant bar magnet. (NOAO)

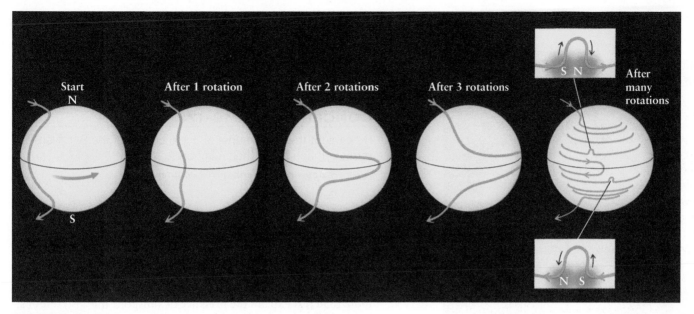

FIGURE 16-24

Babcock's Magnetic Dynamo Model Magnetic field lines tend to move along with the plasma in the Sun's outer layers. Because the Sun rotates faster at the equator than near the poles, a field line that starts off running from the Sun's north magnetic pole (N) to its south magnetic pole (S) ends up wrapped around the Sun like twine wrapped around a ball. The insets on the far right show how sunspot groups appear where the concentrated magnetic field rises through the photosphere.

north magnetic pole, the preceding members of all sunspot groups have north magnetic polarity. In the opposite hemisphere, where the Sun has its south magnetic pole, the preceding members all have south magnetic polarity.

Along with his colleague Seth B. Nicholson, Hale also discovered that the Sun's polarity pattern completely reverses itself every 11 years—the same interval as the time from one solar maximum to the next. The hemisphere that has preceding north magnetic poles during one 11-year sunspot cycle will have preceding south magnetic poles during the next 11-year cycle, and vice versa. The north and south magnetic poles of the Sun itself also reverse every 11 years. Thus, the Sun's magnetic pattern repeats itself only after two sunspot cycles, which is why astronomers speak of a **22-year solar cycle.**

The Magnetic-Dynamo Model

In 1960, the American astronomer Horace Babcock proposed a description that seems to account for many features of this 22-year solar cycle. Babcock's scenario, called a **magnetic-dynamo model,** makes use of two basic properties of the Sun's photosphere—differential rotation and convection. Differential rotation causes the magnetic field in the photosphere to become wrapped around the Sun (Figure 16-24). As a result, the magnetic field becomes concentrated at certain latitudes on either side of the solar equator. Convection in the photosphere creates tangles in the concentrated magnetic field, and "kinks" erupt through the solar surface. Sunspots appear where the magnetic field protrudes through the photosphere. The theory suggests that sunspots should appear first at northern and southern latitudes and later form nearer to the equator, which is just what is observed (see Figure 16-20). Note also that as shown on the far right in Figure 16-24, the preceding member of a sunspot group has the same polarity (N or S) as the Sun's magnetic pole in that hemisphere, which is just as Hale observed.

Differential rotation eventually undoes the twisted magnetic field. The preceding members of sunspot groups move toward the Sun's equator, while the following members migrate toward the poles. Because the preceding members from the two hemispheres have opposite

> The Sun's differential rotation makes the magnetic field twist like a rubber band

magnetic polarities, their magnetic fields cancel each other out when they meet at the equator. The following members in each hemisphere have the opposite polarity to the Sun's pole in that hemisphere; hence, when they converge on the pole, the following members first cancel out and then reverse the Sun's overall magnetic field. The fields are now completely relaxed. Once again, differential rotation begins to twist the Sun's magnetic field, but now with all magnetic polarities reversed. In this way, Babcock's model helps to explain the change in field direction every 11 years.

Recent discoveries in helioseismology (Section 16-3) offer new insights into the Sun's magnetic field. By comparing the speeds of sound waves that travel with and against the Sun's rotation, helioseismologists have been able to determine the Sun's rotation rate at different depths and latitudes. As shown in Figure 16-25, the Sun's surface pattern of differential rotation persists through the convective zone. Farther in, within the radiative zone, the Sun seems to rotate like a rigid object with a period of 27 days at all latitudes. Astronomers suspect that the Sun's magnetic field originates in a relatively thin layer where the radiative and convective zones meet and slide past each other due to their different rotation rates.

One dilemma about sunspots is that compressed magnetic fields tend to push themselves apart, which means that sunspots should dissipate rather quickly. Yet observations show that sunspots can persist for many weeks. The resolution of this paradox may have been found using helioseismology (see Section 16-3). Analysis of the vibrations of the Sun around sunspots shows that beneath the surface of the photosphere, the gases surrounding each sunspot are circulating at high speed—rather like a hurricane as large as Earth. The circulation of charged gases around the magnetic field holds the fields in place, thus stabilizing the sunspot.

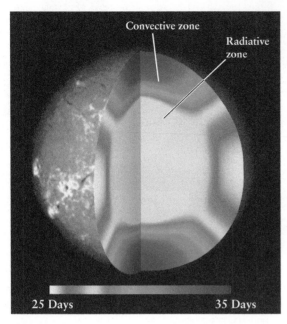

FIGURE 16-25

Rotation of the Solar Interior This cutaway picture of the Sun shows how the solar rotation period (shown by different colors) varies with depth and latitude. The surface and the convective zone have differential rotation (a short period at the equator and longer periods near the poles). Deeper within the Sun, the radiative zone seems to rotate like a rigid sphere.

(Courtesy of K. Libbrecht, Big Bear Solar Observatory)

Helioseismology—the analysis of solar sound waves—can also take advantage of the relationship between sunspots and the solar magnetic field. Compared to typical solar material, sunspots absorb solar sound waves more strongly. When sunspots occur on the *opposite* side of the Sun, their increased absorption means that interior sound waves do not reflect back as strongly through the body of the Sun. Amazingly, this effect is used to estimate sunspot activity on the *opposite side* of the Sun. Knowing the number of sunspots facing away from Earth is actually useful: Added to the sunspots facing Earth, the *total* sunspot activity is continuously monitored to help forecast **space weather**—these are variations in the solar wind and magnetic field that can affect satellites and astronauts.

Much about sunspots and solar activity remains mysterious. There are perplexing irregularities in the solar cycle. For example, the overall reversal of the Sun's magnetic field is often piecemeal and haphazard. One pole may reverse polarity long before the other. For several weeks the Sun's surface may have two north magnetic poles and no south magnetic pole at all.

Furthermore, there seem to be times when all traces of sunspots and the sunspot cycle vanish for many years. For example, virtually no sunspots were seen from 1645 through 1715. Curiously, during these same years Europe experienced record low temperatures, often referred to as the Little Ice Age, whereas the western United States was subjected to severe drought. By contrast, there was apparently a period of increased sunspot activity during the eleventh and twelfth centuries, during which Earth was warmer than it is today. Thus, variations in solar activity appear to affect climates on Earth. The origin of this Sun-Earth connection is a topic of ongoing research.

Is the sunspot cycle an 11-year cycle or a 22-year cycle?

How might the Sun's sunspot cycle change if the Sun were rotating much faster than it is now?

Answers appear at the end of the chapter.

16-10 The Sun's magnetic field heats the corona, produces flares, and causes massive eruptions

Astronomers now understand that the Sun's magnetic field does more than just explain the presence of sunspots. It is also responsible for the existence of spicules, as well as a host of other dramatic phenomena in the chromosphere and corona.

Magnetic Reconnection

In a plasma, magnetic field lines and the material of the plasma tend to move together. Their moving together means that as convection pushes material toward the edge of a supergranule, it pushes magnetic field lines as well. The result is that vertical magnetic field lines pile up around a supergranule. Plasma that "sticks" to these magnetic field lines thus ends up lifted upward, forming a spicule (see Figures 16-12 and 16-13).

The tendency of plasma to follow the Sun's magnetic field can also explain why the temperature of the chromosphere and corona is so high. Spacecraft observations show magnetic field arches extending tens of thousands of kilometers into the corona, with streamers of electrically charged particles moving along each arch (Figure 16-26a). If the magnetic fields of two arches come into proximity, their magnetic fields can rearrange in a process called **magnetic reconnection** (Figure 16-26b). The tremendous amount of energy stored in the magnetic field is then released into the solar atmosphere. (A single arch contains as much energy as a hydroelectric power plant would generate in a million years.) The amount of energy released in this way is thought to help maintain the temperatures of the chromosphere and corona.

ANALOGY The idea that a magnetic field can heat gases has applications on Earth as well as on the Sun. In an automobile engine's ignition system, an electric current is set up in a coil of wire, which produces a magnetic field. When the current is shut off, the magnetic field collapses and its energy is directed to a spark plug in one of the engine's cylinders. The released energy heats the mixture of air and gasoline around the plug, causing the mixture to ignite. This drives the piston in that cylinder and makes the automobile go.

VIDEO 16-6 Magnetic heating can also explain why the parts of the corona that lie on top of sunspots are often the most prominent in ultraviolet images. (Some examples are the bright regions in Figure 16-16.) The intense magnetic field of the sunspots helps trap and compress hot coronal gas, giving it such a high temperature that it emits copious amounts of high-energy ultraviolet photons and even more energetic X-ray photons.

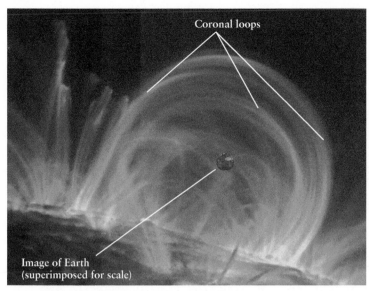

Coronal loops

Image of Earth
(superimposed for scale)

(a)

FIGURE 16-26   R I V U X G

Magnetic Arches and Magnetic Reconnection **(a)** This false-color
ultraviolet image from the *TRACE* spacecraft *(Transition Region and Coronal
Explorer)* shows magnetic field loops suspended high above the solar surface.

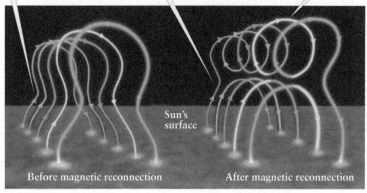

1. If magnetic field loops begin to pinch together...

2. ...the field lines of adjacent loops can reconnect, causing a release of energy.

3. The upper helix or "coil" of magnetic field can break loose, carrying material with it into space.

Sun's surface

Before magnetic reconnection   After magnetic reconnection

(b)

The loops are made visible by the glowing gases trapped within them.
(b) When the magnetic fields in these loops change their arrangement, a
tremendous amount of energy is released and solar material can be ejected
upward. (a: Stanford-Lockheed Institute for Space Research; *TRACE;* and NASA)

CONCEPTCHECK 16-18

Why does glowing plasma on the Sun appear to arch up above
the Sun's photosphere?

Answer appears at the end of the chapter.

Prominences, Flares, and Coronal Mass Ejections

Spicules and coronal heating occur even when the Sun is quiet.
But magnetic fields can also explain many aspects of the active
Sun in addition to sunspots. **Figure 16-27** is an image of the chromosphere made with an H_α filter during a sunspot maximum. The
bright areas are called **plages** (from the French word for "beach").
These plages are bright, hot regions in the chromosphere that tend
to form just before the appearance of new sunspots. They are
probably created by magnetic fields that push upward from the
Sun's interior, compressing and heating a portion of the chromosphere. The dark streaks, called **filaments**, are relatively cool and
dense parts of the chromosphere that have been pulled along with
magnetic field lines as they arch to high altitudes. Images with an
H_α filter are made from visible light, and images using ultraviolet
light can highlight filaments and sunspots (**Figure 16-28**).

When seen from the side, so that they are viewed against the
dark background of space, filaments appear as bright, arching columns of gas called **prominences** (**Figure 16-29**). They can extend
for tens of thousands of kilometers above the photosphere. Some
prominences last for only a few hours, while others persist for
many months. The most energetic prominences break free of the
magnetic fields that confined them and burst into space.

Magnetic reconnection also causes violent, eruptive events
on the Sun, called **solar flares**. Rooted to the same magnetic phenomena, the solar flares occur in complex sunspot groups.
Within only a few minutes, temperatures in a compact region may
soar to 5×10^6 K and vast quantities of particles and radiation—
including as much material as is in the prominence shown in Figure
16-29—are blasted out into space. These eruptions can also cause

disturbances that spread outward in the solar atmosphere, like the
ripples that appear when you drop a rock into a pond.

While the external effects of the Sun's magnetism are on full
display with solar flares, the advancing magnetic fields can also
be discerned below the surface. Observations from NASA's Solar
Dynamics Observatory employed helioseismology to detect magnetic fields about 65,000 km below the surface (that distance is

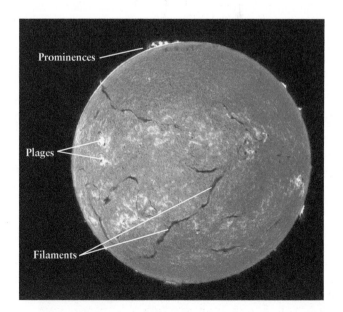

Prominences

Plages

Filaments

FIGURE 16-27   R I V U X G

The Active Sun Seen through an H_α Filter This image was made using
a red filter that only passes light at a wavelength of 656 nm. The spectrum
of the photosphere has an absorption line at this wavelength and so appears
dark. Hence, this filter reveals the chromosphere. Prominences, plages, and
filaments are associated with strong magnetic fields. (NASA)

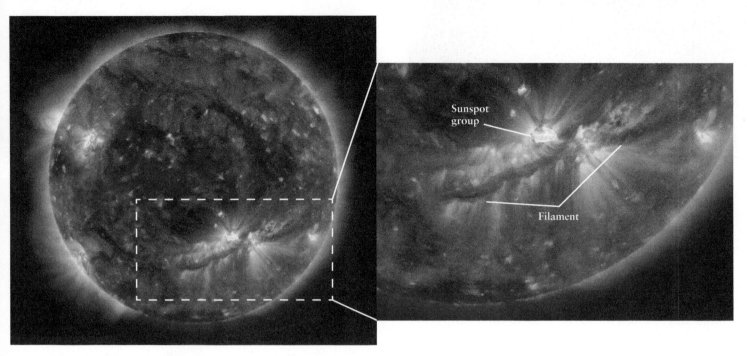

FIGURE 16-28   R I V Ⓤ X G

The Sun in Ultraviolet Light This false-color image was made in ultraviolet light by the Solar Dynamics Observatory. A very long filament is seen cutting across the Sun's southern hemisphere; it is longer than the distance from Earth to the Moon! The filament is confined by a strong magnetic field, and above the filament is a bright sunspot group. (NASA)

about 2½ times Earth's circumference). As predicted, the effects of these magnetic fields emerged several days later as sunspots.

The most energetic flares carry as much as 10^{30} joules of energy, equivalent to 10^{14} one-megaton nuclear weapons being exploded at once! However, the energy of a solar flare does not come from thermonuclear fusion in the solar atmosphere; instead, it appears to be released from the intense magnetic field around a sunspot group.

> A solar flare can have as much energy as 100 trillion nuclear bombs

As energetic as solar flares are, they are dwarfed by **coronal mass ejections.** One such event is shown in Figure 16-30. If directed toward Earth, these events are large enough to affect us directly (Figure 16-31). In a coronal mass ejection, more than 10^{12} kilograms of high-temperature coronal gas is blasted into space at speeds of hundreds of kilometers per second. That is about 200 times the mass of the largest Egyptian pyramid, and a typical coronal mass ejection lasts a few hours. These explosive events seem to be related to large-scale alterations in the Sun's magnetic field, like the magnetic reconnection shown in Figure 16-26b. Coronal mass ejections occur every few months; smaller eruptions may occur almost daily.

If a solar flare or coronal mass ejection happens to be aimed toward Earth, a stream of high-energy electrons and nuclei reaches us a few days later (Figure 16-31b); this is part of space weather (Section 16-9). When this plasma arrives, it can interfere with satellites, pose a health hazard to astronauts in orbit, and disrupt electrical and communications equipment on Earth's surface. Telescopes on Earth and on board spacecraft now monitor the Sun continuously to provide warnings of dangerous levels of solar particles. By design, the walls of the International Space Station are thick enough to protect astronauts from material ejected during regular solar activity, but extra precautions are needed during a stronger solar event.

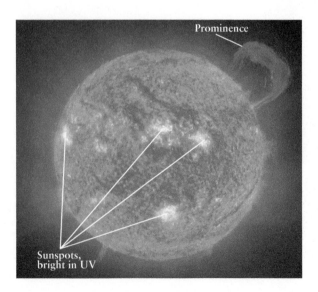

FIGURE 16-29   R I V Ⓤ X G

A Solar Prominence A huge prominence arches above the solar surface in this ultraviolet image from the *SOHO* spacecraft. The image was recorded using ultraviolet light at a wavelength of 30.4 nm, emitted by singly ionized helium atoms at a temperature of about 60,000 K. By comparison, the material within the arches in Figure 16-25 reaches temperatures in excess of 2×10^6 K. (NASA)

FIGURE 16-30 R I V ⊌ X G

Coronal Mass Ejection The Sun produced a huge coronal mass ejection over several hours on April 16, 2012. Not all of the gas was shot into the solar system, and some material from this event was observed crashing back onto the solar surface. (This ejection was not directed toward Earth.) (NASA)

The numbers of plages, filaments, solar flares, and coronal mass ejections all vary with the same 11-year cycle as sunspots. But unlike sunspots, coronal mass ejections never completely cease, even when the Sun is at its quietest. Astronomers are devoting substantial effort to understanding these and other aspects of our dynamic Sun.

CONCEPTCHECK 16-19

Which of the following phenomena is the most energetic: prominences, solar flares, or coronal mass ejections?

Answer appears at the end of the chapter.

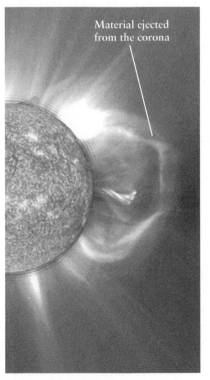

(a) A coronal mass ejection

(b) Two to four days later

FIGURE 16-31 R I V ⊌ X G

A Sun-Earth Connection **(a)** *SOHO* recorded this coronal mass ejection in an X-ray image. (The image of the Sun itself was made at ultraviolet wavelengths.) **(b)** In this artist's illustration we see that within two to four days the fastest-moving ejected material reaches a distance of 1 AU from the Sun. Most particles are deflected by Earth's magnetosphere, but some are able to reach Earth where they can damage satellites and our electrical power grid. (The ejection shown in (a) was not aimed toward Earth and did not affect us.) (SOHO/EIT/LASCO/ESA/NASA)

KEY IDEAS

Hydrogen Fusion in the Sun's Core: The Sun's energy is produced by hydrogen fusion, a sequence of thermonuclear reactions in which four hydrogen nuclei combine to produce a single helium nucleus.

• The energy released in a nuclear reaction corresponds to a slight reduction of mass according to Einstein's equation $E = mc^2$.

• Thermonuclear fusion occurs only at very high temperatures; for example, hydrogen fusion occurs only at temperatures in excess of about 10^7 K. In the Sun, fusion occurs only in the dense, hot core.

Models of the Sun's Interior: A theoretical description of a star's interior can be calculated using the laws of physics.

• The standard model of the Sun suggests that hydrogen fusion takes place in a core extending from the Sun's center to about 0.25 solar radius.

• The core is surrounded by a radiative zone extending to about 0.71 solar radius. In this zone, energy travels outward through radiative diffusion.

• The radiative zone is surrounded by a rather opaque convective zone of gas at relatively low temperature and pressure. In this zone, energy travels outward primarily through convection.

Solar Neutrinos and Helioseismology: Conditions in the solar interior can be inferred from measurements of solar neutrinos and of solar vibrations.

• Neutrinos emitted in thermonuclear reactions in the Sun's core have been detected, but in smaller numbers than expected. Recent neutrino experiments explain why this is so.

• Helioseismology is the study of how the Sun vibrates. These vibrations have been used to infer pressures, densities, chemical compositions, and rotation rates within the Sun.

The Sun's Atmosphere: The Sun's atmosphere has three main layers: the photosphere, the chromosphere, and the corona. Everything below the solar atmosphere is called the solar interior.

• The visible surface of the Sun, the photosphere, is the lowest layer in the solar atmosphere. Its spectrum is similar to that of a blackbody at a temperature of 5800 K. Convection in the photosphere produces granules.

• Above the photosphere is a layer of less dense but higher-temperature gases called the chromosphere. Spicules extend upward from the chromosphere into the corona.

• The outermost layer of the solar atmosphere, the corona, is made of very high-temperature gases at extremely low density. Activity in the corona includes coronal mass ejections and coronal holes. The solar corona blends into the solar wind at great distances from the Sun.

The Active Sun: The Sun's surface features vary in an 11-year cycle. This is related to a 22-year cycle in which the surface magnetic field increases, decreases, and then increases again with the opposite polarity.

• Sunspots are relatively cool regions produced by local concentrations of the Sun's magnetic field. The average number of sunspots increases and decreases in a regular cycle of approximately 11 years, with reversed magnetic polarities from one 11-year cycle to the next. Two such cycles make up the 22-year solar cycle.

• The magnetic-dynamo model suggests that many features of the solar cycle are due to changes in the Sun's magnetic field. These changes are caused by convection and the Sun's differential rotation.

• A solar flare is a brief eruption of hot, ionized gases from a sunspot group. A coronal mass ejection is a much larger eruption that involves immense amounts of gas from the corona.

QUESTIONS

Review Questions

1. What is meant by the luminosity of the Sun?

2. What is Kelvin-Helmholtz contraction? Why is it ruled out as a source of the present-day Sun's energy?

3. **TUTORIAL 16-1** Why is it impossible for the burning of substances like coal to be the source of the Sun's energy?

4. What is hydrogen fusion? Why is hydrogen fusion fundamentally unlike the burning of a log in a fireplace?

5. If the electric force between protons were somehow made stronger, what effect would this have on the temperature required for thermonuclear fusion to take place?

6. Why do thermonuclear reactions occur only in the Sun's core, not in its outer regions?

7. Describe how the net result of the reactions shown in the *Cosmic Connections* figure (Section 16-1) is the conversion of four protons into a single helium nucleus. What other particles are produced in this process? How many of each particle are produced?

8. **TUTORIAL 16-2** Give an everyday example of hydrostatic equilibrium. Give an example of thermal equilibrium. Explain how these equilibrium conditions apply to each example.

9. If thermonuclear fusion in the Sun were suddenly to stop, what would eventually happen to the overall radius of the Sun? Justify your answer using the ideas of hydrostatic equilibrium and thermal equilibrium.

10. Give some everyday examples of conduction, convection, and radiative diffusion.

11. Describe the Sun's interior. Include references to the main physical processes that occur at various depths within the Sun.

12. Suppose thermonuclear fusion in the Sun stopped abruptly. Would the intensity of sunlight decrease just as abruptly? Why or why not?

13. Explain how studying the oscillations of the Sun's surface can give important, detailed information about physical conditions deep within the Sun.

14. What is a neutrino? Why is it useful to study neutrinos coming from the Sun? What do they tell us that cannot be learned from other avenues of research?

15. Unlike all other types of telescopes, neutrino detectors are placed deep underground. Why?

16. What was the solar neutrino problem? What solution to this problem was suggested by the results from the Sudbury Neutrino Observatory?

17. Describe the dangers in attempting to observe the Sun. How have astronomers learned to circumvent these observational problems?

18. Briefly describe the three layers that make up the Sun's atmosphere. In what ways do they differ from one another?

19. What is solar granulation? Describe how convection gives rise to granules.

20. High-resolution spectroscopy of the photosphere reveals that absorption lines are blueshifted in the spectrum of the central, bright regions of granules but are redshifted in the spectrum of the dark boundaries between granules. Explain how these observations show that granulation is due to convection.

21. What is the difference between granules and supergranules?

22. What are spicules? Where are they found? How can you observe them? What causes them?

23. How do astronomers know that the temperature of the corona is so high?

24. ⟳ _TUTORIAL 16-3_ How do astronomers know when the next sunspot maximum and minimum will occur?

25. Why do astronomers say that the solar cycle is really 22 years long, even though the number of sunspots varies over an 11-year period?

26. Explain how the magnetic-dynamo model accounts for the solar cycle.

27. Describe one explanation for why the corona has a higher temperature than the chromosphere.

28. Why should solar flares and coronal mass ejections be a concern for businesses that use telecommunication satellites?

Advanced Questions

Questions preceded by an asterisk (*) involve the topic discussed in Box 16-1.

> **Problem-solving tips and tools**
>
> You may have to review Wien's law and the Stefan-Boltzmann law, which are the subjects of Section 5-4. Section 5-5 discusses the properties of photons. As we described in Box 5-2, you can simplify calculations by taking ratios, such as the ratio of the flux from a sunspot to the flux from the undisturbed photosphere. When you do this, all the cumbersome constants cancel out. Figure 5-7 shows the various parts of the electromagnetic spectrum. We introduced the Doppler effect in Section 5-9 and Box 5-6. For information about the planets, see Table 7-1.

29. Calculate how much energy would be released if each of the following masses were converted _entirely_ into their equivalent energy: (**a**) a carbon atom with a mass of 2×10^{-26} kg, (**b**) 1 kg, and (**c**) a planet as massive as Earth (6×10^{24} kg).

30. Use the luminosity of the Sun (given in Table 16-1) and the answers to the previous question to calculate how long the Sun must shine in order to release an amount of energy equal to that produced by the complete mass-to-energy conversion of (**a**) a carbon atom, (**b**) 1 km, and (**c**) Earth.

31. Assuming that the current rate of hydrogen fusion in the Sun remains constant, what fraction of the Sun's mass will be converted into helium over the next 5 billion years? How will this affect the overall chemical composition of the Sun?

32. (**a**) Estimate how many kilograms of hydrogen the Sun has consumed over the past 4.56 billion years, and estimate the amount of mass that the Sun has lost as a result. Assume that the Sun's luminosity has remained constant during that time. (**b**) In fact, the Sun's luminosity when it first formed was only about 70% of its present value. With this in mind, explain whether your answers to part (a) are an overestimate or an underestimate.

*33. To convert 1 kg of hydrogen (^{1}H) into helium (^{4}He) as described in Box 16-1, you must start with 1.5 kg of hydrogen. Explain why, and explain what happens to the other 0.5 kg. (_Hint:_ How many ^{1}H nuclei are used to make the two ^{3}He nuclei shown in the _Cosmic Connections_ figure in Section 16-1? How many of these ^{1}H nuclei end up being incorporated into the ^{4}He nucleus shown in the figure?)

*34. (**a**) A positron has the same mass as an electron (see Appendix 7). Calculate the amount of energy released by the annihilation of an electron and positron. (**b**) The products of this annihilation are two photons, each of equal energy. Calculate the wavelength of each photon, and confirm from Figure 5-7 that this wavelength is the gamma-ray range.

35. Sirius is the brightest star in the night sky. It has a luminosity of 23.5 $L_\odot$, that is, it is 23.5 times as luminous as the Sun and burns hydrogen at a rate 23.5 times greater than the Sun. How many kilograms of hydrogen does Sirius convert into helium each second?

36. (Refer to the preceding question.) Sirius has 2.3 times the mass of the Sun. Do you expect that the lifetime of Sirius will be longer, shorter, or the same length as that of the Sun? Explain your reasoning.

37. (**a**) If the Sun were not in a state of hydrostatic equilibrium, would its diameter remain the same? Explain your reasoning. (**b**) If the Sun were not in a state of thermal equilibrium, would its luminosity remain the same? What about its surface temperature? Explain your reasoning.

38. Using the mass and size of the Sun given in Table 16-1, verify that the average density of the Sun is 1410 kg/m³. Compare your answer with the average densities of the Jovian planets.

39. Use the data in Table 16-2 to calculate the average density of material within 0.1 $R_\odot$ of the center of the Sun. (You will need to use the mass and radius of the Sun as a whole, given in Table 16-1.) Explain why your answer is not the same as the density at 0.1 $R_\odot$ given in Table 16-2.

40. In a typical solar oscillation, the Sun's surface moves up or down at a maximum speed of 0.1 m/s. An astronomer sets out to measure this speed by detecting the Doppler shift of an

absorption line of iron with wavelength 557.6099 nm. What is the maximum wavelength shift that she will observe?

41. Explain why the results from the Sudbury Neutrino Observatory (SNO) only provide an answer to the solar neutrino problem for relatively high-energy neutrinos. (*Hint:* Can SNO detect solar neutrinos of all energies?)

42. The amount of energy required to dislodge the extra electron from a negative hydrogen ion is 1.2×10^{-19} J. (a) The extra electron can be dislodged if the ion absorbs a photon of sufficiently short wavelength. (Recall from Section 5-5 that the higher the energy of a photon, the shorter its wavelength.) Find the longest wavelength (in nm) that can accomplish this. (b) In what part of the electromagnetic spectrum does this wavelength lie? (c) Would a photon of visible light be able to dislodge the extra electron? Explain. (d) Explain why the photosphere, which contains negative hydrogen ions, is quite opaque to visible light but is less opaque to light with wavelengths longer than the value you calculated in (a).

43. Astronomers often use an H_α filter to view the chromosphere. Explain why this can also be accomplished with filters that are transparent only to the wavelengths of the H and K lines of ionized calcium. (*Hint:* The H and K lines are dark lines in the spectrum of the photosphere.)

44. Calculate the wavelengths at which the photosphere, chromosphere, and corona emit the most radiation. Explain how the results of your calculations suggest the best way to observe these regions of the solar atmosphere. (*Hint:* Treat each part of the atmosphere as a perfect blackbody. Assume average temperatures of 50,000 K and 1.5×10^6 K for the chromosphere and corona, respectively.)

45. On November 15, 1999, the planet Mercury passed in front of the Sun as seen from Earth. The *TRACE* spacecraft made these time-lapse images of this event using ultraviolet light (top) and visible light (bottom). (Mercury moved from left to right in these images. The time between successive views of Mercury is 6 to 9 minutes.) Explain why the Sun appears somewhat larger in the ultraviolet image than in the visible-light image.

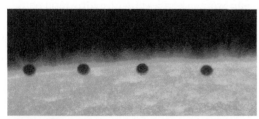

R I V U X G

R I V U X G

(K. Schrijver, Stanford-Lockheed Institute for Space Research, TRACE, and NASA)

46. The moving images on a television set that uses a tube (as opposed to a LCD or plasma flat-screen TV) are made by a fast-moving electron beam that sweeps over the back of the screen. Explain why placing a strong magnet next to the screen distorts the picture. (*Caution:* Do not try this with your television; it can cause permanent damage.)

47. Find the wavelength of maximum emission of the umbra of a sunspot and the wavelength of maximum emission of a sunspot's penumbra. In what part of the electromagnetic spectrum do these wavelengths lie?

48. (a) Find the ratio of the energy flux from a patch of a sunspot's penumbra to the energy flux from an equally large patch of undisturbed photosphere. Which patch is brighter? (b) Find the ratio of the energy flux from a patch of a sunspot's penumbra to the energy flux from an equally large patch of umbra. Again, which patch is brighter?

49. Suppose that you want to determine the Sun's rotation rate by observing its sunspots. Is it necessary to take Earth's orbital motion into account? Why or why not?

50. (a) Using a ruler to make measurements of Figure 16-26, determine how far the arches in that figure extend above the Sun's surface. The diameter of Earth is 12,756 km. (b) In Figure 16-26 (an ultraviolet image) the photosphere appears dark compared to the arches. Explain why.

51. The amount of visible light emitted by the Sun varies only a little over the 11-year sunspot cycle. But the number of X-rays emitted by the Sun can be 10 times greater at solar maximum than at solar minimum. Explain why these two types of radiation should be so different in their variability.

Discussion Questions

52. Discuss the extent to which cultures around the world have worshipped the Sun as a deity throughout history. Why do you suppose there has been such widespread veneration?

53. In the movie *Star Trek IV: The Voyage Home,* the starship *Enterprise* flies on a trajectory that passes close to the Sun's surface. What features should a real spaceship have to survive such a flight? Why?

54. Discuss some of the difficulties in correlating solar activity with changes in Earth's climate.

55. Describe some of the advantages and disadvantages of observing the Sun (a) from space and (b) from Earth's south pole. What kinds of phenomena and issues might solar astronomers want to explore from these locations?

Web/eBook Questions

56. Search the World Wide Web for information about features in the solar atmosphere called *sigmoids*. What are they? What causes them? How might they provide a way to predict coronal mass ejections?

57. **Determining the Lifetime of a Solar Granule.** Access and view the video "Granules on the Sun's Surface" in Chapter 16 of the *Universe* Web site or eBook. Your task is to determine the approximate lifetime of a solar granule on the photosphere. Select an area, then slowly and

rhythmically repeat "start, stop, start, stop" until you can consistently predict the appearance and disappearance of granules. While keeping your rhythm, move to a different area of the video and continue monitoring the appearance and disappearance of granules. When you are confident you have the timing right, move your eyes (or use a partner) to look at the clock shown in the video. Determine the length of time between the appearance and disappearance of the granules and record your answer.

ACTIVITIES

Observing Projects

Observing tips and tools

At the risk of repeating ourselves, we remind you to **never look directly at the Sun, because it can easily cause permanent blindness.** You can view the Sun safely without a telescope just by using two pieces of white cardboard. First, use a pin to poke a small hole in one piece of cardboard; this will be your "lens," and the other piece of cardboard will be your "viewing screen." Hold the "lens" piece of cardboard so that it is face-on to the Sun and sunlight can pass through the hole. With your other hand, hold the "viewing screen" so that the sunlight from the "lens" falls on it. Adjust the distance between the two pieces of cardboard so that you see a sharp image of the Sun on the "viewing screen." This image is perfectly safe to view. It is actually possible to see sunspots with this low-tech apparatus.

For a better view, use a telescope with a solar filter that fits on the front of the telescope. A standard solar filter is a piece of glass coated with a thin layer of metal to give it a mirror-like appearance. This coating reflects almost all the sunlight that falls on it, so that only a tiny, safe amount of sunlight enters the telescope. An Hα filter, which looks like a red piece of glass, keeps the light at a safe level by admitting only a very narrow range of wavelengths. (Filters that fit on the back of the telescope are *not* recommended. The telescope focuses concentrated sunlight on such a filter, heating it and making it susceptible to cracking—and if the filter cracks when you are looking through it, your eye will be ruined instantly and permanently.)

To use a telescope with a solar filter, first aim the telescope away from the Sun, then put on the filter. Keep the lens cap on the telescope's secondary wide-angle "finder scope" (if it has one), because the heat of sunlight can fry the finder scope's optics. Next, aim the telescope toward the Sun, using the telescope's shadow to judge when you are pointed in the right direction. You can then safely look through the telescope's eyepiece. When you are done, make sure you point the telescope away from the Sun before removing the filter and storing the telescope.

Note that the amount of solar activity that you can see (sunspots, filaments, flares, prominences, and so on) will depend on where the Sun is in its 11-year sunspot cycle.

58. Use a telescope with a solar filter to observe the surface of the Sun. Do you see any sunspots? Sketch their appearance. Can you distinguish between the umbrae and penumbrae of the sunspots? Can you see limb darkening? Can you see any granulation?

59. If you have access to an Hα filter attached to a telescope especially designed for viewing the Sun safely, use this instrument to examine the solar surface. How does the appearance of the Sun differ from that in white light? What do sunspots look like in Hα Can you see any prominences? Can you see any filaments? Are the filaments in the Hα image near any sunspots seen in white light? (Note that the amount of activity that you see will be much greater at some times during the solar cycle than at others.)

60. Use the *Starry Night*™ program to examine simulations of various features that appear on the surface of the Sun. Select **Favourites > Explorations > Sun** to show a simulated view of the visible surface of the Sun as it might appear from a spacecraft. **Stop** time flow and use the **Location Scroller** to examine this surface. (**a**) Which layer of the Sun's atmosphere is shown in this part of the simulation? (**b**) List the different features that are visible in this view of the Sun's surface. (**c**) Click and hold the **Decrease current elevation** button in the toolbar to move to a location on the surface of the Sun, from which you can look out into the chromosphere. (The **Viewing Location** panel will indicate the location on the Sun's surface.) This simulated view of the chromosphere is at the color of the wavelength of hydrogen light. The opacity of the gas at this wavelength means that you can see the structure of the hot chromosphere that lies above the visible surface. Use the hand tool or cursor keys to change the gaze direction to view different features of the Sun, zooming in when necessary for a closer look at features on the horizon. (**d**) Provide a detailed description of the various features visible in this simulation of the Sun's surface. You can see current solar images from both ground and space-based solar telescopes by opening the **LiveSky** pane if you have an Internet connection on your computer.

61. Use *Starry Night*™ to measure the Sun's rotation. Select **Favourites > Explorations > Solar Rotation** to display the Sun as seen from about 0.008 AU above its surface, well inside the orbit of Mercury. Use the time controls to stop the Sun's rotation at a time when a line of longitude on the Sun makes a straight line between the solar poles, preferably a line crossing a recognizable solar feature. Note the date and time. **Run Time Forward** and adjust the date and time to place this selected meridian in this position again. (**a**) What is the rotation rate of the Sun as shown in *Starry Night*™? This demonstration does not show one important feature of the Sun, namely its differential rotation, where the equator of this fluid body rotates faster than the polar regions. (**b**) To which region of the Sun does your measured rotation rate refer? It is this differential rotation that is thought to generate the magnetic fields and active regions that make the Sun an active star. Occasional emission of high-energy particles from these active regions can disturb Earth's environment and disrupt electrical transmission systems. Examine this image of the Sun and compare it to real images seen in textbooks, the

Internet, or from the links in the **Solar Images** layer in the **LiveSky** pane. (c) In particular, how does the distribution of sunspots and active regions on this image compare to the distribution of these regions on the real Sun?

Collaborative Exercises

62. Figure 16-20 shows variations in the average latitude of sunspots. Estimate the average latitude of sunspots in the year you were born and estimate the average latitude on your twenty-first birthday. Make rough sketches of the Sun during those years to illustrate your answers.

63. Create a diagram showing a sketch of how limb darkening on the Sun would look different if the Sun had either a thicker or thinner photosphere. Be sure to include a caption explaining your diagram.

64. Solar granules, shown in Figure 16-9, are about 1000 km across. What city is about that distance away from where you are right now? What city is that distance from the birthplace of each group member?

65. Magnetic arches in the corona are shown in Figure 16-26a. How many Earths high are these arches, and how many Earths could fit inside one arch?

ANSWERS

ConceptChecks

ConceptCheck 16-1: The Sun emits most of its energy in the form of visible light.

ConceptCheck 16-2: At the extremely high temperatures and pressures existing in the Sun's core, hydrogen nuclei can move fast enough to overcome the repulsive force from their positive electric charges and fuse together.

ConceptCheck 16-3: When 1 kg of hydrogen combines to form helium, the vast majority of the mass ends up in helium atoms, with only 0.7% of the original mass converted into energy.

ConceptCheck 16-4: Astronomers use the current energy output of the Sun to estimate how fast the Sun is consuming its usable fuel (hydrogen). Then, the amount of hydrogen fuel remaining in the Sun's core indicates how much longer the Sun will shine.

ConceptCheck 16-5: Because pressure in the Sun's core is due to the downward pushing weight of the overlying mass of material, having less mass pressing down would result in a lower pressure at the core.

ConceptCheck 16-6: The energy transport process of conduction occurs when energy moves through a relatively dense material by hot material transferring its kinetic energy to cooler material through direct contact. The Sun's density is simply too low for conduction to be an important process in transferring energy from one part of the Sun to another part.

ConceptCheck 16-7: Only the temperature decreases with increasing distance from the Sun's central core.

ConceptCheck 16-8: As shown in Figure 16-5, sound waves do penetrate into the Sun, which allows helioseismologists to study the solar interior.

ConceptCheck 16-9: By carefully monitoring how sound waves move through the Sun, astronomers are able to deduce the amount of helium in the Sun's core and convective zone, and the thickness of various zones.

ConceptCheck 16-10: Kamiokande could only detect one of three possible types of neutrinos, and this was the type emitted by the Sun. However, along the way, two-thirds of these neutrinos transformed into the other types and were undetectable by Kamiokande. SNO, on the other hand, was able to detect all three types of neutrinos.

ConceptCheck 16-11: The energy transport process of convection causes warmer material to rise to the surface and cooler material to sink back into the photosphere, giving the photosphere a granule-like appearance.

ConceptCheck 16-12: Figure 16-11 is taken with an H_α filter that only allows red light to pass through and form the image. This method is used to block out the Sun's other light that would overwhelm the camera. Since the H_α line is red (at 656.3 nm), the spicules appear red, but they emit other wavelengths as well.

ConceptCheck 16-13: The corona has an extremely low density and can only be observed when the more dominant photosphere is blocked, such as during a solar eclipse.

ConceptCheck 16-14: The solar wind must escape through the corona, and this occurs in greater abundance through holes in the corona that contain less gas.

ConceptCheck 16-15: The photosphere surrounding a sunspot has a temperature of about 5800 K, which far outshines the relatively cooler sunspot region, resulting in the sunspots appearing to be quite dark in comparison.

ConceptCheck 16-16: The length of time between large numbers of sunspots to few numbers of sunspots and back to large numbers of sunspots averages about 11 years. However, the magnetic character of sunspots flips every 11-year cycle, suggesting an overarching 22-year sunspot cycle.

ConceptCheck 16-17: According to the magnetic-dynamo model, the sunspot cycle is a result of twisting magnetic fields. In the event the Sun was turning faster, the twisting would occur more quickly and the length of the sunspot cycle would decrease.

ConceptCheck 16-18: The glowing plasma tends to follow the pathways created by the Sun's invisible magnetic field lines, which form curved arches above the Sun's surface.

ConceptCheck 16-19: Coronal mass ejections are many times more energetic than any other event on the Sun and, when directed at Earth, can cause severe problems with telecommunications, electrical power distribution, and radiation health hazards for astronauts working in space.

CalculationChecks

CalculationCheck 16-1: According to Einstein's equation that $E = mc^2$, a mass of 5 kg is equivalent to 5 kg $\times$ $(3 \times 10^8 \text{ m/s})^2 = 45 \times 10^{16}$ joules, which is equivalent to burning about 100,000 metric tons of coal!

CalculationCheck 16-2: According to Figure 16-3, the Sun's core temperature is about 16 million Kelvins. At a distance of 50% of the Sun's radius, the Sun's temperature has dropped to 4 million Kelvins, which is a drop of about 75%.

An "alien" life-form on Earth: pink, eyeless worms in an underwater mound of solid methane.
(Courtesy of Okeanos Explorer Program/NOAA)

R I **V** U X G

The Search for Extraterrestrial Life

LEARNING GOALS

By reading the sections of this chapter, you will learn

27-1 How comets and meteorites could have aided the origin of life on Earth

27-2 On which other worlds of our solar system life might have evolved

27-3 About the controversy over proposed fossilized life from Mars

27-4 How astronomers estimate the number of other civilizations in the Galaxy

27-5 Why radio telescopes may be a useful tool for contacting alien civilizations

27-6 How astronomers hope to detect Earthlike planets orbiting other stars

One of the most compelling questions in science is also one of the simplest: Are we alone? That is, does life exist beyond Earth? As yet, we have no definitive answer to this question. None of our spacecraft has found life elsewhere in the solar system, and radio telescopes have yet to detect signals of intelligent origin coming from space. Claims of aliens visiting our planet and abducting humans make compelling science fiction, but none of these stories has ever been verified.

Yet there are reasons to suspect that life might indeed exist beyond Earth. One is that biologists find living organisms in some of the most "unearthly" environments on our planet. An example (shown on this page) is at the bottom of the Gulf of Mexico, where the crushing pressure and low temperature cause methane—normally a gas—to form solid, yellowish mounds. Amazingly, these mounds teem with colonies of pink, eyeless, alien-looking worms the size of your thumb. If life can flourish here, might it not also flourish in the seemingly hostile conditions found on other worlds?

In this chapter we will look for places in our solar system where life may once have originated, and where it may exist today. We will see how scientists estimate the chances of finding life beyond our solar system, and how they search for signals from other intelligent species. And we will learn how a new generation of telescopes may make it possible to detect the presence of even single-celled organisms on worlds many light-years away.

27-1 The chemical building blocks of life are found in space

Suppose you were the first visitor to a new and alien planet. If you saw a three-headed lizard running by, you would be sure it was an alien life-form. But how can you distinguish an alien microbe—or even just a fossil-like remnant of a microbe—from a dust grain? What might alien life look like? Questions such as these are central to **astrobiology**, the study of life in the universe. Most astrobiologists suspect that if we find living organisms on other worlds, they will be "life as we know it"—that is, their biochemistry will be based on the unique properties of the carbon atom, as is the case for all life on Earth.

Organic Molecules in the Universe

Why carbon? The reason is that carbon has the most versatile chemistry of any element. Carbon atoms can form chemical bonds to create especially long and complex molecules (Figure 27-1). These carbon-based compounds, called **organic molecules**, include all the molecules of which living organisms are made. (Silicon has some chemical similarities to carbon, and it can also form complex molecules. But as Figure 27-2 shows, complex silicon molecules do not have the right properties to make up complex systems such as living organisms.)

Organic molecules can be linked together to form elaborate structures, such as chains, lattices, and fibers. Some of these structures are capable of complex, self-regulating chemical reactions. Furthermore, the primary constituents of organic molecules—carbon, hydrogen, nitrogen, oxygen, sulfur, and phosphorus—are among the most abundant elements in the universe. The versatility and abundance of carbon suggest that extraterrestrial life is also likely to be based on organic chemistry.

If life is based on organic molecules, then these molecules must initially be present on a planet in order for life to arise from

$$Z - X - X - X - X - Y$$

(a)   Linear molecule

(b)   Glucose

FIGURE 27-1

Complex Molecules and Carbon **(a)** Atoms that can bond to only two other atoms, like the atoms denoted X shown here, can form a chain of atoms called a linear molecule. The chain stops where we introduce an atom, such as those labeled Y and Z, that can bond to only one other atom. **(b)** A carbon atom (denoted C) can bond with up to four other atoms. Hence, carbon atoms can form more complex, nonlinear molecules like glucose. All organic molecules that are found in living organisms have backbones of carbon atoms.

nonliving matter. We now understand that many carbon-based molecules originate from nonbiological processes in interstellar space. One such molecule is carbon monoxide (CO), which is made when a carbon atom and an oxygen atom collide and bond together. Carbon monoxide is found in abundance within giant interstellar clouds that lie along the spiral arms of our Milky Way Galaxy (see Figure 1-7) as well as in other galaxies (see Figure 1-9). Carbon atoms have also combined with other elements to produce an impressive array of interstellar organic molecules, including ethyl alcohol (CH_3CH_2OH), formaldehyde (H_2CO), methylcyanoacetylene (CH_3C_3N), and acetaldehyde (CH_3CHO). Radio astronomers

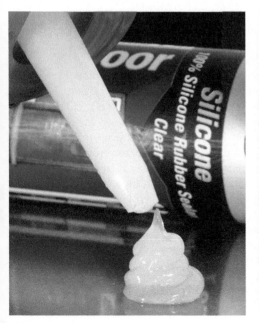

(a)

(b)

FIGURE 27-2 R I V U X G

Why Silicon Is Unsuitable for Making Living Organisms Like carbon, silicon atoms can bond with up to four other atoms. However, the resulting compounds are either too soft or too hard, or react too much or too little, to be suitable for use in living organisms. **(a)** Silicone has a backbone of silicon and oxygen atoms. The molecules form a gel or liquid rather than a solid, and react too slowly to undergo the rapid chemical changes required of molecules in organisms. **(b)** Molecules can also be made with a silicon-carbon-oxygen backbone, but the results (like this quartz crystal) are too rigid for use in organisms. (a: Richard Megna/Fundamental Photographs; b: Ispace/Shutterstock)

FIGURE 27-3   R I ☑ U X G

A Carbonaceous Chondrite Carbonaceous chondrites are primitive meteorites that date back to the very beginning of the solar system. This sample is a piece of the Allende meteorite, a large carbonaceous chondrite that fell in Mexico in 1969. Chemical analyses of newly fallen specimens disclose that they are rich in organic molecules, many of which are the chemical building blocks of life. (Detlev van Ravenswaay/Science Source)

have detected these molecules by looking for the telltale microwave emission lines of carbon-based chemicals in interstellar clouds.

The planets of our solar system formed out of interstellar material (see Section 8-5), and some of the organic molecules in that material must have ended up on the planets' surfaces. Evidence for this comes from meteorites called **carbonaceous chondrites,** like the one shown in Figure 27-3. These meteorites are ancient, date from the formation of the solar system, and are often found to contain a variety of carbon-based molecules. The Murchison meteorite, which fell on Australia in 1969, contains more than 70 amino acids, and these organic molecules are some of the building blocks of life.

The spectra of comets (see Section 7-5)—which are also among the oldest objects in the solar system—show that they, too, contain an assortment of organic compounds. In 2006, NASA's *Stardust* mission returned samples from the atmosphere (or coma) of Comet Wild 2. The samples were very limited, but one type of amino acid was found. In 2014, a probe from the European Space Agency is scheduled to land on a comet's surface for direct samples of its nucleus. Astrobiologists are hoping the lander finds a variety of amino acids, and the results will help them estimate how much of Earth's organic building blocks came from comets.

Comets and meteoroids were much more numerous in the early solar system than they are today, and they were correspondingly more likely to collide with a planet. These collisions would have seeded the planets with organic compounds from the very beginning of our solar system's history. Organic compounds are also found in interplanetary dust particles (see Section 15-5), which continually rain down on the planets. Once meteorites, comets, and interplanetary dust particles bring simple organic chemicals to a planet's surface, additional chemical reactions can produce an even wider range of the complex organic compounds needed for life. Similar processes are thought to take place in other planetary systems, which are thought to form in basically the same way as did our own (see Figure 8-13 and the image that opens Chapter 8).

CONCEPTCHECK 27-1

Why would life-forms throughout the cosmos likely be based on carbon?

CONCEPTCHECK 27-2

How can organic molecules end up on the surfaces of planets?

Answers appear at the end of the chapter.

The Miller-Urey Experiment

Comets and meteorites would not have been the only sources of organic material on the young planets of our solar system. In 1952, the American chemists Stanley Miller and Harold Urey demonstrated that under conditions that are thought to have prevailed on the young Earth, simple chemicals can combine to form the chemical building blocks of life. In a closed container, they prepared a sample of "atmosphere": a mixture of hydrogen (H_2), ammonia (NH_3), methane (CH_4), and water vapor (H_2O), the most common molecules in the solar system. Miller and Urey then exposed this mixture of gases to an electric arc (to simulate atmospheric lightning) for a week. At the end of this period, the inside of the container had become coated with a reddish-brown substance rich in amino acids and other compounds essential to life.

Since Miller and Urey's original experiment, most scientists have come to the conclusion that Earth's primordial atmosphere was composed of carbon dioxide (CO_2), nitrogen (N_2), and water vapor outgassed from volcanoes, along with some hydrogen. Modern versions of the Miller-Urey experiment (Figure 27-4) using these common gases have also succeeded in synthesizing a wide variety of organic compounds. The combination of comets and meteorites falling from space and chemical synthesis in

> Organic molecules do not have to come from living organisms—they can also be synthesized in nature from simple chemicals

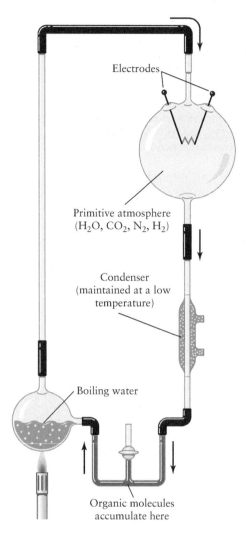

FIGURE 27-4

An Updated Miller-Urey Experiment Modern versions of this classic experiment prove that numerous organic compounds important to life can be synthesized from gases that were present in Earth's primordial atmosphere. This experiment supports the hypothesis that life on Earth arose as a result of ordinary chemical reactions.

the atmosphere could have made the chemical building blocks of life available in substantial quantities on the young Earth.

CAUTION! It is important to emphasize that scientists have *not* created life in a test tube. While organic molecules may have been available on the ancient Earth, biologists have yet to figure out how these molecules gathered themselves into cells and developed systems for self-replication. Nevertheless, because so many chemical components of life are so easily synthesized under conditions that simulate the primordial Earth, it is reasonable to suppose that life could have originated as the result of chemical processes. Furthermore, because the molecules that combine to form these compounds are rather common, it seems equally reasonable that life could have originated in the same way on other planets.

Organic building blocks are commonplace throughout the universe, but their abundance does not guarantee that life is equally commonplace. If a planet's environment is hostile, life may never get started or may quickly be extinguished. But we now have evidence that nearly Earth-sized planets orbit other stars (see Section 8-7) and that additional planetary systems are forming around young stars (see Section 8-4, especially Figure 8-8). It seems increasingly likely that Earthlike planets will be found orbiting other stars, and that conditions on some of these worlds may be suitable for life as we know it. The moons of some planets, in our solar system or associated with other stars, might also have environments suitable for life.

CONCEPTCHECK **27-3**

What did Miller and Urey create when they passed electricity through their sample of "atmosphere" containing a mixture of hydrogen (H_2), ammonia (NH_3), methane (CH_4), and water vapor (H_2O)?

Answer appears at the end of the chapter.

27-2 Water and the potential for life

If life evolved on Earth from nonliving organic molecules, might the same process have taken place elsewhere in our solar system? Scientists are carefully scrutinizing the planets and their moons in an attempt to answer this question.

The Importance of Liquid Water

One major constraint is that liquid water is essential for the survival of life as we know it. However, the water need not be pleasant by human standards—terrestrial organisms have been found in water that is boiling hot, fiercely acidic, or ice cold (see the image that opens this chapter)—but it must be liquid. Organisms living in such extreme conditions are called **extremophiles**, such as the thermophiles that thrive at high temperatures (**Figure 27-5**). Other extremophiles include microorganisms that are resistant to high levels of nuclear radiation, and those that live *within* the tiny porous spaces inside of rocks. All known life-forms require at least some liquid water, and even the extremophile microbes living within ice contain a form of antifreeze to keep their internal water liquid.

In order for water on a planet's surface to remain liquid, the temperature cannot be too hot or too cold. Furthermore, there must be a relatively thick atmosphere to provide enough pressure to keep liquid water from evaporating. Of all the worlds in our present-day solar system, only Earth has the right conditions for water to remain liquid on its surface.

The Oceans of Europa and Enceladus

There is now compelling evidence that Europa, one of the large satellites of Jupiter (see Table 7-2 and Figure 7-4), has an ocean of water *beneath* its icy surface. As it orbits Jupiter, Europa is caught in a tug-of-war between gravitational forces from Jupiter and Jupiter's other large satellites. These forces flex the interior of Europa, and this flexing generates enough heat to keep subsurface water from freezing. Images indicate that chunks of ice have

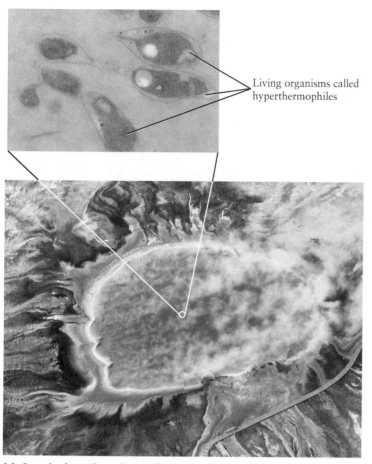

Living organisms called hyperthermophiles

(a) Grand prismatic spring, Yellowstone National Park.

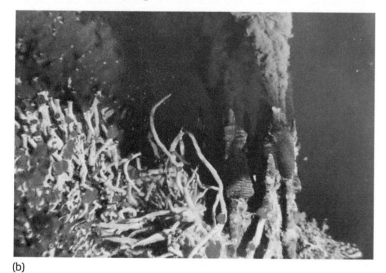

(b)

FIGURE 27-5   R I V U X G

Extremophiles Can Take the Heat **(a)** These microscopic thermophiles (heat-loving organisms) live in water that is between 80°C and 100°C (85°F–140°F). **(b)** Tube worms (light-green tubes) with hemoglobin-rich red plumes. They reside around black smokers—vents in the ocean bottom that are in the same temperature range as the hot springs shown in (a). These vents are more than 3 km (2 mi) under water. (a: Jim Peaco; July 2001 Yellowstone National Park image by NPS Photo; inset: NASA; b: Verena Tunnicliffe, University of Victoria, School of Earth & Ocean Sciences, Department of Biology)

floated around on the surface, and magnetic field measurements reveal that Europa has an underground ocean (see Section 13-6).

No one knows whether life exists in Europa's ocean. But interest in this exotic little world is great, and scientists have proposed several missions to explore Europa in more detail. Unfortunately, the ice layer is probably too thick (several kilometers) for a robotic probe to penetrate the ice and sample the underlying ocean anytime soon. However, just as microorganisms on Earth can survive within ice, perhaps life on Europa has spread beyond its ocean toward the surface.

A mission to search Europa's surface ice is still very challenging, as high levels of radiation on the atmosphere-free moon would likely destroy organisms within a few meters of the surface. A thermal probe designed to simply melt its way down a few meters also faces challenges because of the fine dust (regolith) that is expected to exist near the surface but does not melt. To search for life about 10 m down, scientists are testing a number of ideas, including a thermal probe with a drill tip to get through any patches of dust.

Saturn's moon Enceladus has revealed a large subsurface ocean through salty water spewing out in plumes from ice volcanoes (Figure 27-6 and Figure 13-29). The salts, which are composed of a variety of dissolved minerals, form when the water makes extensive contact with the rocky interior. This evidence points strongly toward a large interior ocean. Enceladus's ocean of liquid water with dissolved minerals, along with some organic material detected in its plumes, suggests that its ocean could make a suitable environment for life.

While Enceladus's ocean could be kilometers beneath the surface and difficult to explore, its ice volcanoes offer a location where microorganisms seeping up from the ocean might be ejected from the surface. In fact, after initially favoring Europa's deep ocean as the primary target in the search for extraterrestrial life,

FIGURE 27-6   R I V U X G

Subsurface Ocean on Enceladus Plumes containing water vapor and dissolved minerals indicate a subsurface ocean on Saturn's moon Enceladus. Organic compounds, in concentrations 20 times higher than expected, were also detected in the plumes. As on Earth, liquid water with dissolved minerals and organic compounds could be a suitable setting for life. (NASA/JPL/Space Science Institute)

many scientists now favor Enceladus. Without the challenges of drilling or melting through ice, as on Europa, a probe can pass right through the plumes of Enceladus's ice volcanoes and test samples for life.

Even if there is no life in the Enceladean ocean, the ice volcanoes themselves might harbor life. The ice volcanoes contain liquid water, organic compounds, and are about 200 K warmer than the surrounding terrain. Within these ice volcanoes, water, food, and energy could make a hospitable environment for life.

CONCEPTCHECK 27-4

Why do dissolved salts (or minerals) ejected from Enceladus suggest a large ocean beneath its surface?

Answer appears at the end of the chapter.

Life in Similar Environments on Earth

Here on Earth, there are environments somewhat similar to those expected in the subsurface oceans of Europa and Enceladus. By investigating these environments, we learn if life might be possible in such extremes. One such place is Lake Vida in Antarctica. Its liquid water is covered by 20 m of frozen ice. At that depth virtually no sunlight reaches the water. The water is also 6 times saltier than seawater, which is too salty for most organisms. However, the high concentration of salt also prevents the water from freezing, where it drops to temperatures of −13.5°C. Incredibly, in 2012, researchers found that bacteria are thriving in these subfreezing waters.

The most extreme liquid water environment known on Earth is Lake Vostok in Antarctica (Figure 27-7). This enormous subsurface lake is completely buried by a whopping 2.2 miles of glacial ice. With this lake so cut off from its surroundings, its water has probably been isolated for more than 15 million years! In 2012, Russian scientists very carefully drilled more than 2 miles into Lake Vostok, piercing its icy depths to sample the waters beneath. When water was finally reached, it rushed up and gushed out of the borehole. In 2013, pristine samples of the lake water were collected and are being investigated for signs of life. Thus, while astronomers look to the skies, they also keep one eye on Lake Vostok.

Searching for Life on Mars

Another possibility for the existence of life in our solar system is Mars. The present-day Martian atmosphere is so thin that water can exist only as ice or as a vapor. However, images made from Martian orbit show dried-up rivers, and the *Curiosity* rover has confirmed that one riverbed must have held water for thousands to millions of years (see Section 11-8 and the *Scientific American* article "Reading the Red Planet" at the end of Chapter 11). These features are evidence that the Martian atmosphere was once thicker and that liquid water once coursed over the planet's surface. Could life have evolved on Mars during its "wet" period? If so, could life—in the form of extremophile microorganisms—have survived as the Martian atmosphere thinned and the surface water either froze or evaporated? While *Curiosity* is designed to find suitable habitats where life may have existed, the rover is not equipped to test for the presence of life itself.

> Some areas on Mars were covered with liquid water for extended periods

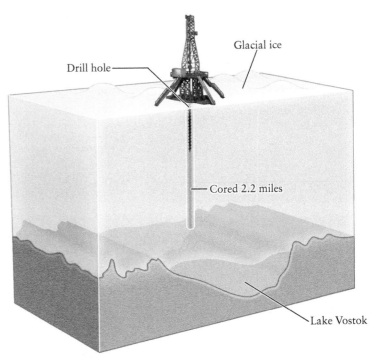

FIGURE 27-7

Exploring Lake Vostok Russian scientists search for life in an environment that might be similar to the oceans of Europa or Enceladus. The coldest temperature recorded on Earth was above Lake Vostok, at −89°C (−128°F). This is similar to typical Martian surface temperatures. However, the lake's liquid water below is much warmer at only about −3°C. This is encouraging because thick ice can act as insulation to keep in a planet or moon's internal heat. (Adapted from Nicolle Rager-Fuller/NSF)

Fortunately, some tests for Martian life have already been carried out, and future tests for life will likely involve similar experiments.

In 1976, the *Viking Lander 1* and *Viking Lander 2* landed on different parts of Mars. Each spacecraft carried a scoop at the end of a mechanical arm to retrieve surface samples (Figure 27-8). These samples were deposited into a compact on-board biological laboratory that carried out three different tests for Martian microorganisms.

1. The *gas-exchange experiment* was designed to detect gases released due to chemical processes involving nutrients. For example, to extract energy from food, many organisms produce carbon dioxide (which is why we exhale this gas in our breath). The experiment involved a surface sample that was placed in a sealed container along with a controlled amount of gas and nutrient-rich water. The gases in the container were then monitored to check for any microbial *exhalation*.

2. The *labeled-release experiment* was designed to detect metabolic processes. A sample was moistened with nutrients containing low levels of radioactive carbon atoms. If any organisms in the sample consumed the nutrients, their waste products should include gases containing the easily detectable radioactive carbon.

3. The *pyrolytic-release experiment* was designed to detect photosynthesis, the biological process by which terrestrial plants use solar energy to help synthesize organic compounds from carbon dioxide. A surface sample was placed in a container along with slightly radioactive carbon dioxide and exposed

FIGURE 27-8 R I V̄ U X G

Digging in the Martian Surface This view from the *Viking Lander 1* spacecraft shows the mechanical arm with its small scoop against the backdrop of the Martian terrain. The scoop was able to dig about 30 cm (12 in.) beneath the surface. (NASA)

to artificial sunlight. If plantlike photosynthesis occurred, microorganisms in the sample would take in some of the radioactive carbon from the gas.

The first data returned from these experiments caused great excitement, for in almost every case, rapid and extensive changes were detected inside the sealed containers. Further analysis of the data, however, led to the conclusion that these changes were due solely to nonbiological chemical processes. It appears that the Martian surface is rich in unstable chemicals that react with water to release oxygen gas. Because the present-day surface of Mars is bone-dry, these chemicals had nothing to react with until they were placed inside the moist interior of the *Viking Lander* laboratory.

At best, the results from the *Viking Lander* biological experiments were inconclusive. Perhaps life never existed on Mars at all. Or perhaps it did originate there, but failed to survive the thinning of the Martian atmosphere, the unstable chemistry of the planet's surface, and exposure to ultraviolet radiation from the Sun. (Unlike Earth, Mars has no ozone layer to block ultraviolet rays.) Another possibility is that Martian microorganisms have survived only in certain locations that the *Viking Landers* did not sample, such as isolated spots on the surface or beneath the ground. And yet another option is that there is life on Mars, but the experimental apparatus on board the *Viking Lander* spacecraft was not sophisticated enough to detect it.

The Power of Measuring Carbon Isotopes

An entirely different set of biological experiments were designed for the British spacecraft *Beagle 2*, which landed on Mars in December 2003. To test for life on Mars, *Beagle 2* was to sample an ancient lakebed region potentially containing the carbon-rich molecules of organisms–*dead or alive.* Then, signs of past or present life might be found by measuring how many of the organic molecules contain the isotope ^{12}C, which appears preferentially in biological molecules, and how many contain ^{13}C (which does not). (See Box 5-5 for a description of isotopes.) Thus, not only can ^{12}C isotopes indicate the existence of life, but this test works even when the organisms have died long ago.

Sadly, scientists on Earth were unable to establish contact with *Beagle 2* after its descent through the Martian atmosphere. However, the *Curiosity* rover presently exploring Mars can also measure the abundance of carbon isotopes and will search for areas of excess ^{12}C. One experiment hopes to examine the carbon in methane gas (CH_4), previously discovered in the Martian atmosphere. Microorganisms on Earth can gain energy by converting carbon dioxide to methane, and if Martian microorganisms do the same, it could explain the unexpected presence of Martian methane. Once it appears, methane rapidly decomposes in the Martian atmosphere, so we know that Mars presently has a source of methane. What we do not know is if this source of methane is geochemical or biological, and *Curiosity* can potentially answer this question by finding an excess of ^{12}C. As of this writing, *Curiosity* has yet to detect Martian methane.

CONCEPTCHECK 27-5

By measuring the abundance of the carbon isotope ^{12}C, how can *Curiosity* gather clues that life once existed on Mars?

Answer appears at the end of the chapter.

The Martian Civilization That Never Was

In 1976, while the *Viking Landers* were carrying out their biological experiments on the Martian surface, the companion *Viking Orbiter* spacecraft photographed some surface features that at first glance seemed to have been crafted by *intelligent* life on Mars. The *Viking Orbiter 1* image in Figure 27-9a shows what appears to be a humanlike face, perhaps the product of an advanced and artistic civilization. However, when the more advanced *Mars Global Surveyor* spacecraft viewed the surface in 1998 using a superior camera (Figure 27-9b), it found no evidence for facial features.

Scientists are universally convinced that the "face" and other apparent patterns in the *Viking Orbiter* images were created by shadows on windblown hills. Microscopic life may once have existed on Mars, and may yet exist today, but there is no evidence that the red planet has ever been the home of intelligent beings.

27-3 Meteorites from Mars have been scrutinized for life-forms

While spacecraft can carry biological experiments to other worlds such as Mars, many astrobiologists look forward to the day when a spacecraft will return Martian samples to laboratories on Earth. Until that day arrives, we have the next best thing: More than a hundred meteorites that appear to have formed on Mars have been found at a variety of locations on Earth.

The feature that identifies meteorites as having come from Mars is the chemical composition of trace amounts of gas trapped within them. This composition is very different from that of Earth's atmosphere, but is a nearly perfect match to the composition of the Martian atmosphere found by the *Viking Landers*.

How could a rock have traveled from Mars to Earth? When an asteroid collides with a planet's surface and forms an impact crater, most of the material thrown upward by the impact falls back onto the planet's surface. But some extraordinarily powerful

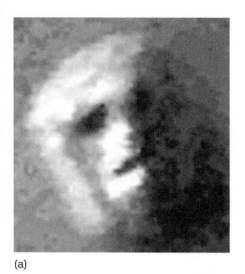

(a)

(b)

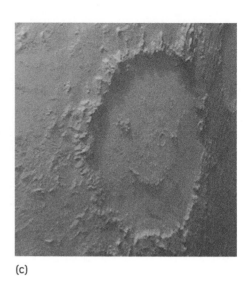

(c)

FIGURE 27-9 R I V U X G

A "Face" on Mars? **(a)** This 1976 image from *Viking Orbiter 1* shows a Martian surface feature that resembles a human face. Some suggested that this feature might have been made by intelligent beings. **(b)** This 1998 *Mars Global Surveyor* (MGS) image, made under different lighting conditions with a far superior camera, reveals the "face" to be just an eroded hill. **(c)** This MGS image shows features of natural origin within a 215-km (134-mi) wide crater on Mars. Can you see this "face"? (a: NSSDC/NASA and Dr. Michael H. Carr; b, c: Malin Space Science Systems/NASA)

impacts have produced large craters on Mars—roughly 100 km in diameter or larger. These tremendous impacts eject some rocks with such speed that they escape the planet's gravitational attraction and fly off into space.

There are numerous large craters on Mars, so a good number of Martian rocks have probably been blasted into space over the planet's history. These ejected rocks then go into elliptical orbits around the Sun. Many such rocks will have orbits that put them on a collision course with Earth, and these are the ones that scientists find as meteorites from Mars. In fact, estimates are that several tons of Martian rocks land on Earth *each year* (equaling around a cubic meter of rock).

Using the radioactive age-dating technique (see Section 8-3), scientists find that most meteorites from Mars are between 200 million and 1.3 billion years old, much younger than the 4.56-billion-year age of the solar system. But one meteorite from Mars, denoted by the serial number ALH 84001 and found in Antarctica in 1984, was found to be 4.5 billion years old (Figure 27-10a). Thus, ALH 84001 is a truly ancient piece of Mars. Further analysis of its radioactivity suggests that ALH 84001 was ejected from Mars by an impact about 16 million years ago and landed in Antarctica a mere 13,000 years ago.

ALH 84001 was on Mars during the era when liquid water existed on the planet's surface. Scientists have therefore investigated this rock carefully for clues about Martian water and possible life. One such clue is the presence of rounded grains of minerals called *carbonates*. Analysis of these carbonates indicates that they formed in liquid water (at a comfortable temperature of about 64°F).

In 1996, David McKay and Everett Gibson of the NASA Johnson Space Center, along with several collaborators, began reporting the results from studies of the carbonate grains in ALH 84001. They provided several pieces of evidence that life may have once existed within the rock's cracks while it was still on Mars:

1. There are large numbers of elongated, tubelike structures around the carbonate grains that could be fossilized microorganisms (Figure 27-10b).

2. Microscopic cracks contain organic molecules—consistent with what is expected from the decay of microorganisms.

3. The carbonate grains contain small magnetic crystals called magnetite, which are similar to magnetite produced by bacteria on Earth.

Are McKay and Gibson's conclusions correct? Their claims of ancient life on Mars are extraordinary, and they require extraordinary proof.

> Claims that scientists have found Martian microorganisms are intriguing but very controversial

Further studies found that the organic compounds detected are not a very good match for decayed organisms. Also, it was later found that the tubelike structures and magnetite crystals found in ALH 84001, while consistent with life, could have been formed through geological processes on Mars. Thus, while some of the evidence is consistent with Martian life, it is far from compelling.

CONCEPTCHECK 27-6

If bacteria on Earth are known to create magnetite crystals like those from Mars, why isn't this proof of life on Mars?

Answer appears at the end of the chapter.

Possible Microbial Tunnels

The most recent indications of possible Martian microbial remnants involve microscopic tunnels found in meteorites from Mars. In 2006, Martin Fisk from Oregon State University led a team that discovered small tunnels in a Martian meteorite named Nakhla. These tunnels are similar to structures found in rocks on Earth that are

(a)

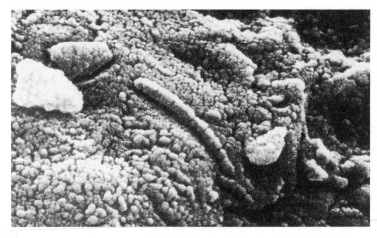

(b)

FIGURE 27-10   R I V U X G

A Meteorite from Mars **(a)** This 1.9-kg meteorite, known as ALH 84001, formed on Mars some 4.5 billion years ago. About 16 million years ago a massive impact blasted it into space, where it drifted in orbit around the Sun until landing in Antarctica 13,000 years ago. The small cube at lower right is 1 cm (0.4 in.) across. **(b)** This electron microscope image, magnified some

100,000 times, shows tubular structures about 100 nanometers (10^{-7} m) in length found within the Martian meteorite ALH 84001. One controversial interpretation is that these are the fossils of microorganisms that lived on Mars billions of years ago. (a: NASA Johnson Space Center; b: *Science,* NASA)

thought to be produced by rock-eating microbes. The Nakhla meteorite is not the only one with tunnels; they are also found in ALH 84001, and a 2013 study at the Johnson Space Center led by Lauren White finds them in the Martian meteorite Y000593 (Figure 27-11).

Could the tunnels have been formed after the meteorites hit Earth? That is possible for ALH 84001, and Y000593, which sat on Earth for thousands of years before being collected. But the Nakhla meteorite was observed falling on June 28, 1911, and its debris was collected too quickly for microbes on Earth to have created its tunnels. Are the microtunnels produced by Martian microbes? As in the case of the tubelike structures and magnetite crystals from ALH 84001, features that *appear* biological in origin might simply

occur naturally through Martian geology. Thus, the Martian tunnels lead us back to Earth, where we must determine if these tunnels can instead be created through geological processes. For now, the origin of these tunnels is uncertain. Though they only provide weak evidence at this early stage of analysis, these microscopic tunnels open up a whole new line of inquiry in the search for Martian life.

CONCEPTCHECK 27-7

Why are scientists convinced that these meteorites actually came from Mars and not from our Moon?

Answer appears at the end of the chapter.

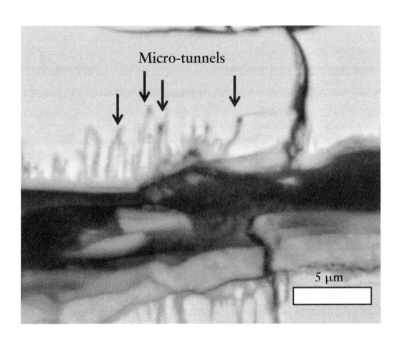

Micro-tunnels

5 μm

FIGURE 27-11   R I V U X G

Microscopic Tunnels in Martian Meteorite This figure shows microtunnels in the Martian meteorite Y000593. Although not common, very similar microscopic tunnels are found in rocks on Earth that contain DNA and are thought to be produced by rock-eating microbes. Even though Y000593 sat on Earth for thousands of years before being collected, the tunnels themselves were likely created on Mars: the tunnels in Y000593 are filled with Martian clay as determined by an analysis of the clay's composition. The microtunnels emanate from larger fractures in the rock and are about 1–5 micrometers long. (Lauren White)

27-4 The Drake equation helps scientists estimate how many civilizations may inhabit our Galaxy

TUTORIAL 27-1 We have seen that only a few locations in our solar system may have been suitable for the origin of

> Are we alone, or does the Galaxy teem with intelligent life? Or is the truth somewhere in between?

life. But what about planets and moons around other stars? The existence of life on Earth seems to suggest that extraterrestrial life, possibly including intelligent species, might evolve on terrestrial planets around other stars, given sufficient time and hospitable conditions. How can we learn whether such worlds exist, given the tremendous distances that separate us from them? This is the great challenge facing the **search for extraterrestrial intelligence, or SETI.**

Close Encounters Versus Remote Communication

A common belief is that alien civilizations do exist and that their spacecraft have visited Earth. Indeed, surveys show that between one-third and one-half of all Americans believe in unidentified flying objects (UFOs) of alien origin. A somewhat smaller percentage believes that aliens have landed on Earth. But, in fact, there is absolutely *no* scientifically verifiable evidence of alien visitations. As an example, many UFO proponents believe that the U.S. government is hiding evidence of an alien spacecraft that crashed near Roswell, New Mexico, in 1947. However, the bits of "spacecraft wreckage" found near Roswell turned out to be nothing more than remnants of an unmanned research balloon. Atomic isotopes are one way to tell if a material comes from Earth or not, because we know that the abundances of many isotopes vary in our solar system and beyond, yet no alien spacecraft debris with a non-Earth isotope signature has been found.

While there is no evidence of alien visits to Earth, alien civilizations could exist around some of the billions of stars in our Galaxy. To find real evidence for intelligent civilizations on other worlds, we must look to the stars.

With our present technology, sending even a small unmanned spacecraft to another star requires a flight time of tens of thousands of years. Speculative design studies have been made for unmanned probes that could reach other stars within a century or less, but these probes are prohibitively expensive. Instead, many astronomers hope to discover extraterrestrial civilizations by detecting radio transmissions from them. Radio waves are a logical choice for interstellar communication because they can travel immense distances without being significantly degraded by the interstellar medium, the thin gas and dust found between the stars (see Section 8-1).

Over the past several decades, astronomers have proposed various ways to search for alien radio transmissions, and several searches have been undertaken. In 1960, Frank Drake first used a radio telescope at the National Radio Astronomy Observatory in West Virginia to listen to two Sunlike stars, Tau Ceti and Epsilon Eridani, without success. Since then, many SETI searches have taken place using radio telescopes around the world. Occasionally,

a search has detected an unusual or powerful signal. But none has ever repeated, as a signal of intelligent origin might be expected to do. To date, we have no confirmed evidence of radio transmissions from another world.

CONCEPTCHECK 27-8

Why might the use of radio waves for exploration for life in the Galaxy be more fruitful than using unmanned interstellar spaceships?

Answer appears at the end of the chapter.

Is There Anybody Out There?

Should we be discouraged by this failure to make contact? What are the chances that a radio astronomer might someday detect radio signals from an extraterrestrial civilization? The first person to tackle this issue was Frank Drake, who proposed that the number of technologically advanced civilizations in the Galaxy could be estimated by a simple equation. This is now called the **Drake equation:**

Drake equation

$$N = R_* \, f_p \, n_e \, f_l \, f_i \, f_c \, L$$

N = number of technologically advanced civilizations in the Galaxy whose messages we might be able to detect

R_* = the rate at which solar-type stars form in the Galaxy

f_p = the fraction of stars that have planets

n_e = the number of planets per solar system that are Earthlike (that is, suitable for life)

f_l = the fraction of those Earthlike planets on which life actually arises

f_i = the fraction of those life-forms that evolve into intelligent species

f_c = the fraction of those species that develop adequate technology and then choose to send messages out into space

L = the lifetime of a technologically advanced civilization

The Drake equation is enlightening because it expresses the number of extraterrestrial civilizations in a simple series of terms. We can estimate some of these terms from what we know about stars, stellar evolution, and planetary orbits. The *Cosmic Connections* figure depicts some of these considerations.

For example, the first two factors, R_* and f_p, can be determined by observation. In estimating R_*, we should probably exclude stars with masses greater than about 1.5 times that of the Sun. These more massive stars use up the hydrogen in their cores in 3 billion (3×10^9) years or less. On Earth, by contrast, human intelligence developed only within the last million years or so, some 4.56 billion years after the formation of the solar system. If that is typical of the time needed to evolve higher life-forms, then a star of 1.5 solar masses or more probably fades away or explodes into a supernova before intelligent creatures like us can evolve on any of that star's planets.

COSMIC CONNECTIONS

Habitable Zones for Life

Intelligent civilizations in our Milky Way Galaxy can evolve only in a certain region called the galactic habitable zone. In that zone, a suitable planet must lie within the planetary habitable zone of its parent star. (After C. H. Lineweaver, Y. Fenner, and B. K. Gibson)

Galactic habitable zone

Too close to the center of the Milky Way Galaxy:
• The distances between stars are small, so there can be close encounters between stars that would disrupt a planetary system.
• There are also frequent outbursts of potentially lethal radiation from supernovae and from the supermassive black hole at the very center of the Galaxy.

Too far from the center of the Milky Way Galaxy:
• Stars are deficient in elements heavier than hydrogen and helium, so they lack both the materials needed to form Earthlike planets and the chemical substances required for life as we know it.

Planetary habitable zone

Inside the galactic habitable zone:
• The star must have a mass that is neither too large nor too small.
• If the star's mass is too large, it will use up its hydrogen fuel so rapidly that it will burn out before life can evolve on any of its planets.
• If the star's mass is too small, it will be too dim to provide the warmth needed for life.

The Neighborhood:
• There needs to be one or more large Jovian planets whose gravitational forces will clear away comets and meteors.

The Planet:
• Must be a terrestrial planet with a solid surface.
• Must have enough mass to provide the gravity needed to retain an atmosphere and oceans.
• Must be at a comfortable distance from the star so that the temperatures are neither too high nor too low.
• Must be in a stable, nearly circular orbit. (A highly elliptical orbit would cause excessively large temperature swings as the planet moved toward and away from the star.)

Although stars less massive than the Sun have much longer life-times, they also seem unsuited for life because they are so dim. Only planets very near a low-mass star would be sufficiently warm for life as we know it, and a planet that close is subject to strong tidal forces from its star. We saw in Section 4-8 how Earth's tidal forces keep the Moon locked in synchronous rotation, with one face continually facing Earth. In the same way, a planet that orbits too close to its star would have one hemisphere that always faced the star, while the other hemisphere would be in perpetual, frigid darkness.

These considerations leave us with stars not too different from the Sun. (Like Goldilocks sampling the three bears' porridge, we must have a star that is not too hot and not too cold but just right.) Based on statistical studies of star formation in the Milky Way, some astronomers estimate that roughly one of these Sunlike stars forms in the Galaxy each year in the galactic **habitable zone** (see the *Cosmic Connections* figure). This result sets R_* at 1 per year.

As we saw in Sections 8-4 and 8-5, the planets in our solar system formed as a natural consequence of the birth of the Sun. We have also seen evidence suggesting that planetary formation may be commonplace around single stars (see Figure 8-8). Many astronomers suspect that most Sunlike stars probably have planets, and so they give f_p a value of 1.

Unfortunately, the rest of the terms in the Drake equation are very uncertain. Let's play with some hypothetical values. The chances that a planetary system has an Earthlike world suitable for life are not known. (But observations of planets around other stars made by the *Kepler* spacecraft will help determine this number in the coming years.) Were we to consider our own solar system as representative, we could put n_e at 1. Let's be more conservative, however, and suppose that 1 in 10 solar-type stars is orbited by a habitable planet, making $n_e = 0.1$.

What about the fraction of Earthlike planets that develop life? Given the existence of life on Earth, we might assume that, given appropriate conditions, the development of life is quite likely, and we could set $f_l = 1$. However, it is also possible that life on Earthlike planets is rare. With only one Earthlike planet (our own Earth) to observe, there is no way to deduce the probability of Earthlike planets in general to form life. Furthermore, biochemists do not understand the detailed steps of life's origins and therefore cannot predict the probability for life to form. This is a topic of intense interest to astrobiologists, but we will be optimistic and assume here that $f_l = 1$.

For the sake of argument, we might also assume that evolution might naturally lead to the development of intelligence (a conjecture that is hotly debated) and also make $f_i = 1$. It is anyone's guess as to whether these intelligent extraterrestrial beings would attempt communication with other civilizations in the Galaxy, but were we to assume that they would, f_c would also be put at 1.

The last variable, L, involving the longevity of an advanced civilization, might be the most uncertain of all. Looking at our own example, we see a planet whose atmosphere and oceans are increasingly polluted by creatures that possess nuclear weapons. If we are typical, perhaps L is as short as 100 years. Putting all these numbers together, we arrive at

$$N = 1/\text{year} \times 1 \times 0.1 \times 1 \times 1 \times 1 \times 100 \text{ years} = 10$$

In other words, out of the hundreds of billions of stars in the Galaxy, we would estimate that there are only 10 tech-nologically advanced civilizations from which we might receive communications.

A wide range of values has been proposed for the terms in the Drake equation, and these various guesses produce vastly different estimates of N. Some scientists argue that there is exactly one advanced civilization in the Galaxy and that we are it. Others speculate that there may be hundreds or thousands of planets inhabited by intelligent creatures. If we wish to know whether our Galaxy is devoid of other intelligence, teeming with civilizations, or something in between, we must keep searching the skies.

CONCEPTCHECK **27-9**

What makes the longevity of an advanced civilization, L, so difficult to estimate?

Answer appears at the end of the chapter.

27-5 Radio searches for alien civilizations are underway

Even if only a few alien civilizations are scattered across the Galaxy, we have the technology to detect radio transmissions from them. On the one hand, the signals would be too weak to detect if they were similar in strength to the radio waves we use for communication and radio stations here on Earth. Waves like this are like ripples on a pond that diminish as they spread outward. On the other hand, if an advanced civilization wanted to make its presence known, it *might* sweep the skies with a focused beam of radio waves, which is much easier to detect. Even if we could not decode such a signal, it should be easy to tell that it results from an advanced technology.

But if other civilizations are trying to communicate with us using radio waves, what frequency are they using? This is an important question, because if we fail to tune our radio telescopes to the right frequency, we might never know whether the aliens are out there.

A reasonable choice would be a frequency that is fairly free of interference from extraneous sources. SETI pioneer Bernard Oliver was the first to draw attention to a range of relatively noise-free frequencies in the neighborhood of the microwave emission lines of hydrogen (H) and hydroxide (OH) (Figure 27-12). This region of the spectrum can be called microwave or radio emission and is called the **water hole,** because H and OH together make H_2O, or water.

In 1989, NASA began work on the High Resolution Microwave Survey (HRMS), an ambitious project to scan the entire sky at frequencies spanning the water hole from 10^3 to 10^4 MHz. HRMS would have observed more than 800 nearby solar-type stars, but, unfortunately, its funding was cut just one year after becoming operational.

Even though NASA no longer funds searches for extraterrestrial intelligence, several teams of scientists remain actively involved in SETI programs. Funding for these projects has come from nongovernmental organizations such as the Planetary Society and from private individuals. In 1995 the SETI Institute in California began Project Phoenix, the direct successor to HRMS. In 2004, after careful observation of the nearest 800 solar-type stars, scientists for Project Phoenix announced that they had not found any promising signals.

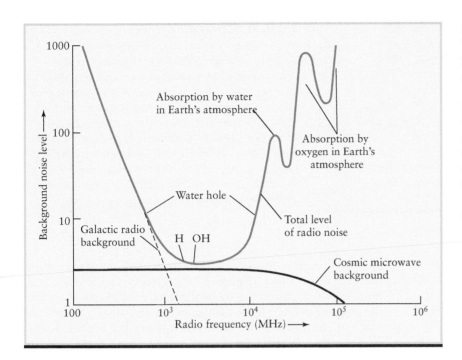

FIGURE 27-12

The Water Hole This graph shows the background noise level from the sky at various radio and microwave frequencies. The so-called water hole is a range of radio frequencies from about 10^3 to 10^4 megahertz (MHz) in which there is little noise and little absorption by Earth's atmosphere. Some scientists suggest that this noise-free region would be well suited for interstellar communication. Within the water hole itself, the principal source of noise is the afterglow of the Big Bang, called the cosmic microwave background. To put this graph in perspective, a frequency of 100 MHz corresponds to "100" on a FM radio, and 10^3 MHz is a frequency used for various types of radar. (Adapted from C. Sagan and F. Drake)

A major challenge facing SETI is the tremendous amount of computer time needed to analyze the mountains of data returned by radio searches. To this end, scientists at the University of California, Berkeley, have recruited more than 5 million personal computer users to participate in a project called SETI@home. Each user receives actual data from a detector called SERENDIP IV (Search for *Extraterrestrial Radio Emissions from Nearby, Developed, Intelligent Populations*) and a data analysis program that also acts as a screensaver. When the computer's screensaver is on, the program runs, the data are analyzed, and the results are reported via the Internet to the researchers at Berkeley. The program then downloads new data to be analyzed. Since 1999, SETI@home users have provided the equivalent of more than 3 million years of computing time!

In 2007, the SETI Institute began operation of a large radio telescope dedicated to the search for intelligent signals. This telescope, called the Allen Telescope Array (ATA), is actually hundreds of relatively small and inexpensive radio dishes working together. Designed for rapid and sensitive searches, the ATA hopes to search the nearest 100,000 solar-type stars. If astronomers knew which stars had Earthlike planets, SETI's telescopes could focus their attention on these objects, and the effort to find these stars is underway.

> Millions of personal computers have helped scan for alien radio transmissions

CONCEPTCHECK 27-10

Why should the range of frequencies in the water hole of Figure 27-12 be an attractive range of frequencies for an alien civilization to send out messages?

Answer appears at the end of the chapter.

27-6 Telescopes have begun searching for Earthlike planets

Although no longer involved in SETI, NASA is carrying out a major effort to search for Earthlike planets suitable for the evolution of an advanced civilization. Such a search poses a major challenge. One problem is that Earth-sized planets are too dim to be seen in visible light against the glare of their parent star. Using a different approach, astronomers have discovered many Jupiter-sized planets by detecting the "wobble" that these planets produce in their parent star (see Section 8-7). But a planet the size of Earth exerts only a weak gravitational force on its parent star. In 2009, the first Earth-sized planet (at 1.9 Earth masses) was discovered through a tiny wobble. However, this planet, named Gliese 581 e, orbits too close to its parent star to allow for liquid water. Furthermore, the wobble is so tiny that few Earth-sized planets are expected to be found using this technique.

Searching for Planetary Transits

An alternative technique was being used by an orbiting telescope called *Kepler,* which began detecting planets in 2009, and discovered hundreds of Earth-sized planets. Recall from Section 8-7, that if a star is orbited by a planet whose orbital plane is oriented edge-on to our line of sight, once per orbit the planet will pass in front of the star in an event called a *transit* (Figure 8-19). By blocking some of the star's light, this transit causes a temporary dimming of the light we see from that star.

Once a transiting planet is detected, astronomers determine the transiting planet's size (inferred from how much dimming takes place), as well as the size of its orbit. Recall from Johannes Kepler's third law that the size of the orbit can be calculated using the orbital period of the planet, which is the same as the time interval

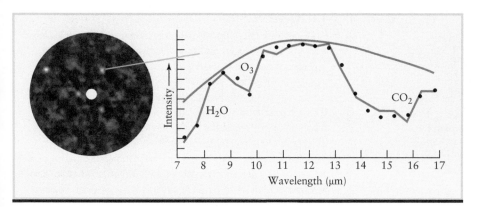

FIGURE 27-13

The Infrared Spectrum of a Simulated Planet The image on the left is a simulation of what a space-based infrared telescope might see once one is launched. The white dot at the center is a nearby Sunlike star, and the smaller dots around it are planets orbiting the star. On the right is the simulated infrared spectrum of one of the planets, showing broad absorption lines of water vapor (H_2O), ozone (O_3), and carbon dioxide (CO_2). Measuring the abundances of these gases can tell us about the planet's conditions and if it might be harboring oxygen-producing life. (Jet Propulsion Laboratory/NASA)

between successive transits. Given the distance from the planet to its parent star and the star's luminosity, astronomers will even be able to estimate the planet's average temperature. However, this estimated temperature would not include the warming effects of an atmosphere without additional observations.

To determine a planet's atmosphere, detailed spectra are required and these observations are only just beginning. It is important to keep in mind that just because a planet is Earth-sized and at a certain distance from its sun, it does not mean it is necessarily Earthlike. Finding a truly Earthlike planet with liquid water is difficult but could be discovered with the current generation of telescopes, or the next generation designed specifically for this goal.

CONCEPTCHECK 27-11

What exactly was the *Kepler* telescope watching for as it searched for planets around stars?

CONCEPTCHECK 27-12

Which planets were the *Kepler* telescope not able to find, even if quite nearby?

Answers appear at the end of the chapter.

Searches Using Images and Spectra

The results from *Kepler* may help SETI and other astronomers select stars to study in more detail. But what comes next? NASA and the European Space Agency have studied how the next generation of telescopes might continue investigating Earthlike planets. These space agencies have proposed the Terrestrial Planet Finder (a system of telescopes) and *Darwin* (a spacecraft mission), but there is no funding for these missions as of this writing.

Studies suggest that a future telescope might search for Earthlike planets by detecting their infrared radiation. The rationale is that stars like the Sun emit much less infrared radiation than visible light, while planets are relatively strong emitters of infrared. Hence, observing in the infrared makes it less difficult (although still technically challenging) to detect planets orbiting a star.

This next generation of telescopes might also be able to measure a "biosignature" of life in a planet's atmosphere. By analyzing the infrared spectra of potentially Earthlike planets, the telescopes could reveal the atmospheric composition from the unique absorption features of gases such as ozone, carbon dioxide, and water vapor (Figure 27-13). Ozone (O_3) is made from ultraviolet sunlight acting on oxygen (O_2), which could indicate the presence of life. This is significant not because oxygen *sustains* many life-forms on Earth (although it does), but because Earth's oxygen-rich atmosphere was *produced* by living organisms, and does not seem attainable through nonbiological means. Therefore, oxygen is a key atmospheric biosignature that will be the target of future studies.

Sometime during the twenty-first century, we will probably answer the question "Are there worlds like Earth orbiting other stars?" If the answer is yes, then searches for life around other stars will gain even more impetus.

> Future telescope technology may make it possible to resolve details on planets orbiting other stars

The potential rewards from such searches are great. Detecting a message from an alien civilization could dramatically change the course of our own civilization, through the sharing of scientific information with another species or an awakening of social or humanistic enlightenment. In only a few years our technology, industry, and social structure might advance the equivalent of centuries into the future. Such changes would touch every person on Earth. Mindful of these profound implications, scientists push ahead with the search for extraterrestrial intelligence.

CONCEPTCHECK 27-13

How does ozone act as a biosignature for life?

CONCEPTCHECK 27-14

Why not look for oxygen gas (O_2) directly? Consult Figure 27-13.

Answers appear at the end of a chapter.

KEY WORDS

KEY IDEAS

Organic Molecules in the Universe: All life on Earth, and presumably on other worlds, depends on organic (carbon-based) molecules. These molecules occur naturally throughout interstellar space.

• The organic molecules needed for life to originate were probably brought to the young Earth by comets, meteorites, and interplanetary dust particles. Another likely source for organic molecules is chemical reactions in Earth's primitive atmosphere. Similar processes may occur on other worlds.

Life in the Solar System: Besides Earth, other worlds in our solar system—the planet Mars, Jupiter's satellite Europa, and Saturn's satellite Enceladus—may have had the right conditions for the origin of life.

• Europa appears to have extensive liquid water beneath its icy surface. Future missions may search for the presence of life there. Enceladus also appears to contain a large subsurface ocean, and some of this material is ejected through ice volcanoes for easier sampling.

• Mars once had liquid water on its surface, though it has none today. Life may have originated on Mars during the liquid water era.

• The *Viking Lander* spacecraft searched for microorganisms on the Martian surface, but found no conclusive sign of their presence. The Mars Science Laboratory is designed to assess the past and present suitability of Mars for microbial life.

• An ancient Martian rock that came to Earth as a meteorite shows features that have been suggested as remnants of Martian life. Most of these features can be explained by geologic processes.

Radio Searches for Extraterrestrial Intelligence: Astronomers have carried out a number of searches for radio signals from other stars. No signs of intelligent life have yet been detected, but searches are continuing and use increasingly sophisticated techniques.

• The Drake equation is a tool for estimating the number of intelligent, communicative civilizations in our Galaxy.

Telescope Searches for Earthlike Planets: A future generation of orbiting telescopes may be able to detect numerous terrestrial planets around nearby stars. If such planets are found, their infrared spectra may reveal the presence or absence of life.

QUESTIONS

Review Questions

1. Why are extreme life-forms on Earth, such as those shown in the photograph that opens this chapter, of interest to astrobiologists?

2. What is meant by "life as we know it"? Why do astrobiologists suspect that extraterrestrial life is likely to be of this form?

3. How have astronomers discovered organic molecules in interstellar space? Does this discovery mean that life of some sort exists in the space between the stars?

4. Mercury, Venus, and the Moon are all considered unlikely places to find life. Suggest why this should be.

5. Why do astronomers think Enceladus could support life? What would be the easiest way to search for this life?

6. Why are scientists interested in searching for life in Lake Vostok, Antarctica?

7. Many science-fiction stories and movies—including *The War of the Worlds, Invaders from Mars, Mars Attacks!,* and *Martians, Go Home*—involve invasions of Earth by intelligent beings from Mars. Why Mars rather than any of the other planets?

8. Summarize the three tests performed by the *Viking Landers* to search for Martian microorganisms.

9. What arguments can you give against the idea that the "face" on Mars (Figure 27-9) is of intelligent origin? What arguments can you give in favor of this idea?

10. Suppose the *Curiosity* rover measured an enhanced abundance of the carbon isotope ^{12}C in an ancient riverbed. What might this tell us?

11. Suppose someone brought you a rock that he claimed was a Martian meteorite. What scientific tests would you recommend be done to test this claim?

12. The Martian meteorite ALH 84001 contains magnetite, which can be produced by microorganisms on Earth. Why isn't this good evidence for Martian life?

13. Why are astronomers interested in the microtunnels within several Martian meteorites? Why don't these prove the existence of Martian life?

14. Why are most searches for extraterrestrial intelligence made using radio telescopes? Why are most of these carried out at frequencies between 10^3 MHz and 10^4 MHz?

15. Explain why telescopes that would look for the infrared spectra of other planets need to be placed in space (*Hint:* See spectrum in Figure 27-13).

Advanced Questions

Problem-solving tips and tools

The small-angle formula, discussed in Box 1-1, will be useful. Section 5-2 gives the relationship between wavelength and frequency, while Section 5-9 and Box 5-6 discuss the Doppler effect. Section 6-3 gives the relationship between the angular resolution of a telescope, the telescope diameter, and the wavelength used. You will find useful data about the planets in Appendix 1.

16. In 1802, when it seemed likely to many scholars that there was life on Mars, the German mathematician Karl Friedrich Gauss proposed that we signal the Martian inhabitants by drawing huge geometric patterns in the snows of Siberia. His plan was never carried out. (a) Suppose patterns had been drawn that were 1000 km across. What minimum diameter would the objective of a Martian telescope need to have to

be able to resolve these patterns? Assume that the observations are made at a wavelength of 550 nm, and assume that Earth and Mars are at their minimum separation. (**b**) Ideally, the patterns used would be ones that could not be mistaken for natural formations. They should also indicate that they were created by an advanced civilization. What sort of patterns would you have chosen?

17. *TUTORIAL 27.1* Assume that all the terms in the Drake equation have the values given in the text, except for N and L. (**a**) If there are 1000 civilizations in the Galaxy today, what must be the average lifetime of a technological civilization? (**b**) What if there are a million such civilizations?

18. (**a**) Of the visually brightest stars in the sky listed in Appendix 5, which might be candidates for having Earthlike planets on which intelligent civilizations have evolved? Explain your selection criteria. (**b**) Repeat part (a) for the nearest stars, listed in Appendix 4.

19. It has been suggested that extraterrestrial civilizations would choose to communicate at a wavelength of 21 cm. Hydrogen atoms in interstellar space naturally emit at this wavelength, so astronomers studying the distribution of hydrogen around the Galaxy would already have their radio telescopes tuned to receive extraterrestrial signals. (**a**) Calculate the frequency of this radiation in megahertz. Is this inside or outside the water hole? (**b**) Discuss the merits of this suggestion.

20. Imagine that a civilization in another planetary system is sending a radio signal toward Earth. As our planet moves in its orbit around the Sun, the wavelength of the signal we receive will change due to the Doppler effect. This gives SETI scientists a way to distinguish stray signals of terrestrial origin (which will not show this kind of wavelength change) from interstellar signals. (**a**) Use the data in Appendix 1 to calculate the speed of Earth in its orbit. For simplicity, assume the orbit is circular. (**b**) If the alien civilization is transmitting at a frequency of 3000 MHz, what wavelength (in meters) would we receive if Earth were moving neither toward nor away from their planet? (**c**) The maximum Doppler shift occurs if Earth's orbital motion takes it directly toward or directly away from the alien planet. How large is that maximum wavelength shift? Express your answer both in meters and as a percentage of the unshifted wavelength you found in (b). (**d**) Discuss why it is important that SETI radio receivers be able to measure frequency and wavelength to very high precision.

21. Astronomers have proposed using interferometry to make an extremely high-resolution telescope. This proposal involves placing a number of infrared telescopes in space, separating them by thousands of kilometers, and combining the light from the individual telescopes. One design of this kind has an effective diameter of 6000 km and uses infrared radiation with a wavelength of 10 μm. If it is used to observe an Earthlike planet orbiting the star Epsilon Eridani, 3.22 parsecs (10.5 light-years) from Earth, what is the size of the smallest detail that this system will be able to resolve on the face of that planet? Give your answer in kilometers.

Discussion Questions

22. Suppose someone told you that the *Viking Landers* failed to detect life on Mars simply because the tests were designed to detect terrestrial life-forms, not Martian life-forms. How would you respond?

23. Science-fiction television shows and movies often depict aliens as looking very much like humans. Discuss the likelihood that intelligent creatures from another world would have (**a**) a biochemistry similar to our own, (**b**) two legs and two arms, and (**c**) about the same dimensions as a human.

24. The late, great science-fiction editor John W. Campbell exhorted his authors to write stories about organisms that think as well as humans but not *like* humans. Discuss the possibility that an intelligent being from another world might be so alien in its thought processes that we could not communicate with it.

25. If a planet always kept the same face toward its star, just as the Moon always keeps the same face toward Earth, most of the planet's surface would be uninhabitable. Discuss why.

26. How do you think our society would respond to the discovery of intelligent messages coming from a civilization on a planet orbiting another star? Explain your reasoning.

27. What do you think will set the limit on the lifetime of our technological civilization? This factor is L in the Drake equation. Explain your reasoning.

28. The first of all Earth spacecraft to venture into interstellar space were *Pioneer 10* and *Pioneer 11*, which were launched in 1972 and 1973, respectively. Their missions took them past Jupiter and Saturn and eventually beyond the solar system. Both spacecraft carry a metal plaque with artwork (reproduced below) that shows where the spacecraft is from and what sort of creatures designed it. If an alien civilization were someday to find one of these spacecraft, which of the features on the plaque do you think would be easily understandable to them? Explain your reasoning.

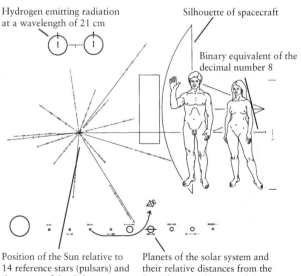

Hydrogen emitting radiation at a wavelength of 21 cm

Silhouette of spacecraft

Binary equivalent of the decimal number 8

Position of the Sun relative to 14 reference stars (pulsars) and the center of the Galaxy

Planets of the solar system and their relative distances from the Sun (shown in binary code)

29. Any living creatures in the subsurface ocean of Europa would have to survive without sunlight. Instead, they might obtain energy from Europa's inner heat. Search the World Wide Web for information about "black smokers," which are associated with high-temperature vents at the bottom of Earth's oceans. What kind of life is found around black smokers? How do these life-forms differ from the more familiar organisms found in the upper levels of the ocean?

30. Search the World Wide Web for information about the *Mars Express* orbiter and the *Spirit* and *Opportunity* rovers. What discoveries have these missions made about water on Mars? Have they found any evidence that liquid water has existed on Mars in the recent past? Describe the evidence, if any.

31. Like other popular media, the World Wide Web is full of claims of the existence of "extraterrestrial intelligence," namely, UFO sightings and alien abductions. (**a**) Choose a Web site of this kind and analyze its content using the idea of Occam's razor, the principle that if there is more than one viable explanation for a phenomenon, one should choose the simplest explanation that fits all the observed facts. (**b**) Read what a skeptical Web site has to say about UFO sightings. A good example is the Web site of the Committee for Skeptical Inquiry, or CSI. After considering what you have read on both sides of the UFO debate, discuss your opinions about whether intelligent aliens really have landed on Earth.

ACTIVITIES

Observing Projects

32. Use the *Starry Night™* program to view Earth from space. Click the **Home** button in the toolbar to move your viewing location to your home town or city. Click the **Increase current elevation** button beneath the **Viewing Location** panel in the toolbar to raise your position to about 11,000 km above your home location. Use the hand tool to move Earth into the center of the view. Locate the position of your home and zoom in on it, using the **Zoom** tool on the right of the toolbar to set the field of view to about 16° × 11°. Use the location scroller to move over various regions of Earth's surface. (**a**) Can you see any evidence of the presence of life or of man-made objects? (**b**) **Right-click** on Earth (**Ctrl-click** on a Mac) and select **Google Maps** from the contextual menu to examine regions of Earth in great detail on these images from an orbiting satellite. (You will need to have access to the Internet in order to use Google Maps.) You can attempt to locate your own home on these maps. As an exercise, find the country of Panama. **Zoom** in progressively upon the Panama Canal and search for the ships traversing this incredible waterway and the massive locks that lift these ships into the upper waterways. What does this suggest about the importance of sending spacecraft to planets to explore their surfaces at close range?

33. Use *Starry Night™* to examine the planet Mars. Select **Favourites > Explorations > Mars. Zoom** in or out on Mars and use the location scroller to examine the planet's surface. Based on what you observe, where on the Martian surface would you choose to land a spacecraft to search for the presence of life? Explain the reasons for your choice.

34. Use the *Starry Night™* program to examine Jupiter's moon Europa. Select **Favourites > Explorations > Europa** to view this enigmatic moon of our largest planet. Use the location scroller cursor and zoom controls to examine the surface of this moon. Are there any surface features that suggest that this moon might harbor life? Is there any other evidence, perhaps not visible in this simulation, that suggests the possibility that life exists there?

35. Use *Starry Night™* to investigate the likelihood of the existence of life in the universe beyond that found upon Earth. Select **Favourites > Explorations > Atlas** and then open the **Options** pane. Expand the **Stars** layer and the **Stars** heading and click the checkbox to turn on the **Mark stars with extrasolar planets** option. Then use the hand tool or cursor keys to look around the sky. Each marked star in this view has at least one planet orbiting around it. Click and hold the **Increase current elevation** button in the toolbar until the viewpoint is about 1000 ly from Earth. Note that the stars with extrasolar planets are contained in a small knot of stars in the region of our Sun, our solar neighborhood. This is because present techniques place a limit upon the distance to which we can find extrasolar planets in space. There is no logical reason why the same proportion of stars in the rest of the Milky Way, or indeed the rest of the universe, should not have companion planets. **Increase current elevation** again to about 90,000 ly from Earth to see a view of the Milky Way galaxy. Use the location scroller to view the Milky Way face-on and compare the clump of stars representing the solar neighborhood to the size of the Milky Way Galaxy. Click the **Increase current elevation** button again until the **Viewing Location** panel indicates a distance to Earth of about 0.300 Mly (300,000 light-years). In this view, each point of light represents a separate galaxy containing hundreds of billions of stars. Now, gradually increase current elevation to about 1500 Mly (1.5 billion light-years from Earth) to see the entire database of 28,000 galaxies included in the *Starry Night™* data bank. The borders of this cube of galaxies reaching out to about 500 Mly from Earth is defined by the limits of the methods for determining distance to galaxies in deep space. In practice, this volume of space represents less than 5% of the size of the observable universe and galaxies abound beyond the limits of this view. To view images of some of these very distant galaxies, open the view named **Hubble Deep Field** from the **Explorations** folder in the **Favourites** pane. This view from the center of Earth is centered upon a seemingly empty patch of sky. Zoom in to see this image of more than 1500 very distant galaxies that were found by the Hubble Space Telescope in a small region of our sky. Then open the view named **Hubble Ultra Deep Field** and zoom in on an image of over 10,000 galaxies that extends to the limits of our observable universe. (**a**) Do these views influence your thoughts on the likelihood that life, including intelligent life, exists elsewhere in the universe other than on Earth? Explain your reasoning using the Drake equation as a guide. (**b**) Why is it unlikely that we will be able to communicate with life forms that might exist on planets in these very distant regions of space?

Collaborative Exercise

36. Imagine that astronomers have discovered intelligent life in a nearby star system. Your group is submitting a proposal for who on Earth should speak for the planet and what 50-word message should be conveyed. Prepare a maximum one-page proposal that states (**a**) who should speak for Earth and why; (**b**) what this person should say in 50 words; and (**c**) why this message is the most important compared to other things that could be said. Only serious responses receive full credit.

ANSWERS

ConceptChecks

ConceptCheck 27-1: Carbon atoms can combine with other elements to form an impressively wide array of different molecules, and this complexity is needed for even the smallest living organisms. Furthermore, carbon is produced in stars and is widely available.

ConceptCheck 27-2: Organic molecules are commonly found in meteorites and comets, which frequently crashed onto planet surfaces during the early formation of the solar system. Also, interplanetary dust grains containing organic compounds steadily fall to a planet's surface.

ConceptCheck 27-3: They created amino acids and other compounds that living organisms need—they did not create living entities.

ConceptCheck 27-4: The salts dissolve in water when the water is in contact with rock. Rather than a small pocket of water within ice, the salts suggest a much larger ocean is in contact with a rocky mantle.

ConceptCheck 27-5: Living organisms that depend on photosynthesis concentrate ^{12}C compared to their surroundings. This increased abundance can remain in the environment long after the organisms are dead.

ConceptCheck 27-6: Geological processes (involving no biology) can also produce this type of magnetite, so there is no compelling reason to assume that the magnetite was produced by Martian microbes. (*Note:* The claims that this magnetite was of biological origin came *before* it was figured out that this magnetite can be produced through geology.)

ConceptCheck 27-7: The Martian meteorites have trace amounts of gas trapped within them that are nearly identical to the unique Martian atmosphere.

ConceptCheck 27-8: At the speed of light, radio waves travel at more than 100,000 km every second, whereas our fastest spacecraft can only travel at tens of thousands of kilometers every hour. Therefore, the vast distances between the stars make it seemingly impossible for spacecraft to carry out exploration of the stars.

ConceptCheck 27-9: L is particularly difficult to estimate because at about the same time the technology to communicate beyond one's own planet is developed, the technology of weapons of mass destruction can develop, which could end the civilization just as it starts (alternatively, it can learn to live without war and exist for a very long time).

ConceptCheck 27-10: At other frequencies, there is more background noise, and more energy would need to be used to reliably transmit the same message across the Galaxy. Any advanced civilization could figure this out and would see the benefits of sending messages in this narrow range.

ConceptCheck 27-11: The *Kepler* telescope was looking for an ever-so-slight dimming of a star as an orbiting planet passes in front of the star, partially blocking the star from our view. This is called a transit.

ConceptCheck 27-12: If the orbital plane of a planet around its star is tilted up or down from our perspective, then the planet will not block out any of the star's light during its orbit. Since *Kepler* looked for the dimming of stars as they were blocked by planets, *Kepler* would not have detected a planet with this kind of orbit.

ConceptCheck 27-13: Ozone comes from oxygen gas, O_2, which is produced by living organisms on Earth. Scientists do not know of any plausible nonbiological way to create a dense oxygen atmosphere.

ConceptCheck 27-14: In the infrared where a planet's spectrum can be more easily measured, an absorption line from oxygen gas is not present, but absorption from ozone (O_3) is.

A Biologist's View of Astrobiology by Kevin W. Plaxco

In the early 1960s, flush with the excitement of Sputnik and the first manned space missions, the Nobel Prize-winning biologist Joshua Lederberg coined the word "exobiology" to describe the scientific study of extraterrestrial life. But after a brief flurry of popularity in the 1960s and 1970s (due partly to the pioneering research and public outreach work of Carl Sagan), exobiology fell out of fashion among scientists; it was, after all, the only field of scientific study without any actual subject material to research.

In the last decade, however, the field of astrobiology has filled the void left by exobiology's demise by being simultaneously more encompassing and more practical. Astrobiology removes exobiology's distinction between life on Earth and life "out there" and focuses on broader, but more tractable, questions regarding the relationship between life (any life, anywhere) and the universe. In a nutshell, astrobiology uses our significant (if incomplete) knowledge of life on Earth to address three broad questions about life in the universe:

- Which of the physical attributes of our universe were necessary in order for it to support the origins and further evolution of life?

- How did life arise and evolve on Earth, and how might it have arisen and evolved elsewhere?

- Where else might life have arisen in our universe, and how do we best search for it?

Much of the worth and appeal of astrobiology lie in the fact that these questions, which are among the most fundamental posed by science, address the most profound issues regarding our origins and our nature.

Perhaps not surprisingly, some aspects of astrobiology are far better understood than others. We understand, for example, the broad details of how the almost metal-free hydrogen and helium left over after the Big Bang was enriched in the metallic elements by fusion reactions occurring in the more massive stars and how these elements—absolutely critical for biology because only they support the complex chemistry required for life—were then returned to the cosmos in supernova explosions. Similarly, we understand in detail how rocky planets—which we see as potentially habitable—form out of the dust and gas of prestellar nebulae. In contrast, however, much of the *biology* in astrobiology remains speculative at best. There is, for example, nothing even remotely resembling consensus regarding how life first arose on Earth much less how—or if—it might have arisen elsewhere in the cosmos.

Perhaps equally unsurprisingly, some branches of the astrobiology community are more "optimistic" than others. Astronomers generally view astrobiology in terms of the processes by which potentially habitable planets form. As these processes are reasonably well understood, astronomers tend to be rather upbeat regarding the events surrounding life's origins. The radio astronomer Frank Drake, who pioneered the search for extraterrestrial intelligence (SETI) in the 1960s, is a relatively extreme example of this hopefulness: When trying to estimate the number of intelligent, "communicating" civilizations in the Galaxy, he assumed that rocky, water-soaked planets almost invariably give rise to life. Biochemists, in contrast, tend to view the key issues in astrobiology as relating to the first steps in the origins of life—that is, the detailed processes by which inanimate chemicals first combine to form a living, evolving, self-replicating system. To date, about a half dozen theories have been put forth to explain how this might have occurred on Earth, but all are fraught with enormous scientific difficulties. Specifically, every single theory of the origins of life postulated to date either utterly lacks experimental evidence to back it up or requires events that are so improbable they make even the "astronomical" number of planets in the universe look small by comparison. When faced with what is, in effect, a complete unknown, biochemists tend to be much more cautious about estimating the probability of life arising and the chances of our being alone in the cosmos. To quote Francis Crick, codiscoverer of the structure of DNA, "We cannot decide whether the origin of life on Earth was an extremely unlikely event or almost a certainty, or any possibility in between these two extremes."

Irrespective of our degree of personal optimism or pessimism, though, no one would be more thrilled than us biologists by the detection of life that had arisen independently of life on Earth. Such a find would revolutionize our field by providing the first, tangible evidence of whether there are viable routes to the evolution of life other than the DNA/proteins/cells path that it took on Earth. But even barring such a momentous discovery as the detection of extraterrestrial life, astrobiology still provides a significant service to biology: By placing life into a broader context, astrobiology provides the perspective that we need in order to truly understand who we are and where we come from.

Prior to becoming a professor of biochemistry at the University of California at Santa Barbara, Kevin Plaxco earned his BS in chemistry from the University of California at Riverside and his Ph.D. in molecular biology from the California Institute of Technology. While his research focuses predominantly on the study of proteins, the "macromolecular machines" that perform the lion's share of the action in our cells, he has also performed NASA-funded research into methods for the detection of extraterrestrial life. In addition, he teaches a course in astrobiology and has published a popular science book on the subject.

Appendices

The Planets: Orbital Data

Planet	Semimajor axis (10⁶ km)	Semimajor axis (AU)	Sidereal period (years)	Sidereal period (days)	Synodic period (days)	Average orbital speed (km/s)	Orbital eccentricity	Inclination of orbit to ecliptic (°)
Mercury	57.9	0.387	0.241	87.969	115.88	47.9	0.206	7.00
Venus	108.2	0.723	0.615	224.70	583.92	35.0	0.007	3.39
Earth	149.6	1.000	1.000	365.256	—	29.79	0.017	0.00
Mars	227.9	1.524	1.88	686.98	779.94	24.1	0.093	1.85
Jupiter	778.3	5.203	11.86		398.9	13.1	0.048	1.30
Saturn	1429	9.554	29.46		378.1	9.64	0.053	2.48
Uranus	2871	19.194	84.10		369.7	6.83	0.043	0.77
Neptune	4498	30.066	164.86		367.5	5.5	0.010	1.77

The Planets: Physical Data

Planet	Equatorial diameter (km)	Equatorial diameter (Earth = 1)	Mass (kg)	Mass (Earth = 1)	Average density (kg/m³)	Rotation period* (solar days)	Inclination of equator to orbit (°)	Surface gravity Earth = 1	Albedo	Escape speed (km/s)
Mercury	4880	0.383	3.302×10^{23}	0.0553	5430	58.646	0.5	0.38	0.12	4.3
Venus	12,104	0.949	4.868×10^{24}	0.8149	5243	243.01^R	177.4	0.91	0.59	10.4
Earth	12,756	1.000	5.974×10^{24}	1.000	5515	0.997	23.45	1.000	0.39	11.2
Mars	6794	0.533	6.418×10^{23}	0.107	3934	1.026	25.19	0.38	0.15	5.0
Jupiter	142,984	11.209	1.899×10^{27}	317.8	1326	0.414	3.12	2.36	0.44	59.5
Saturn	120,536	9.449	5.685×10^{26}	95.16	687	0.444	26.73	1.1	0.47	35.5
Uranus	51,118	4.007	8.682×10^{25}	14.53	1318	0.718^R	97.86	0.92	0.56	21.3
Neptune	49,528	3.883	1.024×10^{26}	17.15	1638	0.671	29.56	1.1	0.51	23.5

*For Jupiter, Saturn, Uranus, and Neptune, the internal rotation period is given. A superscript R means that the rotation is retrograde (opposite the planet's orbital motion).

Appendix 3 | Major Satellites of the Planets by Mass

Planet	Satellite	Date of discovery	Average distance from center of planet (km)	Orbital (sidereal) period (days)*	Orbital eccentricity	Size of satellite (km)**	Mass (kg)
EARTH	Moon	—	384,400	27.322	0.0549	3476	7.349×10^{22}
MARS	Phobos	1877	9378	0.319	0.01	$28 \times 23 \times 20$	1.1×10^{22}
	Deimos	1877	23,460	1.263	0.00	$16 \times 12 \times 10$	1.1×10^{22}
JUPITER	Ganymede	1610	1,070,000	7.155	0.0011	5268	1.482×10^{23}
	Callisto	1610	1,883,000	16.689	0.0074	4800	1.077×10^{23}
	Io	1610	421,600	1.769	0.0041	3642	8.932×10^{22}
	Europa	1610	670,900	3.551	0.0094	3120	4.791×10^{22}
	Himalia	1904	11,461,000	250.56	0.1623	184	6.7×10^{18}
	Amalthea	1892	181,400	0.498	0.0031	$270 \times 200 \times 155$	2.1×10^{18}
	Elara	1905	11,741,000	259.64	0.2174	78	8.7×10^{17}
	Thebe	1979	221,900	0.675	0.0177	98	1.5×10^{18}
	Pasiphaë	1908	23,624,000	743.63^R	0.4090	58	3.0×10^{17}
	Carme	1938	23,404,000	734.17^R	0.2533	46	1.3×10^{17}
SATURN	Titan	1655	1,221,870	15.95	0.0288	5150	1.34×10^{23}
	Rhea	1672	527,070	4.518	0.001	1528	2.3×10^{21}
	Iapetus	1671	3,560,840	79.33	0.0283	1472	2.0×10^{21}
	Dione	1684	377,420	2.737	0.0022	1123	1.1×10^{21}
	Tethys	1684	294,670	1.888	0.0001	1066	6.2×10^{20}
	Enceladus	1789	238,040	1.37	0.0047	504	1.1×10^{20}
	Mimas	1789	185,540	0.942	0.0196	397	3.8×10^{19}
	Phoebe	1898	12,947,780	550.31^R	0.1635	$230 \times 220 \times 210$	8.3×10^{18}
	Hyperion	1848	1,500,880	21.28	0.0274	$360 \times 280 \times 225$	5.7×10^{18}
	Janus	1980	151,460	0.695	0.0068	$193 \times 173 \times 137$	1.9×10^{18}
URANUS	Titania	1787	436,300	8.706	0.0011	1578	3.53×10^{21}
	Oberon	1787	583,500	13.46	0.0014	1522	3.01×10^{21}
	Ariel	1851	190,900	2.52	0.0012	1158	1.35×10^{21}
	Umbriel	1851	266,000	4.144	0.0039	1169	1.2×10^{21}
	Miranda	1948	129,900	1.413	0.0013	471	6.59×10^{19}
	Puck	1985	86,000	0.762	0.0001	162	2.9×10^{18}
	Sycorax	1997	12,179,000	1288.3^R	0.5224	150	5.4×10^{18}
	Portia	1986	66,100	0.513	0.0001	156×126	1.7×10^{18}
	Juliet	1986	64,400	0.493	0.0007	150×74	5.6×10^{17}
	Belinda	1986	75,300	0.624	0.0001	128×64	3.6×10^{17}

Appendix 3   Major Satellites of the Planets by Mass

Planet	Satellite	Date of discovery	Average distance from center of planet (km)	Orbital (sidereal) period (days)*	Orbital eccentricity	Size of satellite (km)**	Mass (kg)
NEPTUNE	Triton	1846	354,800	5.877^R	0	2706	2.15×10^{22}
	Proteus	1989	117,647	1.122	0.0005	$440 \times 416 \times 404$	$4.43\ 10^{19}$
	Nereid	1949	5,513,400	360.14	0.7512	340	3.1×10^{19}
	Larissa	1989	73,548	0.555	0.0014	$216 \times 204 \times 164$	4.2×10^{18}
	Galatea	1989	61,953	0.429	0	$204 \times 184 \times 144$	2.1×10^{18}
	Despina	1989	52,526	0.335	0.0002	$180 \times 150 \times 130$	2.1×10^{18}
	Thalassa	1989	50,075	0.311	0.0002	$108 \times 100 \times 52$	3.5×10^{17}
	Naiad	1989	48,227	0.294	0.0004	$96 \times 60 \times 52$	1.9×10^{17}
	Halimede	2002	15,728,000	1879.71^R	0.5711	62	1.8×10^{17}
	Neso	2002	22,422,000	2914.07	0.2931	44	6.3×10^{16}

This table was compiled from data provided by the Jet Propulsion Laboratory.
A superscript R means that the satellite orbits in a retrograde direction (opposite to the planet's rotation).

**The size of a spherical satellite is equal to its diameter.*

Appendix 4 The Nearest Stars

Name	Parallax (arcsec)	Distance (parsecs)	Distance (light-years)	Spectral Type	Proper motion (arcsec/yr)	Apparent visual magnitude	Absolute visual magnitude	Mass (Sun = 1)
Proxima Centauri	0.772	1.30	4.22	M5.5 V	3.853	+11.09	+15.53	0.107
Alpha Centauri A	0.747	1.34	4.36	G2 V	3.710	−0.01	+4.36	1.144
Alpha Centauri B	0.747	1.34	4.36	K0 V	3.724	+1.34	+5.71	0.916
Barnard's Star	0.547	1.83	5.96	M4.0 V	10.358	+9.53	+13.22	0.166
Wolf 359	0.419	2.39	7.78	M6.0 V	4.696	+13.44	+16.55	0.092
Lalande 21185	0.393	2.54	8.29	M2.0 V	4.802	+7.47	+10.44	0.464
Sirius A	0.380	2.63	8.58	A1 V	1.339	−1.43	+1.47	1.991
Sirius B	0.380	2.63	8.58	white dwarf	1.339	+8.44	+11.34	0.500
UV Ceti	0.374	2.68	8.73	M5.5 V	3.368	+12.54	+15.40	0.109
BL Ceti	0.374	2.68	8.73	M6.0 V	3.368	+12.99	+15.85	0.102
Ross 154	0.337	2.97	9.68	M3.5 V	0.666	+10.43	+13.07	0.171
Ross 248	0.316	3.16	10.32	M5.5 V	1.617	+12.29	+14.79	0.121
Epsilon Eridani	0.310	3.23	10.52	K2 V	0.977	+3.73	+6.19	0.850
Lacailee 9352	0.304	3.29	10.74	M1.5 V	6.896	+7.34	+9.75	0.529
Ross 128	0.299	3.35	10.92	M4.0 V	1.361	+11.13	+13.51	0.156
EZ Aquarii A	0.290	3.45	11.27	M5.0 V	3.254	+13.33	+15.64	0.105
EZ Aquarii B	0.290	3.45	11.27	—	3.254	+13.27	+15.58	0.106
EZ Aquarii C	0.290	3.45	11.27	—	3.254	+14.03	+16.34	0.095
Procyon A	0.286	3.50	11.40	F5 IV-V	1.259	+0.38	+2.66	1.569
Procyon B	0.286	3.50	11.40	white dwarf	1.259	+10.70	+12.98	0.500
61 Cygni A	0.286	3.50	11.40	K5.0 V	5.281	+5.21	+7.49	0.703
61 Cygni B	0.286	3.50	11.40	K7.0 V	5.172	+6.03	+8.31	0.630
GJ725 A	0.283	3.53	11.53	M3.0 V	2.238	+8.90	+11.16	0.351
GJ725 B	0.283	3.53	11.53	M3.5 V	2.313	+9.69	+11.95	0.259
GX Andromedae	0.281	3.56	11.62	M1.5 V	2.918	+8.08	+10.32	0.486
GQ Andromedae	0.281	3.56	11.62	M3.5 V	2.918	+11.06	+13.30	0.163
Epsilon Indi A	0.276	3.63	11.82	K5 V	4.704	+4.69	+6.89	0.766
Epsilon Indi B	0.276	3.63	11.82	T1.0	4.823			0.044
Epsilon Indi C	0.276	3.63	11.82	T6.0	4.823			0.028
DX Cancri	0.276	3.63	11.83	M6.5 V	1.290	+14.78	+16.98	0.087
Tau Ceti	0.274	3.64	11.89	G8 V	1.922	+3.49	+5.68	0.921
RECONS 1	0.272	3.68	11.99	M5.5 V	0.814	+13.03	+15.21	0.113
YZ Ceti	0.269	3.72	12.13	M4.5 V	1.372	+12.02	+14.17	0.136
Luyten's Star	0.264	3.79	12.37	M3.5 V	3.738	+9.86	+11.97	0.257
Kapteyn's Star	0.255	3.92	12.78	M1.5 V	8.670	+8.84	+10.87	0.393
AX Microscopium	0.253	3.95	12.87	M0.0 V	3.455	+6.67	+8.69	0.600

This table, compiled from data reported by the Research Consortium on Nearby Stars, lists all known stars within 4.00 parsecs (13.05 light-years).

Stars that are components of multiple star systems are labeled A, B, and C.

Appendix 5 The Visually Brightest Stars

Name	Designation	Distance (parsecs)	Distance (light-years)	Spectral Type	Radial velocity (km/s)*	Proper motion (arcsec/yr)	Apparent visual magnitude	Apparent visual brightness (Sirius = 1)**	Mass (Sun = 1)
Sirius A	α CMa A	2.63	8.58	A1 V	−7.6	1.34	−1.43	1.000	+1.46
Canopus	α Car	95.9	313	F0 II	+20.5	0.03	−0.72	0.520	−5.63
Arcturus	α Boo	11.3	36.7	K1.5 III	−5.2	2.28	−0.04	0.278	−0.30
Alpha Centauri A	α Cen A	1.34	4.36	G2 V	−25	3.71	−0.01	0.270	+4.36
Vega	α Lyr	7.76	25.3	A0 V	−13.9	0.35	+0.03	0.261	+0.58
Capella	α Aur	12.9	42.2	G5 III	+30.2	0.43	+0.08	0.249	−0.48
Rigel	α Ori A	237	773	B8 Ia	+20.7	0.002	+0.12	0.240	−6.75
Procyon	α CMi A	3.50	11.4	F5 IV-V	−3.2	1.26	+0.34	0.196	+2.62
Achernar	α Eri	44.1	144	B3 V	+16	0.10	+0.50	0.169	−2.72
Betelgeuse	α Ori	131	427	M1 Iab	+21	0.03	+0.58	0.157	−5.01
Hadar	β Cen	161	525	B1 III	+5.9	0.04	+0.60	0.154	−5.43
Altair	α Aql	51.4	168	A7 V	−26.1	0.66	+0.77	0.132	−2.79
Aldebaran	α Tau A	20.0	65.1	K5 III	+54.3	0.20	+0.85	0.122	−0.65
Spica	α Vir	80.4	262	B1 III-IV	+1	0.05	+1.04	0.103	−3.49
Antares	α Sco A	185	604	M1.5 Iab	−3.4	0.03	+1.09	0.098	−5.25
Pollux	β Gem	10.3	33.7	K0 IIIb	+3.3	0.63	+1.15	0.093	+1.08
Fomalhaut	α PsA	7.69	25.1	A3 V	+6.5	0.37	+1.16	0.092	+1.73
Deneb	α Cyg	990	3230	A2 Ia	−4.5	0.002	+1.25	0.085	−8.73
Mimosa	β Cru	108	353	B0.5 IV	+15.6	0.05	+1.297	0.081	−3.87
Regulus	α Leo A	23.8	77.5	B7 V	+5.9	0.25	+1.35	0.077	−0.53

Data in this table were compiled from SIMBAD database operated at the Centre de Données Astronomiques de Strasbourg, France.

*A positive radial velocity means that the star is receding; a negative radial velocity means that the star is approaching.

**This is a ratio of the star's apparent brightness to that of Sirius, the brightest star in the night sky.

Note: Acrux, or α Cru (the brightest star in Crux, the Southern Cross) appears to the naked eye as a star of apparent magnitude +0.87, the same as Aldebaran. However, it does not appear in this table because Acrux is actually a binary star system. The blue-white component stars of this binary system have apparent magnitudes of +1.4 and +1.9, and so they are dimmer than any of the stars listed here.

Appendix 6 — Some Important Astronomical Quantities

Astronomical Unit	1 AU $= 1.4960 \times 10^{11}$ m $= 1.4960 \times 10^8$ km
Light-year	1 pc $= 3.2616$ ly $= 63,240$ AU
Parsec	1 pc $= 3.2616$ ly $= 3.0857 \times 10^{16}$ m $= 206,265$ AU
Year	1 y $= 365.2564$ days $= 3.156 \times 10^7$ s
Solar Mass	1 $M_\odot = 1.989 \times 10^{30}$ kg
Solar Radius	1 $R_\odot = 6.9599 \times 10^8$ m
Solar Luminosity	1 $L_\odot = 3.90 \times 10^{26}$ W

Appendix 7 — Some Important Physical Constants

Speed of light	$c = 2.9979 \times 10^8$ m/s
Gravitational constant	$G = 6.673 \times 10^{-11}$ N m^2/kg^2
Planck's constant	$h = 6.6261 \times 10^{-34}$ J s $= 4.1357 \times 10^{-15}$ eV s
Boltzmann constant	$k = 1.3807 \times 10^{-23}$ J/K $= 8.6173 \times 10^{-5}$ eV/K
Stefan-Boltzmann constant	$\sigma = 5.6704 \times 10^{-8}$ W m^{-2} K^{-4}
Mass of electron	$m_e = 9.1094 \times 10^{-31}$ kg
Mass of proton	$m_p = 1.6726 \times 10^{-27}$ kg
Mass of neutron	$m_n = 1.6749 \times 10^{-27}$ kg
Mass of hydrogen atom	$m_H = 1.6735 \times 10^{-27}$ kg
Rydberg constant	$R = 1.0973 \times 10^{-7}$ kg m^{-1}
Electron volt	1 eV $= 1.6022 \times 10^{-19}$ J

Appendix 8 — Some Useful Mathematics

Area of a rectangle of sides a and b	$A = ab$
Volume of a rectangular solid of sides a, b, and c	$V = abc$
Hypotenuse of a right triangle whose other sides are a and b	$c = \sqrt{a^2 + b^2}$
Circumference of a circle of radius r	$C = 2\pi r$
Area of a circle of radius r	$A = \pi r^2$
Surface areas of a sphere of radius r	$A = 4\pi r^2$
Volume of a sphere of radius r	$V = 4\pi r^3/3$
Value of π	$\pi = 3.1415926536$

Glossary

You can find more information about each term in the indicated chapter or chapters. See the Key Words section in each chapter for the specific page on which the meaning of a given term is described.

A ring One of three prominent rings encircling Saturn. (Chapter 12)

absolute magnitude The apparent magnitude that a star would have if it were at a distance of 10 parsecs. (Chapter 17)

absolute zero A temperature of −273°C (or 0 K), at which all molecular motion stops; the lowest possible temperature. (Chapter 5)

absorption line spectrum Dark lines superimposed on a continuous spectrum. (Chapter 5)

acceleration The rate at which an object's velocity changes due to a change in speed, a change in direction, or both. (Chapter 4)

accretion The gradual accumulation of matter in one location, typically due to the action of gravity. (Chapter 8, Chapter 18, Chapter 24)

accretion disk A disk of gas orbiting a star or black hole. (Chapter 24)

active galactic nucleus (AGN) The center of an active galaxy. (Chapter 24)

active galaxy A galaxy that is emitting exceptionally large amounts of energy; a Seyfert galaxy, radio galaxy, or quasar. (Chapter 24)

active optics A technique for improving a telescopic image by altering the telescope's optics to compensate for variations in air temperature or flexing of the telescope mount. (Chapter 6)

adaptive optics A technique for improving a telescopic image by altering the telescope's optics in a way that compensates for distortion caused by Earth's atmosphere. (Chapter 6)

aerosol Tiny droplets of liquid dispersed in a gas. (Chapter 13)

AGB star See *asymptotic giant branch star.*

AGN See *active galactic nucleus.*

air drag A force that opposes the motion of an object through air. A portion of the object's energy is transferred to the air molecules. (Chapter 4)

albedo The fraction of sunlight that a planet, asteroid, or satellite reflects. (Chapter 9)

alpha particle The nucleus of a helium atom, consisting of two protons and two neutrons. (Chapter 19)

amino acids The chemical building blocks of proteins. (Chapter 15)

angle The opening between two lines that meet at a point. (Chapter 1)

angular diameter The angle subtended by the diameter of an object. (Chapter 1)

angular distance The angle between two points in the sky. (Chapter 1)

angular momentum See *conservation of angular momentum.*

angular resolution The angular size of the smallest feature that can be distinguished with a telescope. (Chapter 6)

angular size See *angular diameter.*

anisotropic Possessing unequal properties in different directions; not isotropic. (Chapter 23)

annihilation The process by which the masses of a particle and antiparticle are converted into energy. (Chapter 26)

annular eclipse An eclipse of the Sun in which the Moon is too distant to cover the Sun completely so that a ring of sunlight is seen around the Moon at mid-eclipse. (Chapter 3)

anorthosite Rock commonly found in ancient, cratered highlands on the Moon. (Chapter 10)

Antarctic Circle A circle of latitude 23½° north of Earth's south pole. (Chapter 2)

antimatter Matter consisting of antiparticles such as antiprotons, antineutrons, and positrons. (Chapter 26)

antiparticle A particle with the same mass as an ordinary particle but with other properties, such as electric charge, reversed. (Chapter 21, Chapter 26)

antipode The point on a planet exactly opposite a specific location. (Chapter 11)

antiproton A particle with the same mass as a proton but with a negative electric charge. (Chapter 26)

aphelion The point in its orbit where a planet is farthest from the Sun. (Chapter 4)

apogee The point in its orbit where a satellite or the Moon is farthest from Earth. (Chapter 3)

apparent brightness The flux of a star's light arriving at Earth. (Chapter 17)

apparent magnitude A measure of the brightness of light from a star or other object as measured from Earth. (Chapter 17)

apparent solar day The interval between two successive transits of the Sun's center across the local meridian. (Chapter 2)

apparent solar time Time reckoned by the position of the Sun in the sky. (Chapter 2)

arcminute One-sixtieth (1/60) of a degree, designated by the symbol ′. (Chapter 1)

arcsecond One-sixtieth (1/60) of an arcminute or 1/3600 of a degree, designated by the symbol ″. (Chapter 1)

Arctic Circle A circle of latitude 23½° south of Earth's north pole. (Chapter 2)

asteroid One of tens of thousands of small, rocky, planetlike objects in orbit about the Sun. Also called minor planet. (Chapter 7, Chapter 15)

asteroid belt A region between the orbits of Mars and Jupiter that encompasses the orbits of many asteroids. (Chapter 7, Chapter 15)

asthenosphere A warm, plastic layer of the mantle beneath the lithosphere of Earth. (Chapter 9)

astrobiology The study of life in the universe. (Chapter 27)

astrometric method A technique for detecting extrasolar planets by looking for stars that "wobble" periodically. (Chapter 8)

astronomical unit (AU) The semimajor axis of Earth's orbit; the average distance between Earth and the Sun. (Chapter 1)

asymptotic giant branch A region on the Hertzsprung-Russell diagram occupied by stars in their second and final red-giant phase. (Chapter 20)

asymptotic giant branch star (AGB star) A red giant star that has ascended the giant branch for the second and final time. (Chapter 20)

atmosphere (atm) A unit of atmospheric pressure. (Chapter 9)

atmospheric pressure The force per unit area exerted by a planet's atmosphere. (Chapter 9)

atom The smallest particle of an element that has the properties characterizing that element. (Chapter 5)

atomic number The number of protons in the nucleus of an atom of a particular element. (Chapter 5, Chapter 8)

AU See *astronomical unit.*

aurora (plural **aurorae**) Light radiated by atoms and ions in Earth's upper atmosphere, mostly in the polar regions. (Chapter 9)

aurora australis Aurorae seen from southern latitudes; the southern lights. (Chapter 9)

aurora borealis Aurorae seen from northern latitudes; the northern lights. (Chapter 9)

autumnal equinox The intersection of the ecliptic and the celestial equator where the Sun crosses the equator from north to south. Also used to refer to the date on which the Sun passes through this intersection. (Chapter 2)

average density The mass of an object divided by its volume. (Chapter 7)

average density of matter In cosmology, the average density of all the material substance in the universe. (Chapter 25)

B ring One of three prominent rings encircling Saturn. (Chapter 12)

Balmer line An emission or absorption line in the spectrum of hydrogen caused by an electron transition between the second and higher energy levels. (Chapter 5)

Balmer series The entire pattern of Balmer lines. (Chapter 5)

Barnard object One of a class of dark nebulae discovered by E. E. Barnard. (Chapter 18)

barred spiral galaxy A spiral galaxy in which the spiral arms begin from the ends of a "bar" running through the nucleus rather than from the nucleus itself. (Chapter 23)

baseline In interferometry, the distance between two telescopes whose signals are combined to give a higher-resolution image. (Chapter 6)

belt A dark band in Jupiter's atmosphere. (Chapter 12)

Big Bang An explosion of all space that took place roughly 13.7 billion years ago and that marks the beginning of the universe. (Chapter 1, Chapter 25)

binary star (also **binary star system** or **binary**) Two stars orbiting each other. (Chapter 17)

biosphere The layer of soil, water, and air surrounding Earth in which living organisms thrive. (Chapter 9)

bipolar outflow Oppositely directed jets of gas expelled from a young star. (Chapter 18)

black hole An object whose gravity is so strong that the escape speed exceeds the speed of light. (Chapter 1, Chapter 25)

black hole evaporation The process by which black holes emit particles. (Chapter 21)

blackbody A hypothetical perfect radiator that absorbs and re-emits all radiation falling upon it. (Chapter 5)

blackbody curve The intensity of radiation emitted by a blackbody plotted as a function of wavelength or frequency. (Chapter 5)

blackbody radiation The radiation emitted by a perfect blackbody. (Chapter 5)

blazar A type of active galaxy whose nucleus has a featureless spectrum without prominent spectral lines. (Chapter 24)

blueshift A decrease in the wavelength of photons emitted by an approaching source of light. (Chapter 5)

Bohr orbits In the model of the atom described by Niels Bohr, the only orbits in which electrons are allowed to move about the nucleus. (Chapter 5)

Bok globule A small, roundish, dark nebula. (Chapter 18)

bow shock A region where charged particles in the solar wind abruptly slow down as they encounter a planet's magnetosphere. (Chapter 12)

bright terrain (on Ganymede) Young, reflective, relatively crater-free terrain on the surface of Ganymede. (Chapter 13)

brightness See *apparent brightness.*

brown dwarf A starlike object that is not massive enough to sustain hydrogen fusion in its core. (Chapter 17)

brown ovals Elongated, brownish features usually seen in Jupiter's northern hemisphere. (Chapter 12)

C ring One of three prominent rings encircling Saturn. (Chapter 12)

capture theory The hypothesis that the Moon was gravitationally captured by Earth. (Chapter 10)

carbon fusion The thermonuclear fusion of carbon to produce heavier nuclei. (Chapter 20)

carbon star A peculiar red giant star whose spectrum shows strong absorption lines of carbon and carbon compounds. (Chapter 20)

carbonaceous chondrite A type of meteorite that has a high abundance of carbon and volatile compounds. (Chapter 15, Chapter 27)

Cassegrain focus An optical arrangement in a reflecting telescope in which light rays are reflected by a secondary mirror to a focus behind the primary mirror. (Chapter 6)

Cassini division An apparent gap between Saturn's A and B rings. (Chapter 12)

CCD See *charge-coupled device.*

celestial equator A great circle on the celestial sphere 90° from the celestial poles. (Chapter 2)

celestial sphere An imaginary sphere of very large radius centered on an observer; the apparent sphere of the sky. (Chapter 2)

center of mass The point between a star and a planet, or between two stars, around which both objects orbit. (Chapter 8, Chapter 10, Chapter 17)

central bulge (of a galaxy) A spherical distribution of stars around the nucleus of a spiral galaxy. (Chapter 22)

Cepheid variable A type of yellow, supergiant, pulsating star. (Chapter 19)

Cerenkov radiation The radiation emitted by a particle traveling through a substance at a speed greater than the speed of light in that substance. (Chapter 20)

Chandrasekhar limit The maximum mass of a white dwarf. (Chapter 20)

charge-coupled device (CCD) A type of solid-state device designed to detect photons. (Chapter 6)

chemical composition A description of which chemical substances make up a given object. (Chapter 7)

chemical differentiation The process by which the heavier elements in a planet sink toward its center while lighter elements rise toward its surface. (Chapter 8)

chemical element See *element.*

chondrule A glassy, roughly spherical blob found within meteorites. (Chapter 8)

chromatic aberration An optical defect whereby different colors of light passing through a lens are focused at different locations. (Chapter 6)

chromosphere A layer in the atmosphere of the Sun between the photosphere and the corona. (Chapter 16)

circumpolar A term describing a star that neither rises nor sets but appears to rotate around one of the celestial poles. (Chapter 2)

circumstellar accretion disk An accretion disk that surrounds a protostar. (Chapter 18)

close binary A binary star system in which the stars are separated by a distance roughly comparable to their diameters. (Chapter 19)

closed universe A universe with positive curvature, so that its geometry is analogous to that of the surface of a sphere. (Chapter 25)

cluster (of galaxies) A collection of galaxies containing a few to several thousand member galaxies. (Chapter 23)

cluster (of stars) A group of stars that formed together and that have remained together because of their mutual gravitational attraction. (Chapter 18)

CNO cycle A series of nuclear reactions in which carbon is used as a catalyst to transform hydrogen into helium. (Chapter 16)

cocoon nebula The nebulosity surrounding a protostar. (Chapter 18)

co-creation theory The hypothesis that Earth and the Moon formed at the same time from the same material. (Chapter 10)

cold dark matter Slowly moving, weakly interacting particles presumed to constitute the bulk of matter in the universe. (Chapter 26)

collapsar A proposed type of supernova in which a black hole forms at the center of a dying star before the outer layers of the star have time to collapse. (Chapter 21)

collisional ejection theory The hypothesis that the Moon formed from material ejected from Earth by the impact of a large asteroid. (Chapter 10)

color ratio The ratio of the apparent brightness of a star measured in one spectral region to its brightness measured in a different region. (Chapter 17)

color-magnitude diagram A plot of the apparent magnitudes (that is, apparent brightnesses) of stars in a cluster versus their color indices (a measure of their surface temperatures). (Chapter 19)

coma (of a comet) The diffuse gaseous component of the head of a comet. (Chapter 15)

coma (optical) The distortion of off-axis images formed by a parabolic mirror. (Chapter 6)

combined average mass density The average density in the observable universe of all forms of matter and energy, measured in units of mass per volume. (Chapter 25)

comet A small body of ice and dust in orbit about the Sun. While passing near the Sun, a comet's vaporized ices give rise to a coma and tail. (Chapter 7, Chapter 15)

compound A substance consisting of two or more chemical elements in a definite proportion. (Chapter 5)

compression A region in a sound wave where the medium carrying the wave is compressed. (Chapter 25)

condensation temperature The temperature at which a particular substance in a low-pressure gas condenses into a solid. (Chapter 8)

conduction The transfer of heat by directly passing energy from atom to atom. (Chapter 16)

conic section The curve of intersection between a circular cone and a plane; this curve can be a circle, ellipse, parabola, or hyperbola. (Chapter 4)

conjunction The geometric arrangement of a planet in the same part of the sky as the Sun, so that the planet is at an elongation of 0°. (Chapter 4)

conservation of angular momentum A law of physics stating that in an isolated system, the total amount of angular momentum—a measure of the amount of rotation—remains constant. (Chapter 8)

conservation of energy A law of physics stating that energy can change forms and be transferred from one object to another but can't be created or destroyed. (Chapter 4)

constellation A configuration of stars in the same region of the sky. (Chapter 2)

contact binary A binary star system in which both members fill their Roche lobes. (Chapter 19)

continuous spectrum A spectrum of light over a range of wavelengths without any spectral lines. (Chapter 5)

convection The transfer of energy by moving currents of fluid or gas containing that energy. (Chapter 9, Chapter 16)

convection cell A circulating loop of gas or liquid that transports heat from a warm region to a cool region. (Chapter 9)

convection current The pattern of motion in a gas or liquid in which convection is taking place. (Chapter 9)

convective zone The region in a star where convection is the dominant means of energy transport. (Chapter 16)

core (of Earth) The iron-rich inner region of Earth's interior. (Chapter 9)

core accretion model The hypothesis that each of the Jovian planets formed by accretion of gas onto a rocky core. (Chapter 8)

core helium fusion The thermonuclear fusion of helium at the center of a star. (Chapter 19, Chapter 20)

core hydrogen fusion The thermonuclear fusion of hydrogen at the center of a star. (Chapter 19)

core-collapse supernova A supernova that occurs at the end of a massive star's lifetime when the star's core collapses to high density and the star's outer layers are expelled into space. (See Type Ib supernova, Type Ic supernova, and Type II supernova) (Chapter 20)

corona (of the Sun) The Sun's outer atmosphere, which has a high temperature and a low density. (Chapter 16)

coronal hole A region in the Sun's corona that is deficient in hot gases. (Chapter 16)

coronal mass ejection An event in which billions of tons of gas from the Sun's corona is suddenly blasted into space at high speed. (Chapter 16)

cosmic background radiation See *cosmic microwave background (CMB)*.

cosmic light horizon An imaginary sphere, centered on Earth, whose radius equals the distance light has traveled since the Big Bang. (Chapter 25, Chapter 26)

cosmic microwave background (CMB) An isotropic radiation field with a blackbody temperature of about 2.725 K that permeates the entire universe. (Chapter 25)

cosmic rays Fast-moving subatomic particles that enter our solar system from interstellar space. (Chapter 11)

cosmological constant (L) In the equations of general relativity, a quantity signifying a pressure throughout all space that helps the universe expand; a type of dark energy. (Chapter 25)

cosmological principle The assumption that the universe is homogeneous and isotropic on the largest scale. (Chapter 25)

cosmological redshift A redshift that is caused by the expansion of the universe. (Chapter 25)

cosmology The study of the structure and evolution of the universe. (Chapter 25)

coudé focus An optical arrangement with a reflecting telescope. A series of mirrors is used to direct light to a remote focus away from the moving parts of the telescope. (Chapter 6)

crater See *impact crater.*

critical density The value of the combined average mass density throughout the universe for which space would be flat. (Chapter 25)

crust (of a planet) The surface layer of a terrestrial planet. (Chapter 9)

crustal dichotomy (on Mars) The contrast between the young northern lowlands and older southern highlands on Mars. (Chapter 11)

cryovolcano An ice-volcano that erupts or oozes water instead of lava. (Chapter 13)

crystal A material in which atoms are arranged in orderly rows. (Chapter 9)

D ring One of several faint rings encircling Saturn. (Chapter 12)

dark ages The era between the formation of the first atoms (when the universe was about 380,000 years old) and the formation of the first stars (when the universe was about 400 million years old). (Chapter 26)

dark energy A form of energy that appears to pervade the universe and causes the expansion of the universe to accelerate but has no discernible gravitational effect. (Chapter 25)

dark energy density parameter (V_L) The ratio of the average mass density of dark energy in the universe to the critical density. (Chapter 25)

dark matter Nonluminous matter that is the dominant form of matter in galaxies and throughout the universe. (Chapter 22)

dark nebula A cloud of interstellar gas and dust that obscures the light of more distant stars. (Chapter 18)

dark terrain (on Ganymede) Older, heavily cratered, dark-colored terrain on the surface of Ganymede. (Chapter 13)

dark-energy–dominated universe A universe in which the mass density of dark energy exceeds both the average density of matter and the mass density of radiation. (Chapter 25)

dark-matter problem The enigma that most of the matter in the universe seems not to emit radiation of any kind and is detectable by its gravity alone. (Chapter 23)

dead quasar A supermassive black hole that has already consumed the gas and dust that previously provided fuel for quasar activity. Any supermassive black hole in the center of a galaxy is presumed to have gone through a brief period when it exhibited quasar activity. (Chapter 24)

dead quasar flare The flare that results if a star wanders too close to a supermassive black hole in the center of a galaxy (which is presumably a dead quasar); gravitational tidal forces can tear some matter away from the star and provide a small source of fuel to briefly "light up" the dead quasar with an approximately year-long flare of light. (Chapter 24)

declination Angular distance of a celestial object north or south of the celestial equator. (Chapter 2)

deferent A stationary circle in the Ptolemaic system along which another circle (an epicycle) moves, carrying a planet, the Sun, or the Moon. (Chapter 4)

degeneracy The phenomenon, due to quantum mechanical effects, whereby the pressure exerted by a gas does not depend on its temperature. (Chapter 19)

degenerate-electron (-neutron) pressure The pressure exerted by degenerate electrons (neutrons). (Chapter 19, Chapter 20)

degree A basic unit of angular measure, designated by the symbol °. (Chapter 1)

degree Celsius A basic unit of temperature, designated by the symbol °C and used on a scale where water freezes at 0° and boils at 100°. (Chapter 5)

degree Fahrenheit A basic unit of temperature, designated by the symbol °F and used on a scale where water freezes at 32° and boils at 212°. (Chapter 5)

density See *average density.*

density fluctuation A variation of density from one place to another. (Chapter 26)

density parameter (V_0) The ratio of the average density of matter and energy in the universe to the critical density. (Chapter 25)

density wave In a spiral galaxy, a localized region in which matter piles up as it orbits the center of the galaxy; this is a proposed explanation of spiral structure. (Chapter 22)

detached binary A binary star system in which neither star fills its Roche lobe. (Chapter 19)

deuterium bottleneck The situation during the first 3 minutes after the Big Bang when deuterium (an isotope of hydrogen whose nucleus contains one proton and one neutron) inhibited the formation of heavier elements. (Chapter 26)

differential rotation The rotation of a nonrigid object in which parts adjacent to each other at a given time do not always stay close together. (Chapter 12, Chapter 16)

differentiated asteroid An asteroid in which chemical differentiation has taken place so that denser material is toward the asteroid's center. (Chapter 15)

differentiation See *chemical differentiation.*

diffraction The spreading out of light passing through an aperture or opening in an opaque object. (Chapter 6)

diffraction grating An optical device, consisting of thousands of closely spaced lines etched in glass or metal, that disperses light into a spectrum. (Chapter 6)

direct motion The apparent eastward movement of a planet seen against the background stars. (Chapter 4)

disk (of a galaxy) The disk-shaped distribution of Population I stars that dominates the appearance of a spiral galaxy. (Chapter 22)

distance ladder The sequence of techniques used to determine the distances to very remote galaxies. (Chapter 23)

distance modulus The difference between the apparent and absolute magnitudes of an object. (Chapter 17)

diurnal motion Any apparent motion in the sky that repeats on a daily basis, such as the rising and setting of stars. (Chapter 2)

Doppler effect The apparent change in wavelength of radiation due to relative motion between the source and the observer along the line of sight. (Chapter 5)

double star A pair of stars located at nearly the same position in the night sky. Some, but not all, double stars are binary stars. (Chapter 17)

Drake equation An equation used to estimate the number of intelligent civilizations in the Galaxy with whom we might communicate. (Chapter 27)

dredge-up The transporting of matter from deep within a star to its surface by convection. (Chapter 20)

dust devil Whirlwind found in dry or desert areas on both Earth and Mars. (Chapter 11)

dust grain A microscopic bit of solid matter found in interplanetary or interstellar space. (Chapter 18)

dust tail The tail of a comet that is composed primarily of dust grains. (Chapter 15)

dwarf elliptical galaxy A low-mass elliptical galaxy that contains only a few million stars. (Chapter 23)

dwarf planet A solar system body that is large enough to be spherical in shape and have a circular orbit around the Sun, but not large enough to clear its own path of other bodies. The term is used for Ceres, Pluto, and Eris. (Chapter 14)

dynamo The mechanism whereby electric currents within an astronomical body generate a magnetic field. Also referred to as a magnetic dynamo. (Chapter 7, Chapter 8)

E ring A very broad, faint ring encircling Saturn. (Chapter 12)

earthquake A sudden vibratory motion of Earth's surface. (Chapter 9)

eccentricity A number between 0 and 1 that describes the shape of an ellipse. (Chapter 4)

eclipse path The track of the tip of the Moon's shadow along Earth's surface during a total or annular solar eclipse. (Chapter 3)

eclipse The cutting off of part or all the light from one celestial object by another. (Chapter 3)

eclipse year The interval between successive passages of the Sun through the same node of the Moon's orbit. (Chapter 3)

eclipsing binary A binary star system in which, as seen from Earth, stars periodically pass in front of each other. (Chapter 17)

ecliptic The apparent annual path of the Sun on the celestial sphere. (Chapter 2)

ecliptic plane The plane of Earth's orbit around the Sun. (Chapter 2)

Eddington limit A constraint on the rate at which a black hole can accrete matter. (Chapter 24)

electromagnetic radiation Radiation consisting of oscillating electric and magnetic fields. Examples include gamma rays, X rays, visible light, ultraviolet and infrared radiation, radio waves, and microwaves. (Chapter 5)

electromagnetic spectrum The entire array of electromagnetic radiation. (Chapter 5)

electromagnetism Electric and magnetic phenomena, including electromagnetic radiation. (Chapter 5)

electron A subatomic particle with a negative charge and a small mass, usually found in orbits about the nuclei of atoms. (Chapter 5)

electron volt (eV) The energy acquired by an electron accelerated through an electric potential of one volt. (Chapter 5)

electroweak force A unification of the electromagnetic and weak forces that occurs at high energies. (Chapter 26)

element A chemical that cannot be broken down into more basic chemicals. (Chapter 5)

elementary particle physics The branch of quantum mechanics that deals with individual subatomic particles and their interactions. (Chapter 26)

ellipse A conic section obtained by cutting completely through a circular cone with a plane. (Chapter 4)

elliptical galaxy A galaxy with an elliptical shape and no conspicuous interstellar material. (Chapter 23)

elongation The angular distance between a planet and the Sun as viewed from Earth. (Chapter 4)

emission line spectrum A spectrum that contains bright emission lines. (Chapter 5)

emission nebula A glowing gaseous nebula whose spectrum has bright emission lines. (Chapter 18)

Encke gap A narrow gap in Saturn's A ring. (Chapter 12)

energy flux The rate of energy flow, usually measured in joules per square meter per second. (Chapter 5)

energy level In an atom, a particular amount of energy possessed by an atom above the atom's least energetic state. (Chapter 5)

energy-level diagram A diagram showing the arrangement of an atom's energy levels. (Chapter 5)

epicenter The location on Earth's surface directly over the focus of an earthquake. (Chapter 9)

epicycle A moving circle in the Ptolemaic system about which a planet revolves. (Chapter 4)

epoch The date used to define the coordinate system for objects on the sky. (Chapter 2)

equal areas, law of See *Kepler's second law.*

equinox One of the intersections of the ecliptic and the celestial equator. Also used to refer to the date on which the Sun passes through such an intersection. (Chapter 2) See also *autumnal equinox* and *vernal equinox.*

equivalence principle In the general theory of relativity, the principle that in a small volume of space, the downward pull of gravity can be accurately and completely duplicated by an upward acceleration of the observer. (Chapter 21)

era of recombination The moment, approximately 380,000 years after the Big Bang, when the universe had cooled sufficiently to permit the formation of hydrogen atoms. (Chapter 25)

ergoregion The region of space immediately outside the event horizon of a rotating black hole where it is impossible to remain at rest. (Chapter 21)

escape speed The speed needed by an object (such as a spaceship) to leave a second object (such as a planet or star) permanently and to escape into interplanetary space. (Chapter 7)

eV See *electron volt.*

event horizon The location around a black hole where the escape speed equals the speed of light; the surface of a black hole. (Chapter 21)

evolutionary track The path on an H-R diagram followed by a star as it evolves. (Chapter 18)

excited state A state of an atom, ion, or molecule with a higher energy than the ground state. (Chapter 5)

exoplanet Planets orbiting a star other than our Sun. (Chapter 8)

exotic dark matter Dark matter that is not made up of regular matter. Regular matter is made of protons, neutrons, and electrons, including all atoms and molecules. There is evidence that the vast majority of dark matter is not made of regular matter, but little else is known about what it might be. (Chapter 23)

exponent A number placed above and after another number to denote the power to which the latter is to be raised, as n in 10^n. (Chapter 1)

extinction (interstellar) See *interstellar extinction.*

extrasolar planet A planet orbiting a star other than the Sun. (Chapter 8)

extremophile Microorganisms that live in extreme conditions such as hot or cold temperatures, high acidity, or high radioactivity. (Chapter 27)

eyepiece lens A magnifying lens used to view the image produced at the focus of a telescope. (Chapter 6)

F ring A thin, faint ring encircling Saturn just beyond the A ring. (Chapter 12)

false color In astronomical images, color used to denote different values of intensity, temperature, or other quantities. (Chapter 6)

false vacuum Empty space with an abnormally high energy density that is thought to have filled the universe immediately after the Big Bang. (Chapter 26)

far side (of the Moon) he side of the Moon that faces perpetually away from Earth. (Chapter 10)

far-infrared The part of the infrared spectrum most different in wavelength from visible light. (Chapter 22)

filament A portion of the Sun's chromosphere that arches to high altitudes. (Chapter 16)

first quarter moon The phase of the Moon that occurs when the Moon is 90° east of the Sun. (Chapter 3)

fission theory The hypothesis that the Moon was pulled out of a rapidly rotating proto-Earth. (Chapter 10)

flake tectonics A model of a planetary interior, particularly Venus, in which a thin crust remains stationary but wrinkles and flakes in response to interior convection currents. (Chapter 11)

flare See *solar flare*.

flat space Space that is not curved; space with zero curvature. (Chapter 25)

flatness problem The dilemma posed by the fact that the combined average mass density of the universe is very nearly equal to the critical density. (Chapter 26)

flocculent spiral galaxy A spiral galaxy with fuzzy, poorly defined spiral arms. (Chapter 22)

fluorescence A process in which high-energy ultraviolet photons are absorbed and the absorbed energy is radiated as lower-energy photons of visible light. (Chapter 18)

focal length The distance from a lens or mirror to the point where converging light rays meet. (Chapter 6)

focal plane The plane in which a lens or mirror forms an image of a distant object. (Chapter 6)

focal point The point at which a lens or mirror forms an image of a distant point of light. (Chapter 6)

focus (of a lens or mirror) The point to which light rays converge after passing through a lens or being reflected from a mirror. (Chapter 6)

focus (of an ellipse) (plural foci) One of two points inside an ellipse such that the combined distance from the two foci to any point on the ellipse is a constant. (Chapter 4)

force A push or pull that acts on an object. (Chapter 4)

frequency The number of crests or troughs of a wave that cross a given point per unit time. Also, the number of vibrations per unit time. (Chapter 5)

full moon A phase of the Moon during which its full daylight hemisphere can be seen from Earth. (Chapter 3)

fundamental plane A relationship among the size of an elliptical galaxy, the average motions of its stars, and how the galaxy's brightness appears distributed over its surface. (Chapter 23)

fusion crust The coating on a stony meteorite caused by the heating of the meteorite as it descended through Earth's atmosphere. (Chapter 15)

G ring A thin, faint ring encircling Saturn. (Chapter 12)

galactic cannibalism A collision between two galaxies of unequal mass and size in which the smaller galaxy seems to be absorbed by the larger galaxy. (Chapter 23)

galactic cluster See *open cluster*.

galactic nucleus The center of a galaxy. (Chapter 22)

galaxy A large assemblage of stars, nebulae, and interstellar gas and dust. (Chapter 1, Chapter 22)

Galilean satellites The four large moons of Jupiter. (Chapter 13)

gamma rays The most energetic form of electromagnetic radiation. (Chapter 5)

gamma-ray bursters Objects found in all parts of the sky that emit a one-time intense burst of high-energy radiation. (Chapter 21)

general theory of relativity A description of gravity formulated by Albert Einstein. It states that gravity affects the geometry of space and the flow of time. (Chapter 21)

geocentric model An Earth-centered theory of the universe. (Chapter 4)

giant A star whose diameter is typically 10 to 100 times that of the Sun and whose luminosity is roughly that of 100 Suns. (Chapter 17)

giant elliptical galaxy A large, massive, elliptical galaxy containing many billions of stars. (Chapter 23)

giant molecular cloud A large cloud of interstellar gas and dust in which temperatures are low enough and densities high enough for atoms to form into molecules. (Chapter 18)

global warming The upward trend of Earth's average temperature caused by increased amounts of greenhouse gases in the atmosphere. (Chapter 9)

globular cluster A large spherical cluster of stars, typically found in the outlying regions of a galaxy. (Chapter 19, Chapter 22)

gluon A particle that is exchanged between quarks. (Chapter 26)

Grand Tack model A model in which the Jovian planets migrate toward the Sun in the first few million years of the solar system and then move back outward. This model can help explain Mars's small mass and icy objects in the asteroid belt. (Chapter 8)

grand unified theory (GUT) A theory that describes and explains the four physical forces. (Chapter 26)

grand-design spiral galaxy A galaxy with well-defined spiral arms. (Chapter 22)

granulation The rice grain-like structure found in the solar photosphere. (Chapter 16)

granule A convective cell in the solar photosphere. (Chapter 16)

grating See *diffraction grating*.

gravitational force See *gravity*.

gravitational lens A massive object that deflects light rays from a remote source, forming an image much as an ordinary lens does. (Chapter 23)

gravitational potential energy Energy associated with Earth's pull on an object toward Earth's center. This energy increases with an object's distance from Earth. (Chapter 4)

gravitational radiation (gravitational waves) Oscillations of space produced by changes in the distribution of matter. (Chapter 21)

gravitational redshift The increase in the wavelength of a photon as it climbs upward in a gravitational field. (Chapter 21)

graviton The particle that is responsible for the gravitational force. (Chapter 26)

gravity The force with which all matter attracts all other matter. (Chapter 4)

Great Red Spot A prominent high-pressure system in Jupiter's southern hemisphere. (Chapter 12)

greatest eastern elongation The configuration of an inferior planet at its greatest angular distance east of the Sun. (Chapter 4, Chapter 11)

greatest western elongation The configuration of an inferior planet at its greatest angular distance west of the Sun. (Chapter 4, Chapter 11)

greenhouse effect The trapping of infrared radiation near a planet's surface by the planet's atmosphere. (Chapter 9)

greenhouse gas A substance whose presence in a planet's atmosphere enhances the greenhouse effect. (Chapter 9)

ground state The state of an atom, ion, or molecule with the least possible energy. (Chapter 5)

group (of galaxies) A poor cluster of galaxies. (Chapter 23)

GUT See *grand unified theory*.

H I Neutral, unionized hydrogen. (Chapter 22)

H II region A region of ionized hydrogen in interstellar space. (Chapter 18)

habitable zone Regions of a galaxy or around a star in which conditions may be suitable for life to have developed. (Chapter 8, Chapter 27)

half-life The time required for one-half of a quantity of a radioactive substance to decay. (Chapter 8)

halo (of a galaxy) A spherical distribution of globular clusters and Population II stars that surround a spiral galaxy. (Chapter 22)

Hawking radiation Light and particles made of matter and antimatter that emanate away from a region just outside a black hole's event horizon. The radiation occurs when one member of a virtual particle pair is captured by the black hole; the other member of the pair becomes a real particle and carries away some of the black hole's energy (and therefore, some of its mass). This loss of energy leads to black hole evaporation. (Chapter 21)

Heisenberg uncertainty principle A principle of quantum mechanics that places limits on the precision of simultaneous measurements. (Chapter 21, Chapter 26)

heliocentric model A Sun-centered theory of the universe. (Chapter 4)

helioseismology The study of the vibrations of the Sun as a whole. (Chapter 16)

helium flash The nearly explosive beginning of helium fusion in the dense core of a red giant star. (Chapter 19)

helium fusion The thermonuclear fusion of helium to form carbon and oxygen. (Chapter 19)

helium shell flash A brief thermal runaway that occurs in the helium-fusing shell of a red supergiant. (Chapter 20)

Herbig-Haro object A small, luminous nebula associated with the end point of a jet emanating from a young star. (Chapter 18)

Hertzsprung-Russell (H-R) diagram A plot of the luminosity (or absolute magnitude) of stars against their surface temperature (or spectral type). (Chapter 17)

highlands (on Mars) See *southern highlands*.

highlands (on the Moon) See *lunar highlands*.

Hirayama family A group of asteroids that have nearly identical orbits about the Sun. (Chapter 15)

homogeneous Having the same property in one region as in every other region. (Chapter 25)

horizon problem See *isotropy problem*.

horizontal branch A group of stars on the color-magnitude diagram of a typical globular cluster that lies near the main sequence and has roughly constant luminosity. (Chapter 20)

horizontal-branch star A low-mass, post–helium-flash star on the horizontal branch. (Chapter 19)

hot dark matter Dark matter consisting of low-mass particles moving at high speeds. (Chapter 26)

hot Jupiter A Jupiter-sized planet orbiting a star other than our Sun with an orbital distance similar to Mercury's, and thus a very hot planet. (Chapter 8)

hot-spot volcanism Volcanic activity that occurs over a hot region buried deep within a planet. (Chapter 11)

H-R diagram See *Hertzsprung-Russell diagram*.

Hubble classification A method of classifying galaxies as spirals, barred spirals, ellipticals, or irregulars according to their appearance. (Chapter 23)

Hubble constant (H_0) In the Hubble law, the constant of proportionality between the recessional velocities of remote galaxies and their distances. (Chapter 23)

Hubble flow The recessional motions of remote galaxies caused by the expansion of the universe. (Chapter 23)

Hubble law The empirical relationship stating that the redshifts of remote galaxies are directly proportional to their distances from Earth. (Chapter 23)

Hubble time A fairly accurate estimate of the age of the universe. It is equal to the inverse of the Hubble constant. This age estimate makes a simplifying assumption that the expansion of the universe has occurred at precisely the same rate since the Big Bang, but the expansion rate was slightly different in the past. (Chapter 25)

hydrocarbon Any one of a variety of chemical compounds composed of hydrogen and carbon. (Chapter 13)

hydrogen envelope A huge, tenuous sphere of gas surrounding the head of a comet. (Chapter 15)

hydrogen fusion The thermonuclear conversion of hydrogen into helium. (Chapter 16)

hydrostatic equilibrium A balance between the weight of a layer in a star and the pressure that supports it. (Chapter 16)

hyperbola A conic section formed by cutting a circular cone with a plane at an angle steeper than the sides of the cone. (Chapter 4)

hyperbolic space Space with negative curvature. (Chapter 25)

hypothesis An idea or collection of ideas that seems to explain a specified phenomenon; a conjecture. (Chapter 1)

ice rafts (Europa) Segments of Europa's icy crust that have been moved by tectonic disturbances. (Chapter 13)

ices Solid materials with low condensation temperatures, including ices of water, methane, and ammonia. (Chapter 7)

ideal gas A gas in which the pressure is directly proportional to both the density and the temperature of the gas; an idealization of a real gas. (Chapter 19)

igneous rock A rock that formed from the solidification of molten lava or magma. (Chapter 9)

imaging The process of recording the image made by a telescope of a distant object. (Chapter 6)

impact breccia A type of rock formed from other rocks that were broken apart, mixed, and fused together by a series of meteoritic impacts. (Chapter 10)

impact crater A circular depression on a planet or satellite caused by the impact of a meteoroid. (Chapter 7, Chapter 10)

induced electric current Electric current resulting from a magnetic field sweeping over an electrically conductive material. (Chapter 13)

induced magnetic field A magnetic field resulting from an induced electric current. (Chapter 13)

inertia, law of See *Newton's first law of motion.*

inferior conjunction The configuration when an inferior planet is between the Sun and Earth. (Chapter 4)

inferior planet A planet that is closer to the Sun than Earth is. (Chapter 4)

inflation A sudden expansion of space. (Chapter 26)

inflationary epoch A brief period shortly after the Big Bang during which the scale of the universe increased very rapidly. (Chapter 26)

infrared radiation Electromagnetic radiation of wavelength longer than visible light but shorter than radio waves. (Chapter 5)

inner core (of Earth) The solid innermost portion of Earth's iron-rich core. (Chapter 9)

inner Lagrangian point The point between the two stars constituting a binary system where their Roche lobes touch; the point across which mass transfer can occur. (Chapter 19)

instability strip A region of the H-R diagram occupied by pulsating stars. (Chapter 19)

interferometry A technique of combining the observations of two or more telescopes to produce images better than one telescope alone could make. (Chapter 6)

intermediate vector boson The particle that is responsible for the weak force. (Chapter 26)

intermediate-mass black hole See *mid-mass black hole.*

interstellar extinction The dimming of starlight as it passes through the interstellar medium. (Chapter 18, Chapter 22)

interstellar matter Diffuse gas and dust spread out in the galaxy between the stars. Interstellar matter is mostly hydrogen. (Chapter 22)

interstellar medium Gas and dust in interstellar space. (Chapter 8, Chapter 18)

interstellar reddening The reddening of starlight passing through the interstellar medium as a result of blue light being scattered more than red. (Chapter 18)

intracluster gas Gas found between the galaxies that make up a cluster of galaxies. (Chapter 23)

inverse-square law The statement that the apparent brightness of a light source varies inversely with the square of the distance from the source. (Chapter 17)

Io torus A doughnut-shaped ring of gas circling Jupiter at the distance of Io's orbit. (Chapter 13)

ion tail The relatively straight tail of a comet produced by the solar wind acting on ions. (Chapter 15)

ionization The process by which a neutral atom becomes an electrically charged ion through the loss or gain of electrons. (Chapter 5)

iron meteorite A meteorite composed primarily of iron. (Chapter 15)

irregular cluster (of galaxies) A sprawling collection of galaxies whose overall distribution in space does not exhibit any noticeable spherical symmetry. (Chapter 23)

irregular galaxy An asymmetrical galaxy having neither spiral arms nor an elliptical shape. (Chapter 23)

isotope Any of several forms for the same chemical element whose nuclei all have the same number of protons but different numbers of neutrons. (Chapter 5)

isotropic Having the same property in all directions. (Chapter 23, Chapter 25)

isotropy problem The dilemma posed by the fact that the cosmic microwave background is isotropic; also called the horizon problem. (Chapter 26)

Jeans length The smallest scale over which a density fluctuation in a medium will contract to form a gravitationally bound object. (Chapter 26)

joule (J) A unit of energy. (Chapter 5)

Jovian planet Low-density planet composed primarily of hydrogen and helium; the Jovian planets include Jupiter, Saturn, Uranus, and Neptune. (Chapter 7)

Jupiter-family comet A comet with an orbital period of less than 20 years. (Chapter 15)

kelvin (K) A unit of temperature on the Kelvin temperature scale, equivalent to a degree Celsius. (Chapter 5)

Kelvin-Helmholtz contraction The contraction of a gaseous body, such as a star or nebula, during which gravitational energy is transformed into thermal energy. (Chapter 8)

Kepler's first law The statement that each planet moves around the Sun in an elliptical orbit with the Sun at one focus of the ellipse. (Chapter 4)

Kepler's second law The statement that a planet sweeps out equal areas in equal times as it orbits the Sun; also called the law of equal areas. (Chapter 4)

Kepler's third law A relationship between the period of an orbiting object and the semimajor axis of its elliptical orbit. (Chapter 4)

kiloparsec (kpc) One thousand parsecs; about 3260 light-years. (Chapter 1)

kinetic energy The energy possessed by an object because of its motion. (Chapter 4, Chapter 7)

Kirchhoff's laws Three statements about circumstances that produce absorption lines, emission lines, and continuous spectra. (Chapter 5)

Kirkwood gaps Gaps in the spacing of asteroid orbits, discovered by Daniel Kirkwood. (Chapter 15)

Kuiper belt A region that extends from around the orbit of Pluto to about 500 AU from the Sun where many icy objects orbit the Sun. (Chapter 7, Chapter 14, Chapter 15)

Lamb shift A tiny shift in the spectral lines of hydrogen caused by the presence of virtual particles. (Chapter 26)

Late Heavy Bombardment A brief, intense period of large impacts on all bodies of the solar system about 600 million years after formation of the solar system. (Chapter 8)

lava Molten rock flowing on the surface of a planet. (Chapter 9)

law of equal areas See *Kepler's second law.*

law of inertia See *Newton's first law of motion.*

law of universal gravitation A formula deduced by Isaac Newton that expresses the strength of the force of gravity that two masses exert on each other. (Chapter 4)

laws of physics A set of physical principles with which we can understand natural phenomena and the nature of the universe. (Chapter 1)

length contraction In the special theory of relativity, the shrinking of an object's length along its direction of motion. (Chapter 21)

lens A piece of transparent material (usually glass) that can bend light and bring it to a focus. (Chapter 6)

lenticular galaxy A galaxy with a central bulge and a disk but no spiral structure; an S0 galaxy. (Chapter 23)

libration An apparent rocking of the Moon whereby an Earth-based observer can, over time, see slightly more than one-half the Moon's surface. (Chapter 10)

light curve A graph that displays how the brightness of a star or other astronomical object varies over time. (Chapter 17)

light pollution Light from cities and towns that degrades telescope images. (Chapter 6)

light scattering The process by which light bounces off particles in its path. (Chapter 5, Chapter 12)

light-gathering power A measure of the amount of radiation brought to a focus by a telescope. (Chapter 6)

light-year (ly) The distance light travels in a vacuum in one year. (Chapter 1)

limb darkening The phenomenon whereby the Sun looks darker near its apparent edge, or limb, than near the center of its disk. (Chapter 16)

line of nodes The line where the plane of Earth's orbit intersects the plane of the Moon's orbit. (Chapter 3)

liquid metallic hydrogen Hydrogen compressed to such a density that it behaves like a liquid metal. (Chapter 7, Chapter 12)

lithosphere The solid, upper layer of Earth; essentially Earth's crust. (Chapter 9)

Local Bubble A large cavity in the interstellar medium within which the Sun and nearby stars are located. (Chapter 22)

Local Group The cluster of galaxies of which our Galaxy is a member. (Chapter 23)

local meridian See *meridian.*

long-period comet A comet that takes hundreds of thousands of years or more to complete one orbit of the Sun. (Chapter 15)

long-period variable A variable star with a period longer than about 100 days. (Chapter 19)

lookback time How far into the past we are looking when we see a particular object. (Chapter 25)

Lorentz transformations Equations that relate the measurements of different observers who are moving relative to each other at high speeds. (Chapter 21)

lower meridian The half of the meridian that lies below the horizon. (Chapter 2)

lowlands (on Mars) See *northern lowlands.*

luminosity The rate at which electromagnetic radiation is emitted from a star or other object. (Chapter 5, Chapter 16, Chapter 17)

luminosity class A classification of a star of a given spectral type according to its luminosity. (Chapter 17)

luminosity function The numbers of stars of differing brightness per cubic parsec. (Chapter 17)

lunar eclipse An eclipse of the Moon by Earth; a passage of the Moon through Earth's shadow. (Chapter 3)

lunar highlands Ancient, high-elevation, heavily cratered terrain on the Moon. (Chapter 10)

lunar month See *synodic month.*

lunar phase The appearance of the illuminated area of the Moon as seen from Earth. (Chapter 3)

Lyman series A series of spectral lines of hydrogen produced by electron transitions to and from the lowest energy state of the hydrogen atom. (Chapter 5)

MACHO See *massive compact halo object.*

magma Molten rock beneath a planet's surface. (Chapter 9)

magnetic axis A line connecting the north and south magnetic poles of a planet or star possessing a magnetic field. (Chapter 14)

magnetic dynamo The mechanism whereby electric currents within an astronomical body generate a magnetic field. Sometimes simply referred to as a dynamo. (Chapter 9)

magnetic reconnection An event where two oppositely directed magnetic fields approach and cancel, thus releasing energy. (Chapter 16)

magnetic resonance imaging (MRI) A technique for viewing the interior of the human body that makes use of the way protons respond to magnetic fields. (Chapter 22)

magnetic-dynamo model A theory that explains the solar cycle as a result of the Sun's differential rotation acting on the Sun's magnetic field. (Chapter 16)

magnetogram An image of the Sun that shows regions of different magnetic polarity. (Chapter 16)

magnetometer A device for measuring magnetic fields. (Chapter 7)

magnetopause That region of a planet's magnetosphere where the magnetic field counterbalances the pressure from the solar wind. (Chapter 9)

magnetosphere The region around a planet occupied by its magnetic field. (Chapter 9)

magnification The factor by which the apparent angular size of an object is increased when viewed through a telescope. (Chapter 6)

magnifying power See *magnification.*

magnitude scale A system for denoting the brightnesses of astronomical objects. (Chapter 17)

main sequence A grouping of stars on the Hertzsprung-Russell diagram extending diagonally across the graph from hot, luminous stars to cool, dim stars. (Chapter 17)

main-sequence lifetime The total time that a star spends fusing hydrogen in its core, and hence the total time that it will spend as a main-sequence star. (Chapter 19)

main-sequence star A star whose luminosity and surface temperature place it on the main sequence on an H-R diagram; a star that derives its energy from core hydrogen fusion. (Chapter 17)

major axis (of an ellipse) The longest diameter of an ellipse. (Chapter 4)

mantle (of a planet) That portion of a terrestrial planet located between its crust and core. (Chapter 9)

mare (plural maria) Latin for "sea"; a large, relatively crater-free plain on the Moon. (Chapter 10)

mare basalt A type of lunar rock commonly found in the mare basins. (Chapter 10)

maser An interstellar cloud in which water molecules emit intensely at microwave wavelengths. (Chapter 23)

mass A measure of the total amount of material in an object. (Chapter 4)

mass density of radiation The energy possessed by a radiation field per unit volume divided by the square of the speed of light. (Chapter 25)

mass loss A process by which a star gently loses matter. (Chapter 19)

mass transfer The flow of gases from one star in a binary system to the other. (Chapter 19)

massive compact halo object (MACHO) A dim star or low-mass black hole that may constitute part of the unseen dark matter. (Chapter 22)

mass-luminosity relation A relationship between the masses and luminosities of main-sequence stars. (Chapter 17)

mass-radius relation A relationship between the masses and radii of white dwarf stars. (Chapter 20)

matter density parameter (V_m) The ratio of the average density of matter in the universe to the critical density. (Chapter 25)

matter-dominated universe A universe in which the average density of matter exceeds both the mass density of radiation and the mass density of dark energy. (Chapter 25)

mean solar day The interval between successive meridian passages of the mean Sun; the average length of a solar day. (Chapter 2)

mean sun A fictitious object that moves eastward at a constant speed along the celestial equator, completing one circuit of the sky with respect to the vernal equinox in one tropical year. (Chapter 2)

medium (plural media) A material through which light travels. (Chapter 6)

medium, interstellar See *interstellar medium.*

megaparsec (Mpc) One million parsecs. (Chapter 1)

melting point The temperature at which a substance changes from solid to liquid. (Chapter 9)

meridian (or local meridian) The great circle on the celestial sphere that passes through an observer's zenith and the north and south celestial poles. (Chapter 2)

meridian transit The crossing of the meridian by any astronomical object. (Chapter 2)

mesosphere A layer in Earth's atmosphere above the stratosphere. (Chapter 9)

metal In reference to stars and galaxies, any element other than hydrogen and helium; in reference to planets, a material such as iron, silver, and aluminum that is a good conductor of electricity and of heat. (Chapter 17)

metal-poor star A star that, compared to the Sun, is deficient in elements heavier than helium; also called a Population II star. (Chapter 19)

metal-rich star A star whose abundance of heavy elements is roughly comparable to that of the Sun; also called a Population I star. (Chapter 19)

metamorphic rock A rock whose properties and appearance have been transformed by the action of pressure and heat beneath Earth's surface. (Chapter 9)

meteor The luminous phenomenon seen when a meteoroid enters Earth's atmosphere; a "shooting star." (Chapter 15)

meteor shower Many meteors that seem to radiate from a common point in the sky. (Chapter 15)

meteorite A fragment of a meteoroid that has survived passage through Earth's atmosphere. (Chapter 1, Chapter 8, Chapter 15)

meteoritic swarm A collection of meteoroids moving together along an orbit about the Sun. (Chapter 15)

meteoroid A small rock in interplanetary space. (Chapter 7, Chapter 15)

microlensing A phenomenon in which a compact object such as a MACHO acts as a gravitational lens, focusing the light from a distant star. (Chapter 8, Chapter 22)

microwaves Short-wavelength radio waves. (Chapter 5)

mid-mass black hole A black hole with a mass of hundreds of Suns. (Chapter 21)

migration Changes in a planet's average orbital distance from the Sun. (Chapter 8)

Milky Way Galaxy Our Galaxy; the band of faint stars seen from Earth in the plane of our Galaxy's disk. (Chapter 22)

mineral A naturally occurring solid composed of a single element or chemical combination of elements, often in the form of crystals. (Chapter 9)

minor planet See *asteroid.*

minute of arc See *arcminute.*

model (1) A hypothesis that has withstood experimental or observational tests. (2) The results of a theoretical calculation that gives the values of temperature, pressure, density, and so forth throughout the interior of an object such as a planet or star. (Chapter 1)

molecule A combination of two or more atoms. (Chapter 5)

moonquake Sudden, vibratory motion of the Moon's surface. (Chapter 10)

MRI See *magnetic resonance imaging.*

M-theory A theory that attempts to describe fundamental particles as 11-dimensional membranes. (Chapter 26)

nanometer (nm) One billionth of a meter: 1 nm $= 10^{-9}$ meter $= 10^{-6}$ millimeter $= 10^{-3}$ μm. (Chapter 5)

neap tide An ocean tide that occurs when the Moon is near first-quarter or third-quarter phase. (Chapter 4)

near-Earth object (NEO) An asteroid whose orbit lies wholly or partly within the orbit of Mars. (Chapter 15)

near-infrared The part of the infrared spectrum closest in wavelength to visible light. (Chapter 22)

nebula (plural nebulae) A cloud of interstellar gas and dust. (Chapter 1, Chapter 18)

nebular hypothesis The idea that the Sun and the rest of the solar system formed from a cloud of interstellar material. (Chapter 8)

nebulosity See *nebula.*

negative curvature The curvature of a surface or space in which parallel lines diverge and the sum of the angles of a triangle is less than 180°. (Chapter 25)

negative hydrogen ion A hydrogen atom that has acquired a second electron. (Chapter 16)

NEO See *near-Earth object.*

neon fusion The thermonuclear fusion of neon to produce heavier nuclei. (Chapter 20)

neutrino A subatomic particle with no electric charge and very little mass, yet one that is important in many nuclear reactions. (Chapter 16)

neutrino oscillation The spontaneous transformation of one type of neutrino into another type. (Chapter 16)

neutron A subatomic particle with no electric charge and with a mass nearly equal to that of the proton. (Chapter 5)

neutron capture The buildup of neutrons inside nuclei. (Chapter 20)

neutron star A very compact, dense star composed almost entirely of neutrons. (Chapter 20)

new moon The phase of the Moon when the dark hemisphere of the Moon faces Earth. (Chapter 3)

Newton's first law of motion The statement that a body remains at rest, or moves in a straight line at a constant speed, unless acted upon by a net outside force; the law of inertia. (Chapter 4)

Newton's form of Kepler's third law A relationship between the period of two objects orbiting each other, the semimajor axis of their orbit, and the masses of the objects. (Chapter 4)

Newton's second law of motion A relationship between the acceleration of an object, the object's mass, and the net outside force acting on the mass. (Chapter 4)

Newton's third law of motion The statement that whenever one body exerts a force on a second body, the second body exerts an equal and opposite force on the first body. (Chapter 4)

Newtonian mechanics The branch of physics based on Newton's laws of motion. (Chapter 1, Chapter 4)

Newtonian reflector A reflecting telescope that uses a small mirror to deflect the image to one side of the telescope tube. (Chapter 6)

Nice model A model in which the Jovian planets began in a more compact configuration in the early solar system and then migrated outward. This model can explain aspects of the Kuiper belt, the Oort cloud, and the Late Heavy Bombardment. (Chapter 8)

noble gas An element whose atoms do not combine into molecules. (Chapter 12)

node See *line of nodes.*

no-hair theorem A statement of the simplicity of black holes. (Chapter 21)

north celestial pole The point directly above Earth's north pole where Earth's axis of rotation, if extended, would intersect the celestial sphere. (Chapter 2)

northern lights See *aurora borealis.*

northern lowlands (on Mars) Relatively young and crater-free terrain in the Martian northern hemisphere. (Chapter 11)

nova (plural novae) A star that experiences a sudden outburst of radiant energy, temporarily increasing its luminosity roughly a thousandfold. (Chapter 20)

nuclear density The density of matter in an atomic nucleus; about $4 \cdot 10^{17}$ kg/m^3. (Chapter 20)

nuclear force The force that holds protons and neutrons together in an atom. It arises from the strong force. It is strong enough to hold protons closely together despite their electrical repulsion. (Chapter 26)

nucleosynthesis The process of building up nuclei such as deuterium and helium from protons and neutrons. (Chapter 26)

nucleus (of an atom) The massive part of an atom, composed of protons and neutrons, about which electrons revolve. (Chapter 5)

nucleus (of a comet) A collection of ices and dust that constitute the solid part of a comet. (Chapter 15)

nucleus (of a galaxy) See *galactic nucleus.*

OB association A grouping of hot, young, massive stars, predominantly of spectral types O and B. (Chapter 18)

OBAFGKM The temperature sequence (from hot to cold) of spectral classes. (Chapter 17)

objective lens The principal lens of a refracting telescope. (Chapter 6)

objective mirror The principal mirror of a reflecting telescope. (Chapter 6)

oblate Flattened at the poles. (Chapter 12)

oblateness A measure of how much a flattened sphere (or spheroid) differs from a perfect sphere. (Chapter 12)

observable universe That portion of the universe inside our cosmic light horizon. (Chapter 25, Chapter 26)

Occam's razor The notion that a straightforward explanation of a phenomenon is more likely to be correct than a convoluted one. (Chapter 4)

occultation The eclipsing of an astronomical object by the Moon or a planet. (Chapter 13, Chapter 14)

oceanic rift A crack in the ocean floor that exudes lava. (Chapter 9)

Olbers's paradox The dilemma associated with the fact that the night sky is dark. (Chapter 25)

1-to-1 spin-orbit coupling See *synchronous rotation.*

Oort cloud A presumed accumulation of comets and cometary material surrounding the Sun at distances of roughly 50,000 AU. (Chapter 7, Chapter 15)

opaque When light cannot travel through an object without being scattered many times or absorbed. Unlike glass or our atmosphere, an opaque material is not transparent. (Chapter 5)

open cluster A loose association of young stars in the disk of our Galaxy; a galactic cluster. (Chapter 18, Chapter 19)

open universe A universe with negative curvature, so that its geometry is analogous to a saddle-shaped hyperbolic surface. (Chapter 25)

opposition The configuration of a planet when it is at an elongation of 180° and thus appears opposite the Sun in the sky. (Chapter 4)

optical double star Two stars that lie along nearly the same line of sight but are actually at very different distances from us. (Chapter 17)

optical telescope A telescope designed to detect visible light. (Chapter 6)

optical window The range of visible wavelengths to which Earth's atmosphere is transparent. (Chapter 6)

orbital energy The net energy of an orbiting object due to its kinetic and gravitational potential energy. The farther a planet orbits from the Sun, the greater its orbital energy. (Chapter 4)

orbital resonance The resonance that occurs whenever the orbital periods of two objects are related by a ratio of small integers—for example, when one moon has twice the orbital period of another. The rhythmic gravitational interactions in an orbital resonance can have either a stabilizing or a destabilizing effect. (Chapter 12, Chapter 13, Chapter 15)

organic molecules Molecules containing carbon, some of which are the molecules of which living organisms are made. (Chapter 27)

outer core (of Earth) The outer, molten portion of Earth's iron-rich core. (Chapter 9)

outgassing The release of gases into a planet's atmosphere by volcanic activity. (Chapter 9)

overcontact binary A close binary system in which the two stars share a common atmosphere. (Chapter 19)

oxygen fusion The thermonuclear fusion of oxygen to produce heavier nuclei. (Chapter 20)

ozone A type of oxygen whose molecules contain three oxygen atoms. (Chapter 9)

ozone hole A region of Earth's atmosphere over Antarctica where the concentration of ozone is abnormally low. (Chapter 9)

ozone layer A layer in Earth's upper atmosphere where the concentration of ozone is high enough to prevent much ultraviolet light from reaching the surface. (Chapter 9)

P wave One of three kinds of seismic waves produced by an earthquake; a primary wave. (Chapter 9)

pair production The creation of a particle and its antiparticle from energy. (Chapter 26)

parabola A conic section formed by cutting a circular cone at an angle parallel to one of the sides of the cone. (Chapter 4)

parallax The apparent displacement of an object due to the motion of the observer. (Chapter 4, Chapter 17)

parsec (pc) A unit of distance; 3.26 light-years. (Chapter 1, Chapter 17)

partial lunar eclipse A lunar eclipse in which the Moon does not appear completely covered. (Chapter 3)

partial solar eclipse A solar eclipse in which the Sun does not appear completely covered. (Chapter 3)

Paschen series A series of spectral lines of hydrogen produced by electron transitions between the third and higher energy levels. (Chapter 5)

Pauli exclusion principle A principle of quantum mechanics stating that no two electrons can have the same position and momentum. (Chapter 19)

penumbra (of a shadow) (plural penumbrae) The portion of a shadow in which only part of the light source is covered by an opaque body. (Chapter 3)

penumbral eclipse A lunar eclipse in which the Moon passes only through Earth's penumbra. (Chapter 3)

perigee The point in its orbit where a satellite or the Moon is nearest Earth. (Chapter 3)

perihelion The point in its orbit where a planet or comet is nearest the Sun. (Chapter 4)

period (of a planet) The interval of time between successive geometric arrangements of a planet and an astronomical object, such as the Sun. (Chapter 4)

periodic table A listing of the chemical elements according to their properties, invented by Dmitri Mendeleev. (Chapter 5)

period-luminosity relation A relationship between the period and average density of a pulsating star. (Chapter 18)

photodisintegration The breakup of nuclei by high-energy gamma rays. (Chapter 20)

photometry The measurement of light intensities. (Chapter 6, Chapter 17)

photon A discrete unit of electromagnetic energy. (Chapter 5)

photosphere The region in the solar atmosphere from which most of the visible light escapes into space. (Chapter 16)

photosynthesis A biochemical process in which solar energy is converted into chemical energy, carbon dioxide and water are absorbed, and oxygen is released. (Chapter 9)

physics, laws of See *laws of physics*.

pixel A picture element. (Chapter 6)

plage A bright region in the solar atmosphere as observed in the monochromatic light of a spectral line. (Chapter 16)

Planck time A fundamental interval of time defined by basic physical constants. (Chapter 25)

Planck's law The relationship between the energy of a photon and its wavelength or frequency; $E = hc/\lambda = h\nu$. (Chapter 5)

planetary nebula A luminous shell of gas ejected from an old, low-mass star. (Chapter 20)

planetesimal One of many small bodies of primordial dust and ice that combined to form the planets. (Chapter 8)

plasma A hot ionized gas. (Chapter 13, Chapter 16, Chapter 25)

plastic The attribute of being nearly solid yet able to flow. (Chapter 9)

plate A large section of Earth's lithosphere that moves as a single unit. (Chapter 9)

plate tectonics The motions of large segments (plates) of Earth's surface over the underlying mantle. (Chapter 7, Chapter 9)

plutino One of about 100 objects in the Kuiper belt that orbit the Sun with nearly the same semimajor axis as Pluto. (Chapter 14)

poor cluster (of galaxies) A cluster of galaxies with very few members; a group of galaxies. (Chapter 23)

Population I star A star whose spectrum exhibits spectral lines of many elements heavier than helium; a metal-rich star. (Chapter 19)

Population II star A star whose spectrum exhibits comparatively few spectral lines of elements heavier than helium; a metal-poor star. (Chapter 19)

Population III star One of the first stars to form after the Big Bang, composed of only hydrogen, helium, and tiny amounts of lithium and beryllium; a "zeroth-generation" star. (Chapter 26)

positional astronomy The study of the apparent positions of the planets and stars and how those positions change. (Chapter 2)

positive curvature The curvature of a surface or space in which parallel lines converge and the sum of the angles of a triangle is greater than 180°. (Chapter 25)

positron An electron with a positive rather than negative electric charge; the antiparticle of the electron. (Chapter 16, Chapter 26)

power of ten The exponent n in 10^n. (Chapter 1)

powers-of-ten notation A shorthand method of writing numbers, involving 10 followed by an exponent. (Chapter 1)

precession (of Earth) A slow, conical motion of Earth's axis of rotation caused by the gravitational pull of the Moon and Sun on Earth's equatorial bulge. (Chapter 2)

precession of the equinoxes The slow westward motion of the equinoxes along the ecliptic due to precession of Earth. (Chapter 2)

primary mirror See *objective mirror*.

prime focus The point in a telescope where the objective focuses light. (Chapter 6)

primitive asteroid See *undifferentiated asteroid*.

primordial black hole A type of black hole that may have formed in the very early universe. (Chapter 21)

primordial fireball The extremely hot gas that filled the universe immediately following the Big Bang. (Chapter 25)

principle of equivalence See *equivalence principle*.

progenitor star A star that later explodes into a supernova. (Chapter 20)

prograde orbit An orbit of a satellite around a planet that is in the same direction as the rotation of the planet. (Chapter 13)

prograde rotation A situation in which an object (such as a planet) rotates in the same direction that it orbits around another object (such as the Sun). (Chapter 11)

projection The projection of a point on Earth is made by an imaginary line from the point, extending perpendicular to the surface of Earth, until the line intersects the imaginary celestial sphere. (Chapter 2)

prominence Flamelike protrusions seen near the limb of the Sun and extending into the solar corona. (Chapter 16)

proper distance See *proper length*.

proper length A length measured by a ruler at rest with respect to an observer. (Chapter 21)

proper motion The angular rate of change in the location of a star on the celestial sphere, usually expressed in arcseconds per year. (Chapter 17)

proper time A time interval measured with a clock at rest with respect to an observer. (Chapter 21)

proplyd See *protoplanetary disk*.

proton A heavy, positively charged subatomic particle that is one of two principal constituents of atomic nuclei. (Chapter 5)

proton-proton chain A sequence of thermonuclear reactions by which hydrogen nuclei are built up into helium nuclei. (Chapter 16)

protoplanet A Moon-sized object formed by the coalescence of planetesimals. (Chapter 8)

protoplanetary disk (proplyd) A disk of material encircling a protostar or a newborn star. (Chapter 8, Chapter 18)

protostar A star in its earliest stages of formation. (Chapter 18)

protosun The part of the solar nebula that eventually developed into the Sun. (Chapter 8)

Ptolemaic system The definitive version of the geocentric cosmogony of ancient Greece. (Chapter 4)

pulsar A pulsating radio source thought to be associated with a rapidly rotating neutron star. (Chapter 1, Chapter 20)

pulsating variable star A star that pulsates in size and luminosity. (Chapter 19)

quantum electrodynamics A theory that describes details of how charged particles interact by exchanging photons. (Chapter 26)

quantum mechanics The branch of physics dealing with the structure and behavior of atoms and their constituents as well as their interaction with light. (Chapter 5, Chapter 26)

quark confinement The permanent bonding of quarks to form particles like protons and neutrons. (Chapter 26)

quark One of several particles thought to be the internal constituents of certain heavy subatomic particles such as protons and neutrons. (Chapter 26)

quasar A very luminous object with a very large redshift and a starlike appearance. (Chapter 1, Chapter 24)

radial velocity That portion of an object's velocity parallel to the line of sight. (Chapter 5, Chapter 17)

radial velocity curve A plot showing the variation of radial velocity with time for a binary star or variable star. (Chapter 17)

radial velocity method A technique used to detect extrasolar planets by observing Doppler shifts in the spectrum of the planet's star. (Chapter 8)

radiation darkening The darkening of methane ice by electron impacts. (Chapter 14)

radiation pressure Pressure exerted on an object by radiation falling on the object. (Chapter 15)

radiation-dominated universe A universe in which the mass density of radiation exceeds both the average density of matter and the mass density of dark energy. (Chapter 25)

radiative diffusion The random migration of photons from a star's center toward its surface. (Chapter 16)

radiative zone A region within a star where radiative diffusion is the dominant mode of energy transport. (Chapter 16)

radio galaxy A galaxy that emits an unusually large amount of radio waves. (Chapter 24)

radio telescope A telescope designed to detect radio waves. (Chapter 6)

radio waves The longest-wavelength electromagnetic radiation. (Chapter 5)

radio window The range of radio wavelengths to which Earth's atmosphere is transparent. (Chapter 6)

radioactive dating A technique for determining the age of a rock sample by measuring the radioactive elements and their decay products in the sample. (Chapter 8)

radioactive decay The process whereby certain atomic nuclei spontaneously transform into other nuclei. (Chapter 8)

rarefaction A region in a sound wave where the medium carrying the wave is spread out or rarefied. (Chapter 25)

recombination The process in which an electron combines with a positively charged ion. (Chapter 18)

red dwarf A main-sequence star with a mass between about 0.08 $M_\odot$ and 0.4 $M_\odot$ that has a fully convective interior and that never goes through a red giant stage. (Chapter 19)

red giant A large, cool star of high luminosity. (Chapter 17, Chapter 19)

reddening (interstellar) See *interstellar reddening.*

red-giant branch The region of an H-R diagram occupied by stars in their first red-giant phase. (Chapter 20)

redshift The shifting to longer wavelengths of the light from remote galaxies and quasars; the Doppler shift of light from a receding source. (Chapter 5, Chapter 23)

reflecting telescope A telescope in which the principal optical component is a concave mirror. (Chapter 6)

reflection The return of light rays by a surface. (Chapter 6)

reflection nebula A comparatively dense cloud of dust in interstellar space that is illuminated by a star. (Chapter 18)

reflector A reflecting telescope. (Chapter 6)

refracting telescope A telescope in which the principal optical component is a lens. (Chapter 6)

refraction The bending of light rays when they pass from one transparent medium to another. (Chapter 6)

refractor A refracting telescope. (Chapter 6)

regolith The layer of rock fragments covering the surface of the Moon. (Chapter 10)

regular cluster (of galaxies) A spherical cluster of galaxies. (Chapter 23)

reionization The process whereby high-energy photons emitted by the most massive of the first stars (Population III stars) ionized most of the atoms in the universe. (Chapter 26)

relativistic cosmology A cosmology based on the general theory of relativity. (Chapter 25)

residual polar cap An ice-covered polar region on Mars that does not completely evaporate during the Martian summer. (Chapter 11)

respiration A biological process that produces energy by consuming oxygen and releasing carbon dioxide. (Chapter 9)

retrograde motion The apparent westward motion of a planet with respect to background stars. (Chapter 4)

retrograde orbit An orbit of a satellite around a planet that is in the direction opposite to which the planet rotates. (Chapter 13)

retrograde rotation A situation in which an object (such as a planet) rotates in the direction opposite to which it orbits around another object (such as the Sun). (Chapter 11)

rich cluster (of galaxies) A cluster of galaxies containing many members. (Chapter 23)

ridge push A force on the oceanic plate that occurs when molten rock emerging from an oceanic rift is pulled by gravity down the ridge's slopes. This segment of ocean seafloor on the slope pushes on the larger seafloor plate and is one of the main forces producing plate tectonics. (Chapter 9)

right ascension A coordinate for measuring the east-west positions of objects on the celestial sphere. (Chapter 2)

ring particles Small particles that constitute a planetary ring. (Chapter 12)

ringlet One of many narrow bands of particles of which Saturn's ring system is composed. (Chapter 12)

Roche limit The smallest distance from a planet or other object at which a second object can be held together by purely gravitational forces. (Chapter 12)

Roche lobe A teardrop-shaped volume surrounding a star in a binary inside which gases are gravitationally bound to that star. (Chapter 19)

rock A mineral or combination of minerals. (Chapter 9)

rotation curve A plot of the orbital speeds of stars and nebulae in a galaxy versus distance from the center of the galaxy. (Chapter 22)

RR Lyrae variable A type of pulsating star with a period of less than one day. (Chapter 19, Chapter 22)

runaway greenhouse effect A greenhouse effect in which the temperature continues to increase. (Chapter 11)

runaway icehouse effect A situation in which a decrease in atmospheric temperature causes a further decrease in temperature. (Chapter 11)

S wave One of three kinds of seismic waves produced by an earthquake; a secondary wave. (Chapter 9)

Sagittarius A* A powerful radio source at the center of our Galaxy. (Chapter 22)

saros A particular cycle of similar eclipses that recur about every 18 years. (Chapter 3)

scarp A line of cliffs formed by the faulting or fracturing of a planet's surface. (Chapter 11)

scattering of light See *light scattering*.

Schwarzschild radius (R_{Sch}) The distance from the singularity to the event horizon in a nonrotating black hole. (Chapter 21)

scientific method The basic procedure used by scientists to investigate phenomena. (Chapter 1)

seafloor spreading The separation of plates under the ocean due to lava emerging in an oceanic rift. (Chapter 9)

search for extraterrestrial intelligence (SETI) The scientific search for evidence of intelligent life on other planets. (Chapter 27)

second of arc See *arcsecond*.

sedimentary rock A rock that is formed from material deposited on land by rain or winds, or on the ocean floor. (Chapter 9)

seeing disk The angular diameter of a star's image. (Chapter 6)

seismic wave A vibration traveling through a terrestrial planet, usually associated with earthquake-like phenomena. (Chapter 9)

seismograph A device used to record and measure seismic waves, such as those produced by earthquakes. (Chapter 9)

selection effect The bias that occurs when an observational technique is more sensitive to one type of system than another. For example, the radial velocity method for detecting exoplanets can more easily detect large planets orbiting closely to their stars. This method is unlikely to detect many planets with Earth's characteristics, but it does not mean they are not there. (Chapter 8)

self-propagating star formation The process by which the formation of stars in one location in a galaxy stimulates the formation of stars in a neighboring location. (Chapter 22)

semidetached binary A binary star system in which one star fills its Roche lobe. (Chapter 19)

semimajor axis One-half of the major axis of an ellipse. For an orbiting object, the semimajor axis is also equal to the average orbital distance. (Chapter 4)

SETI See *search for extraterrestrial intelligence*.

Seyfert galaxy A spiral galaxy with a bright nucleus whose spectrum exhibits emission lines. (Chapter 24)

shell helium fusion The thermonuclear fusion of helium in a shell surrounding a star's core. (Chapter 20)

shell hydrogen fusion The thermonuclear fusion of hydrogen in a shell surrounding a star's core. (Chapter 19)

shepherd satellite A satellite whose gravity restricts the motions of particles in a planetary ring, preventing them from dispersing. (Chapter 12)

shield volcano A volcano with long, gently sloping sides. (Chapter 11)

shock wave An abrupt, localized region of compressed gas caused by an object traveling through the gas at a speed greater than the speed of sound. (Chapter 9)

short-period comet A comet that originated in the Kuiper belt and has a period of less than 200 years. (Chapter 15)

SI units The International System of Units, based on the meter (m), the second (s), and the kilogram (kg). (Chapter 1)

sidereal clock A clock that measures sidereal time. (Chapter 2)

sidereal day The interval between successive meridian passages of the vernal equinox. (Chapter 2)

sidereal month The period of the Moon's revolution about Earth with respect to the stars. (Chapter 3)

sidereal period The orbital period of one object about another as measured with respect to the stars. (Chapter 4)

sidereal time Time reckoned by the location of the vernal equinox. (Chapter 2)

sidereal year The orbital period of Earth about the Sun with respect to the stars. (Chapter 2)

silicon fusion The thermonuclear fusion of silicon to produce heavier elements, especially iron. (Chapter 20)

singularity A place of infinite space-time curvature; the center of a black hole. (Chapter 21)

slab pull A force on seafloor plates that occurs when a portion (or slab) of the seafloor plate sinks down into the mantle at a subduction zone. Gravity pulls on this higher-density slab more than on the surrounding mantle material. The pull on this slab results in a pull on the entire seafloor plate, which is the main force driving plate tectonics. (Chapter 9)

small-angle formula A relationship between the angular and linear sizes of a distant object. (Chapter 1)

snow line The distance from the Sun at which water vapor solidifies into ice or frost. Formation of this solid material beyond the snow line helped to build the large Jovian planets. (Chapter 8)

solar constant The average amount of energy received from the Sun per square meter per second, measured just above Earth's atmosphere. (Chapter 5)

solar corona Hot, faintly glowing gases seen around the Sun during a total solar eclipse; the uppermost regions of the solar atmosphere. (Chapter 3)

solar cycle See *22-year solar cycle.*

solar eclipse An eclipse of the Sun by the Moon; a passage of Earth through the Moon's shadow. (Chapter 3)

solar flare A sudden, temporary outburst of light from an extended region of the solar surface. (Chapter 16)

solar nebula The cloud of gas and dust from which the Sun and solar system formed. (Chapter 8)

solar neutrino A neutrino emitted from the core of the Sun. (Chapter 16)

solar neutrino problem The discrepancy between the predicted and observed numbers of solar neutrinos. (Chapter 16)

solar system The Sun, planets and their satellites, asteroids, comets, and related objects that orbit the Sun. (Chapter 1)

solar wind An outward flow of particles (mostly electrons and protons) from the Sun. (Chapter 9, Chapter 16)

south celestial pole The point directly above Earth's south pole where Earth's axis of rotation, if extended, would intersect the celestial sphere. (Chapter 2)

southern highlands (on Mars) Older, cratered terrain in the Martian southern hemisphere. (Chapter 11)

southern lights See *aurora australis.*

space velocity The speed and direction in which a star moves through space. (Chapter 17)

space weather Variations in the solar wind and magnetic field, which can affect satellites and astronauts. (Chapter 16)

spacetime A four-dimensional combination of time and the three dimensions of space. (Chapter 21)

special theory of relativity A description of mechanics and electromagnetic theory formulated by Albert Einstein, which explains that measurements of distance, time, and mass are affected by the observer's motion. (Chapter 21)

spectral analysis The identification of chemical substances from the patterns of lines in their spectra. (Chapter 5)

spectral class A classification of stars according to the appearance of their spectra. (Chapter 17)

spectral line In a spectrum, an absorption or emission feature that is at a particular wavelength. (Chapter 5)

spectral type A subdivision of a spectral class. (Chapter 17)

spectrograph An instrument for photographing a spectrum. (Chapter 6)

spectroscopic binary A binary star system whose binary nature is deduced from the periodic Doppler shifting of lines in its spectrum. (Chapter 17)

spectroscopic parallax The distance to a star derived by comparing its apparent brightness to a luminosity inferred from the star's spectrum. (Chapter 17)

spectroscopy The study of spectra and spectral lines. (Chapter 5, Chapter 6, Chapter 7)

spectrum (plural spectra) **The result of dispersing a beam of electromagnetic radiation so that components with different wavelengths are separated in space. (Chapter 5)**

spectrum binary A binary star whose binary nature is deduced from the presence of two sets of incongruous spectral lines. (Chapter 17)

speed Distance traveled divided by the time elapsed to cover that distance. (Chapter 4)

spherical aberration The distortion of an image formed by a telescope due to differing focal lengths of the optical system. (Chapter 6)

spherical space Space with positive curvature. (Chapter 25)

spicule A narrow jet of rising gas in the solar chromosphere. (Chapter 16)

spin (of a particle) The intrinsic angular momentum possessed by certain particles. (Chapter 22)

spin-flip transition A transition in the ground state of the hydrogen atom, which occurs when the orientation of the electron's spin changes. (Chapter 22)

spin-orbit coupling See *1-to-1 spin-orbit coupling* and *3-to-2 spin-orbit coupling.*

spiral arms Lanes of interstellar gas, dust, and young stars that wind outward in a plane from the central regions of a galaxy. (Chapter 22)

spiral galaxy A flattened, rotating galaxy with pinwheel-like spiral arms winding outward from the galaxy's nucleus. (Chapter 23)

spontaneous symmetry breaking A process by which certain symmetries in the mathematics of particle physics are suddenly altered to produce new particles and forces. (Chapter 26)

spring tide An ocean tide that occurs at new moon and full moon phases. (Chapter 4)

stable Lagrange points Locations along Jupiter's orbit where the combined gravitational effects of the Sun and Jupiter cause asteroids to collect. (Chapter 15)

standard candle An astronomical object of known intrinsic brightness that can be used to determine the distances to other galaxies. (Chapter 23)

Standard Model The detailed theory, verified by experiments, that explains all of the known forces between all of the known particles (not including the force of gravity). Phenomena such as dark matter and dark energy may or may not be related to the Standard Model. (Chapter 26)

star cluster See *cluster (of stars)*.

starburst galaxy A galaxy that is experiencing an exceptionally high rate of star formation. (Chapter 23)

stationary absorption line An absorption line in the spectrum of a binary star that does not show the same Doppler shift as other lines, indicating that it originates in the interstellar medium. (Chapter 18)

Stefan-Boltzmann law A relationship between the temperature of a blackbody and the rate at which it radiates energy. (Chapter 5)

stellar association A loose grouping of young stars. (Chapter 18)

stellar evolution The changes in size, luminosity, temperature, and so forth that occur as a star ages. (Chapter 18)

stellar parallax The apparent displacement of a star due to Earth's motion around the Sun. (Chapter 17)

stellar-mass black hole A black hole with a mass comparable to that of a star. (Chapter 21)

stony iron meteorite A meteorite composed of both stone and iron. (Chapter 15)

stony meteorite A meteorite composed of stone. (Chapter 15)

stratosphere A layer in Earth's atmosphere directly above the troposphere. (Chapter 9)

string theory The currently favored theory for combining all the known forces in nature. (Chapter 26)

strong force The force that binds protons and neutrons together in nuclei. (Chapter 26)

subduction zone A location where colliding tectonic plates cause Earth's crust to be pulled down into the mantle. (Chapter 9)

subtend To extend over an angle. (Chapter 1)

summer solstice The point on the ecliptic where the Sun is farthest north of the celestial equator. Also used to refer to the date on which the Sun passes through this point. (Chapter 2)

sunspot A temporary cool region in the solar photosphere. (Chapter 16)

sunspot cycle The semiregular 11-year period with which the number of sunspots fluctuates. (Chapter 16)

sunspot maximum/minimum That time during the sunspot cycle when the number of sunspots is highest/lowest. (Chapter 16)

supercluster A collection of clusters of galaxies. (Chapter 23)

supergiant A very large, extremely luminous star of luminosity class I. (Chapter 17, Chapter 20)

supergrand unified theory A complete description of all forces and particles, as well as the structure of space and time; a "theory of everything" (TOE). (Chapter 26)

supergranule A large convective feature in the solar atmosphere, usually outlined by spicules. (Chapter 16)

superior conjunction The configuration of a planet being behind the Sun as viewed from Earth. (Chapter 4)

superior planet A planet that is more distant from the Sun than is Earth. (Chapter 4)

superluminal motion Motion that appears to involve speeds greater than the speed of light. (Chapter 24)

supermassive black hole A black hole with a mass of a million or more Suns. (Chapter 21, Chapter 24)

supernova (plural **supernovae**) A stellar outburst during which a star suddenly increases its brightness roughly a millionfold. (Chapter 1, Chapter 15, Chapter 20)

supernova remnant The gases ejected by a supernova. (Chapter 18, Chapter 20)

supersonic Faster than the speed of sound. (Chapter 18)

surface wave A type of seismic wave that travels only over Earth's surface. (Chapter 9)

synchronous rotation The rotation of a body with a period equal to its orbital period; also called 1-to-1 spin-orbit coupling. (Chapter 3, Chapter 10)

synodic month The period of revolution of the Moon with respect to the Sun; the length of one cycle of lunar phases. Also called the lunar month. (Chapter 3)

synodic period The interval between successive occurrences of the same configuration of a planet. (Chapter 4)

T Tauri stars Young variable stars associated with interstellar matter that show erratic changes in luminosity. (Chapter 18)

tail (of a comet) Gas and dust particles from a comet's nucleus that have been swept away from the comet's head by the radiation pressure of sunlight and the solar wind. (Chapter 15)

tangential velocity That portion of an object's velocity perpendicular to the line of sight. (Chapter 17)

temperature See *degree Celsius, degree Fahrenheit,* and *kelvin.*

terminator The line dividing day and night on the surface of the Moon or a planet; the line of sunset or sunrise. (Chapter 10)

terrae Cratered lunar highlands. (Chapter 10)

terrestrial planet High-density worlds with solid surfaces, including Mercury, Venus, Earth, and Mars. (Chapter 7)

theory A hypothesis that has withstood experimental or observational tests. (Chapter 1)

theory of everything (TOE) See *supergrand unified theory.*

thermal energy An energy that arises from the kinetic energy of the individual atoms and molecules in a substance. The hotter an object, the greater its thermal energy. (Chapter 5)

thermal equilibrium A balance between the input and outflow of heat in a system. (Chapter 16, Chapter 26)

thermal pulse A brief burst in energy output from the helium-fusing shell of an aging low-mass star. (Chapter 20)

thermonuclear Describes nuclear reactions that fuse atoms together as a result of very high temperatures. (Chapter 20)

thermonuclear fusion The combining of nuclei under conditions of high temperature in a process that releases substantial energy. (Chapter 16)

thermonuclear supernova A supernova that occurs when a white dwarf in a close binary system accretes so much matter from its companion star that the white dwarf explodes in a blast of thermonuclear fusion; a Type Ia supernova. (Chapter 20)

thermosphere A region in Earth's atmosphere between the mesosphere and the exosphere. (Chapter 9)

third quarter moon The phase of the Moon that occurs when the Moon is 90° west of the Sun. (Chapter 3)

3-to-2 spin-orbit coupling The rotation of Mercury, which makes three complete rotations on its axis for every two complete orbits around the Sun. (Chapter 11)

threshold temperature The temperature above which photons spontaneously produce particles and antiparticles of a particular type. (Chapter 26)

tidal force A gravitational force whose strength and/or direction varies over a body and thus tends to deform the body. (Chapter 4, Chapter 12)

tidal heating The heating of the interior of a satellite by continually varying tidal stresses. (Chapter 13)

time dilation The slowing of time due to relativistic motion. (Chapter 21)

time zone A region on Earth where, by agreement, all clocks have the same time. (Chapter 2)

TOE See *supergrand unified theory.*

total lunar eclipse A lunar eclipse during which the Moon is completely immersed in Earth's umbra. (Chapter 3)

total solar eclipse A solar eclipse during which the Sun is completely hidden by the Moon. (Chapter 3)

totality (of lunar eclipse) The period during a total lunar eclipse when the Moon is entirely within Earth's umbra. (Chapter 3)

totality (of solar eclipse) The period during a total solar eclipse when the disk of the Sun is completely hidden. (Chapter 3)

transient lunar phenomena A flash of light that can occur when meteorites as small as 10 cm hit the lunar surface. Enough heat is generated in the impact to produce light visible from Earth by the naked eye. (Chapter 10)

transit An event in which an astronomical body moves in front of another. See also *meridian transit.* (Chapter 8)

transit method A method for detecting extrasolar planets that come between us and their parent star, dimming the star's light. (Chapter 8)

Trans-Neptunian object Any small body of rock and ice that orbits the Sun within the solar system, but beyond the orbit of Neptune. (Chapter 7, Chapter 14)

triple alpha process A sequence of two thermonuclear reactions in which three helium nuclei combine to form one carbon nucleus. (Chapter 19)

Trojan asteroid One of several asteroids that share Jupiter's orbit about the Sun. (Chapter 15)

Tropic of Cancer A circle of latitude 23½° north of Earth's equator. (Chapter 2)

Tropic of Capricorn A circle of latitude 23½° south of Earth's equator. (Chapter 2)

tropical year The period of revolution of Earth about the Sun with respect to the vernal equinox. (Chapter 2)

troposphere The lowest level in Earth's atmosphere. (Chapter 9)

Tully-Fisher relation A correlation between the width of the 21-cm line of a spiral galaxy and the total luminosity of that galaxy. (Chapter 23)

tuning fork diagram A diagram that summarizes Edwin Hubble's classification scheme for spiral, barred spiral, and elliptical galaxies. (Chapter 23)

turnoff point The point on an H-R diagram where the stars in a cluster are leaving the main sequence. (Chapter 19)

21-cm radio emission Radio radiation emitted by neutral hydrogen atoms in interstellar space. (Chapter 22)

22-year solar cycle The semiregular 22-year interval between successive appearances of sunspots at the same latitude and with the same magnetic polarity. (Chapter 16)

Type I supernova A supernova whose spectrum lacks hydrogen lines. Type I supernovae are further classified as Type Ia, Ib, or Ic. (Chapter 20)

Type Ia supernova A supernova whose spectrum lacks hydrogen lines but has a strong absorption line of ionized silicon. (Chapter 20)

Type Ib supernova A supernova whose spectrum lacks hydrogen and silicon lines but has a strong helium absorption line. (Chapter 20)

Type Ic supernova A supernova whose spectrum is almost devoid of emission or absorption lines. (Chapter 20)

Type II supernova A supernova with hydrogen emission lines in its spectrum, caused by the explosion of a massive star. (Chapter 20)

UBV photometry A system for determining the surface temperature of a star by measuring the star's brightness in the ultraviolet (U), blue (B), and visible (V) spectral regions. (Chapter 17)

ultraviolet radiation Electromagnetic radiation of wavelengths shorter than those of visible light but longer than those of X rays. (Chapter 5)

umbra (of a shadow) (plural **umbrae**) The central, completely dark portion of a shadow. (Chapter 3)

undifferentiated asteroid An asteroid within which chemical differentiation did not occur. (Chapter 15)

unified model Describes a physical system at the core of active galactic nuclei (AGN) that appears very different depending on the line of sight with which it viewed from Earth. The basic model involves a bright accretion disk that may or may not be obscured by a thick dusty torus. (Chapter 24)

universal constant of gravitation (G) The constant of proportionality in Newton's law of gravitation. (Chapter 4)

upper meridian The half of the meridian that lies above the horizon. (Chapter 2)

Van Allen belts Two doughnut-shaped regions around Earth where many charged particles (protons and electrons) are trapped by Earth's magnetic field. (Chapter 9)

velocity The speed and direction of an object's motion. (Chapter 4)

vernal equinox The point on the ecliptic where the Sun crosses the celestial equator from south to north. Also used to refer to the date on which the Sun passes through this intersection. (Chapter 2)

very-long-baseline interferometry (VLBI) A method of connecting widely separated radio telescopes to make very high-resolution observations. (Chapter 6)

virtual pair (also **virtual particle pair**) A particle and antiparticle that exist for such a brief interval that they cannot be observed. (Chapter 21, Chapter 25, Chapter 26)

virtual particle pair See *virtual pair.*

visible light Electromagnetic radiation detectable by the human eye. (Chapter 5)

visual binary A binary star in which the two components can be resolved through a telescope. (Chapter 17)

VLBI See *very-long-baseline interferometry.*

void A large volume of space, typically 30 to 120 Mpc (100 to 400 million light-years) in diameter, that contains very few galaxies. (Chapter 23)

volatile element An element with low melting and boiling points. (Chapter 10, Chapter 11)

waning crescent moon The phase of the Moon that occurs between third quarter and new moon. (Chapter 3)

waning gibbous moon The phase of the Moon that occurs between full moon and third quarter. (Chapter 3)

water hole A range of frequencies in the microwave spectrum suitable for interstellar radio communication. (Chapter 27)

watt A unit of power equal to one joule of energy per second. (Chapter 5)

wavelength of maximum emission The wavelength at which a heated object emits the greatest intensity of radiation. (Chapter 5)

wavelength The distance between two successive wave crests. (Chapter 5)

waxing crescent moon The phase of the Moon that occurs between new moon and first quarter. (Chapter 3)

waxing gibbous moon The phase of the Moon that occurs between first quarter and full moon. (Chapter 3)

weak force The short-range force that is responsible for transforming certain particles into other particles, such as the decay of a neutron into a proton. (Chapter 26)

weakly interacting massive particle (WIMP) A hypothetical massive particle that may make up part of the unseen dark matter. (Chapter 22)

weight The force with which gravity acts on a body. (Chapter 4)

white dwarf A low-mass star that has exhausted all its thermonuclear fuel and contracted to a size roughly equal to the size of Earth. (Chapter 17, Chapter 20)

white ovals Round, whitish features usually seen in Jupiter's southern hemisphere. (Chapter 12)

Widmanstätten patterns Crystalline structure seen in certain types of meteorites. (Chapter 15)

Wien's law A relationship between the temperature of a blackbody and the wavelength at which it emits the greatest intensity of radiation. (Chapter 5)

WIMP See *weakly interacting massive particle.*

winding dilemma The problem that the spiral arms of a galaxy like the Milky Way do not disappear. (Chapter 22)

winter solstice The point on the ecliptic where the Sun reaches its greatest distance south of the celestial equator. Also used to refer to the date on which the Sun passes through this point. (Chapter 2)

X-ray burster A nonperiodic X-ray source that emits powerful bursts of X rays. (Chapter 20)

X-rays Electromagnetic radiation whose wavelength is between that of ultraviolet light and gamma rays. (Chapter 5)

ZAMS See *zero-age main sequence.*

Zeeman effect A splitting or broadening of spectral lines due to a magnetic field. (Chapter 16)

zenith The point on the celestial sphere directly overhead an observer. (Chapter 2)

zero curvature The curvature of a surface or space in which parallel lines remain parallel and the sum of the angles of a triangle is exactly 180°. (Chapter 25)

zero-age main sequence The main sequence of young stars that have just begun to burn hydrogen at their cores. (Chapter 19)

zero-age main-sequence star A newly formed star that has just arrived on the main sequence. (Chapter 19)

zodiac A band of 12 constellations around the sky centered on the ecliptic. (Chapter 2)

zonal winds The pattern of alternating eastward and westward winds found in the atmospheres of Jupiter and Saturn. (Chapter 12)

zone A light-colored band in Jupiter's atmosphere. (Chapter 12)

Answers to Selected Questions

CAUTION! Only mathematical answers are given, not answers that require interpretation or discussion. Your instructor will expect you to show the steps required to reach each mathematical answer.

Chapter 1

24. 8.5×10^3 km **25.** 2.8×10^7 Suns **26.** About 3×10^{36} times larger **27.** 8.94×10^{56} hydrogen atoms **28.** (a) 1.581×10^{-5} ly (b) 4.848×10^{-6} pc **29.** 4.99×10^2 s **30.** 4.3×10^9 km **31.** (a) 1.59×10^{14} km (b) 16.8 years **32.** 4.32×10^{17} s **34.** (a) 1.5 m (b) 89 m (c) 5.4×10^3 m **35.** 6.9 m **36.** 3.4×10^3 km **37.** 3.7 km **38.** 0.320 arcmin

Chapter 2

28. Around 8:02 P.M. **33.** (a) About 9 hours **46.** October 25, 1917 **49.** 50° **50.** 3:50 A.M. local time **52.** (a) 6:00 P.M. (b) September 21

Chapter 3

31. (a) 0.91 hour (b) 6.6° **32.** 49 arcsec

Chapter 4

18. Semimajor axis = 0.25 AU, period = 0.125 year **19.** Average = 25 AU, farthest = 50 AU **23.** 6 newtons, 4 m/s² **26.** 1/9 as strong **38.** 87.97 days **43.** (a) 4 AU (b) 8 years **44.** (a) 16.0 AU (b) 0.5 AU **45.** (a) 1.26 AU (b) 258 days **46.** Earth exerts a force of 1.98×10^{20} newtons on the Moon. **47.** Forces are approximately the same; Earth's acceleration is 100 times greater. **48.** ¼ as much as on Earth **49.** 62 newtons, 0.13 **53.** 0.5 year **54.** (a) 24 hours (b) 42,375 km **55.** 119 minutes **57.** (a) 5 years (b) 2.92 AU **58.** (a) 3.43×10^{-5} newton (b) 3.21×10^{-5} newton (c) 2.2×10^{-6} newton

Chapter 5

2. 7.5 times **3.** 500 s **7.** 0.340 m **8.** 2.61×10^{14} Hz **17.** 135 nm **18.** 3400 K **27.** 9400 nm **28.** 9980°F **29.** About 10 μm **31.** 2.9 nm **32.** 3.9×10^{26} W **33.** 190 times more **34.** (a) 4890 nm = 4.89 μm (b) 540 times more **35.** (a) 5.75×10^8 W/m (b) 10,000 K **36.** 2.43×10^{-3} nm **37.** (a) 1005 nm **41.** Possible transitions: energy = 1 eV, λ = 1240 nm; energy = 2 eV, λ = 620 nm; energy = 3 eV, λ = 414 nm **42.** 8.6×10^4 km/s **43.** Coming toward us at 13.0 km/s **44.** −85,714 km/s

Chapter 6

31. $^1/_{25}$ = 0.04 **33.** (a) 222× (b) 100× (c) 36× (d) 0.75 arcsec **36.** 300 km (Hubble Space Telescope, Jupiter's moons); 110 km (human eye, our Moon) **37.** (a) 34 ly (b) 37 km **40.** (a) 5.39×10^{-4} m (c) 0.56 m

Chapter 7

23. Mass = 6.4×10^{23} kg, average density = 3900 kg/m³ **25.** (a) 1.0×10^{13} kg (b) 1.2 m/s **26.** (a) 3.3×10^{21} J (b) Equivalent to 3.9×10^7 Hiroshima-type weapons **27.** (a) 2.02 km/s (b) 19.4 km/s **28.** 12 km/s **29.** (a) 618 km/s **31.** 1.76 years **33.** (a) 1000 years (b) 17 days **35.** In 100 years, probability is 8.3×10^{-8} (one chance in 12 million); in 10^6 years, probability is 8.3×10^{-4} (one chance in 1200)

Chapter 8

41. 0.40 kg after 1.3 billion years; 0.20 kg after 2.6 billion years; 0.10 kg after 3.9 billion years **31.** 2.6 billion years **44.** (a) About 180 AU **45.** 2.9×10^7 AU = 140 pc = 460 ly **46.** (a) About 600 AU (b) About 5×10^{40} cubic meters (c) About 10^{55} atoms (d) About 3×10^{14} atoms per cubic meter

Chapter 9

31. (a) 5.4×10^{16} W (b) 1.75×10^{17} W (c) 343 watts per square meter (d) 246 K = 6°C **32.** 4 km **33.** Core: 17%; mantle: 82%; crust: 1% **34.** 0.020 (2% of the total mass) **35.** (b) About 15,000 kg/m³

Chapter 10

26. (a) 4671 km (b) 1707 km below the surface (c) 449 km **30.** 130 newtons on the Moon, 780 newtons on Earth **32.** (a) 5.78×10^{-5} newton (b) 4.15×10^{-5} newton (c) 1.39 **36.** 2.56 seconds **38.** (b) 10^6 times greater

Chapter 11

44. 13.0 arcsec **46.** (a) 0.16 AU **48.** (a) 3.03 m/s (b) 1.26 nm **49.** 0.615 AU **51.** For T = 460°C, λ_{max} = 4.0 μm **58.** (a) 3770 km (b) 370 km **59.** Difference between the round-trip times is 2.0×10^{-4} s **62.** 920 m **73.** Phobos: about 16 arcminutes; Deimos: about 2.7 arcminutes **70.** (a) Radius = 20,400 km, altitude = 17,000 km

Chapter 12

35. 12.7 km/s **38.** 8.5×10^{53} hydrogen atoms, 7.1×10^{52} helium atoms **39.** Roughly 600 km/h **41.** 127 K **43.** 59.5 km/s **44.** 8300 newtons **45.** 7.1×10^4 kg/m³ **49.** (a) 14.4 hours for inner edge of A ring, 7.9 hours for inner edge of B ring

Chapter 13

43. 760 km **47.** 2.8×10^{11} (280 billion) years **48.** About 2.35×10^{-7} (one part in 3×10^{10}) **49.** About 8.1 minutes **52.** Escape speed = 2.6 km/s; mass of molecule = 2.0×10^{-26} kg, corresponding to a molecular weight of 12 **55.** 0.09 arcsec **56.** (a) 65.7 hours (b) 27.6 arcsec

Chapter 14

29. Sun-Uranus force = 1.39×10^{11} newtons, Neptune-Uranus force = 2.24×10^{17} newtons; Neptune reduces the sunward gravitational pull on Uranus by 1.61×10^4, or 0.0161% **31.** (a) 2900 kg/m³ **32.** (a) 8.4 hours **35.** (a) 1.13×10^{-3} as bright as on Earth (b) 4.10×10^{-4} as bright as on Earth (c) 2.77 times as bright **36.** About 7200 km **37.** 0.95 arcsec **42.** Without the boost, a one-way trip would take 30.5 years

Chapter 15

33. 2.82 AU **35.** 30 km **37.** (a) 3.6×10^{12} kg (b) 0.83 m/s
39. About 4% **40.** 3.6×10^7 km $= 0.24$ AU **41.** (a) Period $=$
350 years, lifetime $= 3.5 \times 10^4$ years (b) Period $= 1.1 \times 10^4$ years,
lifetime $= 1.1 \times 10^6$ years (c) Period $= 3.5 \times 10^5$ years, lifetime $=$
3.5×10^7 years (d) Period $= 1.1 \times 10^7$ years, lifetime $= 1.1 \times 10^9$
years **44.** (a) 10^{15} kg (b) 10^{-16} kg/m^3

Chapter 16

29. (a) 1.8×10^{-9} J (b) 9.0×10^{16} J (c) 5.4×10^{41} J **30.** (a) 4.6×10^{-36} s
(b) 2.3×10^{10} s (c) 1.4×10^{15} s $= 4.4 \times 10^7$ years **31.** 0.048 (4.8%) of
the Sun's mass will be converted from hydrogen to helium; chemical
composition of the Sun (by mass) will be 69% hydrogen, 30% helium

32. (a) 8.8×10^{29} kg of hydrogen consumed, 6.2×10^{26} kg lost
34. (a) 1.64×10^{-13} J (b) 2.43×10^{-3} nm **35.** 1.4×10^{13} kg/s
39. 98,600 kg/m^3 **40.** 1.9×10^{-7} nm **42.** (a) 1700 nm **44.** For the
photosphere, 500 nm; for the chromosphere, 58 nm; for the corona,
1.9 nm **47.** For the umbra, 670 nm; for the penumbra, 580 nm
48. (a) (Flux from patch of penumbra)/(flux from patch of
photosphere) $= 0.55$ (b) (Flux from patch of penumbra)/(flux from
patch of umbra) $= 1.8$

Chapter 27

16. (a) 3.7 cm **17.** (a) 10,000 years (b) 10 million years
19. (a) 1430 MHz **20.** (a) 29.8 km/s (b) 0.10 m (c) 9.9×10^{-6} m, or
9.9×10^{-3} % of the unshifted wavelength **21.** 200 km

Index

Page numbers in *italics* indicate figures and captions.

Star Charts

The following set of star charts, one for each month of the year, are from *Griffith Observer* magazine. They are useful in the northern hemisphere only. For a set of star charts suitable for use in the southern hemisphere, see the *Universe* Web site (www.whfreeman.com/universe).

To use these charts, first select the chart that best corresponds to the date and time of your observations. Hold the chart vertically as shown in the illustration below and turn it so that the direction you are facing is shown at the bottom.

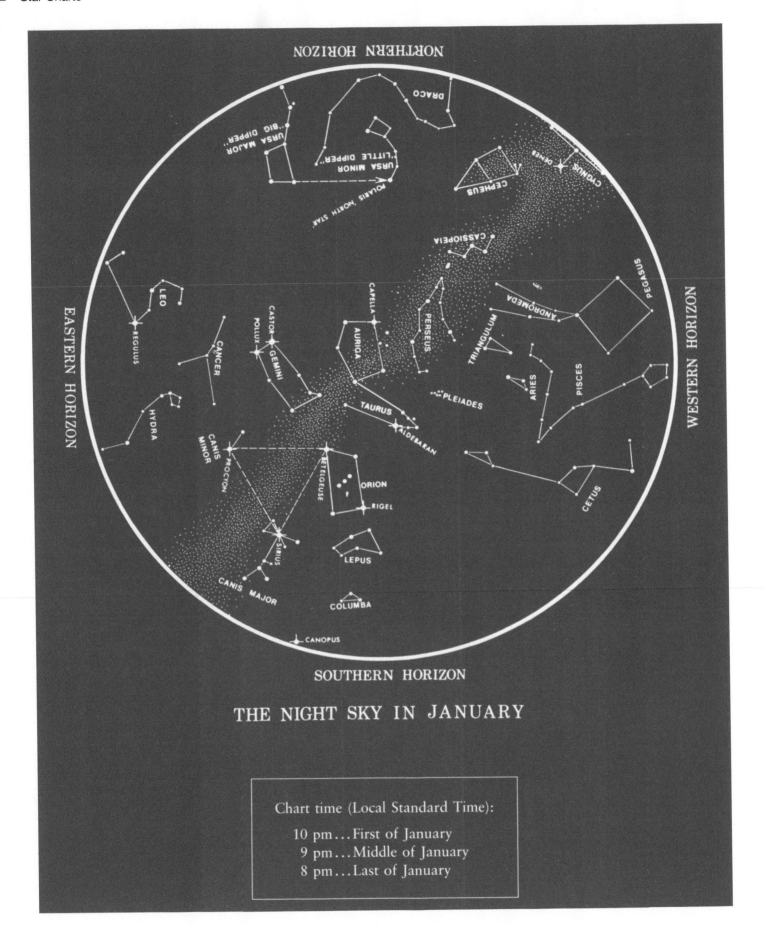

THE NIGHT SKY IN JANUARY

Chart time (Local Standard Time):

10 pm...First of January
9 pm...Middle of January
8 pm...Last of January

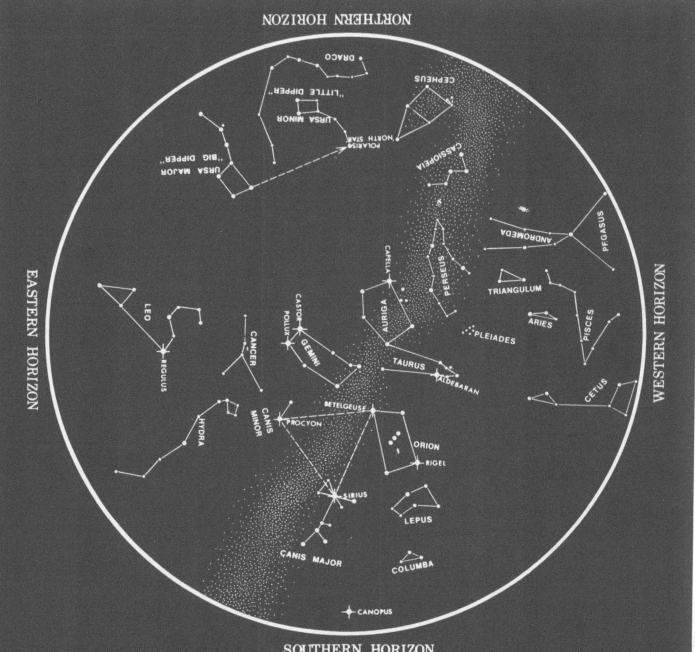

THE NIGHT SKY IN FEBRUARY

Chart time (Local Standard Time):

10 pm...First of February
9 pm...Middle of February
8 pm...Last of February

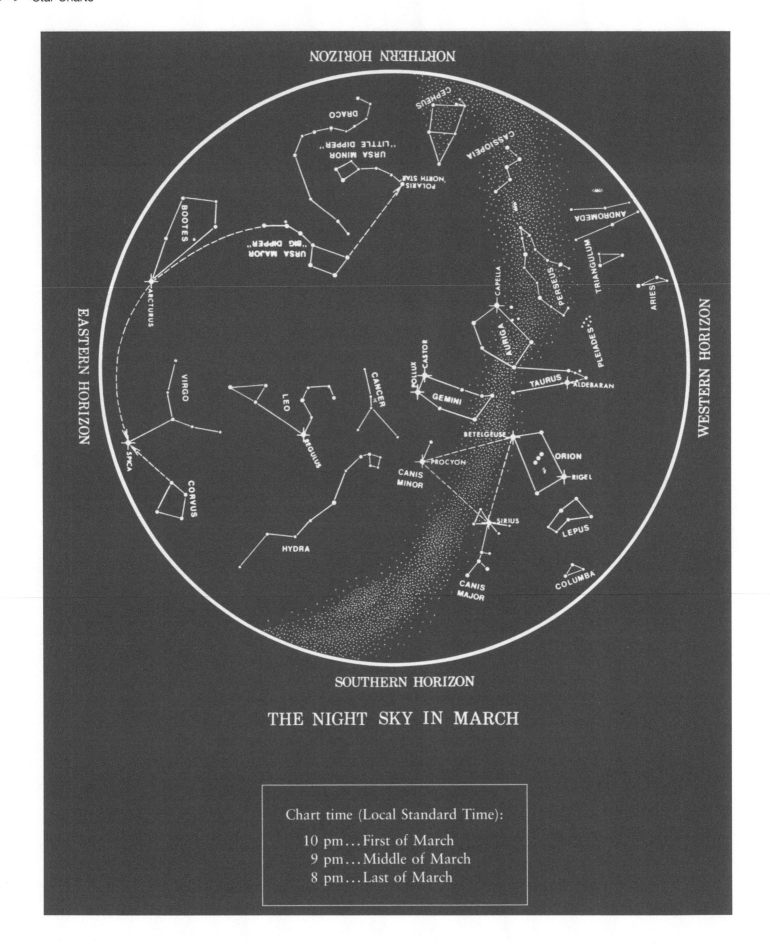

THE NIGHT SKY IN MARCH

Chart time (Local Standard Time):

10 pm...First of March
9 pm...Middle of March
8 pm...Last of March

THE NIGHT SKY IN APRIL

Chart time (Daylight Saving Time):

11 pm...First of April
10 pm...Middle of April
9 pm...Last of April

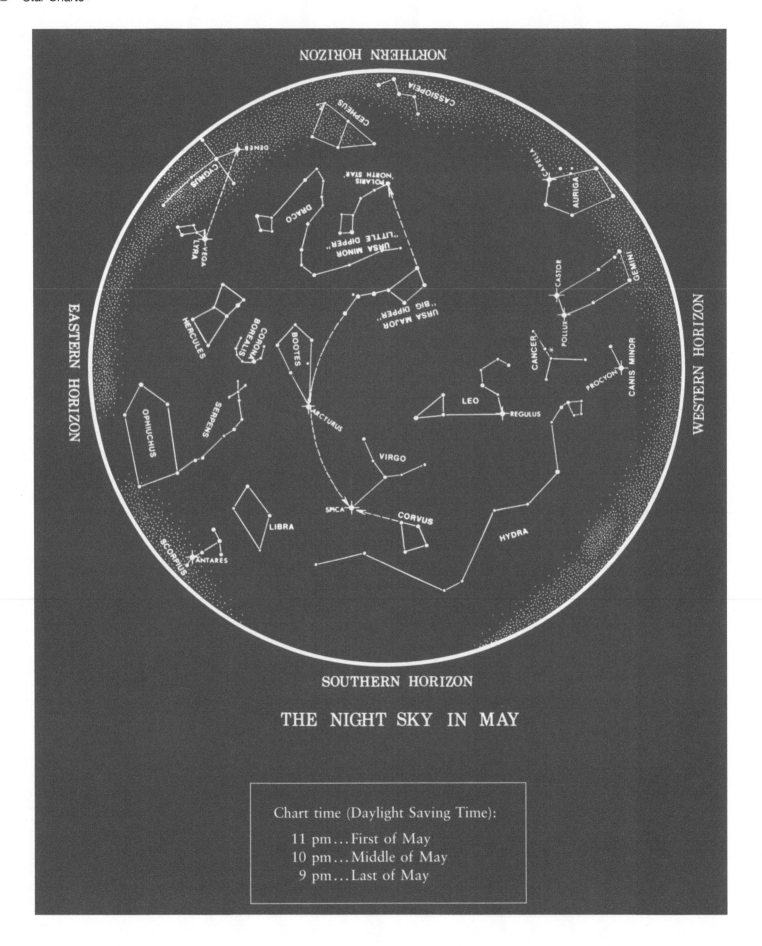

THE NIGHT SKY IN MAY

Chart time (Daylight Saving Time):

11 pm...First of May
10 pm...Middle of May
9 pm...Last of May

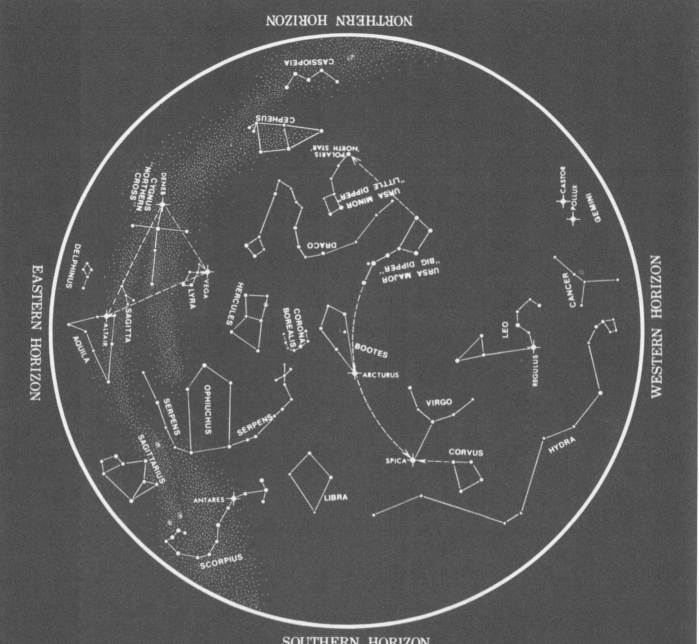

THE NIGHT SKY IN JUNE

Chart time (Daylight Saving Time):

11 pm...First of June
10 pm...Middle of June
9 pm...Last of June

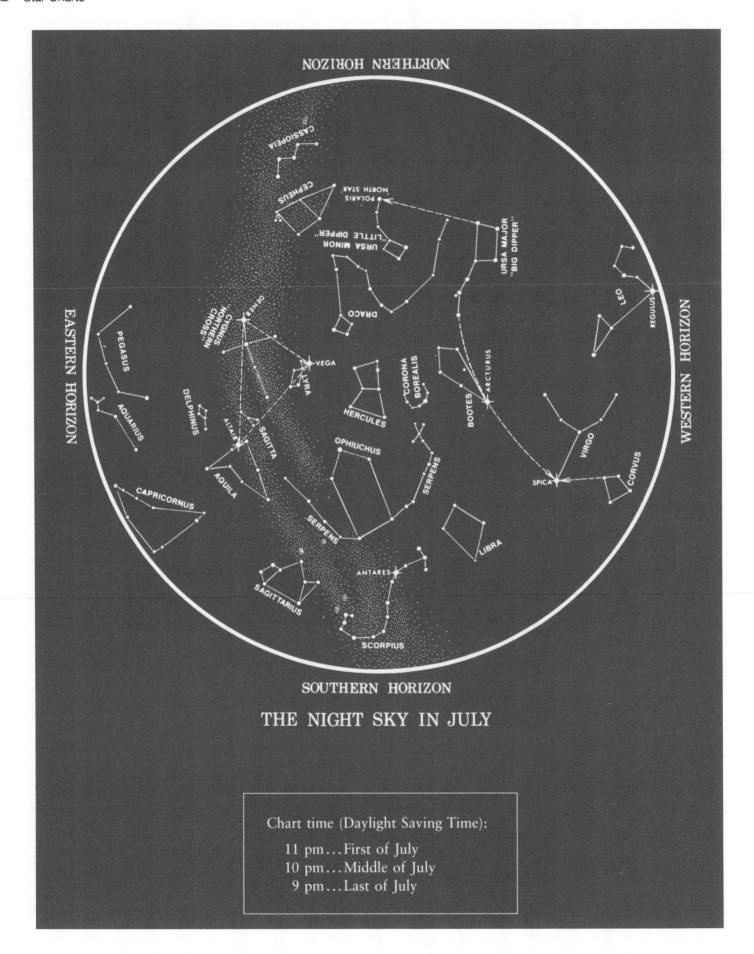

THE NIGHT SKY IN JULY

Chart time (Daylight Saving Time):

11 pm...First of July
10 pm...Middle of July
9 pm...Last of July

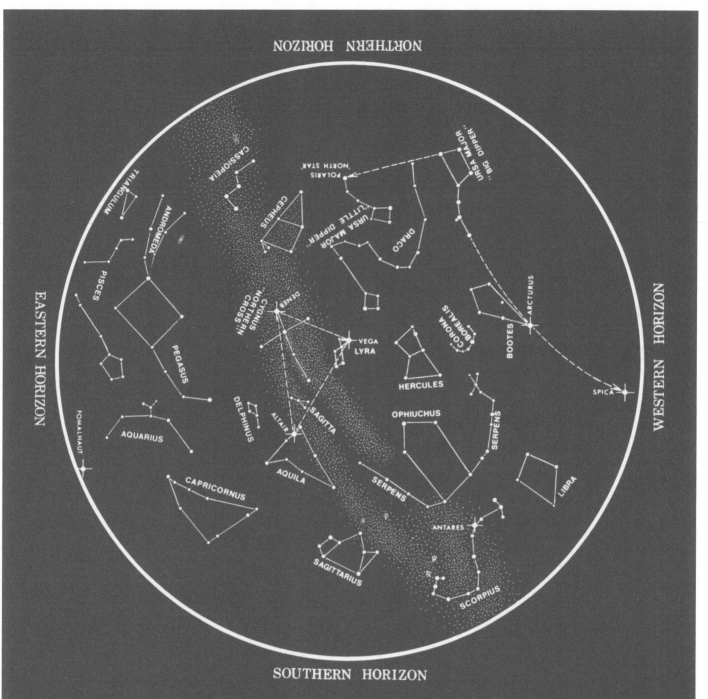

THE NIGHT SKY IN AUGUST

Chart time (Daylight Saving Time):

11 pm...First of August
10 pm...Middle of August
9 pm...Last of August

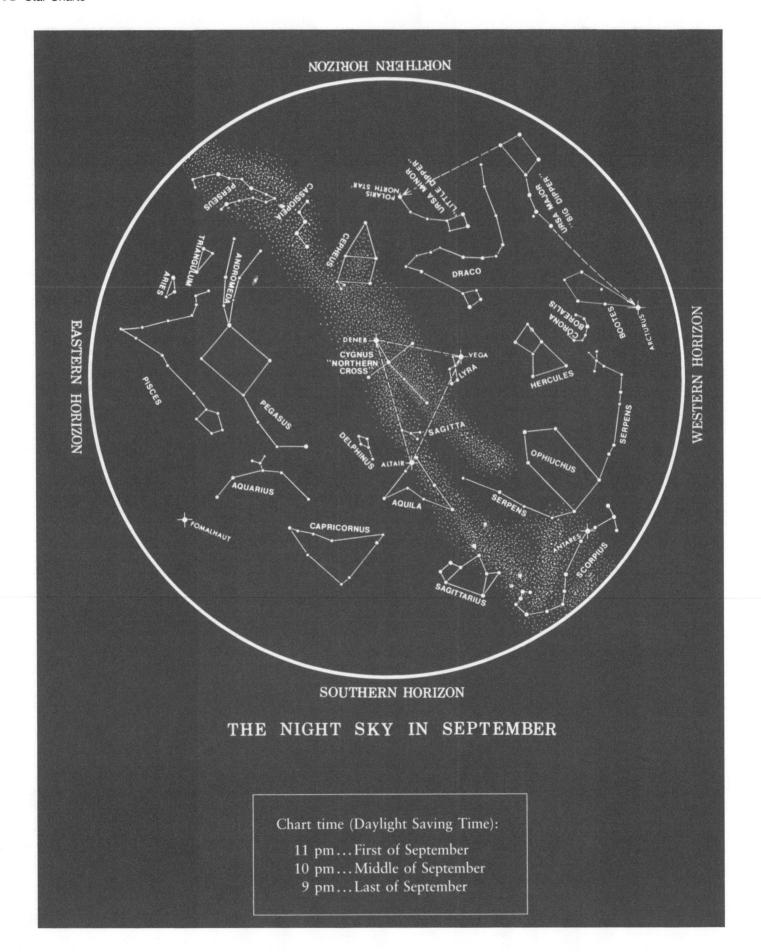

THE NIGHT SKY IN SEPTEMBER

Chart time (Daylight Saving Time):

11 pm...First of September
10 pm...Middle of September
9 pm...Last of September

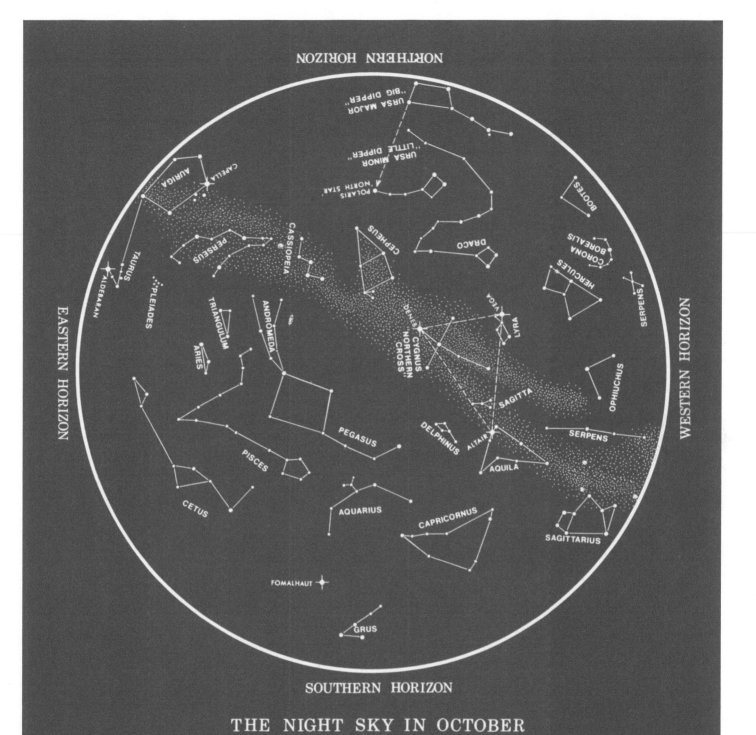

THE NIGHT SKY IN OCTOBER

Chart time (Daylight Saving Time):

11 pm...First of October
10 pm...Middle of October
9 pm...Last of October

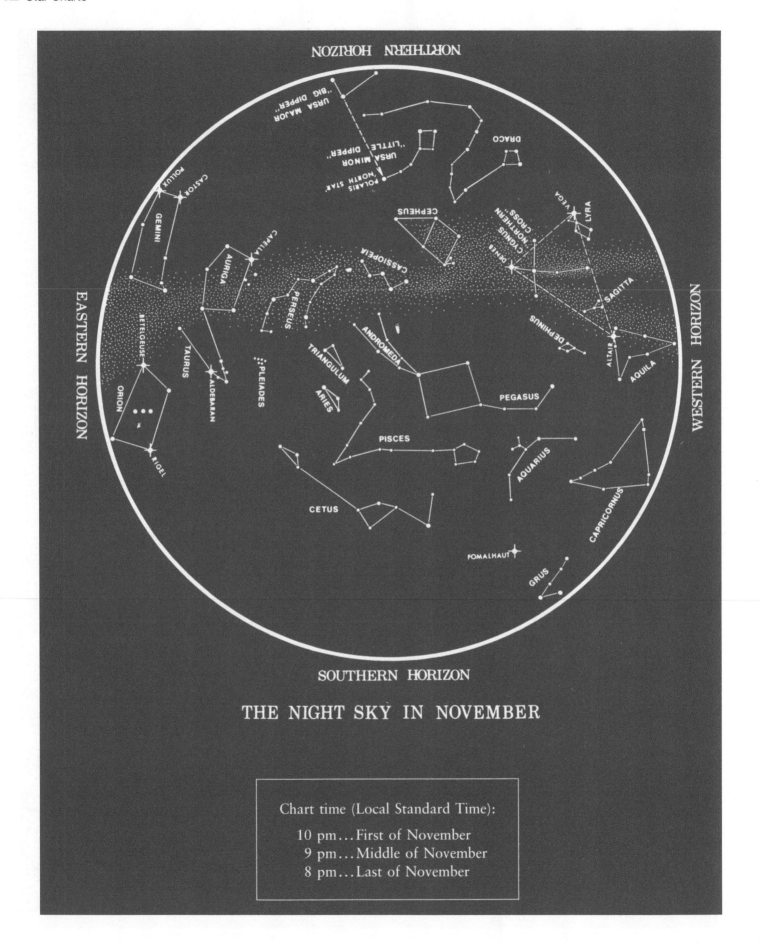

THE NIGHT SKY IN NOVEMBER

Chart time (Local Standard Time):

10 pm...First of November
9 pm...Middle of November
8 pm...Last of November

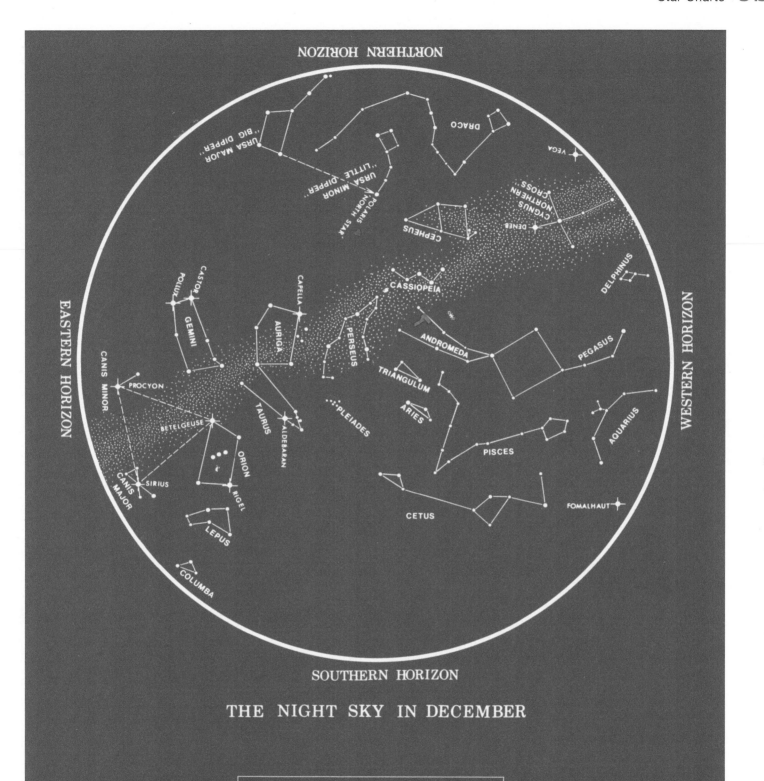

THE NIGHT SKY IN DECEMBER

Chart time (Local Standard Time):

10 pm...First of December
9 pm...Middle of December
8 pm...Last of December